A Study in Derived Algebraic Geometry

Volume I: Correspondences and Duality

Mathematical
Surveys
and
Monographs

Volume 221

A Study in Derived Algebraic Geometry

Volume I: Correspondences and Duality

Dennis Gaitsgory
Nick Rozenblyum

AMS AMERICAN MATHEMATICAL SOCIETY
Providence, Rhode Island

2010 *Mathematics Subject Classification.* Primary 14-02, 14A20, 14F05, 18G55, 18D05.

Library of Congress Cataloging-in-Publication Data

Names: Gaitsgory, D. (Dennis) | Rozenblyum, Nick, 1984-
Title: A study in derived algebraic geometry/ Dennis Gaitsgory, Nick Rozenblyum.
Description: Providence, Rhode Island : American Mathematical Society, 2017 | Series: Mathematical surveys and monographs ; volume 221.1 | Includes bibliographical references and index. Contents: volume 1. Correspondences and duality – volume 2. Deformations, Lie theory, and formal geometry.
Identifiers: LCCN 2016054809 | Volume I ISBN 9781470435691 (alk. paper) | Volume II ISBN 9781470435707 (alk. paper)
Subjects: LCSH:Geometry, Algebraic. | Duality theory (Mathematics) | Lie algebras. | Geometry. | AMS: Algebraic geometry – Research exposition (monographs, survey articles).msc | Algebraic geometry – Foundations – Generalizations (algebraic spaces, stacks).msc | Algebraic geometry – (Co)homology theory – Sheaves, derived categories of sheaves and related constructions. msc | Algebraic geometry – Families, fibrations – Formal methods; deformations. msc | Algebraic geometry – (Co)homology theory – Differentials and other special sheaves; D-modules; Bernstein-Sato ideals and polynomials. msc | Category theory; homological algebra – Homological algebra – Homotopical algebra. msc | Category theory; homological algebra – Categories with structure – Double categories, 2-categories, bicategories and generalizations.msc
Classification: LCC QA564.G35 2017 | DDC 516.3/5–dc23 LC record available at https://lccn.loc.gov/2016054809

♾ The paper used in this book is acid-free and falls within the guidelines established to ensure permanence and durability.
Visit the AMS home page at `http://www.ams.org/`

10 9 8 7 6 5 4 3 2 1 22 21 20 19 18 17

Contents

Preface

Кто я? Не каменщик прямой,
Не кровельщик, не корабельщик, –
Двурушник я, с двойной душой,
Я ночи друг, я дня застрельщик.

О. Мандельштам. Грифельная ода.

Who am I? Not a straightforward mason,
Not a roofer, not a shipbuilder, –
I am a double agent, with a duplicitous soul,
I am a friend of the night, a skirmisher of the day.

O. Mandelshtam. The Graphite Ode.

1. What is the object of study in this book?

The main unifying theme of the two volumes of this book is the notion of *ind-coherent sheaf*, or rather, categories of such on various geometric objects. In this section we will try to explain what ind-coherent sheaves are and why we need this notion.

1.1. Who are we? Let us start with a disclosure: this book is *not* really about algebraic geometry.

Or, rather, in writing this book, its authors do not act as real algebraic geometers. This is because the latter are ultimately interested in geometric objects that are constrained/enriched by the algebraicity requirement.

We, however, use algebraic geometry as a tool: this book is written with a view toward applications to *representation theory*.

It just so happens that algebraic geometry is a very (perhaps, even the most) convenient way to formulate representation-theoretic problems of categorical nature. This is not surprising, since, after all, algebraic groups are themselves objects of algebraic geometry.

The most basic example of how one embeds representation theory into algebraic geometry is this: take the category $\text{Rep}(G)$ of algebraic representations of a linear algebraic group G. Algebraic geometry allows us to define/interpret $\text{Rep}(G)$ as the category of quasi-coherent sheaves on the classifying stack BG.

The advantage of this point of view is that many natural constructions associated with the category of representations are already contained in the package of 'quasi-coherent sheaves on stacks'. For example, the functors of restriction and coinduction[1] along a group homomorphism $G' \to G$ are interpreted as the functors of inverse and direct image along the map of stacks

$$BG' \to BG.$$

But what is the advantage of this point of view? Why not stick to the explicit constructions of all the required functors within representation theory?

The reason is that 'explicit constructions' involve 'explicit formulas', and once we move to the world of higher categories (which we inevitably will, in order to meet the needs of modern representation theory), we will find ourselves in trouble: constructions in higher category theory are intolerant of explicit formulas (for an example of a construction that uses formulas see point (III) in Sect. 1.5 below). Rather, when dealing with higher categories, there is a fairly limited package of constructions that we are allowed to perform (see Chapter 1, Sects. 1 and 2 where some of these constructions are listed), and algebraic geometry seems to contain a large chunk (if not all) of this package.

1.2. A stab in the back. Jumping ahead slightly, suppose for example that we want to interpret algebro-geometrically the category $\mathfrak{g}$-mod of modules over a Lie algebra $\mathfrak{g}$.

The first question is: why would one want to do that? Namely, take the universal enveloping algebra $U(\mathfrak{g})$ and interpret $\mathfrak{g}$-mod as modules over $U(\mathfrak{g})$. Why should one mess with algebraic geometry if all we want is the category of modules over an associative algebra?

But let us say that we have already accepted the fact that we want to interpret $\mathrm{Rep}(G)$ as $\mathrm{QCoh}(BG)$. If we now want to consider restriction functor

$$\mathrm{Rep}(G) \to \mathfrak{g}\text{-mod}, \tag{1.1}$$

(where $\mathfrak{g}$ is the Lie algebra of G), we will need to give an algebro-geometric interpretation of $\mathfrak{g}$-mod as well.

If $\mathfrak{g}$ is a usual (=classical) Lie algebra, one can consider the associated formal group, denoted in the book $\exp(\mathfrak{g})$, and one can show (see Volume II, Chapter 7, Sect. 5) that the category $\mathfrak{g}$-mod is canonically equivalent to $\mathrm{QCoh}(B(\exp(\mathfrak{g})))$, the category of quasi-coherent sheaves on the classifying stack[2] of $\exp(\mathfrak{g})$. With this interpretation of $\mathfrak{g}$-mod, the functor (1.1) is simply the pullback functor along the map

$$B(\exp(\mathfrak{g})) \to BG,$$

induced by the (obvious) map $\exp(\mathfrak{g}) \to G$.

Let us now be given a homomorphism of Lie algebras $\alpha : \mathfrak{g}' \to \mathfrak{g}$. The functor of restriction $\mathfrak{g}\text{-mod} \to \mathfrak{g}'\text{-mod}$ still corresponds to the pullback functor along the corresponding morphism

$$B(\exp(\mathfrak{g}')) \xrightarrow{f_\alpha} B(\exp(\mathfrak{g})). \tag{1.2}$$

---

[1] What we call 'coinduction' is the functor right adjoint to restriction, i.e., it is the usual representation-theoretic operation.

[2] One can (reasonably) get somewhat uneasy from the suggestion to consider the category of quasi-coherent sheaves on the *classifying stack of a formal group*, but, in fact, this is a legitimate operation.

Note, however, that when we talk about representations of Lie algebras, the natural functor in the opposite direction is *induction*, i.e., the *left* adjoint to restriction. And being a left adjoint, it *cannot* correspond to the direct image along (1.2) (whatever the functor of direct image is, it is the *right* adjoint of pullback).

This inconsistency leads to the appearance of *ind-coherent sheaves.*

1.3. The birth of IndCoh.

What happens is that, although we can interpret $\mathfrak{g}$-mod as $\text{QCoh}(B(\exp(\mathfrak{g})))$, a more natural interpretation is as $\text{IndCoh}(B(\exp(\mathfrak{g})))$. The symbol 'IndCoh' will of course be explained in the sequel. It just so happens that for a classical Lie algebra, the categories $\text{QCoh}(B(\exp(\mathfrak{g})))$ and $\text{IndCoh}(B(\exp(\mathfrak{g})))$ are equivalent (as $\text{QCoh}(BG)$ is equivalent to $\text{IndCoh}(BG)$).

Now, the functor of restriction along the homomorphism α will be given by the functor

$$(f_\alpha)^! : \text{IndCoh}(B(\exp(\mathfrak{g}'))) \to \text{IndCoh}(B(\exp(\mathfrak{g})));$$

this is the !-pullback functor, which is the *raison d'être* for the theory of IndCoh.

However, the functor of induction $\mathfrak{g}'$-mod $\to$ $\mathfrak{g}$-mod will be the functor of *IndCoh direct image*

$$(f_\alpha)_*^{\text{IndCoh}} : \text{IndCoh}(B(\exp(\mathfrak{g}'))) \to \text{IndCoh}(B(\exp(\mathfrak{g}))), \tag{1.3}$$

which is the *left* adjoint of $(f_\alpha)^!$. This adjunction is due to the fact that the morphism f_α is, in an appropriate sense, proper.

Now, even though, as was mentioned above, for a usual Lie algebra $\mathfrak{g}$, the categories

$$\text{QCoh}(B(\exp(\mathfrak{g}))) \text{ and } \text{IndCoh}(B(\exp(\mathfrak{g})))$$

are equivalent, the functor $(f_\alpha)_*^{\text{IndCoh}}$ of (1.3) is as different as can be from the functor

$$(f_\alpha)_* : \text{QCoh}(B(\exp(\mathfrak{g}'))) \to \text{QCoh}(B(\exp(\mathfrak{g})))$$

(the latter is quite ill-behaved).

For an analytically minded reader let us also offer the following (albeit somewhat loose) analogy: QCoh(−) behaves more like functions on a space, while IndCoh(−) behaves more like measures on the same space.

1.4. What can we do with ind-coherent sheaves? As we saw in the example of Lie algebras, the kind of geometric objects on which we will want to consider IndCoh (e.g., $B(\exp(\mathfrak{g}))$) are quite a bit more general than the usual objects on which we consider quasi-coherent sheaves, the latter being schemes (or algebraic stacks).

A natural class of algebro-geometric objects for which IndCoh is defined is that of *inf-schemes*, introduced and studied in Volume II, Part I of the book. This class includes all schemes, but also formal schemes, as well as classifying spaces of formal groups, etc. In addition, if X is a scheme, its de Rham prestack[3] X_{dR} is an inf-scheme, and ind-coherent sheaves on X_{dR} will be the same as *crystals* (a.k.a. D-modules) on X.

[3]The de Rham prestack of a given scheme X is obtained by 'modding' out X by the groupoid of its infinitesimal symmetries, see Volume II, Chapter 4, Sect. 1.1.1 for a precise definition.

Thus, for any inf-scheme $\mathcal{X}$ we have a well-defined category $\text{IndCoh}(\mathcal{X})$. For any map of inf-schemes $f : \mathcal{X}' \to \mathcal{X}$ we have functors

$$f_*^{\text{IndCoh}} : \text{IndCoh}(\mathcal{X}') \to \text{IndCoh}(\mathcal{X})$$

and

$$f^! : \text{IndCoh}(\mathcal{X}) \to \text{IndCoh}(\mathcal{X}').$$

Moreover, if f is proper[4], then the functors $(f_*^{\text{IndCoh}}, f^!)$ form an adjoint pair.

Why should we be happy to have this? The reason is that this is exactly the kind of operations one needs in geometric representation theory.

1.5. Some examples of what we can do.

(I) Take $\mathcal{X}'$ to be a scheme X and $\mathcal{X} = X_{\text{dR}}$, with f being the canonical projection $X \to X_{\text{dR}}$. Then the adjoint pair

$$f_*^{\text{IndCoh}} : \text{IndCoh}(X) \rightleftarrows \text{IndCoh}(X_{\text{dR}}) : f^!$$

identifies with the pair

$$\mathbf{ind}_{\text{D-mod}} : \text{IndCoh}(X) \rightleftarrows \text{D-mod}(X) : \mathbf{ind}_{\text{D-mod}},$$

corresponding to forgetting and inducing the (right) D-module structure (as we shall see shortly in Sect. 2.3, for a scheme X, the category $\text{IndCoh}(X)$ is only slightly different from the usual category of quasi-coherent sheaves $\text{QCoh}(X)$).

(II) Suppose we have a morphism of schemes $g : Y \to X$ and set

$$Y_{\text{dR}} \overset{f:=g_{\text{dR}}}{\longrightarrow} X_{\text{dR}}.$$

The corresponding functors

$$f_*^{\text{IndCoh}} : \text{IndCoh}(Y_{\text{dR}}) \to \text{IndCoh}(X_{\text{dR}}) \text{ and } f^! : \text{IndCoh}(X_{\text{dR}}) \to \text{IndCoh}(Y_{\text{dR}})$$

identify with the functors

$$g_{*,\text{Dmod}} : \text{Dmod}(Y) \to \text{Dmod}(X) \text{ and } g^!_{\text{Dmod}} : \text{Dmod}(X) \to \text{Dmod}(Y)$$

of D-module (a.k.a. de Rham) push-forward and pullback, respectively.

Note that while the operation of pullback of (right) D-modules corresponds to !-pullback on the underlying $\mathcal{O}$-module, the operation of D-module push-forward is less straightforward as it involves taking fiber-wise de Rham cohomology. So, the operation of the IndCoh direct image does something quite non-trivial in this case.

(III) Suppose we have a Lie algebra $\mathfrak{g}$ that acts (by vector fields) on a scheme X. In this case we can create a diagram

$$B(\exp(\mathfrak{g})) \overset{f_1}{\longleftarrow} B_X(\exp(\mathfrak{g})) \overset{f_2}{\longrightarrow} X_{\text{dR}},$$

where $B_X(\exp(\mathfrak{g}))$ is an inf-scheme, which is the quotient of X by the action of $\mathfrak{g}$.

Then the composite functor

$$(f_2)_*^{\text{IndCoh}} \circ (f_1)^! : \text{IndCoh}(B(\exp(\mathfrak{g}))) \to \text{IndCoh}(X_{\text{dR}})$$

identifies with the *localization functor*

$$\mathfrak{g}\text{-mod} \to \text{Dmod}(X).$$

[4]Properness means the following: to every inf-scheme there corresponds its underlying reduced scheme, and a map between inf-schemes is proper if and only if the map of the underlying reduced schemes is proper in the usual sense.

This third example should be a particularly convincing one: the localization functor, which is usually defined by an explicit formula

$$M \mapsto D_X \underset{U(\mathfrak{g})}{\otimes} M,$$

is given here by the general formalism.

2. How do we do we construct the theory of IndCoh?

Whatever inf-schemes are, for an individual inf-scheme $\mathcal{X}$, the category $\text{IndCoh}(\mathcal{X})$ is bootstrapped from the corresponding categories for schemes by the following procedure:

$$\text{IndCoh}(\mathcal{X}) = \lim_{Z \to \mathcal{X}} \text{IndCoh}(Z). \tag{2.1}$$

Some explanations are in order.

2.1. What do we mean by limit?

(a) In formula (2.1), the symbol 'lim' appears. This is the limit of categories, but not quite. If we were to literally take the limit in the category of categories, we would obtain utter nonsense. This is a familiar phenomenon: the (literally understood) limit of, say, triangulated categories is not well-behaved. A well-known example of this is that the derived category of sheaves on a space cannot be recovered from the corresponding categories on an open cover. However, this can be remedied if instead of the triangulated categories we consider their higher categorical enhancements, i.e., the corresponding ∞-categories.

So, what we actually mean by 'limit', is the limit taken in the *∞-category* of *∞-categories*. That is, in the preceding discussion, all our $\text{IndCoh}(-)$ are actually ∞-categories. In our case, they have a bit more structure: they are k-linear over a fixed ground field k; we call them *DG categories*, and denote the ∞-category of such by DGCat.

Thus, ∞-categories inevitably appear in this book.

(b) The indexing (∞)-category appearing in the expression (2.1) is the (∞)-category opposite to that of schemes Z equipped with a map $Z \to \mathcal{X}$ to our inf-scheme $\mathcal{X}$. The transition functors are given by

$$(Z' \xrightarrow{f} Z) \in \text{Sch}_{/\mathcal{X}} \rightsquigarrow \text{IndCoh}(Z) \xrightarrow{f^!} \text{IndCoh}(Z').$$

So, in order for the expression in (2.1) to make sense we need to make the assignment

$$Z \rightsquigarrow \text{IndCoh}(Z), \quad (Z' \xrightarrow{f} Z) \rightsquigarrow (\text{IndCoh}(Z) \xrightarrow{f^!} \text{IndCoh}(Z')) \tag{2.2}$$

into a *functor of ∞-categories*

$$\text{IndCoh}^!_{\text{Sch}} : (\text{Sch})^{\text{op}} \to \text{DGCat}. \tag{2.3}$$

To that end, before we proceed any further, we need to explain what the DG category $\text{IndCoh}(Z)$ is for a scheme Z.

For a scheme Z, the category $\text{IndCoh}(Z)$ will be almost the same as $\text{QCoh}(Z)$. The former is obtained from the latter by a *renormalization* procedure, whose nature we shall now explain.

2.2. Why renormalize? Keeping in mind the examples of $\mathrm{Rep}(G)$ and $\mathfrak{g}$-mod, it is natural to expect that the assignment (2.2) (for schemes, and then also for inf-schemes) should have the following properties:

(i) For every scheme Z, the DG category $\mathrm{IndCoh}(Z)$ should contain infinite direct sums;

(ii) For a map $Z' \xrightarrow{f} Z$, the functor $\mathrm{IndCoh}(Z) \xrightarrow{f^!} \mathrm{IndCoh}(Z')$ should preserve infinite direct sums.

This means that the functor (2.3) takes values in the subcategory of DGCat, where we allow as objects only DG categories satisfying (i)[5] and as 1-morphisms only functors that satisfy (ii)[6].

Let us first try to make this work with the usual QCoh. We refer the reader to Chapter 3, where the DG category $\mathrm{QCoh}(\mathcal{X})$ is introduced for an arbitrary *prestack*, and in particular a scheme. However, for a scheme Z, whatever the DG category $\mathrm{QCoh}(Z)$ is, its homotopy category (which is a triangulated category) is the usual (unbounded) derived category of quasi-coherent sheaves on Z.

Suppose we have a map of schemes $Z' \xrightarrow{f} Z$. The construction of the !-pullback functor

$$f^! : \mathrm{QCoh}(Z) \to \mathrm{QCoh}(Z')$$

is quite complicated, except when f is proper. In the latter case, $f^!$, which from now on we will denote by $f^{!,\mathrm{QCoh}}$, is defined to be the *right adjoint* of

$$f_* : \mathrm{QCoh}(Z') \to \mathrm{QCoh}(Z).$$

The only problem is that the above functor $f^{!,\mathrm{QCoh}}$ *does not* preserve infinite direct sums. The simplest example of a morphism for which this happens is

$$f : \mathrm{Spec}(k) \to \mathrm{Spec}(k[t]/t^2)$$

(or the embedding of a *singular* point into any scheme).

The reason for the failure to preserve infinite direct sums is this: the left adjoint of $f^{!,\mathrm{QCoh}}$, i.e., f_*, *does not* preserve compactness. Indeed, f_* does not necessarily send *perfect complexes* on Z' to perfect complexes on Z, unless f is of finite Tor-dimension[7].

So, our attempt with QCoh fails (ii) above.

2.3. Ind-coherent sheaves on a scheme. The nature of the renormalization procedure that produces $\mathrm{IndCoh}(Z)$ out of $\mathrm{QCoh}(Z)$ is to force (ii) from Sect. 2.2 'by hand'.

[5]Such DG categories are called *cocomplete*.

[6]Such functors are called *continuous*.

[7]We remark that a similar phenomenon, where instead of the category $\mathrm{QCoh}(\mathrm{Spec}(k[t]/t^2)) = k[t]/t^2$-mod we have the category of representations of a finite group, leads to the notion of Tate cohomology: the trivial representation on $\mathbb{Z}$ is *not* a compact object in the category of representations.

As we just saw, the problem with $f^{!,\mathrm{QCoh}}$ was that its left adjoint f_* did not send the corresponding subcategories of perfect complexes to one another. However, f_* sends the subcategory

$$\mathrm{Coh}(Z') \subset \mathrm{QCoh}(Z')$$

to

$$\mathrm{Coh}(Z) \subset \mathrm{QCoh}(Z),$$

where Coh(−) denotes the subcategory of bounded complexes, whose cohomology sheaves are *coherent* (as opposed to quasi-coherent).

The category $\mathrm{IndCoh}(Z)$ is defined as the *ind-completion* of $\mathrm{Coh}(Z)$ (see Chapter 1, Sect. 7.2 for what this means). The functor f_* gives rise to a functor $\mathrm{Coh}(Z') \to \mathrm{Coh}(Z)$, and ind-extending we obtain a functor

$$f_*^{\mathrm{IndCoh}} : \mathrm{IndCoh}(Z') \to \mathrm{IndCoh}(Z).$$

Its right adjoint, denoted $f^! : \mathrm{IndCoh}(Z) \to \mathrm{IndCoh}(Z')$ satisfies (ii) from Sect. 2.2.

Are we done? Far from it. First, we need to define the functor

$$f_*^{\mathrm{IndCoh}} : \mathrm{IndCoh}(Z') \to \mathrm{IndCoh}(Z)$$

for a morphism f that is not necessarily proper. This will not be difficult, and will be done by appealing to t-structures, see Sect. 2.4 below.

What is much more serious is to define $f^!$ for any f. More than that, we need $f^!$ not just for an individual f, but we need the data of (2.2) to be a functor of ∞-categories as in (2.3). Roughly a third of the work in this book goes into the construction of the functor (2.3); we will comment on the nature of this work in Sect. 2.5 and then in Sect. 3 below.

2.4. In what sense is IndCoh a 'renormalization' of QCoh? The tautological embedding $\mathrm{Coh}(Z) \hookrightarrow \mathrm{QCoh}(Z)$ induces, by ind-extension, a functor

$$\Psi_Z : \mathrm{IndCoh}(Z) \to \mathrm{QCoh}(Z).$$

The usual t-structure on the DG category $\mathrm{Coh}(Z)$ induces one on $\mathrm{IndCoh}(Z)$. The key feature of the functor Ψ_Z is that it is *t-exact*. Moreover, for every fixed n, the resulting functor

$$\mathrm{IndCoh}(Z)^{\geq -n} \to \mathrm{QCoh}(Z)^{\geq -n}$$

is an *equivalence*[8]. The reason for this is that any coherent complex can be approximated by a perfect one up to something in $\mathrm{Coh}(Z)^{<-n}$ for any given n.

In other words, the difference between $\mathrm{IndCoh}(Z)$ and $\mathrm{QCoh}(Z)$ occurs 'somewhere at $-\infty$'. So, this difference can only become tangible in the finer questions of homological algebra (such as convergence of spectral sequences).

However, we do need to address such questions adequately if we want to have a functioning theory, and for the kind of applications we have in mind (see Sect. 1.5 above) this necessitates working with IndCoh rather than QCoh.

As an illustration of how the theory of IndCoh takes something very familiar and unravels it to something non-trivial, consider the IndCoh direct image functor.

[8]But the functor Ψ_Z is an equivalence on all of $\mathrm{IndCoh}(Z)$ if *and only if* Z is smooth.

In the case of schemes, for a morphism $f : Z' \to Z$, the functor

$$f_*^{\mathrm{IndCoh}} : \mathrm{IndCoh}(Z') \to \mathrm{IndCoh}(Z)$$

does 'little new' as compared to the usual

$$f_* : \mathrm{QCoh}(Z') \to \mathrm{QCoh}(Z).$$

Namely, f_*^{IndCoh} is the unique functor that preserves infinite direct sums and makes the diagram

$$\begin{array}{ccc} \mathrm{IndCoh}(Z')^{\geq -n} & \xrightarrow[\sim]{\Psi_{Z'}} & \mathrm{QCoh}(Z')^{\geq -n} \\ {\scriptstyle f_*^{\mathrm{IndCoh}}}\big\downarrow & & \big\downarrow{\scriptstyle f_*} \\ \mathrm{IndCoh}(Z)^{\geq -n} & \xrightarrow[\sim]{\Psi_{Z}} & \mathrm{QCoh}(Z)^{\geq -n} \end{array}$$

commute for every n.

However, as was already mentioned, once we extend the formalism of IndCoh direct image to inf-schemes, we will in particular obtain the de Rham direct image functor. So, it is in the world of inf-schemes that IndCoh shows its full strength.

2.5. Construction of the !-pullback functor. As has been mentioned already, a major component of work in this book is the construction of the functor

$$\mathrm{IndCoh}^!_{\mathrm{Sch}} : (\mathrm{Sch})^{\mathrm{op}} \to \mathrm{DGCat}$$

of (2.3).

We already know what $\mathrm{IndCoh}(Z)$ is for an individual scheme. We now need to extend it to morphisms.

For a morphism $f : Z' \to Z$, we can factor it as

$$Z' \overset{f_1}{\to} \overline{Z}' \overset{f_2}{\to} Z, \tag{2.4}$$

where f_1 is an open embedding and f_2 is proper. We then define

$$f^! : \mathrm{IndCoh}(Z) \to \mathrm{IndCoh}(Z')$$

to be

$$f_1^! \circ f_2^!,$$

where

(i) $f_2^!$ is the *right adjoint* of $(f_2)_*^{\mathrm{IndCoh}}$;

(ii) $f_1^!$ is the *left adjoint* of $(f_1)_*^{\mathrm{IndCoh}}$.

Of course, in order to have $f^!$ as a well-defined functor, we need to show that its definition is independent of the factorization of f as in (2.4). Then we will have to show that the definition is compatible with compositions of morphisms. But this is only the tip of the iceberg.

Since we want to have a functor between ∞-categories, we need to supply the assignment

$$f \rightsquigarrow f^!$$

with a *homotopy-coherent* system of compatibilities for n-fold compositions of morphisms, a task which appears infeasible to do 'by hand'.

What we do instead is we prove an existence and uniqueness theorem... not for (2.3), but rather for a more ambitious piece of structure. We refer the reader to Chapter 5, Proposition 2.1.4 for the precise formulation. Here we will only say that, in addition to (2.3), this structure contains the data of a functor

$$\text{IndCoh} : \text{Sch} \to \text{DGCat}, \tag{2.5}$$

$$Z \rightsquigarrow \text{IndCoh}(Z), \quad (Z' \xrightarrow{f} Z) \rightsquigarrow (\text{IndCoh}(Z') \xrightarrow{f_*^{\text{IndCoh}}} \text{IndCoh}(Z)),$$

as well as *compatibility* between (2.3) and (2.5).

The latter means that whenever we have a Cartesian square

$$\begin{array}{ccc} Z_1' & \xrightarrow{g'} & Z' \\ {\scriptstyle f_1}\downarrow & & \downarrow{\scriptstyle f} \\ Z_1 & \xrightarrow{g} & Z \end{array} \tag{2.6}$$

there is a canonical isomorphism of functors, called base change:

$$(f_1)_*^{\text{IndCoh}} \circ (g')^! \simeq g^! \circ f_*^{\text{IndCoh}}. \tag{2.7}$$

2.6. Enter DAG. The appearance of the Cartesian square (2.6) heralds another piece of 'bad news'. Namely, Z_1' must be the fiber product

$$Z_1 \underset{Z}{\times} Z'.$$

But what category should we take this fiber product in? If we look at the example

$$\begin{array}{ccc} \text{pt} \underset{\mathbb{A}^1}{\times} \text{pt} & \longrightarrow & \text{pt} \\ \downarrow & & \downarrow \\ \text{pt} & \longrightarrow & \mathbb{A}^1, \end{array}$$

(here $\text{pt} = \text{Spec}(k)$, $\mathbb{A}^1 = \text{Spec}(k[t])$), we will see that the fiber product $\text{pt} \underset{\mathbb{A}^1}{\times} \text{pt}$ *cannot* be taken to be the point-scheme, i.e., it cannot be the fiber product in the category of usual (=classical) schemes. Rather, we need to take

$$\text{pt} \underset{\mathbb{A}^1}{\times} \text{pt} = \text{Spec}(k \underset{k[t]}{\otimes} k),$$

where the tensor product is understood in the *derived* sense, i.e.,

$$k \underset{k[t]}{\otimes} k = k[\epsilon], \quad \deg(\epsilon) = -1.$$

This is to say that in building the theory of IndCoh, we cannot stay with classical schemes, but rather need to enlarge our world to that of *derived algebraic geometry.*

So, unless the reader has already guessed this, in all the previous discussion, the word 'scheme' had to be understood as 'derived scheme'[9] (although in the main body of the book we say just 'scheme', because everything is derived).

However, this is not really 'bad news'. Since we are already forced to work with ∞-categories, passing from classical algebraic geometry to DAG does not add

[9]Technically, for whatever has to do with IndCoh, we need to add the adjective 'laft'='locally almost of finite type', see Chapter 2, Sect. 3.5 for what this means.

a new level of complexity. But it does add *a lot* of new techniques, for example in anything that has to do with deformation theory (see Volume II, Chapter 1).

Moreover, many objects that appear in geometric representation theory naturally belong to DAG (e.g., Springer fibers, moduli of local systems on a curve, moduli of vector bundles on a surface). That is, these objects are *not* classical, i.e., we cannot ignore their derived structure if we want to study their scheme-theoretic (as opposed to topological) properties. So, we would have wanted to do DAG in any case.

Here are two particular examples:

(I) Consider the category of D-modules (resp., perverse) sheaves on the double quotient

$$I\backslash G((t))/I,$$

where G is a connected reductive group, $G((t))$ is the corresponding loop group (considered as an ind-scheme) and $I \subset G((t))$ is the Iwahori subgroup. Then Bezrukavnikov's theory (see [**Bez**]) identifies this category with the category of ad-equivariant ind-coherent (resp., coherent) sheaves on the *Steinberg scheme* (for the Langlands dual group). But what do we mean by the Steinberg scheme? By definition, this is the fiber product

$$\widetilde{\mathcal{N}} \underset{\mathfrak{g}}{\times} \widetilde{\mathcal{N}}, \tag{2.8}$$

where $\widetilde{\mathcal{N}}$ is the Springer resolution of the nilpotent cone. However, in order for this equivalence to hold, the fiber product in (2.8) needs be understood in the *derived* sense.

(II) Let X be a smooth and complete curve. Let $\mathrm{Pic}(X)$ be the *Picard stack* of X, i.e., the stack parameterizing line bundles on X. Let $\mathrm{LocSys}(X)$ be the stack parameterizing 1-dimensional local systems on X. The Fourier-Mukaï-Laumon transform defines an equivalence

$$\mathrm{Dmod}(\mathrm{Pic}(X)) \simeq \mathrm{QCoh}(\mathrm{LocSys}(X)).$$

However, in order for this equivalence to hold, we need to understand $\mathrm{LocSys}(X)$ as a *derived stack.*

2.7. Back to inf-schemes. The above was a somewhat lengthy detour into the constructions of the theory of IndCoh on schemes. Now, if $\mathcal{X}$ is an inf-scheme, the category $\mathrm{IndCoh}(\mathcal{X})$ is defined by the formula (2.1).

Thus, informally, an object $\mathcal{F} \in \mathrm{IndCoh}(\mathcal{X})$ is a family of assignments

$$(Z \overset{x}{\to} \mathcal{X}) \rightsquigarrow \mathcal{F}_{Z,x} \in \mathrm{IndCoh}(Z)$$

(here Z is a scheme) plus

$$(Z' \overset{f}{\to} Z) \in \mathrm{Sch}_{/\mathcal{X}} \rightsquigarrow f^!(\mathcal{F}_{Z,x}) \simeq \mathcal{F}_{Z',x'},$$

along with a homotopy-coherent compatibility data for compositions of morphisms.

For a map $g : \mathcal{X}' \to \mathcal{X}$, the functor

$$g^! : \mathrm{IndCoh}(\mathcal{X}) \to \mathrm{IndCoh}(\mathcal{X}')$$

is essentially built into the construction. Recall, however, that our goal is to also have the functor

$$g_*^{\mathrm{IndCoh}} : \mathrm{IndCoh}(\mathcal{X}') \to \mathrm{IndCoh}(\mathcal{X}).$$

The construction of the latter requires some work (which occupies most of Volume II, Chapter 3). What we show is that there exists a unique system of such functors such that for every commutative (but not necessarily Cartesian) diagram

$$\begin{array}{ccc} Z' & \xrightarrow{i'} & \mathcal{X}' \\ {\scriptstyle f}\downarrow & & \downarrow{\scriptstyle g} \\ Z & \xrightarrow{i} & \mathcal{X} \end{array}$$

with Z, Z' being schemes and the morphisms i, i' proper, we have an isomorphism

$$g_*^{\mathrm{IndCoh}} \circ (i')_*^{\mathrm{IndCoh}} \simeq i_*^{\mathrm{IndCoh}} \circ f^{\mathrm{IndCoh}},$$

where i_*^{IndCoh} (resp., $(i')_*^{\mathrm{IndCoh}}$) is the left adjoint of $i^!$ (resp., $(i')^!$).

Amazingly, this procedure contains the de Rham push-forward functor as a particular case.

3. What is actually done in this book?

This book consists of two volumes. The first volume consists of three parts and an appendix and the second volume consists of two parts. Each part consists of several chapters. The chapters are designed so that they can be read independently from one another (in a sense, each chapter is structured as a separate paper with its own introduction that explains what this particular chapter does).

Below we will describe the contents of the different parts and chapters from several different perspectives: (a) goals and role in the overall project; (b) practical implications; (c) nature of work; (d) logical dependence.

3.1. The contents of the different parts.

Volume I, Part I is called 'preliminaries', and it is really preliminaries.

Volume I, Part II builds the theory of IndCoh on schemes.

Volume I, Part III develops the formalism of *categories of correspondences*; it is used as a 'black box' in the key constructions in Volume I, Part II and Volume II, Part I: this is our tool of bootstrapping the theory of IndCoh out of a much smaller amount of data.

Volume I, Appendix provides a sketch of the theory of $(\infty, 2)$-categories, which, in turn, is crucially used in Volume I, Part III.

Volume II, Part I defines the notion of *inf-scheme* and extends the formalism of IndCoh from schemes to inf-schemes, and in that it achieves one of the two main goals of this book.

Volume II, Part II consists of applications of the theory of IndCoh: we consider formal moduli problems, Lie theory and infinitesimal differential geometry; i.e., exactly the things one needs for geometric representation theory. Making these constructions available is the second of our main goals.

3.2. Which chapters should a practically minded reader be interested in? Not all the Chapters in this book make an enticing read; some are downright technical and tedious. Here is, however, a description of the 'cool' things that some of the Chapters do:

None of the material in Volume I, Part I alters the pre-existing state of knowledge.

Volume I, Chapters 4 and 5 should not be a difficult read. They construct the theory of IndCoh on schemes (the hard technical work is delegated to Volume I, Chapter 7). The reader cannot avoid reading these chapters if he/she is interested in the applications of IndCoh: one has to have an idea of what IndCoh is in order to use it.

Volume I, Chapters 6 is routine. The only really useful thing from it is the functor

$$\Upsilon_Z : \mathrm{QCoh}(Z) \to \mathrm{IndCoh}(Z),$$

given by tensoring an object of QCoh(Z) with the dualizing complex

$$\omega_Z \in \mathrm{IndCoh}(Z).$$

Extract this piece of information from Sects. 3.2-3.3 and move on.

Volume I, Chapter 7 introduces the formalism of correspondences. The idea of the category of correspondences is definitely something worth knowing. We recommend the reader to read Sect. 1 in its entirety, then understand the universal property stated in Sect. 3, and finally get an idea about the two extension theorems, proved in Sects. 4 and 5, respectively. These extension theorems are the mechanism by means of which we construct IndCoh as a functor out of the category of correspondences in Volume I, Chapter 5.

Volume I, Chapter 8 proves a rather technical extension theorem, stated in Sect. 1; we do not believe that the reader will gain much by studying its proof. This theorem is key to the extension of IndCoh from schemes to inf-schemes in Volume II, Chapter 3.

Volume I, Chapter 9 is routine, except for one observation, contained in Sects. 2.2-2.3: the natural involution on the category of correspondences encodes *duality*. In fact, this is how we construct Serre duality on IndCoh(Z) and Verdier duality on Dmod(Z) where Z is a scheme (or inf-scheme), see Chapter 5, Sect. 4.2, Volume II, Chapter 3, Sect. 6.2, and Volume II, Chapter 4, Sect. 2.2, respectively.

Volume I, Chapter 10 introduces the notion of $(\infty, 2)$-category and some basic constructions in the theory of $(\infty, 2)$-categories. This Chapter is not very technical (mainly because it omits most proofs) and might be of independent interest.

Volume I, Chapter 11 does a few more technical things in the theory of $(\infty, 2)$-categories. It introduces the $(\infty, 2)$-*category of* $(\infty, 2)$-*categories*, denoted 2-Cat. We then discuss the straightening/unstraightening procedure in the $(\infty, 2)$-categorical context and the $(\infty, 2)$-categorical Yoneda lemma. The statements of the results from this Chapter may be of independent interest.

Volume I, Chapter 12 discusses the notion of adjunction in the context of $(\infty, 2)$-categories. The main theorem in this Chapter explicitly constructs the *universal adjointable functor* (and its variants), and we do believe that this is of interest beyond the particular goals of this book.

Volume II, Chapter 1 is background on deformation theory. The reason it is included in the book is that the notion of inf-scheme is based on deformation theory. However, the reader may find the material in Sects. 1-7 of this Chapter useful without any connection to the contents of the rest of the book.

Volume II, Chapter 2 introduces inf-schemes. It is quite technical. So, the practically minded reader should just understand the definition (Sect. 3.1) and move on.

Volume II, Chapter 3 bootstraps the theory of IndCoh from schemes to inf-schemes. It is not too technical, and should be read (for the same reason as Volume I, Chapters 4 and 5). The hard technical work is delegated to Volume I, Chapter 8.

Volume II, Chapter 4 explains how the theory of crystals/D-modules follows from the theory of IndCoh on inf-schemes. Nothing in this Chapter is very exciting, but it should not be a difficult read either.

Volume II, Chapter 5 is about formal moduli problems. It proves a pretty strong result, namely, the equivalence of categories between formal groupoids acting on a given prestack $\mathcal{X}$ (assumed to admit deformation theory) and formal moduli problems *under* $\mathcal{X}$.

Volume II, Chapter 6 is a digression on the general notion of Lie algebra and Koszul duality in a symmetric monoidal DG category. It gives a nice interpretation of the universal enveloping algebra of a Lie algebra of $\mathfrak{g}$ as the homological Chevalley complex of the Lie algebra obtained by *looping* $\mathfrak{g}$. The reader may find this Chapter useful and independently interesting.

Volume II, Chapter 7 develops Lie theory in the context of inf-schemes. Namely, it establishes an equivalence of categories between group inf-schemes (over a given base $\mathcal{X}$) and Lie algebras in IndCoh($\mathcal{X}$). One can regard this result as one of the main applications of the theory developed hereto.

Volume II, Chapters 8 and 9 use the theory developed in the preceding Chapters for 'differential calculus' in the context of DAG. We discuss Lie algebroids and their universal envelopes, the procedure of deformation to the normal cone, etc. For example, the notion of n-th infinitesimal neighborhood developed in Volume II, Chapter 9 gives rise to the Hodge filtration.

3.3. The nature of the technical work. The substance of mathematical thought in this book can be roughly split into three modes of cerebral activity: (a) making constructions; (b) overcoming difficulties of homotopy-theoretic nature; (c) dealing with issues of *convergence*.

Mode (a) is hard to categorize or describe in general terms. This is what one calls 'the fun part'.

Mode (b) is something much better defined: there are certain constructions that are obvious or easy for ordinary categories (e.g., define categories or functors by an explicit procedure), but require some ingenuity in the setting of higher categories. For many readers that would be the least fun part: after all it is clear that the thing should work, the only question is how to make it work without spending another 100 pages.

Mode (c) can be characterized as follows. In low-tech terms it consists of showing that certain spectral sequences converge. In a language better adapted for our needs, it consists of proving that in some given situation we can swap a limit and a colimit (the very idea of IndCoh was born from this mode of thinking). One can say that mode (c) is a sort of analysis within algebra. Some people find it fun.

Here is where the different Chapters stand from the point of view of the above classification:

Volume I, Chapter 1 is (b) and a little of (c).

Volume I, Chapter 2 is (a) and a little of (c).

Volume I, Chapter 3 is (c).

Volume I, Chapter 4 is (a) and (c).

Volume I, Chapter 5 is (a).

Volume I, Chapter 6 is (b).

Volume I, Chapters 7-9 are (b).

Volume I, Chapters 10-12 are (b).

Volume II, Chapter 1 is (a) and a little of (c).

Volume II, Chapter 2 is (a) and a little of (c).

Volume II, Chapter 3 is (a).

Volume II, Chapter 4 is (a).

Volume II, Chapter 5 is (a).

Volume II, Chapter 6 is (c) and a little of (b).

Volume II, Chapter 7 is (c) and a little of (a).

Volume II, Chapters 8 and 9 are (a).

3.4. Logical dependence of chapters. This book is structured so that Volume I prepares the ground and Volume II reaps the fruit. However, below is a scheme of the logical dependence of chapters, where we allow a 5% *skip margin* (by which we mean that the reader skips certain things[10] and comes back to them when needed).

3.4.1. Volume I, Chapter 1 reviews ∞-categories and higher algebra. Read it only if you have no prior knowledge of these subjects. In the latter case, here is what you will need in order to understand the constructions in the main body of the book:

Read Sects. 1-2 to get an idea of how to operate with ∞-categories (this is a basis for everything else in the book).

Read Sects. 5-7 for a summary of *stable* ∞-categories: this is what our QCoh$(-)$ and IndCoh$(-)$ are; forget on the first pass about the additional structure of k-linear DG category (the latter is discussed in Sect. 10).

[10] These are things that can be taken on faith without compromising the overall understanding of the material.

Read Sects. 3-4 for a summary of monoidal structures and duality in the context of higher category theory. You will need it for this discussion of Serre duality and for Volume I, Chapter 6.

Sects. 8-9 are about algebra in (symmetric) monoidal stable ∞-categories. You will need it for Volume II, Part II of the book.

Volume I, Chapter 2 introduces DAG proper. If you have not seen any of it before, read Sect. 1 for the (shockingly general, yet useful) notion of prestack. Every category of geometric objects we will encounter in this book (e.g., (derived) schemes, Artin stacks, inf-schemes, etc.) will be a full subcategory of the ∞-category of prestacks. Proceed to Sect. 3.1 for the definition of derived schemes. Skip all the rest.

Volume I, Chapter 3 introduces QCoh on prestacks. Even though the main focus of this book is the theory of ind-coherent sheaves, the latter theory takes a significant input and interacts with that of quasi-coherent sheaves. If you have not seen this before, read Sect. 1 and then Sects. 3.1-3.2.

3.4.2. In Volume I, Chapter 4 we develop the elementary aspects of the theory of IndCoh on schemes: we define the DG category $\text{IndCoh}(Z)$ for an individual scheme Z, construct the IndCoh direct image functor, and also the !-pullback functor for proper morphisms. This Chapter uses the material from Volume I, Part I mentioned above. You will need the material from this chapter in order to proceed with the reading of the book.

Volume I, Chapter 5 builds on Volume 1, Chapter 4, and accomplishes (modulo the material delegated to Volume I, Chapter 7) one of the main goals of this book. We construct IndCoh as a *functor out of the category of correspondences*. In particular, we construct the functor (2.3). The material from this Chapter is also needed for the rest of the book.

In Volume I, Chapter 6 we study the interaction between IndCoh and QCoh. For an individual scheme Z we have an action of $\text{QCoh}(Z)$ (viewed as a monoidal category) on $\text{IndCoh}(Z)$. We study how this action interacts with the formalism of correspondences from Volume I, Chapter 5, and in particular with the operation of !-pullback. The material in this Chapter uses the formalism of monoidal categories and modules over them from Volume I, Chapter 1, as well as the material from Volume I, Chapter 5. Skipping Volume I, Chapter 6 will not impede your understanding of the rest of the book, so it might be a good idea to do so on the first pass.

3.4.3. Volume I, Part II develops the theory of categories of correspondences. It plays a service role for Volume I, Chapter 6 and Volume II, Chapter 3, and relies on the theory of $(\infty, 2)$-categories, developed in Volume I, Appendix.

3.4.4. Volume I, Appendix develops the theory of $(\infty, 2)$-categories. It plays a service role for Volume I, Part III.

Volume I, Chapters 11 and 12 rely on Volume I, Chapter 10, but can be read independently of one another.

3.4.5. Volume II, Chapter 1 introduces deformation theory. It is needed for the definition of inf-schemes and, therefore, for proofs of any results about inf-schemes (that is, for Volume II, Chapter 2). We will also need it for the discussion of formal moduli problems in Volume II, Chapter 5. The prerequisites for Volume II, Chapter 1 are Volume I, Chapters 2 and 3, so it is (almost)[11] independent of the material from Volume I, Part II.

In Volume II, Chapter 2 we introduce inf-schemes and some related notions (ind-schemes, ind-inf-schemes). The material here relies in that of Volume II, Chapter 1, and will be needed in Volume II, Chapter 3.

In Volume II, Chapter 3 we construct the theory of IndCoh on inf-schemes. The material here relies on that from Volume I, Chapter 5 and Volume II, Chapter 2 (and also a tedious general result about correspondences from Volume I, Chapter 8). Thus, Volume II, Chapter 3 achieves one of our goals, the later being making the theory of IndCoh on inf-schemes available. The material from Volume II, Chapter 3 will (of course) be used when will apply the theory of IndCoh, in Volume II, Chapter 4 and 7–9.

In Volume II, Chapter 4 we apply the material from Volume II, Chapter 3 in order to develop a proper framework for crystals (=D-modules), together with the forgetful/induction functors that related D-modules to $\mathcal{O}$-modules. The material from this Chapter will not be used later, except for the extremely useful notion of the de Rham prestack construction $\mathcal{X} \rightsquigarrow \mathcal{X}_{\mathrm{dR}}$.

3.4.6. In Volume II, Chapter 5 we prove a key result that says that in the category of prestacks that admit deformation theory, the operation of taking the quotient with respect to a formal groupoid is well-defined. The material here relies on that from Volume II, Chapter 1 (at some point we appeal to a proposition from Volume II, Chapter 3, but that can be avoided). So, the main result from Volume II, Chapter 5 is independent of the discussion of IndCoh.

Volume II, Chapter 6 is about Lie algebras (or more general operad algebras) in symmetric monoidal DG categories. It only relies on the material from Volume I, Chapter 1, and is independent of the preceding Chapters of the book (no DAG, no IndCoh). The material from this Chapter will be used for the subsequent Chapters in Volume II, Part II.

3.4.7. *A shortcut.* As has been mentioned earlier, Volume II, Chapters 7–9 are devoted to applications of IndCoh to 'differential calculus'. This 'differential calculus' occurs on prestacks that admit deformation theory.

If one really wants to use arbitrary such prestacks, one needs the entire machinery of IndCoh provided by Volume II, Chapter 3. However, if one is content with

[11]Whenever we want to talk about *tangent* (as opposed to *cotangent*) spaces, we have to use IndCoh rather than QCoh, and these parts in Volume II, Chapter 1 use the material from Volume I, Chapter 5.

working with inf-schemes (which would suffice for the majority of applications), much less machinery would suffice:

The cofinality result from Volume II, Chapter 3, Sect. 4.3 implies that we can bypass the entire discussion of correspondences, and only use the material from Volume I, Chapter 4, i.e., IndCoh on schemes and !-pullbacks for proper (in fact, finite) morphisms.

3.4.8. Volume II, Chapters 7-9 form a logical succession. As input from the preceding chapters they use Volume II, Chapter 3 (resp., Volume I, Chapter 5 (see Sect. 3.4.7 above), Volume II, Chapter 1 and Volume II, Chapters 5–6.

Acknowledgements

In writing this book, we owe a particular debt of gratitude to Jacob Lurie for teaching us both the theory of higher categories and derived algebraic geometry. Not less importantly, some of the key constructions in this book originated from his ideas; among them is the concept of the category of correspondences.

We would like to thank V. Drinfeld, J. Francis and S. Raskin for numerous illuminating discussions and suggestions related to this book.

We are grateful to A. Khan and S. Lysenko for pointing out multiple errors as this book was evolving.

Introduction

In describing the contents of Volume I we will use some terminology pertaining to higher category theory, derived algebraic geometry and the (derived) category of quasi-coherent sheaves. The reader is referred to the part of this book, called Preliminaries, where the relevant notions are surveyed.

1. Ind-coherent sheaves

The goal of Volume I is to set up the machinery of ind-coherent sheaves on (derived) schemes, in order to apply it in Volume II and describe algebro-geometrically categories and functors that naturally arise in representation theory.

1.1. How do ind-coherent sheaves arise? We start the development of the theory of ind-coherent sheaves in Chapter 4. The idea is the following:

Given a (derived) scheme X (assumed *almost of finite type*), the usual DG category $\mathrm{QCoh}(X)$ of quasi-coherent sheaves on X can be realized as the ind-completion of its full subcategory $\mathrm{QCoh}(X)^{\mathrm{perf}}$ of *perfect* objects.

The category $\mathrm{IndCoh}(X)$ is defined to be the ind-completion of another subcategory of $\mathrm{QCoh}(X)$, namely $\mathrm{Coh}(X)$ that consists of objects that are cohomologically bounded (i.e., have non-zero cohomologies only in finitely many degrees) and all of whose cohomologies are *coherent* as sheaves on the classical scheme underlying X.

The first question is: why should we consider such a thing? In the next few subsections we will try to provide an answer.

1.1.1. One motivation for the theory of ind-coherent sheaves is to have a robust theory of Grothendieck-Serre duality. In particular, we would like to have a well behaved exceptional inverse image functor for arbitrary maps of schemes (as well as more general prestacks). Moreover, for the needs of representation theory, we would like to study $\mathcal{O}$-modules on algebro-geometric objects much more general than schemes. In fact, we want to consider all prestacks (locally almost of finite type). A particularly important class of prestacks is that of inf-schemes, a notion that will be introduced in Volume II. The theory of ind-coherent sheaves addresses both of these concerns.

To simplify the discussion, let $\mathcal{X}$ be an *ind-scheme*, i.e., a filtered colimit of schemes

$$\mathcal{X} = \underset{i}{\mathrm{colim}}\, X_i,$$

where the transition maps $X_i \xrightarrow{f_{i,j}} X_j$ are closed embeddings.

We would like to have a version of the category of $\mathcal{O}$-modules on $\mathcal{X}$ which is the *colimit* of the corresponding categories on the X_i's, where the transition functors are given by taking *direct images* with respect to the $f_{i,j}$'s. I.e., morally, an $\mathcal{O}$-module on $\mathcal{X}$ is a union of its submodules supported on the X_i's.

Let us try to interpret the category of $\mathcal{O}$-modules as QCoh(−) and see what we get. If we apply the definition, we obtain

$$\mathrm{QCoh}(\mathcal{X}) \simeq \lim_i \mathrm{QCoh}(X_i), \quad (i \to j) \rightsquigarrow (\mathrm{QCoh}(X_j) \xrightarrow{f_{i,j}^*} \mathrm{QCoh}(X_i)),$$

i.e., our category on $\mathcal{X}$ is the *limit* of the corresponding categories on the X_i's with respect to pullbacks (rather than the *colimit* of the same categories with respect to pushforwards).

So, the category that we seek on $\mathcal{X}$ is not the usual $\mathrm{QCoh}(\mathcal{X})$. Let us, however, try something else: let us try to 'force' the definition as a colimit, while still using QCoh(−) on the X_i's as building blocks. I.e., consider the category

$$\operatorname*{colim}_i \mathrm{QCoh}(X_i), \quad (i \to j) \rightsquigarrow (\mathrm{QCoh}(X_i) \xrightarrow{(f_{i,j})_*} \mathrm{QCoh}(X_j)). \tag{1.1}$$

The above gives a well-defined category, but the problem is that it may be quite ill-behaved. Namely, one can formally rewrite the above *colimit* as a *limit*,

$$\lim_i \mathrm{QCoh}(X_i), \quad (i \to j) \rightsquigarrow (\mathrm{QCoh}(X_j) \xrightarrow{f_{i,j}^{!,\mathrm{QCoh}}} \mathrm{QCoh}(X_i)),$$

where $f_{i,j}^{!,\mathrm{QCoh}}$ is the functor *right* adjoint to $(f_{i,j})_*$. The problem is caused by the potential bad behavior of the functors $f_{i,j}^{!,\mathrm{QCoh}}$.

1.1.2. Let us isolate the problem. Let $f : X \to Y$ be a closed embedding (or, more generally, a proper map). We have the usual direct image functor

$$f_* : \mathrm{QCoh}(X) \to \mathrm{QCoh}(Y),$$

and it follows formally from Lurie's Adjoint Functor Theorem that this functor admits a right adjoint, denoted

$$f^{!,\mathrm{QCoh}} : \mathrm{QCoh}(Y) \to \mathrm{QCoh}(X).$$

The trouble is, however, that the above functor $f^{!,\mathrm{QCoh}}$ may be ill-behaved. Technically, 'ill-behaved' means that it may fail to be *continuous* (i.e., preserve colimits).

One can ask further: why is non-contuniuty a problem? The answer to this is that the world that we would like to work in is that of DG categories that are *cocomplete*, and continuous functors between them. The reason for the latter is that in this world we have a well-defined operation of tensor product of DG categories

$$\mathbf{C}, \mathbf{D} \rightsquigarrow \mathbf{C} \otimes \mathbf{D}. \tag{1.2}$$

I.e., this is the world in which we can really 'do algebra', which is exactly what we want to do in Volume II, with a view to applications to representation theory.

In addition, it is this world in which it is most convenient to talk about duality, which will be discussed in the sequel.

1.1.3. Now, the obstruction to the functor $f^{!,\mathrm{QCoh}}$ being continuous is that its left adjoint, namely, f_* does not *preserve compactness*, i.e., it does not necessarily send $\mathrm{QCoh}(X)^{\mathrm{perf}}$ to $\mathrm{QCoh}(Y)^{\mathrm{perf}}$. However, it does send $\mathrm{Coh}(X)$ to $\mathrm{Coh}(Y)$, by virtue of properness. Therefore, the right adjoint to the corresponding functor

$$f_*^{\mathrm{IndCoh}} : \mathrm{IndCoh}(X) \to \mathrm{IndCoh}(Y),$$

denoted

$$f^! : \mathrm{IndCoh}(Y) \to \mathrm{IndCoh}(X)$$

is continuous.

So, replacing $\mathrm{QCoh}(-)$ by $\mathrm{IndCoh}(-)$ fixes the bug of non-continuity of $f^{!,\mathrm{QCoh}}$.

1.1.4. In particular, returning to the case of an ind-scheme $\mathcal{X}$, we can define $\mathrm{IndCoh}(X)$ as

$$\underset{i}{\mathrm{colim}}\, \mathrm{IndCoh}(X_i), \quad (i \to j) \rightsquigarrow (\mathrm{IndCoh}(X_i) \overset{(f_{i,j})_*^{\mathrm{IndCoh}}}{\longrightarrow} \mathrm{IndCoh}(X_j)), \tag{1.3}$$

and thus obtain a reasonable category, which we can also write as

$$\underset{i}{\lim}\, \mathrm{QCoh}(X_i), \quad (i \to j) \rightsquigarrow (\mathrm{IndCoh}(X_j) \overset{f_{i,j}^!}{\longrightarrow} \mathrm{IndCoh}(X_i)), \tag{1.4}$$

The above category matches exactly the needs of representation theory and one that we will use.

1.2. What does the theory of ind-coherent sheaves consist of? Let us now take $\mathcal{X}$ to be a general prestack (locally almost of finite type). We would like to define the category $\mathrm{IndCoh}(\mathcal{X})$ that reproduces the answer given above in the case when $\mathcal{X}$ is an ind-scheme. However, we no longer expect that $\mathrm{IndCoh}(\mathcal{X})$ could be written as a *colimit*. But we can try to approach $\mathrm{IndCoh}(\mathcal{X})$ as a *limit*, so that in the case of ind-schemes, we recover (1.4).

Thus, we would like to define

$$\mathrm{IndCoh}(\mathcal{X}) := \underset{X_i \to \mathcal{X}}{\lim}\, \mathrm{IndCoh}(X_i), \tag{1.5}$$

where the limit is taken over the category of *all* schemes (almost of finite type) mapping to $\mathcal{X}$, and where the transition functors $\mathrm{IndCoh}(X_j) \to \mathrm{IndCoh}(X_i)$ are given by

$$(X_i \overset{f_{i,j}}{\to} X_j) \rightsquigarrow (\mathrm{IndCoh}(X_j) \overset{f_{i,j}^!}{\longrightarrow} \mathrm{IndCoh}(X_j).$$

But we now face a new problem: the maps $f_{i,j}$ are no longer closed embeddings (or proper); they are arbitrary maps between schemes (almost of finite type). So, we need the definition of the functor

$$f^! : \mathrm{IndCoh}(Y) \to \mathrm{IndCoh}(X)$$

in the case of an arbitrary map $X \overset{f}{\to} Y$.

Moreover, in order for the limit (1.5) to make sense in the world of higher categories, we need the assignment

$$X \rightsquigarrow \mathrm{IndCoh}(X), \ (X \overset{f}{\to} Y) \rightsquigarrow (\mathrm{IndCoh}(Y) \overset{f^!}{\to} \mathrm{IndCoh}(X))$$

be a functor from the category opposite to that of schemes almost of finite type to that of DG categories and continuous functors:

$$\text{IndCoh}^!_{\text{Sch}_{\text{aft}}} : (\text{Sch}_{\text{aft}})^{\text{op}} \to \text{DGCat}_{\text{cont}} . \tag{1.6}$$

1.2.1. The problem is that for an arbitrary map f, the functor $f^!$ is not adjoint to anything. However, we can factor f as a composition $f_1 \circ f_2$, where f_2 is an open embedding, and f_1 is a proper morphism, and define

$$f^! := f_2^! \circ f_1^!,$$

where $f_1^!$ is the right adjoint to $(f_1)_*^{\text{IndCoh}}$, and $f_2^!$ is just restriction.

If we were to realize this idea, we would have to show that the above definition of $f^!$ does not depend on the factorization f as $f_1 \circ f_2$, and moreover that it upgrades to a functor (1.6). This can be handled explicitly if our target was an ordinary category (rather than $\text{DGCat}_{\text{cont}}$), but in the world of higher categories we will have to extract (1.6) using the (somewhat constrained) toolbox of constructions that produce functors from already existing ones.

1.2.2. Suppose, nevertheless, that we have constructed the functor (1.6). We may (and do) want more, however: for a *schematic* map between prestacks $g : \mathcal{X} \to \mathcal{Y}$ we want to have the direct image functor

$$g_*^{\text{IndCoh}} : \text{IndCoh}(\mathcal{X}) \to \text{IndCoh}(\mathcal{Y}),$$

determined by the requirement that for a scheme Y and a map $f_Y : Y \to \mathcal{Y}$, for the Cartesian square

$$\begin{array}{ccc} X & \xrightarrow{f_X} & \mathcal{X} \\ {\scriptstyle g'}\downarrow & & \downarrow{\scriptstyle g} \\ Y & \xrightarrow{f_Y} & \mathcal{Y}, \end{array}$$

we have an isomorphism of functors

$$f_Y^! \circ g_*^{\text{IndCoh}} \simeq (g')_*^{\text{IndCoh}} \circ f_X^!. \tag{1.7}$$

In order for this to happen, at the very least, we need an analogous property for maps between schemes. I.e., we want that for a Cartesian diagram of schemes

$$\begin{array}{ccc} X' & \xrightarrow{f_X} & X \\ {\scriptstyle g'}\downarrow & & \downarrow{\scriptstyle g} \\ Y' & \xrightarrow{f_Y} & Y, \end{array}$$

there exists a canonical isomorphism of functors

$$f_Y^! \circ g_*^{\text{IndCoh}} \simeq (g')_*^{\text{IndCoh}} \circ f_X^!. \tag{1.8}$$

However, in order for the isomorphisms (1.8) to give rise to (1.7), the isomorphisms (1.8) themselves must be functorial with respect to compositions of the maps f and g.

Such a functoriality is easy to spell out in the world of ordinary categories, but it becomes a non-trivial problem when we are dealing with higher categories (more precisely, when the target category, which in our case is $\text{DGCat}_{\text{cont}}$, is a higher category).

1.2.3. This brings us to the idea of the *category of correspondences*, discussed below, and to which we devoted Part III of this volume.

In Chapter 5 we prove the existence and uniqueness of IndCoh as a functor out of the category of correspondences. Moreover, it turns out that this extended formalism (rather than just the functor (1.6)) is a very natural way to construct the functor (1.6) itself.

So, the formalism of correspondences is not only needed in order to extend IndCoh to prestacks, but a necessity for the construction of the !-pullback on schemes.

1.3. Before we pass to the discussion of the formalism of correspondences, let us mention the role of Chapter 6 of this volume. In that Chapter, we undertake a systematic study of the relationship between QCoh(−) and IndCoh(−), when both are viewed as functors

$$(\mathrm{Sch}_{\mathrm{aft}})^{\mathrm{op}} \to \mathrm{DGCat}_{\mathrm{cont}}.$$

The upshot is that both have a natural symmetric monoidal structure, where the symmetric monoidal structure on $(\mathrm{Sch}_{\mathrm{aft}})^{\mathrm{op}}$ is given by the Cartesian product of schemes, and on $\mathrm{DGCat}_{\mathrm{cont}}$ by (1.2).

Moreover, there is a natural transformation between the above two functors, denoted Υ. For an individual scheme X, the corresponding functor

$$\Upsilon_X : \mathrm{QCoh}(X) \to \mathrm{IndCoh}(X),$$

given by tensoring the dualizing object $\omega_X \in \mathrm{IndCoh}(X)$ by an object of $\mathrm{QCoh}(X)$.

2. Correspondences and the six functor formalism

For our purposes, the main function of the category of correspondences is to encode all of the data of Grothendieck's six functor formalism.

Let $\mathbf{C}$ be an $(\infty,1)$-category with Cartesian product, and let $\mathbb{S}$ be a target $(\infty,1)$-category. The role of the category of correspondences $\mathrm{Corr}(\mathbf{C})$ is to encode a 'bivariant' functor from $\mathbf{C}$ to $\mathbb{S}$. The example that one should keep in mind is $\mathbf{C} = \mathrm{Sch}_{\mathrm{aft}}$, $\mathbb{S} = \mathrm{DGCat}_{\mathrm{cont}}$ and the functor in question is IndCoh, where we consider both the !-pullback and *-push forward. Namely, we will see that functors $\mathrm{Corr}(\mathbf{C}) \to \mathbb{S}$ will exactly correspond to such 'bivariant' functors.

The Cartesian product on the category $\mathbf{C}$ induces a symmetric monoidal structure on the category $\mathrm{Corr}(\mathbf{C})$. Moreover, the data of a (right-lax) symmetric monoidal structure on a functor $\Phi_{\mathrm{corr}} : \mathrm{Corr}(\mathbf{C}) \to \mathbb{S}$, in particular induces a commutative algebra structure on $\Phi_{\mathrm{corr}}(c)$ for every object $\mathbf{c} \in \mathbf{C}$. In the example where $\mathbf{C} = \mathrm{Sch}_{\mathrm{aft}}$, $\mathbb{S} = \mathrm{DGCat}_{\mathrm{cont}}$ and the functor in question is IndCoh, this gives $\mathrm{IndCoh}(X)$ a symmetric monoidal structure for each scheme X. Moreover, as will be explained in Sect. 2.4, one also recovers Grothendieck-Serre duality for $\mathrm{IndCoh}(X)$ from the symmetric monoidal structure on the functor from correspondences.

Additionally, the $(\infty,2)$-categorical enhancement $\mathrm{Corr}(\mathbf{C})^{2\text{-Cat}}$ of the category of correspondences is used to encode the various relations between the pullback and pushforward functors as well as the tensor structure, such as the fact that for IndCoh, the !-pullback is the right adjoint to *-pushforward for a proper map. In this way, the $(\infty,2)$-category of correspondences encodes the full six functor formalism (the other three functors, when they exist, are adjoint to the pullback, pushforward and tensor functors). The reader is referred to Part III, Introduction, Sect. 2 for a general discussion of the six functor formalism and its relation to the category of correspondences. For the time being, we will focus on the parts relevant to IndCoh.

2.1. Why do correspondences arise? Suppose that we are given functors

$$\Phi : \mathbf{C} \to \mathbb{S} \text{ and } \Phi^! : \mathbf{C}^{\text{op}} \to \mathbb{S},$$

(not necessarily related to each other by any sort of adjunction) that agree on objects, and for every Cartesian square

$$\begin{array}{ccc} \mathbf{c}'_0 & \xrightarrow{\alpha_0} & \mathbf{c}_0 \\ {\scriptstyle\beta'}\downarrow & & \downarrow{\scriptstyle\beta} \\ \mathbf{c}'_1 & \xrightarrow{\alpha_1} & \mathbf{c}_1 \end{array} \tag{2.1}$$

we are given an isomorphism of maps $\Phi(\mathbf{c}_0) \to \Phi(\mathbf{c}'_1)$

$$\Phi^!(\alpha_1) \circ \Phi(\beta) \simeq \Phi(\beta') \circ \Phi^!(\alpha_0).$$

We want to encode this data by a functor $\text{Corr}(\mathbf{C}) \to \mathbb{S}$.

2.1.1. If $\mathbf{C}$ is an ordinary category, it is easy to say what $\text{Corr}(\mathbf{C})$ should be. Namely, its objects are the same as those of $\mathbf{C}$, but now morphisms from $\mathbf{c}_0$ to $\mathbf{c}_1$ are diagrams

$$\begin{array}{ccc} \mathbf{c}_{0,1} & \xrightarrow{g} & \mathbf{c}_0 \\ {\scriptstyle f}\downarrow & & \\ \mathbf{c}_1, & & \end{array} \tag{2.2}$$

and the compositions are given as follows: the composition of (2.2) and

$$\begin{array}{ccc} \mathbf{c}_{1,2} & \longrightarrow & \mathbf{c}_1 \\ \downarrow & & \\ \mathbf{c}_2, & & \end{array}$$

is the diagram

$$\begin{array}{ccc} \mathbf{c}_{0,2} & \longrightarrow & \mathbf{c}_0 \\ \downarrow & & \\ \mathbf{c}_2, & & \end{array}$$

where $\mathbf{c}_{0,2} = \mathbf{c}_{1,2} \underset{\mathbf{c}_1}{\times} \mathbf{c}_{0,1}$.

A bivariant functor as above defines a functor out of the category $\text{Corr}(\mathbf{C})$ by setting

$$\Phi_{\text{corr}}(\mathbf{c}) := \Phi(\mathbf{c}) = \Phi^!(\mathbf{c})$$

at the level of objects, and for a morphism (2.2), the corresponding map

$$\Phi(\mathbf{c}_0) \to \Phi(\mathbf{c}_1)$$

is given by $\Phi(\beta) \circ \Phi^!(\alpha)$. The isomorphisms (2.1) ensure that Φ_{corr} respects compositions.

2.1.2. However, if $\mathbf{C}$ is a higher category, we cannot just define $\mathrm{Corr}(\mathbf{C})$ by specifying the objects, morphisms and compositions. Instead, we need to invent a device which would produce the desired $(\infty,1)$-category from the (rather limited) list of procedures that produce $(\infty,1)$-categories from the existing ones. Moreover, the $(\infty,1)$-category $\mathrm{Corr}(\mathbf{C})$ should exactly encode what it means for the isomorphisms (2.1) to be compatible with the compositions of vertical and horizontal morphisms.

We introduce and study such a device in Chapter 7 of this volume.

2.2. How to construct functors out of a category of correspondences? Once the category $\mathrm{Corr}(\mathbf{C})$ is constructed, we would like to describe a mechanism that produces functors out of it (and thereby gives rise to bi-variant functors with all the necessary compatibilities).

2.2.1. Here is a construction, a generalization of which will be one of our basic tools. Let us start with a functor

$$\Phi : \mathbf{C} \to \mathbb{S},$$

where $\mathbb{S}$ is the $(\infty,1)$-category 1-Cat. Suppose that for every 1-morphism $\mathbf{c}_0 \xrightarrow{f} \mathbf{c}_1$ in $\mathbf{C}$, the corresponding map in 1-Cat, i.e., a *functor between* $(\infty,1)$*-categories*,

$$\Phi(\mathbf{c}_0) \to \Phi(\mathbf{c}_1),$$

admits a right adjoint.

Then the operation of *passage to the right adjoint* adjoint defines a functor

$$\Phi^! : \mathbf{C}^{\mathrm{op}} \to \mathbb{S}.$$

Suppose now that the following condition holds: for a Cartesian diagram (2.1), the natural transformation

$$\Phi(\beta') \circ \Phi^!(\alpha_0) \to \Phi^!(\alpha_1) \circ \Phi(\beta)$$

that arises by adjunction from the isomorphism

$$\Phi(\alpha_1) \circ \Phi(\beta') \simeq \Phi(\beta) \circ \Phi(\alpha_0),$$

is an isomorphism.

In this case we do expect that the functors $(\Phi, \Phi^!)$ comprise the datum of a functor

$$\Phi_{\mathrm{corr}} : \mathrm{Corr}(\mathbf{C}) \to \mathbb{S}.$$

And this turns out to indeed be the case.

Let us denote the subcategory of functors $\mathrm{Funct}(\mathbf{C}, \mathbb{S})$ satisfying the above properties by $\mathrm{Funct}(\mathbf{C}, \mathbb{S})^{\mathrm{BC}}$ (here 'BC' stands either for 'Beck-Chevalley' or 'base change'). Thus, we obtain a functor

$$\mathrm{Funct}(\mathbf{C}, \mathbb{S})^{\mathrm{BC}} \to \mathrm{Funct}(\mathrm{Corr}(\mathbf{C}), \mathbb{S}), \quad \mathbb{S} = 1\text{-Cat}. \tag{2.3}$$

However, the functor (2.3) is not an equivalence, and it is not quite adequate for our purposes, for two reasons.

2.2.2. For one thing, we would like to 'upgrade' (2.3) (by modifying the right-hand side) to make it an equivalence, in order to be more robust and suitable for applications.

But more importantly, for now, the above is just wishful thinking: we would not even be able to construct the functor (2.3) unless we make a sharper claim.

2.3. The 2-categorical enhancement. To make the sought-for sharper claim, we notice that our discussion was specific to the target $(\infty, 1)$-category being 1-Cat, in that we used the notion of 'adjoint' 1-morphism.

However, this is not specific just to 1-Cat, but rather is an artifact of a richer structure on the totality of $(\infty, 1)$-categories: namely that 1-Cat is the $(\infty, 1)$-category underlying a canonically defined $(\infty, 2)$-category, denoted $\mathbf{1}\text{-}\mathbf{Cat}$.

Thus, one expects to find a construction analogous to (2.3), where $\mathbb{S}$ is (the $(\infty, 1)$-category underlying) an $(\infty, 2)$-category. And such a construction is indeed possible, and can be sharpened to an equivalence, once we understand $\mathrm{Corr}(\mathbf{C})$ differently:

Namely, we should enhance $\mathrm{Corr}(\mathbf{C})$ itself to an $(\infty, 2)$-category, denoted $\mathrm{Corr}(\mathbf{C})^{\text{2-Cat}}$.

2.3.1. If $\mathbf{C}$ was an ordinary category, then $\mathrm{Corr}(\mathbf{C})^{\text{2-Cat}}$ would be an ordinary 2-category, where we introduce 2-morphisms as follows:

For a morphism $\mathbf{c}_0 \to \mathbf{c}_1$, given by (2.2), and another one, given by

$$\begin{array}{ccc} \mathbf{c}'_{0,1} & \xrightarrow{\alpha'} & \mathbf{c}_0 \\ {\scriptstyle \beta'}\downarrow & & \\ \mathbf{c}_1, & & \end{array}$$

the set of maps between them is that of commutative diagrams

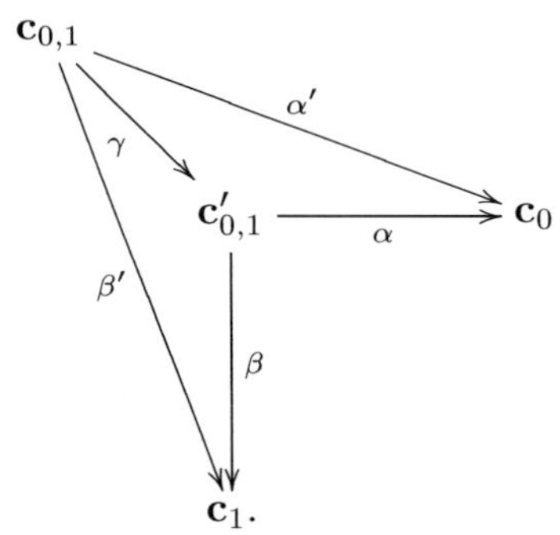

When $\mathbf{C}$ is a genuine $(\infty, 1)$-category, we construct $\mathrm{Corr}(\mathbf{C})^{\text{2-Cat}}$ using a device that we call 'Segal categories'.

2.3.2. Thus, in order to have an adequate theory of categories of correspondences, one has to venture into the (so far, not so well explored) world of $(\infty, 2)$-categories. Once we do this, we will have a naturally defined map

$$\mathrm{Funct}(\mathrm{Corr}(\mathbf{C})^{\text{2-Cat}}, \mathbb{S}) \to \mathrm{Funct}(\mathbf{C}, \mathbb{S}),$$

whose essential image is $\mathrm{Funct}(\mathbf{C}, \mathbb{S})^{\mathrm{BC}}$, i.e., we obtain the sought-for equivalence

$$\mathrm{Funct}(\mathbf{C}, \mathbb{S})^{\mathrm{BC}} \simeq \mathrm{Funct}(\mathrm{Corr}(\mathbf{C})^{\text{2-Cat}}, \mathbb{S}), \tag{2.4}$$

refining (2.3).

2.3.3. In addition to defining the $(\infty,1)$-category $\mathrm{Corr}(\mathbf{C})^{2\text{-Cat}}$ and also the $(\infty,2)$-category $\mathrm{Corr}(\mathbf{C})^{2\text{-Cat}}$, in Chapter 7 we prove an extension theorem, that allows us to construct IndCoh as a functor out of the category of correspondences to $\mathrm{DGCat}_{\mathrm{cont}}$.

In Chapter 8 we prove two more extension theorems that allow us to extend a functor from one category of correspondences to a larger one. These theorems will be applied in Volume II to extending IndCoh from schemes to *inf-schemes*.

2.4. Correspondences and duality. The formalism of functors out of $\mathrm{Corr}(\mathbf{C})$ is also an efficient way of encoding the idea of duality.

2.4.1. Note that the Cartesian product on $\mathbf{C}$ makes $\mathrm{Corr}(\mathbf{C})$ into a symmetric monoidal category. Moreover, every object $\mathbf{c} \in \mathrm{Corr}(\mathbf{C})$ is canonically self-dual. Namely, the unit and co-unit maps are given by the diagrams

$$\begin{array}{ccc} \mathbf{c} & \longrightarrow & * \\ \downarrow & & \\ \mathbf{c}\times\mathbf{c} & & \end{array}$$

and

$$\begin{array}{ccc} \mathbf{c} & \longrightarrow & \mathbf{c}\times\mathbf{c} \\ \downarrow & & \\ *, & & \end{array}$$

where $*$ denotes the final object of $\mathbf{C}$, and $\mathbf{c} \to \mathbf{c}\times\mathbf{c}$ is the diagonal map.

2.4.2. Suppose that we are given a functor

$$\Phi_{\mathrm{corr}} : \mathrm{Corr}(\mathbf{C}) \to \mathbb{S},$$

where both $\mathbb{S}$ and Φ_{corr} are equipped with symmetric monoidal structures.

Then we obtain that for any $\mathbf{c} \in \mathbf{C}$, the corresponding object $\Phi_{\mathrm{corr}}(\mathbf{c}) \in \mathbb{S}$ is canonically self-dual.

2.4.3. Applying this observation to the IndCoh functor, we will obtain that Serre duality is a formal consequence of the existence of IndCoh as a functor out of the category of correspondences. Namely, we obtain that for $X \in \mathrm{Sch}_{\mathrm{aft}}$, the DG category $\mathrm{IndCoh}(X)$ is equipped with a canonical identification

$$\mathbf{D}^{\mathrm{Serre}} : \mathrm{IndCoh}(X) \simeq \mathrm{IndCoh}(X)^{\vee}, \tag{2.5}$$

where $\mathrm{IndCoh}(X)^{\vee}$ is the dual category, i.e., $\mathrm{Funct}_{\mathrm{cont}}(\mathrm{IndCoh}(X), \mathrm{Vect})$.

At the level of compact objects, the equivalence (2.5) gives rise to an equivalence

$$\mathbb{D}^{\mathrm{Serre}} : \mathrm{Coh}(X)^{\mathrm{op}} \simeq \mathrm{Coh}(X),$$

which is the usual Serre duality.

A similar reasoning leads to Verdier duality for D-modules, which will be developed in Volume II.

3. The appendix on $(\infty, 2)$-categories

As mentioned above, in order to establish, or even formulate, the isomorphism (2.4), one needs to venture into the world of $(\infty, 2)$-categories.

Some of the foundations of the theory of $(\infty, 2)$-categories can be found in the existing literature. However, the theory developed so far does quite meet our needs. For this reason, we have decided to include an appendix consisting of Chapters 10–12, which lays out this theory the way we would like to see it (albeit, omitting some proofs).

3.1. Setting up the theory. We approach $(\infty, 2)$-categories by imitating the complete Segal space approach to $(\infty, 1)$-categories.

3.1.1. Namely, we recall that the datum of an $(\infty, 1)$-category $\mathbf{C}$ is completely recovered from the datum of the simplicial space $\mathrm{Seq}_\bullet(\mathbf{C})$ that sends $[n] \in \mathbf{\Delta}$ to the space of strings of objects of $\mathbf{C}$

$$\mathbf{c}_0 \to \mathbf{c}_1 \to \ldots \to \mathbf{c}_n. \tag{3.1}$$

Thus, we obtain a functor

$$\mathrm{Seq}_\bullet : 1\text{-Cat} \to \mathrm{Spc}^{\mathbf{\Delta}^{\mathrm{op}}},$$

which is fully faithful, and one can explicitly describe its essential image.

3.1.2. We would like to define $(\infty, 2)$-categories similarly. Namely, we wish to define the $(\infty, 1)$-category 2-Cat as a certain full subcategory in $1\text{-Cat}^{\mathbf{\Delta}^{\mathrm{op}}}$. However, one immediately runs into the following dilemma: if $\mathbb{S}$ is an $(\infty, 2)$-category (whatever this notion is), there are two possibilities of what the $(\infty, 1)$-category of length n strings objects could be.

In both cases, the objects of our category are strings as in (3.1), where the arrows are 1-morphisms. But there is a choice involved in how we define morphisms between such objects. In one case, we ask for diagrams

(3.2)

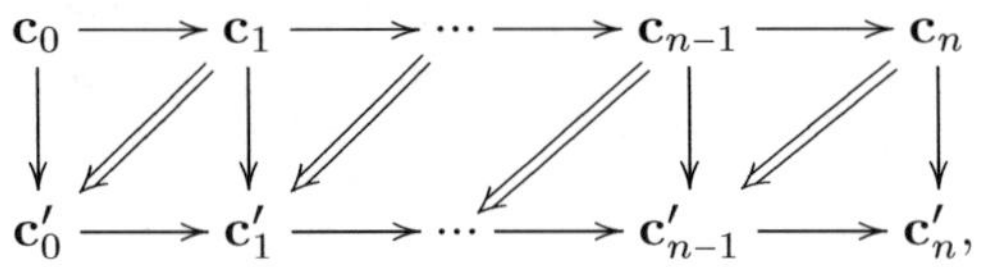

where the slanted arrows stand for 2-morphisms.

In the other case, we ask for diagrams

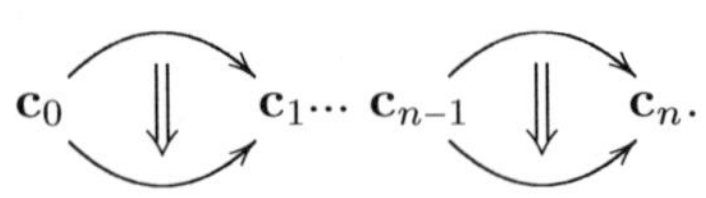

I.e., these are the same as diagrams (3.2), but with the vertical 1-morphisms being isomorphisms.

3.1.3. What we obtain is that, whatever the $(\infty,1)$-category 2-Cat is, it is equipped with two functors

$$\mathrm{Seq}_\bullet : 2\text{-Cat} \to 1\text{-Cat}^{\mathbf{\Delta}^{\mathrm{op}}}$$

(corresponding to the second kind of 1-morphisms on n-simplices), and

$$\mathrm{Seq}^{\mathrm{ext}}_\bullet : 2\text{-Cat} \to 1\text{-Cat}^{\mathbf{\Delta}^{\mathrm{op}}},$$

(corresponding to the first kind of 1-morphisms on n-simplices), both of which are supposed to be fully faithful, with an explicitly described essential image.

We take the first realization (i.e., one with $\mathrm{Seq}_\bullet$) as the definition of 2-Cat, which by [**BarS**] is known to be equivalent to various other notions of $(\infty,2)$-category that appear in the literature. We prove that the other realization (i.e., one with $\mathrm{Seq}^{\mathrm{ext}}_\bullet$) has the expected properties. This done in Chapter 10 of this volume.

In turns out that the first realization is more convenient for taking the theory off the ground, while the second one is necessary for our treatment of adjunctions, as described below.

3.1.4. In Chapter 11 we study some basic constructions associated with $(\infty,2)$-categories, namely, the straightening/unstraightening equivalence (generalizing the familiar construction in the context of $(\infty,1)$-categories), and the Yoneda embedding.

3.2. Adjunctions. The main reason we need to develop the theory of $(\infty,2)$-categories is to have a theory of adjunctions, adequate for establishing the equivalence (2.4).

3.2.1. Let $\mathbb{S}$ an $(\infty,2)$-category. Then for a 1-morphism $\alpha : \mathbf{s}_0 \to \mathbf{s}_1$, there is a notion of what it means to *admit a right adjoint.*

If $\mathbb{S}$ is an ordinary category, if a right adjoint of a 1-morphism exists, it is uniquely defined up to a canonical isomorphism. More generally, if $F : \mathbf{I} \to \mathbb{S}$ is a functor, and if for every arrow $i_0 \to i_1$ in $\mathbf{I}$, the corresponding 1-morphism $F(i_0) \to F(i_1)$ admits a right adjoint, we can canonically construct a functor

$$G : \mathbf{I}^{\mathrm{op}} \to \mathbb{S},$$

which is the same as F at the level of objects, but which at the level of morphisms is obtained from F by replacing each $F(i_0) \to F(i_1)$ by its right adjoint. In this case we will say that *G is obtained from F by passing to right adjoints.*

Let $\mathrm{Funct}(\mathbf{I},\mathbb{S})^L$ be the full subcategory of $\mathrm{Funct}(\mathbf{I},\mathbb{S})$ consisting of those functors $F : \mathbf{I} \to \mathbb{S}$ such that for every 1-morphism $i_0 \to i_1$ in $\mathbf{I}$, the corresponding 1-morphism $F(i_0) \to F(i_1)$ admits a right adjoint. Let $\mathrm{Funct}(\mathbf{I}^{\mathrm{op}},\mathbb{S})^R$ be the corresponding full subcategory of $\mathrm{Funct}(\mathbf{I}^{\mathrm{op}},\mathbb{S})$ (replace 'right' by 'left'). Then the assignment $F \mapsto G$ defines an equivalence

$$\mathrm{Funct}(\mathbf{I},\mathbb{S})^L \simeq \mathrm{Funct}(\mathbf{I}^{\mathrm{op}},\mathbb{S})^R. \tag{3.3}$$

3.2.2. The same assertions – canonicity of the adjoint for an individual morphism and the equivalence (3.3) – remain true in the context of $(\infty, 2)$-categories, but it is a non-trivial task to formulate, and subsequently prove them.

We develop the theory of adjunctions in Chapter 12. We show that for any $\mathbf{I}$ there exists an $(\infty, 2)$-category, $\mathbf{I}^R$, equipped with a pair of functors

$$\mathbf{I} \to \mathbf{I}^R \leftarrow \mathbf{I}^{\mathrm{op}},$$

such that for any target $(\infty, 2)$-category $\mathbb{S}$, restrictions along the above functors define equivalences

$$\mathrm{Funct}(\mathbf{I}, \mathbb{S})^R \overset{\sim}{\leftarrow} \mathrm{Funct}(\mathbf{I}^R, \mathbb{S}) \overset{\sim}{\to} \mathrm{Funct}(\mathbf{I}^{\mathrm{op}}, \mathbb{S})^L. \tag{3.4}$$

Moreover, it turns out that the equivalences (3.4) are precisely suited for establishing the equivalence (2.4).

The construction of $\mathbf{I}^R$ is based on the 'second' realization of 2-Cat as a full subcategory of $1\text{-Cat}^{\mathbf{\Delta}^{\mathrm{op}}}$, i.e., one using the functor $\mathrm{Seq}_\bullet^{\mathrm{ext}}$.

Part I

Preliminaries

Introduction

Why do we need these preliminaries?

0.3. None of the contents of Part I is original mathematics.

Chapter I.1 is a review of higher category theory and higher algebra, mostly following [**Lu1**] and [**Lu2**].

Chapter I.2 is a review of the basic definitions of derived algebraic geometry (derived schemes, Artin stack and general prestacks), mostly following [**TV1, TV2**].

Chapter I.3 is a review of the basics of quasi-coherent sheaves (there are no deep theorems there, so one can say that it is mostly folklore).

0.4. We wish to emphasize that by no means do these chapters supply a self-contained exposition of elements of the theory required for the rest of the book. Our goal is rather to give the reader a concise account of the most ubiquitous structures, in order to enable him/her to start reading the subsequent chapters.

Our hope is that once he/she gets started, he/she will gradually acquire the ability to look up or reconstruct the necessary bits of foundational material.

1. ∞-categories and higher algebra

1.1. Let us accept the inevitable: when we talk about algebraic geometry, we need to speak in the language of categories.

For one thing, geometric objects (such as schemes and their generalizations) form a category. But even more importantly, the flora to be found on these geometric objects (sheaves of various sorts) consists of categories: there is no way to develop the theory of sheaves without using categories.

Since the introduction of the categorical language to the study of algebraic geometry by Grothendieck in the 1950's, and up until the late 2000's, the methods of *usual* (=ordinary) category theory sufficed for most purposes. People used either the *abelian* category of quasi-coherent sheaves or its derived category, which is a *triangulated category*.

However, there are some instances where triangulated categories are not enough. Perhaps the main example of this is the failure of gluing: one cannot glue the derived category of quasi-coherent sheaves on a scheme from just knowing it on an open cover.

Now, the problem of inadequacy of triangulated categories becomes even more acute in the context of *derived algebraic geometry* (DAG). So, having accepted the inevitability of categories for usual algebraic geometry, we now have no choice but accept the inevitability of ∞-categories if we want to work in DAG. This is further reinforced by the fact that the geometric objects themselves (derived schemes or, more generally, prestacks) now form an ∞-category.

1.2. In Chapter 1, Sects. 1 and 2 we give a concise review of the basics of ∞-categories.

We mostly focus on the syntax: how to use the language of ∞-categories. In other words, the reader does not have to be familiar with a particular model for ∞-categories, be it topological categories, simplicial categories, or the model that is now most widely used in the literature – Joyal's quasi-categories, animated by Lurie in [Lu0].

We introduce the key notions of Cartesian/coCartesian fibration, Yoneda, limit/colimit, cofinality, left/right Kan extension, adjunction for functors.

1.3. In Chapter 1, Sects. 3 and 4 we give the first taste of *higher algebra*. We introduce the notions of *monoidal ∞-category* and of associative algebra inside a monoidal ∞-category. We also introduce the corresponding commutative notions.

We also introduce the corresponding notions of module (that is, a module category for a given monoidal ∞-category, and the notion of module for an associative algebra).

We then proceed to the discussion of duality. We discuss the notion of left/right dualizability of an object in a monoidal ∞-category, and the related notion of dualizability of left/right module over an algebra.

1.4. In Chapter 1, Sects. 5, 6, 7 we discuss the notion of *stable ∞-category.*

Stable ∞-categories are the higher categorical replacement of triangulated categories, i.e., this is where we really do algebra.

An operation that will play a key role in the book is that of the *Lurie tensor product* of (cocomplete) stable ∞-categories, which gives the totality of the latter, denoted, $1\text{-Cat}^{\text{St,cocmpl}}_{\text{cont}}$ a structure of symmetric monoidal ∞-category.

1.5. In Chapter 1, Sects. 8 and 9 we supply a framework for "really doing algebra":

We talk about (symmetric) monoidal stable ∞-categories, i.e., associative (resp., commutative) algebra objects in the symmetric monoidal category $1\text{-Cat}^{\text{St,cocmpl}}_{\text{cont}}$.

1.6. Finally, in Chapter 1, Sect. 10 we introduce the notion of *DG category*, i.e., a differential graded category that is equipped with a linear structure over a fixed ground field k of characteristic 0.

2. Basics of derived algebraic geometry

In Chapter 2 we begin to discuss derived algebraic geometry proper, i.e., we introduce the ∞-category of the corresponding geometric objects.

2.1. We start with the category of (derived)[1] affine schemes over our ground field k, denoted $\mathrm{Sch}^{\mathrm{aff}}$, which is, by definition, the category opposite to that of *connective commutative DG algebras over* k.

In Chapter 2, Sect. 1 we introduce the most general class of geometric objects: prestacks. The ∞-category of the latter is simply that of functors

$$(\mathrm{Sch}^{\mathrm{aff}})^{\mathrm{op}} \to \mathrm{Spc},$$

where Spc is the ∞-category of spaces (a.k.a. ∞-groupoids). I.e., a prestack is just something that has a Grothendieck functor of points.

All other geometric objects that we will consider (schemes, Artin stacks, etc.) will be prestacks. For example, a scheme (resp., Artin stack) will be a prestack with certain properties (as opposed to additional pieces of structure).

In later Chapters of the book we will be interested in yet another particular class of prestacks, namely, inf-schemes.

2.2. In Chapter 2, Sect. 2 we will introduce the descent condition with respect to the Zariski, étale or faithfully flat topology. We call prestacks that satisfy the descent condition *stacks*.

We study how the descent condition interacts with the basic properties that a prestack can possess (such as being locally of finite type).

2.3. In Chapter 2, Sect. 3 we introduce what is, arguably, the main object of study in derived algebraic geometry: (derived) schemes.

According to what was said above, we *do not* introduce schemes as locally ringed spaces. Rather, we define schemes as prestacks that satisfy a certain condition. The condition in question is that to admit an open covering by affine schemes.

2.4. In Chapter 2, Sect. 4 we introduce the hierarchy of k-Artin stacks, $k \geq 0$. We should say that we call a k-Artin stack for a particular k may diverge from elsewhere in the literature (for example, for us, a 0-Artin stack is a stack that is a (possibly infinite) disjoint of affine schemes). However, the union over all k produces the same class of objects. The advantage of our particular system of definitions is that it makes inductive proofs of various properties of k-Artin stacks very simple.

We should also point out that from the point of view of our hierarchy of k-Artin stacks, schemes are a red herring. They are more general than 0-Artin stacks, but are a tiny particular case of 1-Artin stacks.

3. Quasi-coherent sheaves

In Chapter 3 we introduce what is perhaps the main object of study of *categorical* (derived) algebraic geometry: quasi-coherent sheaves.

[1] Henceforth the adjective 'derived' will be dropped, because everything will be derived.

3.1. In Chapter 3, Sect. 1 we start with the functor

$$\mathrm{QCoh}^*_{\mathrm{Sch}^{\mathrm{aff}}} : (\mathrm{Sch}^{\mathrm{aff}})^{\mathrm{op}} \to \mathrm{DGCat}_{\mathrm{cont}}, \quad S = \mathrm{Spec}(A) \rightsquigarrow A\text{-mod}, \quad (S' \xrightarrow{f} S) \rightsquigarrow f^*.$$

We apply the procedure of *right Kan extension* along the (Yoneda) embedding $\mathrm{Sch}^{\mathrm{aff}} \hookrightarrow \mathrm{PreStk}$ and thus obtain a functor

$$\mathrm{QCoh}^*_{\mathrm{PreStk}} : \mathrm{PreStk}^{\mathrm{op}} \to \mathrm{DGCat}_{\mathrm{cont}}.$$

Thus, for any prestack $\mathcal{Y}$ we have a well-defined DG category $\mathrm{QCoh}(\mathcal{Y})$ and for a morphism $f : \mathcal{Y}' \to \mathcal{Y}$ we have a pullback functor $f^* : \mathrm{QCoh}(\mathcal{Y}) \to \mathrm{QCoh}(\mathcal{Y}')$.

3.2. Note, in particular, that if Z is a scheme, we obtain a category $\mathrm{QCoh}(Z)$. This definition of QCoh of a scheme is equivalent to any other (reasonable) definition. However, we note that we *do not* approach it via first considering *all sheaves of $\mathcal{O}$-modules in Zariski topology*, and then passing to a subcategory. Instead, we directly glue $\mathrm{QCoh}(Z)$ from affines.

A similar feature of our approach to the definition of $\mathrm{QCoh}(-)$ is also present in the case of Artin stacks.

3.3. In Chapter 3, Sect. 2 we study the functor of *direct image* for quasi-coherent sheaves

$$f_* : \mathrm{QCoh}(\mathcal{Y}') \to \mathrm{QCoh}(\mathcal{Y})$$

for a morphism $f : \mathcal{Y}' \to \mathcal{Y}$. By definition, f_* is the right adjoint of f^*, and it exists for abstract reasons (the Adjoint Functor Theorem).

For a general morphism f, the functor f_* is very badly behaved. For example, it fails to satisfy base change. However, by imposing some additional assumptions on f one can ensure that it is reasonable. One such assumption is that f is schematic quasi-compact.

3.4. In Chapter 3, Sect. 3 we study the natural *right lax* symmetric monoidal structure on the functor $\mathrm{QCoh}^*_{\mathrm{PreStk}}$. Concretely, this structure amounts to (a compatible family of) functors

$$\mathrm{QCoh}(\mathcal{Y}_1) \otimes \mathrm{QCoh}(\mathcal{Y}_2) \to \mathrm{QCoh}(\mathcal{Y}_1 \times \mathcal{Y}_2), \quad \mathcal{Y}_1, \mathcal{Y}_2 \in \mathrm{PreStk}.$$

We study the question of when the above functor is an equivalence.

The symmetric monoidal structure on $\mathrm{QCoh}^*_{\mathrm{PreStk}}$ induces a symmetric monoidal structure on the category $\mathrm{QCoh}(\mathcal{Y})$ for an individual prestack $\mathcal{Y} \in \mathrm{PreStk}$.

We study how various properties of a prestack $\mathcal{Y}$ reflect in properties of $\mathrm{QCoh}(\mathcal{Y})$ (compact generation, dualizability, rigidity).

CHAPTER 1

Some higher algebra

Introduction

This Chapter is meant to provide some background on ∞-categories and *higher algebra* (the latter includes the notions of (symmetric) monoidal ∞-category, (commutative) algebras in a (symmetric) monoidal ∞-category, and modules over such algebras).

0.1. Why $(\infty, 1)$-categories? At this point in the development of mathematics one hardly needs to make a case for ∞-categories. Nonetheless, in this subsection we explain why they necessarily appear in this book. I.e., why we cannot remain in the world of, say, triangulated categories (if we talk about 'linear' categories).

0.1.1. In fact, there are two separate (but related) reasons that force one to work with ∞-categories, rather than triangulated ones: *extrinsic* and *intrinsic*.

The extrinsic reason has to do with the behavior of the *totality* of ∞/triangulated categories, and the intrinsic reason has to do with what is going on within a given ∞/triangulated category.

We begin by discussing the extrinsic reason, which we believe is more fundamental.

0.1.2. The extrinsic reason has to do the with the operation of *limit* of a diagram of ∞ (resp., triangulated) categories.

An example of a limit is *gluing*: imagine that you want to glue the ∞/triangulated category of quasi-coherent sheaves on a scheme X from an affine cover.

Below we will explain why the above operation of gluing along an open cover unavoidably appears in the theory that we are trying to build. However, taking that on faith, we arrive at the necessity to work with ∞-categories: it is well-known that triangulated categories do not glue well.

For example, given a scheme/topological space X with an action of an algebraic/compact group G, there is no known way to define the G-equivariant derived category of sheaves on X while only using the *derived* category of sheaves as an input: all the existing definitions appeal to constructions that take place at the *chain level*.

But once we put ourselves in the context of ∞-categories, everything works as expected. For example, given a prestack $\mathcal{Y}$ (i.e., an arbitrary functor from the category of affine schemes to that of ∞-groupoids), one can define the category $\mathrm{Shv}(\mathcal{Y})$ of sheaves on $\mathcal{Y}$ as the limit

$$\lim_{S, y: S \to \mathcal{Y}} \mathrm{Shv}(S),$$

where the limit is taken over the category of affine schemes over $\mathcal{Y}$. As a tiny particular case of this, we can take $\mathcal{Y} = X/G$ and recover the G-equivariant derived category on X.

0.1.3. We now explain the *intrinsic* reason why one is often forced to work with ∞-categories.

It also has to do with...–*suspense*–...the operation of taking a limit (or colimit), but now within our given ∞/triangulated category $\mathbf{C}$.

But here, in a sense, we will not say anything new. A basic example of limit is the operation of fiber product of objects

$$\mathbf{c}_1 \underset{\mathbf{c}}{\times} \mathbf{c}_2.$$

When working in a triangulated category, we usually want to interpret the latter as the (shifted by $[-1]$) *cone* of the map

$$\mathbf{c}_1 \oplus \mathbf{c}_2 \to \mathbf{c},$$

and we arrive to the familiar problem that cones are not well-defined (or, rather, that they do not have a functorial description).

Of course, one can say that cones exist, even though they are not canonical. But this non-canonicity prevents one from defining more general *homotopy colimits*, e.g., geometric realizations of simplicial objects, and without that, one cannot really do algebra, of the kind that we will be doing in Volume II, Part II of this book (operads, Lie algebras and Lie algebroids, etc.)

For example, in a monoidal triangulated category, one cannot form the tensor product of a right module and a left module over an associative algebra.

0.2. The emergence of derived algebraic geometry and why we need gluing. We shall first explain why derived algebraic geometry enters our game (that is, even if, at the start, one tries to work in the world of usual schemes).

We will then see how objects of derived algebraic geometry necessitate a gluing procedure.

0.2.1. Let us consider the pattern of *base change* for the derived category of quasi-coherent sheaves. Let us be given a Cartesian diagram of (usual) schemes

$$\begin{array}{ccc} X_1 & \xrightarrow{g_X} & X_2 \\ {\scriptstyle f_1}\downarrow & & \downarrow{\scriptstyle f_2} \\ Y_1 & \xrightarrow{g_Y} & Y_2. \end{array}$$

Then from the isomorphism of functors

$$(f_2)_* \circ (g_X)_* \simeq (g_Y)_* \circ (f_1)_*$$

one obtains by adjunction the natural transformation

$$(g_Y)^* \circ (f_2)_* \to (f_1)_* \circ (g_X)^*. \tag{0.1}$$

The base change theorem says that (0.1) is an isomorphism. The only problem is that this theorem is *false*.

More precisely, it is false if X_1 is taken to be the fiber product $Y_1 \underset{Y_2}{\times} X_2$ in the category of usual schemes: consider the case when all schemes are affine; $X_i = \mathrm{Spec}(A_i)$ and $Y_i = \mathrm{Spec}(B_i)$, and apply (0.1) to

$$A_2 \in A_2\text{-mod} = \mathrm{QCoh}(X_2).$$

We obtain the map

$$B_1 \underset{B_2}{\otimes} A_2 \to A_1,$$

where the tensor product is understood in the *derived* sense, while the right hand side is its top (i.e., 0-th) cohomology.

To remedy this problem, we need to take A_1 to be the full derived tensor product $B_1 \underset{B_2}{\otimes} A_2$, so that A_1 is no longer a plain commutative algebra, but what one calls a *connective commutative DG algebra*. The spectrum of such a thing is, by definition, an affine derived scheme.

0.2.2. Thus, we see that affine derived schemes are necessary if we wish to have base change. And if we want to do algebraic geometry (i.e., consider not just affine schemes), we need to introduce the notion of general derived scheme.

We will delay the discussion of what derived schemes actually are until Chapter 2. However, whatever they are, a derived scheme X is glued from an open cover of affine derived schemes $U_i = \mathrm{Spec}(A_i)$, and let us try to imagine what the ∞/triangulated category $\mathrm{QCoh}(X)$ of quasi-coherent sheaves on X should be.

By definition for each element of the cover we have

$$\mathrm{QCoh}(U_i) = A_i\text{-mod},$$

i.e., this is the category of A_i-modules. Now, whatever X is, the ∞/triangulated category $\mathrm{QCoh}(X)$ should be obtained as a gluing of $\mathrm{QCoh}(U_i)$, i.e., as the limit of the diagram of categories

$$\mathrm{QCoh}(U_{i_0} \cap \ldots \cap U_{i_n}), \quad n = 0, 1, \ldots$$

Thus, we run into the problem of taking the limit of a diagram of categories, and as we said before, in order to take this limit, we should understand the above diagram as one of ∞-categories rather than triangulated ones.

0.3. What is done in this Chapter? This Chapter (with the exception of Sect. 9) contains no original mathematics; it is mostly a review of the foundational works of J. Lurie, [**Lu1**] and [**Lu2**].

Thematically, it can be split into the following parts:

0.3.1. Sects. 1-2 are a review of higher category theory, i.e., the theory of $(\infty,1)$-categories, following [**Lu1**, Chapters 1-5].

In Sect. 1 we introduce the basic words of the vocabulary of $(\infty,1)$-categories. In Sect. 2 we discuss some of the most frequent manipulations that one performs with $(\infty,1)$-categories.

0.3.2. In Sects. 3-4 we review the basics of higher algebra, following [**Lu2**, Chapter 4].

In Sect. 3 we discuss the notions of (symmetric) monoidal $(\infty,1)$-category, the notion of associative/commutative algebra in a given (symmetric) monoidal $(\infty,1)$-category, and the notion of module over an algebra.

In Sect. 4 we discuss the pattern of duality in higher algebra.

0.3.3. In Sects. 5-7, we discuss *stable* $(\infty,1)$-categories, following [**Lu2**, Sects. 1.1, 1.4 and 4.8] with an incursion into [**Lu1**, Sect. 5.4].

In Sect. 5 we introduce the notion of *stable* $(\infty,1)$-category.

In Sect. 6 we discuss the operation of *Lurie tensor product* on *cocomplete stable $(\infty,1)$-categories.*

In Sect. 7 we discuss the notions of compactness, compact generation and ind-completion (we do this in the context of stable categories, even though these notions make sense more generally, see [**Lu1**, Sect. 5.3]).

0.3.4. In Sects. 8-10 we start discussing *algebra.*

In Sect. 8 we specialize the general concepts of higher algebra to the case of stable categories. I.e., we will discuss stable (symmetric) monoidal categories, module categories over them, duality for such, etc.

In Sect. 9 (which is the only section that contains some original mathematics) we introduce the notion of *rigid* monoidal category. By a loose analogy, one can think of rigid monoidal categories as Frobenius algebras in the world of stable categories. These stable monoidal categories exhibit particularly strong *extrinsic* finiteness properties: i.e., properties of module categories over them.

Finally, in Sect. 10 we introduce the notion of *DG category.* This will be the world in which we will do algebra in the main body of this book.

0.4. What do we have to say about the theory of ∞-categories? The theory of ∞-categories, in the form that is amenable for use by non-experts, has been constructed by J. Lurie in [**Lu1**]. It is based on the model of ∞-categories as *quasi-categories* (a.k.a., *weak Kan simplicial sets*), developed in the foundational work of A. Joyal, [**Jo**].

0.4.1. The remarkable thing about this theory is that one does *not* really need to know the contents of [**Lu1**] in order to apply it.

What Lurie's book provides is a syntax of allowed words and sentences in the the theory of ∞-categories, and ensures that this syntax can be realized in the model of quasi-categories.

0.4.2. In Sect. 1 we make an attempt to summarize this syntax. However, we are not making a mathematical assertion here: our grammar is incomplete and suffers from circularity (e.g., we appeal to fiber products before introducing limits).

The task of actually writing down such a syntax appears to be a non-trivial problem on its own. It seems likely, however, that in order to do that, one has to completely disengage oneself from viewing objects of a (higher) category as a set.

The latter would be desirable in any case: the simplical set underlying an ∞-category is a phantom; indeed, we never use it for any 'yes or no' questions or when we need to compute something.

0.4.3. In Sect. 2, we assume that we know how to speak the language of ∞-categories, and we introduce some basic tools that one uses to create new ∞-categories from existing ones, and similarly for functors.

These have to do with the operation of taking limits and colimits (within a given ∞-category or the totality of such), and the procedure of Kan extension.

0.4.4. Our tool-kit regarding ∞-categories is far from complete.

For example, we do not define what filtered/sifted ∞-categories are.

And, quite possibly, there are multiple other pieces of terminology, common in the theory of ∞-categories, that we use without being aware of not having introduced them. Whenever this happens, the reader should go back to [**Lu1**], and find the definition therein.

0.4.5. *What about set theory?* As is written in [**Lu1**, Sect. 1.2.15], one needs to make a decision on how one treats the sizes of our categories, i.e., the distinction between 'large' and 'small' categories.

Our policy is option (3) from *loc.cit.*, i.e., we just *ignore* these issues.

One reason for this is that the mention of cardinals when stating lemmas and theorems clutters the exposition.

Another reason is that it is very difficult to make a mistake of set-theoretic nature, unless one makes a set-theoretic argument (which we never do).

So, we will assume that our reader will not be conflicted about cutting his/her own hair, and live in the happy cardinal-free world.

0.5. What do we have to say about higher algebra? Nothing, in fact, beyond what is written in [**Lu2**, Chapter 4]. But we need much less (e.g., we do need general operads), so we decided to present a concise summary of things that we will actually use.

0.5.1. We start by discussing associative and commutative structures, i.e., monoidal/symmetric monoidal ∞-categories and associative/commutative algebras in them. In fact, it all boils down to the notion of monoid/commutative monoid in a given ∞ category.

The remarkable thing is that it is easy to encode monoids/commutative monoids using functors between ∞-categories. This idea originated in Segal's foundational work [**Seg**], and was implemented in the present context in [**Lu2**, Chapter 4].

Namely, the datum of a monoid in an ∞-category $\mathbf{C}$ is encoded by a functor

$$F : \mathbf{\Delta}^{\mathrm{op}} \to \mathbf{C},$$

that satisfies the following condition: $F([0]) = *$, and for every $n = 1, ..., $ the map

$$F([n]) \to \underset{i=1,\ldots,n}{\Pi} F([1])$$

is an isomorphism, where the i-th map $F([n]) \to F([1])$ corresponds to the map

$$[1] \to [n], \quad 0 \mapsto i-1, 1 \mapsto i.$$

For example, the binary operation is the map

$$F([2]) \to F([1])$$

that corresponds to the map

$$[1] \to [2], \quad 0 \mapsto 0, 1 \mapsto 2.$$

Similarly, the datum of a commutative monoid is encoded by a functor

$$F : \mathrm{Fin}_* \to \mathbf{C},$$

where Fin_* is the category of *pointed finite sets.*

Once we take this point of view, the basic definitions of higher algebra roll out quite easily. This is what is done in Sect. 3.

0.5.2. In Sect. 4 we discuss the notion of duality. It appears in several flavors: the notion of left/right dual of an object in a monoidal ∞-category; the notion of dual of module over an algebra; and also as the notion of adjoint functor.

It is easy to define what it means for an object to be *dualizable.*

However, the question of *canonicity* of the dual is trickier: in what sense is the dual uniquely defined? I.e., what kind of duality datum specifies it uniquely (i.e., *up to a contractible space of choices*)?

In fact, this question can be answered precisely, but for this one needs to work in the context of $(\infty, 2)$-categories. And we actually do this, in Chapter 12, in the framework of discussing the notion of adjoint 1-morphism in a given $(\infty, 2)$-category.

The upshot of *loc.cit.* is that the dual is canonically defined, and one can specify (albeit not too explicitly) the data that fixes it uniquely.

0.6. Stable ∞-categories. In the main body of the book we will be doing algebra in DG categories (over a field k of characteristic 0). There are (at least) two routes to set this theory up.

0.6.1. One route would be to proceed directly by working with (ordinary) categories enriched over the category of complexes of vector spaces over k.

In fact, this way of approaching DG categories has been realized in [**Dr**]. However, one of the essential ingredients of a functioning theory is that the totality of DG categories should itself be endowed with a structure of ∞-category (in order to be able to take limits). But since the paper [**Dr**] appeared before the advent of the language of ∞-categories, some amount of work would be needed to explain how to organize DG categories into an ∞-category.

The situation with the operation of *tensor product* of DG categories is similar. It had been developed in [**FG**], prior to the appearance of [**Lu2**]. However, this structure had not been formulated as a symmetric monoidal ∞-category in language that we use today.

So, instead of trying to rewrite the constructions of [**Dr**] and [**FG**] in the language of (symmetric monoidal) ∞-categories, we decided to abandon this approach, and access DG categories via a more robust (=automatic, tautological) approach using the general notion of stable ∞-category and the symmetric monoidal structure on such, developed in [**Lu2**].

0.6.2. The definition of stable ∞-categories given in [**Lu2**] has the following huge advantage: being stable is *not* an additional piece of structure, but a *property* of an ∞-category.

As a consequence of this, we do not have to labor to express the fact that any stable ∞-category is *enriched* over the ∞-category Sptr of spectra (i.e., that mapping spaces in a stable ∞-category naturally lift to objects of Sptr): whatever meaning we assign to this phrase, this structure is automatic from the definition (see Sect. 0.6.5).

0.6.3. Given a stable ∞-category, one can talk about t-structures on it. We count on the reader's familiarity with this notion: a t-structure on a stable category is the same as the t-structure on the associated triangulated category.

In terms of notation, given a stable ∞-category $\mathbf{C}$ with a t-structure, we let

$$\mathbf{C}^{\leq 0} \subset \mathbf{C} \supset \mathbf{C}^{\geq 0}$$

the corresponding full subcategories of connective/coconnective objects (so that $\mathbf{C}^{>0}$ is the *right* orthogonal to $\mathbf{C}^{\leq 0}$). We let

$$\mathbf{C}^{\leq 0} \overset{\tau^{\leq 0}}{\longleftarrow} \mathbf{C} \text{ and } \mathbf{C} \overset{\tau^{\geq 0}}{\longrightarrow} \mathbf{C}^{\geq 0}$$

be the corresponding right and left adjoints (i.e., the truncation functors).

We let

$$\mathbf{C}^{\heartsuit} = \mathbf{C}^{\leq 0} \cap \mathbf{C}^{\geq 0}$$

denote the heart of the t-structure; this is an abelian category.

We will also use the notation

$$\mathbf{C}^{-} = \underset{n\geq 0}{\cup} \mathbf{C}^{\leq n} \text{ and } \mathbf{C}^{+} = \underset{n\geq 0}{\cup} \mathbf{C}^{\geq -n}.$$

We will refer to $\mathbf{C}^{-}$ (resp., $\mathbf{C}^{+}$) as the *bounded above* or *eventually connective* (resp., *bounded below* or *eventually coconnective*) subcategory of $\mathbf{C}$.

0.6.4. *The Lurie tensor product.* One of the key features of the ∞-category of stable categories $1\text{-Cat}^{\mathrm{St,cocmpl}}_{\mathrm{cont}}$ (here we restrict objects to be *cocomplete* stable categories, and morphisms to be colimit-preserving functors) is that it carries a symmetric monoidal structure[1], which we call the *Lurie tensor product.*

Another huge advantage of the way this theory is set up in [**Lu2**, Sect. 4.8] is that the definition of this structure is automatic (=obtained by passing to appropriate full subcategories) from the Cartesian symmetric monoidal structure on the ∞-category 1-Cat of all ∞-categories.

The intuitive idea behind the Lurie tensor product is this: if A and B are associative algebras, then the tensor product of A-mod and B-mod should be $(A \otimes B)$-mod.

[1]The existence of the Lurie tensor product is yet another advantage of working with stable ∞-categories rather than triangulated ones: one cannot define the tensor product for the latter.

0.6.5. *Spectra.* The symmetric monoidal structure on $1\text{-Cat}^{\mathrm{St,cocmpl}}_{\mathrm{cont}}$ leads to a very concise definition of the ∞-category Sptr of spectra. Namely, this is the unit object in $1\text{-Cat}^{\mathrm{St,cocmpl}}_{\mathrm{cont}}$ with respect to its symmetric monoidal structure.

In particular, every (cocomplete) stable ∞-category $\mathbf{C}$ is automatically a module over Sptr. Thus, for any two objects $\mathbf{c}_0, \mathbf{c}_1 \in \mathbf{C}$, we can consider their *relative internal Hom*

$$\underline{\mathrm{Hom}}_{\mathbf{C},\mathrm{Sptr}}(\mathbf{c}_0, \mathbf{c}_1) \in \mathrm{Sptr}\,.$$

This is the enrichment structure on $\mathbf{C}$ with respect to Sptr, mentioned earlier.

0.6.6. In Sect. 7 we study a class of cocomplete stable categories that are particularly amenable to calculations: these are the compactly generated stable categories. This material is covered by [**Lu1**, Sect. 5.3], and a parallel theory in the framework of DG categories can be found in [**Dr**].

The main point is that a compactly generated stable category $\mathbf{C}$ can be obtained as the *ind-completion*[2] of its full subcategory $\mathbf{C}^c$ of compact objects. The ind-completion procedure can be thought of as formally adjoining to $\mathbf{C}^c$ all filtered colimits. However, we can also define it explicitly as the category of all exact functors

$$(\mathbf{C}^c)^{\mathrm{op}} \to \mathrm{Sptr}\,.$$

The advantage of compactly generated stable categories is that the data involved in describing colimit-preserving functors out of them is manageable: for a compactly generated $\mathbf{C}$ and an arbitrary cocomplete $\mathbf{D}$ we have

$$\mathrm{Funct}_{\mathrm{ex,cont}}(\mathbf{C}, \mathbf{D}) \simeq \mathrm{Funct}_{\mathrm{ex}}(\mathbf{C}^c, \mathbf{D}).$$

0.6.7. As in a symmetric monoidal ∞-category, given an object $\mathbf{C} \in 1\text{-Cat}^{\mathrm{St,cocmpl}}_{\mathrm{cont}}$ we can ask about its dualizability.

It is a basic fact that if $\mathbf{C}$ is compactly generated, then it is dualizable. Moreover, its dual can be described very explicitly: it is the ind-completion of $(\mathbf{C}^c)^{\mathrm{op}}$.

In other words, $\mathbf{C}^\vee$ is also compactly generated and we have a canonical equivalence

$$(\mathbf{C}^\vee)^c \simeq (\mathbf{C}^c)^{\mathrm{op}}. \tag{0.2}$$

0.6.8. *Categorical meaning of Verdier duality.* The equivalence (0.2) is key to the categorical understanding of such phenomena as Verdier duality. Indeed, let X be a scheme (of finite type), and consider the cocomplete stable ∞-category $\mathrm{Dmod}(X)$ of D-modules on X.

The subcategory $(\mathrm{Dmod}(X))^c$ consists of those objects that have finitely many cohomologies (with respect to the usual t-structure) all of which are *coherent* D-modules. Denote this subcategory by $\mathrm{Dmod}(X)_{\mathrm{coh}}$.

The usual Verdier duality for D-modules defines a contravariant auto-equivalence

$$\mathbb{D}_X^{\mathrm{Verdier}} : (\mathrm{Dmod}(X)_{\mathrm{coh}})^{\mathrm{op}} \simeq \mathrm{Dmod}(X)_{\mathrm{coh}}.$$

[2]Ind-completion is another operation that requires having a stable category, rather than a triangulated one.

Now, the above description of duality for compactly generated stable ∞-categories implies that we can perceive Verdier duality as an equivalence

$$\mathbf{D}_X^{\text{Verdier}} : (\text{Dmod}(X))^\vee \simeq \text{Dmod}(X),$$

which reduces to $\mathbb{D}_X^{\text{Verdier}}$ at the level of compact objects.

We also obtain a more functorial understanding of expressions such as "the Verdier conjugate of the *-direct image is the !-direct image". The categorical formulation of this is the fact that for a morphism of schemes $f : X \to Y$, the functors

$$f_{\text{dR},*} : \text{Dmod}(X) \to \text{Dmod}(Y) \text{ and } f^! : \text{Dmod}(Y) \to \text{Dmod}(X)$$

are each other's *duals* in terms of the identifications $\mathbf{D}_X^{\text{Verdier}}$ and $\mathbf{D}_Y^{\text{Verdier}}$, see Proposition 7.3.5.

0.6.9. In Sect. 8 we discuss stable monoidal ∞-categories, and algebras in them. This consists of studying the interaction of the concepts introduced in Sect. 3 with the Lurie tensor product.

Let us give one example. Let $\mathbf{A}$ be a stable monoidal ∞-category, and let $\mathbf{M}$ be a stable module category over $\mathbf{A}$. Let $\mathcal{A}$ be an algebra object in $\mathbf{A}$.

On the one hand we can consider the (stable) ∞-category $\mathcal{A}\text{-mod}(\mathbf{M})$ of $\mathcal{A}$-modules in $\mathbf{M}$. On the other hand, we can consider $\mathbf{A}$ as acting on itself on the left, and thus consider

$$\mathcal{A}\text{-mod} := \mathcal{A}\text{-mod}(\mathbf{A}).$$

The action of $\mathbf{A}$ on itself on the right makes $\mathcal{A}$-mod into a right $\mathbf{A}$-module category.

Now, the claim is (this is Corollary 8.5.7) that there is a canonical equivalence

$$\mathcal{A}\text{-mod}(\mathbf{M}) \simeq \mathcal{A}\text{-mod} \underset{\mathbf{A}}{\otimes} \mathbf{M}.$$

0.6.10. *Rigid monoidal categories.* In Sect. 9 we discuss a key technical notion of stable *rigid* monoidal ∞-category.

If a stable monoidal ∞-category $\mathbf{A}$ is compactly generated, then being rigid is equivalent to the combination of the following conditions:

- (i) the unit object in $\mathbf{A}$ is compact;
- (ii) the monoidal operation on $\mathbf{A}$ preserves compactness;
- (iii) every compact object of $\mathbf{A}$ admits a left and a right dual.

For example, the category of modules over a commutative algebra has this property.

From the point of view of its intrinsic properties, a rigid monoidal category can be as badly behaved as any other category. However, the ∞-category of its module categories satisfies very strong finiteness conditions.

For example, given a rigid symmetric monoidal ∞-category $\mathbf{A}$, we have:

(i) Any functor between $\mathbf{A}$-module categories that is lax-compatible with $\mathbf{A}$-actions, is actually strictly compatible;

(ii) The tensor product of $\mathbf{A}$-module categories is equivalent to the co-tensor product;

(iii) An **A**-module category is dualizable as such if and only if it is dualizable as a plain stable category, and the duals in both senses are isomorphic.

0.6.11. *DG categories.* We can now spell out our definition of DG categories:

Let Vect be the ∞-category of chain complexes of k-vector spaces. It is stable and cocomplete, and carries a symmetric monoidal structure. We define the ∞-category $\mathrm{DGCat}_{\mathrm{cont}}$ to be that of Vect-modules in $1\text{-}\mathrm{Cat}^{\mathrm{St,cocmpl}}_{\mathrm{cont}}$.

The stable monoidal category Vect is rigid (see above), and this ensures the good behavior of $\mathrm{DGCat}_{\mathrm{cont}}$.

1. $(\infty, 1)$-categories

In this section we make an attempt to write down a user guide to the theory of $(\infty, 1)$-categories. In that, the present section may be regarded as a digest of [**Lu1**, Chapters 1-3 and Sect. 5.2], with a view to applications (i.e., we will not be interested in how to construct the theory of $(\infty, 1)$-categories, but, rather, what one needs to know in order to use it).

The main difference between this section and the introductory Sect. 1 of [**Lu1**] is the following. In *loc.cit.* it is explained how to use quasi-categories (i.e., weak Kan simplicial sets) to capture the structures of higher category theory, the point of departure being that Kan simplicial sets incarnate *spaces*.

By contrast, we take the basic concepts of $(\infty, 1)$-categories on faith, and try to show how to use them to construct further notions. In that respect we try to stay *model independent*, i.e., we try to avoid, as much as possible, referring to simplicial sets that realize our $(\infty, 1)$-categories.

The reader familiar with [**Lu1**] can safely skip this section.

1.1. The basics. In most of the practical situations, when working with $(\infty, 1)$-categories, one *does not need to know* what they actually are, i.e., how exactly one defines the notion of $(\infty, 1)$-category.

What one does use is the syntax: one believes that the notion of $(\infty, 1)$-category exists, and all one needs to know is how to use the words correctly.

Below is the summary of the few basic words of the vocabulary. However, as was mentioned in Sect. 0.4.2, this vocabulary is flawed and incomplete. So, strictly speaking, what follows does nothing more than introduce notation, because circularity appears from the start (e.g., we talk about full subcategories and adjoints).

The reference for the material here is [**Lu1**, Sect. 1.2].

1.1.1. We let 1-Cat denote the $(\infty, 1)$-category of $(\infty, 1)$-categories.

1.1.2. We let Spc denote the $(\infty, 1)$-category of spaces. We have a canonical fully faithful embedding

$$\mathrm{Set} \hookrightarrow \mathrm{Spc},$$

which admits a left adjoint, denoted

$$\mathcal{S} \mapsto \pi_0(\mathcal{S}).$$

In particular, for any $\mathcal{S} \in \mathrm{Spc}$, we have a canonical map of spaces $\mathcal{S} \to \pi_0(\mathcal{S})$.

We denote by $* \in \mathrm{Set} \subset \mathrm{Spc}$ the point space.

1.1.3. We will regard Spc as a full subcategory of 1-Cat. In particular, we will regard a space as an $(\infty, 1)$-category, and maps between spaces as functors between the corresponding $(\infty, 1)$-categories.

We will refer to objects of the $(\infty, 1)$-category corresponding to a space $\mathcal{S}$ as *points* of $\mathcal{S}$.

The inclusion $\mathrm{Spc} \hookrightarrow 1\text{-Cat}$ admits a right adjoint, denoted $\mathbf{C} \mapsto \mathbf{C}^{\mathrm{Spc}}$; it is usually referred to as 'discarding non-invertible morphisms'.

1.1.4. For an $(\infty, 1)$-category $\mathbf{C}$, and objects $\mathbf{c}_0, \mathbf{c}_1 \in \mathbf{C}$, we denote by

$$\mathrm{Maps}_{\mathbf{C}}(\mathbf{c}_0, \mathbf{c}_1) \in \mathrm{Spc}$$

the corresponding *mapping space*.

1.1.5. We let $1\text{-Cat}_{\mathrm{ordn}}$ denote the full subcategory of 1-Cat formed by ordinary categories.

This inclusion admits a left adjoint, denoted $\mathbf{C} \mapsto \mathbf{C}^{\mathrm{ordn}}$. (Sometimes, $\mathbf{C}^{\mathrm{ordn}}$ is called the *homotopy category of* $\mathbf{C}$ and denoted $\mathrm{Ho}(\mathbf{C})$.) The objects of $\mathbf{C}^{\mathrm{ordn}}$ are the same as those of $\mathbf{C}$, and we have

$$\mathrm{Hom}_{\mathbf{C}^{\mathrm{ordn}}}(\mathbf{c}_0, \mathbf{c}_1) = \pi_0\left(\mathrm{Maps}_{\mathbf{C}}(\mathbf{c}_0, \mathbf{c}_1)\right).$$

Warning: we need to distinguish the $(\infty, 1)$-category $1\text{-Cat}_{\mathrm{ordn}}$ (which is in fact a $(2, 1)$-category) from the ordinary category $(1\text{-Cat})^{\mathrm{ordn}} = \mathrm{Ho}(1\text{-Cat})$.

1.1.6. A map $\phi : \mathbf{c}_0 \to \mathbf{c}_1$ in $\mathbf{C}$ (i.e., a point in $\mathrm{Maps}_{\mathbf{C}}(\mathbf{c}_0, \mathbf{c}_1)$) is said to be an *isomorphism* if the corresponding map in $\mathbf{C}^{\mathrm{ordn}}$, i.e., the image of ϕ under the projection

$$\mathrm{Maps}_{\mathbf{C}}(\mathbf{c}_0, \mathbf{c}_1) \to \pi_0\left(\mathrm{Maps}_{\mathbf{C}}(\mathbf{c}_0, \mathbf{c}_1)\right) = \mathrm{Hom}_{\mathbf{C}^{\mathrm{ordn}}}(\mathbf{c}_0, \mathbf{c}_1),$$

is an isomorphism.

1.1.7. For a pair of $(\infty, 1)$-categories $\mathbf{C}$ and $\mathbf{D}$, we denote by $\mathrm{Funct}(\mathbf{D}, \mathbf{C})$ the $(\infty, 1)$ category of functors $\mathbf{D} \to \mathbf{C}$.

We have

$$\mathrm{Funct}(*, \mathbf{C}) \simeq \mathbf{C}$$

and

$$\mathrm{Maps}_{1\text{-Cat}}(\mathbf{D}, \mathbf{C}) = (\mathrm{Funct}(\mathbf{D}, \mathbf{C}))^{\mathrm{Spc}}.$$

A functor $F : \mathbf{C} \to \mathbf{D}$ is said to be an *equivalence* if it is an isomorphism in 1-Cat, i.e., if it induces an isomorphism in $(1\text{-Cat})^{\mathrm{ordn}}$ (which implies, but is much stronger than asking that $F^{\mathrm{ordn}} : \mathbf{C}^{\mathrm{ordn}} \to \mathbf{D}^{\mathrm{ordn}}$ be an isomorphism in $1\text{-Cat}_{\mathrm{ordn}}$).

1.1.8. For a diagram of categories

$$\mathbf{C}' \to \mathbf{C} \leftarrow \mathbf{C}'',$$

we can form their *fiber product*

$$\mathbf{C}' \underset{\mathbf{C}}{\times} \mathbf{C}'' \in 1\text{-Cat}.$$

For $\mathbf{C}' = \mathbf{C}'' = *$, $\mathbf{C} = \mathcal{S} \in \text{Spc}$, and the maps $* \to \mathcal{S} \leftarrow *$ corresponding to a particular point $s \in \mathcal{S}$, we will denote by $\Omega(\mathcal{S})$ the loop space of $\mathcal{S}$ with base point s,

$$\Omega(\mathcal{S}) = * \underset{\mathcal{S}}{\times} *.$$

For $(\mathcal{S}, s)$ as above, the homotopy groups $\pi_i(\mathcal{S}, s)$ are defined inductively by

$$\pi_i(\mathcal{S}, s) = \pi_{i-1}(\Omega(\mathcal{S})).$$

1.1.9. The $(\infty, 1)$-category 1-Cat carries a canonical involutive auto-equivalence

$$\mathbf{C} \mapsto \mathbf{C}^{\mathrm{op}}. \tag{1.1}$$

1.1.10. For $n = 0, 1, \ldots$ we let $[n]$ denote the ordinary category $0 \to 1 \to \ldots \to n$. We have $[0] = *$, this is the point-category.

We let $\mathbf{\Delta}$ denote the full subcategory of $1\text{-Cat}_{\mathrm{ordn}}$, spanned by the objects $[n]$.

The category $\mathbf{\Delta}$ carries a canonical involutive auto-equivalence, denoted rev: it acts as reversal on each $[n]$, i.e.,

$$rev : i \mapsto n - i.$$

(Note that rev acts as the identity on objects of $\mathbf{\Delta}$.)

1.2. Some auxiliary notions. In this subsection we introduce some terminology and notation to be used throughout the book.

1.2.1. A functor between $(\infty, 1)$-categories $F : \mathbf{D} \to \mathbf{C}$ is said to be *fully faithful* if for for every $\mathbf{d}_1, \mathbf{d}_2 \in \mathbf{D}$ the map

$$\mathrm{Maps}_{\mathbf{D}}(\mathbf{d}_1, \mathbf{d}_2) \to \mathrm{Maps}_{\mathbf{C}}(F(\mathbf{d}_1), F(\mathbf{d}_2))$$

is a *isomorphism* in Spc.

A map of spaces $F : \mathcal{S}_0 \to \mathcal{S}_1$ is said to be a *monomorphism* if it is fully faithful as a functor, when $\mathcal{S}_0$ and $\mathcal{S}_1$ are regarded as $(\infty, 1)$-categories.

Concretely, F is a monomorphism if $\pi_0(F)$ is injective, and for every point $s_0 \in S_0$, the induced map $\pi_i(S_0, s_0) \to \pi_i(S_1, F(s_0))$ is an isomorphism for all $i > 0$.

1.2.2. Let $\mathbf{C}$ be an $(\infty, 1)$-category. Then to every full subcategory $\overline{\mathbf{C}}'$ of $\mathbf{C}^{\mathrm{ordn}}$ one can attach an ∞-category $\mathbf{C}'$. It has the same objects as $\overline{\mathbf{C}}'$ and for $\mathbf{c}_1, \mathbf{c}_2 \in \mathbf{C}'$, we have

$$\mathrm{Maps}_{\mathbf{C}'}(\mathbf{c}_1, \mathbf{c}_2) = \mathrm{Maps}_{\mathbf{C}}(\mathbf{c}_1, \mathbf{c}_2).$$

We shall refer to $(\infty, 1)$-categories arising in the way as *full subcategories of* $\mathbf{C}$.

A fully faithful functor is an equivalence onto a full subcategory.

1.2.3. A *full subspace* of a space $\mathcal{S}$ is the same as a full subcategory of $\mathcal{S}$, considered as an $(\infty, 1)$-category. Those are in bijection with subsets of $\pi_0(\mathcal{S})$.

A connected component of $\mathcal{S}$ is a full subspace that projects to a single point in $\pi_0(\mathcal{S})$.

1.2.4. A functor between $(\infty, 1)$-categories $F : \mathbf{D} \to \mathbf{C}$ is said to be *1-fully faithful* if for for every $\mathbf{d}_1, \mathbf{d}_2 \in \mathbf{D}$ the map

$$\mathrm{Maps}_{\mathbf{D}}(\mathbf{d}_1, \mathbf{d}_2) \to \mathrm{Maps}_{\mathbf{C}}(F(\mathbf{d}_1), F(\mathbf{d}_2))$$

is a *monomorphism* in Spc.

If $\mathbf{D}$ and $\mathbf{C}$ are ordinary categories, a functor between them is 1-fully faithful if and only if it induces an *injection* on Hom sets.

1.2.5. A functor between $(\infty, 1)$-categories $F : \mathbf{D} \to \mathbf{C}$ is said to be *1-replete* if it is 1-fully faithful, and for every $\mathbf{d}_1, \mathbf{d}_2 \in \mathbf{D}$, the connected components of $\mathrm{Maps}_{\mathbf{C}}(F(\mathbf{d}_1), F(\mathbf{d}_2))$ *that correspond to isomorphisms* are in the image of

$$\mathrm{Maps}_{\mathbf{D}}(\mathbf{d}_1, \mathbf{d}_2).$$

It is not difficult to show that a functor is 1-replete if and only if it is 1-fully faithful and $\mathbf{D}^{\mathrm{Spc}} \to \mathbf{C}^{\mathrm{Spc}}$ is a monomorphism.

1.2.6. Let $\overline{\mathbf{C}}$ be an ordinary category. By a *1-full subcategory* we shall mean the category obtained by choosing a sub-class $\overline{\mathbf{C}}'$ of objects in $\overline{\mathbf{C}}$, and for every $\mathbf{c}_1, \mathbf{c}_2 \in \overline{\mathbf{C}}'$ a subset $\mathrm{Hom}_{\overline{\mathbf{C}}'}(\mathbf{c}_1, \mathbf{c}_2) \subset \mathrm{Hom}_{\overline{\mathbf{C}}}(\mathbf{c}_1, \mathbf{c}_2)$, such that $\mathrm{Hom}_{\overline{\mathbf{C}}'}(\mathbf{c}_1, \mathbf{c}_2)$ contains all isomorphisms and is closed under compositions.

Let $\mathbf{C}$ be an $(\infty, 1)$-category. Then to every 1-full subcategory $\overline{\mathbf{C}}'$ of $\mathbf{C}^{\mathrm{ordn}}$ one can attach an $(\infty, 1)$-category $\mathbf{C}'$. It has the same objects as $\overline{\mathbf{C}}'$. For $\mathbf{c}_1, \mathbf{c}_2 \in \mathbf{C}'$, we have

$$\mathrm{Maps}_{\mathbf{C}'}(\mathbf{c}_1, \mathbf{c}_2) = \mathrm{Maps}_{\mathbf{C}}(\mathbf{c}_1, \mathbf{c}_2) \underset{\mathrm{Hom}_{\mathbf{C}^{\mathrm{ordn}}}(\mathbf{c}_1, \mathbf{c}_2)}{\times} \mathrm{Hom}_{\overline{\mathbf{C}}'}(\mathbf{c}_1, \mathbf{c}_2).$$

We shall refer to $(\infty, 1)$-categories arising in the way as *1-full subcategories of* $\mathbf{C}$.

1.2.7. In the above situation, for any $\mathbf{D} \in 1\text{-Cat}$, the resulting functor

$$\mathrm{Funct}(\mathbf{D}, \mathbf{C}') \to \mathrm{Funct}(\mathbf{D}, \mathbf{C})$$

is 1-replete. I.e., if a functor $\mathbf{D} \to \mathbf{C}$ can be factored through $\mathbf{C}'$, it can be done in an essentially unique way.

Vice versa, if $F : \mathbf{D} \to \mathbf{C}$ is a functor and the corresponding functor $\mathbf{D}^{\mathrm{ordn}} \to \mathbf{C}^{\mathrm{ordn}}$ factors (automatically uniquely) through a functor $\mathbf{D}^{\mathrm{ordn}} \to \overline{\mathbf{C}}'$, then F gives rise to a well-defined functor $\mathbf{D} \to \mathbf{C}'$.

In the above situation, the functor $\mathbf{D} \to \mathbf{C}'$ is an equivalence if and only if F is 1-replete and $\mathbf{D}^{\mathrm{ordn}} \to \overline{\mathbf{C}}'$ is an equivalence.

In particular, a 1-replete functor is an equivalence onto a uniquely defined 1-full subcategory.

1.2.8. A functor $F : \mathbf{D} \to \mathbf{C}$ is said to be conservative if for a morphism $\alpha \in \mathrm{Maps}_{\mathbf{D}}(\mathbf{d}_0, \mathbf{d}_1)$ the fact that $F(\alpha)$ is an isomorphism implies that α itself is an isomorphism.

1.3. Cartesian and coCartesian fibrations. Now that we have the basic words of the vocabulary, we want to take the theory of $(\infty,1)$-categories off the ground. Here are two basic things that one would want to do:

(1) For an $(\infty,1)$-category $\mathbf{C}$, define the Yoneda functor $\mathbf{C} \times \mathbf{C}^{\mathrm{op}} \to \mathrm{Spc}$.

(2) For a functor $F : \mathbf{D} \to \mathbf{C}$ we would like to talk about its left or right adjoint.

It turns out that this is much easier said than done: the usual way of going about this in ordinary category theory uses the construction of functors by specifying what they do on objects and morphisms, something that is not allowed in higher category theory.

To overcome this, we will use the device of straightening/unstraightening, described in the next subsection. In order to explain it, we will first need to introduce the key notion of Cartesian/coCartesian fibration.

The reference for the material in this subsection is [**Lu1**, Sect. 2.4].

1.3.1. *Cartesian arrows.* Let $F : \mathbf{D} \to \mathbf{C}$ be a functor between $(\infty,1)$-categories. We shall say that a morphism $\mathbf{d}_0 \overset{\alpha}{\to} \mathbf{d}_1$ in $\mathbf{D}$ is *Cartesian* over $\mathbf{C}$ if for every $\mathbf{d}' \in \mathbf{D}$, the map

$$\mathrm{Maps}_{\mathbf{D}}(\mathbf{d}',\mathbf{d}_0) \to \mathrm{Maps}_{\mathbf{D}}(\mathbf{d}',\mathbf{d}_1) \underset{\mathrm{Maps}_{\mathbf{C}}(F(\mathbf{d}'),F(\mathbf{d}_1))}{\times} \mathrm{Maps}_{\mathbf{C}}(F(\mathbf{d}'),F(\mathbf{d}_0))$$

is an isomorphism in Spc.

1.3.2. *Cartesian fibrations.* A functor $F : \mathbf{D} \to \mathbf{C}$ is said to be a *Cartesian fibration* if for every morphism $\mathbf{c}_0 \to \mathbf{c}_1$ in $\mathbf{C}$ and an object $\mathbf{d}_1 \in \mathbf{D}$ equipped with an isomorphism $F(\mathbf{d}_1) \simeq \mathbf{c}_1$, there *exists a Cartesian* morphism $\mathbf{d}_0 \to \mathbf{d}_1$ that fits into a commutative diagram

$$\begin{array}{ccc} F(\mathbf{d}_0) & \longrightarrow & F(\mathbf{d}_1) \\ {\scriptstyle\sim}\downarrow & & \downarrow{\scriptstyle\sim} \\ \mathbf{c}_0 & \longrightarrow & \mathbf{c}_1. \end{array}$$

1.3.3. *Cartesian fibrations in spaces.* We shall say that a functor $F : \mathbf{D} \to \mathbf{C}$ is a *Cartesian fibration in spaces* if it is a Cartesian fibration and for every $\mathbf{c} \in \mathbf{C}$, the $(\infty,1)$-category

$$\mathbf{D}_{\mathbf{c}} := \mathbf{D} \underset{\mathbf{C}}{\times} \{\mathbf{c}\}$$

is a space.

An alternative terminology for 'Cartesian fibration in spaces' is *right fibration*, see [**Lu1**, Sect. 2.1].

1.3.4. *CoCartesian counterparts.* Inverting the arrows, one obtains the parallel notions of coCartesian morphism, coCartesian fibrations and coCartesian fibrations in spaces (a.k.a. *left fibration*).

1.3.5. *Over- and under-categories.* Given a functor $F : \mathbf{I} \to \mathbf{C}$ consider the corresponding over-category and under-category

$$\mathbf{C}_{/F} := \mathbf{C} \underset{\mathrm{Funct}(\mathbf{I},\mathbf{C})}{\times} \mathrm{Funct}([1] \times \mathbf{I}, \mathbf{C}) \underset{\mathrm{Funct}(\mathbf{I},\mathbf{C})}{\times} \{F\}$$

and

$$\mathbf{C}_{F/} := \{F\} \underset{\mathrm{Funct}(\mathbf{I},\mathbf{C})}{\times} \mathrm{Funct}([1] \times \mathbf{I}, \mathbf{C}) \underset{\mathrm{Funct}(\mathbf{I},\mathbf{C})}{\times} \mathbf{C},$$

where the functors

$$\mathrm{Funct}([1] \times \mathbf{I}, \mathbf{C}) \to \mathrm{Funct}(\mathbf{I}, \mathbf{C}),$$

are given by evaluation at the objects 1 and 0 in $[1]$, respectively, and the functor

$$\mathbf{C} \to \mathrm{Funct}(\mathbf{I}, \mathbf{C})$$

corresponds to

$$\mathbf{C} \simeq \mathrm{Funct}(*, \mathbf{C}) \to \mathrm{Funct}(\mathbf{I}, \mathbf{C}).$$

For future use we mention that when $\mathbf{I} = *$ and F is given by an object $\mathbf{c} \in \mathbf{C}$, we will simply write $\mathbf{C}_{/\mathbf{c}}$ for $\mathbf{C}_{/F}$ and $\mathbf{C}_{\mathbf{c}/}$ for $\mathbf{C}_{/F}$, respectively.

The forgetful functors

$$\mathbf{C}_{/F} \to \mathbf{C} \text{ and } \mathbf{C}_{F/} \to \mathbf{C}$$

are a Cartesian and a coCartesian fibrations in spaces, respectively.

1.3.6. Note that we have the following canonical isomorphism of spaces: for $\mathbf{c}_0, \mathbf{c}_1 \in \mathbf{C}$

$$\mathrm{Funct}([1], \mathbf{C}) \underset{\mathbf{C}\times\mathbf{C}}{\times} \{\mathbf{c}_0, \mathbf{c}_1\} \simeq \mathrm{Maps}_{\mathbf{C}}(\mathbf{c}_0, \mathbf{c}_1), \tag{1.2}$$

where the left-hand side, although defined to be an $(\infty,1)$-category, is actually a space.

In particular, for $\mathbf{c}, \mathbf{c}' \in \mathbf{C}$, we have

$$(\mathbf{C}_{/\mathbf{c}})_{\mathbf{c}'} \simeq \mathrm{Maps}_{\mathbf{C}}(\mathbf{c}', \mathbf{c}) \text{ and } (\mathbf{C}_{\mathbf{c}/})_{\mathbf{c}'} \simeq \mathrm{Maps}_{\mathbf{C}}(\mathbf{c}, \mathbf{c}').$$

(we remind that the superscript $\mathbf{c}'$ means taking the fiber over $\mathbf{c}'$).

Taking $\mathbf{C} = 1\text{-Cat}$, from (1.2) we obtain

$$\mathrm{Funct}([1], 1\text{-Cat}) \underset{1\text{-Cat}\times 1\text{-Cat}}{\times} \{\mathbf{C}_0, \mathbf{C}_1\} \simeq (\mathrm{Funct}(\mathbf{C}_0, \mathbf{C}_1))^{\mathrm{Spc}}. \tag{1.3}$$

1.3.7. For future reference we introduce the following notation. For a functor $F : [1] \to \mathbf{C}$ that sends $0 \mapsto \mathbf{c}_0$ and $1 \mapsto \mathbf{c}_1$ we we will denote by

$$\mathbf{C}_{\mathbf{c}_0//\mathbf{c}_1}$$

the fiber product

$$\mathrm{Funct}([2], \mathbf{C}) \underset{\mathrm{Funct}([1],\mathbf{C})}{\times} *,$$

where $* \to \mathrm{Funct}([1], \mathbf{C})$ corresponds to the initial functor F, and $\mathrm{Funct}([2], \mathbf{C}) \to \mathrm{Funct}([1], \mathbf{C})$ is given by precomposition with

$$[1] \to [2], \quad 0 \mapsto 0, \ 1 \mapsto 2.$$

This is the $(\infty,1)$-category, whose objects are diagrams

$$\mathbf{c}_0 \to \mathbf{c} \to \mathbf{c}_1,$$

where the composition is the map $\mathbf{c}_0 \to \mathbf{c}_1$, specified by F.

1.4. Straightening/unstraightening. Straightening, also known as the Grothendieck construction, is the higher-categorical counterpart to the fact that the datum of a Cartesian (resp., coCartesian) fibration of ordinary categories $\mathbf{D} \to \mathbf{C}$ is equivalent to the datum of a functor from $\mathbf{C}^{\mathrm{op}}$ (resp., $\mathbf{C}$) to the category of categories.

It is hard to overestimate the importance of this assertion in higher category theory: it paves a way to constructing functors $\mathbf{C} \to 1\text{-Cat}$.

The reason being that it is usually easier to exhibit a functor $\mathbf{D} \to \mathbf{C}$ and then *check its property* of being a Cartesian/coCartesian fibration, than to construct a functor $\mathbf{C} \to 1\text{-Cat}$.

1.4.1. Fix an $(\infty,1)$-category $\mathbf{C}$. Consider the category

$$1\text{-Cat}_{/\mathbf{C}}.$$

Note that its objects are pairs $(\mathbf{D}; \mathbf{D} \overset{F}{\to} \mathbf{C})$.

Let $\mathrm{coCart}_{/\mathbf{C}}$ (resp., $0\text{-coCart}_{/\mathbf{C}}$) be the full subcategory of $1\text{-Cat}_{/\mathbf{C}}$ whose objects are those $(\mathbf{D}, F)$, for which F is a coCartesian fibration (resp., coCartesian fibration in spaces).

Let $(\mathrm{coCart}_{/\mathbf{C}})_{\mathrm{strict}}$ be the 1-full subcategory of $\mathrm{coCart}_{/\mathbf{C}}$, where we allow as 1-morphisms those functors $\mathbf{D}_1 \to \mathbf{D}_2$ over $\mathbf{C}$ that send coCartesian arrows to coCartesian arrows. We note that the inclusion

$$(\mathrm{coCart}_{/\mathbf{C}})_{\mathrm{strict}} \cap 0\text{-coCart}_{/\mathbf{C}} \hookrightarrow 0\text{-coCart}_{/\mathbf{C}}$$

is an equivalence.

1.4.2. *Straightening/unstraightening for coCartesian fibrations.* The following is the basic feature of coCartesian fibrations (see [**Lu1**, Sect. 3.2]):

There is a canonical equivalence between $(\mathrm{coCart}_{/\mathbf{C}})_{\mathrm{strict}}$ *and* $\mathrm{Funct}(\mathbf{C}, 1\text{-Cat})$.

Under the above equivalence, the full subcategory

$$0\text{-coCart}_{/\mathbf{C}} \subset (\mathrm{coCart}_{/\mathbf{C}})_{\mathrm{strict}}$$

corresponds to the full subcategory

$$\mathrm{Funct}(\mathbf{C}, \mathrm{Spc}) \subset \mathrm{Funct}(\mathbf{C}, 1\text{-Cat}).$$

1.4.3. Explicitly, for a coCartesian fibration $\mathbf{D} \to \mathbf{C}$, the value of the corresponding functor $\mathbf{C} \to 1\text{-Cat}$ on $\mathbf{c} \in \mathbf{C}$ equals the fiber $\mathbf{D}_{\mathbf{c}}$ of $\mathbf{D}$ over $\mathbf{c}$.

Vice versa, given a functor

$$\Phi : \mathbf{C} \to 1\text{-Cat}, \quad \mathbf{c} \mapsto \Phi(\mathbf{c}), \quad (\mathbf{c}_0 \overset{f}{\to} \mathbf{c}_1) \mapsto \Phi(\mathbf{c}_0) \overset{\Phi_f}{\longrightarrow} \Phi(\mathbf{c}_1),$$

the objects of the corresponding coCartesian fibration $\mathbf{D} \to \mathbf{C}$ are pairs $(\mathbf{c} \in \mathbf{C}, \mathbf{d} \in \Phi(\mathbf{c}))$, and morphisms

$$\mathrm{Maps}_{\mathbf{D}}((\mathbf{c}_0, \mathbf{d}_0 \in \Phi(\mathbf{c}_0)), (\mathbf{c}_1, \mathbf{d}_1 \in \Phi(\mathbf{c}_1)))$$

are pairs consisting of $f \in \mathrm{Maps}_{\mathbf{C}}(\mathbf{c}_0, \mathbf{c}_1)$ and $g \in \mathrm{Maps}_{\Phi(\mathbf{c}_1)}(\Phi_f(\mathbf{d}_0), \mathbf{d}_1)$.

1.4.4. One defines the $(\infty, 1)$-categories

$$0\text{-Cart}_{/\mathbf{C}} \subset (\text{Cart}_{/\mathbf{C}})_{\text{strict}} \subset \text{Cart}_{/\mathbf{C}} \subset 1\text{-Cat}_{/\mathbf{C}}$$

in a similar way.

Note that the involution (1.1) defines an equivalence $1\text{-Cat}_{/\mathbf{C}} \to 1\text{-Cat}_{/\mathbf{C}^{\text{op}}}$ that identifies

$$0\text{-coCart}_{/\mathbf{C}} \simeq 0\text{-Cart}_{/\mathbf{C}^{\text{op}}}, \quad (\text{coCart}_{/\mathbf{C}})_{\text{strict}} \simeq (\text{Cart}_{/\mathbf{C}^{\text{op}}})_{\text{strict}}$$

and

$$\text{coCart}_{/\mathbf{C}} \simeq \text{Cart}_{/\mathbf{C}^{\text{op}}}.$$

1.4.5. *Straightening/unstraightening for Cartesian fibrations.* From Sect. 1.4.2, and using the involution (1.1) on 1-Cat, one obtains:

There is a canonical equivalence between $(\text{Cart}_{/\mathbf{C}})_{\text{strict}}$ *and* $\text{Funct}(\mathbf{C}^{\text{op}}, 1\text{-Cat})$.

Under the above equivalence, the full subcategory

$$0\text{-Cart}_{/\mathbf{C}} \subset (\text{Cart}_{/\mathbf{C}})_{\text{strict}}$$

corresponds to the full subcategory

$$\text{Funct}(\mathbf{C}^{\text{op}}, \text{Spc}) \subset \text{Funct}(\mathbf{C}^{\text{op}}, 1\text{-Cat}).$$

Explicitly, for a Cartesian fibration $\mathbf{D} \to \mathbf{C}$, the value of the corresponding functor

$$\mathbf{C}^{\text{op}} \to 1\text{-Cat}$$

on $\mathbf{c} \in \mathbf{C}$ still equals the fiber $\mathbf{D}_{\mathbf{c}}$ of $\mathbf{D}$ over $\mathbf{c}$.

1.5. Yoneda. In this subsection we will illustrate how one uses straightening/unstraightening by constructing the various incarnations of the Yoneda functor.

1.5.1. For an $(\infty, 1)$-category $\mathbf{C}$, consider the $(\infty, 1)$-category $\text{Funct}([1], \mathbf{C})$, equipped with the functor

$$\text{Funct}([1], \mathbf{C}) \to \text{Funct}(*, \mathbf{C}) \times \text{Funct}(*, \mathbf{C}) \simeq \mathbf{C} \times \mathbf{C}, \tag{1.4}$$

given by evaluation on $0, 1 \in [1]$.

We can view the above functor as a morphism in the category $(\text{Cart}_{/\mathbf{C}})_{\text{strict}}$ with respect to the projection on the first factor.

1.5.2. Applying straightening, the above morphism gives rise to a morphism in the $(\infty, 1)$-category $\text{Funct}(\mathbf{C}^{\text{op}}, 1\text{-Cat})$ from the functor

$$\mathbf{c} \mapsto \mathbf{C}_{\mathbf{c}/}$$

to the functor with constant value $\mathbf{C} \in 1\text{-Cat}$.

1.5.3. For any triple of $(\infty, 1)$-categories we have a canonical isomorphism

$$\text{Funct}(\mathbf{E}, \text{Funct}(\mathbf{E}', \mathbf{D})) \simeq \text{Funct}(\mathbf{E} \times \mathbf{E}', \mathbf{D}) \simeq \text{Funct}(\mathbf{E}', \text{Funct}(\mathbf{E}, \mathbf{D})).$$

In particular, taking $\mathbf{E}' = [1]$ and a fixed $F : \mathbf{E} \to \mathbf{D}$ and $\mathbf{d} \in \mathbf{D}$, using (1.2), we obtain that the datum of a morphism in $\text{Funct}(\mathbf{E}, \mathbf{D})$ from F to the constant functor with value $\mathbf{d}$ is equivalent to the datum of a map

$$\mathbf{E} \to \mathbf{D}_{/\mathbf{d}},$$

whose composition with the projection $\mathbf{D}_{/\mathbf{d}} \to \mathbf{D}$, is identified with F.

1.5.4. Thus (taking $\mathbf{E} = \mathbf{C}^{\mathrm{op}}$ and $\mathbf{D} = 1\text{-Cat}$), we can view the datum of the morphism in Sect. 1.5.2 as a functor from $\mathbf{C}^{\mathrm{op}}$ to $1\text{-Cat}_{/\mathbf{C}}$.

It is easy to check that the latter functor factors through

$$0\text{-coCart}_{/\mathbf{C}} \subset 1\text{-Cat}_{/\mathbf{C}}.$$

1.5.5. Applying straightening again, we thus obtain a functor

$$\mathbf{C}^{\mathrm{op}} \to \mathrm{Funct}(\mathbf{C}, \mathrm{Spc}),$$

hence a functor

$$\mathrm{Yon}_{\mathbf{C}} : \mathbf{C}^{\mathrm{op}} \times \mathbf{C} \to \mathrm{Spc}.$$

1.6. Enhanced version of straightening/unstraightening. In this subsection we will discuss a version of the straightening/unstraightening equivalence that takes into account functoriality in the base $(\infty,1)$-category $\mathbf{C}$.

1.6.1. Consider the $(\infty,1)$-category $\mathrm{Funct}([1], 1\text{-Cat})$. Note that its objects are triples

$$\mathbf{D} \overset{F}{\to} \mathbf{C}.$$

Let

$$\mathrm{Funct}^{\mathrm{coCart}}([1], 1\text{-Cat}) \subset \mathrm{Funct}([1], 1\text{-Cat})$$

be the full subcategory whose objects are those $\mathbf{D} \overset{F}{\to} \mathbf{C}$ that are coCartesian fibrations.

Let

$$\mathrm{Funct}^{\mathrm{coCart}}([1], 1\text{-Cat})_{\mathrm{strict}} \subset \mathrm{Funct}^{\mathrm{coCart}}([1], 1\text{-Cat})$$

be the 1-full subcategory, where we only allow as morphisms those commutative diagrams

$$\begin{array}{ccc} \mathbf{D}_1 & \xrightarrow{G_{\mathbf{D}}} & \mathbf{D}_2 \\ {\scriptstyle F_1}\downarrow & & \downarrow{\scriptstyle F_2} \\ \mathbf{C}_1 & \xrightarrow{G_{\mathbf{C}}} & \mathbf{C}_2 \end{array}$$

for which the functor $G_{\mathbf{D}}$ sends morphisms in $\mathbf{D}_1$ coCartesian over $\mathbf{C}_1$ to morphisms in $\mathbf{D}_2$ coCartesian over $\mathbf{C}_2$.

Evaluation on $1 \in [1]$ defines a functor

$$\mathrm{Funct}^{\mathrm{coCart}}([1], 1\text{-Cat})_{\mathrm{strict}} \to 1\text{-Cat}. \tag{1.5}$$

The functor (1.5) is a Cartesian fibration.

1.6.2. An enhanced version of the straightening/unstraightening equivalence says:

The functor $1\text{-Cat}^{\mathrm{op}} \to 1\text{-Cat}$ *corresponding to the Cartesian fibration* (1.5) *is canonically isomorphic to the functor*

$$\mathbf{C} \mapsto \mathrm{Funct}(\mathbf{C}, 1\text{-Cat}).$$

1.6.3. Again, by applying the involution $\mathbf{D} \mapsto \mathbf{D}^{\mathrm{op}}$, we obtain a counterpart of Sect. 1.6.2 for Cartesian fibrations:

The functor $1\text{-Cat}^{\mathrm{op}} \to 1\text{-Cat}$ *corresponding to the Cartesian fibration*

$$\mathrm{Funct}^{\mathrm{Cart}}([1], 1\text{-Cat})_{\mathrm{strict}} \to 1\text{-Cat}$$

is canonically isomorphic to the functor

$$\mathbf{C} \mapsto \mathrm{Funct}(\mathbf{C}^{\mathrm{op}}, 1\text{-Cat}).$$

1.7. Adjoint functors. In this subsection we will finally introduce the notion of *adjoint functor*, following [**Lu1**, Sect. 5.2.1].

However, this will not be the end of the story. We will *not* describe the datum of an adjunction as pair of a unit map and a co-unit map that satisfy some natural conditions (because in the context of higher categories, there is an infinite tail of these conditions).

We will return to the latter approach to adjunction in Sect. 4.4, and more fundamentally in Chapter 12: it turns out that it is most naturally described in the context of $(\infty, 2)$-categories.

1.7.1. Let $F : \mathbf{C}_0 \to \mathbf{C}_1$ be a functor. Using (1.3), we can view F as a functor $[1] \to 1\text{-Cat}$. We now apply unstraightening and regard it as a coCartesian fibration

$$\widetilde{\mathbf{C}} \to [1]. \tag{1.6}$$

We shall say that F *admits a right adjoint* if the above functor (1.6) is a *bi-Cartesian fibration*, i.e., if it happens to be a Cartesian fibration, in addition to being a coCartesian one.

In this case, viewing (1.6) as a *Cartesian fibration* and applying straightening, we transform (1.6) into to a functor

$$[1]^{\mathrm{op}} \to 1\text{-Cat}. \tag{1.7}$$

The resulting functor $\mathbf{C}_1 \to \mathbf{C}_0$ (obtained by applying the equivalence (1.3) to the functor (1.7)) is called the *right adjoint* of F, and denoted F^R. By construction, F^R is uniquely determined by F.

1.7.2. Inverting the arrows, we obtain the notion of a functor $G : \mathbf{D}_0 \to \mathbf{D}_1$, *admitting a left adjoint*. We denote the left adjoint of G by G^L.

By construction, the data of realizing G as a right adjoint of F is equivalent to the data of realizing F as a left adjoint of G: both are encoded by a bi-Cartesian fibration

$$\mathbf{E} \to [1].$$

By construction, for $\mathbf{c}_0 \in \mathbf{C}_0$ and $\mathbf{c}_1 \in \mathbf{C}_1$ we have a canonical isomorphism in Spc

$$\mathrm{Maps}_{\mathbf{C}_0}(\mathbf{c}_0, F^R(\mathbf{c}_1)) \simeq \mathrm{Maps}_{\mathbf{C}_1}(F(\mathbf{c}_0), \mathbf{c}_1).$$

1.7.3. Let us be in the situation Sect. 1.7.1, but *without* assuming that (1.6) is bi-Cartesian. Let $\mathbf{C}'_1 \subset \mathbf{C}_1$ be the full subcategory consisting of those objects $\mathbf{c}_1 \in \mathbf{C}_1$, for which there exists a *Cartesian* morphism

$$\mathbf{c}_0 \to \mathbf{c}_1$$

in $\widetilde{\mathbf{C}}$, covering the morphism $0 \to 1$ in $[1]$.

Let $\widetilde{\mathbf{C}}' \subset \widetilde{\mathbf{C}}$ be the corresponding full subcategory of $\widetilde{\mathbf{C}}$, so that

$$\widetilde{\mathbf{C}}'_0 = \mathbf{C}_0 \text{ and } \widetilde{\mathbf{C}}'_1 = \mathbf{C}'_1.$$

The functor

$$\widetilde{\mathbf{C}}' \to [1]$$

is now a Cartesian fibration. Applying straightening, we obtain a functor $F'^R : \mathbf{C}'_1 \to \mathbf{C}_0$.

1.7.4. We will refer to F'^R as the *partially defined right adjoint of* F. By construction, we have a canonical isomorphism

$$\mathrm{Maps}_{\mathbf{C}_1}(F(\mathbf{c}_0), \mathbf{c}_1) \simeq \mathrm{Maps}_{\mathbf{C}_0}(\mathbf{c}_0, F'^R(\mathbf{c}_1)), \quad \mathbf{c}_0 \in \mathbf{C}_0, \quad \mathbf{c}_1 \in \mathbf{C}'_1.$$

The original functor F admits a right adjoint if and only if $\mathbf{C}'_1 = \mathbf{C}_1$.

1.7.5. Inverting the arrows, in a similar way we define the notion of *partially defined left adjoint of* F.

2. Basic operations with $(\infty, 1)$-categories

In this section we will assume that we 'know' what $(\infty, 1)$-categories are, as well as the basic rules of the syntax of operating with them. I.e., we know the 'theory', but what we need now is 'practice'.

Here are some of the primary practical questions that one needs to address:

(Q1) How do we produce 'new' $(\infty, 1)$-categories?

(Q2) How do we construct functors between two given $(\infty, 1)$-categories?

Of course, there are some cheap answers: for (Q1) take a full subcategory of an existing $(\infty, 1)$-category; for (Q2) compose two existing functors, or pass to the adjoint of a given functor. But in this way, we will not get very far.

Here are, however, some additional powerful tools:

(A1) Start a diagram of existing ones and take its limit.

(A2) Start with a given functor, and apply the procedure of Kan extension.

These answers entail the next question: how to we calculate limits when we need to?

This circle of ideas is the subject of the present section. The material here can be viewed as a user guide to (some parts of) [**Lu1**, Chapter 4 and Sect. 5.5].

2.1. Left and right Kan extensions. Let us say that at this point we have convinced ourselves that we should work with $(\infty,1)$-categories. But here comes a question: how do we ever construct functors between two given $(\infty,1)$-categories?

The difficulty is that, unlike ordinary categories, we cannot simply specify what a functor does on objects and morphisms: we would need to specify an infinite tail of compatibilities for multi-fold compositions. (Rigorously, we would have to go to the model of $(\infty,1)$-categories given by quasi-categories, and specify a map of the underlying simplicial sets, which, of course, no one wants to do in a practical situation.)

Here to our rescue comes the operation *Kan extension*: given a functor $\Phi : \mathbf{D} \to \mathbf{E}$ and a functor $F : \mathbf{D} \to \mathbf{C}$, we can (sometimes) canonically construct a functor from $\mathbf{C} \to \mathbf{E}$.

A particular case of this operation leads to the notion of limit/colimit of a functor $\mathbf{D} \to \mathbf{E}$ (we can think of such a functor as a diagram of objects in $\mathbf{E}$, parameterized by $\mathbf{D}$).

By taking $\mathbf{E}$ to be 1-Cat, we arrive to the notion of *limit of* $(\infty,1)$*-categories*, which in itself is a *key tool of constructing* $(\infty,1)$*-categories.*

The reference for this material is [**Lu1**, Sect. 4.3].

2.1.1. Let $F : \mathbf{D} \to \mathbf{C}$ be a functor between $(\infty,1)$-categories. For a (target) $(\infty,1)$-category $\mathbf{E}$, consider the functor

$$\mathrm{Funct}(\mathbf{C},\mathbf{E}) \mapsto \mathrm{Funct}(\mathbf{D},\mathbf{E}),$$

given by restriction along F (i.e., composition with F).

Its partially defined left (resp., right) adjoint is called the functor of *left* (resp., *right*) *Kan extension along* F, and denoted LKE_F (resp., RKE_F).

2.1.2. If $\mathbf{C} = *$, the corresponding left and right Kan extension functors are the functors of *colimit* (resp., *limit*):

$$\underset{\mathbf{D}}{\mathrm{colim}} : \mathrm{Funct}(\mathbf{D},\mathbf{E}) \to \mathbf{E} \text{ and } \underset{\mathbf{D}}{\lim} : \mathrm{Funct}(\mathbf{D},\mathbf{E}) \to \mathbf{E}.$$

We record the following piece of terminology: colimits over the category $\mathbf{\Delta}^{\mathrm{op}}$ are called *geometric realizations*, and limits over the category $\mathbf{\Delta}$ are called *totalizations.*

2.1.3. In general, for $\Phi : \mathbf{D} \to \mathbf{E}$, suppose that for every given $\mathbf{c} \in \mathbf{C}$, the colimit

$$\underset{\mathbf{D}\underset{\mathbf{C}}{\times}\mathbf{C}_{/\mathbf{c}}}{\mathrm{colim}} \Phi \tag{2.1}$$

exists. Then $\mathrm{LKE}_F(\Phi)$ exists and (2.1) calculates its value on $\mathbf{c}$.

Similarly, suppose that for every given $\mathbf{c}$, the limit

$$\underset{\mathbf{D}\underset{\mathbf{C}}{\times}\mathbf{C}_{\mathbf{c}/}}{\lim} \Phi, \tag{2.2}$$

exists. Then $\mathrm{RKE}_F(\Phi)$ exists and (2.2) calculates its value on $\mathbf{c}$.

2.1.4. Note that by transitivity,

$$\operatorname*{colim}_{\mathbf{D}} \Phi \simeq \operatorname*{colim}_{\mathbf{C}} \mathrm{LKE}_F(\Phi)$$

and

$$\lim_{\mathbf{D}} \Phi \simeq \lim_{\mathbf{C}} \mathrm{RKE}_F(\Phi).$$

2.1.5. To an $(\infty, 1)$-category $\mathbf{C}$ one attaches the space

$$|\mathbf{C}| := \operatorname*{colim}_{\mathbf{C}} *, \tag{2.3}$$

where $*$ is the functor $\mathbf{C} \to \mathrm{Spc}$ with constant value $*$.

The assignment

$$\mathbf{C} \mapsto |\mathbf{C}|$$

is the functor *left adjoint* to the inclusion $\mathrm{Spc} \hookrightarrow 1\text{-Cat}$. This procedure is usually referred to as *inverting all morphisms.* In particular, for $\mathcal{S} \in \mathrm{Spc} \subset 1\text{-Cat}$, we have a canonical isomorphism in Spc

$$|\mathcal{S}| \simeq \mathcal{S}.$$

An $(\infty, 1)$-category $\mathbf{C}$ is said to be *contractible* if $|\mathbf{C}|$ is isomorphic to $*$.

2.1.6. Let $\mathbf{C}$ be an $(\infty, 1)$-category and let $\Phi : \mathbf{C} \to \mathrm{Spc}$ be a functor. Then it follows from Sect. 1.4.2 that there is a canonical equivalence

$$\operatorname*{colim}_{\mathbf{C}} \Phi \simeq |\widetilde{\mathbf{C}}_\Phi|,$$

where $\widetilde{\mathbf{C}}_\Phi \to \mathbf{C}$ is the coCartesian fibration corresponding to Φ.

2.1.7. Here is a typical application of the procedure of left Kan extension:

Let $\mathbf{C}$ be an arbitrary $(\infty, 1)$-category that contains colimits. We have:

LEMMA 2.1.8. *Restriction and left Kan extension along* $* \hookrightarrow \mathrm{Spc}$ *define an equivalence between the subcategory of* $\mathrm{Funct}(\mathrm{Spc}, \mathbf{C})$ *consisting of colimit-preserving functors and* $\mathrm{Funct}(*, \mathbf{C}) \simeq \mathbf{C}$.

We note that the inverse functor in Lemma 2.1.8 is explicitly given as follows: it sends $\mathbf{c} \in \mathbf{C}$ to the functor $\mathrm{Spc} \to \mathbf{C}$, given by

$$(\mathcal{S} \in \mathrm{Spc}) \mapsto (\operatorname*{colim}_{\mathcal{S}} \mathbf{c}_{\mathcal{S}} \in \mathbf{C}),$$

where $\mathbf{c}_{\mathcal{S}}$ denotes the constant functor $\mathcal{S} \to \mathbf{C}$ with value $\mathbf{c}$, where $\mathcal{S}$ is considered as an $(\infty, 1)$-category.

2.2. Cofinality. Many of the actual calculations that one performs in higher category theory amount to calculating limits and colimits. How does one ever do this?

A key tool here is the notion of *cofinality* that allows to replace the limit/colimit over a given index $(\infty, 1)$-category, by the limit/colimit over another one, which is potentially simpler.

Iterating this procedure, one eventually arrives to a limit/colimit that can be evaluated 'by hand'. Sometimes, at the end our limit/colimit will be given just by evaluation (or a manageable fiber product/push-out). Sometimes, it will still be a limit/colimit, but in the world of ordinary categories.

The reference for the material here is [**Lu1**, Sect. 4.1].

2.2.1. A functor $F : \mathbf{D} \to \mathbf{C}$ is said to be cofinal if for any $\mathbf{c} \in \mathbf{C}$, the category

$$\mathbf{D} \underset{\mathbf{C}}{\times} \mathbf{C}_{\mathbf{c}/}$$

is contractible.

We have:

LEMMA 2.2.2. *The following are equivalent:*

(i) $F : \mathbf{D} \to \mathbf{C}$ *is cofinal;*

(ii) *For any* $\Phi : \mathbf{C} \to \mathbf{E}$, *the natural map*

$$\underset{\mathbf{D}}{\operatorname{colim}}\, \Phi \circ F \to \underset{\mathbf{C}}{\operatorname{colim}}\, \Phi$$

is an isomorphism, whenever either side is defined;

(ii') *Same as* (ii), *but we take* $\mathbf{E} = \operatorname{Spc}$ *(in which case, the colimits are always defined);*

(ii") *Same as* (ii'), *but we only consider the Yoneda functors* $\mathbf{c} \mapsto \operatorname{Maps}_{\mathbf{C}}(\mathbf{c}_0, \mathbf{c})$ *for* $\mathbf{c}_0 \in \mathbf{C}$*;*

(iii) *For any functor* $\Phi : \mathbf{C}^{\mathrm{op}} \to \mathbf{E}$, *the map*

$$\lim_{\mathbf{C}^{\mathrm{op}}} \Phi \to \lim_{\mathbf{D}^{\mathrm{op}}} \Phi \circ F^{\mathrm{op}}$$

is an isomorphism, whenever either side is defined;

(iii') *Same as* (iii), *but we take* $\mathbf{E} = \operatorname{Spc}$ *(in which case, the limits are always defined);*

(iv) *For any* $\Phi : \mathbf{C} \to \mathbf{E}$ *and any functor* $\Phi' : \mathbf{C} \to \mathbf{E}$ that sends all morphisms to isomorphisms, *the map*

$$\operatorname{Maps}_{\operatorname{Funct}(\mathbf{C},\mathbf{E})}(\Phi, \Phi') \to \operatorname{Maps}_{\operatorname{Funct}(\mathbf{D},\mathbf{E})}(\Phi \circ F, \Phi' \circ F)$$

is an isomorphism.

2.2.3. For example, any functor that admits a left adjoint is cofinal. Indeed, in this case, the category $\mathbf{D} \underset{\mathbf{C}}{\times} \mathbf{C}_{\mathbf{c}/}$ admits an initial object, given by

$$\mathbf{c} \mapsto F \circ F^L(\mathbf{c}).$$

2.2.4. Let $\mathbf{D} \to \mathbf{C}$ be a coCartesian fibration. We note that in this case for any $\mathbf{c} \in \mathbf{C}$, the functor

$$\mathbf{D}_{\mathbf{c}} \to \mathbf{D} \underset{\mathbf{C}}{\times} \mathbf{C}_{/\mathbf{c}}$$

is cofinal. Hence, we obtain that for $\Phi : \mathbf{D} \to \mathbf{E}$, the value of $\operatorname{LKE}_F(\Phi)$ at $\mathbf{c} \in \mathbf{C}$ is canonically isomorphic to

$$\underset{\mathbf{D}_{\mathbf{c}}}{\operatorname{colim}}\, \Phi.$$

I.e., instead of computing the colimit over the slice category, we can do so over the fiber.

Similarly, if $\mathbf{D} \to \mathbf{C}$ is a Cartesian fibration, then for $\Phi : \mathbf{D} \to \mathbf{E}$, the value of $\operatorname{RKE}_F(\Phi)$ at $\mathbf{c} \in \mathbf{C}$ is canonically isomorphic to

$$\lim_{\mathbf{D}_{\mathbf{c}}} \Phi.$$

2.3. Contractible functors. The contents of this subsection can be skipped on the first pass. It is included in order to address the following question that arises naturally after introducing the notion of cofinality:

Let $F : \mathbf{D} \to \mathbf{C}$ be a functor. For $\Phi, \Phi' \in \mathrm{Funct}(\mathbf{C}, \mathbf{E})$, consider the restriction map

$$\mathrm{Maps}_{\mathrm{Funct}(\mathbf{C},\mathbf{E})}(\Phi, \Phi') \to \mathrm{Maps}_{\mathrm{Funct}(\mathbf{D},\mathbf{E})}(\Phi \circ F, \Phi' \circ F). \tag{2.4}$$

The condition that (2.4) be an isomorphism for any Φ and Φ' that take all morphisms to isomorphisms is equivalent to the map

$$|\mathbf{D}| \to |\mathbf{C}|$$

being an isomorphism in Spc.

According to Lemma 2.2.2 the condition that (2.4) be an isomorphism for any Φ' that takes all morphisms to isomorphisms is equivalent to F being cofinal.

We will now formulate the condition that (2.4) be an isomorphism for all pairs Φ, Φ'. I.e., that the restriction functor

$$\mathrm{Funct}(\mathbf{C}, \mathbf{E}) \to \mathrm{Funct}(\mathbf{D}, \mathbf{E})$$

be fully faithful.

2.3.1. For $\mathbf{c}, \mathbf{c}' \in \mathbf{C}$ and a morphism $\mathbf{c} \overset{\alpha}{\to} \mathbf{c}'$, consider the $(\infty, 1)$-category $\mathrm{Factor}_{\mathbf{D}}(\alpha)$:

$$\left((\mathbf{C}_{\mathbf{c}/} \underset{\mathbf{C}}{\times} \mathbf{D}) \underset{\mathbf{D}}{\times} (\mathbf{D} \underset{\mathbf{C}}{\times} \mathbf{C}_{/\mathbf{c}'}) \right) \underset{\mathrm{Maps}_{\mathbf{C}}(\mathbf{c},\mathbf{c}')}{\times} \{\alpha\}.$$

I.e., this is the category, whose objects are

$$(\widetilde{\mathbf{d}} \in \mathbf{D}, \mathbf{c} \overset{\beta}{\to} F(\widetilde{\mathbf{d}}) \overset{\gamma}{\to} \mathbf{c}', \gamma \circ \beta \sim \alpha).$$

We shall say that F is *contractible* if for any $\mathbf{c} \overset{\alpha}{\to} \mathbf{c}'$, the category $\mathrm{Factor}_{\mathbf{D}}(\alpha)$ is contractible.

2.3.2. We have:

LEMMA 2.3.3. *The following conditions are equivalent:*

(i) *F is contractible;*

(i') *$F^{\mathrm{op}} : \mathbf{D}^{\mathrm{op}} \to \mathbf{C}^{\mathrm{op}}$ is contractible;*

(ii) *For any $\mathbf{E}$, the restriction functor*

$$\mathrm{Funct}(\mathbf{C}, \mathbf{E}) \to \mathrm{Funct}(\mathbf{D}, \mathbf{E})$$

is fully faithful;

(ii') *Same as* (ii) *but $\mathbf{E} = \mathrm{Spc}$;*

(iii) *The unit of the adjunction*

$$\Phi \to \mathrm{RKE}_F(\Phi \circ F), \quad \Phi \in \mathrm{Funct}(\mathbf{C}, \mathbf{E})$$

is an isomorphism for any $\mathbf{E}$ and Φ;

(iii') *Same as (iii), but $\mathbf{E} = \mathrm{Spc}$;*

(iv) *The counit of the adjunction*

$$\mathrm{LKE}_F(\Phi \circ F) \to \Phi, \quad \Phi \in \mathrm{Funct}(\mathbf{C}, \mathbf{E})$$

is an isomorphism for any $\mathbf{E}$ and Φ;

(iv') *Same as (iv), but* $\mathbf{E} = \mathrm{Spc}$;

(iv") *Same as (iv'), but* Φ *are taken to be the Yoneda functors* $\mathbf{c} \mapsto \mathrm{Maps}_{\mathbf{C}}(\mathbf{c}_0, \mathbf{c})$.

2.3.4. We also note:

LEMMA 2.3.5. *Let* $F : \mathbf{D} \to \mathbf{C}$ *be a Cartesian or coCartesian fibration. Then it is contractible if and only if it has contractible fibers.*

2.4. The operation of 'passing to adjoints'. Let

$$i \mapsto \mathbf{C}_i, \quad i \in \mathbf{I}$$

be an $\mathbf{I}$-diagram of $(\infty,1)$-categories. In this subsection we will discuss the procedure of creating a new diagram, parameterized by $\mathbf{I}^{\mathrm{op}}$, that still sends i to $\mathbf{C}_i$, but replaces the transition functors by their adjoints.

This procedure generalizes the situation of Sect. 1.7.1: in the latter our index category $\mathbf{I}$ was simply $[1]$.

2.4.1. Let

$$\mathbf{C}_{\mathbf{I}} : \mathbf{I} \to 1\text{-Cat}, \quad i \mapsto \mathbf{C}_i, \quad (i_0 \overset{\alpha}{\to} i_1) \mapsto \left(\mathbf{C}_{i_0} \overset{F_\alpha}{\longrightarrow} \mathbf{C}_{i_1}\right)$$

be a functor, where $\mathbf{I} \in 1\text{-Cat}$. Let

$$\widetilde{\mathbf{C}} \to \mathbf{I} \tag{2.5}$$

be the coCartesian fibration corresponding to $\mathbf{C}_{\mathbf{I}}$.

Assume that for every morphism $(i_0 \overset{\alpha}{\to} i_1) \in \mathbf{I}$, the resulting functor $\mathbf{C}_{i_0} \overset{F_\alpha}{\longrightarrow} \mathbf{C}_{i_1}$ admits a right adjoint. In this case the coCartesian fibration (2.5) is bi-Cartesian.

2.4.2. Viewing (2.5) as a Cartesian fibration, and applying straightening, we transform (2.5) into a functor

$$\mathbf{C}^R_{\mathbf{I}^{\mathrm{op}}} : \mathbf{I}^{\mathrm{op}} \to 1\text{-Cat}.$$

On this case, we shall say that $\mathbf{C}^R_{\mathbf{I}^{\mathrm{op}}}$ is obtained from $\mathbf{C}_{\mathbf{I}}$ *by passing to right adjoints.*

By construction, the value of $\mathbf{C}^R_{\mathbf{I}^{\mathrm{op}}}$ on $i \in \mathbf{I}$ is still $\mathbf{C}_i$. However, for a morphism $i_0 \overset{\alpha}{\to} i_1$ in $\mathbf{I}$, viewed as a morphism $i_1 \to i_0$ in $\mathbf{I}^{\mathrm{op}}$, the corresponding functor

$$\mathbf{C}_{i_1} \to \mathbf{C}_{i_0}$$

is $(F_\alpha)^R$.

2.4.3. Similarly, by inverting the arrows, we talk about a functor

$$\mathbf{D}^L_{\mathbf{J}^{\mathrm{op}}} : \mathbf{J}^{\mathrm{op}} \to 1\text{-Cat}$$

being obtained from a functor $\mathbf{D}_{\mathbf{J}} : \mathbf{J} \to 1\text{-Cat}$ by *passing to left adjoints.*

For $\mathbf{J} = \mathbf{I}^{\mathrm{op}}$, the datum of realizing $\mathbf{D}_{\mathbf{J}}$ as obtained from $\mathbf{C}_{\mathbf{I}}$ by passing to right adjoints is equivalent to the datum of realizing $\mathbf{C}_{\mathbf{I}}$ as obtained from $\mathbf{D}_{\mathbf{J}}$ by passing to left adjoints: both are encoded by a bi-Cartesian fibration

$$\mathbf{E} \to \mathbf{I}.$$

2.5. Colimits in presentable $(\infty,1)$-categories. As was mentioned in the introduction, the primary reason for working with $(\infty,1)$-categories is the fact that the operation of *limit of a diagram of* $(\infty,1)$*-categories* is well-behaved (as opposed to one within the world of triangulated categories).

But here comes a problem: while limits are, by definition, adjusted to mapping *to* them, how do we ever construct a functor *out of* an $(\infty,1)$-category, defined as a limit? However, quite an amazing thing happens: in a wide class of situations, a *limit* in 1-Cat happens to also be the *colimit* (taken in a slightly different category).

The pattern of how this happens will be described in this subsection.

2.5.1. We let $1\text{-Cat}_{\mathrm{Prs}} \subset 1\text{-Cat}$ be the 1-full subcategory whose objects are *presentable* $(\infty,1)$-categories contain colimits[3], and where we allow as morphisms functors that preserve colimits.

We have the following basic fact:

LEMMA 2.5.2 ([**Lu1**], Proposition 5.5.3.13).

(a) *The* $(\infty,1)$*-category* $1\text{-Cat}_{\mathrm{Prs}}$ *contains limits and colimits.*

(b) *The inclusion functor*

$$1\text{-Cat}_{\mathrm{Prs}} \to 1\text{-Cat}$$

preserves limits.

2.5.3. Here is a version of the Adjoint Functor Theorem:

THEOREM 2.5.4 ([**Lu1**], Corollary 5.5.2.9).

(a) *Any morphism in* $1\text{-Cat}_{\mathrm{Prs}}$*, viewed as a functor between* $(\infty,1)$*-categories, admits a right adjoint.*

(b) *If* $\mathbf{C}$ *and* $\mathbf{D}$ *are objects in* $1\text{-Cat}_{\mathrm{Prs}}$*, and* $G:\mathbf{D}\to\mathbf{C}$ *is a functor that preserves* limits[4]*, then this functor admits a left adjoint, which is a morphism in* $1\text{-Cat}_{\mathrm{Prs}}$.

2.5.5. Let

$$\mathbf{C}_{\mathbf{I}}:\mathbf{I}\to 1\text{-Cat}_{\mathrm{Prs}}$$

be a functor.

By the Adjoint Functor Theorem and Sect. 2.4.1, there exists a canonically defined functor

$$\mathbf{C}^R_{\mathbf{I}^{\mathrm{op}}}:\mathbf{I}^{\mathrm{op}}\to 1\text{-Cat},$$

obtained from the composition

$$\mathbf{I}\xrightarrow{\mathbf{C}_{\mathbf{I}}} 1\text{-Cat}_{\mathrm{Prs}} \hookrightarrow 1\text{-Cat}$$

by passing to right adjoints.

[3]Presentability is a technical condition of set-theoretic nature (see [**Lu1**, Sect. 5.5.]), which is necessary for the Adjoint Functor Theorem to hold. However, following our conventions (see Sect. 0.4.5), we will omit the adjective 'presentable' even when it should properly be there.

[4]One also needs to impose a condition of set-theoretic nature that G be *accessible*, see [**Lu1**, Defn. 5.4.2.5] for what this means.

2.5.6. Let $\mathbf{C}_*$ denote the *colimit* of $\mathbf{C}_{\mathbf{I}}$ in $1\text{-Cat}_{\text{Prs}}$.

Let $\mathbf{I}'$ be the category obtained from $\mathbf{I}$ by adjoining a final object $*$. The functor $\mathbf{C}_{\mathbf{I}}$ canonically extents to a functor

$$\mathbf{C}_{\mathbf{I}'} : \mathbf{I}' \to 1\text{-Cat}_{\text{Prs}},$$

whose value on $*$ is $\mathbf{C}_*$.

By the Adjoint Functor Theorem and Sect. 2.4.1, there exists a canonically defined functor

$$\mathbf{C}^R_{\mathbf{I}'^{\text{op}}} : \mathbf{I}'^{\text{op}} \to 1\text{-Cat},$$

obtained from the composition

$$\mathbf{I}' \xrightarrow{\mathbf{C}_{\mathbf{I}'}} 1\text{-Cat}_{\text{Prs}} \hookrightarrow 1\text{-Cat}$$

by passing to right adjoints, and whose restriction to $\mathbf{I}^{\text{op}}$ is the functor $\mathbf{C}^R_{\mathbf{I}^{\text{op}}}$.

Note that the category $\mathbf{I}'^{\text{op}}$ is obtained from $\mathbf{I}^{\text{op}}$ by adjoining an initial object.

In particular, we obtain a canonically defined functor

$$\mathbf{C}_* \to \lim_{\mathbf{I}^{\text{op}}} \mathbf{C}^R_{\mathbf{I}^{\text{op}}}. \tag{2.6}$$

We have the following fundamental fact, which follows from [**Lu1**, Corollary 5.5.3.4]:

PROPOSITION 2.5.7. *The functor* (2.6) *is an equivalence.*

The equivalence of Proposition 2.5.7 will be used all the time in this book. We emphasize that it states the equivalence

$$\underset{i\in\mathbf{I}}{\text{colim}}\, \mathbf{C}_i \simeq \lim_{i\in\mathbf{I}^{\text{op}}} \mathbf{C}_i,$$

where the colimit in the left-hand side is taken in $1\text{-Cat}_{\text{Prs}}$, and the limit in the right-hand side is taken in 1-Cat.

2.5.8. In the setting of Proposition 2.5.7, for $i \in \mathbf{I}$, we will denote by ins_i the tautological functor

$$\mathbf{C}_i \to \mathbf{C}_*.$$

In terms of the identification

$$\mathbf{C}_* \simeq \lim_{\mathbf{I}^{\text{op}}} \mathbf{C}^R_{\mathbf{I}^{\text{op}}},$$

the functor ins_i is the left adjoint of the tautological evaluation functor

$$\text{ev}_i : \lim_{\mathbf{I}^{\text{op}}} \mathbf{C}^R_{\mathbf{I}^{\text{op}}} \to \mathbf{C}_i.$$

Thus, we can restate Proposition 2.5.7 by saying that each of the functors ev_i admits a left adjoint, and the resulting family of functors

$$(\text{ev}_i)^L : \mathbf{C}_i \to \lim_{\mathbf{I}^{\text{op}}} \mathbf{C}^R_{\mathbf{I}^{\text{op}}}$$

gives rise to an equivalence

$$\underset{\mathbf{I}}{\text{colim}}\, \mathbf{C}_{\mathbf{I}} \xrightarrow{\sim} \lim_{\mathbf{I}^{\text{op}}} \mathbf{C}^R_{\mathbf{I}^{\text{op}}},$$

where the colimit in the left-hand side is taken in $1\text{-Cat}_{\text{Prs}}$.

2.6. Limits and adjoints. In this subsection we will discuss two general results about the interaction of limits of $(\infty, 1)$-categories with adjunctions and with limits *within* a given $(\infty, 1)$-category. We will use them in multiple places in the book.

2.6.1. Let

$$\mathbf{I} \to 1\text{-Cat}, \quad i \mapsto \mathbf{C}_i$$

be a diagram of $(\infty, 1)$-categories. Set

$$\mathbf{C} := \lim_{i \in \mathbf{I}} \mathbf{C}_i.$$

Let

$$A \to \mathbf{C}, \quad a \mapsto \mathbf{c}^a$$

be a functor, where A is some other index category. Consider the corresponding functors

$$A \to \mathbf{C} \xrightarrow{\text{ev}_i} \mathbf{C}_i, \quad a \mapsto \mathbf{c}_i^a.$$

Suppose that for each i, the limit

$$\lim_{a \in A} \mathbf{c}_i^a =: \mathbf{c}_i \in \mathbf{C}_i$$

exists. Assume also that for every 1-morphism $i \to j$ in $\mathbf{I}$, the corresponding map $F_{i,j}(\mathbf{c}_i) \to \mathbf{c}_j$ happens to be an isomorphism.

We claim:

LEMMA 2.6.2. *Under the above circumstances, the limit*

$$\lim_{a \in A} \mathbf{c}^a =: \mathbf{c} \in \mathbf{C}$$

exists and the natural maps $\text{ev}_i(\mathbf{c}) \to \mathbf{c}_i$ *are isomorphisms.*

2.6.3. Let now $\mathbf{I}$ be an $(\infty, 1)$-category of indices, and let be given a functor

$$\mathbf{I} \to \text{Funct}([1], 1\text{-Cat}), \quad i \mapsto (\mathbf{D}_i \xrightarrow{\Phi_i} \mathbf{C}_i).$$

Assume that for every i, the corresponding functor Φ_i admits a right adjoint. Assume also that for every map $i \to j$ in I the natural transformation

$$F_{i,j}^{\mathbf{D}} \circ (\Phi_i)^R \to (\Phi_j)^R \circ F_{i,j}^{\mathbf{C}}$$

is an isomorphism, where $F_{i,j}^{\mathbf{D}}$ (resp., $F_{i,j}^{\mathbf{C}}$) denotes the transition functor $\mathbf{C}_i \to \mathbf{C}_j$ (resp., $\mathbf{D}_i \to \mathbf{D}_j$).

Set

$$\mathbf{D} := \lim_{i \in I} \mathbf{D}_i \text{ and } \mathbf{C} := \lim_{i \in I} \mathbf{C}_i.$$

We claim:

LEMMA 2.6.4. *The resulting functor* $\Phi : \mathbf{D} \to \mathbf{C}$ *admits a right adjoint, and for every* i *the natural transformation*

$$\text{ev}_i^{\mathbf{D}} \circ \Phi^R \to \Phi_i^R \circ \text{ev}_i^{\mathbf{C}}$$

is an isomorphism.

3. Monoidal structures

This section is meant to be a user guide to some aspects of Higher Algebra, roughly Sects. 4.1-4.3, 4.5 and 4.7 of [**Lu2**].

We discuss the notion of monoidal $(\infty, 1)$-category; the notion of module over a given monoidal category, the notion of associative algebra in a given monoidal category, and the notion of module over an algebra in a given module category.

At the end of this section, we discuss monads and the Barr-Beck-Lurie theorem.

A reader who is familiar with [**Lu2**] can safely skip this section.

3.1. The notion of monoidal $(\infty, 1)$-category. In this subsection we introduce the notion of monoidal $(\infty, 1)$-category. The idea is very simple: a monoidal $(\infty, 1)$-category will be encoded by a functor from the category $\mathbf{\Delta}^{\mathrm{op}}$ to 1-Cat.

3.1.1. Recall the category $\mathbf{\Delta}$, see Sect. 1.1.10.

We define a monoidal $(\infty, 1)$-category to be a functor

$$\mathbf{A}^{\otimes} : \mathbf{\Delta}^{\mathrm{op}} \to 1\text{-Cat},$$

subject to the following conditions:

- $\mathbf{A}^{\otimes}([0]) = *$;
- For any n, the functor, given by the n-tuple of maps in $\mathbf{\Delta}$

$$[1] \to [n], \quad 0 \mapsto i, 1 \mapsto i+1, \quad i = 0, ..., n-1, \tag{3.1}$$

  defines an equivalence

$$\mathbf{A}^{\otimes}([n]) \to \mathbf{A}^{\otimes}([1]) \times ... \times \mathbf{A}^{\otimes}([1]).$$

3.1.2. If $\mathbf{A}^{\otimes}$ is a monoidal $(\infty, 1)$-category, we shall denote by $\mathbf{A}$ the *underlying* $(\infty, 1)$-category, i.e., $\mathbf{A}^{\otimes}([1])$.

Sometimes, we will abuse the notation and say that "$\mathbf{A}$ is a monoidal $(\infty, 1)$-category". Whenever we say this we will mean that $\mathbf{A}$ is obtained in the above way from a functor $\mathbf{A}^{\otimes}$.

3.1.3. The map

$$[1] \to [2], \quad 0 \mapsto 0, 1 \mapsto 2$$

defines a functor

$$\mathbf{A} \times \mathbf{A} \to \mathbf{A}.$$

This functor is the monoidal operation on $\mathbf{A}$, corresponding to $\mathbf{A}^{\otimes}$. Unless a confusion is likely to occur, we denote the above functor by

$$\mathbf{a}_1, \mathbf{a}_2 \mapsto \mathbf{a}_1 \otimes \mathbf{a}_2.$$

The map $[1] \to [0]$ defines a functor $* \to \mathbf{A}$; the corresponding object is the unit of the monoidal structure $\mathbf{1}_{\mathbf{A}} \in \mathbf{A}$.

3.1.4. We let $1\text{-}\mathrm{Cat}^{\mathrm{Mon}}$ denote the $(\infty,1)$-category of monoidal $(\infty,1)$-categories, which is by definition a full subcategory in $\mathrm{Funct}(\boldsymbol{\Delta}^{\mathrm{op}}, 1\text{-Cat})$.

The involution $(-)^{\mathrm{op}}$ of (1.1) on 1-Cat induces one on $\mathrm{Funct}(\boldsymbol{\Delta}^{\mathrm{op}}, 1\text{-Cat})$, and the latter preserves the full subcategory $1\text{-}\mathrm{Cat}^{\mathrm{Mon}}$. At the level of underlying $(\infty,1)$-categories, this involution acts as $\mathbf{A} \mapsto \mathbf{A}^{\mathrm{op}}$.

In other words, the opposite of a monoidal $(\infty,1)$-category carries a natural monoidal structure.

Recall the involution rev on the category $\boldsymbol{\Delta}$; see Sect. 1.1.10. This involution also induces one on $\mathrm{Funct}(\boldsymbol{\Delta}^{\mathrm{op}}, 1\text{-Cat})$, and the latter preserves also preserves $1\text{-}\mathrm{Cat}^{\mathrm{Mon}}$.

This is the operation of passing to the monoidal $(\infty,1)$-category with the *reversed multiplication*,

$$\mathbf{A} \mapsto \mathbf{A}^{\text{rev-mult}}.$$

3.1.5. *An example: endo-functors.* Let $\mathbf{C}$ be an $(\infty,1)$-category. We claim that the $(\infty,1)$-category $\mathrm{Funct}(\mathbf{C},\mathbf{C})$ acquires a natural monoidal structure. Indeed, we define the functor

$$\mathrm{Funct}(\mathbf{C},\mathbf{C})^{\otimes} : \boldsymbol{\Delta}^{\mathrm{op}} \to 1\text{-Cat} \tag{3.2}$$

as follows: it sends $[n]$ to

$$\mathrm{Cart}_{/[n]^{\mathrm{op}}} \underset{1\text{-Cat}\times\ldots\times 1\text{-Cat}}{\times} \{\mathbf{C}\times\ldots\times\mathbf{C}\},$$

where the functor

$$\mathrm{Cart}_{/[n]^{\mathrm{op}}} \to \mathrm{Cart}_{/*\sqcup\ldots\sqcup*} \simeq 1\text{-Cat}\times\ldots\times 1\text{-Cat}$$

is given by restriction along

$$(*\sqcup\ldots\sqcup*) = ([n]^{\mathrm{op}})^{\mathrm{Spc}} \to [n]^{\mathrm{op}}.$$

By Chapter 10, Corollary 2.4.4, we have that

$$\mathrm{Cart}_{/[1]^{\mathrm{op}}} \underset{1\text{-Cat}\times 1\text{-Cat}}{\times} \{\mathbf{C}\times\mathbf{C}\} \simeq \mathrm{Funct}(\mathbf{C},\mathbf{C}),$$

and the functor (3.2) is easily seen to satisfy the conditions of Sect. 3.1.1.

REMARK 3.1.6. The fact that (3.2) is well-defined as a functor follows from the *enhanced straightening procedure*, see Sect. 1.6.2.

3.1.7. Unstraightening defines a fully faithful embedding

$$1\text{-}\mathrm{Cat}^{\mathrm{Mon}} \hookrightarrow (\mathrm{coCart}_{/\boldsymbol{\Delta}^{\mathrm{op}}})_{\mathrm{strict}},$$

denoted

$$\mathbf{A}^{\otimes} \mapsto \mathbf{A}^{\otimes,\boldsymbol{\Delta}^{\mathrm{op}}}.$$

Its essential image is singled out by the condition in Sect. 3.1.1.

3.2. Lax functors and associative algebras. In this subsection we introduce the notion of associative algebra in a given monoidal $(\infty,1)$-category.

The method by which we will do it (following [**Lu2**, Sect. 4.2]) will exhibit the power of the idea of unstraightening.

3.2.1. We introduce another $(\infty,1)$-category, denoted $(1\text{-}\mathrm{Cat}^{\mathrm{Mon}})_{\text{right-lax}_{\text{non-untl}}}$. It will have the same objects as $1\text{-}\mathrm{Cat}^{\mathrm{Mon}}$, and will contain the latter as a 1-full subcategory.

The idea of the category $(1\text{-}\mathrm{Cat}^{\mathrm{Mon}})_{\text{right-lax}_{\text{non-untl}}}$ is that we now allow functors $\mathbf{A}_0 \to \mathbf{A}_1$ such that the diagrams

$$\begin{array}{ccc} \mathbf{A}_0 \times \mathbf{A}_0 & \longrightarrow & \mathbf{A}_0 \\ \downarrow & & \downarrow \\ \mathbf{A}_1 \times \mathbf{A}_1 & \longrightarrow & \mathbf{A}_1 \end{array}$$

no longer commute, but do so up to a natural transformation.

3.2.2. Namely, we let $(1\text{-}\mathrm{Cat}^{\mathrm{Mon}})_{\text{right-lax}_{\text{non-untl}}}$ be the 1-full subcategory of $\mathrm{coCart}_{/\mathbf{\Delta}^{\mathrm{op}}}$, whose objects are those lying in the essential image of $1\text{-}\mathrm{Cat}^{\mathrm{Mon}}$, and where we allow as morphisms functors

$$\mathbf{A}_0^{\otimes,\mathbf{\Delta}^{\mathrm{op}}} \to \mathbf{A}_1^{\otimes,\mathbf{\Delta}^{\mathrm{op}}}$$

that map morphisms in $\mathbf{A}_0^{\otimes,\mathbf{\Delta}^{\mathrm{op}}}$ that are coCartesian over morphisms in $\mathbf{\Delta}^{\mathrm{op}}$ of the form (3.1) to morphisms in $\mathbf{A}_1^{\otimes,\mathbf{\Delta}^{\mathrm{op}}}$ with the same property.

Such a functor will be called a *right-lax monoidal functor.*

3.2.3. Passing to the opposite categories, one obtains the notion of *left-lax monoidal functor.* The next assertion follows by unwinding the definitions:

LEMMA 3.2.4. *Let $\mathbf{A}_1^{\otimes}$ and $\mathbf{A}_2^{\otimes}$ be a pair of monoidal $(\infty,1)$-categories, and let*

$$F : \mathbf{A}_1 \rightleftarrows \mathbf{A}_2 : G$$

be a pair of adjoint functors of the underlying plain $(\infty,1)$-categories. Then the datum of left-lax monoidal functor on F is equivalent to the datum of right-lax monoidal functor on G.

3.2.5. Let $*^{\otimes}$ be the point category, equipped with a natural monoidal structure, i.e., $*^{\otimes}([n]) = *$ for any n.

Given a monoidal $(\infty,1)$-category $\mathbf{A}^{\otimes}$, we define the notion of associative algebra in $\mathbf{A}^{\otimes}$ to be a right-lax monoidal functor

$$\mathcal{A}^{\otimes,\mathbf{\Delta}^{\mathrm{op}}} : \mathbf{\Delta}^{\mathrm{op}} = *^{\otimes,\mathbf{\Delta}^{\mathrm{op}}} \to \mathbf{A}^{\otimes,\mathbf{\Delta}^{\mathrm{op}}}.$$

We denote the $(\infty,1)$-category of associative algebras in $\mathbf{A}^{\otimes}$ by $\mathrm{AssocAlg}(\mathbf{A})$ (suppressing the $\otimes$ superscript). We let

$$\mathbf{oblv}_{\mathrm{Assoc}} : \mathrm{AssocAlg}(\mathbf{A}) \to \mathbf{A}$$

denote the tautological forgetful functor.

Given $\mathcal{A}^{\otimes,\mathbf{\Delta}^{\mathrm{op}}} \in \mathrm{AssocAlg}(\mathbf{A})$, we denote by $\mathcal{A}$ its *underlying* object of $\mathbf{A}$, i.e., the value of $\mathcal{A}^{\otimes,\mathbf{\Delta}^{\mathrm{op}}}$ on the object $[1] \in \mathbf{\Delta}^{\mathrm{op}}$.

3.2.6. Let $\mathcal{A}$ be an associative algebra in $\mathbf{A}$. Then we obtain, tautologically, an associative algebra $\mathcal{A}^{\text{rev-mult}}$ in $\mathbf{A}^{\text{rev-mult}}$, with the same underlying object of $\mathbf{A}$ as a plain $(\infty,1)$-category.

3.3. The symmetric(!) monoidal case. In this subsection we explain the modifications necessary in order to talk about *symmetric* monoidal $(\infty,1)$-categories, and commutative algebras inside them.

3.3.1. The definitions involving monoidal categories and associative algebras in them can be rendered into the world of *symmetric monoidal* $(\infty,1)$-categories and *commutative algebras*, by replacing the category $\mathbf{\Delta}^{\mathrm{op}}$ by that of *finite pointed sets*, denoted Fin_*. We replace the condition in Sect. 3.1.1 by the following one:

- $\mathbf{A}^{\otimes}(\{*\}) = *$;
- For any finite pointed set $(* \in I)$ and any $i \in I - \{*\}$, we have the map $(* \in I) \to (* \in \{* \cup i\})$ given by $i \mapsto i$ and $j \mapsto *$ for $j \neq i$. We require that the induced map
$$\mathbf{A}^{\otimes}(* \in I) \to \prod_{i \in I - \{*\}} \mathbf{A}^{\otimes}(* \in \{* \cup i\}).$$
be an equivalence.

We let $1\text{-}\mathrm{Cat}^{\mathrm{SymMon}}$ denote the $(\infty,1)$-category of symmetric monoidal $(\infty,1)$-categories.

Given $\mathbf{A} \in 1\text{-}\mathrm{Cat}^{\mathrm{SymMon}}$, we let $\mathrm{ComAlg}(\mathbf{A})$ denote the $(\infty,1)$-category of commutative algebras in $\mathbf{A}$. We let

$$\mathbf{oblv}_{\mathrm{Com}} : \mathrm{Com}(\mathbf{A}) \to \mathbf{A}$$

denote the tautological forgetful functor.

3.3.2. Note that we have a canonically defined functor

$$\mathbf{\Delta}^{\mathrm{op}} \to \mathrm{Fin}_* . \tag{3.3}$$

At the level of objects this functor sends $[n] \mapsto (0 \in \{0,...,n\})$. At the level of morphisms, it sends a non-decreasing map $\phi : [m] \to [n]$ to the map $\psi : \{0,...,n\} \to \{0,...,m\}$ defined as follows:

For $i \in \{0,...,n\}$ we set $\psi(i) = j$ if there exists (an automatically unique) $j \in \{0,...,m\}$ such that $\phi(j-1) < i \leq \phi(j)$, and $\phi(i) = 0$ otherwise.

Using the functor (3.3) we obtain that any object of commutative nature (e.g., symmetric monoidal $(\infty,1)$-category or a commutative algebra in one such) gives rise to the corresponding associative one (monoidal $(\infty,1)$-category or associative algebra in one such).

3.3.3. Any $(\infty,1)$-category $\mathbf{C}$ that admits Cartesian products[5] has a canonically defined (symmetric) monoidal structure. Namely, we start with the functor

$$(\mathrm{Fin}_*)^{\mathrm{op}} \to 1\text{-}\mathrm{Cat}, \tag{3.4}$$

given by

$$(* \in I) \mapsto \mathrm{Funct}(I, \mathbf{C}) \underset{\mathrm{Funct}(\{*\},\mathbf{C})}{\times} \{*\},$$

where $\{*\} \to \mathrm{Funct}(\{*\}, \mathbf{C})$ is given by the functor that maps $*$ to the final object.

Now, the condition that $\mathbf{C}$ admits Cartesian products implies that the functor (3.4) satisfies the assumption of Sect. 2.4.1. Hence, we obtain a well-defined functor

$$\mathrm{Fin}_* \to 1\text{-}\mathrm{Cat}, \tag{3.5}$$

[5]Including the empty Cartesian product, i.e., a final object.

obtained from (3.4) by passing to *right adjoints*. It is easy to see that the functor (3.4) satisfies the assumptions of Sect. 3.3.1, thereby giving rise to a symmetric monoidal structure on $\mathbf{C}$.

In particular, we can talk about commutative (and if we regard $\mathbf{C}$ just as a monoidal category, also associative) algebras in $\mathbf{C}$. These objects are called *commutative monoids* (resp., just *monoids*). We denote the corresponding categories by

$$\text{ComMonoid}(\mathbf{C}) \text{ and } \text{Monoid}(\mathbf{C}),$$

respectively.

Dually, if $\mathbf{C}$ admits coproducts, it has a *coCartesian* symmetric monoidal structure.

3.3.4. In particular, we can consider the $(\infty,1)$-category 1-Cat equipped with the Cartesian symmetric monoidal structure.

Commutative (resp., associative) algebras in 1-Cat with respect to the Cartesian structure, i.e., commutative monoids (resp., just monoids) in 1-Cat are the same as symmetric monoidal (resp., monoidal) $(\infty,1)$-categories, see Chapter 9, Sect. 1.3.3.

3.3.5. Let $\mathbf{A}$ be a symmetric monoidal $(\infty,1)$-category. In this case, the $(\infty,1)$-category $\text{AssocAlg}(\mathbf{A})$ acquires a symmetric monoidal structure, compatible with the forgetful functor $\text{AssocAlg}(\mathbf{A}) \to \mathbf{A}$, see [**Lu2**, Proposition 3.2.4.3 and Example 3.2.4.4].

3.3.6. Furthermore, the $(\infty,1)$-category $\text{ComAlg}(\mathbf{A})$ also acquires a symmetric monoidal structure, and this symmetric monoidal structure equals the *coCartesian* symmetric monoidal structure on $\text{ComAlg}(\mathbf{A})$, see [**Lu2**, Proposition 3.2.4.7].

In particular, every object $\mathcal{A} \in \text{ComAlg}(\mathbf{A})$ has a natural structure of commutative algebra in $\text{ComAlg}(\mathbf{A})$, and hence also in $\text{AssocAlg}(\mathbf{A})$.

3.4. Module categories. In this section we extend the definition of monoidal $(\infty,1)$-categories to the case of modules.

3.4.1. Let $\mathbf{\Delta}^+$ be the 1-full subcategory of $1\text{-Cat}_{\text{ordn}}$, where we allow as objects categories of the form

$$[n] = (0 \to 1 \to \dots \to n), \quad n = 0,1,\dots$$

and

$$[n]^+ = (0 \to 1 \to \dots \to n \to +), \quad n = 0,1.\dots$$

As 1-morphisms we allow:

- Arbitrary functors $[n] \to [m]$;
- Functors $[n] \to [m]^+$, whose essential image does not contain $+$;
- Functors $[n]^+ \to [m]^+$ that send $+$ to $+$ and such that the preimage of $+$ is $+$.

3.4.2. Given a monoidal $(\infty,1)$-category $\mathbf{A}^{\otimes}$, a module for it is a datum of extension of the functor

$$\mathbf{A}^{\otimes} : \mathbf{\Delta}^{\text{op}} \to 1\text{-Cat},$$

to a functor

$$\mathbf{A}^{+,\otimes} : \mathbf{\Delta}^{+,\text{op}} \to 1\text{-Cat},$$

such that the following condition holds:

- For any $n \geq 0$, the functor
$$\mathbf{A}^{+,\otimes}([n]^+) \to \mathbf{A}^{\otimes}([n]) \times \mathbf{A}^{+,\otimes}([0]^+),$$
given by the morphisms
$$[n] \to [n]^+, \quad i \mapsto i \text{ and } [0]^+ \to [n]^+, \quad 0 \mapsto n, + \mapsto +,$$
is an equivalence.

3.4.3. We will think of the $(\infty,1)$-category
$$\mathbf{M} := \mathbf{A}^{+,\otimes}([0]^+)$$
as the $(\infty,1)$-*category underlying the module.*

Note that $\mathbf{A}^{+,\otimes}([1]^+)$ identifies with $\mathbf{A} \times \mathbf{M}$. The map
$$[0]^+ \to [1]^+, \quad 0 \mapsto 0, + \mapsto +,$$
defines a functor
$$\mathbf{A} \times \mathbf{M} \to \mathbf{M},$$
which is the functor of action of $\mathbf{A}$ on $\mathbf{M}$.

Unless a confusion is likely to occur, we denote the above functor by
$$\mathbf{a}, \mathbf{m} \mapsto \mathbf{a} \otimes \mathbf{m}.$$

3.4.4. We let $1\text{-}\mathrm{Cat}^{\mathrm{Mon}^+}$ denote the $(\infty,1)$-category of pairs of a monoidal $(\infty,1)$-category equipped with a module, which is a full subcategory in
$$\mathrm{Funct}(\mathbf{\Delta}^{+,\mathrm{op}}, 1\text{-Cat}).$$

For a fixed $\mathbf{A}^{\otimes} \in 1\text{-}\mathrm{Cat}^{\mathrm{Mon}}$, we let
$$\mathbf{A}\text{-}\mathbf{mod} := 1\text{-}\mathrm{Cat}^{\mathrm{Mon}^+} \underset{1\text{-}\mathrm{Cat}^{\mathrm{Mon}}}{\times} \{\mathbf{A}^{\otimes}\}.$$

This is the $(\infty,1)$-category of (left) $\mathbf{A}$-module categories.

3.4.5. Replacing $\mathbf{A}$ by $\mathbf{A}^{\text{rev-mult}}$ we obtain the $(\infty,1)$-category of right $\mathbf{A}$-module categories, denoted $\mathbf{A}\text{-}\mathbf{mod}^r$.

If $\mathbf{C}$ is an $(\infty,1)$-category with a structure of $\mathbf{A}$-module category, then $\mathbf{C}^{\mathrm{op}}$ acquires a structure of $\mathbf{A}^{\mathrm{op}}$-module category.

3.4.6. Let $\mathbf{C}$ be an $(\infty,1)$-category. Recall that $\mathrm{Funct}(\mathbf{C},\mathbf{C})$ acquires a natural monoidal structure (see Sect. 3.1.5). The same construction as in *loc.cit.* shows that $\mathbf{C}$ is naturally a module category for $\mathrm{Funct}(\mathbf{C},\mathbf{C})$.

In addition, for any $\mathbf{D}$, the category $\mathrm{Funct}(\mathbf{D},\mathbf{C})$ (resp., $\mathrm{Funct}(\mathbf{C},\mathbf{D})$) is naturally a left (resp., right) module over $\mathrm{Funct}(\mathbf{C},\mathbf{C})$.

3.5. Modules for algebras. In this subsection we will explain that, given a monoidal $(\infty,1)$-category $\mathbf{A}$, an $\mathbf{A}$-module $\mathbf{M}$ and $\mathcal{A} \in \mathrm{AssocAlg}(\mathbf{A})$, we can talk about $\mathcal{A}$-modules in $\mathbf{M}$.

The idea is the same as that giving rise to the definition of associative algebras: we will use unstraightening.

3.5.1. Parallel to Sect. 3.2.2, we define the $(\infty,1)$-category

$$(1\text{-}\mathrm{Cat}^{\mathrm{Mon}^+})_{\text{right-lax}_{\text{non-untl}}}.$$

Thus, given two pairs $(\mathbf{A}_1,\mathbf{M}_1)$, $(\mathbf{A}_2,\mathbf{M}_2)$ we can talk about a pair of functors

$$F_{\mathrm{Alg}}:\mathbf{A}_1\to\mathbf{A}_2 \text{ and } F_{\mathrm{mod}}:\mathbf{M}_1\to\mathbf{M}_2,$$

where F_{Alg} is a right-lax monoidal functor, and F_{mod} is right-lax compatible with actions.

In particular, for a fixed $\mathbf{A}$, and $\mathbf{M},\mathbf{N}\in\mathbf{A}\text{-}\mathbf{mod}$ we can talk about *right-lax functors* $\mathbf{M}\to\mathbf{N}$ of $\mathbf{A}$-modules.

3.5.2. Passing to opposite categories, we obtain the corresponding notion of left-lax functor. The following is not difficult to obtain from the definitions (see also [**Lu2**, Corollary 7.3.2.7]):

LEMMA 3.5.3. *Let* $\mathbf{A}$ *be a monoidal* $(\infty,1)$*-category, and let* $\mathbf{M},\mathbf{N}\in\mathbf{A}\text{-}\mathbf{mod}$. *Let*

$$F:\mathbf{M}\rightleftarrows\mathbf{N}:G$$

be a pair of adjoint functors as plain $(\infty,1)$*-categories. Then the structure on* F *of left-lax functor of* $\mathbf{A}$*-modules is equivalent to the structure on* G *of right-lax functor of* $\mathbf{A}$*-modules.*

3.5.4. Consider the point-object

$$*^{+,\otimes}\in 1\text{-}\mathrm{Cat}^{\mathrm{Mon}^+}.$$

Given $\mathbf{A}^{+,\otimes}\in 1\text{-}\mathrm{Cat}^{\mathrm{Mon}^+}$ with the corresponding $\mathbf{A},\mathbf{M}$ we let

$$\mathrm{AssocAlg}{+}\mathrm{mod}(\mathbf{A},\mathbf{M})$$

denote the resulting category of right-lax functors

$$*^{+,\otimes}\to\mathbf{A}^{+,\otimes}.$$

This is, by definition, the category of pairs

$$\mathcal{A}\in\mathrm{AssocAlg}(\mathbf{A}) \text{ and } \mathcal{M}\in\mathcal{A}\text{-mod}(\mathbf{M}).$$

The fiber of the forgetful functor

$$\mathrm{AssocAlg}+\mathrm{mod}(\mathbf{A},\mathbf{M})\to\mathrm{AssocAlg}(\mathbf{A}) \tag{3.6}$$

over a given $\mathcal{A}\in\mathrm{AssocAlg}(\mathbf{A})$ is the category of $\mathcal{A}$-modules in $\mathbf{M}$, denoted $\mathcal{A}\text{-mod}(\mathbf{M})$.

3.5.5. The forgetful functor (3.6) is a Cartesian fibration via the operation of *restricting the module structure.*

If $\mathbf{M}$ admits geometric realizations, then the functor (3.6) is also a coCartesian fibration via the operation of *inducing the module structure.*

3.5.6. Note that we have a naturally defined functor

$$\mathbf{\Delta}^+ \to \mathbf{\Delta}, \quad [n] \mapsto [n], [n]^+ \mapsto [n+1].$$

Restriction along this functor shows that for any $\mathbf{A}^\otimes \in 1\text{-Cat}^{\mathrm{Mon}}$, the underlying $(\infty,1)$-category $\mathbf{A}$ is naturally a module for $\mathbf{A}^\otimes$.

Thus, we can talk about the category

$$\mathcal{A}\text{-mod} := \mathcal{A}\text{-mod}(\mathbf{A})$$

of $\mathcal{A}$-modules in $\mathbf{A}$ itself.

3.5.7. For example, taking $\mathbf{A}$ equal to 1-Cat with the Cartesian monoidal structure, and $\mathcal{A}$ being an associative algebra object in 1-Cat, i.e., a monoidal $(\infty,1)$-category $\mathbf{O}$, the resulting $(\infty,1)$-category

$$\mathbf{O}\text{-mod} = \mathbf{O}\text{-mod}(1\text{-Cat})$$

is the same thing as what we denoted earlier by $\mathbf{O}$-mod, i.e., this is the $(\infty,1)$-category of $\mathbf{O}$-module categories.

3.5.8. Similarly, we obtain the $(\infty,1)$-category $\mathcal{A}\text{-mod}^r$ of *right* $\mathcal{A}$-modules, i.e.,

$$\mathcal{A}\text{-mod}^r := \mathcal{A}^{\text{rev-mult}}\text{-mod}(\mathbf{A}^{\text{rev-mult}}).$$

Tensor product *on the right* makes $\mathcal{A}$-mod into a right $\mathbf{A}$-module category, and tensor product *on the left* makes $\mathcal{A}\text{-mod}^r$ into a left $\mathbf{A}$-module category.

3.5.9. By a pattern similar to Sect. 3.5.6, for $\mathcal{A} \in \mathrm{AssocAlg}(\mathbf{A})$, the object $\mathcal{A} \in \mathbf{A}$ has a natural structure of an object of $\mathcal{A}$-mod (resp., $\mathcal{A}\text{-mod}^r$).

3.6. The relative inner Hom.

3.6.1. Let $\mathbf{A}$ be a monoidal $(\infty,1)$-category, and let $\mathbf{M}$ an $\mathbf{A}$-module $(\infty,1)$-category.

Given two objects $\mathbf{m}_0, \mathbf{m}_1 \in \mathbf{M}$, consider the functor

$$\mathbf{A}^{\mathrm{op}} \to \mathrm{Spc}, \quad \mathbf{a} \mapsto \mathrm{Maps}_{\mathbf{M}}(\mathbf{a} \otimes \mathbf{m}_0, \mathbf{m}_1).$$

If this functor is representable, we will denote the representing object by

$$\underline{\mathrm{Hom}}_{\mathbf{A}}(\mathbf{m}_0, \mathbf{m}_1) \in \mathbf{A}.$$

This is the *relative inner* Hom.

3.6.2. In particular we can take $\mathbf{M} = \mathbf{A}$, regarded as a module over itself. In this case, for $\mathbf{a}_0, \mathbf{a}_1 \in \mathbf{A}$, we obtain the notion of usual inner Hom

$$\underline{\mathrm{Hom}}_{\mathbf{A}}(\mathbf{a}_0, \mathbf{a}_1) \in \mathbf{A}.$$

3.6.3. For example, let us take $\mathbf{A} = 1\text{-Cat}$, equipped with the Cartesian monoidal structure. Then for $\mathbf{C}_0, \mathbf{C}_1 \in 1\text{-Cat}$, the resulting object

$$\underline{\mathrm{Hom}}_{1\text{-Cat}}(\mathbf{C}_0, \mathbf{C}_1) \in 1\text{-Cat}$$

identifies with $\mathrm{Funct}(\mathbf{C}_0, \mathbf{C}_1)$.

3.6.4. Let $\mathcal{A} \in \mathbf{A}$ be an associative algebra. Following Sect. 3.5.8, we consider the $(\infty,1)$-category $\mathcal{A}\text{-mod}^r$ as a (left) module category over $\mathbf{A}$.

Thus, for two objects $\mathcal{M}_0, \mathcal{M}_1 \in \mathcal{A}\text{-mod}^r$, it makes sense to ask about the existence of their inner Hom as an object of $\mathbf{A}$. We shall denote it by

$$\underline{\mathrm{Hom}}_{\mathbf{A},\mathcal{A}}(\mathcal{M}_0, \mathcal{M}_1).$$

3.6.5. Assume now that $\mathbf{A}$ is symmetric monoidal, and that $\mathcal{A}$ is a commutative algebra in $\mathbf{A}$. In this case, for $\mathcal{M}, \mathcal{N} \in \mathcal{A}\text{-mod}$, the above object

$$\underline{\mathrm{Hom}}_{\mathbf{A},\mathcal{A}}(\mathcal{M}, \mathcal{N}) \in \mathbf{A}$$

naturally acquires a structure of $\mathcal{A}$-module.

3.6.6. Let $\mathbf{A}$ be again a monoidal $(\infty,1)$-category, and let $\mathbf{M}$ be an $\mathbf{A}$-module $(\infty,1)$-category.

Let $\mathbf{m} \in \mathbf{M}$ be an object. Suppose that the relative inner Hom object

$$\underline{\mathrm{Hom}}_{\mathbf{A}}(\mathbf{m}, \mathbf{m}) \in \mathbf{A}$$

exists.

Then $\underline{\mathrm{Hom}}_{\mathbf{A}}(\mathbf{m}, \mathbf{m})$ has a natural structure of associative algebra in $\mathbf{A}$. This is the unique algebra structure, for which the tautological map

$$\underline{\mathrm{Hom}}_{\mathbf{A}}(\mathbf{m}, \mathbf{m}) \otimes \mathbf{m} \to \mathbf{m}$$

extends to a structure of $\underline{\mathrm{Hom}}_{\mathbf{A}}(\mathbf{m}, \mathbf{m})$-module on $\mathbf{m}$, see [**Lu2**, Sect. 4.7.2].

3.7. Monads and Barr-Beck-Lurie.

3.7.1. Let $\mathbf{C}$ be an $(\infty,1)$-category. Recall that $\mathrm{Funct}(\mathbf{C}, \mathbf{C})$ has a natural structure of monoidal category, and $\mathbf{C}$ that of $\mathrm{Funct}(\mathbf{C}, \mathbf{C})$-module, see Sects. 3.1.5 and 3.4.6.

By definition, a *monad* acting on $\mathbf{C}$ is an associative algebra $\mathcal{A} \in \mathrm{Funct}(\mathbf{C}, \mathbf{C})$.

3.7.2. Given a monad $\mathcal{A}$, we can consider the category $\mathcal{A}\text{-mod}(\mathbf{C})$. We denote by

$$\mathbf{oblv}_{\mathcal{A}} : \mathcal{A}\text{-mod}(\mathbf{C}) \to \mathbf{C}$$

the tautological forgetful functor.

The functor $\mathbf{oblv}_{\mathcal{A}}$ admits a left adjoint, denoted

$$\mathbf{ind}_{\mathcal{A}} : \mathbf{C} \to \mathcal{A}\text{-mod}(\mathbf{C}).$$

The composite functor

$$\mathbf{oblv}_{\mathcal{A}} \circ \mathbf{ind}_{\mathcal{A}} : \mathbf{C} \to \mathbf{C}$$

identifies with the functor

$$\mathbf{c} \mapsto \mathcal{A}(\mathbf{c}),$$

where we view $\mathcal{A}$ as an endo-functor of $\mathbf{C}$, see [**Lu2**, Corollary 4.2.4.8].

3.7.3. Recall now that for any $(\infty,1)$-category $\mathbf{D}$, the $(\infty,1)$-category $\mathrm{Funct}(\mathbf{D},\mathbf{C})$ is also a module over $\mathrm{Funct}(\mathbf{C},\mathbf{C})$, see Sect. 3.4.6.

One can deduce from the construction that for a given $G \in \mathrm{Funct}(\mathbf{D},\mathbf{C})$, a structure on G of $\mathcal{A}$-module, i.e., that of object in

$$\mathcal{A}\text{-mod}(\mathrm{Funct}(\mathbf{D},\mathbf{C}))$$

is equivalent to that of factoring G as

$$\mathbf{D} \to \mathcal{A}\text{-mod}(\mathbf{C}) \overset{\mathbf{oblv}_{\mathcal{A}}}{\to} \mathbf{C},$$

see [**Lu2**, Sect. 4.2.4].

3.7.4. Let G be a functor $\mathbf{D} \to \mathbf{C}$. It is easy to see that if G admits a left adjoint, then the inner Hom object

$$\underline{\mathrm{Hom}}_{\mathrm{Funct}(\mathbf{C},\mathbf{C})}(G,G) \in \mathrm{Funct}(\mathbf{C},\mathbf{C})$$

exists and identifies with $G \circ G^L$ (see [**Lu2**, Lemma 4.2.4.1].).

Note that according to Sect. 3.6.6,

$$\mathcal{A} := G \circ G^L \in \mathrm{Funct}(\mathbf{C},\mathbf{C})$$

acquires a structure of associative algebra.

By the above, the functor G canonically factors as

$$\mathbf{D} \overset{G^{\mathrm{enh}}}{\longrightarrow} \mathcal{A}\text{-mod}(\mathbf{C}) \overset{\mathbf{oblv}_{\mathcal{A}}}{\to} \mathbf{C}.$$

DEFINITION 3.7.5. *We shall say that G is* monadic *if the above functor*

$$G^{\mathrm{enh}} : \mathbf{D} \to \mathcal{A}\text{-mod}(\mathbf{C})$$

is an equivalence.

3.7.6. Here is the statement of a simplified version of the Barr-Beck-Lurie theorem (see [**Lu2**, Theorem 4.7.4.5] for the general statement):

PROPOSITION 3.7.7. *Suppose that in the above situation both categories $\mathbf{C}$ and $\mathbf{D}$ contain geometric realizations. Then the functor G is monadic provided that the following two conditions hold:*

(1) *G is conservative;*
(2) *G preserves geometric realizations.*

PROOF. By assumption, the functor G^{enh} is conservative. Hence, it suffices to show that G^{enh} admits a left adjoint, to be denoted F^{enh}, and that the natural transformation

$$\mathbf{oblv}_{\mathcal{A}} \to \mathbf{oblv}_{\mathcal{A}} \circ G^{\mathrm{enh}} \circ F^{\mathrm{enh}} \simeq G \circ F^{\mathrm{enh}} \tag{3.7}$$

is an isomorphism.

It is clear that the (a priori partially defined) left adjoint F^{enh} is defined on objects of the form $\mathbf{ind}_{\mathcal{A}}(\mathbf{c})$ for $\mathbf{c} \in \mathbf{C}$, and by transitivity $F^{\mathrm{enh}} \circ \mathbf{ind}_{\mathcal{A}} = G^L$. The corresponding map

$$\mathbf{oblv}_{\mathcal{A}} \circ \mathbf{ind}_{\mathcal{A}} \to G \circ F^{\mathrm{enh}} \circ \mathbf{ind}_{\mathcal{A}} \tag{3.8}$$

is the tautological isomorphism $\mathbf{oblv}_{\mathcal{A}} \circ \mathbf{ind}_{\mathcal{A}} \to G \circ G^L$.

Now, every object of $\mathcal{A}\text{-mod}(\mathbf{C})$ can be obtained as a geometric realization of a simplicial object, whose terms are of the form $\mathbf{ind}_{\mathcal{A}}(\mathbf{c})$ for $\mathbf{c} \in \mathbf{C}$. Hence, the fact that F^{enh} is defined on such objects implies that it is defined on all of $\mathcal{A}\text{-mod}(\mathbf{C})$. Given that (3.8) is an isomorphism, in order to deduce the corresponding fact for (3.7), it suffices to show that both sides in (3.7) preserve geometric realizations.

This is clear for the right-hand side in (3.7), since G preserves geometric realizations. The fact that $\mathbf{oblv}_{\mathcal{A}}$ preserves geometric realizations follows from the fact that the functor

$$\mathcal{A} \otimes - \simeq G \circ G^L$$

has this property.

□

3.7.8. Here is a typical situation in which Proposition 3.7.7 applies. Let $\mathbf{A}$ be a monoidal $(\infty,1)$-category, $\mathcal{A} \in \text{Assoc}(\mathbf{A})$, and $\mathbf{M} \in \mathbf{A}\text{-}\mathbf{mod}$. Then the forgetful functor

$$\mathbf{oblv}_{\mathcal{A}} : \mathcal{A}\text{-mod}(\mathbf{M}) \to \mathbf{M}$$

is monadic, and the corresponding monad on $\mathbf{M}$ is given by

$$\mathbf{m} \mapsto \mathcal{A} \otimes \mathbf{m}.$$

Consistently with Sect. 3.7.2, we denote the corresponding left adjoint $\mathbf{M} \to \mathcal{A}\text{-mod}(\mathbf{M})$ by $\mathbf{ind}_{\mathcal{A}}$.

4. Duality

In this section we will discuss the general pattern of duality. It will apply to the notion of dualizable object in a monoidal $(\infty,1)$-category, dualizable module over an algebra, and also to that of adjoint functor.

The material in this section can be viewed as a user guide to (some parts) of [**Lu2**, Sects. 4.4 and 4.6].

4.1. Dualizability.

In this subsection we introduce the notion of dualizability of an object in a monoidal $(\infty,1)$-category.

4.1.1. Let $\mathbf{A}$ be a monoidal $(\infty,1)$-category. We shall say that an object $\mathbf{a} \in \mathbf{A}$ is *right-dualizable* if it is so as an object of $\mathbf{A}^{\text{ordn}}$.

I.e., $\mathbf{a}$ admits a *right dual* if there exists an object $\mathbf{a}^{\vee,R} \in \mathbf{A}$ equipped with 1-morphisms

$$\mathbf{a} \otimes \mathbf{a}^{\vee,R} \overset{\text{co-unit}}{\longrightarrow} \mathbf{1}_{\mathbf{A}} \text{ and } \mathbf{1}_{\mathbf{A}} \overset{\text{unit}}{\longrightarrow} \mathbf{a}^{\vee,R} \otimes \mathbf{a},$$

such that the composition

$$\mathbf{a} \overset{\text{id}\otimes\text{unit}}{\longrightarrow} \mathbf{a} \otimes \mathbf{a}^{\vee,R} \otimes \mathbf{a} \overset{\text{co-unit}\otimes\text{id}}{\longrightarrow} \mathbf{a} \tag{4.1}$$

projects to the identity element in $\pi_0(\text{Maps}_{\mathbf{A}}(\mathbf{a},\mathbf{a}))$, and the composition

$$\mathbf{a}^{\vee,R} \overset{\text{unit}\otimes\text{id}}{\longrightarrow} \mathbf{a}^{\vee,R} \otimes \mathbf{a} \otimes \mathbf{a}^{\vee,R} \overset{\text{id}\otimes\text{co-unit}}{\longrightarrow} \mathbf{a}^{\vee,R} \tag{4.2}$$

projects to the identity element in $\pi_0(\text{Maps}_{\mathbf{A}}(\mathbf{a}^{\vee,R},\mathbf{a}^{\vee,R}))$.

Similarly, one defines the notion of being left-dualizable.

If $\mathbf{A}$ is *symmetric* monoidal, then there is no difference between being right or left dualizable.

4.1.2. Let us be given $\mathbf{a} \in \mathbf{A}$ that admits a right dual. Consider the corresponding data

$$(\mathbf{a}^{\vee,R}, \mathbf{a} \otimes \mathbf{a}^{\vee,R} \overset{\text{co-unit}}{\longrightarrow} \mathbf{1}_{\mathbf{A}}). \tag{4.3}$$

We obtain that for any $\mathbf{a}' \in \mathbf{A}$, the composite map

$$\mathrm{Maps}_{\mathbf{A}}(\mathbf{a}', \mathbf{a}^{\vee,R}) \to \mathrm{Maps}_{\mathbf{A}}(\mathbf{a} \otimes \mathbf{a}', \mathbf{a} \otimes \mathbf{a}^{\vee,R}) \overset{\text{co-unit}}{\longrightarrow} \mathrm{Maps}_{\mathbf{A}}(\mathbf{a} \otimes \mathbf{a}', \mathbf{1}_{\mathbf{A}}).$$

is an isomorphism.

From here, we obtain that the data of (4.3) is *uniquely defined.*

Similarly, the data of

$$(\mathbf{a}^{\vee,R}, \mathbf{1}_{\mathbf{A}} \overset{\text{unit}}{\longrightarrow} \mathbf{a}^{\vee,R} \otimes \mathbf{a}) \tag{4.4}$$

is uniquely defined.

Furthermore, we can fix both (4.3) and (4.4) uniquely by choosing a path between (4.1) with $\mathrm{id}_{\mathbf{a}}$ *or* a path between (4.2) with $\mathrm{id}_{\mathbf{a}^{\vee,R}}$.

4.1.3. A convenient framework for viewing the notions of right or left dual is that of adjunction of 1-morphisms in an $(\infty,2)$-category, developed in Chapter 12: the datum of a monoidal $(\infty,1)$-category is equivalent to that of an $(\infty,2)$-category with a single object.

In particular, it follows from Chapter 12, Sect. 1, that given an object $\mathbf{a} \in \mathbf{A}$ that admits a right dual there exists a canonically defined $\mathbf{a}^{\vee,R}$, equipped with the data of (4.3) and (4.4), as well as paths connecting (4.1) with $\mathrm{id}_{\mathbf{a}}$ *and* (4.2) with $\mathrm{id}_{\mathbf{a}^{\vee,R}}$. These data are fixed uniquely by requiring that they satisfy a certain *infinite* set of compatibility conditions, specified in *loc.cit.*

Thus, we can talk about *the* right dual of an object $\mathbf{a} \in \mathbf{A}$.

A similar discussion applies to the word 'right' replaced by 'left'. By construction, the datum of making $\mathbf{a}'$ the right dual of $\mathbf{a}$ is equivalent to the datum of making $\mathbf{a}$ the left dual of $\mathbf{a}'$.

4.1.4. Let $\mathbf{A}^{\text{right-dualizable}}$ (resp., $\mathbf{A}^{\text{left-dualizable}}$) denote the full subcategory spanned by right (resp., left) dualizable objects. Applying Chapter 12, Corollary 1.3.6 we obtain that dualization defines an equivalence of monoidal $(\infty,1)$-categories

$$(\mathbf{A}^{\text{right-dualizable}})^{\text{op}} \simeq (\mathbf{A}^{\text{left-dualizable}})^{\text{rev-mult}}.$$

For a morphism $\phi : \mathbf{a}_1 \to \mathbf{a}_2$ we denote by $\phi^{\vee,R}$ (resp., $\phi^{\vee,L}$) the corresponding morphism $\mathbf{a}_2^{\vee,R} \to \mathbf{a}_1^{\vee,R}$ (resp., $\mathbf{a}_2^{\vee,L} \to \mathbf{a}_1^{\vee,L}$).

If $\mathbf{A}$ is symmetric monoidal we denote

$$\mathbf{A}^{\text{right-dualizable}} =: \mathbf{A}^{\text{dualizable}} := \mathbf{A}^{\text{left-dualizable}}.$$

For a morphism $\phi : \mathbf{a}_1 \to \mathbf{a}_2$ we let $\phi^{\vee} : \mathbf{a}_2^{\vee} \to \mathbf{a}_1^{\vee}$ denote its dual.

4.1.5. Consider $\mathbf{A}$ as a module over itself, and for two objects $\mathbf{a}_1, \mathbf{a}_2 \in \mathbf{A}$ recall the notation

$$\underline{\mathrm{Hom}}_{\mathbf{A}}(\mathbf{a}_1, \mathbf{a}_2) \in \mathbf{A}$$

(see Sect. 3.6.1). I.e., this is an object of $\mathbf{A}$ (if it exists) such that

$$\mathrm{Maps}_{\mathbf{A}}(\mathbf{a}', \underline{\mathrm{Hom}}_{\mathbf{A}}(\mathbf{a}_1, \mathbf{a}_2)) \simeq \mathrm{Maps}_{\mathbf{A}}(\mathbf{a}' \otimes \mathbf{a}_1, \mathbf{a}_2).$$

Assume that $\mathbf{a}_1 \in \mathbf{A}$ is left dualizable. Then it is easy to see that $\underline{\mathrm{Hom}}_{\mathbf{A}}(\mathbf{a}_1, \mathbf{a}_2)$ exists and we have a canonical isomorphism

$$\underline{\mathrm{Hom}}_{\mathbf{A}}(\mathbf{a}_1, \mathbf{a}_2) := \mathbf{a}_2 \otimes \mathbf{a}_1^{\vee, L}.$$

LEMMA 4.1.6. *Let $\mathbf{A}$ be a monoidal $(\infty, 1)$-category.*

(a) *Suppose that the functor $\mathrm{Maps}_{\mathbf{A}}(\mathbf{1}_{\mathbf{A}}, -)$ is conservative. Then if $\mathbf{a} \in \mathbf{A}$ is right dualizable, then the functor $\mathbf{a}' \mapsto \mathbf{a}' \otimes \mathbf{a}$ commutes with limits.*

(b) *Let $\mathbf{I}$ be an index category, and suppose that the functor $\otimes : \mathbf{A} \times \mathbf{A} \to \mathbf{A}$ preserves colimits in each variable indexed by $\mathbf{I}$. Assume also that the functor $\mathrm{Maps}_{\mathbf{A}}(\mathbf{1}_{\mathbf{A}}, -)$ preserves colimits indexed by $\mathbf{I}$. Then for any $\mathbf{a} \in \mathbf{A}$ that is left or right dualizable, the functor $\mathrm{Maps}_{\mathbf{A}}(\mathbf{a}, -)$ preserves colimits indexed by $\mathbf{I}$.*

PROOF. For (a), we rewrite the functor $\mathbf{a}' \mapsto \mathbf{a}' \otimes \mathbf{a}$ as $\mathbf{a}' \mapsto \underline{\mathrm{Hom}}_{\mathbf{A}}(\mathbf{a}^{\vee, R}, \mathbf{a}')$, so it is sufficient to show that the latter functor preserves limits. Since the functor $\mathrm{Maps}_{\mathbf{A}}(\mathbf{1}_{\mathbf{A}}, -)$ commutes with limits and is conservative (by assumption), it is enough to show that the functor

$$\mathbf{a}' \mapsto \mathrm{Maps}_{\mathbf{A}}(\mathbf{1}_{\mathbf{A}}, (\underline{\mathrm{Hom}}_{\mathbf{A}}(\mathbf{a}^{\vee, R}, \mathbf{a}')))$$

preserves limits. However, the latter functor is isomorphic to $\mathrm{Maps}_{\mathbf{A}}(\mathbf{a}^{\vee, R}, \mathbf{a}')$.

For (b) we give a proof when $\mathbf{a}$ is left-dualizable. Indeed, the functor

$$\mathbf{a}' \mapsto \mathrm{Maps}_{\mathbf{A}}(\mathbf{a}, \mathbf{a}')$$

is the composition of the functor

$$\mathbf{a}' \mapsto \underline{\mathrm{Hom}}_{\mathbf{A}}(\mathbf{a}, \mathbf{a}') \simeq \mathbf{a}' \otimes \mathbf{a}^L,$$

followed by the functor $\mathrm{Maps}_{\mathbf{A}}(\mathbf{1}_{\mathbf{A}}, -)$.

□

4.1.7. Let $\mathcal{A}$ be an associative algebra in $\mathbf{A}$, and $\mathbf{a} \in \mathbf{A}$ an $\mathcal{A}$-module, which is left-dualizable as a plain object of $\mathbf{A}$.

In this case, the left dual $\mathbf{a}^{\vee, L}$ of $\mathbf{a}$ acquires a natural structure of right $\mathcal{A}$-module.

The corresponding action map $\mathbf{a}^{\vee, L} \otimes \mathcal{A} \to \mathbf{a}^{\vee, L}$ is explicitly given by

$$\mathbf{a}^{\vee, L} \otimes \mathcal{A} \overset{\mathrm{id}_{\mathbf{a}^{\vee, L}} \otimes \mathrm{id}_{\mathcal{A}} \otimes \mathrm{unit}}{\longrightarrow} \mathbf{a}^{\vee, L} \otimes \mathcal{A} \otimes \mathbf{a} \otimes \mathbf{a}^{\vee, L} \to \mathbf{a}^{\vee, L} \otimes \mathbf{a} \otimes \mathbf{a}^{\vee, L} \overset{\text{co-unit} \otimes \mathrm{id}_{\mathbf{a}^{\vee, L}}}{\longrightarrow} \mathbf{a}^{\vee, L},$$

where the middle arrow is given by the action map $\mathcal{A} \otimes \mathbf{a} \to \mathbf{a}$.

More generally, for any $\mathcal{A}$-module $\mathbf{a}$ and $\mathbf{a}' \in \mathbf{A}$ for which $\underline{\mathrm{Hom}}_{\mathbf{A}}(\mathbf{a}, \mathbf{a}') \in \mathbf{A}$ exists, the object $\underline{\mathrm{Hom}}_{\mathbf{A}}(\mathbf{a}, \mathbf{a}')$ is naturally a right $\mathcal{A}$-module.

4.2. Tensor products of modules. In this subsection we will make a digression and discuss the operation of tensor product of modules over an associative (resp., commutative) algebra.

4.2.1. Assume now that $\mathbf{A}$ contains geometric realizations that distribute over the monoidal operation in $\mathbf{A}$. We claim that in this case there exists a canonically defined functor

$$\mathcal{A}\text{-mod}^r \times \mathcal{A}\text{-mod} \to \mathbf{A}, \quad \mathcal{N}, \mathcal{M} \mapsto \mathcal{N} \underset{\mathcal{A}}{\otimes} \mathcal{M},$$

see [**Lu2**, Sect. 4.4].

Indeed, it is *uniquely* defined by the following conditions:

- It preserves geometric realizations in each variable;
- It is a functor of $\mathbf{A}$-bimodule categories;
- It sends

$$(\mathcal{A} \times \mathcal{A} \in \mathcal{A}\text{-mod}^r \times \mathcal{A}\text{-mod}) \mapsto (\mathcal{A} \in \mathbf{A}),$$

in a way compatible with the homomorphisms

$$\mathcal{A}\times\mathcal{A}^{\text{rev-mult}} \to \underline{\mathrm{Hom}}_{\mathbf{A}\times\mathbf{A}^{\text{rev-mult}}}(\mathcal{A}\times\mathcal{A}, \mathcal{A}\times\mathcal{A}) \text{ and } \mathcal{A}\times\mathcal{A}^{\text{rev-mult}} \to \underline{\mathrm{Hom}}_{\mathbf{A}\times\mathbf{A}^{\text{rev-mult}}}(\mathcal{A}, \mathcal{A}).$$

4.2.2. Let now $\mathbf{A}$ be a symmetric monoidal $(\infty,1)$-category. In this case, the $(\infty,1)$-category

$$\mathrm{AssocAlg} + \mathrm{mod}(\mathbf{A}) := \mathrm{AssocAlg} + \mathrm{mod}(\mathbf{A}, \mathbf{A})$$

has a natural symmetric monoidal structure, so that the forgetful functor

$$\mathrm{AssocAlg} + \mathrm{mod}(\mathbf{A}) \to \mathrm{AssocAlg}(\mathbf{A}) \tag{4.5}$$

is symmetric monoidal, see [**Lu2**, Proposition 3.2.4.3].

4.2.3. Assume now that $\mathbf{A}$ contains geometric realizations that distribute over the monoidal operation in $\mathbf{A}$. In this case (4.5) is a coCartesian fibration.

4.2.4. Let now $\mathcal{A}$ be a commutative algebra in $\mathbf{A}$, viewed as a commutative algebra object in $\mathrm{AssocAlg}(\mathbf{A})$, see Sect. 3.3.6.

Combining with the above, we obtain that the $(\infty,1)$-category $\mathcal{A}$-mod acquires a canonically defined symmetric monoidal structure (thought of as given by tensor product over $\mathcal{A}$), see [**Lu2**, Theorem 4.5.2.1].

4.3. Duality for modules over an algebra. In this subsection we will discuss the notion of duality between left and right modules over a given associative algebra.

4.3.1. Let $\mathbf{A}$ be a monoidal $(\infty,1)$-category, and $\mathcal{A}$ an associative algebra in $\mathbf{A}$. Let $\mathcal{N}$ and $\mathcal{M}$ be a right and left $\mathcal{A}$-modules in $\mathbf{A}$. A duality datum between $\mathcal{N}$ and $\mathcal{M}$ is a pair of morphisms

$$\mathrm{unit} : \mathbf{1}_{\mathbf{A}} \to \mathcal{N} \underset{\mathcal{A}}{\otimes} \mathcal{M}$$

and

$$\text{co-unit} : \mathcal{M} \otimes \mathcal{N} \to \mathcal{A},$$

the latter being a map of $\mathcal{A} \otimes \mathcal{A}^{\text{rev-mult}}$-modules, such that the composition

$$\mathcal{M} \overset{\mathrm{id}\otimes\mathrm{unit}}{\longrightarrow} \mathcal{M} \otimes (\mathcal{N} \underset{\mathcal{A}}{\otimes} \mathcal{M}) \simeq (\mathcal{M} \otimes \mathcal{N}) \underset{\mathcal{A}}{\otimes} \mathcal{M} \overset{\text{co-unit}\otimes\mathrm{id}}{\longrightarrow} \mathcal{A} \underset{\mathcal{A}}{\otimes} \mathcal{M} \simeq \mathcal{M}$$

projects to the identity element in $\pi_0(\mathrm{Maps}_{\mathbf{A}}(\mathcal{M},\mathcal{M}))$, and the composition

$$\mathcal{N} \overset{\mathrm{unit}\otimes\mathrm{id}}{\longrightarrow} (\mathcal{N} \underset{\mathcal{A}}{\otimes} \mathcal{M}) \otimes \mathcal{N} \simeq \mathcal{N} \underset{\mathcal{A}}{\otimes} (\mathcal{M} \otimes \mathcal{N}) \overset{\mathrm{id}\otimes\mathrm{co\text{-}unit}}{\longrightarrow} \mathcal{N} \underset{\mathcal{A}}{\otimes} \mathcal{A} \simeq \mathcal{N}$$

projects to the identity element in $\pi_0(\mathrm{Maps}_{\mathbf{A}}(\mathcal{N},\mathcal{N}))$.

Thus, it makes sense to talk about dualizable left or right $\mathcal{A}$-modules.

The discussion in Sect. 4.1.2 regarding the canonicity of the dual and the duality data applies *mutatis mutandis* to the present setting.

4.3.2. Consider $\mathcal{A}\text{-mod}^r$ as a $\mathbf{A}$-module category. Let $\mathcal{M}$ and $\mathcal{N}$ be a pair of objects of $\mathcal{A}\text{-mod}^r$. Assume that $\mathcal{M}$ is dualizable, and let $\mathcal{M}^\vee \in \mathcal{A}\text{-mod}$ denote its dual. In this case, it is easy to see that the object

$$\underline{\mathrm{Hom}}_{\mathbf{A},\mathcal{A}}(\mathcal{M},\mathcal{N}) \in \mathbf{A}$$

exists and identifies canonically with $\mathcal{N} \underset{\mathcal{A}}{\otimes} \mathcal{M}^\vee$.

In particular, we obtain that in the situation of Lemma 4.1.6(a), the functor

$$\mathcal{N} \mapsto \mathcal{N} \underset{\mathcal{A}}{\otimes} \mathcal{M}^\vee$$

preserves $\mathbf{I}$-limits.

4.3.3. Assume now that $\mathbf{A}$ is symmetric monoidal and $\mathcal{A}$ is commutative. Recall that in this case, the $(\infty,1)$-category $\mathcal{A}$-mod itself carries a symmetric monoidal structure.

In this case, the duality datum between $\mathcal{M}$ and $\mathcal{N}$ in the sense of Sect. 4.3.1 is equivalent to the duality datum between them as objects of $\mathcal{A}$-mod as a symmetric monoidal $(\infty,1)$-category.

Furthermore, in this case, if $\mathcal{M}$ is dualizable, the isomorphism

$$\underline{\mathrm{Hom}}_{\mathbf{A},\mathcal{A}}(\mathcal{M},\mathcal{N}) \simeq \mathcal{N} \underset{\mathcal{A}}{\otimes} \mathcal{M}^\vee$$

upgrades to one in the category $\mathcal{A}$-mod.

4.4. Adjoint functors, revisited. We will now make a digression and discuss the point of view on the notion of adjoint functor parallel to that of the dual object. (This is, of course, more than an analogy: the two are part of the same paradigm–the notion of adjunction for 1-morphisms in an $(\infty,2)$-category.)

The reference for the material here is [**Lu1**, Sect. 5.2].

4.4.1. Let

$$F : \mathbf{C}_0 \rightleftarrows \mathbf{C}_1 : G$$

be a pair of functors between $(\infty,1)$-categories.

An adjunction datum between F and G is the datum of natural transformations

$$\mathrm{unit} : \mathrm{Id}_{\mathbf{C}_0} \to G \circ F \text{ and } \mathrm{co\text{-}unit} : F \circ G \to \mathrm{Id}_{\mathbf{C}_1}, \tag{4.6}$$

such that the composition

$$F \overset{\mathrm{id}\circ\mathrm{unit}}{\longrightarrow} F \circ G \circ F \overset{\mathrm{co\text{-}unit}\circ\mathrm{id}}{\longrightarrow} F \tag{4.7}$$

maps to the identity element in $\pi_0(\mathrm{Maps}_{\mathrm{Funct}(\mathbf{C}_0,\mathbf{C}_1)}(F,F))$, and the composition

$$G \overset{\text{unit}\circ\text{id}}{\longrightarrow} G\circ F\circ G \overset{\text{id}\circ\text{co-unit}}{\longrightarrow} F \tag{4.8}$$

maps to the identity element in $\pi_0(\mathrm{Maps}_{\mathrm{Funct}(\mathbf{C}_1,\mathbf{C}_0)}(G,G))$.

4.4.2. Given F (resp., G) that can be complemented to an adjunction datum, the discussion in Sect. 4.1.2 applies as to the canonicity of the data of $(G, \text{unit}, \text{co-unit})$ (resp., $(F, \text{unit}, \text{co-unit})$).

4.4.3. Suppose that F and G are mutually adjoint in the sense of Sect. 1.7.1. Then F and G can be *canonically* equipped with the adjunction datum; moreover there exists a canonical choice for a path between (4.7) (resp., (4.8)) and the identity endomorphism of F (resp., G).

Vice versa, a functor F (resp., G) that can be complemented to an adjunction datum admits a right (resp., left) adjoint in the sense of Sect. 1.7.1.

5. Stable $(\infty, 1)$-categories

In this section we study the notion of *stable* $(\infty, 1)$-category. This is the ∞-categorical enhancement of the notion of triangulated category.

The main point of difference between these two notions is that stable categories are *much better behaved* when it comes to such operations as taking the limit of a diagram of categories.

Related to this is the fact that given a pair of stable categories, we can form their *tensor product*, discussed in the next section.

5.1. The notion of stable category. In this subsection we define the notion of stable $(\infty, 1)$-category.

In a way parallel to abelian categories, the additive structure carried by $(\infty, 1)$-categories is in fact *not* an additional piece of structure, but rather a property of an $(\infty, 1)$-category.

The material here follows [**Lu2**, Sect. 1.1].

5.1.1. Let $\mathbf{C}$ be an $(\infty, 1)$-category. We say that $\mathbf{C}$ is *stable* if:

- It contains fiber products and push-outs[6];
- The map from the initial object to the final object is an isomorphism; we will henceforth denote it by 0;
- A diagram

$$\begin{array}{ccc} \mathbf{c}_0 & \longrightarrow & \mathbf{c}_1 \\ \downarrow & & \downarrow \\ \mathbf{c}_2 & \longrightarrow & \mathbf{c}_3 \end{array}$$

  is a pullback square if and only if it is a push-out square.

Clearly, $\mathbf{C}$ is stable if and only if $\mathbf{C}^{\mathrm{op}}$ is.

[6]Including the empty ones, i.e., a final and and an initial objects.

5.1.2. Let $\mathbf{C}$ be a stable category. For $\mathbf{c} \in \mathbf{C}$ we will use the short-hand notation $\mathbf{c}[-1]$ and $\mathbf{c}[1]$ for

$$\Omega(\mathbf{c}) := 0 \underset{\mathbf{c}}{\times} 0 \text{ and } \Sigma(\mathbf{c}) := 0 \underset{\mathbf{c}}{\sqcup} 0,$$

respectively. It follows from the axioms that the functors $[1]$ and $[-1]$, which are a priori are mutually adjoint, are actually mutually inverse.

Consider the homotopy category $\mathrm{Ho}(\mathbf{C})$ of $\mathbf{C}$, i.e., in our notation $\mathbf{C}^{\mathrm{ordn}}$. Then $\mathbf{C}^{\mathrm{ordn}}$ has a structure of triangulated category: its distinguished triangles are images of fiber sequences

$$\mathbf{c}_1 \to \mathbf{c}_2 \to \mathbf{c}_3,$$

i.e.,

$$\mathbf{c}_1 \simeq 0 \underset{\mathbf{c}_3}{\times} \mathbf{c}_2.$$

The map $\mathbf{c}_3[-1] \to \mathbf{c}_1$ comes from the tautological map

$$0 \underset{\mathbf{c}_3}{\sqcup} 0 \to \mathbf{c}_1 \simeq 0 \underset{\mathbf{c}_3}{\times} \mathbf{c}_2.$$

5.1.3. A functor between stable categories is said to be *exact* if it preserves pullbacks (equivalently, push-outs)[7].

We let $1\text{-Cat}^{\mathrm{St}}$ denote the 1-full subcategory of 1-Cat, whose objects are stable categories and whose morphisms are exact functors.

It is clear that the inclusion functor

$$1\text{-Cat}^{\mathrm{St}} \to 1\text{-Cat}$$

preserves limits.

5.1.4. For a pair of stable categories $\mathbf{C}$ and $\mathbf{D}$, let

$$\mathrm{Funct}_{\mathrm{ex}}(\mathbf{C}, \mathbf{D})$$

denote the full subcategory of $\mathrm{Funct}(\mathbf{C}, \mathbf{D})$ spanned by exact functors. We have

$$(\mathrm{Funct}_{\mathrm{ex}}(\mathbf{C}, \mathbf{D}))^{\mathrm{Spc}} = \mathrm{Maps}_{1\text{-Cat}^{\mathrm{St}}}(\mathbf{C}, \mathbf{D}).$$

The $(\infty, 1)$-category $\mathrm{Funct}_{\mathrm{ex}}(\mathbf{C}, \mathbf{D})$ is itself stable.

5.1.5. We shall say that a stable category is *cocomplete*[8] if it contains filtered colimits. This condition is equivalent to the (seemingly stronger) condition of containing arbitrary colimits, and also to the (seemingly weaker) condition of containing direct sums.

We let $1\text{-Cat}^{\mathrm{St,cocmpl}} \subset 1\text{-Cat}^{\mathrm{St}}$ be the full subcategory of $1\text{-Cat}^{\mathrm{St}}$ spanned by cocomplete stable categories.

[7]Including the empty one, i.e. the 0 object.

[8]When talking about cocomplete categories we will always assume that they are presentable.

5.1.6. Let $\mathbf{C}$ and $\mathbf{D}$ be a pair of cocomplete stable categories, and let $F : \mathbf{D} \to \mathbf{C}$ be an exact functor.

We shall say that F is *continuous* if it preserves filtered colimits. This condition is equivalent to the (seemingly stronger) condition of preserving arbitrary colimits, and also to the (seemingly weaker) condition of preserving direct sums.

We let

$$1\text{-Cat}^{\mathrm{St,cocmpl}}_{\mathrm{cont}} \subset 1\text{-Cat}^{\mathrm{St,cocmpl}}$$

denote the 1-full subcategory where we restrict morphisms to continuous functors.

5.1.7. Let $\mathbf{C}$ and $\mathbf{D}$ be a pair of stable categories. Consider the stable category $\mathrm{Funct}_{\mathrm{ex}}(\mathbf{C},\mathbf{D})$. If $\mathbf{D}$ is cocomplete, then $\mathrm{Funct}_{\mathrm{ex}}(\mathbf{C},\mathbf{D})$ is also cocomplete (this follows from the definition of cocompleteness via direct sums).

Assume now that $\mathbf{C}$ and $\mathbf{D}$ are cocomplete. We let

$$\mathrm{Funct}_{\mathrm{ex,cont}}(\mathbf{C},\mathbf{D}) \subset \mathrm{Funct}_{\mathrm{ex}}(\mathbf{C},\mathbf{D})$$

be the full subcategory spanned by continuous functors. We have

$$(\mathrm{Funct}_{\mathrm{ex,cont}}(\mathbf{C},\mathbf{D}))^{\mathrm{Spc}} = \mathrm{Maps}_{1\text{-Cat}^{\mathrm{St,cocmpl}}_{\mathrm{cont}}}(\mathbf{C},\mathbf{D}).$$

The $(\infty,1)$-category $\mathrm{Funct}_{\mathrm{ex,cont}}(\mathbf{C},\mathbf{D})$ is stable and cocomplete. The inclusion

$$\mathrm{Funct}_{\mathrm{ex,cont}}(\mathbf{C},\mathbf{D}) \hookrightarrow \mathrm{Funct}_{\mathrm{ex}}(\mathbf{C},\mathbf{D})$$

is continuous.

5.1.8. Let $\mathbf{C}$ be any $(\infty,1)$-category that contains coproducts. We equip $\mathbf{C}$ with the *coCartesian* symmetric monoidal structure. Then the forgetful functor

$$\mathrm{ComAlg}(\mathbf{C}) \overset{\mathbf{oblv}_{\mathrm{ComAlg}}}{\longrightarrow} \mathbf{C}$$

is an equivalence, see [**Lu2**, Corollary 2.4.3.10]. (Informally, every object $\mathbf{c} \in \mathbf{C}$ has a uniquely defined structure of commutative algebra, given by $\mathbf{c} \sqcup \mathbf{c} \to \mathbf{c}$).

Let now $\mathbf{C}$ be stable. In this case, the *coCartesian* symmetric monoidal structure coincides with the Cartesian one. Hence, we obtain that the forgetful functor

$$\mathrm{ComMonoid}(\mathbf{C}) \overset{\mathbf{oblv}_{\mathrm{ComMonoid}}}{\longrightarrow} \mathbf{C} \tag{5.1}$$

is an equivalence, where the notation $\mathrm{ComMonoid}(-)$ is as in Sect. 3.3.3.

Let $\mathrm{ComGrp}(\mathbf{C}) \subset \mathrm{ComMonoid}(\mathbf{C})$ be the full subcategory of group-like objects[9]. The following assertion is immediate (it happens at the level of the underlying triangulated category):

LEMMA 5.1.9. *For a stable category* $\mathbf{C}$*, the inclusion*

$$\mathrm{ComGrp}(\mathbf{C}) \hookrightarrow \mathrm{ComMonoid}(\mathbf{C})$$

is an equivalence.

[9]A (commutative) monoid in an $(\infty,1)$-category is said to be *group-like*, if it is is such in the corresponding ordinary category.

5.1.10. Let now F be an arbitrary functor $\mathbf{C} \to \mathbf{D}$, where $\mathbf{C}$ is stable and $\mathbf{D}$ is an $(\infty, 1)$-category with Cartesian products. We obtain that F canonically factors as

$$\mathbf{C} \to \mathrm{ComGrp}(\mathbf{D}) \overset{\mathbf{oblv}_{\mathrm{ComGrp}}}{\longrightarrow} \mathbf{D},$$

where $\mathbf{oblv}_{\mathrm{ComGrp}}$ denotes the tautological forgetful functor.

5.2. The 2-categorical structure. In the later chapters in this book (specifically, for the formalism of IndCoh as a functor out of the category of correspondences), we will need to consider the $(\infty, 2)$-categorical enhancement of the totality of stable categories.

We refer the reader to Chapter 10, Sect. 2, where the notion of $(\infty, 2)$-category is introduced, along with the corresponding terminology.

This subsection could (and, probably, should) be skipped on the first pass.

5.2.1. The structure of $(\infty, 1)$-category on 1-Cat naturally upgrades to a structure of $(\infty, 2)$-category, denoted $\mathbf{1}\text{-}\mathbf{Cat}$, see Chapter 10, Sect. 2.4.

We let $\mathbf{1}\text{-}\mathbf{Cat}^{\mathrm{St}}$ be the 1-full subcategory of $\mathbf{1}\text{-}\mathbf{Cat}$, where we restrict objects to stable categories, and 1-morphisms to exact functors.

We let

$$\mathbf{1}\text{-}\mathbf{Cat}^{\mathrm{St,cocmpl}} \subset \mathbf{1}\text{-}\mathbf{Cat}^{\mathrm{St}}$$

be the full subcategory where we restrict objects to be cocomplete stable categories.

Let

$$\mathbf{1}\text{-}\mathbf{Cat}^{\mathrm{St,cocmpl}}_{\mathrm{cont}} \subset \mathbf{1}\text{-}\mathbf{Cat}^{\mathrm{St,cocmpl}}$$

be the 1-full subcategory, where we restrict 1-morphisms to exact functors that are continuous.

5.2.2. Explicitly, the $(\infty, 2)$-category $\mathbf{1}\text{-}\mathbf{Cat}$ is defined in Chapter 10, Sect. 2.4 as follows:

The simplicial $(\infty, 1)$-category $\mathrm{Seq}_\bullet(\mathbf{1}\text{-}\mathbf{Cat})$ is defined so that each $\mathrm{Seq}_n(\mathbf{1}\text{-}\mathbf{Cat})$ is the 1-full subcategory of $\mathrm{Cart}_{/[n]^{\mathrm{op}}}$, where we restrict 1-morphisms to those functors between $(\infty, 1)$-categories over $[n]^{\mathrm{op}}$ that induce an equivalence on the fiber over each $i \in [n]^{\mathrm{op}}$.

Then

$$\mathrm{Seq}_n(\mathbf{1}\text{-}\mathbf{Cat}^{\mathrm{St,cocmpl}}_{\mathrm{cont}}) \subset \mathrm{Seq}_n(\mathbf{1}\text{-}\mathbf{Cat}^{\mathrm{St,cocmpl}}) \subset \mathrm{Seq}_n(\mathbf{1}\text{-}\mathbf{Cat}^{\mathrm{St}})$$

are the full subcategories of $\mathrm{Seq}_n(\mathbf{1}\text{-}\mathbf{Cat})$ defined by the following conditions:

We take those $(\infty, 1)$-categories $\mathbf{C}$ equipped with a Cartesian fibration over $[n]^{\mathrm{op}}$ for which:

- In all three cases, we require that for every $i = 0, ..., n$, the $(\infty, 1)$-category $\mathbf{C}_i$ be stable, and in the case of $\mathbf{1}\text{-}\mathbf{Cat}^{\mathrm{St,cocmpl}}$ and $\mathbf{1}\text{-}\mathbf{Cat}^{\mathrm{St,cocmpl}}_{\mathrm{cont}}$ that it be cocomplete;
- In all three cases, we require that for every $i = 1, ..., n$ the corresponding functor $\mathbf{C}_{i-1} \to \mathbf{C}_i$ be exact, and in the case of $\mathbf{1}\text{-}\mathbf{Cat}^{\mathrm{St,cocmpl}}_{\mathrm{cont}}$ that it be continuous.

5.2.3. By construction, we have:

$$\mathbf{Maps}_{\mathbf{1}\text{-}\mathbf{Cat}^{\mathrm{St,cocmpl}}_{\mathrm{cont}}}(\mathbf{D},\mathbf{C}) = \mathrm{Funct}_{\mathrm{ex,cont}}(\mathbf{D},\mathbf{C}), \quad \mathbf{C},\mathbf{D} \in 1\text{-}\mathrm{Cat}^{\mathrm{St,cocmpl}}_{\mathrm{cont}}$$

and

$$\mathbf{Maps}_{\mathbf{1}\text{-}\mathbf{Cat}^{\mathrm{St,}}}(\mathbf{D},\mathbf{C}) = \mathrm{Funct}_{\mathrm{ex}}(\mathbf{D},\mathbf{C}), \quad \mathbf{C},\mathbf{D} \in 1\text{-}\mathrm{Cat}^{\mathrm{St}}.$$

5.3. Some residual 2-categorical features. The $(\infty,2)$-categories introduced in Sect. 5.2.1 allow to assign an intrinsic meaning to the notion of adjunction of (various classes of) functors between (various classes of) stable categories.

We will exploit this in the present subsection.

We note, however, that, unlike Sect. 5.2, the constructions here are not esoteric, but are of direct practical import (e.g., the notion of exact monad).

5.3.1. According to Chapter 12, Sect. 1, it makes sense to ask whether a 1-morphism $F:\mathbf{C}\to\mathbf{D}$ in $\mathbf{1}\text{-}\mathbf{Cat}^{\mathrm{St}}$ (resp., $\mathbf{1}\text{-}\mathbf{Cat}^{\mathrm{St,cocmpl}}$, $\mathbf{1}\text{-}\mathbf{Cat}^{\mathrm{St,cocmpl}}_{\mathrm{cont}}$) admits a right adjoint 1-morphism within the corresponding $(\infty,2)$-category.

The following results from Theorem 2.5.4(a):

LEMMA 5.3.2. *Let $F:\mathbf{C}\to\mathbf{D}$ be a morphism in $\mathbf{1}\text{-}\mathbf{Cat}^{\mathrm{St,cocmpl}}_{\mathrm{cont}}$.*

(a) *The right adjoint of F always exists as a 1-morphism in $\mathbf{1}\text{-}\mathbf{Cat}^{\mathrm{St,cocmpl}}$.*

(b) *The right adjoint from* (a)*, when viewed as a functor between plain $(\infty,1)$-categories, is the right adjoint F^R of F, when the latter is viewed also as a functor between plain $(\infty,1)$-categories.*

(c) *The right adjoint of F exists in $\mathbf{1}\text{-}\mathbf{Cat}^{\mathrm{St,cocmpl}}_{\mathrm{cont}}$ if and only if F^R preserves filtered colimits (equivalently, all colimits or direct sums).*

(d) *If a 1-morphism in $\mathbf{1}\text{-}\mathbf{Cat}^{\mathrm{St,cocmpl}}$ admits a left adjoint (as a plain functor), then this left adjoint is automatically a 1-morphism in $\mathbf{1}\text{-}\mathbf{Cat}^{\mathrm{St,cocmpl}}_{\mathrm{cont}}$.*

5.3.3. In particular, to any functor

$$\mathbf{C}_{\mathbf{I}}:\mathbf{I}\to 1\text{-}\mathrm{Cat}^{\mathrm{St,cocmpl}}_{\mathrm{cont}}$$

one can associate a functor

$$\mathbf{C}^R_{\mathbf{I}^{\mathrm{op}}}:\mathbf{I}^{\mathrm{op}}\to 1\text{-}\mathrm{Cat}^{\mathrm{St,cocmpl}},$$

obtained by passing to right adjoints.

The following is a formal consequence of Lemma 2.5.2 and Proposition 2.5.7:

COROLLARY 5.3.4.

(a) *The $(\infty,1)$-category $1\text{-}\mathrm{Cat}^{\mathrm{St,cocmpl}}_{\mathrm{cont}}$ contains limits and colimits, and the functor*

$$1\text{-}\mathrm{Cat}^{\mathrm{St,cocmpl}}_{\mathrm{cont}}\to 1\text{-}\mathrm{Cat}^{\mathrm{St,cocmpl}}$$

preserves limits.

(b) *Let $\mathbf{C}_{\mathbf{I}}:\mathbf{I}\to 1\text{-}\mathrm{Cat}^{\mathrm{St,cocmpl}}_{\mathrm{cont}}$ be a functor and let $\mathbf{C}_*$ denote its colimit in $1\text{-}\mathrm{Cat}^{\mathrm{St,cocmpl}}_{\mathrm{cont}}$. Let $\mathbf{C}^R_{\mathbf{I}^{\mathrm{op}}}:\mathbf{I}^{\mathrm{op}}\to 1\text{-}\mathrm{Cat}^{\mathrm{St,cocmpl}}$ be the functor obtained from $\mathbf{C}_{\mathbf{I}}$ by passing to right adjoints. Then the resulting map in $1\text{-}\mathrm{Cat}^{\mathrm{St,cocmpl}}$*

$$\mathbf{C}_*\to\lim_{\mathbf{I}^{\mathrm{op}}}\mathbf{C}^R_{\mathbf{I}^{\mathrm{op}}}$$

is an isomorphism.

5.3.5. Given $\mathbf{C} \in 1\text{-Cat}^{\text{St,cocmpl}}_{\text{cont}}$ we can consider the monoidal $(\infty,1)$-category

$$\mathbf{Maps}_{1\text{-Cat}^{\text{St,cocmpl}}_{\text{cont}}}(\mathbf{C},\mathbf{C}), \tag{5.2}$$

which is equipped with an action on

$$\mathbf{Maps}_{1\text{-Cat}^{\text{St,cocmpl}}_{\text{cont}}}(\mathbf{D},\mathbf{C}),$$

for any $\mathbf{D} \in 1\text{-Cat}^{\text{St,cocmpl}}_{\text{cont}}$, see Chapter 9, Sect. 4.1.1 where the general paradigm is explained.

In particular, we can talk about the $(\infty,1)$-category of *exact continuous* monads acting on $\mathbf{C}$, which are by definition associative algebra objects in the monoidal $(\infty,1)$-category (5.2).

5.3.6. Given a monad $\mathcal{A}$, we can consider the $(\infty,1)$-category

$$\mathcal{A}\text{-mod}(\mathbf{C})$$

in the sense of Sect. 3.7.2.

The category $\mathcal{A}\text{-mod}(\mathbf{C})$ is itself an object of $1\text{-Cat}^{\text{St,cocmpl}}_{\text{cont}}$ and the adjoint pair

$$\mathbf{ind}_{\mathcal{A}} : \mathbf{C} \rightleftarrows \mathcal{A}\text{-mod}(\mathbf{C}) : \mathbf{oblv}_{\mathcal{A}}$$

takes place in $\mathbf{1}\text{-}\mathbf{Cat}^{\text{St,cocmpl}}_{\text{cont}}$.

For a 1-morphism $G \in \mathbf{Maps}_{1\text{-Cat}^{\text{St,cocmpl}}_{\text{cont}}}(\mathbf{D},\mathbf{C})$ the datum of action of $\mathcal{A}$ on G is equivalent to that of factoring G as

$$\mathbf{oblv}_{\mathcal{A}} \circ G^{\text{enh}}, \quad G^{\text{enh}} \in \mathbf{Maps}_{1\text{-Cat}^{\text{St,cocmpl}}_{\text{cont}}}(\mathbf{D},\mathcal{A}\text{-mod}(\mathbf{C})).$$

5.3.7. Let $G : \mathbf{D} \to \mathbf{C}$ be as above, and assume that it admits a left adjoint G^L as a plain functor. Recall (see Lemma 5.3.2) that G^L is then automatically a 1-morphism in $\mathbf{1}\text{-}\mathbf{Cat}^{\text{St,cocmpl}}_{\text{cont}}$. Consider the corresponding monad

$$\mathcal{A} := G \circ G^L,$$

see Sect. 3.7.4, so that G gives rise to a 1-morphism in $\mathbf{1}\text{-}\mathbf{Cat}^{\text{St,cocmpl}}_{\text{cont}}$:

$$G^{\text{enh}} : \mathbf{D} \to \mathcal{A}\text{-mod}(\mathbf{C}).$$

The following is an immediate consequence of Proposition 3.7.7:

COROLLARY 5.3.8. *Suppose that G does not send non-zero objects to zero. Then G^{enh} is an equivalence.*

5.4. Generation.

5.4.1. Let $\mathbf{C}$ be an object of $1\text{-Cat}^{\text{St,cocmpl}}$. A collection of objects $\{\mathbf{c}_\alpha\}$ is said to *generate* if

$$\text{Maps}_{\mathbf{C}}(\mathbf{c}_\alpha[-i],\mathbf{c}) = *, \ \forall \alpha, \ \forall i = 0,1,... \Rightarrow \mathbf{c} = 0,$$

where $[-i]$ denotes the shift functor on $\mathbf{C}$, i.e., the i-fold loop functor Ω^i.

5.4.2. The following is tautological:

LEMMA 5.4.3. *Let $\mathbf{D}$ be a (not necessarily stable) $(\infty,1)$-category, and let $F : \mathbf{D} \to \mathbf{C}$ be a functor that admits a right adjoint, and whose essential image is preserved by the loop functor. Then the essential image of F generates $\mathbf{C}$ if and only if its right adjoint F^R is conservative.*

5.4.4. We have the following basic statement:

PROPOSITION 5.4.5. *A collection* $\{\mathbf{c}_\alpha\}$ *of objects generates* $\mathbf{C}$ *if and only if* $\mathbf{C}$ *does not properly contain a cocomplete stable subcategory that contains all* $\mathbf{c}_\alpha$.

PROOF. Let $\mathbf{C}'$ be the smallest cocomplete stable full subcategory of $\mathbf{C}$ that contains the objects of the form $\mathbf{c}_\alpha$. The inclusion $\iota : \mathbf{C}' \hookrightarrow \mathbf{C}$ admits an (*a priori* non-continuous) right adjoint, denoted ι^R. Set $\mathbf{C}'' := \ker(\iota^R)$.

The inclusion

$$\mathbf{C} \overset{\jmath}{\hookleftarrow} \mathbf{C}''$$

admits a left adjoint $\jmath^L$, given by

$$\mathbf{c} \mapsto \text{coFib}(\iota \circ \iota^R(\mathbf{c}) \to \mathbf{c}).$$

By definition,

$$\mathbf{c} \in \mathbf{C}'' \Leftrightarrow \mathbf{c} \in (\mathbf{C}')^\perp \Leftrightarrow \text{Maps}_{\mathbf{C}}(\mathbf{c}_\alpha[-i], \mathbf{c}) = 0, \ \forall \alpha, \ \forall i = 0, 1, \ldots.$$

Now it is clear that the inclusion ι is an equivalence if and only if $\jmath^L$ is zero if and only if $\mathbf{C}'' = 0$.

□

5.4.6. Finally, we have:

PROPOSITION 5.4.7. *Let* $F : \mathbf{D} \to \mathbf{C}$ *be a continuous functor. Then its essential image generates the target (i.e.,* $\mathbf{C}$*) if and only if for any continuous functor* $G : \mathbf{C} \to \mathbf{C}'$ *with* $G \circ F = 0$ *we have* $G = 0$.

PROOF. We first prove the 'only if' direction. Assume that the essential image of F generates $\mathbf{C}$, and let $G : \mathbf{C} \to \mathbf{C}'$ be such that $F \circ G = 0$. Since G is continuous it admits a (possibly discontinuous) right adjoint G^R, and it suffices to show that $G^R = 0$. Since F^R is conservative, it suffices to show that $F^R \circ G^R = 0$. However, the latter identifies with $(G \circ F)^R$, which vanishes by assumption.

We now prove the 'if' direction. Let $\mathbf{C}' \subset \mathbf{C}$ be the full subcategory, generated by the essential image of F (i.e., the smallest stable cocomplete subcategory of $\mathbf{C}$ that contains the essential image of F). Let $(\mathbf{C}'', \imath, \jmath)$ be as in the proof of Proposition 5.4.5.

Being a left adjoint, $\jmath^L$ preserves colimits. Hence, the fact that $\mathbf{C}$ is cocomplete implies that $\mathbf{C}''$ is cocomplete (and $\jmath^L$ is continuous).

Now, by the construction of $\mathbf{C}'$, the composition $F \circ \jmath^L$ is zero. Hence, $\jmath^L = 0$, i.e., $\imath$ is an equivalence.

□

6. The symmetric monoidal structure on $1\text{-Cat}^{\text{St,cocmpl}}_{\text{cont}}$

In this section we will discuss some of the key features of the $(\infty, 1)$-category $1\text{-Cat}^{\text{St,cocmpl}}_{\text{cont}}$: the symmetric monoidal structure, given by *tensor product of stable categories*, which we call the *Lurie tensor product*, and the notion of dualizable stable category.

In the process we will encounter the most basic stable category–that of *spectra*.

This section can be regarded as a user guide to [**Lu2**, 1.4 and 4.8].

6.1. The Lurie tensor product. In this subsection we introduce the Lurie tensor product.

It is quite remarkable that one does not have to work very hard in order to characterize it uniquely: for a pair of stable categories $\mathbf{C}_1$ and $\mathbf{C}_2$ and a third one $\mathbf{D}$, the space of exact continuous functors

$$\mathbf{C}_1 \otimes \mathbf{C}_2 \to \mathbf{D}$$

is a full subspace in

$$\text{Maps}_{1\text{-Cat}}(\mathbf{C}_1 \times \mathbf{C}_2, \mathbf{D})$$

that consists of functors that are exact and continuous in each variable.

I.e., one does not need to introduce any additional pieces of structure, but rather impose conditions.

6.1.1. Consider the coCartesian fibration

$$1\text{-Cat}^{\times,\text{Fin}_*} \to \text{Fin}_*,$$

corresponding to the Cartesian symmetric monoidal structure on 1-Cat.

We let

$$(1\text{-Cat}^{\text{St,cocmpl}}_{\text{cont}})^{\otimes,\text{Fin}_*} \subset 1\text{-Cat}^{\times,\text{Fin}_*}$$

be the 1-full subcategory, where:

- We restrict objects to those
$$(I,*), \quad (i \in I - \{*\}) \mapsto (\mathbf{C}_i \in 1\text{-Cat}),$$
where each $\mathbf{C}_i$ is stable and cocomplete;
- We restrict morphisms to those
$$\phi : (I,*) \to (J,*), \quad (j \in J - \{*\}) \mapsto (\underset{i \in \phi^{-1}(j)}{\Pi} \mathbf{C}_i \overset{F_j}{\to} \mathbf{C}_j),$$
where each F_j is exact and continuous in each variable.

THEOREM 6.1.2. *The composite functor* $(1\text{-Cat}^{\text{St,cocmpl}}_{\text{cont}})^{\otimes,\text{Fin}_*} \to \text{Fin}_*$ *is a coCartesian fibration, that lies in the essential image of the fully faithful functor*

$$1\text{-Cat}^{\text{SymMon}} \to (\text{coCart}_{/\text{Fin}_*})_{\text{strict}}.$$

This theorem is a combination of [**Lu2**, Propositions 4.8.1.3, 4.8.1.14 and 4.8.1.18].

6.1.3. It follows from Theorem 6.1.2 that the $(\infty,1)$-category $1\text{-Cat}^{\text{St,cocmpl}}_{\text{cont}}$ of stable categories acquires a symmetric monoidal structure. We will refer to it as the *Lurie symmetric monoidal structure.*

The corresponding monoidal operation, denoted

$$(\mathbf{C}_i, i \in I) \mapsto \underset{i \in I}{\otimes} \mathbf{C}_i$$

is the *Lurie tensor product.*

6.1.4. By construction, for $\mathbf{D} \in 1\text{-Cat}^{\text{St,cocmpl}}_{\text{cont}}$ the space of exact continuous functors

$$\underset{i \in I}{\otimes} \mathbf{C}_i \to \mathbf{D}$$

is the full subspace in the space of functors

$$\underset{i \in I}{\Pi} \mathbf{C}_i \to \mathbf{D}$$

that are exact and continuous in each variable.

It follows from the above description and Proposition 2.5.7 that the monoidal operation:

$$\{\mathbf{C}_i\} \mapsto \underset{i \in I}{\otimes} \mathbf{C}_i$$

preserves colimits (taken in $1\text{-Cat}^{\text{St,cocmpl}}_{\text{cont}}$) in each variable.

REMARK 6.1.5. A remarkable aspect of this theory is that Theorem 6.1.2 is not very hard. The existence of the tensor product $\underset{i \in I}{\otimes} \mathbf{C}_i$ follows from the Adjoint Functor Theorem. The fact tat the canonical functor

$$\mathbf{C}_1 \otimes (\mathbf{C}_2 \otimes \mathbf{C}_3) \to \mathbf{C}_1 \otimes \mathbf{C}_2 \otimes \mathbf{C}_3$$

follows by interpreting exact continuous functors

$$\mathbf{C}_1 \otimes (\mathbf{C}_2 \otimes \mathbf{C}_3) \to \mathbf{D}$$

as exact continuous functors

$$\mathbf{C}_2 \otimes \mathbf{C}_3 \to \text{Funct}_{\text{ex,cont}}(\mathbf{C}_1, \mathbf{D}).$$

6.1.6. By construction, we have a tautological functor

$$\underset{i \in I}{\Pi} \mathbf{C}_i \to \underset{i \in I}{\otimes} \mathbf{C}_i, \tag{6.1}$$

which is exact and continuous in each variable.

For $\mathbf{c}_i \in \mathbf{C}_i$, we let

$$\underset{i}{\boxtimes} \mathbf{c}_i \in \underset{i \in I}{\otimes} \mathbf{C}_i$$

denote the image of the object $(\underset{i}{\times} \mathbf{c}_i) \in \underset{i \in I}{\Pi} \mathbf{C}_i$ under the functor (6.1).

6.1.7. Note that for $\mathbf{C}, \mathbf{D} \in 1\text{-Cat}^{\text{St,cocmpl}}_{\text{cont}}$ the object

$$\text{Funct}_{\text{ex,cont}}(\mathbf{D}, \mathbf{C}) \in 1\text{-Cat}^{\text{St,cocmpl}}_{\text{cont}}$$

(see Sect. 5.1.7) identifies with the inner Hom object

$$\underline{\text{Hom}}_{1\text{-Cat}^{\text{St,cocmpl}}_{\text{cont}}}(\mathbf{D}, \mathbf{C}).$$

I.e., for $\mathbf{E} \in 1\text{-Cat}^{\text{St,cocmpl}}_{\text{cont}}$ we have a canonical isomorphism

$$\text{Maps}_{1\text{-Cat}^{\text{St,cocmpl}}_{\text{cont}}}(\mathbf{E} \otimes \mathbf{D}, \mathbf{C}) \simeq \text{Maps}_{1\text{-Cat}^{\text{St,cocmpl}}_{\text{cont}}}(\mathbf{E}, \text{Funct}_{\text{ex,cont}}(\mathbf{D}, \mathbf{C})).$$

6.1.8. *The symmetric monoidal structure on* $\mathbf{1\text{-}Cat}^{\text{St,cocmpl}}_{\text{cont}}$. For future use, we note that the structure of symmetric monoidal $(\infty,1)$-category on $1\text{-Cat}^{\text{St,cocmpl}}_{\text{cont}}$ canonically extends to that of symmetric monoidal $(\infty,2)$-category on the 2-categorical enhancement of $1\text{-Cat}^{\text{St,cocmpl}}_{\text{cont}}$, i.e., $\mathbf{1\text{-}Cat}^{\text{St,cocmpl}}_{\text{cont}}$ (see Chapter 9, Sect. 1.4 for the notion of symmetric monoidal structure on an $(\infty,2)$-category).

Indeed, we start with the 2-coCartesian fibration $\mathbf{1\text{-}Cat}^{\otimes,\text{Fin}_*} \to \text{Fin}_*$ that defines the Cartesian symmetric monoidal structure on $\mathbf{1\text{-}Cat}$.

Note that

$$\left(\mathbf{1\text{-}Cat}^{\otimes,\text{Fin}_*}\right)^{1\text{-Cat}} \simeq 1\text{-Cat}^{\otimes,\text{Fin}_*}.$$

We let

$$\left(\mathbf{1\text{-}Cat}^{\text{St,cocmpl}}_{\text{cont}}\right)^{\otimes,\text{Fin}_*}$$

be the 1-full subcategory of $\mathbf{1\text{-}Cat}^{\otimes,\text{Fin}_*}$ that corresponds to

$$\left(1\text{-Cat}^{\text{St,cocmpl}}_{\text{cont}}\right)^{\otimes,\text{Fin}_*} \subset 1\text{-Cat}^{\otimes,\text{Fin}_*}.$$

One checks that the composite functor

$$\left(\mathbf{1\text{-}Cat}^{\text{St,cocmpl}}_{\text{cont}}\right)^{\otimes,\text{Fin}_*} \to \mathbf{1\text{-}Cat}^{\otimes,\text{Fin}_*} \to \text{Fin}_*$$

is a 2-coCartesian fibration, and as such defines a symmetric monoidal structure on the $(\infty,2)$-category $\mathbf{1\text{-}Cat}^{\text{St,cocmpl}}_{\text{cont}}$.

6.2. The $(\infty,1)$-category of spectra. The symmetric monoidal structure on $1\text{-Cat}^{\text{St,cocmpl}}_{\text{cont}}$ leads to a concise definition of the $(\infty,1)$-category of spectra, along with (some of) its key features.

6.2.1. The $(\infty,1)$-category Sptr of spectra can be defined as the unit object in the symmetric monoidal $(\infty,1)$-category $1\text{-Cat}^{\text{St,cocmpl}}_{\text{cont}}$.

Let $\mathbf{1}_{\text{Sptr}}$ denote the unit object in Sptr. This is the *sphere spectrum*.

6.2.2. Recall the setting of Lemma 2.1.8. We obtain that the object $\mathbf{1}_{\text{Sptr}} \in \text{Sptr}$ gives rise to a functor

$$\text{Spc} \to \text{Sptr}.$$

We denote this functor by Σ^∞.

6.2.3. The functor Σ^∞ has the following universal property (see [**Lu2**, Corollary 1.4.4.5]):

LEMMA 6.2.4. *For* $\mathbf{C} \in 1\text{-Cat}^{\text{St,cocmpl}}_{\text{cont}}$, *restriction and left Kan extension along* Σ^∞ *define an equivalence between* $\text{Funct}_{\text{ex,cont}}(\text{Sptr},\mathbf{C})$ *and the full subcategory of* $\text{Funct}(\text{Spc},\mathbf{C})$ *consisting of colimit-preserving functors.*

The above lemma expresses the universal property of the category Sptr as the *stabilization* of Spc.

6.2.5. Combining Lemmas 6.2.4 and 2.1.8 we obtain:

COROLLARY 6.2.6. *For* $\mathbf{C} \in 1\text{-Cat}^{\text{St,cocmpl}}_{\text{cont}}$*, restriction and left Kan extension along*

$$\{\mathbf{1}_{\text{Sptr}}\} \hookrightarrow \text{Sptr}$$

define an equivalence

$$\text{Funct}_{\text{ex,cont}}(\text{Sptr}, \mathbf{C}) \simeq \mathbf{C}.$$

6.2.7. The functor Σ^∞ admits a right adjoint, denoted Ω^∞. By Sect. 5.1.10, the functor Ω^∞ canonically factors via a functor

$$\text{Sptr} \to \text{ComGrp}(\text{Spc}), \tag{6.2}$$

followed by the forgetful functor

$$\text{ComGrp}(\text{Spc}) \overset{\mathbf{oblv}_{\text{ComGrp}}}{\longrightarrow} \text{Spc}.$$

The functor Ω^∞ preserves filtered colimits.

6.2.8. The stable category Sptr has a t-structure, uniquely determined by the condition that an object $\mathcal{S} \in \text{Sptr}$ is strictly *coconnective*, i.e., belongs to $\text{Sptr}^{>0}$, if and only if $\Omega^\infty(\mathcal{S}) = *$; see [**Lu2**, Proposition 1.4.3.6].

The t-structure on Sptr is both left and right complete. This means that for $\mathcal{S} \in \text{Sptr}$ the canonical maps

$$\mathcal{S} \mapsto \lim_n \tau^{\geq -n}(\mathcal{S}) \text{ and } \operatorname{colim}_n \tau^{\leq n}(\mathcal{S}) \to \mathcal{S}$$

are isomorphisms.

6.2.9. The restriction of the functor (6.2) to the full subcategory of *connective spectra*

$$\text{Sptr}^{\leq 0} \subset \text{Sptr}$$

defines an equivalence

$$\text{Sptr}^{\leq 0} \to \text{ComGrp}(\text{Spc}); \tag{6.3}$$

see [**Lu2**, Theorem 5.2.6.10] (this statement goes back to [**May**] and [**BoV**]).

6.2.10. Let $\mathbf{C}$ be an object of $1\text{-Cat}^{\text{St,cocmpl}}_{\text{cont}}$. Since Sptr is the unit object in the symmetric monoidal category $1\text{-Cat}^{\text{St,cocmpl}}_{\text{cont}}$, our $\mathbf{C}$ has a canonical structure of Sptr-module category.

For $\mathbf{c}_0, \mathbf{c}_1 \in \mathbf{C}$, consider the corresponding relative inner Hom object

$$\underline{\text{Hom}}_{\text{Sptr}}(\mathbf{c}_0, \mathbf{c}_1) \in \text{Sptr}$$

see Sect. 3.6.1.

We will also use the notation

$$\mathcal{M}aps_{\mathbf{C}}(\mathbf{c}_0, \mathbf{c}_1) := \underline{\text{Hom}}_{\text{Sptr}}(\mathbf{c}_0, \mathbf{c}_1) \in \text{Sptr}.$$

I.e., for $\mathcal{S} \in \text{Sptr}$ we have

$$\text{Maps}_{\text{Sptr}}(\mathcal{S}, \mathcal{M}aps_{\mathbf{C}}(\mathbf{c}_0, \mathbf{c}_1)) \simeq \text{Maps}_{\mathbf{C}}(\mathcal{S} \otimes \mathbf{c}_0, \mathbf{c}_1).$$

By adjunction, we have

$$\text{Maps}_{\mathbf{C}}(\mathbf{c}_0, \mathbf{c}_1) \simeq \Omega^\infty(\mathcal{M}aps_{\mathbf{C}}(\mathbf{c}_0, \mathbf{c}_1)).$$

6.3. Duality of stable categories. Since $1\text{-Cat}^{\mathrm{St,cocmpl}}_{\mathrm{cont}}$ has a symmetric monoidal structure, we can talk about dualizable objects in it, see Sect. 4.1.1. Thus, we arrive at the notion of *dualizable cocomplete stable category*. In the same vein, we can talk about the datum of duality between two objects of $1\text{-Cat}^{\mathrm{St,cocmpl}}_{\mathrm{cont}}$.

These notions turn out to be immensely useful in practice.

6.3.1. By definition, a duality datum between $\mathbf{C}$ and $\mathbf{D}$ is the datum of a morphism

$$\epsilon : \mathbf{C} \otimes \mathbf{D} \to \mathrm{Sptr} \text{ and } \mu : \mathrm{Sptr} \to \mathbf{D} \otimes \mathbf{C},$$

such that the composition

$$\mathbf{C} \overset{\mathrm{Id}_{\mathbf{C}} \otimes \mu}{\longrightarrow} \mathbf{C} \otimes \mathbf{D} \otimes \mathbf{C} \overset{\epsilon \otimes \mathrm{Id}_{\mathbf{C}}}{\longrightarrow} \mathbf{C}$$

is isomorphic to $\mathrm{Id}_{\mathbf{C}}$, and the composition

$$\mathbf{D} \overset{\mu \otimes \mathrm{Id}_{\mathbf{D}}}{\longrightarrow} \mathbf{D} \otimes \mathbf{C} \otimes \mathbf{D} \overset{\mathrm{Id}_{\mathbf{C}} \otimes \epsilon}{\longrightarrow} \mathbf{D}$$

is isomorphic to $\mathrm{Id}_{\mathbf{D}}$.

6.3.2. Let $\mathbf{C}$ and $\mathbf{D}$ be dualizable objects in $1\text{-Cat}^{\mathrm{St,cocmpl}}_{\mathrm{cont}}$, and let $\mathbf{C}^{\vee}$ and $\mathbf{D}^{\vee}$ denote their respective duals.

For a continuous functor $F : \mathbf{C} \to \mathbf{D}$, we denote by $F^{\vee} : \mathbf{D}^{\vee} \to \mathbf{C}^{\vee}$ the dual functor (see Sect. 4.1.4). Explicitly, $F^{\vee}$ is given as the composition

$$\mathbf{D}^{\vee} \overset{\mu_{\mathbf{C}} \otimes \mathrm{Id}_{\mathbf{D}^{\vee}}}{\longrightarrow} \mathbf{C}^{\vee} \otimes \mathbf{C} \otimes \mathbf{D}^{\vee} \overset{\mathrm{Id}_{\mathbf{C}^{\vee}} \otimes F \otimes \mathrm{Id}_{\mathbf{D}^{\vee}}}{\longrightarrow} \mathbf{C}^{\vee} \otimes \mathbf{D} \otimes \mathbf{D}^{\vee} \overset{\mathrm{Id}_{\mathbf{C}^{\vee}} \otimes \epsilon_{\mathbf{D}}}{\longrightarrow} \mathbf{C}^{\vee}.$$

By Sect. 4.1.5 and Sect. 6.1.7, we have the canonical isomorphisms

$$\mathrm{Funct}_{\mathrm{ex,cont}}(\mathbf{C}, \mathbf{D}) \simeq \mathbf{C}^{\vee} \otimes \mathbf{D} \simeq \mathrm{Funct}_{\mathrm{ex,cont}}(\mathbf{D}^{\vee}, \mathbf{C}^{\vee}).$$

6.3.3. Let us again be in the situation of Sect. 5.3.3. Assume that all the objects $\mathbf{C}_i$ are dualizable, and that the right adjoints of the transition functors $\mathbf{C}_i \to \mathbf{C}_j$ are continuous.

By applying the dualization functor (see Sect. 4.1.4), from

$$\mathbf{C}_{\mathbf{I}} : \mathbf{I} \to 1\text{-Cat}^{\mathrm{St,cocmpl}}_{\mathrm{cont}},$$

we obtain another functor, denoted

$$\mathbf{C}^{\vee}_{\mathbf{I}^{\mathrm{op}}} : \mathbf{I}^{\mathrm{op}} \to 1\text{-Cat}^{\mathrm{St,cocmpl}}_{\mathrm{cont}}, \quad i \mapsto \mathbf{C}^{\vee}_i.$$

We claim:

PROPOSITION 6.3.4. *Under the above circumstances, the object $\mathbf{C}_*$ is dualizable, and the dual of the colimit diagram*

$$\operatorname*{colim}_{\mathbf{I}} \mathbf{C}_{\mathbf{I}} \to \mathbf{C}_*$$

is a limit diagram, i.e., the map

$$(\mathbf{C}_*)^{\vee} \to \lim_{\mathbf{I}^{\mathrm{op}}} \mathbf{C}^{\vee}_{\mathbf{I}^{\mathrm{op}}}$$

is an equivalence.

The rest of this subsection is devoted to the proof of this proposition.

6.3.5. We will construct the duality datum between $\underset{\mathbf{I}}{\operatorname{colim}}\, \mathbf{C}_{\mathbf{I}}$ and $\underset{\mathbf{I}^{\mathrm{op}}}{\lim}\, \mathbf{C}^{\vee}_{\mathbf{I}^{\mathrm{op}}}$.

The functor

$$\epsilon : \left(\underset{\mathbf{I}}{\operatorname{colim}}\, \mathbf{C}_{\mathbf{I}}\right) \otimes \left(\underset{\mathbf{I}^{\mathrm{op}}}{\lim}\, \mathbf{C}^{\vee}_{\mathbf{I}^{\mathrm{op}}}\right) \to \mathrm{Sptr}$$

is given as follows:

Since

$$\left(\underset{i\in\mathbf{I}}{\operatorname{colim}}\, \mathbf{C}_i\right) \otimes \left(\underset{j\in\mathbf{I}^{\mathrm{op}}}{\lim}\, \mathbf{C}^{\vee}_j\right) \simeq \underset{i\in\mathbf{I}}{\operatorname{colim}} \left(\mathbf{C}_i \otimes \left(\underset{j\in\mathbf{I}^{\mathrm{op}}}{\lim}\, \mathbf{C}^{\vee}_j\right)\right),$$

the datum of ϵ is equivalent to a compatible family of functors

$$\mathbf{C}_i \otimes \left(\underset{j\in\mathbf{I}^{\mathrm{op}}}{\lim}\, \mathbf{C}^{\vee}_j\right) \to \mathrm{Sptr}, \quad i \in \mathbf{I}.$$

The latter are given by

$$\mathbf{C}_i \otimes \left(\underset{j\in\mathbf{I}^{\mathrm{op}}}{\lim}\, \mathbf{C}^{\vee}_j\right) \overset{\mathrm{Id}_{\mathbf{C}_i} \otimes \mathrm{ev}_i}{\longrightarrow} \mathbf{C}_i \otimes \mathbf{C}^{\vee}_i \overset{\epsilon_{\mathbf{C}_i}}{\to} \mathrm{Sptr},$$

where ev_i denotes the evaluation functor $\underset{j\in\mathbf{I}^{\mathrm{op}}}{\lim}\, \mathbf{C}^{\vee}_j \to \mathbf{C}^{\vee}_i$.

6.3.6. Let us now construct the functor

$$\mu : \mathrm{Sptr} \to \left(\underset{\mathbf{I}^{\mathrm{op}}}{\lim}\, \mathbf{C}^{\vee}_{\mathbf{I}^{\mathrm{op}}}\right) \otimes \left(\underset{\mathbf{I}}{\operatorname{colim}}\, \mathbf{C}_{\mathbf{I}}\right).$$

For this we note that the functor $\mathbf{C}^{\vee}_{\mathbf{I}^{\mathrm{op}}}$ is obtained by passing to *right adjoints* from a functor $\mathbf{I} \to 1\text{-}\mathrm{Cat}^{\mathrm{St,cocmpl}}_{\mathrm{cont}}$, which in turn is given by passing to right adjoints in $\mathbf{C}_{\mathbf{I}}$, and then passing to the duals.

Hence, by Corollary 5.3.4, the limit $\underset{\mathbf{I}^{\mathrm{op}}}{\lim}\, \mathbf{C}^{\vee}_{\mathbf{I}^{\mathrm{op}}}$ can be rewritten as a colimit. Hence, for any $\mathbf{D} \in 1\text{-}\mathrm{Cat}^{\mathrm{St,cocmpl}}_{\mathrm{cont}}$ the natural map

$$\mathbf{D} \otimes \left(\underset{j\in\mathbf{I}^{\mathrm{op}}}{\lim}\, \mathbf{C}^{\vee}_j\right) \to \underset{j\in\mathbf{I}^{\mathrm{op}}}{\lim} \left(\mathbf{D} \otimes \mathbf{C}^{\vee}_j\right)$$

is an equivalence.

Hence,

$$\left(\underset{j\in\mathbf{I}^{\mathrm{op}}}{\lim}\, \mathbf{C}^{\vee}_j\right) \otimes \left(\underset{i\in\mathbf{I}}{\operatorname{colim}}\, \mathbf{C}_i\right) \simeq \underset{j\in\mathbf{I}^{\mathrm{op}}}{\lim} \left(\mathbf{C}^{\vee}_j \otimes \left(\underset{i\in\mathbf{I}}{\operatorname{colim}}\, \mathbf{C}_i\right)\right).$$

Therefore, the datum of μ amounts to a compatible family of functors

$$\mathrm{Sptr} \to \mathbf{C}^{\vee}_j \otimes \left(\underset{i\in\mathbf{I}}{\operatorname{colim}}\, \mathbf{C}_i\right), \quad j \in \mathbf{I}.$$

The latter are given by

$$\mathrm{Sptr} \overset{\mu_{\mathbf{C}_j}}{\longrightarrow} \mathbf{C}^{\vee}_j \otimes \mathbf{C}_j \overset{\mathrm{Id}_{\mathbf{C}_j} \otimes \mathrm{ins}_j}{\longrightarrow} \mathbf{C}^{\vee}_j \otimes \left(\underset{i\in\mathbf{I}}{\operatorname{colim}}\, \mathbf{C}_i\right),$$

where ins_j denotes the insertion functor $\mathbf{C}_j \to \underset{i\in\mathbf{I}}{\operatorname{colim}}\, \mathbf{C}_i$.

6.3.7. The fact that the functors ϵ and μ constructed above satisfy the adjunction identities is a straightforward verification.

6.4. Generation of tensor products.

6.4.1. Let $\mathbf{C}_1$ and $\mathbf{C}_2$ be objects of $1\text{-Cat}^{\text{St,cocmpl}}$, and consider their tensor product

$$\mathbf{C}_1 \otimes \mathbf{C}_2.$$

We have the following basic fact:

PROPOSITION 6.4.2. *Let $F_i : \mathbf{D}_i \to \mathbf{C}_i$, $i = 1, 2$ be continuous functors, such that their respective essential images generate the target. The essential image of the tautological functor*

$$\otimes : \mathbf{C}_1 \times \mathbf{C}_2 \to \mathbf{C}_1 \otimes \mathbf{C}_2, \quad \mathbf{c}_1 \times \mathbf{c}_2 \mapsto \mathbf{c}_1 \boxtimes \mathbf{c}_2$$

generates the target.

PROOF. Let $\mathbf{C}'$ be the smallest cocomplete stable full subcategory of $\mathbf{C} := \mathbf{C}_1 \otimes \mathbf{C}_2$ that contains the objects of the form $\mathbf{c}_1 \boxtimes \mathbf{c}_2$. Recall the notations in the proof of Proposition 5.4.5.

Being a left adjoint, $\jmath^L$ preserves colimits. Hence, the fact that $\mathbf{C}$ is cocomplete implies that $\mathbf{C}''$ is cocomplete.

We need to show that $\mathbf{C}'' = 0$, which is equivalent to the functor $\jmath^L$ being zero. By the universal property of $\mathbf{C}_1 \otimes \mathbf{C}_2$, the latter is equivalent to the fact that the composition

$$\mathbf{C}_1 \times \mathbf{C}_2 \to \mathbf{C}_1 \otimes \mathbf{C}_2 = \mathbf{C} \xrightarrow{\jmath^L} \mathbf{C}''$$

maps to the zero object of $\mathbf{C}''$.

However, the latter composition factors as

$$\mathbf{C}_1 \times \mathbf{C}_2 \to \mathbf{C}' \overset{\iota}{\hookrightarrow} \mathbf{C} \xrightarrow{\jmath^L} \mathbf{C}'',$$

while $\jmath^L \circ \iota$ is tautologically 0. □

In addition, we have:

PROPOSITION 6.4.3. *Let $F_i : \mathbf{D}_i \to \mathbf{C}_i$, $i = 1, 2$ be continuous functors, such that their respective essential images generate the target. Then the same is true for*

$$F_1 \otimes F_2 : \mathbf{D}_1 \otimes \mathbf{D}_2 \to \mathbf{C}_1 \otimes \mathbf{C}_2.$$

PROOF. By Proposition 5.4.7, it is enough to show that for a continuous functor

$$G : \mathbf{C}_1 \otimes \mathbf{C}_2 \to \mathbf{C}',$$

if the composition $G \circ (F_1 \otimes F_2)$ is zero, then $G = 0$. Thus, we have to show that for a fixed $\mathbf{c}_1 \in \mathbf{C}_1$, the functor

$$G(\mathbf{c}_1 \otimes -) : \mathbf{C}_2 \to \mathbf{C}'$$

is zero. By Proposition 5.4.7, it suffices to show that

$$G(\mathbf{c}_1 \otimes F_2(\mathbf{d}_2)) = 0$$

for any $\mathbf{d}_2 \in \mathbf{D}_2$. I.e., it suffices to show that the functor

$$G(- \otimes F_2(\mathbf{d}_2)) : \mathbf{C}_1 \to \mathbf{C}'$$

is zero (for a fixed $\mathbf{d}_2 \in \mathbf{D}_2$).

Applying Proposition 5.4.7 again, we obtain that it suffices to show that

$$G(F_1(\mathbf{d}_1) \otimes F_2(\mathbf{d}_2))$$

is zero for any $\mathbf{d}_1 \in \mathbf{D}_1$. However, the latter is just the assumption that $G \circ (F_1 \otimes F_2) = 0$.

□

6.4.4. Consider the following situation: let $\mathbf{C}_i$, $i = 1, 2$ be objects of $1\text{-Cat}^{\text{St,cocmpl}}_{\text{cont}}$, and let

$$\mathcal{A}_i \in \text{AssocAlg}(\text{Funct}_{\text{ex,cont}}(\mathbf{C}_i, \mathbf{C}_i))$$

be a monad acting on $\mathbf{C}$.

Consider the monad

$$\mathcal{A}_1 \otimes \mathcal{A}_2 \in \text{AssocAlg}(\text{Funct}_{\text{ex,cont}}(\mathbf{C}_1 \otimes \mathbf{C}_2, \mathbf{C}_1 \otimes \mathbf{C}_2)).$$

The tautological action of $\mathcal{A}_i$ on $\mathbf{oblv}_{\mathcal{A}_i} \in \text{Funct}_{\text{ex,cont}}(\mathcal{A}_i\text{-mod}(\mathbf{C}_i), \mathbf{C}_i)$ induces an action of $\mathcal{A}_1 \otimes \mathcal{A}_2$ on

$$\mathbf{oblv}_{\mathcal{A}_1} \otimes \mathbf{oblv}_{\mathcal{A}_2} \in \\ \in \text{Funct}_{\text{ex,cont}}(\mathcal{A}_1\text{-mod}(\mathbf{C}_1) \otimes \mathcal{A}_2\text{-mod}(\mathbf{C}_2), \mathcal{A}_1\text{-mod}(\mathbf{C}_1) \otimes \mathcal{A}_2\text{-mod}(\mathbf{C}_2)).$$

Hence, by Sect. 5.3.6, the functor $\mathbf{oblv}_{\mathcal{A}_1} \otimes \mathbf{oblv}_{\mathcal{A}_2}$ upgrades to a functor

$$\mathcal{A}_1\text{-mod}(\mathbf{C}_1) \otimes \mathcal{A}_2\text{-mod}(\mathbf{C}_2) \to (\mathcal{A}_1 \otimes \mathcal{A}_2)\text{-mod}(\mathbf{C}_1 \otimes \mathbf{C}_2). \tag{6.4}$$

We claim:

LEMMA 6.4.5. *The functor* (6.4) *is an equivalence.*

PROOF. The left adjoint of $\mathbf{oblv}_{\mathcal{A}_1} \otimes \mathbf{oblv}_{\mathcal{A}_2}$ is provided by

$$\mathbf{ind}_{\mathcal{A}_1} \otimes \mathbf{ind}_{\mathcal{A}_2},$$

and hence, the canonical map from $\mathcal{A}_1 \otimes \mathcal{A}_2$ to the monad on $\mathbf{C}_1 \otimes \mathbf{C}_2$, corresponding to $\mathbf{oblv}_{\mathcal{A}_1} \otimes \mathbf{oblv}_{\mathcal{A}_2}$, is an isomorphism.

Hence, by Corollary 5.3.8, it suffices to show that the functor $\mathbf{oblv}_{\mathcal{A}_1} \otimes \mathbf{oblv}_{\mathcal{A}_2}$ is conservative. By Lemma 5.4.3, this is equivalent to the fact that the essential image of

$$\mathbf{ind}_{\mathcal{A}_1} \otimes \mathbf{ind}_{\mathcal{A}_2}$$

generates the target. This is true for each $\mathbf{ind}_{\mathcal{A}_i}$ (since $\mathbf{oblv}_{\mathcal{A}_i}$ are conservative), and hence the required assertion follows from Proposition 6.4.2.

□

7. Compactly generated stable categories

Among all stable categories one singles out a class of those that are particularly manageable: these are the *compactly generated stable categories.*

One favorable property of compactly generated stable categories is that they are dualizable with a very explicit description of the dual.

Another is that the tensor product of two compactly generated categories can also be described rather explicitly.

The material in Sects. 7.1 and 7.2 is based on [**Lu1**, Sect. 5.3].

7.1. Compactness. The notion of compactness is key for doing computations in a given stable category: we usually can calculate the mapping spaces out of compact objects.

For a related reason, compactly generated stable categories are those that we know how to calculate functors *from.*

7.1.1. Let $\mathbf{C}$ be an object of $1\text{-Cat}^{\mathrm{St,cocmpl}}$. An object $\mathbf{c} \in \mathbf{C}$ is said to be *compact* if the functor

$$\mathrm{Maps}_{\mathbf{C}}(\mathbf{c}, -) : \mathbf{C} \to \mathrm{Spc}$$

preserves filtered colimits.

Equivalently, $\mathbf{c}$ is compact if the functor

$$\mathcal{M}aps_{\mathbf{C}}(\mathbf{c}, -) : \mathbf{C} \to \mathrm{Sptr}$$

preserves filtered colimits (equivalently, all colimits or direct sums).

We let $\mathbf{C}^c \subset \mathbf{C}$ denote the full subcategory spanned by compact objects. We have $\mathbf{C}^c \in 1\text{-Cat}^{\mathrm{St}}$.

7.1.2. We give the following definition:

DEFINITION 7.1.3. *An object* $\mathbf{C} \in 1\text{-Cat}^{\mathrm{St,cocmpl}}$ *is said to be* compactly generated *of it admits a set of compact generators.*

7.1.4. Let $F : \mathbf{C} \to \mathbf{D}$ be a morphism in $1\text{-Cat}^{\mathrm{St,cocmpl}}_{\mathrm{cont}}$, and assume that $\mathbf{C}$ is compactly generated.

In this case one can give an easy criterion for when the right adjoint F^R of F, which is *a priori* a morphism in $1\text{-Cat}^{\mathrm{St,cocmpl}}$ (see Lemma 5.3.2), is in fact a morphism in $1\text{-Cat}^{\mathrm{St,cocmpl}}_{\mathrm{cont}}$.

Namely, we have the following (almost immediate) assertion:

LEMMA 7.1.5. *Under the above circumstance,* F^R *is continuous if and only if* F *sends* $\mathbf{C}^c$ *to* $\mathbf{D}^c$.

7.2. The operation of ind-completion. In the previous subsection we attached to a cocomplete stable category $\mathbf{C}$ its full subcategory consisting of compact objects.

In this subsection we will discuss the inverse procedure: starting from a non-cocomplete stable category $\mathbf{C}_0$ we will be able to canonically produce a cocomplete one by 'adding all filtered colimits'. This is the operation of *ind-completion.*

7.2.1. Let $\mathbf{C}_0$ be an object of $1\text{-Cat}^{\mathrm{St}}$. Consider the following $(\infty, 1)$-categories:

(1) The full subcategory of $\mathrm{Funct}(\mathbf{C}_0, \mathrm{Spc})$ that consists of functors that preserve fiber products.
(2) The full subcategory of $\mathrm{Funct}(\mathbf{C}_0, \mathrm{ComGrp}(\mathrm{Spc}))$ that consists of functors that preserve fiber products.
(3) The full subcategory of $\mathrm{Funct}(\mathbf{C}_0, \mathrm{Sptr})$ that consists of functors that preserve fiber products, i.e., $\mathrm{Funct}_{\mathrm{ex}}(\mathbf{C}_0, \mathrm{Sptr})$.

The functors

$$\mathrm{Sptr} \xrightarrow{\tau^{\leq 0}} \mathrm{Sptr}^{\leq 0} \simeq \mathrm{ComGrp}(\mathrm{Spc}) \xrightarrow{\mathbf{oblv}_{\mathrm{ComGrp}}} \mathrm{Spc}$$

define functors (3) ⇒ (2) ⇒ (1).

We have (see [**Lu2**, Corollary 1.4.2.23]):

LEMMA 7.2.2. *The above functors (3) ⇒ (2) ⇒ (1) are equivalences.*

7.2.3. For $\mathbf{C}_0 \in 1\text{-}\mathrm{Cat}^{\mathrm{St}}$, we define the $(\infty,1)$-categpry

$$\mathrm{Ind}(\mathbf{C}_0) := \mathrm{Funct}_{\mathrm{ex}}((\mathbf{C}_0)^{\mathrm{op}}, \mathrm{Sptr}).$$

According to Sect. 5.1.7, $\mathrm{Ind}(\mathbf{C}_0)$ is stable and cocomplete. Yoneda defines a fully faithful functor

$$\mathbf{C}_0 \to \mathrm{Ind}(\mathbf{C}_0). \tag{7.1}$$

We have (see [**Lu1**, 5.3.5] and [**Lu2**, Remark 1.4.2.9])

LEMMA 7.2.4.

(1) *The essential image of* (7.1) *is contained in* $\mathrm{Ind}(\mathbf{C}^0)^c$.

(1') *The essential image of* (7.1) *generates* $\mathrm{Ind}(\mathbf{C}^0)$. *Moreover, any object of* $\mathrm{Ind}(\mathbf{C}^0)$ *can be written as a* filtered *colimit of objects from* $\mathbf{C}_0$.

(1") *Any compact object in* $\mathrm{Ind}(\mathbf{C}^0)$ *is a direct summand of one in the essential image of* (7.1).

(2) *For* $\mathbf{C} \in 1\text{-}\mathrm{Cat}^{\mathrm{St,cocmpl}}$, *restriction along* (7.1) *defines an equivalence*

$$\mathrm{Funct}_{\mathrm{ex,cont}}(\mathrm{Ind}(\mathbf{C}^0), \mathbf{C}) \to \mathrm{Funct}_{\mathrm{ex}}(\mathbf{C}^0, \mathbf{C}).$$

(3) *Let* $\mathbf{C}$ *be an object of* $1\text{-}\mathrm{Cat}^{\mathrm{St,cocmpl}}$, *and let* $\mathbf{C}^0 \subset \mathbf{C}^c$ *be a full subcategory that generates* $\mathbf{C}$. *Then the functor* $\mathrm{Ind}(\mathbf{C}^0) \to \mathbf{C}$, *arising from* (2), *is an equivalence.*

(3') *For a compactly generated* $\mathbf{C} \in 1\text{-}\mathrm{Cat}^{\mathrm{St,cocmpl}}$, *the functor* $\mathrm{Ind}(\mathbf{C}^c) \to \mathbf{C}$ *is an equivalence.*

7.2.5. Note that point (2) in Lemma 7.2.4 says that the assignment

$$\mathbf{C}_0 \mapsto \mathrm{Ind}(\mathbf{C}_0) \tag{7.2}$$

provides a functor $1\text{-}\mathrm{Cat}^{\mathrm{St}} \to 1\text{-}\mathrm{Cat}^{\mathrm{St,cocmpl}}_{\mathrm{cont}}$, left adjoint to the inclusion

$$1\text{-}\mathrm{Cat}^{\mathrm{St,cocmpl}}_{\mathrm{cont}} \hookrightarrow 1\text{-}\mathrm{Cat}^{\mathrm{St}}.$$

7.2.6. Let us return to the setting of Sect. 5.3.3. Assume that each of the categories $\mathbf{C}_i$ is compactly generated, and that each of the functors $\mathbf{C}_i \to \mathbf{C}_j$ preserves compactness.

In this case, the functor $\mathbf{C}_\mathbf{I} : \mathbf{I} \to 1\text{-Cat}^{\text{St,cocmpl}}_{\text{cont}}$ gives rise to a functor

$$\mathbf{C}^c_\mathbf{I} : \mathbf{I} \to 1\text{-Cat}^{\text{St}}, \quad i \mapsto \mathbf{C}^c_i.$$

We have a tautological exact functor

$$\underset{\mathbf{I}}{\operatorname{colim}}\, \mathbf{C}^c_\mathbf{I} \to \mathbf{C}_*,$$

where the colimit in the left-hand side is taken in 1-Cat^{St}. Using Lemma 7.2.4(2), we obtain a morphism in $1\text{-Cat}^{\text{St,cocmpl}}_{\text{cont}}$

$$\operatorname{Ind}\left(\underset{\mathbf{I}}{\operatorname{colim}}\, \mathbf{C}^c_\mathbf{I}\right) \to \mathbf{C}_*. \tag{7.3}$$

Since the functor (7.2) is a left adjoint, it preserves colimits. Hence, combining with Lemma 7.2.4(3'), we obtain:

COROLLARY 7.2.7.

(a) *The functor* (7.3) *is an equivalence.*

(b) *The category* $\mathbf{C}_*$ *is compactly generated by the essential images of the functors*

$$\mathbf{C}^c_i \to \mathbf{C}_i \xrightarrow{\text{ins}_i} \mathbf{C}_*.$$

REMARK 7.2.8. Note that in the present situation, the assertion of Corollary 5.3.4 becomes particularly obvious. Namely, we have

$$\begin{aligned}\mathbf{C}_* \simeq \operatorname{Ind}\left(\underset{i\in\mathbf{I}}{\operatorname{colim}}\, \mathbf{C}^c_i\right) = \operatorname{Funct}_{\text{ex}}\left(\left(\underset{i\in\mathbf{I}}{\operatorname{colim}}\, \mathbf{C}^c_i\right)^{\text{op}}, \text{Sptr}\right) \simeq \\ \simeq \operatorname{Funct}_{\text{ex}}\left(\underset{i\in\mathbf{I}}{\operatorname{colim}}\, (\mathbf{C}^c_i)^{\text{op}}, \text{Sptr}\right) \simeq \underset{i\in\mathbf{I}^{\text{op}}}{\lim}\, \operatorname{Funct}_{\text{ex}}((\mathbf{C}^c_i)^{\text{op}}, \text{Sptr}) \simeq \\ \simeq \underset{i\in\mathbf{I}^{\text{op}}}{\lim}\, \operatorname{Ind}(\mathbf{C}^c_i) \simeq \underset{i\in\mathbf{I}^{\text{op}}}{\lim}\, \mathbf{C}_i.\end{aligned}$$

7.3. The dual of a compactly generated category. One of the key features of compactly generated stable categories is that they are dualizable in the sense of the Lurie symmetric monoidal structure. Moreover, the dual can be described very explicitly:

$$(\operatorname{Ind}(\mathbf{C}_0))^\vee \simeq \operatorname{Ind}((\mathbf{C}_0)^\vee).$$

As was mentioned in the introduction, the latter equivalence provides a framework for such phenomena as Verdier duality: rather than talking about a contravariant self-equivalence (on a small category of compact objects), we talk about the datum of self-duality on the entire category.

7.3.1. Let $\mathbf{C} \in 1\text{-Cat}^{\text{St,cocmpl}}$ be of the form $\text{Ind}(\mathbf{C}_0)$ for some $\mathbf{C}_0 \in 1\text{-Cat}^{\text{St}}$. In particular, $\mathbf{C}$ is compactly generated, and any compactly generated object of $1\text{-Cat}^{\text{St,cocmpl}}$ is of this form.

The assignment

$$(\mathbf{c}, \mathbf{c}' \in \mathbf{C}_0) \mapsto \mathcal{M}aps_{\mathbf{C}}(\mathbf{c}, \mathbf{c}')$$

defines a functor

$$(\mathbf{C}_0)^{\text{op}} \times (\mathbf{C}_0) \to \text{Sptr},$$

which is exact in each variable.

Applying left Kan extension along

$$(\mathbf{C}_0)^{\text{op}} \times (\mathbf{C}_0) \hookrightarrow \text{Ind}((\mathbf{C}_0)^{\text{op}}) \times \text{Ind}(\mathbf{C}_0),$$

we obtain a functor

$$\text{Ind}((\mathbf{C}_0)^{\text{op}}) \times \text{Ind}(\mathbf{C}_0) \to \text{Sptr},$$

which is exact and continuous in each variable. Hence, it gives rise to a functor

$$\text{Ind}((\mathbf{C}_0)^{\text{op}}) \otimes \text{Ind}(\mathbf{C}_0) \to \text{Sptr}. \tag{7.4}$$

PROPOSITION 7.3.2. *The functor* (7.4) *provides the co-unit map of an adjunction data, thereby identifying* $\text{Ind}(\mathbf{C}_0)$ *and* $\text{Ind}((\mathbf{C}_0)^{\text{op}})$ *as each other's duals.*

The proof given below essentially copies [**Lu2**, Proposition 4.8.1.16].

PROOF. We have

$$\text{Funct}_{\text{ex,cont}}(\text{Ind}(\mathbf{C}_0), \text{Sptr}) \simeq \text{Funct}_{\text{ex}}(\mathbf{C}_0, \text{Sptr}) \simeq \text{Ind}((\mathbf{C}_0)^{\text{op}}).$$

Hence, it suffices to show that for $\mathbf{D} \in 1\text{-Cat}^{\text{St,cocmpl}}$, the tautological functor

$$\text{Ind}((\mathbf{C}_0)^{\text{op}}) \otimes \mathbf{D} \to \text{Funct}_{\text{ex,cont}}(\text{Ind}(\mathbf{C}_0), \mathbf{D})$$

is an equivalence.

Thus, we need to show that for $\mathbf{E} \in 1\text{-Cat}^{\text{St,cocmpl}}$, the space of continuous functors

$$\text{Funct}_{\text{ex,cont}}(\text{Ind}(\mathbf{C}_0), \mathbf{D}) \to \mathbf{E},$$

which is is the same as the space of continuous functors

$$\text{Funct}_{\text{ex}}(\mathbf{C}_0, \mathbf{D}) \to \mathbf{E},$$

maps isomorphically to the space of continuous functors

$$\text{Ind}((\mathbf{C}_0)^{\text{op}}) \otimes \mathbf{D} \to \mathbf{E},$$

while the latter identifies with the space of exact functors

$$(\mathbf{C}_0)^{\text{op}} \to \text{Funct}_{\text{ex,cont}}(\mathbf{D}, \mathbf{E}).$$

We will use the following observation:

LEMMA 7.3.3. *For* $\mathbf{F}_1, \mathbf{F}_2 \in 1\text{-Cat}^{\text{St,cocmpl}}$, *the passage to the right adjoint functor and the opposite category defines an equivalence*

$$\text{Funct}_{\text{ex,cont}}(\mathbf{F}_1, \mathbf{F}_2) \to \text{Funct}_{\text{ex,cont}}((\mathbf{F}_2)^{\text{op}}, (\mathbf{F}_1)^{\text{op}}).$$

Applying the lemma, we rewrite

$$\begin{aligned}
&\mathrm{Maps}_{1\text{-}\mathrm{Cat}^{\mathrm{St,cocmpl}}_{\mathrm{cont}}}(\mathrm{Funct}_{\mathrm{ex}}(\mathbf{C}_0,\mathbf{D}),\mathbf{E}) \\
&\simeq \mathrm{Maps}_{1\text{-}\mathrm{Cat}^{\mathrm{St,cocmpl}}_{\mathrm{cont}}}(\mathbf{E}^{\mathrm{op}},(\mathrm{Funct}_{\mathrm{ex}}(\mathbf{C}_0,\mathbf{D}))^{\mathrm{op}}) \simeq \\
&\simeq \mathrm{Maps}_{1\text{-}\mathrm{Cat}^{\mathrm{St,cocmpl}}_{\mathrm{cont}}}(\mathbf{E}^{\mathrm{op}},\mathrm{Funct}_{\mathrm{ex}}((\mathbf{C}_0)^{\mathrm{op}},\mathbf{D}^{\mathrm{op}})) \simeq \\
&\simeq \mathrm{Maps}_{1\text{-}\mathrm{Cat}^{\mathrm{St,cocmpl}}}((\mathbf{C}_0)^{\mathrm{op}},\mathrm{Funct}_{\mathrm{ex,cont}}(\mathbf{E}^{\mathrm{op}},\mathbf{D}^{\mathrm{op}})) \simeq \\
&\simeq \mathrm{Maps}_{1\text{-}\mathrm{Cat}^{\mathrm{St,cocmpl}}}((\mathbf{C}_0)^{\mathrm{op}},\mathrm{Funct}_{\mathrm{ex,cont}}(\mathbf{D},\mathbf{E}))\,,
\end{aligned}$$

as required.

□

7.3.4. Let $F_0 : \mathbf{C}_0 \to \mathbf{D}_0$ be an exact functor between stable categories. Set

$$\mathbf{C} := \mathrm{Ind}(\mathbf{C}_0), \quad \mathbf{D} := \mathrm{Ind}(\mathbf{D}_0).$$

Let $F : \mathbf{C} \to \mathbf{D}$ be the left Kan extension along $\mathbf{C}_0 \to \mathbf{C}$ of the composite functor

$$\mathbf{C}_0 \overset{F_0}{\to} \mathbf{D}_0 \to \mathbf{D}.$$

We can also think about F as being obtained from F_0 by applying the functor

$$\mathrm{Ind} : 1\text{-}\mathrm{Cat}^{\mathrm{St}} \to 1\text{-}\mathrm{Cat}^{\mathrm{St,cocmpl}}_{\mathrm{cont}}.$$

Note that according to Proposition 7.3.2, we have a canonical identification

$$\mathbf{C}^\vee \simeq \mathrm{Ind}((\mathbf{C}_0)^{\mathrm{op}}) \text{ and } \mathbf{D}^\vee \simeq \mathrm{Ind}((\mathbf{D}_0)^{\mathrm{op}}).$$

Consider the functor

$$(F_0)^{\mathrm{op}} : (\mathbf{C}_0)^{\mathrm{op}} \to (\mathbf{D}_0)^{\mathrm{op}},$$

and let

$$F^{\text{fake-op}} : \mathbf{C}^\vee \to \mathbf{D}^\vee$$

denote its ind-extension.

PROPOSITION 7.3.5. *The functor $F^{\text{fake-op}}$ is the dual of the right adjoint F^R of F. I.e.,*

$$F^{\text{fake-op}} \simeq (F^R)^\vee.$$

REMARK 7.3.6. The functor $F^{\text{fake-op}}$ is in no sense the opposite of F; the latter would be a (not necessarily continuous) functor

$$\mathbf{C}^{\mathrm{op}} \to \mathbf{D}^{\mathrm{op}}.$$

However, the two agree on the full subcategory

$$\mathbf{C}^{\mathrm{op}} \supset (\mathbf{C}_0)^{\mathrm{op}} \subset \mathbf{C}^\vee,$$

which they both map to

$$\mathbf{D}^{\mathrm{op}} \supset (\mathbf{D}_0)^{\mathrm{op}} \subset \mathbf{D}^\vee.$$

PROOF. We need to show that the functor

$$\mathbf{C}^\vee \times \mathbf{D} \to \mathbf{C}^\vee \otimes \mathbf{D} \overset{F^{\text{fake-op}} \otimes \mathrm{Id}_{\mathbf{D}}}{\longrightarrow} \mathbf{D}^\vee \otimes \mathbf{D} \overset{\epsilon_{\mathbf{D}}}{\longrightarrow} \mathrm{Sptr} \tag{7.5}$$

identifies with

$$\mathbf{C}^\vee \times \mathbf{D} \to \mathbf{C}^\vee \otimes \mathbf{D} \overset{\mathrm{Id}_{\mathbf{C}^\vee} \otimes F^R}{\longrightarrow} \mathbf{C}^\vee \otimes \mathbf{C} \overset{\epsilon_{\mathbf{C}}}{\longrightarrow} \mathrm{Sptr}\,. \tag{7.6}$$

Both functors are left Kan extensions from their respective restrictions to

$$(\mathbf{C}_0)^{\mathrm{op}} \times \mathbf{D}_0 \subset \mathrm{Ind}((\mathbf{C}_0)^{\mathrm{op}}) \times \mathrm{Ind}(\mathbf{D}_0) \simeq \mathbf{C}^{\vee} \times \mathbf{D},$$

and are uniuquely recovered from their respective compositions with $\Omega^\infty : \mathrm{Sptr} \longrightarrow \mathrm{Spc}$.

The functor $(\mathbf{C}_0)^{\mathrm{op}} \times \mathbf{D}_0 \to \mathrm{Spc}$ obtained from (7.5) is

$$(\mathbf{C}_0)^{\mathrm{op}} \times \mathbf{D}_0 \overset{(F_0)^{\mathrm{op}} \times \mathrm{Id}_{\mathbf{D}_0}}{\longrightarrow} (\mathbf{D}_0)^{\mathrm{op}} \times \mathbf{D}_0 \overset{\mathrm{Yon}_{\mathbf{D}_0}}{\longrightarrow} \mathrm{Spc},$$

which we can further rewrite as

$$(\mathbf{C}_0)^{\mathrm{op}} \times \mathbf{D}_0 \to \mathbf{C}^{\mathrm{op}} \times \mathbf{D} \overset{F^{\mathrm{op}} \times \mathrm{Id}_{\mathbf{D}}}{\longrightarrow} \mathbf{D}^{\mathrm{op}} \times \mathbf{D} \overset{\mathrm{Yon}_{\mathbf{D}}}{\longrightarrow} \mathrm{Spc},$$

where the first arrow is obtained from the embeddings $\mathbf{C}_0 \to \mathbf{C}$ and $\mathbf{D}_0 \to \mathbf{D}$.

The functor $(\mathbf{C}_0)^{\mathrm{op}} \times \mathbf{D}_0 \to \mathrm{Spc}$, obtained from (7.6), is

$$(\mathbf{C}_0)^{\mathrm{op}} \times \mathbf{D}_0 \to \mathbf{C}^{\mathrm{op}} \times \mathbf{D} \to \mathrm{Funct}(\mathbf{C}, \mathrm{Spc}) \times \mathbf{D} \to \mathrm{Funct}(\mathbf{C}, \mathrm{Spc}) \times \mathbf{C} \to \mathrm{Spc},$$

where the first arrow is obtained from the embeddings $\mathbf{C}_0 \to \mathbf{C}$ and $\mathbf{D}_0 \to \mathbf{D}$, respectively, the second arrow is obtained from the Yoneda embedding for $\mathbf{C}$, the third arrow from F^R, and the last arrow is evaluation.

Thus, it suffices to see that the functors

$$\mathbf{C}^{\mathrm{op}} \times \mathbf{D} \overset{F^{\mathrm{op}} \times \mathrm{Id}_{\mathbf{D}}}{\longrightarrow} \mathbf{D}^{\mathrm{op}} \times \mathbf{D} \overset{\mathrm{Yon}_{\mathbf{D}}}{\longrightarrow} \mathrm{Spc}$$

and

$$\mathbf{C}^{\mathrm{op}} \times \mathbf{D} \to \mathrm{Funct}(\mathbf{C}, \mathrm{Spc}) \times \mathbf{D} \overset{\mathrm{Id}_{\mathrm{Funct}(\mathbf{C},\mathrm{Spc})} \times F^R}{\longrightarrow} \mathrm{Funct}(\mathbf{C}, \mathrm{Spc}) \times \mathbf{C} \to \mathrm{Spc},$$

are canonically identified. However, the latter fact expresses the adjunction between F and F^R.

□

7.4. Compact generation of tensor products. In this subsection we will discuss a variant of Proposition 6.4.2 in the compactly generated case. This turns out to be a more explicit statement, which will tell us 'what the tensor product actually looks like'.

7.4.1. Let us be given a pair of compactly generated stable categories $\mathbf{C}$ and $\mathbf{D}$. We claim:

PROPOSITION 7.4.2.

(a) *The tensor product* $\mathbf{C} \otimes \mathbf{D}$ *is compactly generated by objects of the form* $\mathbf{c}_0 \boxtimes \mathbf{d}_0$ *with* $\mathbf{c}_0 \in \mathbf{C}^c$ *and* $\mathbf{d}_0 \in \mathbf{D}^c$.

(b) *For* $\mathbf{c}_0, \mathbf{d}_0$ *as above, and* $\mathbf{c} \in \mathbf{C}$, $\mathbf{d} \in \mathbf{D}$, *we have a canonical isomorphism*

$$\mathcal{M}aps_{\mathbf{C}}(\mathbf{c}_0, \mathbf{c}) \otimes \mathcal{M}aps_{\mathbf{D}}(\mathbf{d}_0, \mathbf{d}) \simeq \mathcal{M}aps_{\mathbf{C} \otimes \mathbf{D}}(\mathbf{c}_0 \boxtimes \mathbf{d}_0, \mathbf{c} \boxtimes \mathbf{d}).$$

The rest of this subsection is devoted to the proof of this proposition.

7.4.3. To prove the proposition we will give an alternative description of the tensor product $\mathbf{C} \otimes \mathbf{D}$.

Set

$$\mathbf{C} = \operatorname{Ind}(\mathbf{C}_0) \text{ and } \mathbf{D} = \operatorname{Ind}(\mathbf{D}_0).$$

Note that we have a canonically defined functor

$$\mathbf{C} \otimes \mathbf{D} \to \operatorname{Funct}_{\text{ex,cont}}(\mathbf{C}^\vee \otimes \mathbf{D}^\vee, \operatorname{Sptr}) \to \operatorname{Funct}((\mathbf{C}_0)^{\text{op}} \times (\mathbf{D}_0)^{\text{op}}, \operatorname{Sptr}), \tag{7.7}$$

defined so that the corresponding functor

$$\mathbf{C} \times \mathbf{D} \to \operatorname{Funct}((\mathbf{C}_0)^{\text{op}} \times (\mathbf{D}_0)^{\text{op}}, \operatorname{Sptr})$$

is given by

$$(\mathbf{c}, \mathbf{d}) \mapsto ((\mathbf{c}_0, \mathbf{d}_0) \mapsto \mathcal{M}aps_{\mathbf{C}}(\mathbf{c}_0, \mathbf{c}) \otimes \mathcal{M}aps_{\mathbf{D}}(\mathbf{d}_0, \mathbf{d})).$$

7.4.4. By construction, the essential image of (7.7) is contained in the full subcategory

$$\operatorname{Funct}_{\text{bi-ex}}((\mathbf{C}_0)^{\text{op}} \times (\mathbf{D}_0)^{\text{op}}, \operatorname{Sptr}) \subset \operatorname{Funct}((\mathbf{C}_0)^{\text{op}} \times (\mathbf{D}_0)^{\text{op}}, \operatorname{Sptr})$$

that consists of functors that are exact in each variable.

Denote the resulting functor

$$\mathbf{C} \otimes \mathbf{D} \to \operatorname{Funct}_{\text{bi-ex}}((\mathbf{C}_0)^{\text{op}} \times (\mathbf{D}_0)^{\text{op}}, \operatorname{Sptr}) \tag{7.8}$$

by $h_{\mathbf{C},\mathbf{D}}$.

7.4.5. We claim that the functor $h_{\mathbf{C},\mathbf{D}}$ is an equivalence. Indeed, this follows from the interpretation of (7.8) as the composition

$$\begin{aligned} &\mathbf{C} \otimes \mathbf{D} \simeq \operatorname{Funct}_{\text{ex,cont}}(\mathbf{C}^\vee, \mathbf{D}) \simeq \operatorname{Funct}_{\text{ex}}((\mathbf{C}_0)^{\text{op}}, \mathbf{D}) \simeq \\ &\simeq \operatorname{Funct}_{\text{ex}}((\mathbf{C}_0)^{\text{op}}, \operatorname{Funct}_{\text{ex}}((\mathbf{D}_0)^{\text{op}}, \operatorname{Sptr})) \simeq \operatorname{Funct}_{\text{bi-ex}}((\mathbf{C}_0)^{\text{op}} \times (\mathbf{D}_0)^{\text{op}}, \operatorname{Sptr}). \end{aligned}$$

7.4.6. Now, an analog of Yoneda's lemma for $h_{\mathbf{C},\mathbf{D}}$ says that for $\mathbf{c}_0 \in \mathbf{C}_0$ and $\mathbf{d}_0 \in \mathbf{D}_0$, and any $F \in \operatorname{Funct}_{\text{bi-ex}}((\mathbf{C}_0)^{\text{op}} \times (\mathbf{D}_0)^{\text{op}}, \operatorname{Sptr})$ we have a canonical isomorphism

$$\mathcal{M}aps_{\operatorname{Funct}_{\text{bi-ex}}((\mathbf{C}_0)^{\text{op}} \times (\mathbf{D}_0)^{\text{op}}, \operatorname{Sptr})}(h_{\mathbf{C},\mathbf{D}}(\mathbf{c}_0 \boxtimes \mathbf{d}_0), F) \simeq F(\mathbf{c}_0 \times \mathbf{d}_0).$$

This implies that the objects

$$h_{\mathbf{C},\mathbf{D}}(\mathbf{c}_0 \boxtimes \mathbf{d}_0) \in \operatorname{Funct}_{\text{bi-ex}}((\mathbf{C}_0)^{\text{op}} \times (\mathbf{D}_0)^{\text{op}}, \operatorname{Sptr})$$

are compact, generate $\operatorname{Funct}_{\text{bi-ex}}((\mathbf{C}_0)^{\text{op}} \times (\mathbf{D}_0)^{\text{op}}, \operatorname{Sptr})$, and

$$\begin{aligned} &\mathcal{M}aps_{\operatorname{Funct}_{\text{bi-ex}}((\mathbf{C}_0)^{\text{op}} \times (\mathbf{D}_0)^{\text{op}}, \operatorname{Sptr})}(h_{\mathbf{C},\mathbf{D}}(\mathbf{c}_0 \boxtimes \mathbf{d}_0), h_{\mathbf{C},\mathbf{D}}(\mathbf{c} \boxtimes \mathbf{d})) \simeq \\ &\qquad \simeq \mathcal{M}aps_{\mathbf{C}}(\mathbf{c}_0, \mathbf{c}) \otimes \mathcal{M}aps_{\mathbf{D}}(\mathbf{d}_0, \mathbf{d}). \end{aligned}$$

8. Algebra in stable categories

In this section we apply the theory developed above to study *stable monoidal categories*, which are by definition associative algebra objects in $1\text{-}\mathrm{Cat}^{\mathrm{St,cocmpl}}_{\mathrm{cont}}$.

Our particular points of interest are how the behavior of modules over stable monoidal categories interacts with such notions as the Lurie tensor product, duality and compactness.

8.1. Modules over a stable monoidal category. We consider the symmetric monoidal category $1\text{-}\mathrm{Cat}^{\mathrm{St,cocmpl}}_{\mathrm{cont}}$. Our interest in this and the next few sections are associative and commutative algebra objects $\mathbf{A}$ in $1\text{-}\mathrm{Cat}^{\mathrm{St,cocmpl}}_{\mathrm{cont}}$. We will refer to them as *stable monoidal* (resp., *symmetric monoidal*) *categories*.

In this subsection we summarize and adapt some pieces of notation introduced earlier to the present context.

8.1.1. Note that the 1-fully faithful embedding $1\text{-}\mathrm{Cat}^{\mathrm{St,cocmpl}}_{\mathrm{cont}} \to 1\text{-}\mathrm{Cat}$ induces 1-fully faithful embeddings

$$\mathrm{AssocAlg}(1\text{-}\mathrm{Cat}^{\mathrm{St,cocmpl}}_{\mathrm{cont}}) \to 1\text{-}\mathrm{Cat}^{\mathrm{Mon}} \text{ and } \mathrm{ComAlg}(1\text{-}\mathrm{Cat}^{\mathrm{St,cocmpl}}_{\mathrm{cont}}) \to 1\text{-}\mathrm{Cat}^{\mathrm{SymMon}},$$

respectively.

I.e., a monoidal (resp., symmetric monoidal) cocomplete stable category is a particular case of a monoidal (symmetric monoidal) $(\infty,1)$-category.

So, we can talk about right-lax functors between monoidal (resp., symmetric monoidal) cocomplete stable categories.

In particular, given $\mathbf{A}$, we can talk about associative (resp., commutative) algebras in $\mathbf{A}$.

8.1.2. Given $\mathbf{A}$, following Sect. 3.4.4, we can consider the corresponding $(\infty,1)$-category of $\mathbf{A}$-modules in $1\text{-}\mathrm{Cat}^{\mathrm{St,cocmpl}}_{\mathrm{cont}}$, i.e., $\mathbf{A}\text{-mod}(1\text{-}\mathrm{Cat}^{\mathrm{St,cocmpl}}_{\mathrm{cont}})$, for which we will also use the notation

$$\mathbf{A}\text{-}\mathbf{mod}^{\mathrm{St,cocmpl}}_{\mathrm{cont}}.$$

Note that $\mathbf{A}\text{-}\mathbf{mod}^{\mathrm{St,cocmpl}}_{\mathrm{cont}}$ is a 1-full subcategory in $\mathbf{A}\text{-}\mathbf{mod}$, the latter being the $(\infty,1)$-category of module categories over $\mathbf{A}$, when the latter is considered as a plain monoidal $(\infty,1)$-category.

Namely, an object $\mathbf{M} \in \mathbf{A}\text{-}\mathbf{mod}$ belongs to $\mathbf{A}\text{-}\mathbf{mod}^{\mathrm{St,cocmpl}}_{\mathrm{cont}}$ if and only if $\mathbf{M}$ is a cocomplete stable category, and the action functor

$$\mathbf{A} \times \mathbf{M} \to \mathbf{M}$$

is exact and continuous in each variable.

A morphism $F : \mathbf{M}_0 \to \mathbf{M}_1$ in $\mathbf{A}\text{-}\mathbf{mod}$ belongs to $\mathbf{A}\text{-}\mathbf{mod}^{\mathrm{St,cocmpl}}_{\mathrm{cont}}$ if and only if, when viewed as a plain functor, F is exact and continuous.

8.1.3. We have a pair of adjoint functors

$$\mathbf{ind}_{\mathbf{A}} : 1\text{-}\mathrm{Cat}^{\mathrm{St,cocmpl}}_{\mathrm{cont}} \rightleftarrows \mathbf{A}\text{-}\mathbf{mod}^{\mathrm{St,cocmpl}}_{\mathrm{cont}} : \mathbf{oblv}_{\mathbf{A}},$$

and the corresponding monad on $1\text{-}\mathrm{Cat}^{\mathrm{St,cocmpl}}_{\mathrm{cont}}$ is given by tensor product with $\mathbf{A}$.

The functor $\mathbf{oblv}_{\mathbf{A}}$ preserves limits (being a right adjoint), and also colimits (because $\mathbf{A} \otimes -$ does).

8.1.4. Let $F : \mathbf{M} \to \mathbf{N}$ be a morphism in $\mathbf{A}\text{-}\mathbf{mod}^{\mathrm{St,cocmpl}}_{\mathrm{cont}}$, and suppose that F, *when viewed as a plain functor between* $(\infty,1)$*-categories*, admits a right adjoint, F^R.

Then, according to Lemma 3.5.3, F^R has a natural structure of *right-lax* functor between $\mathbf{A}$-module categories.

8.1.5. According to Sect. 3.5.1, given $\mathbf{A} \in \operatorname{AssocAlg}(1\text{-}\mathrm{Cat}^{\mathrm{St,cocmpl}}_{\mathrm{cont}})$, $\mathbf{M} \in \mathbf{A}\text{-}\mathbf{mod}^{\mathrm{St,cocmpl}}_{\mathrm{cont}}$ and $\mathcal{A} \in \operatorname{AssocAlg}(\mathbf{A})$, we can consider the $(\infty,1)$-category

$$\mathcal{A}\text{-mod}(\mathbf{M}).$$

8.2. Inner Hom and tensor products.

8.2.1. According to Sect. 3.6, for a given $\mathbf{A} \in \operatorname{AssocAlg}(1\text{-}\mathrm{Cat}^{\mathrm{St,cocmpl}}_{\mathrm{cont}})$ and a pair of objects $\mathbf{M},\mathbf{N} \in \mathbf{A}\text{-}\mathbf{mod}^{\mathrm{St,cocmpl}}_{\mathrm{cont}}$, we can consider their relative inner Hom

$$\underline{\mathrm{Hom}}_{1\text{-}\mathrm{Cat}^{\mathrm{St,cocmpl}}_{\mathrm{cont}},\mathbf{A}}(\mathbf{M},\mathbf{N}) \in 1\text{-}\mathrm{Cat}^{\mathrm{St,cocmpl}}_{\mathrm{cont}}.$$

We will use the notation

$$\mathrm{Funct}_{\mathbf{A}}(\mathbf{M},\mathbf{N}) := \underline{\mathrm{Hom}}_{1\text{-}\mathrm{Cat}^{\mathrm{St,cocmpl}}_{\mathrm{cont}},\mathbf{A}}(\mathbf{M},\mathbf{N}) \in 1\text{-}\mathrm{Cat}^{\mathrm{St,cocmpl}}_{\mathrm{cont}}.$$

We have:

$$(\mathrm{Funct}_{\mathbf{A}}(\mathbf{M},\mathbf{N}))^{\mathrm{Spc}} \simeq \mathrm{Maps}_{\mathbf{A}\text{-}\mathbf{mod}^{\mathrm{St,cocmpl}}_{\mathrm{cont}}}(\mathbf{M},\mathbf{N}).$$

8.2.2. By Corollary 6.2.6, evaluation at $\mathbf{1}_{\mathbf{A}}$ defines an equivalence of stable categories.

$$\mathrm{Funct}_{\mathbf{A}}(\mathbf{A},\mathbf{M}) \simeq \mathbf{M}.$$

8.2.3. According to Sect. 3.6.6, in the case $\mathbf{N} = \mathbf{M}$, the object $\mathrm{Funct}_{\mathbf{A}}(\mathbf{M},\mathbf{M})$ has a natural structure of associative algebra, i.e., a structure of stable monoidal category.

8.2.4. According to Sect. 3.6.5, if $\mathbf{A}$ is a stable *symmetric* monoidal category, then for $\mathbf{M},\mathbf{N} \in \mathbf{A}\text{-}\mathbf{mod}^{\mathrm{St,cocmpl}}_{\mathrm{cont}}$ as above, the object $\mathrm{Funct}_{\mathbf{A}}(\mathbf{M},\mathbf{N})$ has a natural structure of $\mathbf{A}$-module in $1\text{-}\mathrm{Cat}^{\mathrm{St,cocmpl}}_{\mathrm{cont}}$, i.e., lifts to $\mathbf{A}\text{-}\mathbf{mod}^{\mathrm{St,cocmpl}}_{\mathrm{cont}}$.

8.2.5. According to Sect. 4.2.1, for a given $\mathbf{A} \in \operatorname{AssocAlg}(1\text{-}\mathrm{Cat}^{\mathrm{St,cocmpl}}_{\mathrm{cont}})$, we have a well-defined functor

$$\mathbf{A}^{\text{rev-mult}}\text{-}\mathbf{mod}^{\mathrm{St,cocmpl}}_{\mathrm{cont}} \times \mathbf{A}\text{-}\mathbf{mod}^{\mathrm{St,cocmpl}}_{\mathrm{cont}} \to 1\text{-}\mathrm{Cat}^{\mathrm{St,cocmpl}}_{\mathrm{cont}}, \quad \mathbf{N},\mathbf{M} \mapsto \mathbf{N} \underset{\mathbf{A}}{\otimes} \mathbf{M}.$$

LEMMA 8.2.6. *For* $\mathbf{M}$ *and* $\mathbf{N}$ *as above, the image of the tautological functor of stable categories*

$$\mathbf{N} \otimes \mathbf{M} \to \mathbf{N} \underset{\mathbf{A}}{\otimes} \mathbf{M}$$

generates the target.

PROOF. The object $\mathbf{N} \underset{\mathbf{A}}{\otimes} \mathbf{M}$ can be calculated as the geometric realization of a simplicial object of $1\text{-}\mathrm{Cat}^{\mathrm{St,cocmpl}}_{\mathrm{cont}}$ with terms given by $\mathbf{N} \otimes \mathbf{A}^{\otimes n} \otimes \mathbf{M}$ (see [**Lu2**, Theorem 4.4.2.8]). By Corollary 5.3.4, this geometric realization can be rewritten as a totalization (taken in $1\text{-}\mathrm{Cat}^{\mathrm{St,cocmpl}}$) of the corresponding co-simplcial object. By Lemma 5.4.3, we need to show that the functor of evaluation on 0-simplices

$$\mathbf{N} \underset{\mathbf{A}}{\otimes} \mathbf{M} \to \mathbf{N} \otimes \mathbf{M}$$

is conservative. However, this follows from the fact that every object in $\mathbf{\Delta}$ admits a morphism from $[0]$. □

8.2.7. According to Sect. 4.2.4, if $\mathbf{A}$ is an object of $\operatorname{ComAlg}(1\text{-}\mathrm{Cat}^{\mathrm{St,cocmpl}}_{\mathrm{cont}})$, then the operation of tensor product of modules extends to a structure of symmetric monoidal $(\infty,1)$-category on $\mathbf{A}\text{-}\mathbf{mod}^{\mathrm{St,cocmpl}}_{\mathrm{cont}}$.

8.2.8. According to Sect. 4.3.1, given a right $\mathbf{A}$-module $\mathbf{M}$ and a left $\mathbf{A}$-module $\mathbf{N}$, we can talk about the data of duality between them.

According to Sect. 4.3.3, if $\mathbf{A}$ is symmetric monoidal, a datum of duality between $\mathbf{M}$ and $\mathbf{N}$ in the above sense is equivalent to that in the sense of objects of $\mathbf{A}\text{-}\mathbf{mod}^{\mathrm{St,cocmpl}}_{\mathrm{cont}}$ as a symmetric monoidal $(\infty,1)$-category.

8.3. The 2-categorical structure. The material in this subsection is an extension of that in Sect. 5.2; it will be needed in later Chapters in the book.

8.3.1. Let $\mathbf{A}$ be a stable monoidal category. We claim that the structure of $(\infty,1)$-category on $\mathbf{A}\text{-}\mathbf{mod}^{\mathrm{St,cocmpl}}_{\mathrm{cont}}$ can be naturally upgraded to that of $(\infty,2)$-category, to be denoted

$$\left(\mathbf{A}\text{-}\mathbf{mod}^{\mathrm{St,cocmpl}}_{\mathrm{cont}}\right)^{2\text{-Cat}}.$$

Namely, we define the corresponding simplicial $(\infty,1)$-category

$$\mathrm{Seq}_\bullet\left(\left(\mathbf{A}\text{-}\mathbf{mod}^{\mathrm{St,cocmpl}}_{\mathrm{cont}}\right)^{2\text{-Cat}}\right)$$

as follows.

We let $\mathrm{Seq}_n\left(\left(\mathbf{A}\text{-}\mathbf{mod}^{\mathrm{St,cocmpl}}_{\mathrm{cont}}\right)^{2\text{-Cat}}\right)$ be the full $(\infty,1)$-category in

$$\mathbf{A}\text{-}\mathbf{mod} \underset{1\text{-Cat}}{\times} \mathrm{Seq}_n(\mathbf{1}\text{-}\mathbf{Cat}) \subset \mathbf{A}\text{-}\mathbf{mod} \underset{1\text{-Cat}}{\times} \mathrm{Cart}_{/[n]^{\mathrm{op}}},$$

singled out by the following conditions:

We take those $(\infty,1)$-categories $\mathbf{C}$, equipped with an action of $\mathbf{A}$ (regarded as a monoidal $(\infty,1)$-category), and a Cartesian fibration $\mathbf{C}\to[n]^{\mathrm{op}}$ for which:

- We require that for every $i=0,...,n$, the $(\infty,1)$-category $\mathbf{C}_i$ be stable and cocomplete;
- For every $i=1,...,n$ the corresponding functor $\mathbf{C}_{i-1}\to\mathbf{C}_i$ be exact and continuous;
- The action functor $\mathbf{A}\times\mathbf{C}\to\mathbf{C}$ should be a morphism in $(\mathrm{Cart}_{/[n]^{\mathrm{op}}})_{\mathrm{strict}}$.

8.3.2. One checks that the object

$$\mathrm{Seq}_\bullet\left(\left(\mathbf{A}\text{-}\mathbf{mod}^{\mathrm{St,cocmpl}}_{\mathrm{cont}}\right)^{2\text{-Cat}}\right)\in 1\text{-Cat}^{\mathbf{\Delta}^{\mathrm{op}}}$$

defined above indeed lies in the essential image of the functor

$$\mathrm{Seq}_\bullet : 2\text{-Cat}\to 1\text{-Cat}^{\mathbf{\Delta}^{\mathrm{op}}}$$

and thus defines an object

$$\left(\mathbf{A}\text{-}\mathbf{mod}^{\mathrm{St,cocmpl}}_{\mathrm{cont}}\right)^{2\text{-Cat}}\in 2\text{-Cat},$$

whose underlying $(\infty,1)$-category is $\mathbf{A}\text{-}\mathbf{mod}^{\mathrm{St,cocmpl}}_{\mathrm{cont}}$.

8.3.3. By construction, for $\mathbf{M},\mathbf{N}\in\mathbf{A}\text{-}\mathbf{mod}^{\mathrm{St,cocmpl}}_{\mathrm{cont}}$, we have

$$\mathbf{Maps}_{\left(\mathbf{A}\text{-}\mathbf{mod}^{\mathrm{St,cocmpl}}_{\mathrm{cont}}\right)^{2\text{-Cat}}}(\mathbf{M},\mathbf{N})=\mathrm{Funct}_{\mathbf{A}}(\mathbf{M},\mathbf{N}).$$

8.3.4. Finally, we note that by repeating Sects. 4.2.4 and 4.2.2, we obtain that if $\mathbf{A}$ is a stable *symmetric* monoidal category, then the $(\infty,2)$-category

$$\left(\mathbf{A}\text{-}\mathbf{mod}^{\mathrm{St,cocmpl}}_{\mathrm{cont}}\right)^{2\text{-Cat}}$$

acquires a natural symmetric monoidal structure.

8.4. Some residual 2-categorical features. The material in this subsection is an extension of that in Sect. 5.3.

8.4.1. Let $F:\mathbf{M}\to\mathbf{N}$ be a morphism in $\mathbf{A}\text{-}\mathbf{mod}^{\mathrm{St,cocmpl}}_{\mathrm{cont}}$, and suppose that F, when viewed as a plain functor between $(\infty,1)$-categories, admits a right adjoint, F^R. According to Sect. 8.1.4, the functor F^R has a natural structure of *right-lax* functor between $\mathbf{A}$-module categories.

It follows from the definitions that F^R is a *strict* functor between $\mathbf{A}$-module categories if and only if F, when viewed as a 1-morphism in $\left(\mathbf{A}\text{-}\mathbf{mod}^{\mathrm{St,cocmpl}}_{\mathrm{cont}}\right)^{2\text{-Cat}}$, admits a right adjoint.

8.4.2. *Limits and colimits.* Let $\mathbf{I}$ be an index category, and let

$$\mathbf{C}_{\mathbf{I}}:\mathbf{I}\to\mathbf{A}\text{-}\mathbf{mod}^{\mathrm{St,cocmpl}}_{\mathrm{cont}}.$$

Denote

$$\mathbf{C}_*:=\operatorname*{colim}_{\mathbf{I}}\mathbf{C}_{\mathbf{I}}\in\mathbf{A}\text{-}\mathbf{mod}^{\mathrm{St,cocmpl}}_{\mathrm{cont}}.$$

Assume that for every arrow $i\to j$ in $\mathbf{I}$, the corresponding 1-morphism $\mathbf{C}_i\to\mathbf{C}_j$ admits a right adjoint. Then, the procedure of passage to right adjoints (see Chapter 12, Corollary 1.3.4) gives rise to a functor

$$\mathbf{C}^R_{\mathbf{I}^{\mathrm{op}}}:\mathbf{I}^{\mathrm{op}}\to\mathbf{A}\text{-}\mathbf{mod}^{\mathrm{St,cocmpl}}_{\mathrm{cont}}.$$

The following results from Proposition 2.5.7:

COROLLARY 8.4.3. *The canonically defined morphism*

$$\mathbf{C}_*\to\lim_{\mathbf{I}^{\mathrm{op}}}\mathbf{C}_{\mathbf{I}^{\mathrm{op}}}$$

is an equivalence, where the above limit is taken in $\mathbf{A}\text{-}\mathbf{mod}^{\mathrm{St,cocmpl}}_{\mathrm{cont}}$.

8.4.4. Let $\mathbf{C}$ be an object of $\mathbf{A}\text{-}\mathbf{mod}^{\mathrm{St,cocmpl}}_{\mathrm{cont}}$. Then in the way parallel to Sect. 5.3.5, we can consider the monoidal $(\infty,1)$-category

$$\mathbf{Maps}_{\mathbf{A}\text{-}\mathbf{mod}^{\mathrm{St,cocmpl}}_{\mathrm{cont}}}(\mathbf{C},\mathbf{C}),\tag{8.1}$$

which is equipped with an action on

$$\mathbf{Maps}_{\mathbf{A}\text{-}\mathbf{mod}^{\mathrm{St,cocmpl}}_{\mathrm{cont}}}(\mathbf{D},\mathbf{C}),$$

for any $\mathbf{D}\in 1\text{-}\mathrm{Cat}^{\mathrm{St,cocmpl}}_{\mathrm{cont}}$.

In particular, we can talk about the $(\infty,1)$-category of $\mathbf{A}$-linear monads acting on $\mathbf{C}$, which are by definition associative algebra objects in the monoidal $(\infty,1)$-category (8.1).

Given an $\mathbf{A}$-linear monad $\mathcal{B}$, we can consider the $(\infty,1)$-category

$$\mathcal{B}\text{-mod}(\mathbf{C})$$

in the sense of Sect. 3.7.2.

The category $\mathcal{B}$-mod$(\mathbf{C})$ is itself an object of $\mathbf{A}\text{-}\mathbf{mod}^{\mathrm{St,cocmpl}}_{\mathrm{cont}}$ and the adjoint pair

$$\mathbf{ind}_{\mathcal{B}} : \mathbf{C} \rightleftarrows \mathcal{B}\text{-mod}(\mathbf{C}) : \mathbf{oblv}_{\mathcal{B}}$$

takes place in $\mathbf{A}\text{-}\mathbf{mod}^{\mathrm{St,cocmpl}}_{\mathrm{cont}}$.

For a 1-morphism $G \in \mathbf{Maps}_{\mathbf{A}\text{-}\mathbf{mod}^{\mathrm{St,cocmpl}}_{\mathrm{cont}}}(\mathbf{D},\mathbf{C})$ the datum of action of $\mathcal{B}$ on G is equivalent to that of factoring G as

$$\mathbf{oblv}_{\mathcal{B}} \circ G^{\mathrm{enh}}, \quad G^{\mathrm{enh}} \in \mathbf{Maps}_{\mathbf{A}\text{-}\mathbf{mod}^{\mathrm{St,cocmpl}}_{\mathrm{cont}}}(\mathbf{D},\mathcal{B}\text{-mod}(\mathbf{C})).$$

Let $G : \mathbf{D} \to \mathbf{C}$ be as above, and assume that it admits a left adjoint G^L. The functor G^L acquires a natural left-lax functor between $\mathbf{A}$-module categories. Assume, however, that this left-lax structure is strict. Then

$$\mathcal{B} := G \circ G^L$$

acquires a natural structure of $\mathbf{A}$-linear monad. The functor G gives rise to a 1-morphism in $\mathbf{A}\text{-}\mathbf{mod}^{\mathrm{St,cocmpl}}_{\mathrm{cont}}$

$$G^{\mathrm{enh}} : \mathbf{D} \to \mathcal{B}\text{-mod}(\mathbf{C}).$$

8.5. Modules over an algebra. In this subsection we will start combining the general features of modules over algebras with the specifics of dealing with cocomplete stable categories.

8.5.1. Let $\mathbf{A}$ be an object of $\mathrm{AssocAlg}(1\text{-}\mathrm{Cat}^{\mathrm{St,cocmpl}}_{\mathrm{cont}})$. Fix also $\mathbf{M} \in \mathbf{A}\text{-}\mathbf{mod}^{\mathrm{St,cocmpl}}_{\mathrm{cont}}$ and $\mathcal{A} \in \mathrm{AssocAlg}(\mathbf{A})$.

Consider the category $\mathcal{A}$-mod$(\mathbf{M})$. Recall that we have a pair of adjoint functors

$$\mathbf{ind}_{\mathcal{A}} : \mathbf{M} \rightleftarrows \mathcal{A}\text{-mod}(\mathbf{M}) : \mathbf{oblv}_{\mathcal{A}},$$

where $\mathbf{oblv}_{\mathcal{A}}$ is monadic, and the corresponding monad on $\mathbf{M}$ is given by

$$\mathbf{m} \mapsto \mathcal{A} \otimes \mathbf{m}.$$

8.5.2. Suppose that we have two such triples $(\mathbf{A}_1, \mathbf{M}_1, \mathcal{A}_1)$ and $(\mathbf{A}_2, \mathbf{M}_2, \mathcal{A}_2)$. By Sect. 4.2.4, we can regard $\mathbf{A}_1 \otimes \mathbf{A}_2$ as an object of $\mathrm{AssocAlg}(1\text{-}\mathrm{Cat}^{\mathrm{St,cocmpl}}_{\mathrm{cont}})$, and

$$\mathcal{A}_1 \boxtimes \mathcal{A}_2 \in \mathbf{A}_1 \otimes \mathbf{A}_2$$

has a natural structure of object in $\mathrm{Assoc}(\mathbf{A}_1 \otimes \mathbf{A}_2)$.

Furthermore, $\mathbf{M}_1 \otimes \mathbf{M}_2$ has a natural structure of object of

$$(\mathbf{A}_1 \otimes \mathbf{A}_2)\text{-}\mathbf{mod}^{\mathrm{St,cocmpl}}_{\mathrm{cont}}.$$

8.5.3. Suppose that in the above situation, the $\mathbf{A}_1$-module structure on $\mathbf{M}_1$ (resp., $\mathbf{A}_2$-module structure on $\mathbf{M}_2$) has been extended to a structure of module over $\mathbf{A}_1 \otimes \mathbf{A}^{\text{rev-mult}}$ (resp., $\mathbf{A}_2 \otimes \mathbf{A}$), where $\mathbf{A}$ is yet another monoidal stable category.

In this case, we can form $\mathbf{M}_1 \underset{\mathbf{A}}{\otimes} \mathbf{M}_2$, which is a module over $\mathbf{A}_1 \otimes \mathbf{A}_2$. In addition, $\mathcal{A}_1$-mod$(\mathbf{M}_1)$ (resp., $\mathcal{A}_2$-mod$(\mathbf{M}_2)$) is a right (resp., left) module over $\mathbf{A}$, so we can form

$$\mathcal{A}_1\text{-mod}(\mathbf{M}_1) \underset{\mathbf{A}}{\otimes} \mathcal{A}_2\text{-mod}(\mathbf{M}_2).$$

We have a canonically defined functor

$$\mathcal{A}_1\text{-mod}(\mathbf{M}_1) \underset{\mathbf{A}}{\otimes} \mathcal{A}_2\text{-mod}(\mathbf{M}_2) \to \mathbf{M}_1 \underset{\mathbf{A}}{\otimes} \mathbf{M}_2,$$

on which $\mathcal{A}_1 \boxtimes \mathcal{A}_2$ acts as a monad. Hence, we obtain a functor

$$\mathcal{A}_1\text{-mod}(\mathbf{M}_1) \underset{\mathbf{A}}{\otimes} \mathcal{A}_2\text{-mod}(\mathbf{M}_2) \to (\mathcal{A}_1 \boxtimes \mathcal{A}_2)\text{-mod}(\mathbf{M}_1 \underset{\mathbf{A}}{\otimes} \mathbf{M}_2). \tag{8.2}$$

PROPOSITION 8.5.4. *The functor* (8.2) *is an equivalence.*

PROOF. Follows in the same way as Lemma 6.4.5 from Corollary 5.3.8. □

Here are some particular cases of Proposition 8.5.4.

8.5.5. First, let us take $\mathbf{A} = \text{Sptr}$. In this case, Proposition 8.5.4 says that the functor

$$\mathcal{A}_1\text{-mod}(\mathbf{M}_1) \otimes \mathcal{A}_2\text{-mod}(\mathbf{M}_2) \to (\mathcal{A}_1 \boxtimes \mathcal{A}_2)\text{-mod}(\mathbf{M}_1 \otimes \mathbf{M}_2)$$

is an equivalence. Note this is also a corollary of Lemma 6.4.5.

8.5.6. Let us now take $\mathbf{A}_1 = \mathbf{A}$, $\mathcal{A}_1 =: \mathcal{A}$ and $\mathbf{M}_1 = \mathbf{A}$ with its natural structure of $\mathbf{A}$-bimodule. Take $\mathbf{A}_2 = \text{Sptr}$, $\mathcal{A}_2 = \mathbf{1}_{\text{Sptr}}$ and $\mathbf{M}_2 =: \mathbf{M} \in \mathbf{A}\text{-}\mathbf{mod}^{\text{St,cocmpl}}_{\text{cont}}$. Thus, from Proposition 8.5.4 we obtain:

COROLLARY 8.5.7. *The functor*

$$\mathcal{A}\text{-mod} \underset{\mathbf{A}}{\otimes} \mathbf{M} \to \mathcal{A}\text{-mod}(\mathbf{M})$$

is an equivalence.

8.5.8. Let now take $(\mathbf{A}_1, \mathcal{A}_1, \mathbf{M}_2) = (\mathbf{A}, \mathcal{A}, \mathbf{A})$ as above, and let

$$(\mathbf{A}_2, \mathcal{A}_2, \mathbf{M}) = (\mathbf{A}^{\text{mult-rev}}, \mathcal{A}^{\text{mult-rev}}, \mathbf{A}^{\text{mult-rev}}).$$

We obtain:

COROLLARY 8.5.9. *The functor*

$$\mathcal{A}\text{-mod} \underset{\mathbf{A}}{\otimes} \mathcal{A}\text{-mod}^r \to (\mathcal{A} \boxtimes \mathcal{A}^{\text{mult-rev}})\text{-mod}(\mathbf{A})$$

is an equivalence.

8.5.10. Let $\mathbf{A}$ be a stable symmetric monoidal category. Recall the Cartesian fibration

$$\text{AssocAlg} + \text{mod}(\mathbf{A}) \to \text{AssocAlg}(\mathbf{A}),$$

of (4.5), and the corresponding functor

$$(\text{AssocAlg}(\mathbf{A}))^{\text{op}} \to 1\text{-Cat}, \quad \mathcal{A} \mapsto \mathcal{A}\text{-mod}. \tag{8.3}$$

It follows from Proposition 8.5.4 that (8.3) upgrades to a symmetric monoidal functor

$$(\text{AssocAlg}(\mathbf{A}))^{\text{op}} \to \mathbf{A}\text{-mod}^{\text{St,cocmpl}}_{\text{cont}}.$$

8.6. Duality for module categories.

8.6.1. Consider $\mathbf{A}$ as an associative algebra object in the monoidal category $1\text{-Cat}^{\text{St,cocmpl}}_{\text{cont}}$. Hence, it makes sense to talk about duality between left and right $\mathbf{A}$-modules, see Sect. 4.3.1.

8.6.2. From Corollary 8.5.9 we will now deduce:

COROLLARY 8.6.3. *The left* $\mathbf{A}$*-module category* $\mathcal{A}\text{-mod}^r$ *is naturally dual to the right* $\mathbf{A}$*-module category* $\mathcal{A}\text{-mod}$.

PROOF. We will construct explicitly the duality datum. The functor

$$\text{co-unit} : \mathcal{A}\text{-mod}^r(\mathbf{A}) \otimes \mathcal{A}\text{-mod}(\mathbf{A}) \to \mathbf{A}$$

corresponds to the functor of tensor product

$$\mathcal{A}\text{-mod}^r(\mathbf{A}) \times \mathcal{A}\text{-mod}(\mathbf{A}) \to \mathbf{A}$$

of Sect. 4.2.1.

The functor

$$\text{unit} : \text{Sptr} \to \mathcal{A}\text{-mod}(\mathbf{A}) \underset{\mathbf{A}}{\otimes} \mathcal{A}\text{-mod}^r(\mathbf{A})$$

is constructed as follows. Under the identification

$$\mathcal{A}\text{-mod}(\mathbf{A}) \underset{\mathbf{A}}{\otimes} \mathcal{A}\text{-mod}^r(\mathbf{A}) \simeq (\mathcal{A} \boxtimes \mathcal{A}^{\text{mult-rev}})\text{-mod}(\mathbf{A})$$

of Corollary 8.5.9, it corresponds to the object

$$\mathcal{A} \in (\mathcal{A} \boxtimes \mathcal{A}^{\text{mult-rev}})\text{-mod}(\mathbf{A}).$$

□

COROLLARY 8.6.4. *For* $\mathbf{M} \in \mathbf{A}\text{-}\mathbf{mod}$ *there is a canonical equivalence*

$$\text{Funct}_{\mathbf{A}}(\mathcal{A}\text{-mod}^r, \mathbf{M}) \simeq \mathcal{A}\text{-mod}(\mathbf{M}).$$

8.7. Compact generation of tensor products.
8.7.1. Let $\mathbf{A}$ be a monoidal stable category, and let $\mathbf{M}$ and $\mathbf{N}$ be a left and a right $\mathbf{A}$-modules, respectively.

Assume that the monoidal operation $\mathbf{A} \otimes \mathbf{A} \to \mathbf{A}$ admits a continuous right adjoint, and that so do the action functors $\mathbf{A} \otimes \mathbf{M} \to \mathbf{M}$ and $\mathbf{N} \otimes \mathbf{A} \to \mathbf{N}$.

We will prove:

PROPOSITION 8.7.2. *Under the above circumstances, the right adjoint to the tautological functor*

$$\mathbf{N} \otimes \mathbf{M} \to \mathbf{N} \underset{\mathbf{A}}{\otimes} \mathbf{M}$$

is continuous.

PROOF. The category $\mathbf{N} \underset{\mathbf{A}}{\otimes} \mathbf{M}$ is given as the geometric realization of the simplicial category

$$i \mapsto \mathbf{N} \otimes \mathbf{A}^{\otimes i} \otimes \mathbf{M}.$$

Hence, applying Corollary 5.3.4, it is enough to show that the functor

$$\mathbf{\Delta}^{\text{op}} \to 1\text{-Cat}^{\text{St,cocmpl}}_{\text{cont}}, \quad [n] \mapsto \mathbf{N} \otimes \mathbf{A}^{\otimes n} \otimes \mathbf{M}$$

has the property that it sends every morphism in $\mathbf{\Delta}^{\text{op}}$ to a 1-morphism in $\mathbf{1}\text{-}\mathbf{Cat}^{\text{St,cocmpl}}_{\text{cont}}$ that admits a continuous right adjoint. However, this follows from the assumption on the monoidal operation on $\mathbf{A}$ and the action functors.

□

8.7.3. Combining with Lemma 8.2.6 and Proposition 7.4.2, we obtain:

COROLLARY 8.7.4. *Assume that $\mathbf{A}$, $\mathbf{M}_1$, $\mathbf{M}_2$ are compactly generated, and that the functors*

$$\mathbf{A} \otimes \mathbf{A} \to \mathbf{A}, \ \mathbf{A} \otimes \mathbf{M} \to \mathbf{M}, \ \mathbf{N} \otimes \mathbf{A} \to \mathbf{N}$$

preserve compact objects. Then the functor

$$\mathbf{N} \otimes \mathbf{M} \to \mathbf{N} \underset{\mathbf{A}}{\otimes} \mathbf{M}$$

sends compact objects to compact ones. In particular, $\mathbf{N} \underset{\mathbf{A}}{\otimes} \mathbf{M}$ is compactly generated.

8.8. Compactness and relative compactness.
8.8.1. Let $\mathbf{A}$ be a stable monoidal category, and let $\mathbf{M}$ be an object of

$$\mathbf{A}\text{-}\mathbf{mod}^{\text{St,cocmpl}}_{\text{cont}}.$$

For an object $\mathbf{m} \in \mathbf{M}$ consider the functor

$$\mathbf{M} \to \mathbf{A}, \quad \mathbf{m}' \mapsto \underline{\mathrm{Hom}}_{\mathbf{A}}(\mathbf{m}, \mathbf{m}'). \tag{8.4}$$

DEFINITION 8.8.2. *We shall say that* $\mathbf{m}$ *is* compact relative to $\mathbf{A}$ *if the functor* (8.4) *preserves filtered colimits (equivalently, all colimits or direct sums).*

8.8.3. The following is immediate:

LEMMA 8.8.4.

(a) *Suppose that* $\mathbf{A}$ *is compactly generated, and that the action functor* $\mathbf{A}\times\mathbf{M}\to\mathbf{M}$ *sends* $\mathbf{A}^c\times\mathbf{M}^c$ *to* $\mathbf{M}^c$. *Then every compact object in* $\mathbf{M}$ *is compact relative to* $\mathbf{A}$.

(b) *Suppose that* $\mathbf{1_A}\in\mathbf{A}$ *is compact. Then every object in* $\mathbf{M}$ *that is compact relative to* $\mathbf{A}$ *is compact.*

8.8.5. Let us now take $\mathbf{M}=\mathbf{A}$. It is clear that if $\mathbf{a}\in\mathbf{A}$ is left-dualizable (see Sect. 4.1.1 for what this means), then it is compact relative to $\mathbf{A}$: indeed

$$\underline{\mathrm{Hom}}_{\mathbf{A}}(\mathbf{a},\mathbf{a}')\simeq\mathbf{a}'\otimes\mathbf{a}^{\vee,L},$$

while the monoidal operation on $\mathbf{A}$ distributes over colimits.

We have the following partial converse to this statement:

LEMMA 8.8.6. *Suppose that* $\mathbf{A}$ *is generated by left-dualizable objects. Then every object of* $\mathbf{A}$ *that is compact relative to* $\mathbf{A}$ *is left-dualizable.*

PROOF. To show that an object $\mathbf{a}\in\mathbf{A}$ is left-dualizable, it suffices to show that for any $\mathbf{a}'\in\mathbf{A}$, the natural map

$$\mathbf{a}'\otimes\underline{\mathrm{Hom}}_{\mathbf{A}}(\mathbf{a},\mathbf{1_A})\to\underline{\mathrm{Hom}}_{\mathbf{A}}(\mathbf{a},\mathbf{a}') \tag{8.5}$$

is an isomorphism.

Let $\mathbf{a}\in\mathbf{A}$ be compact relative to $\mathbf{A}$. By assumption, both sides in (8.5) preserve colimits in $\mathbf{a}'$. Hence, it suffices to show that (8.5) is an isomorphism for $\mathbf{a}'$ taken from a generating collection of objects of $\mathbf{A}$. We take this collection to be that left-dualizable obejcts. However, (8.5) is an isomorphism for any $\mathbf{a}$, provided that $\mathbf{a}'$ is left-dualizable.

□

9. Rigid monoidal categories

This section contains, what probably is, the only piece of original mathematics in this chapter–the notion of *rigid monoidal category*. These are stable monoidal categories with particularly strong finiteness properties.

9.1. The notion of rigid monoidal category.

9.1.1. Let $\mathbf{A}$ be a stable monoidal category. Let $\mathrm{mult}_{\mathbf{A}}$ denote the tensor product functor $\mathbf{A}\otimes\mathbf{A}\to\mathbf{A}$.

DEFINITION 9.1.2. *We shall say that* $\mathbf{A}$ *is* rigid *if the following conditions hold:*

- *The object* $\mathbf{1_A}\in\mathbf{A}$ *is compact;*
- *The right adjoint of* $\mathrm{mult}_{\mathbf{A}}$, *denoted* $(\mathrm{mult}_{\mathbf{A}})^R$, *is continuous;*
- *The functor* $(\mathrm{mult}_{\mathbf{A}})^R:\mathbf{A}\to\mathbf{A}\otimes\mathbf{A}$ *is a functor of* $\mathbf{A}$*-bimodule categories (a priori it is only a right-lax functor);*

A tautological example of a rigid stable monoidal category is $\mathbf{A}=\mathrm{Sptr}$.

9.1.3. *An example.* Let $\mathcal{A}$ be a commutative algebra object in the stable symmetric monoidal category Sptr. Then the stable (symmetric) monoidal category $\mathcal{A}$-mod is rigid.

More generally, let $\mathbf{A}$ be a rigid symmetric monoidal category, and let $\mathcal{A}$ be a commutative algebra in $\mathbf{A}$. Then the stable (symmetric) monoidal category $\mathcal{A}$-mod is rigid.

9.1.4. Here is the link to the more familiar definition of rigidity:

LEMMA 9.1.5. *Suppose that* $\mathbf{A}$ *is compactly generated. Then* $\mathbf{A}$ *is rigid if and only if the following conditions hold:*

- *The object* $\mathbf{1}_{\mathbf{A}}$ *is compact;*
- *The functor* $\mathrm{mult}_{\mathbf{A}}$ *sends* $\mathbf{A}^c \times \mathbf{A}^c$ *to* $\mathbf{A}^c$*;*
- *Every compact object in* $\mathbf{A}$ *admits both a left and a right dual.*

PROOF. First, the fact that $\mathbf{A}$, and hence $\mathbf{A}\otimes\mathbf{A}$, is compactly generated implies that $\mathrm{mult}_{\mathbf{A}}$ preserves compactness if and only if $(\mathrm{mult}_{\mathbf{A}})^R$ is continuous.

Assume that every compact object in $\mathbf{A}$ admits a left dual. We claim that in this case, every right-lax functor between $\mathbf{A}$-module categories $F : \mathbf{M} \to \mathbf{N}$ is strict. Indeed, it suffices to show that for every $\mathbf{m} \in \mathbf{M}$ and $\mathbf{a} \in \mathbf{A}^c$, the map

$$\mathbf{a} \otimes F(\mathbf{m}) \to F(\mathbf{a} \otimes \mathbf{m})$$

is an isomorphism. However, the above map admits an explicit inverse, given by

$$F(\mathbf{a}\otimes\mathbf{m}) \to \mathbf{a}\otimes\mathbf{a}^{\vee,L} \otimes F(\mathbf{a}\otimes\mathbf{m}) \to \mathbf{a}\otimes F(\mathbf{a}^{\vee,L}\otimes\mathbf{a}\otimes\mathbf{m}) \to \mathbf{a}\otimes F(\mathbf{m}).$$

Suppose, vice versa, that $\mathbf{A}$ is rigid. Let us show that every object $\mathbf{a} \in \mathbf{A}^c$ admits a left dual. For that end, it suffices to show that the functor

$$\mathbf{a}' \mapsto \mathbf{a}' \otimes \mathbf{a}, \quad \mathbf{A} \to \mathbf{A}$$

admits a right adjoint, and this right adjoint is a strict (as opposed to right-lax) functor between left $\mathbf{A}$-modules. However, the right adjoint in question is given by

$$\mathbf{A} \overset{(\mathrm{mult}_{\mathbf{A}})^R}{\longrightarrow} \mathbf{A}\otimes\mathbf{A} \overset{\mathrm{Id}\otimes\mathcal{M}aps_{\mathbf{A}}(\mathbf{a},-)}{\longrightarrow} \mathbf{A}\otimes\mathrm{Sptr} \simeq \mathbf{A}.$$

The situation with right duals is similar.

□

9.1.6. As a corollary, we obtain:

COROLLARY 9.1.7. *Let* $\mathbf{A}$ *be rigid and compactly generated. Then an object of* $\mathbf{A}$ *is compact if and only if it is left-dualizable and if and only if it is right-dualizable.*

9.2. Basic properties of rigid monoidal categories. A fundamental property of a rigid monoidal category (and one that entails the multiple properties of its modules) is that it is *canonically self-dual* when viewed as a plain stable category.

Moreover, this self-duality interacts in a very explicit way with many operations (such as the monoidal operation on $\mathbf{A}$ or monoidal functors between rigid monoidal categories).

9.2.1. Suppose that $\mathbf{A}$ is rigid. In this case, it is easy to see that the data of

$$\epsilon : \mathbf{A} \otimes \mathbf{A} \overset{\text{mult}_{\mathbf{A}}}{\longrightarrow} \mathbf{A} \overset{\mathcal{M}aps_{\mathbf{A}}(\mathbf{1}_{\mathbf{A}},-)}{\longrightarrow} \text{Sptr}$$

and

$$\mu : \text{Sptr} \overset{\mathbf{1}_{\mathbf{A}}}{\longrightarrow} \mathbf{A} \overset{(\text{mult}_{\mathbf{A}})^R}{\longrightarrow} \mathbf{A} \otimes \mathbf{A}$$

define an isomorphism

$$\mathbf{A} \to \mathbf{A}^{\vee,R} = \mathbf{A}^{\vee}.$$

We denote the above isomorphism by $\phi_{\mathbf{A}}$.

9.2.2. *An example.* Consider again the example from Sect. 9.1.3. The co-unit of the above self-duality data on $\mathcal{A}$-mod is given by

$$\mathcal{A}\text{-mod} \otimes \mathcal{A}\text{-mod} \overset{\otimes}{\longrightarrow} \mathcal{A}\text{-mod} \overset{\mathcal{M}aps_{\mathbf{A}}(\mathbf{1}_{\mathbf{A}},-)}{\longrightarrow} \text{Sptr}\,.$$

9.2.3. Let us regard $\mathbf{A}$ as a bimodule over itself. Then, according to Sect. 4.1.7, $\mathbf{A}^{\vee}$ also acquires a structure of $\mathbf{A}$-bimodule. It is easy to see that the isomorphism

$$\phi_{\mathbf{A}} : \mathbf{A} \to \mathbf{A}^{\vee}$$

is compatible with the *left* $\mathbf{A}$-module structure.

LEMMA 9.2.4. *Suppose that $\mathbf{A}$ is compactly generated. Then the equivalence $(\mathbf{A}^c)^{\text{op}} \to \mathbf{A}^c$, induced by $\phi_{\mathbf{A}}$, identifies with $\mathbf{a} \mapsto \mathbf{a}^{\vee,R}$.*

PROOF. We need to construct a functorial isomorphism

$$\text{Maps}_{\mathbf{A}}(\phi_{\mathbf{A}}(\mathbf{a}),\mathbf{a}') \simeq \text{Maps}_{\mathbf{A}}(\mathbf{a}^{\vee,R},\mathbf{a}'), \quad \mathbf{a}' \in \mathbf{A}.$$

By definition,

$$\text{Maps}_{\mathbf{A}}(\phi_{\mathbf{A}}(\mathbf{a}),\mathbf{a}') = \epsilon(\mathbf{a}' \boxtimes \mathbf{a}) \simeq \text{Maps}_{\mathbf{A}}(\mathbf{1}_{\mathbf{A}},\mathbf{a}' \otimes \mathbf{a}),$$

while $\text{Maps}_{\mathbf{A}}(\mathbf{a}^{\vee,R},\mathbf{a}')$ also identifies with $\text{Maps}_{\mathbf{A}}(\mathbf{1}_{\mathbf{A}},\mathbf{a}' \otimes \mathbf{a})$, as required. □

9.2.5. The following is obtained by diagram chase:

LEMMA 9.2.6. *Let $\mathbf{A}$ be a rigid monoidal $(\infty,1)$-category. Then:*

(a) *The following diagram commutes:*

$$\begin{array}{ccc} \mathbf{A}^{\vee} & \xrightarrow{(\text{mult}_{\mathbf{A}})^{\vee}} & \mathbf{A}^{\vee} \otimes \mathbf{A}^{\vee} \\ {\scriptstyle \phi_{\mathbf{A}}}\uparrow & & \uparrow{\scriptstyle \phi_{\mathbf{A}} \otimes \phi_{\mathbf{A}}} \\ \mathbf{A} & \xrightarrow{(\text{mult}_{\mathbf{A}})^{R}} & \mathbf{A} \otimes \mathbf{A} \end{array}$$

commutes.

(b) *Let $F : \mathbf{A}_1 \to \mathbf{A}_2$ be a monoidal functor between rigid monoidal $(\infty,1)$-categories. Then its right adjoint F^R is continuous and the following diagram commutes:*

$$\begin{array}{ccc} \mathbf{A}_2^{\vee} & \xrightarrow{F^{\vee}} & \mathbf{A}_1^{\vee} \\ {\scriptstyle \phi_{\mathbf{A}_2}}\uparrow & & \uparrow{\scriptstyle \phi_{\mathbf{A}_1}} \\ \mathbf{A}_2 & \xrightarrow{F^{R}} & \mathbf{A}_1. \end{array}$$

9.2.7. It is clear that $\mathbf{A}$ is rigid if and only if $\mathbf{A}^{\text{rev-mult}}$ is. Reversing the multiplication on $\mathbf{A}$ we obtain another identification $\mathbf{A} \to \mathbf{A}^{\vee}$, denoted $\phi_{\mathbf{A}^{\text{rev-mult}}}$.

We have $\phi_{\mathbf{A}^{\text{rev-mult}}} = \phi_{\mathbf{A}} \circ \varphi_{\mathbf{A}}$, where $\varphi_{\mathbf{A}}$ is an automorphism of $\mathbf{A}$.

It is easy to see, however, that $\varphi_{\mathbf{A}}$ is naturally an automorphism[10] of $\mathbf{A}$ as a monoidal $(\infty, 1)$-category.

If $\mathbf{A}$ is symmetric monoidal, then $\varphi_{\mathbf{A}}$ is canonically isomorphic to the identity functor.

Unwinding the definitions, we obtain:

LEMMA 9.2.8. *Suppose that* $\mathbf{A}$ *is compactly generated. Then* $\varphi_{\mathbf{A}}$ *is induced by the automorphism*

$$\mathbf{a} \mapsto (\mathbf{a}^{\vee,L})^{\vee,L}$$

of $\mathbf{A}^c$.

9.3. Modules over rigid categories. It turns out that modules over rigid monoidal categories exhibit some very special features:

- For a $\mathbf{A}$-module $\mathbf{M}$, the action map $\text{act}_{\mathbf{A},\mathbf{M}} : \mathbf{A} \otimes \mathbf{M} \to \mathbf{M}$ admits a continuous right adjoint, and this right adjoint identifies with the *dual* of $\text{act}_{\mathbf{A},\mathbf{M}}$ with respect to the self-duality on $\mathbf{A}$;
- Any right-lax (or left-lax) functor between $\mathbf{A}$-module categories is *strict*;
- The tensor product of modules over $\mathbf{A}$ is isomorphic to the co-tensor product;
- An $\mathbf{A}$-module is dualizable if and only if it is such as a plain stable category, and the stable category underlying the dual of an $\mathbf{A}$-module $\mathbf{M}$ identifies with the dual of $\mathbf{M}$ as a plain stable category.

9.3.1. Throughout this subsection we let $\mathbf{A}$ be a rigid monoidal category. Let $\mathbf{M}$ be an $\mathbf{A}$-module. Let

$$\text{act}_{\mathbf{A},\mathbf{M}} : \mathbf{A} \otimes \mathbf{M} \to \mathbf{M}$$

denote the action functor.

LEMMA 9.3.2. *The action functor* $\text{act}_{\mathbf{A},\mathbf{M}}$ *admits a continuous right adjoint, which is given by the composition*

$$\mathbf{M} \simeq \text{Sptr} \otimes \mathbf{M} \overset{\mu \otimes \text{Id}_{\mathbf{M}}}{\longrightarrow} \mathbf{A} \otimes \mathbf{A} \otimes \mathbf{M} \overset{\text{Id}_{\mathbf{A}} \otimes \text{act}_{\mathbf{A},\mathbf{M}}}{\longrightarrow} \mathbf{A} \otimes \mathbf{M}. \tag{9.1}$$

PROOF. We construct the adjunction data as follows. The composition

$$\mathbf{M} \overset{(9.1)}{\longrightarrow} \mathbf{A} \otimes \mathbf{M} \overset{\text{act}_{\mathbf{A},\mathbf{M}}}{\longrightarrow} \mathbf{M}$$

identifies with

$$\mathbf{M} \simeq \text{Sptr} \otimes \mathbf{M} \overset{\mathbf{1}_{\mathbf{A}} \otimes \text{Id}_{\mathbf{M}}}{\longrightarrow} \mathbf{A} \otimes \mathbf{M} \overset{(\text{mult}_{\mathbf{A}})^R \otimes \text{Id}_{\mathbf{M}}}{\longrightarrow} \mathbf{A} \otimes \mathbf{A} \otimes \mathbf{M} \overset{\text{mult}_{\mathbf{A}} \otimes \text{Id}_{\mathbf{M}}}{\longrightarrow} \mathbf{A} \otimes \mathbf{M} \overset{\text{act}_{\mathbf{A},\mathbf{M}}}{\longrightarrow} \mathbf{M},$$

which, by virtue of the $(\text{mult}_{\mathbf{A}}, (\text{mult}_{\mathbf{A}})^R)$-adjunction, admits a canonically defined map to

$$\mathbf{M} \simeq \text{Sptr} \otimes \mathbf{M} \overset{\mathbf{1}_{\mathbf{A}} \otimes \text{Id}_{\mathbf{M}}}{\longrightarrow} \mathbf{A} \otimes \mathbf{M} \overset{\text{act}_{\mathbf{A},\mathbf{M}}}{\longrightarrow} \mathbf{M},$$

the latter being the identity map on $\mathbf{M}$.

[10] We are grateful to J. Lurie for pointing this out to us.

The composition

$$\mathbf{A}\otimes\mathbf{M}\overset{\mathrm{act}_{\mathbf{A},\mathbf{M}}}{\longrightarrow}\mathbf{M}\overset{(9.1)}{\longrightarrow}\mathbf{A}\otimes\mathbf{M}$$

identifies with

$$\mathbf{A}\otimes\mathbf{M}\simeq\mathrm{Sptr}\otimes\mathbf{A}\otimes\mathbf{M}\overset{\mathbf{1}_{\mathbf{A}}\otimes\mathrm{Id}_{\mathbf{A}}\otimes\mathrm{Id}_{\mathbf{M}}}{\longrightarrow}\mathbf{A}\otimes\mathbf{A}\otimes\mathbf{M}\overset{(\mathrm{mult}_{\mathbf{A}})^R\otimes\mathrm{Id}_{\mathbf{A}}\otimes\mathrm{Id}_{\mathbf{M}}}{\longrightarrow}$$

$$\to\mathbf{A}\otimes\mathbf{A}\otimes\mathbf{A}\otimes\mathbf{M}\overset{\mathrm{Id}_{\mathbf{A}}\otimes\mathrm{mult}_{\mathbf{A}}\otimes\mathrm{Id}_{\mathbf{M}}}{\longrightarrow}\mathbf{A}\otimes\mathbf{A}\otimes\mathbf{M}\overset{\mathrm{Id}_{\mathbf{A}}\otimes\mathrm{act}_{\mathbf{A},\mathbf{M}}}{\longrightarrow}\mathbf{A}\otimes\mathbf{M},$$

and the latter, in turn identifies with

$$\mathbf{A}\otimes\mathbf{M}\overset{(\mathrm{mult}_{\mathbf{A}})^R\otimes\mathrm{Id}_{\mathbf{M}}}{\longrightarrow}\mathbf{A}\otimes\mathbf{A}\otimes\mathbf{M}\overset{\mathrm{Id}_{\mathbf{A}}\otimes\mathrm{act}_{\mathbf{A},\mathbf{M}}}{\longrightarrow}\mathbf{A}\otimes\mathbf{M},$$

which by adjunction receives a map from

$$\mathbf{A}\otimes\mathbf{M}\simeq\mathbf{A}\otimes\mathrm{Sptr}\otimes\mathbf{M}\overset{\mathrm{Id}_{\mathbf{A}}\otimes\mathbf{1}_{\mathbf{A}}\otimes\mathrm{Id}_{\mathbf{M}}}{\longrightarrow}\mathbf{A}\otimes\mathbf{A}\otimes\mathbf{M}\overset{\mathrm{Id}_{\mathbf{A}}\otimes\mathrm{act}_{\mathbf{A},\mathbf{M}}}{\longrightarrow}\mathbf{A}\otimes\mathbf{M},$$

while the latter is the identity functor on $\mathbf{A}\otimes\mathbf{M}$. □

Combining with Proposition 8.7.2, we obtain:

COROLLARY 9.3.3. *For a left* $\mathbf{A}$*-module* $\mathbf{M}$ *and a right* $\mathbf{A}$*-module* $\mathbf{N}$*, the right adjoint to the tautological functor*

$$\mathbf{N}\otimes\mathbf{M}\to\mathbf{N}\underset{\mathbf{A}}{\otimes}\mathbf{M}$$

is continuous.

Combining with Lemma 8.8.4, we obtain:

COROLLARY 9.3.4. *Let* $\mathbf{M}$ *be an* $\mathbf{A}$*-module category. Then an object* $\mathbf{m}\in\mathbf{M}$ *is compact relative to* $\mathbf{A}$ *if and only if it is compact.*

9.3.5. We also claim:

LEMMA 9.3.6. *Any right-lax or (left-lax) functor between* $\mathbf{A}$*-module categories is strict.*

REMARK 9.3.7. Note that if $\mathbf{A}$ is compactly generated, the assertion of Lemma 9.3.6 has been established in the course of the proof of Lemma 9.1.5.

PROOF. Let $F:\mathbf{M}\to\mathbf{N}$ be a right-lax functor between $\mathbf{A}$-module categories. We need to show that the (given) natural transformation from

$$\mathbf{A}\otimes\mathbf{M}\overset{\mathrm{Id}_{\mathbf{A}}\otimes F}{\longrightarrow}\mathbf{A}\otimes\mathbf{N}\overset{\mathrm{act}_{\mathbf{A},\mathbf{N}}}{\longrightarrow}\mathbf{N}\tag{9.2}$$

to

$$\mathbf{A}\otimes\mathbf{M}\overset{\mathrm{act}_{\mathbf{A},\mathbf{N}}}{\longrightarrow}\mathbf{M}\overset{F}{\longrightarrow}\mathbf{N}\tag{9.3}$$

is an isomorphism. We will construct an explicit inverse natural transformation.

We consider two more functors $\mathbf{A}\otimes\mathbf{M}\to\mathbf{N}$. One is

(9.4)

$$\mathbf{A}\otimes\mathbf{M}\overset{\mathbf{1}_{\mathbf{A}}\otimes\mathrm{Id}_{\mathbf{A}}\otimes\mathrm{Id}_{\mathbf{N}}}{\longrightarrow}\mathbf{A}\otimes\mathbf{A}\otimes\mathbf{M}\overset{(\mathrm{mult}_{\mathbf{A}})^R\otimes\mathrm{Id}_{\mathbf{A}}\otimes\mathrm{Id}_{\mathbf{M}}}{\longrightarrow}\mathbf{A}\otimes\mathbf{A}\otimes\mathbf{A}\otimes\mathbf{M}\overset{\mathrm{Id}_{\mathbf{A}}\otimes\mathrm{Id}_{\mathbf{A}}\otimes\mathrm{act}_{\mathbf{A},\mathbf{M}}}{\longrightarrow}$$

$$\to\mathbf{A}\otimes\mathbf{A}\otimes\mathbf{M}\overset{\mathrm{Id}_{\mathbf{A}}\otimes\mathrm{Id}_{\mathbf{A}}\otimes F}{\longrightarrow}\mathbf{A}\otimes\mathbf{A}\otimes\mathbf{N}\overset{\mathrm{Id}_{\mathbf{A}}\otimes\mathrm{act}_{\mathbf{A},\mathbf{N}}}{\longrightarrow}\mathbf{A}\otimes\mathbf{N}\overset{\mathrm{act}_{\mathbf{A},\mathbf{N}}}{\longrightarrow}\mathbf{N}.$$

The other is

(9.5)

$$\mathbf{A}\otimes\mathbf{M} \overset{\mathbf{1_A}\otimes\mathrm{Id}_\mathbf{A}\otimes\mathrm{Id}_\mathbf{N}}{\longrightarrow} \mathbf{A}\otimes\mathbf{A}\otimes\mathbf{M} \overset{(\mathrm{mult}_\mathbf{A})^R\otimes\mathrm{Id}_\mathbf{A}\otimes\mathrm{Id}_\mathbf{M}}{\longrightarrow} \mathbf{A}\otimes\mathbf{A}\otimes\mathbf{A}\otimes\mathbf{M} \overset{\mathrm{Id}_\mathbf{A}\otimes\mathrm{Id}_\mathbf{A}\otimes\mathrm{act}_{\mathbf{A},\mathbf{M}}}{\longrightarrow}$$

$$\to \mathbf{A}\otimes\mathbf{A}\otimes\mathbf{M} \overset{\mathrm{Id}_\mathbf{A}\otimes\mathrm{act}_{\mathbf{A},\mathbf{M}}}{\longrightarrow} \mathbf{A}\otimes\mathbf{M} \overset{\mathrm{Id}_\mathbf{A}\otimes F}{\longrightarrow} \mathbf{A}\otimes\mathbf{N} \overset{\mathrm{act}_{\mathbf{A},\mathbf{N}}}{\longrightarrow} \mathbf{N}.$$

The unit of the adjunction in Lemma 9.3.2 gives rise to a natural transformation from (9.3) to (9.4). The right-lax structure on F gives rise to a natural transformation from (9.4) to (9.5). Finally, the co-unit of the adjunction in Lemma 9.3.2 gives rise to a natural transformation from (9.5) to (9.3).

□

9.4. Duality for modules over rigid categories–the commutative case. In this subsection we will show that the theory of duality for modules over a rigid category is particularly transparent.

9.4.1. Recall (see Sect. 4.1.7) that if $\mathbf{A}$ is a stable monoidal category, and $\mathbf{M}$ is a left (resp., right) $\mathbf{A}$-module, and $\mathbf{M}$ is dualizable as a plain stable category, then $\mathbf{M}^\vee$ is naturally a right (resp., left) $\mathbf{A}$-module.

More generally, if $\mathbf{M}$ is a left (resp., right) $\mathbf{A}$-module, and $\mathbf{C}$ is a stable category, then $\mathrm{Funct}_{\mathrm{ex,cont}}(\mathbf{M},\mathbf{C})$ is naturally a right (resp., left) $\mathbf{A}$-module.

9.4.2. For the duration of this subsection we let $\mathbf{A}$ be a rigid *symmetric* monoidal category, so that there is no distinction between left and right modules.

Recall that in this case, the automorphism $\varphi_\mathbf{A}$ of $\mathbf{A}$ is canonically the identity map. So, $\mathbf{A}^\vee$ identifies with $\mathbf{A}$ as an $\mathbf{A}$-*bimodule.*

9.4.3. We claim:

PROPOSITION 9.4.4. *Let* $\mathbf{M}$ *be an* $\mathbf{A}$*-module. Then* $\mathbf{M}$ *is dualizable as an* $\mathbf{A}$*-module if and only if* $\mathbf{M}$ *is dualizable as a plain stable category. In this case, the dual of* $\mathbf{M}$ *as a* $\mathbf{A}$*-module identifies canonically with* $\mathbf{M}^\vee$ *with its natural* $\mathbf{A}$*-module structure.*

PROOF. Let first $\mathbf{A}$ be any stable monoidal category such that the underlying stable category is dualizable. We consider $\mathbf{A}^\vee$ equipped with a natural structure of bimodule over $\mathbf{A}$, see Sect. 9.4.1.

Note that if $\mathbf{M}$ is an $\mathbf{A}$-module and $\mathbf{C}\in 1\text{-}\mathrm{Cat}^{\mathrm{St,cocmpl}}_{\mathrm{cont}}$, we have

$$\mathrm{Funct}_\mathbf{A}(\mathbf{M},\mathbf{C}\otimes\mathbf{A}^\vee)\simeq \mathrm{Funct}_{\mathrm{ex,cont}}(\mathbf{M},\mathbf{C}), \tag{9.6}$$

as right $\mathbf{A}$-modules.

Assume that $\mathbf{M}$ is dualizable as a left $\mathbf{A}$-module. In this case, from (9.6) we obtain that for any $\mathbf{C},\mathbf{D}\in 1\text{-}\mathrm{Cat}^{\mathrm{St,cocmpl}}_{\mathrm{cont}}$, the functor

$$\mathbf{D}\otimes\mathrm{Funct}_{\mathrm{ex,cont}}(\mathbf{M},\mathbf{C})\to \mathbf{D}\otimes\mathrm{Funct}_{\mathrm{ex,cont}}(\mathbf{M},\mathbf{D}\otimes\mathbf{C})$$

is an equivalence. Hence, $\mathbf{M}$ is dualizable as a plain stable category.

Let us now restore the assumption that $\mathbf{A}$ be rigid. The dual of $\mathbf{M}$ as a $\mathbf{A}$-module is given by $\mathrm{Funct}_\mathbf{A}(\mathbf{M},\mathbf{A})$. Using the equivalence $\mathbf{A}\simeq\mathbf{A}^\vee$ and (9.6), we obtain the stated description of the dual of $\mathbf{M}$.

It remains to show that if $\mathbf{M}$ is dualizable as a plain stable category, then it is dualizable as an $\mathbf{A}$-module. For that it suffices to show that the functor

$$\mathbf{N} \mapsto \mathrm{Funct}_{\mathbf{A}}(\mathbf{M}, \mathbf{N}), \quad \mathbf{A}\text{-}\mathbf{mod}^{\mathrm{St,cocmpl}}_{\mathrm{cont}} \to 1\text{-}\mathrm{Cat}^{\mathrm{St,cocmpl}}_{\mathrm{cont}}$$

preserves sifted colimits and the operation of tensoring up by an object of $1\text{-}\mathrm{Cat}^{\mathrm{St,cocmpl}}_{\mathrm{cont}}$.

We note that $\mathrm{Funct}_{\mathbf{A}}(\mathbf{M}, \mathbf{N})$ is given as the totalization of the co-simplicial category with terms

$$\mathbf{M}^{\vee} \otimes (\mathbf{A}^{\vee})^{\otimes n} \otimes \mathbf{N}.$$

Now, by Lemma 9.3.2, the transition maps in this cosimplicial category are *continuous* functors, and hence, by Corollary 5.3.4, the above totalization can be rewritten as a geometric realization. This implies the required assertion.

□

9.4.5. *Digression: the co-tensor product.*

Let $\mathcal{O}$ be an associative algebra in a *symmetric* monoidal category $\mathbf{O}$. Note that in this case we can regard $\mathcal{O}^{\text{rev-mult}}$ also as an associative algebra in $\mathbf{O}$. Furthermore, the category of $\mathcal{O}$-bimodules in $\mathbf{O}$ identifies with

$$(\mathcal{O} \otimes \mathcal{O}^{\text{rev-mult}})\text{-mod}.$$

Let $\mathcal{M}$ and $\mathcal{N}$ be a left and a right $\mathcal{O}$-modules in $\mathbf{O}$, respectively. We regard $\mathcal{M} \otimes \mathcal{N}$ as a $(\mathcal{O} \otimes \mathcal{O}^{\text{rev-mult}})$-module in $\mathbf{O}$.

We let

$$\mathcal{M} \overset{\mathcal{O}}{\otimes} \mathcal{N} \in \mathbf{O}$$

denote the object

$$\underline{\mathrm{Hom}}_{\mathbf{O}, \mathcal{O} \otimes \mathcal{O}^{\text{rev-mult}}}(\mathcal{O}, \mathcal{M} \otimes \mathcal{N}),$$

provided that the latter exists.

9.4.6. Applying this to $\mathbf{O} = 1\text{-}\mathrm{Cat}^{\mathrm{St,cocmpl}}_{\mathrm{cont}}$ and

$$\mathcal{O} = \mathbf{A} \in \mathrm{AssocAlg}(1\text{-}\mathrm{Cat}^{\mathrm{St,cocmpl}}_{\mathrm{cont}}),$$

we obtain the notion of *co-tensor* product of $\mathbf{A}$-module categories.

9.4.7. We claim:

PROPOSITION 9.4.8. *Suppose that* $\mathbf{A}$ *is rigid. Then for* $\mathbf{A}$*-modules* $\mathbf{M}$ *and* $\mathbf{N}$*, we have a canonical isomorphism in* $1\text{-}\mathrm{Cat}^{\mathrm{St,cocmpl}}_{\mathrm{cont}}$

$$\mathbf{M} \overset{\mathbf{A}}{\otimes} \mathbf{N} \simeq \mathbf{N} \underset{\mathbf{A}}{\otimes} \mathbf{M}.$$

PROOF. First we note that the assumption that $\mathbf{A}$ is rigid implies that $\mathbf{A} \otimes \mathbf{A}^{\text{rev-mult}}$ is also rigid. Consider $\mathbf{A}$ as a module over $\mathbf{A} \otimes \mathbf{A}^{\text{rev-mult}}$. From Proposition 9.4.4, it follows that the dual of $\mathbf{A}$ as a module over $\mathbf{A} \otimes \mathbf{A}^{\text{rev-mult}}$ identifies with $\mathbf{A}$.

Hence,

$$\mathbf{M} \overset{\mathbf{A}}{\otimes} \mathbf{N} := \mathrm{Funct}_{\mathbf{A} \otimes \mathbf{A}^{\text{rev-mult}}}(\mathbf{A}, \mathbf{M} \otimes \mathbf{N}) \simeq \mathbf{A} \underset{\mathbf{A} \otimes \mathbf{A}^{\text{rev-mult}}}{\otimes} (\mathbf{M} \otimes \mathbf{N}) \simeq \mathbf{N} \underset{\mathbf{A}}{\otimes} \mathbf{M},$$

as required.

□

9.5. Duality for modules over rigid categories–the general case. In this subsection we will explain the (minor) modifications needed to generalize the results from Sect. 9.4 to the case when $\mathbf{A}$ is just a rigid monoidal category (i.e., *not necessarily* symmetric monoidal).

These modifications will amount to a twist by the automorphism $\varphi_{\mathbf{A}}$ of $\mathbf{A}$.

9.5.1. In what follows, for a right $\mathbf{A}$-module $\mathbf{N}$, we denote by $\mathbf{N}_\varphi$ the $\mathbf{A}$-module, with the same underlying category, but where the action of $\mathbf{A}$ is obtained by pre-composing with the *inverse* of the automorphism $\varphi_{\mathbf{A}}$ of $\mathbf{A}$.

9.5.2. Then we have the following variant of Proposition 9.4.4 (with the same proof):

PROPOSITION 9.5.3. *Let $\mathbf{M}$ be a left $\mathbf{A}$-module. Then $\mathbf{M}$ is dualizable as a left $\mathbf{A}$-module if and only if $\mathbf{M}$ is dualizable as a plain stable category. In this case, the dual of $\mathbf{M}$ as a right $\mathbf{A}$-module identifies canonically with $(\mathbf{M}^\vee)_\varphi$.*

As a corollary we obtain:

COROLLARY 9.5.4. *Let $\mathbf{M}$ (resp., $\mathbf{N}$) be a dualizable left (resp., right) $\mathbf{A}$-module. Then $\mathbf{N} \underset{\mathbf{A}}{\otimes} \mathbf{M} \in 1\text{-}\mathrm{Cat}^{\mathrm{St,cocmpl}}_{\mathrm{cont}}$ is dualizable, and its dual is given by*

$$(\mathbf{M}^\vee)_\varphi \underset{\mathbf{A}}{\otimes} \mathbf{N}^\vee.$$

PROOF. Follows from the fact that for any stable monoidal category $\mathbf{A}$, if $\mathbf{M}$ is a dualizable left $\mathbf{A}$-module with dual $\mathbf{L}$, and $\mathbf{N}$ is a right module, dualizable as a plain stable category, then the tensor product $\mathbf{N} \underset{\mathbf{A}}{\otimes} \mathbf{M}$ is dualizable with dual given by

$$\mathbf{L} \underset{\mathbf{A}}{\otimes} \mathbf{N}^\vee.$$

□

9.5.5. We will now consider the co-tensor product of modules over $\mathbf{A}$. We have the following variant of Proposition 9.4.8, with the same proof:

PROPOSITION 9.5.6. *Suppose that $\mathbf{A}$ is rigid. Then for a left $\mathbf{A}$-module $\mathbf{M}$ and a right $\mathbf{A}$-module $\mathbf{N}$, we have a canonical isomorphism in $1\text{-}\mathrm{Cat}^{\mathrm{St,cocmpl}}_{\mathrm{cont}}$*

$$\mathbf{M} \overset{\mathbf{A}}{\otimes} \mathbf{N} \simeq \mathbf{N}_\varphi \underset{\mathbf{A}}{\otimes} \mathbf{M}.$$

10. DG categories

10.1. The $(\infty, 1)$-category of vector spaces.

10.1.1. Throughout this book we will be working over a ground field k of characteristic 0. To k we can attach the $(\infty, 1)$-category Vect of complexes of vector spaces over k.

This is the *derived ∞-category* attached to the abelian category of vector spaces, in the sense of [**Lu2**, Sect. 1.3.2].

This $(\infty, 1)$-category is endowed with a t-structure, and the corresponding abelian category $\mathrm{Vect}^\heartsuit$ is the usual abelian category of vector spaces over k.

The $(\infty, 1)$-category Vect is *stable* and *cocomplete*.

REMARK 10.1.2. Starting from an abelian category with enough projectives $\mathcal{A}$, the definition in [**Lu2**, Sect. 1.3.2.7] produces the 'bounded above' derived $(\infty,1)$-category $\mathcal{D}^-(\mathcal{A})$. In the case of $\mathcal{A} = \text{Vect}^\heartsuit$, one recovers the entire Vect as the *right completion* of Vect^- with respect to its t-structure. I.e., Vect is the unique stable category equipped with a t-structure such that its bounded above part is Vect^- and for any $V \in \text{Vect}$, the tautological map

$$\underset{n}{\text{colim}}\, \tau^{\leq n}(V) \to V$$

is an isomorphism.

The construction of the derived $(\infty,1)$-category $\mathcal{D}^-(\mathcal{A})$ given in [**Lu2**, Sect. 1.3.2.7] appeals to an explicit procedure called 'the differential graded nerve'. We have no desire to reproduce it here because this construction appeals to a particular model of $(\infty,1)$-category (namely, quasi-categories): the explicit knowledge of what it is does not usually add any information of practical import. What is important to know is that the homotopy category of $\mathcal{D}^-(\mathcal{A})$, i.e., $(\mathcal{D}^-(\mathcal{A}))^{\text{ordn}}$, is the usual triangulated bounded above derived category of $\mathcal{A}$.

The good news, however, is that the derived $(\infty,1)$-category $\mathcal{D}^-(\mathcal{A})$ can be characterized by a universal property, see [**Lu2**, Theorem 1.3.3.2] or the less heavy looking [**Lu2**, Proposition 1.3.3.7].

10.1.3. We let

$$\text{Vect}^{\text{f.d.}} \subset \text{Vect},$$

denote the full subcategory of *finite complexes of finite-dimensional vector spaces* over k.

The corresponding abelian category $(\text{Vect}^{\text{f.d.}})^\heartsuit$ is that of usual finite-dimensional vector spaces over k.

We have

$$\text{Vect}^{\text{f.d.}} = \text{Vect}^c,$$

and Vect is compactly generated by $\text{Vect}^{\text{f.d.}}$.

10.1.4. The fact of crucial importance is that the stable category Vect carries a symmetric monoidal structure uniquely characterized by the following conditions ([**Lu2**, Theorems 4.5.2.1 and 7.1.2.13]):

- It is compatible with the (usual) symmetric monoidal structure on $\text{Vect}^\heartsuit \subset \text{Vect}$.
- The monoidal operation $\text{Vect} \times \text{Vect} \to \text{Vect}$ preserves colimits in each variable.

The second of the above conditions means that Vect is a commutative algebra object in $1\text{-Cat}^{\text{St,cocmpl}}_{\text{cont}}$.

10.1.5. The symmetric monoidal structure on Vect induces one on its full subcategory $\text{Vect}^{\text{f.d.}}$.

Every object in the symmetric monoidal category $\mathrm{Vect}^{\mathrm{f.d.}}$ is dualizable. Hence, by Sect. 4.1.4, the functor of dualization defines an equivalence

$$(\mathrm{Vect}^{\mathrm{f.d.}})^{\mathrm{op}} \to \mathrm{Vect}^{\mathrm{f.d.}}.$$

From Lemma 9.1.5 we obtain:

COROLLARY 10.1.6. *The stable symmetric monoidal* $(\infty,1)$*-category* Vect *is rigid.*

10.2. The Dold-Kan functor(s).

10.2.1. Since Sptr is the unit object in the symmetric monoidal $(\infty,1)$-category $1\text{-}\mathrm{Cat}^{\mathrm{St,cocmpl}}_{\mathrm{cont}}$, we have a canonically defined symmetric monoidal functor

$$\mathrm{Sptr} \to \mathrm{Vect}. \tag{10.1}$$

This functor admits a right adjoint, denoted

$$\mathrm{Vect} \overset{\text{Dold-Kan}^{\mathrm{Sptr}}}{\longrightarrow} \mathrm{Sptr}.$$

The functor $\text{Dold-Kan}^{\mathrm{Sptr}}$ is continuous (e.g., by Lemma 9.2.6(b)).

10.2.2. The functor $\text{Dold-Kan}^{\mathrm{Sptr}}$ has the following additional property: it is *t-exact* (i.e., compatible with the t-structures).

In particular, $\text{Dold-Kan}^{\mathrm{Sptr}}$ restricts to a functor

$$\mathrm{Vect}^{\leq 0} \overset{\text{Dold-Kan}^{\mathrm{ComGrp}}}{\longrightarrow} \mathrm{ComGrp}(\mathrm{Spc}),$$

where we recall that ComGrp(Spc) identifies with $\mathrm{Sptr}^{\leq 0}$.

10.2.3. The composition

$$\mathrm{Vect}^{\leq 0} \overset{\text{Dold-Kan}^{\mathrm{ComGrp}}}{\longrightarrow} \mathrm{ComGrp}(\mathrm{Spc}) \overset{\mathbf{oblv}_{\mathrm{ComGrp}}}{\longrightarrow} \mathrm{Spc},$$

or, which is the same

$$\Omega^{\infty} \circ \text{Dold-Kan}^{\mathrm{Sptr}},$$

is the usual Dold-Kan functor

$$\mathrm{Vect}^{\leq 0} \overset{\text{Dold-Kan}}{\longrightarrow} \mathrm{Spc}.$$

The functor Dold-Kan preserves filtered colimits and all limits. In addition, Dold-Kan commutes with *sifted* colimits (because the forgetful functor $\mathbf{oblv}_{\mathrm{ComGrp}}$ does, see Volume II, Chapter 6, Sect. 1.1.3).

For $V \in \mathrm{Vect}^{\leq 0}$ we have

$$\pi_i(\text{Dold-Kan}(V)) = H^{-i}(V), \quad i = 0, 1, \ldots$$

10.2.4. By construction, the functor Dold-Kan is the right adjoint to the composition

$$\mathrm{Spc} \overset{\Sigma^{\infty}}{\longrightarrow} \mathrm{Sptr} \overset{(10.1)}{\longrightarrow} \mathrm{Vect}. \tag{10.2}$$

In terms of the equivalence of Lemma 2.1.8, the above functor (10.2) corresponds to the object $k \in \mathrm{Vect}$, and can be thought of as the functor of *chains with coefficients in* k.

$$\mathcal{S} \mapsto \mathrm{C}_\bullet(\mathcal{S}, k).$$

10.3. The notion of DG category. In the rest of this chapter we will develop the theory of modules (in $1\text{-Cat}^{\mathrm{St}}$) over the (symmetric) monoidal categories $\mathrm{Vect}^{\mathrm{f.d.}}$ and Vect.

However, the entire discussion is equally applicable, when we replace the pair $\mathrm{Vect}^{\mathrm{f.d.}} \subset \mathrm{Vect}$ by

$$(\mathbf{A}^c \subset \mathbf{A}),$$

where $\mathbf{A}$ is a *rigid symmetric monoidal category* that satisfies the equivalent conditions of Lemma 9.1.5.

10.3.1. We let $\mathrm{DGCat}^{\mathrm{non\text{-}cocmpl}}$ denote the full subcategory in the $(\infty,1)$-category

$$\mathrm{Vect}^{\mathrm{f.d.}}\text{-}\mathbf{mod} = \mathrm{Vect}^{\mathrm{f.d.}}\text{-}\mathbf{mod}(1\text{-Cat})$$

(see Sect. 3.5.7 for the notation), consisting of those $\mathrm{Vect}^{\mathrm{f.d.}}$-modules $\mathbf{C}$, for which:

- $\mathbf{C}$ is stable;
- The action functor $\mathrm{Vect}^{\mathrm{f.d.}} \times \mathbf{C} \to \mathbf{C}$ is *exact* in both variables.

10.3.2. The identification $(\mathrm{Vect}^{\mathrm{f.d.}})^{\mathrm{op}} \simeq \mathrm{Vect}^{\mathrm{f.d.}}$ induces an involution

$$\mathbf{C} \mapsto \mathbf{C}^{\mathrm{op}}$$

on $\mathrm{DGCat}^{\mathrm{non\text{-}cocmpl}}$.

10.3.3. We let $\mathrm{DGCat}_{\mathrm{cont}}$ denote the $(\infty,1)$-category

$$\mathrm{Vect}\text{-}\mathbf{mod}^{\mathrm{St,cocmpl}}_{\mathrm{cont}} := \mathrm{Vect}\text{-}\mathbf{mod}(1\text{-Cat}^{\mathrm{St,cocmpl}}_{\mathrm{cont}}).$$

By unwinding the definitions, we obtain:

LEMMA 10.3.4.

(a) *The functor*

$$\mathrm{DGCat}_{\mathrm{cont}} \to \mathrm{DGCat}^{\mathrm{non\text{-}cocmpl}},$$

given by restriction of action along $\mathrm{Vect}^{\mathrm{f.d.}} \hookrightarrow \mathrm{Vect}$ *is* 1-replete, *i.e., is an equivalence on a 1-full subcategory.*

(b) *An object of* $\mathrm{DGCat}^{\mathrm{non\text{-}cocmpl}}$ *lies in the essential image of the functor from* (a) *if and only if the underlying stable category is cocomplete.*

(c) *A morphism in* $\mathrm{DGCat}^{\mathrm{non\text{-}cocmpl}}$ *between objects in the essential image of* $\mathrm{DGCat}_{\mathrm{cont}}$ *comes from a morphism in* $\mathrm{DGCat}_{\mathrm{cont}}$ *if and only if the underlying functor between the corresponding stable categories is continuous.*

The $(\infty,1)$-category $\mathrm{DGCat}_{\mathrm{cont}}$ will be the principal actor in this book.

We introduce one more notion: we let $\mathrm{DGCat} \subset \mathrm{DGCat}^{\mathrm{non\text{-}cocmpl}}$ be the full subcategory equal to the essential image of the functor $\mathrm{DGCat}_{\mathrm{cont}} \to \mathrm{DGCat}^{\mathrm{non\text{-}cocmpl}}$.

Thus, $\mathrm{DGCat}_{\mathrm{cont}}$ is a 1-full subcategory of DGCat with the same class of objects (i.e., cocomplete DG categories), but in the latter we allow non-continuous functors.

10.3.5. Let $\mathbf{C}, \mathbf{D}$ be two objects of $\mathrm{DGCat}^{\text{non-cocmpl}}$. By Sect. 3.6.5, we can associate to them an object

$$\underline{\mathrm{Hom}}_{1\text{-Cat},\mathrm{Vect}^{\mathrm{f.d.}}}(\mathbf{D}, \mathbf{C}) \in \mathrm{Vect}^{\mathrm{f.d.}}\text{-}\mathbf{mod}(1\text{-Cat}).$$

It is easy to see, however, that $\underline{\mathrm{Hom}}_{1\text{-Cat},\mathrm{Vect}^{\mathrm{f.d.}}}(\mathbf{D}, \mathbf{C})$ belongs to the full subcategory

$$\mathrm{DGCat}^{\text{non-cocmpl}} \subset \mathrm{Vect}^{\mathrm{f.d.}}\text{-}\mathbf{mod}(1\text{-Cat}).$$

We will use the notation:

$$\mathrm{Funct}_k(\mathbf{D}, \mathbf{C}) := \underline{\mathrm{Hom}}_{1\text{-Cat},\mathrm{Vect}^{\mathrm{f.d.}}}(\mathbf{D}, \mathbf{C}).$$

This is the DG category of exact k-linear functors from $\mathbf{D}$ to $\mathbf{C}$. We have

$$(\mathrm{Funct}_k(\mathbf{D}, \mathbf{C}))^{\mathrm{Spc}} = \mathrm{Maps}_{\mathrm{DGCat}^{\text{non-cocmpl}}}(\mathbf{D}, \mathbf{C}).$$

Note that if $\mathbf{C}$ is cocomplete, then so is $\mathrm{Funct}_k(\mathbf{D}, \mathbf{C})$, i.e., in this case it is an object of DGCat.

10.3.6. Let now $\mathbf{C}, \mathbf{D}$ be two objects of $\mathrm{DGCat}_{\mathrm{cont}}$. By Sect. 8.2.1, we can consider the object

$$\underline{\mathrm{Hom}}_{1\text{-Cat}^{\mathrm{St,cocmpl}}_{\mathrm{cont}},\mathrm{Vect}}(\mathbf{D}, \mathbf{C}) =: \mathrm{Funct}_{\mathrm{Vect}}(\mathbf{D}, \mathbf{C}) \in \mathrm{DGCat}_{\mathrm{cont}}.$$

We will use the notation:

$$\mathrm{Funct}_{k,\mathrm{cont}}(\mathbf{D}, \mathbf{C}) := \mathrm{Funct}_{\mathrm{Vect}}(\mathbf{D}, \mathbf{C}).$$

This is the DG category of continuous exact k-linear functors $\mathbf{D}$ to $\mathbf{C}$. We have

$$(\mathrm{Funct}_{k,\mathrm{cont}}(\mathbf{D}, \mathbf{C}))^{\mathrm{Spc}} = \mathrm{Maps}_{\mathrm{DGCat}_{\mathrm{cont}}}(\mathbf{D}, \mathbf{C}).$$

By construction, we have a map in $\mathrm{DGCat}_{\mathrm{cont}}$:

$$\mathrm{Funct}_{k,\mathrm{cont}}(\mathbf{D}, \mathbf{C}) \to \mathrm{Funct}_k(\mathbf{D}, \mathbf{C}), \tag{10.3}$$

which is fully faithful at the level of the underlying $(\infty, 1)$-categories.

10.3.7. Let $\mathbf{C}$ be an object of $\mathrm{DGCat}^{\text{non-cocmpl}}$. For a pair of objects $\mathbf{c}_0, \mathbf{c}_1 \in \mathbf{C}$ we introduce the object

$$\mathcal{M}aps_{k,\mathbf{C}}(\mathbf{c}_0, \mathbf{c}_1) \in \mathrm{Vect}$$

by

$$\mathrm{Maps}_{\mathrm{Vect}}(V, \mathcal{M}aps_{k,\mathbf{C}}(\mathbf{c}_0, \mathbf{c}_1)) \simeq \mathrm{Maps}_{\mathbf{C}}(V \otimes \mathbf{c}_0, \mathbf{c}_1), \quad V \in \mathrm{Vect}^{\mathrm{f.d.}}. \tag{10.4}$$

It is easy to see that $\mathcal{M}aps_{k,\mathbf{C}}(\mathbf{c}_0, \mathbf{c}_1)$ always exists.

10.3.8. If $\mathbf{C} \in \mathrm{DGCat}$, then we have a canonical isomorphism

$$\mathcal{M}aps_{k,\mathbf{C}}(\mathbf{c}_0, \mathbf{c}_1) \simeq \underline{\mathrm{Hom}}_{\mathrm{Vect}}(\mathbf{c}_0, \mathbf{c}_1),$$

i.e., the isomorphism (10.4) holds for $V \in \mathrm{Vect}$ (and not just $\mathrm{Vect}^{\mathrm{f.d.}}$).

Finally, we note that we have

$$\mathcal{M}aps_{\mathbf{C}}(\mathbf{c}_0, \mathbf{c}_1) \simeq \text{Dold-Kan}^{\mathrm{Sptr}}\left(\mathcal{M}aps_{k,\mathbf{C}}(\mathbf{c}_0, \mathbf{c}_1)\right).$$

10.3.9. *The 2-categorical structure.* According to Sect. 8.3, the structure of $(\infty, 1)$-category on $\mathrm{DGCat}_{\mathrm{cont}}$ can be naturally upgraded to a structure of $(\infty, 2)$-category.

We denote the resulting $(\infty, 2)$-category by $\mathrm{DGCat}^{2\text{-Cat}}_{\mathrm{cont}}$. By Sect. 8.3.3, we have

$$\mathbf{Maps}_{\mathrm{DGCat}^{2\text{-Cat}}_{\mathrm{cont}}}(\mathbf{D}, \mathbf{C}) \simeq \mathrm{Funct}_{k,\mathrm{cont}}(\mathbf{D}, \mathbf{C}).$$

We also note (see Lemma 9.3.6) that if a morphism in $\mathrm{DGCat}_{\mathrm{cont}}$, when viewed as plain stable categories, admits a continuous right adjoint, then the initial 1-morphism admits a right adjoint in the $(\infty, 2)$-category $\mathrm{DGCat}^{2\text{-Cat}}_{\mathrm{cont}}$.

10.4. The symmetric monoidal structure on DG categories.
10.4.1. According to Sect. 8.2.7, the $(\infty, 1)$-category $\mathrm{DGCat}_{\mathrm{cont}}$ is equipped with a symmetric monoidal structure. We will denote the corresponding monoidal operation by

$$\mathbf{C}, \mathbf{D} \mapsto \mathbf{C} \underset{\mathrm{Vect}}{\otimes} \mathbf{D}.$$

For $\mathbf{c} \in \mathbf{C}$ and $\mathbf{d} \in \mathbf{D}$ we let denote by $\mathbf{c} \underset{k}{\boxtimes} \mathbf{d} \in \mathbf{C} \underset{\mathrm{Vect}}{\otimes} \mathbf{D}$ the image of $\mathbf{c} \times \mathbf{d} \in \mathbf{C} \times \mathbf{D}$ under the tautological functor

$$\mathbf{C} \times \mathbf{D} \to \mathbf{C} \underset{\mathrm{Vect}}{\otimes} \mathbf{D}.$$

10.4.2. In particular, given $\mathbf{C}, \mathbf{D} \in \mathrm{DGCat}_{\mathrm{cont}}$, we can talk about the datum of *duality* between them, the latter being the datum of functors

$$\mu : \mathrm{Vect} \to \mathbf{C} \underset{\mathrm{Vect}}{\otimes} \mathbf{D} \text{ and } \mathbf{D} \underset{\mathrm{Vect}}{\otimes} \mathbf{C} \to \mathrm{Vect}$$

such that the corresponding identities hold.

10.4.3. According to Proposition 9.4.4, a DG category $\mathbf{C}$ is dualizable as an object of $\mathrm{DGCat}_{\mathrm{cont}}$ if and only if it is dualizable as a plain stable category.

Moreover, again by Proposition 9.4.4, the datum of duality between $\mathbf{C}$ and $\mathbf{D}$ as DG categories is equivalent to the datum of duality between $\mathbf{C}$ and $\mathbf{D}$ as plain stable categories.

10.4.4. Explicitly, if

$$\mathbf{C} \underset{\mathrm{Vect}}{\otimes} \mathbf{D} \to \mathrm{Vect}$$

is the co-unit of a duality in $\mathrm{DGCat}_{\mathrm{cont}}$, then the composition

$$\mathbf{C} \otimes \mathbf{D} \to \mathbf{C} \underset{\mathrm{Vect}}{\otimes} \mathbf{D} \to \mathrm{Vect} \overset{\text{Dold-Kan}^{\mathrm{Sptr}}}{\longrightarrow} \mathrm{Sptr}$$

is the co-unit of a duality in $1\text{-}\mathrm{Cat}^{\mathrm{St,cocmpl}}_{\mathrm{cont}}$.

In fact, for any $\mathbf{C} \in \mathrm{DGCat}_{\mathrm{cont}}$, the composed functor

$$\mathrm{Funct}_{k,\mathrm{cont}}(\mathbf{C}, \mathrm{Vect}) \to \mathrm{Funct}_{\mathrm{ex,cont}}(\mathbf{C}, \mathrm{Vect}) \overset{\text{Dold-Kan}^{\mathrm{Sptr}}}{\longrightarrow} \mathrm{Funct}_{\mathrm{ex,cont}}(\mathbf{C}, \mathrm{Sptr})$$

is an equivalence, see the proof of Proposition 9.4.4.

10.4.5. Finally, we mention that according to Sect. 8.3.4, the above symmetric monoidal structure on the $(\infty, 1)$-category $\mathrm{DGCat}_{\mathrm{cont}}$ naturally upgrades to a symmetric monoidal structure on the $(\infty, 2)$-category

$$\mathrm{DGCat}^{2\text{-Cat}}_{\mathrm{cont}}.$$

10.5. Compact objects and ind-completions.

10.5.1. Let $\mathbf{C}$ be an object of DGCat. We note that, according to Corollary 9.3.4, an object $\mathbf{c} \in \mathbf{C}$ is compact if and only if it is compact relative to Vect, i.e., the functor

$$\mathcal{M}aps_{k,\mathbf{C}}(\mathbf{c},-), \quad \mathbf{C} \to \mathrm{Vect}$$

preserves filtered colimits (equivalently, direct sums or all colimits).

Alternatively, the equivalence of the two notions follows from the fact that the functor

$$\text{Dold-Kan}^{\mathrm{Sptr}} : \mathrm{Vect} \to \mathrm{Sptr}$$

is continuous and conservative.

10.5.2. The full subcategory $\mathbf{C}^c \subset \mathbf{C}$ is preserved by the monoidal operation

$$\mathrm{Vect}^{\mathrm{f.d}} \times \mathbf{C} \to \mathbf{C}.$$

Hence, $\mathbf{C}^c$ naturally acquires a structure of object of $\mathrm{DGCat}^{\text{non-cocmpl}}$.

10.5.3. Vice versa, let $\mathbf{C}_0$ be an object of $\mathrm{DGCat}^{\text{non-cocmpl}}$. Consider the corresponding object

$$\mathrm{Ind}(\mathbf{C}_0) = \mathrm{Funct}_{\mathrm{ex}}((\mathbf{C}_0)^{\mathrm{op}}, \mathrm{Sptr}).$$

The action of $\mathrm{Vect}^{\mathrm{f.d.}}$ on $\mathbf{C}_0$ defines an action of $\mathrm{Vect}^{\mathrm{f.d.}}$ on $\mathrm{Ind}(\mathbf{C}_0)$ by Sect. 4.1.7.

By Lemma 10.3.4(b), since $\mathrm{Ind}(\mathbf{C}_0)$ is cocomplete, we obtain that $\mathrm{Ind}(\mathbf{C}_0)$ is an object of DGCat.

10.5.4. By construction, the tautological functor

$$\mathbf{C}_0 \to \mathrm{Ind}(\mathbf{C}_0)$$

is a functor of $\mathrm{Vect}^{\mathrm{f.d.}}$-module categories.

For $\mathbf{C} \in \mathrm{DGCat}_{\mathrm{cont}}$, the equivalence

$$\mathrm{Funct}_{\mathrm{ex,cont}}(\mathrm{Ind}(\mathbf{C}_0), \mathbf{C}) \to \mathrm{Funct}_{\mathrm{ex}}(\mathbf{C}_0, \mathbf{C})$$

is a functor of bimodule categories over $\mathrm{Vect}^{\mathrm{f.d.}}$. Hence, combining with Lemma 10.3.4(c), we obtain an equivalence

$$\mathrm{Funct}_{k,\mathrm{cont}}(\mathrm{Ind}(\mathbf{C}_0), \mathbf{C}) \simeq \mathrm{Funct}_k(\mathbf{C}_0, \mathbf{C}). \tag{10.5}$$

In other words, we obtain that the ind-completion of $\mathbf{C}_0$ as a plain stable category is also the ind-completion as a DG category.

10.5.5. We claim that the DG category $\mathrm{Ind}(\mathbf{C}_0)$ can also be described as

$$\mathrm{Funct}_k((\mathbf{C}_0)^{\mathrm{op}}, \mathrm{Vect}).$$

More precisely:

LEMMA 10.5.6. *The functor*

$$\mathrm{Funct}_k((\mathbf{C}_0)^{\mathrm{op}}, \mathrm{Vect}) \to$$

$$\to \mathrm{Funct}_{\mathrm{ex}}((\mathbf{C}_0)^{\mathrm{op}}, \mathrm{Vect}) \overset{\text{Dold-Kan}^{\mathrm{Sptr}}}{\longrightarrow} \mathrm{Funct}_{\mathrm{ex}}((\mathbf{C}_0)^{\mathrm{op}}, \mathrm{Sptr}) = \mathrm{Ind}(\mathbf{C}_0)$$

is an equivalence.

PROOF. By (10.5), we have

$$\mathrm{Funct}_k((\mathbf{C}_0)^{\mathrm{op}}, \mathrm{Vect}) \simeq \mathrm{Funct}_{k,\mathrm{cont}}(\mathrm{Ind}((\mathbf{C}_0)^{\mathrm{op}}), \mathrm{Vect}),$$

which by Sect. 10.4.3 identifies with

$$(\mathrm{Ind}((\mathbf{C}_0)^{\mathrm{op}}))^{\vee} \simeq \mathrm{Ind}(\mathbf{C}_0),$$

as required.

□

10.5.7. It follows from Corollary 8.7.4 that if $\mathbf{C}$ and $\mathbf{D}$ are compactly generated DG categories, then the same is true for $\mathbf{C} \underset{\mathrm{Vect}}{\otimes} \mathbf{D}$. Moreover, objects of the form

$$\mathbf{c} \underset{k}{\boxtimes} \mathbf{d} \in \mathbf{C} \underset{\mathrm{Vect}}{\otimes} \mathbf{D}, \quad \mathbf{c} \in \mathbf{C}^c, \mathbf{d} \in \mathbf{D}^d$$

are the compact generators of $\mathbf{C} \underset{\mathrm{Vect}}{\otimes} \mathbf{D}$.

In addition, the following is obtained by repeating the proof of Proposition 7.4.2:

PROPOSITION 10.5.8. *For $\mathbf{c}_0, \mathbf{c} \in \mathbf{C}$ and $\mathbf{d}_0, \mathbf{d} \in \mathbf{D}$ with $\mathbf{c}_0, \mathbf{d}_0$ compact, we have a canonical isomorphism*

$$\mathcal{M}aps_{k,\mathbf{C}}(\mathbf{c}_0, \mathbf{c}) \underset{k}{\otimes} \mathcal{M}aps_{k,\mathbf{D}}(\mathbf{d}_0, \mathbf{d}) \simeq \mathcal{M}aps_{k,\mathbf{C} \underset{\mathrm{Vect}}{\otimes} \mathbf{D}}(\mathbf{c}_0 \underset{k}{\boxtimes} \mathbf{d}_0, \mathbf{c} \underset{k}{\boxtimes} \mathbf{d}).$$

10.6. Change of notations. In the main body of the book, the only stable categories that we will ever encounter will be DG categories. For this reason we will simplify our notations as follows:

- For $\mathbf{C} \in \mathrm{DGCat}^{\text{non-cocmpl}}$ and $\mathbf{c}_0, \mathbf{c}_1 \in \mathbf{C}$ we will write $\mathcal{M}aps_{\mathbf{C}}(\mathbf{c}_0, \mathbf{c}_1)$ instead of $\mathcal{M}aps_{k,\mathbf{C}}(\mathbf{c}_0, \mathbf{c}_1)$ (i.e., our $\mathcal{M}aps_{\mathbf{C}}(-,-)$ is an object of Vect, rather than Sptr; the latter is obtained by applying the functor $\text{Dold-Kan}^{\mathrm{Sptr}}$);
- For $\mathbf{C}, \mathbf{D} \in \mathrm{DGCat}^{\text{non-cocmpl}}$ we will write $\mathrm{Funct}(\mathbf{D}, \mathbf{C})$ instead of $\mathrm{Funct}_k(\mathbf{D}, \mathbf{C})$;
- For $\mathbf{C}, \mathbf{D} \in \mathrm{DGCat}$ we will write $\mathrm{Funct}_{\mathrm{cont}}(\mathbf{D}, \mathbf{C})$ instead of $\mathrm{Funct}_{k,\mathrm{cont}}(\mathbf{D}, \mathbf{C})$;
- For $\mathbf{C}, \mathbf{D} \in \mathrm{DGCat}$ we will write $\mathbf{C} \otimes \mathbf{D}$ instead of $\mathbf{C} \underset{\mathrm{Vect}}{\otimes} \mathbf{D}$;
- For $\mathbf{C}, \mathbf{D} \in \mathrm{DGCat}$ and $\mathbf{c} \in \mathbf{C}, \mathbf{d} \in \mathbf{D}$ we will write $\mathbf{c} \boxtimes \mathbf{d}$ instead of $\mathbf{c} \underset{k}{\boxtimes} \mathbf{d}$.

CHAPTER 2

Basics of derived algebraic geometry

Introduction

This Chapter is meant to introduce the basic objects of study in derived algebraic geometry that will be used in the subsequent chapters.

0.1. Why prestacks? The most general (and, perhaps, also the most important) type of algebro-geometric object that we will introduce is the notion of *prestack*.

0.1.1. Arguably, there is an all-pervasive problem with how one introduces classical algebraic geometry. Even nowadays, any introductory book on algebraic geometry defines schemes as *locally ringed spaces*. The problem with this is that a locally ringed space is a lot of structure, so the definition is quite heavy.

However, one does not have to go this way if one adopts Grothendieck's language of points. Namely, whatever the category of schemes is, it embeds fully faithfully into the category of functors

$$(\mathrm{Sch}^{\mathrm{aff}})^{\mathrm{op}} \to \mathrm{Set},$$

where $\mathrm{Sch}^{\mathrm{aff}}$ is the category of affine schemes, i.e., $(\mathrm{Sch}^{\mathrm{aff}})^{\mathrm{op}}$ is the category of commutative rings.

Now, it is not difficult to characterize which functors $(\mathrm{Sch}^{\mathrm{aff}})^{\mathrm{op}} \to \mathrm{Set}$ correspond to schemes: essentially the functor needs to have a Zariski atlas, a notion that has an intrinsic meaning.

0.1.2. This is exactly the point of view that we will adopt in this Chapter and throughout the book, with the difference that instead of classical (=usual=ordinary) affine schemes we consider derived affine schemes, where, by definition, the category of the latter is the one opposite to the category of connective commutative DG algebras (henceforth, when we write $\mathrm{Sch}^{\mathrm{aff}}$ we will mean the derived version, and denote the category of classical affine schemes by ${}^{\mathrm{cl}}\mathrm{Sch}^{\mathrm{aff}}$).

And instead of functors with values in the category Set of sets we consider the category of functors

$$(\mathrm{Sch}^{\mathrm{aff}})^{\mathrm{op}} \to \mathrm{Spc}, \tag{0.1}$$

where Spc is the category of *spaces* (a.k.a. ∞-groupoids).

We denote the category of functors (0.1) by PreStk and call its objects *prestacks*. I.e., a prestack is something that has a Grothendieck functor of points attached to it, with no further conditions or pieces of structure.

Thus, a prestack is the most general kind of space that one can have in algebraic geometry.

All other kinds of algebro-geometric objects that we will encounter will be prestacks, that *have some particular properties* (as opposed to *extra pieces of structure*). This includes schemes (considered in Sect. 3), Artin stacks (considered in Sect. 4), ind-schemes and inf-schemes (considered in Volume II, Chapter 2), formal moduli problems (considered in Volume II, Chapter 5), etc.

0.1.3. However, the utility of the notion of prestack goes beyond being a general concept that contains the other known types of algebro-geometric objects as particular cases.

Namely, there are some algebro-geometric constructions that can be carried out in this generality, and it turns out to be convenient to do so.

The central among these is the assignment to a prestack $\mathcal{Y}$ of the category $\mathrm{QCoh}(\mathcal{Y})$ of quasi-coherent sheaves on $\mathcal{Y}$, considered in the next Chapter, i.e. Chapter 3. In fact, there is a canonically defined functor

$$\mathrm{QCoh}^*_{\mathrm{PreStk}} : (\mathrm{PreStk})^{\mathrm{op}} \to \mathrm{DGCat}_{\mathrm{cont}}, \quad \mathcal{Y} \mapsto \mathrm{QCoh}(\mathcal{Y}).$$

The definition of $\mathrm{QCoh}^*_{\mathrm{PreStk}}$ is actually automatic: it is the *right Kan extension* of the functor

$$\mathrm{QCoh}^*_{\mathrm{Sch}^{\mathrm{aff}}} : (\mathrm{Sch}^{\mathrm{aff}})^{\mathrm{op}} \to 1\text{-}\mathrm{Cat}$$

that attaches to

$$\mathrm{Spec}(A) = S \in \mathrm{Sch}^{\mathrm{aff}}$$

the DG category

$$\mathrm{QCoh}(S) := A\text{-mod}$$

and to a map $f : S' \to S$ the pullback functor $f^* : \mathrm{QCoh}(S) \to \mathrm{Coh}(S')$.

In other words,

$$\mathrm{QCoh}(\mathcal{Y}) = \lim_{(S \xrightarrow{y} \mathcal{Y}) \in ((\mathrm{Sch}^{\mathrm{aff}})_{/\mathcal{Y}})^{\mathrm{op}}} \mathrm{QCoh}(S). \tag{0.2}$$

So an object $\mathcal{F} \in \mathrm{QCoh}(\mathcal{Y})$ is a assignment

$$(S \xrightarrow{y} \mathcal{Y}) \rightsquigarrow \mathcal{F}_{S,y} \in \mathrm{QCoh}(S),$$

$$(S' \xrightarrow{f} S) \rightsquigarrow \mathcal{F}_{S',y\circ f} \simeq f^*(\mathcal{F}_{S,y}),$$

satisfying a *homotopy compatible system of compatibilities* for compositions of morphisms between affine schemes.

Note that the expression in (0.2) involves taking a limit in the ∞-category 1-Cat. Thus, in order to assign a meaning to it (equivalently, the meaning to the expression 'homotopy compatible system of compatibilities') we need to input the entire machinery of ∞-categories, developed in [**Lu1**]. Thus, it is fair to say that Lurie gave us the freedom to consider quasi-coherent sheaves on prestacks.

Note that before the advent of the language of ∞-categories, the definition of the (derived) category of quasi-coherent sheaves on even such benign objects as algebraic stacks was quite awkward (see [**LM**]). Essentially, in the past, each time one needed to construct a triangulated category, one had to start from an abelian category, take its derived category, and then perform some manipulations on it in order to obtained the desired one.

As an application of the assignment

$$\mathcal{Y} \rightsquigarrow \mathrm{QCoh}(\mathcal{Y})$$

we obtain an automatic construction of the category of D-modules/crystals (see Volume II, Chapter 4). Namely,

$$\text{D-mod}(\mathcal{Y}) := \mathrm{QCoh}(\mathcal{Y}_{\mathrm{dR}}),$$

where $\mathcal{Y}_{\mathrm{dR}}$ is the *de Rham prestack* of $\mathcal{Y}$.

0.1.4. Another example of a theory that is convenient to develop in the generality of prestacks is *deformation theory*, considered in Volume II, Chapter 1. Here, too, it is crucial that we work in the context of derived (as opposed to classical) algberaic geometry.

0.1.5. As yet another application of the general notion of prestack is the construction of the *Ran space* of a given scheme, along with its category of quasi-coherent sheaves or D-modules. We will not discuss it explicitly in this book, and refer the reader to, e.g., [**Ga2**].

0.2. What do we say about prestacks? The notion of prestack is so general that it is, of course, impossible to prove anything non-trivial about arbitrary prestacks.

What we do in Sect. 1 is study some very formal properties of prestacks, which will serve us in the later chapters of this book.

0.2.1. *The notion of n-coconnectivity.* As was said before, the category PreStk is that of functors $(\mathrm{Sch}^{\mathrm{aff}})^{\mathrm{op}} \to \mathrm{Spc}$, where

$$(\mathrm{Sch}^{\mathrm{aff}})^{\mathrm{op}} := \mathrm{ComAlg}(\mathrm{Vect}^{\leq 0}).$$

Now, arguably, the category $\mathrm{ComAlg}(\mathrm{Vect}^{\leq 0})$ is complicated, and it is natural to try to approach it via its successive approximations, namely, the categories

$$\mathrm{ComAlg}(\mathrm{Vect}^{\geq -n, \leq 0})$$

of connective commutative DG algebras that live in cohomological degrees $\geq -n$.

We denote the corresponding full subcategory in $\mathrm{Sch}^{\mathrm{aff}}$ by ${}^{\leq n}\mathrm{Sch}^{\mathrm{aff}}$; we call its objects n-coconnective affine schemes. We can consider the corresponding category of functors

$$({}^{\leq n}\mathrm{Sch}^{\mathrm{aff}})^{\mathrm{op}} \to \mathrm{Spc}$$

and denote it by ${}^{\leq n}\mathrm{PreStk}$.

The ∞-categories ${}^{\leq n}\mathrm{PreStk}$ and ${}^{\leq n}\mathrm{Stk}$ are related by a pair of mutually adjoint functors

$$(0.3) \qquad {}^{\leq n}\mathrm{PreStk} \rightleftarrows \mathrm{PreStk},$$

given by restriction and *left Kan extension* along the inclusion ${}^{\leq n}\mathrm{Sch}^{\mathrm{aff}} \hookrightarrow \mathrm{Sch}^{\mathrm{aff}}$, respectively, with the left adjoint in (0.3) being fully faithful.

Thus, we can think of each ${}^{\leq n}\mathrm{PreStk}$ as a full subcategory in PreStk; we referred to its objects as n-coconnective prestacks. Informally, a functor in (0.1) is n-coconnective if it is *completely determined* by its values on n-coconnective affine schemes.

The subcategories ${}^{\leq n}\mathrm{PreStk}$ form a sequence of approximations to PreStk.

0.2.2. *Convergence.* A technically convenient condition that one can impose on a prestack is that of *convergence.* By definition, a functor $\mathcal{Y}$ in (0.1) is convergent if for any $S \in \text{Sch}^{\text{aff}}$ the map

$$\mathcal{Y}(S) \to \lim_n \mathcal{Y}(^{\leq n}S),$$

is an isomorphism, where $^{\leq n}S$ denotes the n-coconnective truncation of S.

Convergence is a necessary condition for a prestack to satisfy in order to *admit deformation theory,* see Volume II, Chapter 1, Sect. 7.1.

0.2.3. *Finite typeness.* Consider the categories $^{\leq n}\text{Sch}^{\text{aff}}$ and $^{\leq n}\text{PreStk}$. We shall say that an object $S \in {}^{\leq n}\text{Sch}^{\text{aff}}$ (resp., $\mathcal{Y} \in {}^{\leq n}\text{PreStk}$) is of *finite type* (resp., *locally of finite type*) if the corresponding functor (0.1) takes *filtered limits of affine schemes* to colimits in Spc.

It follows tautologically that an object $\mathcal{Y} \in {}^{\leq n}\text{PreStk}$ is locally of finite type if and only if the corresponding functor (0.1) is completely determined by its values on affine schemes of finite type.

Now, the point is that, as in the case of classical algebraic geometry, the condition on an object $\text{Spec}(A) = S \in {}^{\leq n}\text{Sch}^{\text{aff}}$ to be of finite type is very explicit: it is equivalent to $H^0(A)$ being finitely generated over our ground field, and each $H^{-i}(A)$ (where i runs from 1 to n) being finitely generated as a module over $H^0(A)$.

Thus, objects of $^{\leq n}\text{PreStk}$ that are locally of finite type are precisely those that can be expressed via affine schemes that are 'finite dimensional'.

0.2.4. *Inserting the word 'almost'.* Consider now the category PreStk.

We shall say that a prestack is *locally almost of finite type* if it is convergent, and for any n, the functor $\leftarrow$ in (0.3) produces from it an object locally of finite type.

The class of prestacks locally almost of finite type will play a central role in this book. Namely, it is for this class of prestacks that we will develop the theory of ind-coherent sheaves and crystals.

0.3. What else is done in this Chapter?

0.3.1. In Sect. 2 we introduce a hierarchy of Grothendieck topologies on Sch^{aff}: flat, ppf, étale, Zariski. Each of the above choices gives rise to a full subcategory

$$\text{Stk} \subset \text{PreStk}$$

consisting of objects that satisfy the corresponding descent condition. We refer to the objects of Stk as *stacks.*

The primary interest in Sect. 2 is how the descent condition interacts with the conditions of n-coconnectivity, convergence and local (almost) finite typeness.

0.3.2. In the rest of this Chapter we discuss two specific classes of stacks: schemes and Artin stacks (the former being a particular case of the latter).

The corresponding sections are essentially a paraphrase of some parts of [**TV1, TV2**] in the language of ∞-categories.

0.3.3. In Sect. 3 we introduce the full subcategory Sch $\subset$ PreStk of (derived) schemes[1].

Essentially, a prestack Z is a scheme if it is a stack and admits a Zariski atlas (i.e., a collection of affine schemes S_i equipped with *open embeddings* $S_i \to \mathcal{Y}$).

We will not go deep into the study of derived schemes, but content ourselves with establishing the properties related to n-coconnectivity and finite typeness. These can be summarized by saying that a scheme is n-coconnective (resp., of finite type) if and only if some (equivalently, any) Zariski atlas consists of affine schemes that are n-coconnective (resp., of finite type).

0.3.4. In Sect. 4 we introduce the hierarchy of k-Artin stacks, $k = 0, 1, 2....$ Our definition is a variation of the notion of a k-geometric stack defined by Simpson in [**Sim**] and developed in the derived context in [**TV2**].

For an individual k, what we call a k-Artin stack may be different from what is accepted elsewhere in the literature (e.g., in our definition, only schemes that are disjoint unions of affines are 0-Artin stacks; all other schemes are 1-Artin stacks). However, the union over all k produces the same class of objects as in other definitions, called Artin stacks.

The definition of k-Artin stacks proceeds by induction on k. By definition, a k-Artin stack is an étale prestack that admits a *smooth* $(k-1)$*-representable atlas* by affine schemes.

As in the case of schemes, we will only discuss the properties of Artin stacks related to n-coconnectivity and finite typeness, with results parallel to those mentioned above: an Artin stack is n-coconnective (resp., of finite type) if and only if some (equivalently, any) smooth atlas consists of affine schemes that are n-coconnective (resp., of finite type).

1. Prestacks

In this section we introduce the principal actors in derived algebraic geometry: prestacks.

We will focus on the very formal aspects of the theory, such as what it means for a prestack to be n-coconnective (for some integer n) or to be locally (almost) of finite type.

1.1. The notion of prestack.

Derived algebraic geometry is 'born' from connective commutative DG algebras, in the same way as classical algebraic geometry (over a given ground field k) is born from commutative algebras. Following Grothendieck, we will think of algebro-geometric objects as *prestacks*, i.e., arbitrary functors from the ∞-category of connective commutative DG algebras to that of spaces.

[1]Henceforth we will drop the adjective 'derived'.

1.1.1. Consider the stable symmetric monoidal Vect, and its full monoidal subcategory $\mathrm{Vect}^{\le 0}$. By a connective commutative DG algebra over k we shall mean a commutative algebra object in $\mathrm{Vect}^{\le 0}$. The totality of such algebras forms an $(\infty,1)$-category, $\mathrm{ComAlg}(\mathrm{Vect}^{\le 0})$.

REMARK 1.1.2. Note that what we call a 'connective commutative DG algebra over k' is really an abstract notion: we are appealing to the general notion of commutative algebra in symmetric monoidal category from Chapter 1, Sect. 3.3.

However, one can show (see [**Lu2**, Proposition 7.1.4.11]) that the homotopy category of the ∞-category $\mathrm{ComAlg}(\mathrm{Vect}^{\le 0})$ is a familiar object: it is obtained from the category of what one classically calls 'commutative differential graded algebras over k concentrated in degrees ≤ 0' by inverting quasi-isomorphisms.

1.1.3. We define the category of (derived) affine schemes over k to be

$$\mathrm{Sch}^{\mathrm{aff}} := (\mathrm{ComAlg}(\mathrm{Vect}^{\le 0}))^{\mathrm{op}}.$$

1.1.4. By a (derived) prestack we shall mean a functor $(\mathrm{Sch}^{\mathrm{aff}})^{\mathrm{op}} \to \mathrm{Spc}$. We let PreStk denote the $(\infty,1)$-category of prestacks, i.e.,

$$\mathrm{PreStk} := \mathrm{Funct}((\mathrm{Sch}^{\mathrm{aff}})^{\mathrm{op}}, \mathrm{Spc}).$$

1.1.5. Yoneda defines a fully faithful embedding

$$\mathrm{Sch}^{\mathrm{aff}} \hookrightarrow \mathrm{PreStk}.$$

For $S \in \mathrm{Sch}^{\mathrm{aff}}$ and $\mathcal{Y} \in \mathrm{PreStk}$ we have, tautologically,

$$\mathrm{Maps}_{\mathrm{PreStk}}(S, \mathcal{Y}) \simeq \mathcal{Y}(S).$$

1.1.6. Let $f : \mathcal{Y}_1 \to \mathcal{Y}_2$ be a map of prestacks. We shall say that f is *affine schematic* if for every $S \in (\mathrm{Sch}^{\mathrm{aff}})_{/\mathcal{Y}_2}$, the fiber product $S \underset{\mathcal{Y}_2}{\times} \mathcal{Y}_1 \in \mathrm{PreStk}$ is representable by an affine scheme.

1.2. Coconnectivity conditions: affine schemes. Much of the analysis in derived algebraic geometry proceeds by induction on how many negative cohomological degrees we allow our DG algebras to live in. We initiate this discussion in the present subsection.

1.2.1. For $n \ge 0$, consider the full subcategory

$$\mathrm{Vect}^{\ge -n, \le 0} \subset \mathrm{Vect}^{\le 0}.$$

This fully faithful embedding admits a left adjoint, given by the truncation functor $\tau^{\ge -n}$. It is clear that if $V_1' \to V_1'$ is a morphism in $\mathrm{Vect}^{\le 0}$ such that $\tau^{\ge -n}(V_1') \to \tau^{\ge -n}(V_1')$ is an isomorphism, then

$$\tau^{\ge -n}(V_1' \otimes V_2) \to \tau^{\ge -n}(V_1' \otimes V_2)$$

is an isomorphism for any $V_2 \in \mathrm{Vect}^{\le 0}$.

This implies that the $(\infty,1)$-category $\mathrm{Vect}^{\ge -n, \le 0}$ acquires a uniquely defined symmetric monoidal structure for which the functor $\tau^{\ge -n}$ is symmetric monoidal.

It follows from the *symmetric* monoidal version of Chapter 1, Lemma 3.2.4 that the embedding

$$\mathrm{Vect}^{\geq -n, \leq 0} \hookrightarrow \mathrm{Vect}^{\leq 0} \tag{1.1}$$

has a natural *right-lax* symmetric monoidal structure.

1.2.2. In particular, the functor (1.1) induces a fully faithful functor

$$\mathrm{ComAlg}(\mathrm{Vect}^{\geq -n, \leq 0}) \to \mathrm{ComAlg}(\mathrm{Vect}^{\leq 0}), \tag{1.2}$$

whose essential image consists of those objects of $\mathrm{ComAlg}(\mathrm{Vect}^{\leq 0})$ that belong to $\mathrm{Vect}^{\geq -n, \leq 0}$ when regarded as plain objects of $\mathrm{Vect}^{\leq 0}$.

The functor (1.2) admits a left adjoint

$$\tau^{\geq -n} : \mathrm{ComAlg}(\mathrm{Vect}^{\leq 0}) \to \mathrm{ComAlg}(\mathrm{Vect}^{\geq -n, \leq 0}) \tag{1.3}$$

that makes the diagram

$$\begin{array}{ccc} \mathrm{ComAlg}(\mathrm{Vect}^{\leq 0}) & \xrightarrow{\tau^{\geq -n}} & \mathrm{ComAlg}(\mathrm{Vect}^{\geq -n, \leq 0}) \\ {\scriptstyle \mathbf{oblv}_{\mathrm{ComAlg}}} \downarrow & & \downarrow {\scriptstyle \mathbf{oblv}_{\mathrm{ComAlg}}} \\ \mathrm{Vect}^{\leq 0} & \xrightarrow{\tau^{\geq -n}} & \mathrm{Vect}^{\geq -n, \leq 0} \end{array}$$

commute.

1.2.3. We shall say that $S \in \mathrm{Sch}^{\mathrm{aff}}$ is n-coconnective if $S = \mathrm{Spec}(A)$ with A lying in the essential image of (1.2). In other words, if $H^{-i}(A) = 0$ for $i > n$.

We shall denote the full subcategory of $\mathrm{Sch}^{\mathrm{aff}}$ spanned by n-coconnective objects by ${}^{\leq n}\mathrm{Sch}^{\mathrm{aff}}$.

1.2.4. For $n = 0$ we recover

$${}^{\mathrm{cl}}\mathrm{Sch}^{\mathrm{aff}} := {}^{\leq 0}\mathrm{Sch}^{\mathrm{aff}},$$

the category of classical affine schemes.

1.2.5. The embedding ${}^{\leq n}\mathrm{Sch}^{\mathrm{aff}} \hookrightarrow \mathrm{Sch}^{\mathrm{aff}}$ admits a fully faithful right adjoint, denoted

$$S \mapsto {}^{\leq n}S,$$

and given at the level of commutative DG algebras by the functor (1.3).

Thus, ${}^{\leq n}\mathrm{Sch}^{\mathrm{aff}}$ is a *colocalization* of $\mathrm{Sch}^{\mathrm{aff}}$. We denote the corresponding colocalization functor

$$\mathrm{Sch}^{\mathrm{aff}} \to {}^{\leq n}\mathrm{Sch}^{\mathrm{aff}} \hookrightarrow \mathrm{Sch}^{\mathrm{aff}}$$

by $S \mapsto \tau^{\leq n}(S)$.

REMARK 1.2.6. We choose to notationally distinguish objects of ${}^{\leq n}\mathrm{Sch}^{\mathrm{aff}}$ and their images in $\mathrm{Sch}^{\mathrm{aff}}$. Doing otherwise would cause notational clashes when talking about descent conditions.

1.2.7. We will say that $S \in \mathrm{Sch}^{\mathrm{aff}}$ is *eventually coconnective* if it belongs to ${}^{\leq n}\mathrm{Sch}^{\mathrm{aff}}$ for some n.

We denote the full subcategory of $\mathrm{Sch}^{\mathrm{aff}}$ spanned by eventually coconnective objects by ${}^{<\infty}\mathrm{Sch}^{\mathrm{aff}}$.

1.3. Coconnectivity conditions: prestacks.

1.3.1. Consider the $(\infty, 1)$-category

$$^{\leq n}\text{PreStk} := \text{Funct}((^{\leq n}\text{Sch}^{\text{aff}})^{\text{op}}, \text{Spc}).$$

Restriction defines a functor

$$\text{PreStk} \to {}^{\leq n}\text{PreStk}, \tag{1.4}$$

that we will denote by $\mathcal{Y} \mapsto {}^{\leq n}\mathcal{Y}$.

1.3.2. The functor (1.4) admits a *fully faithful left adjoint*, given by the left Kan extension

$$\text{LKE}_{^{\leq n}\text{Sch}^{\text{aff}} \hookrightarrow \text{Sch}^{\text{aff}}} : {}^{\leq n}\text{PreStk} \to \text{PreStk}.$$

Thus, $^{\leq n}$PreStk is a colocalization of PreStk. We denote the resulting colocalization functor

$$\text{PreStk} \to {}^{\leq n}\text{PreStk} \to \text{PreStk}$$

by $\mathcal{Y} \mapsto \tau^{\leq n}(\mathcal{Y})$.

REMARK 1.3.3. The usage of the symbol $\tau^{\leq n}$ may diverge from other sources' conventions: the latter use $\tau^{\leq n}$ to denote the corresponding truncation of the Postnikov tower, whereas we denote the latter by the symbol $\mathsf{P}_{\leq n}$, see Sect. 1.8.5 below.

Tautologically, if $\mathcal{Y}$ is representable by an affine scheme $S = \text{Spec}(A)$, then the above two meanings of $\tau^{\leq n}$ coincide: the prestack $\tau^{\leq n}(\mathcal{Y})$ is representable by the affine scheme $\tau^{\leq n}(S)$.

1.3.4. We shall say that $\mathcal{Y} \in \text{PreStk}$ is n-coconnective if it belongs to the essential image of the functor $\text{LKE}_{^{\leq n}\text{Sch}^{\text{aff}} \hookrightarrow \text{Sch}^{\text{aff}}}$.

For example, an affine scheme is n-coconnective in the sense of Sect. 1.2.3 if and only if its image under the Yoneda functor is n-coconnective as a prestack.

We will often identify $^{\leq n}$PreStk with its essential image under the above functor, and thus think of $^{\leq n}$PreStk as a full subcategory of PreStk.

1.3.5. We will say that $\mathcal{Y} \in \text{PreStk}$ is *eventually coconnective* if it is n-coconnective for some n. We shall denote the full subcategory of eventually coconnective objects of PreStk by $^{<\infty}$PreStk.

1.3.6. *Classical prestacks.* Let $n = 0$. We shall call objects of $^{\leq 0}$PreStk 'classical' prestacks, and use for it also the alternative notation ^{cl}PreStk.

We will also denote the corresponding restriction functor $\mathcal{Y} \mapsto {}^{\text{cl}}\mathcal{Y}$, and the corresponding colocalization functor

$$\text{PreStk} \to {}^{\text{cl}}\text{PreStk} \to \text{PreStk}$$

by $\mathcal{Y} \mapsto \tau^{\text{cl}}(\mathcal{Y})$.

1.3.7. *The right Kan extension.* The restriction functor

$$\mathcal{Y} \mapsto {}^{\leq n}\mathcal{Y} : \text{PreStk} \to {}^{\leq n}\text{PreStk}$$

admits also a right adjoint, given by *right Kan extension.*

This functor lacks a clear geometric meaning. However, it can be explicitly described: by adjunction we have

$$(\text{RKE}_{^{\leq n}\text{Sch}^{\text{aff}} \hookrightarrow \text{Sch}^{\text{aff}}}(\mathcal{Y}))\,(S) \simeq \mathcal{Y}(\tau^{\leq n}(S)).$$

1.4. Convergence. The idea of the notion of convergence is that if we perceive a connective commutative DG algebra as built iteratively by adding lower and lower cohomologies, we can ask whether the value of a given prestack on such an algebra to be determined by its values on the above sequence of truncations.

Convergence is a necessary condition if we want to approach our prestack via deformation theory (see Volume II, Chapter 1, Sect. 7.1).

1.4.1. Le S be an object of $\mathrm{Sch}^{\mathrm{aff}}$. Note that the assignment

$$n \mapsto \tau^{\leq n}(S)$$

is naturally a functor

$$\mathbb{Z}^{\geq 0} \to (\mathrm{Sch}^{\mathrm{aff}})_{/S}.$$

1.4.2. Let $\mathcal{Y}$ be a prestack. We say that $\mathcal{Y}$ is convergent if for $S \in \mathrm{Sch}^{\mathrm{aff}}$, the map

$$\mathcal{Y}(S) \to \lim_{n} \mathcal{Y}(\tau^{\leq n}(S))$$

is an isomorphism.

1.4.3. Since for every connective commutative DG algebra A, the map

$$A \to \lim_{n} \tau^{\geq -n}(A)$$

is an isomorphism, we have:

LEMMA 1.4.4. *Any prestack representable by a (derived) affine scheme is convergent.*

REMARK 1.4.5. As we shall see in the sequel, all prestacks 'of geometric nature', such as (derived) schemes and Artin stacks (and also ind-schemes), are convergent.

Here is, however, an example of a non-convergent prestack: consider the prestack that associates to an affine scheme $S = \mathrm{Spec}(A)$ the category $(A\text{-mod})^{\mathrm{Spc}}$, i.e., this is the prestack

$$(\mathrm{Sch}^{\mathrm{aff}})^{\mathrm{op}} \overset{\mathrm{QCoh}^*_{\mathrm{Sch}^{\mathrm{aff}}}}{\longrightarrow} 1\text{-Cat} \overset{\mathbf{C} \mapsto \mathbf{C}^{\mathrm{Spc}}}{\longrightarrow} \mathrm{Spc},$$

where $\mathrm{QCoh}^*_{\mathrm{Sch}^{\mathrm{aff}}}$ is as in Chapter 3, Sect. 1.1.2.

1.4.6. We have:

PROPOSITION 1.4.7. *A prestack $\mathcal{Y}$ is convergent if and only if, when as a functor*

$$(\mathrm{Sch}^{\mathrm{aff}})^{\mathrm{op}} \to \mathrm{Spc},$$

it is the right *Kan extension from the subcategory* ${}^{<\infty}\mathrm{Sch}^{\mathrm{aff}} \subset \mathrm{Sch}^{\mathrm{aff}}$.

PROOF. We claim that the functor of right Kan extension along ${}^{<\infty}\mathrm{Sch}^{\mathrm{aff}} \subset \mathrm{Sch}^{\mathrm{aff}}$ is given by sending

$$\mathcal{Z}' \in \mathrm{Funct}(({}^{<\infty}\mathrm{Sch}^{\mathrm{aff}})^{\mathrm{op}}, \mathrm{Spc}) \mapsto \mathcal{Z} \in \mathrm{Funct}((\mathrm{Sch}^{\mathrm{aff}})^{\mathrm{op}}, \mathrm{Spc}),$$

with

$$\mathcal{Z}(S) = \lim_{n} \mathcal{Z}'(\tau^{\leq n}(S)).$$

Indeed, a priori, the value of $\mathcal{Z}$ on S is given by

$$\lim_{S' \to S} \mathcal{Z}'(S'),$$

where the limit is taken over the category opposite to $({}^{<\infty}\mathrm{Sch}^{\mathrm{aff}})_{/S}$. Now, the assertion follows from the fact that the functor

$$\mathbb{Z}^{\geq 0} \to ({}^{<\infty}\mathrm{Sch}^{\mathrm{aff}})_{/S}, \quad n \mapsto \tau^{\leq n}(S)$$

is cofinal.

□

1.4.8. Let ${}^{\mathrm{conv}}\mathrm{PreStk} \subset \mathrm{PreStk}$ denote the full subcategory of convergent prestacks. This embedding admits a left adjoint, which we call the *convergent completion* and denote by

$$\mathcal{Y} \mapsto {}^{\mathrm{conv}}\mathcal{Y}.$$

According to Proposition 1.4.7, we have:

$${}^{\mathrm{conv}}\mathcal{Y} \simeq \mathrm{RKE}_{{}^{<\infty}\mathrm{Sch}^{\mathrm{aff}} \hookrightarrow \mathrm{Sch}^{\mathrm{aff}}}(\mathcal{Y}|_{{}^{<\infty}\mathrm{Sch}^{\mathrm{aff}}}).$$

Explicitly,

$${}^{\mathrm{conv}}\mathcal{Y}(S) = \lim_n \mathcal{Y}(\tau^{\leq n}(S)).$$

1.4.9. Consider the canonical map

$$\operatorname*{colim}_n \tau^{\leq n}(\mathcal{Y}) \to \mathcal{Y}.$$

Tautologically, $\mathcal{Y}_1 \in \mathrm{PreStk}$ is convergent if and only if for every $\mathcal{Y}$, the map

$$\mathrm{Maps}(\mathcal{Y}, \mathcal{Y}_1) \to \mathrm{Maps}(\operatorname*{colim}_n \tau^{\leq n}(\mathcal{Y}), \mathcal{Y}_1) = \lim_n \mathrm{Maps}(\tau^{\leq n}(\mathcal{Y}), \mathcal{Y}_1)$$

is an isomorphism.

REMARK 1.4.10. Note that the *left* Kan extension functor

$$\mathrm{LKE}_{{}^{\leq n}\mathrm{Sch}^{\mathrm{aff}} \hookrightarrow \mathrm{Sch}^{\mathrm{aff}}} : {}^{\leq n}\mathrm{PreStk} \to \mathrm{PreStk}$$

does *not* map into ${}^{\mathrm{conv}}\mathrm{PreStk}$.

1.5. Affine schemes of finite type (the eventually coconnective case). We will now introduce the notion of what it means for a (derived) affine scheme to be of finite type. This generalizes the usual notion of being of finite type over a field. As in classical algebraic geometry, *finite typeness* puts us in the context of finite-dimensional geometry.

1.5.1. We say that an object $S = \mathrm{Spec}(A) \in {}^{<\infty}\mathrm{Sch}^{\mathrm{aff}}$ is *of finite type* if $H^0(A)$ is of finite type over k, and each $H^{-i}(A)$ is finitely generated as a module over $H^0(A)$.

Let ${}^{<\infty}\mathrm{Sch}^{\mathrm{aff}}_{\mathrm{ft}}$ denote the full subcategory of ${}^{<\infty}\mathrm{Sch}^{\mathrm{aff}}$ consisting of affine schemes of finite type.

1.5.2. Denote by ${}^{\leq n}\mathrm{Sch}^{\mathrm{aff}}_{\mathrm{ft}}$ the intersection ${}^{<\infty}\mathrm{Sch}^{\mathrm{aff}}_{\mathrm{ft}} \cap {}^{\leq n}\mathrm{Sch}^{\mathrm{aff}}$.

The following theorem is proved by induction on n using deformation theory (but we will not do it here, but see [**Lu2**, Proposition 7.2.5.31]):

THEOREM 1.5.3.
(a) *The objects of* $({}^{\leq n}\mathrm{Sch}^{\mathrm{aff}}_{\mathrm{ft}})^{\mathrm{op}}$ *are compact in* $({}^{\leq n}\mathrm{Sch}^{\mathrm{aff}})^{\mathrm{op}}$.
(b) *For every object* $S \in {}^{\leq n}\mathrm{Sch}^{\mathrm{aff}}$, *the category opposite to* $({}^{<n}\mathrm{Sch}^{\mathrm{aff}}_{\mathrm{ft}})_{S/}$ *is filtered, and the map*

$$S \mapsto \lim_{S_0 \in ({}^{\leq n}\mathrm{Sch}^{\mathrm{aff}}_{\mathrm{ft}})_{S/}} S_0$$

is an isomorphism.

REMARK 1.5.4. We note that the filteredness assertion in Theorem 1.5.3(b) is easy: it follows from the fact that the category $({}^{\leq n}\mathrm{Sch}^{\mathrm{aff}}_{\mathrm{ft}})_{S/}$ has fiber products.

1.5.5. By [**Lu1**, Proposition 5.3.5.11], the assertion of Theorem 1.5.3 is equivalent to the following:

COROLLARY 1.5.6. *We have a canonical equivalence:*

$${}^{\leq n}\mathrm{Sch}^{\mathrm{aff}} \simeq \mathrm{Pro}({}^{\leq n}\mathrm{Sch}^{\mathrm{aff}}_{\mathrm{ft}}).$$

1.5.7. Since ${}^{\leq n}\mathrm{Sch}^{\mathrm{aff}}_{\mathrm{ft}}$ is closed under retracts, using [**Lu1**, Lemma 5.4.2.4], from Corollary 1.5.6 we obtain:

COROLLARY 1.5.8. *The inclusion* $({}^{\leq n}\mathrm{Sch}^{\mathrm{aff}}_{\mathrm{ft}})^{\mathrm{op}} \subset (({}^{\leq n}\mathrm{Sch}^{\mathrm{aff}})^{\mathrm{op}})^{c}$ *of Theorem* 1.5.6*(a) is an equality.*

1.6. Prestacks locally of finite type (the eventually coconnective case). In this subsection we will make precise the following idea: a prestack is locally of finite type if and only if it is completely determined by its values on affine schemes of finite type.

1.6.1. Let $\mathcal{Y}$ be an object of ${}^{\leq n}\mathrm{PreStk}$ for some n. We say that it is *locally of finite type* if it is the left Kan extension (of its own restriction) along the embedding

$${}^{\leq n}\mathrm{Sch}^{\mathrm{aff}}_{\mathrm{ft}} \hookrightarrow {}^{\leq n}\mathrm{Sch}^{\mathrm{aff}}.$$

We denote the resulting full subcategory of ${}^{\leq n}\mathrm{PreStk}$ by ${}^{\leq n}\mathrm{PreStk}_{\mathrm{lft}}$.

1.6.2. In other words, we can identify ${}^{<n}\mathrm{PreStk}_{\mathrm{lft}}$ with the category of functors

$$({}^{\leq n}\mathrm{Sch}^{\mathrm{aff}}_{\mathrm{ft}})^{\mathrm{op}} \to \mathrm{Spc},$$

and we have a pair of mutually adjoint functors

$${}^{\leq n}\mathrm{PreStk}_{\mathrm{lft}} \rightleftarrows {}^{\leq n}\mathrm{PreStk},$$

given by restriction and left Kan extension along ${}^{\leq n}\mathrm{Sch}^{\mathrm{aff}}_{\mathrm{ft}} \hookrightarrow {}^{\leq n}\mathrm{Sch}^{\mathrm{aff}}$, respectively, where the left Kan extension functor is fully faithful.

1.6.3. Now, using [**Lu1**, Proposition 5.3.5.10], from Corollary 1.5.6, we obtain:

COROLLARY 1.6.4. *An object* $\mathcal{Y} \in {}^{\leq n}\mathrm{PreStk}$ *belongs to* ${}^{\leq n}\mathrm{PreStk}_{\mathrm{lft}}$ *if and only if it takes filtered limits in* ${}^{\leq n}\mathrm{Sch}^{\mathrm{aff}}$ *to colimits in* Spc.

1.6.5. Combining Corollaries 1.6.4 and 1.5.8, we obtain:

LEMMA 1.6.6. *Let S be an object of* ${}^{\leq n}\mathrm{Sch}^{\mathrm{aff}}$. *It belongs to* ${}^{\leq n}\mathrm{Sch}^{\mathrm{aff}}_{\mathrm{ft}}$ *if and only if the prestack that it represents belongs to* ${}^{\leq n}\mathrm{PreStk}_{\mathrm{lft}}$.

1.6.7. Evidently, the restriction functor ${}^{\leq n}\mathrm{PreStk}_{\mathrm{lft}} \leftarrow {}^{\leq n}\mathrm{PreStk}$ commutes with limits and colimits. The functor

$$\mathrm{LKE}_{{}^{\leq n}\mathrm{Sch}^{\mathrm{aff}}_{\mathrm{ft}} \hookrightarrow {}^{\leq n}\mathrm{Sch}^{\mathrm{aff}}} : {}^{\leq n}\mathrm{PreStk}_{\mathrm{lft}} \to {}^{\leq n}\mathrm{PreStk},$$

being a left adjoint commutes with colimits.

In addition, we have the following:

LEMMA 1.6.8. *The functor* $\mathrm{LKE}_{{}^{\leq n}\mathrm{Sch}^{\mathrm{aff}}_{\mathrm{ft}} \hookrightarrow {}^{\leq n}\mathrm{Sch}^{\mathrm{aff}}}$ *commutes with finite limits.*

PROOF. This follows from Corollary 1.6.4: indeed, the condition of taking filtered limits in ${}^{\leq n}\mathrm{Sch}^{\mathrm{aff}}$ to colimits in Spc is preserved by the operation of taking finite limits of prestacks.

□

1.7. The 'locally almost of finite type' condition. In Sect. 1.6 we introduced the 'locally of finite type' condition for n-coconnective prestacks. In this subsection we will give a definition crucial for the rest of the book: what it means for an object of PreStk to be locally *almost* of finite type (=laft). This will be the class of prestacks for which we will develop the theory of ind-coherent sheaves.

1.7.1. We say that an affine (derived) scheme S is *almost of finite type* if ${}^{\leq n}S$ is of finite type for every n.

I.e., $S = \mathrm{Spec}(A)$ is almost of finite type if $H^0(A)$ is of finite type over k, and each $H^{-i}(A)$ is finitely generated as a module over $H^0(A)$.

Let $\mathrm{Sch}^{\mathrm{aff}}_{\mathrm{aft}}$ denote the full subcategory of $\mathrm{Sch}^{\mathrm{aff}}$ consisting of affine schemes almost of finite type.

1.7.2. We say that $\mathcal{Y} \in \mathrm{PreStk}$ is *locally almost of finite type* if the following conditions hold:

(1) $\mathcal{Y}$ is convergent.
(2) For every n, we have ${}^{\leq n}\mathcal{Y} \in {}^{\leq n}\mathrm{PreStk}_{\mathrm{lft}}$

We denote the corresponding full subcategory by

$$\mathrm{PreStk}_{\mathrm{laft}} \subset \mathrm{PreStk}.$$

By Lemma 1.6.6, we have

$$\mathrm{Sch}^{\mathrm{aff}}_{\mathrm{aft}} = \mathrm{Sch}^{\mathrm{aff}} \cap \mathrm{PreStk}_{\mathrm{laft}}.$$

1.7.3. In particular, if $\mathcal{Y} \in \mathrm{PreStk}_{\mathrm{laft}}$, then ${}^{\mathrm{cl}}\mathcal{Y}$ is an object of ${}^{\mathrm{cl}}\mathrm{PreStk}$ locally of finite type, i.e., it is a classical prestack locally of finite type.

REMARK 1.7.4. Note that by Remark 1.4.10, the left Kan extension functor does *not* send ${}^{\leq n}\mathrm{PreStk}_{\mathrm{lft}}$ to $\mathrm{PreStk}_{\mathrm{laft}}$: the resulting prestack will satisfy the second condition, but in general, not the first one.

However, if $\mathcal{Y} \in \mathrm{PreStk}$ is obtained as a left Kan extension functor of an object of ${}^{\leq n}\mathrm{PreStk}$ that belongs to ${}^{\leq n}\mathrm{PreStk}_{\mathrm{lft}}$, then its convergent completion ${}^{\mathrm{conv}}\mathcal{Y}$ will belong to $\mathrm{PreStk}_{\mathrm{laft}}$, see Corollary 1.7.8 below.

1.7.5. We claim:

PROPOSITION 1.7.6. *Restriction along* ${}^{<\infty}\mathrm{Sch}^{\mathrm{aff}}_{\mathrm{ft}} \hookrightarrow \mathrm{Sch}^{\mathrm{aff}}$ *defines an equivalence*

$$\mathrm{PreStk}_{\mathrm{laft}} \to \mathrm{Funct}(({}^{<\infty}\mathrm{Sch}^{\mathrm{aff}}_{\mathrm{ft}})^{\mathrm{op}}, \mathrm{Spc}).$$

The inverse functor is given by first applying the left *Kan extension along*

$${}^{<\infty}\mathrm{Sch}^{\mathrm{aff}}_{\mathrm{ft}} \hookrightarrow {}^{<\infty}\mathrm{Sch}^{\mathrm{aff}},$$

followed by the right *Kan extension along*

$${}^{<\infty}\mathrm{Sch}^{\mathrm{aff}} \hookrightarrow \mathrm{Sch}^{\mathrm{aff}}.$$

PROOF. By Proposition 1.4.7, it suffices to show that the following conditions on a functor

$${}^{<\infty}\mathrm{Sch}^{\mathrm{aff}} \to \mathrm{Spc}$$

are equivalent:

(i) It is a left Kan extension along ${}^{<\infty}\mathrm{Sch}^{\mathrm{aff}}_{\mathrm{ft}} \to {}^{<\infty}\mathrm{Sch}^{\mathrm{aff}}$;

(ii) Its restriction to any ${}^{\leq n}\mathrm{Sch}^{\mathrm{aff}}$ is a left Kan extension along ${}^{\leq n}\mathrm{Sch}^{\mathrm{aff}}_{\mathrm{ft}} \to {}^{\leq n}\mathrm{Sch}^{\mathrm{aff}}$.

First, it is clear that (i) implies (ii): indeed, the diagram

$$\begin{array}{ccc} \mathrm{Funct}({}^{<\infty}\mathrm{Sch}^{\mathrm{aff}}_{\mathrm{ft}}, \mathrm{Spc}) & \xrightarrow{\mathrm{LKE}} & \mathrm{Funct}({}^{<\infty}\mathrm{Sch}^{\mathrm{aff}}, \mathrm{Spc}) \\ \downarrow & & \downarrow \\ \mathrm{Funct}({}^{\leq n}\mathrm{Sch}^{\mathrm{aff}}_{\mathrm{ft}}, \mathrm{Spc}) & \xrightarrow{\mathrm{LKE}} & \mathrm{Funct}({}^{\leq n}\mathrm{Sch}^{\mathrm{aff}}, \mathrm{Spc}) \end{array}$$

is commutative.

Vice versa, let $\mathcal{Y}$ satisfy (ii). We need to show that for any $S \in {}^{\leq n}\mathrm{Sch}^{\mathrm{aff}}$, the map

$$\operatorname*{colim}_{S \to S'} \mathcal{Y}(S') \to \mathcal{Y}(S) \tag{1.5}$$

is an isomorphism, where the colimit is taken over the index category

$$\left(({}^{<\infty}\mathrm{Sch}^{\mathrm{aff}}_{\mathrm{ft}})_{S/}\right)^{\mathrm{op}}.$$

However, cofinal in the above index category is the full subcategory consisting of those $S \to S'$, for which $S' \in {}^{\leq n}\mathrm{Sch}^{\mathrm{aff}}_{\mathrm{ft}}$; indeed the embedding of this full subcategory admits a left adjoint, given by $S' \mapsto \tau^{\leq n}(S')$.

Hence, the colimit in (1.5) can be replaced by

$$\operatorname*{colim}_{S \to S'} \mathcal{Y}(S')$$

taken over $\left(({}^{\leq n}\mathrm{Sch}^{\mathrm{aff}}_{\mathrm{ft}})_{S/}\right)^{\mathrm{op}}$. However, the latter colimit computes

$$\mathrm{LKE}_{{}^{\leq n}\mathrm{Sch}^{\mathrm{aff}}_{\mathrm{ft}} \hookrightarrow {}^{\leq n}\mathrm{Sch}^{\mathrm{aff}}}({}^{\leq n}\mathcal{Y}).$$

□

1.7.7. We note:

COROLLARY 1.7.8. *The composite functor*

$${}^{\leq n}\mathrm{PreStk}_{\mathrm{lft}} \hookrightarrow {}^{\leq n}\mathrm{PreStk} \overset{\mathrm{LKE}_{{}^{\leq n}\mathrm{Sch}^{\mathrm{aff}} \hookrightarrow \mathrm{Sch}^{\mathrm{aff}}}}{\longrightarrow} \mathrm{PreStk} \overset{\mathcal{Y} \mapsto {}^{\mathrm{conv}}\mathcal{Y}}{\longrightarrow} {}^{\mathrm{conv}}\mathrm{PreStk}$$

takes values in $\mathrm{PreStk}_{\mathrm{laft}}$.

PROOF. By Proposition 1.7.6, it suffices to show that the composition of the functor in the corollary with the identification

$${}^{\mathrm{conv}}\mathrm{PreStk} \simeq \mathrm{Funct}({}^{<\infty}\mathrm{Sch}^{\mathrm{aff}}, \mathrm{Spc})$$

lands in the full subcategory spanned by functors obtained as a left Kan extension from

$${}^{<\infty}\mathrm{Sch}^{\mathrm{aff}}_{\mathrm{ft}} \hookrightarrow {}^{<\infty}\mathrm{Sch}^{\mathrm{aff}}.$$

However, the above composition is given by left Kan extension along

$${}^{\leq n}\mathrm{Sch}^{\mathrm{aff}}_{\mathrm{ft}} \hookrightarrow {}^{<\infty}\mathrm{Sch}^{\mathrm{aff}}.$$

□

1.7.9. By combining Lemma 1.6.8 and Proposition 1.7.6, we obtain:

COROLLARY 1.7.10. *The subcategory* $\mathrm{PreStk}_{\mathrm{laft}} \subset \mathrm{PreStk}$ *is closed under finite limits.*

1.8. Truncatedness.

1.8.1. For $k = 0, 1, \ldots$, let $\mathrm{Spc}_{\leq k} \subset \mathrm{Spc}$ denote the full subcategory of *k-truncated* spaces. I.e., it is spanned by those objects $\mathcal{S} \in \mathrm{Spc}$ such that each connected component $\mathcal{S}'$ of $\mathcal{S}$ satisfies

$$\pi_l(\mathcal{S}') = 0 \text{ for } l > k.$$

For example, for $k = 0$, we have $\mathrm{Spc}_{\leq 0} = \mathrm{Set}$.

1.8.2. The embedding

$$\mathrm{Spc}_{\leq k} \hookrightarrow \mathrm{Spc}$$

admits a left adjoint.

The corresponding localization functor

$$\mathrm{Spc} \to \mathrm{Spc}_{\leq k} \to \mathrm{Spc}$$

will be denoted $\mathsf{P}_{\leq k}$.

REMARK 1.8.3. The $(\infty, 1)$-category $\mathrm{Spc}_{\leq k}$ is actually a $(k+1, 1)$-category. I.e., the mapping spaces between objects are k-truncated.

1.8.4. For $\mathcal{S} \in \text{Spc}$, the assignment $k \mapsto \mathsf{P}_{\leq k}(\mathcal{S})$ is a functor

$$(\mathbb{Z}^{\geq 0})^{\text{op}} \to \text{Spc},$$

called the *Postnikov tower* of $\mathcal{S}$.

It is a basic fact that the natural map

$$\mathcal{S} \to \lim_k \mathsf{P}_{\leq k}(\mathcal{S})$$

is an isomorphism.

1.8.5. For a fixed n, and an integer $k = 0, 1, \ldots$ we will say that $\mathcal{Y} \in {}^{\leq n}\text{PreStk}$ is *k-truncated* if, as a functor

$$({}^{\leq n}\text{Sch}^{\text{aff}})^{\text{op}} \to \text{Spc},$$

it takes values in the full subcategory of $\text{Spc}_{\leq k} \subset \text{Spc}$ of k-truncated spaces.

1.8.6. For example, if $\mathcal{Y} \in {}^{\leq n}\text{PreStk}$ is representable, i.e., is the Yoneda image of $S \in {}^{\leq n}\text{Sch}^{\text{aff}}$, then $\mathcal{Y}$ is n-truncated.

This reflects the fact that $\text{ComAlg}(\text{Vect}^{\geq -n, \leq 0})$ is an $(n+1,1)$-category, which, in turn, formally follows from the fact that $\text{Vect}^{\geq -n, \leq 0}$ is an $(n+1,1)$-category.

REMARK 1.8.7. In the sequel, we will see that for any (derived) scheme, its restriction to ${}^{\leq n}\text{Sch}^{\text{aff}}$ is n-truncated as an object of ${}^{\leq n}\text{PreStk}$.

Similarly, for a k-Artin stack, its restriction to ${}^{\leq n}\text{Sch}^{\text{aff}}$ is $(n+k)$-truncated as an object of ${}^{\leq n}\text{PreStk}$.

1.8.8. *Another example.* To any object $\mathcal{K} \in \text{Spc}$ we can attach the corresponding constant prestack $\underline{\mathcal{K}}$:

$$\underline{\mathcal{K}}(S) := \mathcal{K}, \quad S \in \text{Sch}^{\text{aff}}.$$

If $\mathcal{K}$ is k-truncated, then $\underline{\mathcal{K}}$ is k-truncated.

1.8.9. Let ${}^{\leq n}\text{PreStk}_{\leq k} \subset {}^{\leq n}\text{PreStk}$ denote the full subcategory of k-truncated prestacks. This embedding admits a left adjoint. The corresponding localization functor

$${}^{\leq n}\text{PreStk} \to {}^{\leq n}\text{PreStk}_{\leq k} \to {}^{\leq n}\text{PreStk}$$

will be denoted $\mathsf{P}_{\leq k}$. Explicitly,

$$(\mathsf{P}_{\leq k}(\mathcal{Y}))(S) = \mathsf{P}_{\leq k}(\mathcal{Y}(S)), \quad S \in {}^{\leq n}\text{Sch}^{\text{aff}}.$$

The full subcategory ${}^{\leq n}\text{PreStk}_{\leq k} \subset {}^{\leq n}\text{PreStk}$ is actually a $(k+1,1)$-category.

1.8.10. When $n = 0$ and $k = 0$, the (ordinary) category ${}^{\text{cl}}\text{PreStk}_{\leq 0}$ is that of *presheaves of sets* on ${}^{\text{cl}}\text{Sch}^{\text{aff}}$.

When $n = 0$ and $k = 1$, we shall call objects of ${}^{\text{cl}}\text{PreStk}_{\leq 1}$ 'ordinary classical prestacks'. I.e., ${}^{\text{cl}}\text{PreStk}_{\leq 1}$ is the $(2,1)$-category of functors from the category of classical affine schemes to that of ordinary groupoids.

2. Descent and stacks

The object of study in this section is the notion of *stack*–the result of the interaction of the general notion of prestack with a given Grothendieck topology (flat, ppf, étale or Zariski) on the category of affine schemes; see [**TV2**, Sect. 2.2.2].

Specifically, we will be interested in how the stack condition interacts with n-coconnectivity and the finite typeness.

2.1. Flat morphisms. In this subsection we will introduce the crucial notion of flatness for a morphism between (derived) affine schemes. Knowing what it means to be flat, we will give the definition of what it means to be an open embedding, étale, smooth, ppf, etc.

2.1.1. Let us recall, following [**TV2**], the notion of flatness for a morphism between (derived) affine schemes:

A map $\mathrm{Spec}(B) \to \mathrm{Spec}(A)$ between affine schemes is said to be flat if $H^0(B)$ is flat as a module over $H^0(A)$, plus the following *equivalent* conditions hold:

- The natural map
$$H^0(B) \underset{H^0(A)}{\otimes} H^i(A) \to H^i(B)$$
is an isomorphism for every i.
- For any A-module M, the natural map
$$H^0(B) \underset{H^0(A)}{\otimes} H^i(M) \to H^i(B \underset{A}{\otimes} M)$$
is an isomorphism for every i.
- If an A-module N is concentrated in degree 0 then so is $B \underset{A}{\otimes} N$.

2.1.2. Note in particular that if $S' \to S$ is flat, then
$$S \in {}^{\leq n}\mathrm{Sch}^{\mathrm{aff}} \quad \Rightarrow \quad S' \in {}^{\leq n}\mathrm{Sch}^{\mathrm{aff}}.$$

The following assertion is easily established by induction:

LEMMA 2.1.3. *For a map $S' \to S$ between affine schemes, S' is flat over S if and only if each ${}^{\leq n}S'$ is flat over ${}^{\leq n}S$.*

2.1.4. Let $f : S' \to S$ be a morphism of affine schemes. We shall say that it is ppf[2] (resp., smooth, étale, open embedding, Zariski) if the following conditions hold:

(1) The morphism f is flat (in particular, the base-changed (derived!) affine scheme
$$\tau^{\mathrm{cl}}(S) \underset{S}{\times} S'$$
is classical and thus identifies with $\tau^{\mathrm{cl}}(S')$);

(2) The map of classical affine schemes ${}^{\mathrm{cl}}S' \to {}^{\mathrm{cl}}S$ is of finite presentation (resp., smooth, étale, open embedding, disjoint union of open embeddings).

[2] ppf=plat de présentation finie= flat of finite presentation

For future reference, we quote the following basic fact that can be proved using deformation theory (see [**TV2**, Corollaries 2.2.2.9 and 2.2.2.10]):

LEMMA 2.1.5. *For a given* $S \in \mathrm{Sch}^{\mathrm{aff}}$, *the operation of passage to the underlying classical subscheme defines an equivalence between the full subcategory* $(\mathrm{Sch}^{\mathrm{aff}})_{/S}$ *spanned by* $S' \xrightarrow{f} S$ *with* f *étale and the full subcategory of* $({}^{\mathrm{cl}}\mathrm{Sch}^{\mathrm{aff}})_{/{}^{\mathrm{cl}}S}$ *spanned by* $\widetilde{S}' \xrightarrow{\widetilde{f}} {}^{\mathrm{cl}}S$ *with* $\widetilde{f}$ *étale. Furthermore,* f *is an open embedding (resp., Zariski) if and only if* $\widetilde{f}$ *is.*

2.1.6. We say that a morphism is $f : S' \to S$ is a covering with respect to the flat (resp., ppf, smooth, étale, Zariski) topology, if it is flat (resp., ppf, smooth, étale, Zariski), and the induced map of classical affine schemes ${}^{\mathrm{cl}}S' \to {}^{\mathrm{cl}}S$ is surjective.

Thus, the category $\mathrm{Sch}^{\mathrm{aff}}$ acquires a hierarchy of Grothendieck topologies: flat, ppf, smooth, étale and Zariski.

2.1.7. The property of a morphism to be flat (resp., ppf, smooth, étale, open embedding, Zariski) is obviously stable under base change.

Moreover, the property of a morphism $f : S' \to S$ to be flat (resp., ppf, smooth, étale, open embedding) it itself local with respect to any of the above topologies on S.

In addition, the property of a morphism $f : S' \to S$ to be flat (resp., ppf, smooth, étale, Zariski) is local with respect to the flat (resp., ppf, smooth, étale, Zariski) topology on S'.

REMARK 2.1.8. For obvious reasons, the property of a morphism to be an open embedding is *not* Zariski-local on the source. And the property of a morphism to be Zariski is *not* étale-local on the target.

2.1.9. Let $f : \mathcal{Y}_1 \to \mathcal{Y}_2$ be an *affine schematic* morphism in PreStk (see Sect. 1.1.6 for what this means).

We shall say that it is flat (resp., ppf, smooth, étale, open embedding, Zariski) if for every $S \in (\mathrm{Sch}^{\mathrm{aff}})_{/\mathcal{Y}_2}$, the corresponding map

$$S \underset{\mathcal{Y}_2}{\times} \mathcal{Y}_1 \to S$$

(of affine schemes(!)) is flat (resp., ppf, smooth, étale, open embedding, Zariski).

2.2. Digression: the Čech nerve.

2.2.1. Let Fin denote the category of finite sets.

Let $\mathbf{C}$ be an arbitrary ∞-category with Cartesian products. Then to an object $\mathbf{c} \in \mathbf{C}$ we can attach a functor

$$\mathrm{Fin}^{\mathrm{op}} \to \mathbf{C}, \quad I \mapsto \mathbf{c}^I.$$

In terms of the Yoneda embedding, this functor is uniquely characterized by

$$\mathrm{Maps}_{\mathbf{C}}(\mathbf{c}', \mathbf{c}^I) = \mathrm{Maps}_{\mathrm{Spc}}(I, \mathrm{Maps}_{\mathbf{C}}(\mathbf{c}', \mathbf{c})), \quad \mathbf{c}' \in \mathbf{C}'.$$

Composing with the functor $\mathbf{\Delta} \to \mathrm{Fin}$, we obtain a functor

$$\mathbf{\Delta}^{\mathrm{op}} \to \mathbf{C}.$$

2.2.2. Let now $\mathbf{D}$ be an ∞-category with fiber products, and $\mathbf{d} \in \mathbf{D}$ an object. Set

$$\mathbf{C} := \mathbf{D}_{/\mathbf{d}},$$

so that Cartesian products in $\mathbf{C}$ are the fiber products in $\mathbf{D}$ *over* $\mathbf{d}$.

Given an object $\mathbf{c} \in \mathbf{D}_{/\mathbf{d}}$ we thus obtain a functor

$$\mathbf{\Delta}^{\mathrm{op}} \to \mathbf{D}_{/\mathbf{d}} \to \mathbf{D}.$$

It is called the *Čech nerve* of the morphism $\mathbf{c} \to \mathbf{d}$, and denoted $\mathbf{c}^{\bullet}/\mathbf{d}$.

2.2.3. Thus, we have $\mathbf{c}^0/\mathbf{d} = \mathbf{c}$,

$$\mathbf{c}^1/\mathbf{d} = \mathbf{c} \underset{\mathbf{d}}{\times} \mathbf{c}.$$

In general, the object $\mathbf{c}^{\bullet}/\mathbf{d} \in \mathrm{Funct}(\mathbf{\Delta}^{\mathrm{op}}, \mathbf{D})$ is an example of a *groupoid object* of $\mathbf{D}$; see [**Lu1**, Sect. 6.1.2] for what this means.

2.3. The descent condition. In this subsection we will impose the descent condition that singles out the class of *stacks* among all prestacks.

This discussion here is not specific to the category $\mathrm{Sch}^{\mathrm{aff}}$. It is applicable to any ∞-category (with fiber products) equipped with a Grothendieck topology. So, we can view this subsection is a summary of some results from [**Lu1**, Sect. 6] and [**TV1**].

2.3.1. Let $\mathcal{Y}$ be a prestack. We say that it satisfies flat (resp., ppf, smooth, étale, Zarski) descent if whenever

$$f : S' \to S \in \mathrm{Sch}^{\mathrm{aff}}$$

is a flat covering, the map

$$\mathcal{Y}(S) \to \mathrm{Tot}(\mathcal{Y}(S'^{\bullet}/S))$$

is an isomorphism, where $S'^{\bullet}/S$ is the Čech nerve of the map f.

2.3.2. In what follows we will assume that our topology is chosen to be étale. However, the entire discussion equally applies to the other cases, i.e. flat, ppf, smooth or Zariski.

We shall call prestacks that satisfy the above descent condition *stacks*, and denote the corresponding full subcategory of PreStk by Stk.

As in the case of classical algebraic geometry, one shows that if an object of PreStk satisfies étale descent, then it satisfies smooth descent.

2.3.3. We say that a map $\mathcal{Y}_1 \to \mathcal{Y}_2$ in PreStk is an *étale equivalence* if it induces an isomorphism

$$\mathrm{Maps}(\mathcal{Y}_2, \mathcal{Y}) \to \mathrm{Maps}(\mathcal{Y}_1, \mathcal{Y})$$

whenever $\mathcal{Y} \in \mathrm{Stk}$.

2.3.4. The inclusion

$$\mathrm{Stk} \hookrightarrow \mathrm{PreStk}$$

admits a left adjoint making Stk a localization of PreStk.

Concretely, the functor PreStk → Stk is universal among functors that turn étale equivalences into isomorphisms, see [**Lu1**, Sect. 6.2.1].

We will denote by L the corresponding localization (=sheafification) functor, i.e., the composition

$$\mathrm{PreStk} \to \mathrm{Stk} \to \mathrm{PreStk}.$$

Tautologically, a map $\mathcal{Y}_1 \to \mathcal{Y}_2$ is an étale equivalence if and only if $L(\mathcal{Y}_1) \to L(\mathcal{Y}_2)$ is an isomorphism.

2.3.5. We have the following assertion (see [**Lu1**, Corollary 6.2.1.6 and Proposition 6.2.2.7][3]):

LEMMA 2.3.6. *The functor L is left exact, i.e., commutes with finite limits.*

2.3.7. Let $f : \mathcal{Y}_1 \to \mathcal{Y}_2$ be a morphism in PreStk.

We say that f is an étale surjection if for every $S \in \mathrm{Sch}^{\mathrm{aff}}$ and an object $y_2 \in \mathcal{Y}_2(S)$ there exists an étale cover $\phi : S' \to S$, such that $\phi^*(y_2) \in \mathcal{Y}_2(S')$ belongs to the essential image of $f(S') : \mathcal{Y}_1(S') \to \mathcal{Y}_2(S')$.

The following is [**Lu1**, Corollary 6.2.3.5]:

LEMMA 2.3.8. *Let $\mathcal{Y}_1 \to \mathcal{Y}_2$ be an étale surjection. Then the induced map*

$$|\mathcal{Y}_1^\bullet/\mathcal{Y}_2|_{\mathrm{PreStk}} \to \mathcal{Y}_2$$

is an étale equivalence, where $\mathcal{Y}_1^\bullet/\mathcal{Y}_2$ is the Čech nerve of f, and $|-|_{\mathrm{PreStk}}$ denotes geometric realization taken in the category PreStk.

Note that the assertion of Lemma 2.3.8 can be reformulated as the statement that if $\mathcal{Y}_1 \to \mathcal{Y}_2$ is an étale surjection, then the map

$$|L(\mathcal{Y}_1^\bullet/\mathcal{Y}_2)|_{\mathrm{Stk}} \simeq |L(\mathcal{Y}_1)^\bullet/L(\mathcal{Y}_2)|_{\mathrm{Stk}} \simeq L(|\mathcal{Y}_1^\bullet/\mathcal{Y}_2|_{\mathrm{PreStk}}) \to L(\mathcal{Y}_2)$$

is an isomorphism.

2.3.9. Finally, we have:

LEMMA 2.3.10. *For $\mathcal{Y} \subset$ PreStk, the unit of the adjunction*

$$\mathcal{Y} \to L(\mathcal{Y})$$

is an étale surjection.

2.4. Descent for affine schemes. In this subsection we state (without proof) the standard, but crucial, fact that affine schemes are in fact stacks, and discuss some of its corollaries.

As in the previous subsection, the results stated in this subsection here hold also for the flat, ppf and Zariski topologies.

---

[3]For this proposition the reader should use the version of [**Lu1**] available on Lurie's website rather than the printed version.

2.4.1. We have the following basic fact (see [**TV2**, Lemma 2.2.2.13]):

PROPOSITION 2.4.2.

(a) *The image of the Yoneda embedding* $\mathrm{Sch}^{\mathrm{aff}} \hookrightarrow \mathrm{PreStk}$ *belongs to* Stk.

(b) *Let*

$$\begin{array}{ccc} \mathcal{Y} & \longleftarrow & \mathcal{Y}' \\ \downarrow & & \downarrow \\ S & \xleftarrow{f} & S' \end{array}$$

be a pullback diagram in Stk *with* $S, S' \in \mathrm{Sch}^{\mathrm{aff}}$. *Assume that* $\mathcal{Y}'$ *also belongs to* $\mathrm{Sch}^{\mathrm{aff}} \subset \mathrm{PreStk}$ *and the morphism* f *is an étale covering. Then* $\mathcal{Y} \in \mathrm{Sch}^{\mathrm{aff}}$.

2.4.3. As a corollary, we obtain:

COROLLARY 2.4.4. *Let* $f : \mathcal{Y}_1 \to \mathcal{Y}_2$ *be an affine schematic morphism in* PreStk. *Then the morphism* $L(\mathcal{Y}_1) \to L(\mathcal{Y}_2)$ *is also affine schematic.*

PROOF. We need to show that for $S \in \mathrm{Sch}^{\mathrm{aff}}$ and a map $S \to L(\mathcal{Y}_2)$, the fiber product $S \underset{L(\mathcal{Y}_2)}{\times} L(\mathcal{Y}_1)$ belongs to $\mathrm{Sch}^{\mathrm{aff}}$. By Proposition 2.4.2(b), it suffices to show that that the fiber product $S' \underset{L(\mathcal{Y}_2)}{\times} L(\mathcal{Y}_1)$ belongs to $\mathrm{Sch}^{\mathrm{aff}}$ for some étale covering map $S' \to S$ with $S' \in \mathrm{Sch}^{\mathrm{aff}}$.

However, by Lemma 2.3.10, we can choose $S' \to S$ so that the composition $S' \to S \to L(\mathcal{Y}_2)$ factors as $S' \to \mathcal{Y}_2 \to L(\mathcal{Y}_2)$. Since the functor L commutes with fiber products (by Lemma 2.3.6), we have

$$S' \underset{L(\mathcal{Y}_2)}{\times} L(\mathcal{Y}_1) \simeq L(S' \underset{\mathcal{Y}_2}{\times} \mathcal{Y}_1).$$

Now, by assumption, $S' \underset{\mathcal{Y}_2}{\times} \mathcal{Y}_1 \in \mathrm{Sch}^{\mathrm{aff}}$, and

$$S' \underset{\mathcal{Y}_2}{\times} \mathcal{Y}_1 \to L(S' \underset{\mathcal{Y}_2}{\times} \mathcal{Y}_1)$$

is an isomorphism by Proposition 2.4.2(a)

□

2.4.5. The same proof also gives:

COROLLARY 2.4.6. *Let* $f : \mathcal{Y}_1 \to \mathcal{Y}_2$ *be affine flat (resp., ppf, smooth, étale, open embedding). Then so is* $L(\mathcal{Y}_1) \to L(\mathcal{Y}_2)$.

2.5. Descent and n-coconnectivity. In this subsection we will study how the étale descent condition interacts with the operation of restriction and left Kan extension to the (full) subcategory ${}^{\leq n}\mathrm{Sch}^{\mathrm{aff}} \subset \mathrm{Sch}^{\mathrm{aff}}$.

Again, the entire discussion is applicable when we replace the word 'étale' by 'flat', 'ppf' or 'Zariski'.

2.5.1. Let us denote by ${}^{\leq n}\mathrm{Stk}$ the full subcategory of ${}^{\leq n}\mathrm{PreStk}$ consisting of objects that satisfy descent for étale covers $S_1 \to S_2 \in {}^{\leq n}\mathrm{Sch}^{\mathrm{aff}}$.

We obtain that ${}^{\leq n}\mathrm{Stk}$ is a localization of ${}^{\leq n}\mathrm{PreStk}$. Let ${}^{\leq n}L$ denote the corresponding localization functor

$$ {}^{\leq n}\mathrm{PreStk} \to {}^{\leq n}\mathrm{Stk} \to {}^{\leq n}\mathrm{PreStk}. $$

The analog of Lemma 2.3.6 equally applies in the present context.

2.5.2. The sheafification functor ${}^{\leq n}L$ on *truncated objects* can be described explicitly as follows (see [**Lu1**, Sect. 6.5.3]):

We have the following endo-functor, denoted

$$ \mathcal{Y} \mapsto \mathcal{Y}^{+} \tag{2.1} $$

of ${}^{\leq n}\mathrm{PreStk}$.

Namely, for $\mathcal{Y} \in {}^{\leq n}\mathrm{PreStk}$, the value of $\mathcal{Y}^{+}$ on $S \in {}^{\leq n}\mathrm{Sch}^{\mathrm{aff}}$ is the colimit over all étale covers $S' \to S$ of $\mathcal{Y}(S')$.

Now, if $\mathcal{Y}$ is $k-2$-truncated for $k = 2, 3, ...$, then the value of $L(\mathcal{Y})$ on $S \in {}^{\leq n}\mathrm{Sch}^{\mathrm{aff}}$ is

$$ \mathcal{Y}^{+^{k}}(S), $$

where $\mathcal{Y}^{+^{k}}$ denotes the k-th iteration of the functor (2.1).

In particular, since the colimit involved in its description is filtered, we obtain:

LEMMA 2.5.3. *The functor* ${}^{\leq n}L : {}^{\leq n}\mathrm{PreStk} \to {}^{\leq n}\mathrm{PreStk}$ *sends k-truncated objects to k-truncated ones.*

2.5.4. The following results from the definitions:

LEMMA 2.5.5.

(a) *The restriction functor* $\mathrm{PreStk} \to {}^{\leq n}\mathrm{PreStk}$ *sends* Stk *to* ${}^{\leq n}\mathrm{Stk}$.

(b) *The functor*

$$ \mathrm{LKE}_{{}^{\leq n}\mathrm{Sch}^{\mathrm{aff}} \hookrightarrow \mathrm{Sch}^{\mathrm{aff}}} : {}^{\leq n}\mathrm{PreStk} \to \mathrm{PreStk} $$

sends étale equivalences to étale equivalences.

2.5.6. Note now that the *right* Kan extension functor along ${}^{\leq n}\mathrm{Sch}^{\mathrm{aff}} \hookrightarrow \mathrm{Sch}^{\mathrm{aff}}$:

$$ \mathrm{RKE}_{{}^{\leq n}\mathrm{Sch}^{\mathrm{aff}} \hookrightarrow \mathrm{Sch}^{\mathrm{aff}}} : {}^{\leq n}\mathrm{PreStk} \to \mathrm{PreStk} $$

tautologically sends ${}^{\leq n}\mathrm{Stk}$ to Stk. This implies that the restriction functor $\mathcal{Y} \mapsto {}^{\leq n}\mathcal{Y}$ sends étale equivalences to étale equivalences.

Thus, from Lemma 2.5.5 we obtain:

COROLLARY 2.5.7. *For* $\mathcal{Y} \in \mathrm{PreStk}$ *we have:*

$$ {}^{\leq n}L({}^{\leq n}\mathcal{Y}) \simeq {}^{\leq n}(L(\mathcal{Y})). $$

2.5.8. *Right Kan extensions from* ${}^{<\infty}\mathrm{Sch}^{\mathrm{aff}}$. Let $\mathcal{Y}'$ be a functor

$$({}^{<\infty}\mathrm{Sch}^{\mathrm{aff}})^{\mathrm{op}} \to \mathrm{Spc},$$

which we can think of as a compatible family of objects $\mathcal{Y}'_n \in {}^{\leq n}\mathrm{PreStk}$. Let

$$\mathcal{Y} := \mathrm{RKE}_{{}^{<\infty}\mathrm{Sch}^{\mathrm{aff}} \hookrightarrow \mathrm{Sch}^{\mathrm{aff}}}(\mathcal{Y}') \in \mathrm{PreStk}.$$

LEMMA 2.5.9. *Assume that for all* n, $\mathcal{Y}'_n \in {}^{\leq n}\mathrm{Stk}$. *Then* $\mathcal{Y}$ *belongs to* Stk.

PROOF. Follows from the description of the functor $\mathrm{RKE}_{{}^{<\infty}\mathrm{Sch}^{\mathrm{aff}} \hookrightarrow \mathrm{Sch}^{\mathrm{aff}}}$ given in the proof of Proposition 1.4.7.

□

From here we obtain:

COROLLARY 2.5.10. *Suppose that* $\mathcal{Y} \in \mathrm{PreStk}$ *belongs to* Stk. *Then so does* ${}^{\mathrm{conv}}\mathcal{Y}$.

2.6. The notion of n-coconnective stack.

2.6.1. Note that the functor

$$\mathrm{LKE}_{{}^{\leq n}\mathrm{Sch}^{\mathrm{aff}} \hookrightarrow \mathrm{Sch}^{\mathrm{aff}}} : {}^{\leq n}\mathrm{PreStk} \to \mathrm{PreStk}$$

does *not* send ${}^{\leq n}\mathrm{Stk}$ to Stk. Instead, the left adjoint to the restriction functor ${}^{\leq n}\mathrm{Stk} \leftarrow \mathrm{Stk}$ is given by the composition

$${}^{\leq n}\mathrm{Stk} \hookrightarrow {}^{\leq n}\mathrm{PreStk} \overset{\mathrm{LKE}}{\longrightarrow} \mathrm{PreStk} \overset{L}{\longrightarrow} \mathrm{Stk};$$

we denote this composite functor by ${}^{L}\mathrm{LKE}_{{}^{\leq n}\mathrm{Sch}^{\mathrm{aff}} \hookrightarrow \mathrm{Sch}^{\mathrm{aff}}}$.

2.6.2. The above left adjoint is easily seen to be fully faithful. Hence, we can identify ${}^{\leq n}\mathrm{Stk}$ with a full subcategory of Stk. We shall denote by ${}^{L}\tau^{\leq n} : \mathrm{Stk} \to \mathrm{Stk}$ the resulting colocalization functor

$$\mathcal{Y} \mapsto {}^{L}\mathrm{LKE}_{{}^{\leq n}\mathrm{Sch}^{\mathrm{aff}} \hookrightarrow \mathrm{Sch}^{\mathrm{aff}}}({}^{\leq n}\mathcal{Y}).$$

By definition, ${}^{L}\tau^{\leq n} \simeq L \circ \tau^{\leq n}$.

2.6.3. We shall call objects of Stk that belong to the essential image of

$${}^{L}\mathrm{LKE}_{{}^{\leq n}\mathrm{Sch}^{\mathrm{aff}} \hookrightarrow \mathrm{Sch}^{\mathrm{aff}}}$$

n-coconnective stacks. I.e., $\mathcal{Y} \in \mathrm{Stk}$ is n-coconnective as a stack if and only if the adjunction map

$${}^{L}\tau^{\leq n}(\mathcal{Y}) \to \mathcal{Y}$$

is an isomorphism.

I.e., the functor ${}^{L}\mathrm{LKE}_{{}^{\leq n}\mathrm{Sch}^{\mathrm{aff}} \hookrightarrow \mathrm{Sch}^{\mathrm{aff}}}$ identifies the category ${}^{\leq n}\mathrm{Stk}$ with the full subcategory of PreStk spanned by n-coconnective stacks.

We shall refer to objects of ${}^{\leq 0}\mathrm{Stk} =: {}^{\mathrm{cl}}\mathrm{Stk}$ as 'classical stacks', and also denote ${}^{L}\tau^{\leq 0} =: {}^{L}\tau^{\mathrm{cl}}$.

REMARK 2.6.4. We emphasize again that, as subcategories PreStk, it is *not* true that ${}^{\leq n}\mathrm{Stk}$ is contained in ${}^{\leq n}\mathrm{PreStk}$. That is to say, that a n-coconnective stack is not necessarily n-coconnective as a prestack.

Note, however, that we do have an inclusion

$$\mathrm{Stk} \cap {}^{\leq n}\mathrm{PreStk} \subset {}^{\leq n}\mathrm{Stk}$$

as subcategories of PreStk.

2.6.5. We shall say that a stack is eventually coconnective if it is n-coconnective for some n.

2.7. Descent and the 'locally of finite type' condition. In this subsection we will study how the descent condition interacts with the condition of being of finite type.

The entire discussion is applicable if we replace the étale topology by the ppf, or Zariski one.

However, the flat topology (without the finite type condition) would not do: we need finite typeness for the validity of Lemma 2.8.2.

2.7.1. Let n be a fixed integer. We can consider the étale topology on the category ${}^{\leq n}\mathrm{Sch}^{\mathrm{aff}}_{\mathrm{ft}}$. Thus, we obtain a localization of ${}^{\leq n}\mathrm{PreStk}_{\mathrm{lft}}$ that we denote ${}^{\leq n}\mathrm{NearStk}_{\mathrm{lft}}$.

We shall denote by ${}^{\leq n}L_{\mathrm{ft}}$ the corresponding localization functor

$$ {}^{\leq n}\mathrm{PreStk}_{\mathrm{lft}} \to {}^{\leq n}\mathrm{NearStk}_{\mathrm{lft}} \to {}^{\leq n}\mathrm{PreStk}_{\mathrm{lft}}. $$

As in Lemma 2.5.3, we have:

LEMMA 2.7.2. *The functor* ${}^{\leq n}L_{\mathrm{ft}} : {}^{\leq n}\mathrm{PreStk}_{\mathrm{lft}} \to {}^{\leq n}\mathrm{PreStk}_{\mathrm{lft}}$ *sends k-truncated objects to k-truncated ones.*

2.7.3. Consider the restriction functor for ${}^{\leq n}\mathrm{Sch}^{\mathrm{aff}}_{\mathrm{ft}} \hookrightarrow {}^{\leq n}\mathrm{Sch}^{\mathrm{aff}}$, i.e.,

$$ {}^{\leq n}\mathrm{PreStk}_{\mathrm{lft}} \leftarrow {}^{\leq n}\mathrm{PreStk}. $$

It is clear that it sends ${}^{\leq n}\mathrm{Stk}$ to ${}^{\leq n}\mathrm{NearStk}_{\mathrm{lft}}$. By adjunction, the functor of left Kan extension

$$ \mathrm{LKE}_{{}^{\leq n}\mathrm{Sch}^{\mathrm{aff}}_{\mathrm{ft}} \hookrightarrow {}^{\leq n}\mathrm{Sch}^{\mathrm{aff}}} : {}^{\leq n}\mathrm{PreStk}_{\mathrm{lft}} \to {}^{\leq n}\mathrm{PreStk} $$

sends étale equivalences to étale equivalences.

Moreover, we claim:

LEMMA 2.7.4. *The functor of* right *Kan extension*

$$ \mathrm{RKE}_{{}^{\leq n}\mathrm{Sch}^{\mathrm{aff}}_{\mathrm{ft}} \hookrightarrow {}^{\leq n}\mathrm{Sch}^{\mathrm{aff}}} : {}^{\leq n}\mathrm{PreStk}_{\mathrm{lft}} \to {}^{\leq n}\mathrm{PreStk} $$

sends ${}^{\leq n}\mathrm{NearStk}_{\mathrm{lft}}$ *to* ${}^{\leq n}\mathrm{Stk}$.

PROOF. For $\mathcal{Y} \in {}^{\leq n}\mathrm{PreStk}_{\mathrm{lft}}$ the value of $\mathrm{RKE}_{{}^{\leq n}\mathrm{Sch}^{\mathrm{aff}}_{\mathrm{ft}} \hookrightarrow {}^{\leq n}\mathrm{Sch}^{\mathrm{aff}}}(\mathcal{Y})$ on $S \in {}^{\leq n}\mathrm{Sch}^{\mathrm{aff}}$ is given as

$$ \lim_{S_0 \to S} \mathcal{Y}(S_0), $$

where the limit is taken over the category opposite to $({}^{\leq n}\mathrm{Sch}^{\mathrm{aff}}_{\mathrm{ft}})_{/S}$.

Let $S' \to S$ be an étale cover. We need to show that the map from $\lim\limits_{S_0 \to S} \mathcal{Y}(S_0)$ to the totalization of the cosimplicial space whose n-simplices are given by

$$ \lim_{S_0^n \to (S'^n/S)} \mathcal{Y}(S_0^n), $$

is an isomorphism.

However, this follows from the fact that the functor

$$ (({}^{\leq n}\mathrm{Sch}^{\mathrm{aff}}_{\mathrm{ft}})_{/S})^{\mathrm{op}} \to (({}^{\leq n}\mathrm{Sch}^{\mathrm{aff}}_{\mathrm{ft}})_{/(S'^n/S)})^{\mathrm{op}}, \quad S_0 \mapsto S_0^n := S_0 \underset{S}{\times} (S'^n/S), $$

is cofinal. □

From Lemma 2.7.4 we obtain:

COROLLARY 2.7.5.

(a) *The restriction functor* ${}^{\leq n}\mathrm{PreStk}_{\mathrm{lft}} \leftarrow {}^{\leq n}\mathrm{PreStk}$ *sends étale equivalences to étale equivalences.*

(b) *For* $\mathcal{Y} \in {}^{\leq n}\mathrm{PreStk}$ *we have:*

$$ {}^{\leq n}L(\mathcal{Y})|_{{}^{\leq n}\mathrm{Sch}^{\mathrm{aff}}_{\mathrm{ft}}} \simeq {}^{\leq n}L_{\mathrm{ft}}(\mathcal{Y}|_{{}^{\leq n}\mathrm{Sch}^{\mathrm{aff}}_{\mathrm{ft}}}). $$

2.7.6. Let us return to the functor

$$ \mathrm{LKE}_{{}^{\leq n}\mathrm{Sch}^{\mathrm{aff}}_{\mathrm{ft}} \hookrightarrow {}^{\leq n}\mathrm{Sch}^{\mathrm{aff}}} : {}^{\leq n}\mathrm{PreStk}_{\mathrm{lft}} \to {}^{\leq n}\mathrm{PreStk}. $$

It is not clear, and probably not true, that this functor sends ${}^{\leq n}\mathrm{NearStk}_{\mathrm{lft}}$ to ${}^{\leq n}\mathrm{Stk}$. However, as we have learned from J. Lurie, there is the following partial result, proved below:

PROPOSITION 2.7.7. *Suppose that an object* $\mathcal{Y} \in {}^{\leq n}\mathrm{PreStk}_{\mathrm{lft}}$ *is k-truncated for some k (see Sect.* 1.8.5*), and that* $\mathcal{Y} \in {}^{\leq n}\mathrm{NearStk}_{\mathrm{lft}}$. *Then the object*

$$ \mathrm{LKE}_{{}^{\leq n}\mathrm{Sch}^{\mathrm{aff}}_{\mathrm{ft}} \hookrightarrow {}^{\leq n}\mathrm{Sch}^{\mathrm{aff}}}(\mathcal{Y}) $$

of ${}^{\leq n}\mathrm{PreStk}$ *belongs to* ${}^{\leq n}\mathrm{Stk}$.

2.7.8. In what follows we shall use the notation

$$ {}^{\leq n}\mathrm{Stk}_{\mathrm{lft}} := {}^{\leq n}\mathrm{Stk} \cap {}^{\leq n}\mathrm{PreStk}_{\mathrm{lft}}. $$

We shall refer to objects of the subcategory ${}^{\leq n}\mathrm{Stk}_{\mathrm{lft}}$ of ${}^{\leq n}\mathrm{Stk}$ as 'n-coconnective stacks locally of finite type$n$-coconnective stacks locally of finite type'.

We have the inclusion

$$ {}^{\leq n}\mathrm{Stk}_{\mathrm{lft}} \subset {}^{\leq n}\mathrm{NearStk}_{\mathrm{lft}}. $$

Thus, Proposition 2.7.7 says that the essential image of this inclusion contains all truncated objects.

2.7.9. As a corollary of Proposition 2.7.7 and Lemma 2.7.2, we obtain:

COROLLARY 2.7.10. *For* $\mathcal{Y} \in {}^{\leq n}\mathrm{PreStk}_{\mathrm{lft}}$, *which is truncated, the natural map*

$$ \mathrm{LKE}_{{}^{\leq n}\mathrm{Sch}^{\mathrm{aff}}_{\mathrm{ft}} \hookrightarrow {}^{\leq n}\mathrm{Sch}^{\mathrm{aff}}}\left({}^{\leq n}L_{\mathrm{ft}}(\mathcal{Y})\right) \to {}^{\leq n}L\left(\mathrm{LKE}_{{}^{\leq n}\mathrm{Sch}^{\mathrm{aff}}_{\mathrm{ft}} \hookrightarrow {}^{\leq n}\mathrm{Sch}^{\mathrm{aff}}}(\mathcal{Y})\right) $$

is an isomorphism.

2.8. Proof of Proposition 2.7.7.

2.8.1. The proof will use the following assertion:

Let $f : S_1 \to S_2$ be an étale morphism in ${}^{\leq n}\mathrm{Sch}^{\mathrm{aff}}$. Consider the category of Cartesian diagrams

$$\begin{array}{ccc} S_1 & \longrightarrow & S_1' \\ {\scriptstyle f}\downarrow & & \downarrow{\scriptstyle f'} \\ S_2 & \longrightarrow & S_2' \end{array}$$

with $S_2', S_1' \in {}^{\leq n}\mathrm{Sch}^{\mathrm{aff}}_{\mathrm{ft}}$, and f' is étale. Denote this category by f_{ft}. We have the natural forgetful functors

$$\{S_2 \to S_2', S_2' \in {}^{\leq n}\mathrm{Sch}^{\mathrm{aff}}_{\mathrm{ft}}\} \leftarrow f_{\mathrm{ft}} \to \{S_1 \to S_1', S_1' \in {}^{\leq n}\mathrm{Sch}^{\mathrm{aff}}_{\mathrm{ft}}\}. \tag{2.2}$$

LEMMA 2.8.2. *Both functors opposite to those in* (2.2) *are cofinal.*

PROOF. We first show that the functor opposite to

$$f_{\mathrm{ft}} \to \{S_1 \to S_1', S_1' \in {}^{\leq n}\mathrm{Sch}^{\mathrm{aff}}_{\mathrm{ft}}\}$$

is cofinal.

Both categories in question are filtered: the above categories (before passing to the opposite) admit fiber products. Hence, it is enough to show that for any $S_1 \to S_1'$ with $S_1' \in {}^{\leq n}\mathrm{Sch}^{\mathrm{aff}}_{\mathrm{ft}}$, there *exists* an object of f_{ft} such that the map $S_1 \to S_1'$ factors as $S_1 \to S_1' \to S_1'$. For $n = 0$ this a standard fact in classical algebraic geometry, and for general n, it follows by induction using deformation theory (specifically, Volume II, Chapter 1, Proposition 5.4.2(b)).

To prove the assertion concerning

$$f_{\mathrm{ft}} \to \{S_2 \to S_2', S_2' \in {}^{\leq n}\mathrm{Sch}^{\mathrm{aff}}_{\mathrm{ft}}\},$$

we note that the corresponding fact holds for $n = 0$, i.e., in classical algebraic geometry.

Consider the following diagram

$$\begin{array}{ccc} f_{\mathrm{ft}} & \longrightarrow & {}^{\mathrm{cl}}f_{\mathrm{ft}} \\ \downarrow & & \downarrow \\ \{S_2 \to S_2', S_2' \in {}^{\leq n}\mathrm{Sch}^{\mathrm{aff}}_{\mathrm{ft}}\} & \longrightarrow & \{{}^{\mathrm{cl}}S_2 \to S_{2,0}', S_{2,0}' \in {}^{\mathrm{cl}}\mathrm{Sch}^{\mathrm{aff}}_{\mathrm{ft}}\}. \end{array}$$

By Lemma 2.1.5, this is a pullback diagram. In addition, the bottom horizontal arrow is a Cartesian fibration. Hence, the cofinality of the functor opposite to the right vertical arrow implies the corresponding fact for the left vertical arrow.

□

REMARK 2.8.3. An assertion parallel to Lemma 2.8.2 remains valid if we replace the word 'étale' by 'ppf', but the proof is more involved.

2.8.4. Let $\mathcal{Y}'$ be an object of ${}^{\leq n}\mathrm{NearStk}_{\mathrm{lft}}$, and let $\mathcal{Y}$ be its left Kan extension to an object of ${}^{\leq n}\mathrm{PreStk}$. Let $f : S_1 \to S_2$ be an étale cover. To prove Proposition 2.7.7, we need to check that the map

$$\mathcal{Y}(S_2) \to \mathrm{Tot}(\mathcal{Y}(S_1^\bullet/S_2)) \tag{2.3}$$

is an isomorphism.

For $S \in {}^{\leq n}\mathrm{Sch}^{\mathrm{aff}}$, the value of $\mathcal{Y}$ on S is calculated as

$$\operatorname*{colim}_{S \to S'} \mathcal{Y}'(S'),$$

where the colimit is taken over the category opposite to $({}^{\leq n}\mathrm{Sch}^{\mathrm{aff}}_{\mathrm{ft}})_{S/}$. Recall that according to Theorem 1.5.3(b), the above category is *filtered.* This implies that if $\mathcal{Y}'$ is k-truncated, then so is $\mathcal{Y}$.

Hence, we can replace Tot in (2.3), which is a limit in Spc over the index category $\boldsymbol{\Delta}$, by the corresponding limit, denoted $\mathrm{Tot}^{\leq k}$, in the category $\mathrm{Spc}_{\leq k}$, over the index category $\boldsymbol{\Delta}^{\leq k}$ of finite ordered sets of cardinality $\leq k+1$.

2.8.5. We rewrite the left-hand side in (2.3) as

$$\operatorname*{colim}_{S_2 \to S_2', S_2' \in {}^{\leq n}\mathrm{Sch}^{\mathrm{aff}}_{\mathrm{ft}}} \mathcal{Y}'(S_2').$$

Applying Lemma 2.8.2 for the $\to$ functor, we rewrite the right-hand side in (2.3) as

$$\mathrm{Tot}^{\leq k}\left(\operatorname*{colim}_{(f_{\mathrm{ft}})^{\mathrm{op}}} \mathcal{Y}(S_1'^{\bullet}/S_2')\right).$$

The category $(f_{\mathrm{ft}})^{\mathrm{op}}$ is filtered, as it contains push-outs. Since $\mathrm{Tot}^{\leq k}$ is a *finite* limit, we can commute the limit and the colimit in the above expression, and therefore rewrite it as

$$\operatorname*{colim}_{(f_{\mathrm{ft}})^{\mathrm{op}}} \left(\mathrm{Tot}^{\leq k}(\mathcal{Y}(S_1'^{\bullet}/S_2'))\right).$$

By the descent condition for $\mathcal{Y}'$, the latter expression is isomorphic to $\operatorname*{colim}_{(f_{\mathrm{ft}})^{\mathrm{op}}} \mathcal{Y}(S_2')$. Applying Lemma 2.8.2 for the $\leftarrow$ functor, we obtain that

$$\operatorname*{colim}_{(f_{\mathrm{ft}})^{\mathrm{op}}} \mathcal{Y}(S_2') \simeq \operatorname*{colim}_{S_2 \to S_2', S_2' \in {}^{\leq n}\mathrm{Sch}^{\mathrm{aff}}_{\mathrm{ft}}} \mathcal{Y}'(S_2'),$$

as required.

□

2.9. Stacks locally almost of finite type.

2.9.1. Recall the full subcategory $\mathrm{PreStk}_{\mathrm{laft}} \subset \mathrm{PreStk}$. In this subsection we will perceive it as the category

$$\mathrm{Funct}\left(({}^{<\infty}\mathrm{Sch}^{\mathrm{aff}}_{\mathrm{ft}})^{\mathrm{op}}, \mathrm{Spc}\right),$$

see Proposition 1.7.6.

2.9.2. Consider the étale topology on the category ${}^{<\infty}\mathrm{Sch}^{\mathrm{aff}}_{\mathrm{ft}}$. Thus, we obtain a localization of $\mathrm{PreStk}_{\mathrm{laft}}$ that we denote $\mathrm{NearStk}_{\mathrm{laft}}$.

Let us denote by L_{laft} the corresponding localization functor

$$\mathrm{PreStk}_{\mathrm{laft}} \to \mathrm{NearStk}_{\mathrm{laft}} \to \mathrm{PreStk}_{\mathrm{laft}}.$$

2.9.3. Consider the functor

$$\text{PreStk} \to \text{PreStk}_{\text{laft}}$$

given by restriction along

$$^{<\infty}\text{Sch}^{\text{aff}}_{\text{ft}} \hookrightarrow {}^{<\infty}\text{Sch}^{\text{aff}} \hookrightarrow \text{Sch}^{\text{aff}}.$$

It is clear that this functor sends Stk to $\text{NearStk}_{\text{laft}}$. Moreover, as in Corollary 2.7.5 and Corollary 2.5.7, we obtain:

LEMMA 2.9.4. *For* $\mathcal{Y} \in \text{PreStk}$ *we have:*

$$L(\mathcal{Y})|_{^{<\infty}\text{Sch}^{\text{aff}}_{\text{ft}}} \simeq L_{\text{laft}}(\mathcal{Y}|_{^{<\infty}\text{Sch}^{\text{aff}}_{\text{ft}}}).$$

From Proposition 2.7.7 and Lemma 2.5.9 we obtain:

COROLLARY 2.9.5. *Let* $\mathcal{Y}$ *be an object of* $\text{NearStk}_{\text{laft}}$*, thought of as an object of* PreStk *via*

$$\text{NearStk}_{\text{laft}} \subset \text{PreStk}_{\text{laft}} \subset \text{PreStk}$$

(see Proposition 1.7.6*). Suppose that for each* n*, the restriction* $^{\leq n}\mathcal{Y}$ *of* $\mathcal{Y}$ *to* $^{\leq n}\text{Sch}^{\text{aff}}_{\text{ft}}$ *is* k_n*-truncated for some* $k_n \in \mathbb{N}$*. Then* $\mathcal{Y} \in \text{Stk}$.

2.9.6. In what follows, we will denote the intersection

$$\text{Stk} \cap \text{PreStk}_{\text{laft}}$$

by Stk_{laft}. We shall refer to objects of the subcategory $\text{Stk}_{\text{laft}} \subset \text{Stk}$ as 'stacks locally almost of finite type'.

We have an evident inclusion

$$\text{Stk}_{\text{laft}} \subset \text{NearStk}_{\text{laft}}.$$

Corollary 2.9.5 says that the essential image of Stk_{laft} in $\text{NearStk}_{\text{laft}}$ contains all objects $\mathcal{Y}$, such that for every n, the restriction $^{\leq n}\mathcal{Y}$ of $\mathcal{Y}$ to $^{\leq n}\text{Sch}^{\text{aff}}_{\text{ft}}$ is truncated.

3. (Derived) schemes

In this section we introduce the basic object of study in derived algebraic geometry–the notion of (derived) scheme[4].

We investigate some basic properties of schemes: what it means to be n-coconnective and locally (almost) of finite type.

3.1. The definition of (derived) schemes. Our approach to the definition of (derived) schemes (or more general algebro-geometric objects) is that they are prestacks that have some specific properties. I.e., we never need to introduce additional pieces of structure.

In the case of (derived) schemes, the relevant properties are descent and the existence of a *Zariski atlas.*

[4]In the main body of the text we drop the adjective 'derived': everything is derived unless specified otherwise.

3.1.1. Recall the notion of an affine open embedding, see Sect. 2.1.9.

Following [**TV2**, Sect. 2.2], we say that an object $Z \in \text{PreStk}$ is a *scheme* if:

(1) Z satisfies étale descent;
(2) The diagonal map $Z \to Z \times Z$ is affine schematic, and for every $T \in (\text{Sch}^{\text{aff}})_{/Z\times Z}$, the induced map of classical schemes ${}^{\text{cl}}(T \underset{Z\times Z}{\times} Z) \to {}^{\text{cl}}T$ is a closed embedding;
(3) There exists a collection of affine schemes S_i and maps $f_i : S_i \to Z$ (called a *Zariski atlas*), such that:
   - Each f_i (which is affine schematic by the previous point) is an open embedding;
   - For every $T \in (\text{Sch}^{\text{aff}})_{/Z}$, the images of the maps ${}^{\text{cl}}(T \underset{Z}{\times} S_i) \to {}^{\text{cl}}T$ cover ${}^{\text{cl}}T$.

We shall denote the full subcategory of Stk spanned by schemes by Sch.

REMARK 3.1.2. One can show that the étale descent condition can be replaced by a weaker one: namely, it is sufficient to require that Z satisfy Zariski descent. In addition, it is not difficult to see that schemes as defined above actually satisfy flat descent.

REMARK 3.1.3. Our definition gives what is usually called a *separated* scheme. The non-separated case will be covered under the rubric of Artin stacks, discussed in the next section.

3.1.4. We shall say that a scheme Z is quasi-compact if the classical scheme ${}^{\text{cl}}Z$ is. Equivalently, this means that Z admits a Zariski cover by a finite collection of affine schemes.

3.1.5. It follows from the definition that if $(S_i \overset{f_i}{\to} Z)$ is a Zariski atlas, then the map

$$\underset{i}{\sqcup} S_i \to Z$$

is an étale (and, in fact, Zariski) surjection.

Hence, from Lemma 2.3.8, we obtain:

LEMMA 3.1.6. *Let Z be a scheme. For a given Zariski atlas $\underset{i}{\sqcup} S_i \to Z$, we have $Z \simeq L(|(\underset{i}{\sqcup} S_i)^\bullet/Z|_{\text{PreStk}})$.*

3.1.7. The following results from Lemma 2.1.5:

COROLLARY 3.1.8.

(a) *Given a Zariski morphism of affine schemes $S' \to S$, for $T \to S$, the datum of its lift to a map $T \to S'$ is equivalent to the datum of a lift of ${}^{\text{cl}}T \to {}^{\text{cl}}S$ to a map ${}^{\text{cl}}T \to {}^{\text{cl}}S'$.*

(b) *Let $Z' \to Z$ be an affine Zariski map, where $Z', Z \in \text{Sch}$. Then for $T \to Z$ with $T \in \text{Sch}^{\text{aff}}$, the datum of a lift of f to a map $f' : T \to Z'$ is equivalent to the datum of a lift of ${}^{\text{cl}}f : {}^{\text{cl}}T \to {}^{\text{cl}}Z$ to a map ${}^{\text{cl}}f' : {}^{\text{cl}}T \to {}^{\text{cl}}Z'$.*

REMARK 3.1.9. Both points in Corollary 3.1.8 remain valid if we replace the word 'Zariski' by 'étale'.

3.2. Construction of schemes. In this subsection we will prove an assertion that provides a converse to Lemma 3.1.6.

3.2.1. First, we claim:

PROPOSITION 3.2.2. *Let* Z *be an object of* Stk, *equipped with a collection of affine open embeddings* $S_i \to Z$, *where* $S_i \in \mathrm{Sch}^{\mathrm{aff}}$. *Suppose that* ${}^{\mathrm{cl}}Z$ *is a classical scheme*[5] *and* $\underset{i}{\sqcup}\, {}^{\mathrm{cl}}S_i \to {}^{\mathrm{cl}}Z$ *is its Zariski atlas. Then:*

(a) *Z is a scheme;*

(b) *The maps* $\underset{i \in I}{\sqcup} S_i \to Z$ *form a Zariski atlas of* Z.

PROOF. We only have to show that the diagonal map $Z \to Z \times Z$ is affine schematic. This is equivalent to showing that for any $T, U \in (\mathrm{Sch}^{\mathrm{aff}})_{/Z}$, the fiber product $T \underset{Z}{\times} U$ is an affine scheme.

Consider the fiber products $S_i \underset{Z}{\times} T$. By assumption, these are affine schemes, and the map

$$\underset{i}{\sqcup} S_i \underset{Z}{\times} T \to T$$

is a Zariski covering. Therefore, by Proposition 2.4.2(b), it suffices to show that the fiber products

$$S_i \underset{Z}{\times} T \underset{Z}{\times} U$$

are affine schemes. However,

$$S_i \underset{Z}{\times} T \underset{Z}{\times} U \simeq (S_i \underset{Z}{\times} T) \underset{S_i}{\times} (S_i \underset{Z}{\times} U).$$

□

3.2.3. Let $S^\bullet$ be a groupoid-object of PreStk (see [**Lu1**, Sect. 6.1.2] for what this means).

Denote

$$Z := L(|S^\bullet|).$$

We claim:

PROPOSITION 3.2.4. *Assume that* S^0 *and* S^1 *are of the form*

$$S^0 = \underset{i \in I}{\sqcup} S_i^0 \text{ and } S^1 = \underset{j \in J}{\sqcup} S_j^1,$$

where S_i^0 *and* S_i^1 *are affine schemes, and the maps* $S^0 \rightrightarrows S^1$ *are comprised of open embeddings* $S_i^0 \to S_j^1$. *Assume, moreover, that* ${}^{\mathrm{cl}}Z$ *is a classical scheme and that* $\underset{i}{\sqcup}\, {}^{\mathrm{cl}}S_i^0 \to {}^{\mathrm{cl}}Z$ *is its Zariski atlas. Then:*

(a) *Z is a scheme;*

(b) *The maps* $\underset{i \in I}{\sqcup} S_i^0 \to Z$ *form a Zariski atlas of* Z.

[5]Following our conventions, when talking about classical schemes, we impose the hypothesis that they be separated.

PROOF. By Proposition 3.2.2, it is enough to show that each of the maps $S_i^0 \to Z$ is an affine open embedding. By Corollary 2.4.4, it suffices to show that the each of the maps

$$S_i^0 \to |S^\bullet|$$

is an affine open embedding.

Fix a map $T \to |S^\bullet|$. By definition, such a map factors as $T \to S^0 \to |S^\bullet|$. Hence, we have

$$T \underset{|S^\bullet|}{\times} S_i^0 \simeq T \underset{S^0}{\times} S^0 \underset{|S^\bullet|}{\times} S_i^0.$$

Thus, it suffices to show that each of the maps $S^0 \underset{|S^\bullet|}{\times} S_i^0 \to S^0$ is an affine open embedding.

We have

$$S^0 \underset{|S^\bullet|}{\times} S_i^0 \simeq (S^0 \underset{|S^\bullet|}{\times} S^0) \underset{S^0}{\times} S_i^0.$$

Now,

$$S^0 \underset{|S^\bullet|}{\times} S^0 \simeq S^1,$$

and the assertion follows from the assumption on the map $S^1 \to S^0$. □

3.2.5. Combining Proposition 3.2.4 with Lemma 2.1.5, we obtain:

COROLLARY 3.2.6. *Let Z be a scheme. Then the operation of passage to the underlying classical subscheme defines an equivalence between the full subcategory* $\mathrm{Sch}_{/Z}$ *spanned by* $Z' \xrightarrow{f} Z$ *with f affine Zariski and the full subcategory of* ${}^{\mathrm{cl}}\mathrm{Sch}_{/{}^{\mathrm{cl}}Z}$ *spanned by* $\widetilde{Z}' \xrightarrow{\widetilde{f}} {}^{\mathrm{cl}}Z$ *with $\widetilde{f}$ affine Zariski. Furthermore, f is an open embedding if and only if $\widetilde{f}$ is.*

Further, combining with Proposition 2.4.2(b), we obtain:

COROLLARY 3.2.7. *In the circumstances of Corollary 3.2.6, the scheme Z' is affine if and only if the classical scheme ${}^{\mathrm{cl}}Z'$ is affine.*

And finally:

COROLLARY 3.2.8. *A scheme Z is affine if and only if the classical scheme ${}^{\mathrm{cl}}Z$ is affine.*

3.3. Schemes and n-coconnectivity. In this subsection we study the question of how the notion of scheme interacts with the notion of n-coconnective stack.

3.3.1. Replacing the category PreStk by ${}^{\leq n}\mathrm{PreStk}$ in the definition of the notion of scheme we obtain a category that we denote by ${}^{\leq n}\mathrm{Sch}$.

For $n = 0$ we recover the category of classical (separated) schemes.

3.3.2. We claim:

PROPOSITION 3.3.3. *Any object of* ${}^{\leq n}\mathrm{Sch}$ *is* n*-truncated as an object of* PreStk.

PROOF. Let Z be an object of ${}^{\leq n}\mathrm{Sch}$ and let us be given a map $f_0 : {}^{\mathrm{cl}}T \to Z$, where $T \in {}^{\leq n}\mathrm{Sch}^{\mathrm{aff}}$. We will show that the space of maps $T \to Z$ that restrict to f_0 is n-truncated.

Fix a Zariski atlas $\underset{i}{\sqcup} S_i \to Z$. Consider the induced Zariski cover ${}^{\mathrm{cl}}T \underset{Z}{\times} S_i$ of ${}^{\mathrm{cl}}T$. Since ${}^{\mathrm{cl}}T$ is quasi-compact, we can replace the initial index set by its finite subset, denoted I, so that

$$\underset{i\in I}{\sqcup} {}^{\mathrm{cl}}T \underset{Z}{\times} S_i \to {}^{\mathrm{cl}}T$$

is still a cover.

By Lemma 2.1.5, there exists a canonically defined Zariski cover $\underset{i\in I}{\sqcup} T_i = T' \to T$ such that

$$\underset{i\in I}{\sqcup} {}^{\mathrm{cl}}T \underset{Z}{\times} S_i = {}^{\mathrm{cl}}T'.$$

Now, the datum of a map $f : T \to Z$ that restricts to f_0 is equivalent to the datum of a point of

$$\mathrm{Tot}(\mathrm{Maps}(T'^{\bullet}/T, Z) \underset{\mathrm{Maps}({}^{\mathrm{cl}}T'^{\bullet}/{}^{\mathrm{cl}}T, Z)}{\times} \{f_0\}|_{{}^{\mathrm{cl}}T'^{\bullet}/{}^{\mathrm{cl}}T}).$$

We now claim that the above cosimplicial space is n-truncated simplex-wise.

Indeed, by Corollary 3.1.8(b), for every $m \geq 0$, the corresponding space of m-simplices is the product over the set of $(m+1)$-tuples $(i_0, ..., i_m)$ of elements of I of

$$\mathrm{Maps}(T_{i_0} \underset{T}{\times} ... \underset{T}{\times} T_{i_m}, S_{i_0}) \underset{\mathrm{Maps}({}^{\mathrm{cl}}T_{i_0} \underset{{}^{\mathrm{cl}}T}{\times} ... \underset{{}^{\mathrm{cl}}T}{\times} {}^{\mathrm{cl}}T_{i_m}, S_{i_0})}{\times} \{f_0\}|_{{}^{\mathrm{cl}}T_{i_0} \underset{{}^{\mathrm{cl}}T}{\times} ... \underset{{}^{\mathrm{cl}}T}{\times} {}^{\mathrm{cl}}T_{i_m}}.$$

Now, the assertion follows from the fact that mapping spaces in ${}^{\leq n}\mathrm{Sch}^{\mathrm{aff}}$ are n-truncated, by Sect. 1.8.6.

□

3.3.4. It is easy to see that the restriction functor for ${}^{\leq n}\mathrm{Sch}^{\mathrm{aff}} \hookrightarrow \mathrm{Sch}^{\mathrm{aff}}$ sends Sch to ${}^{\leq n}\mathrm{Sch}$ (replace the original Zariski cover S_i by ${}^{\leq n}S_i$).

We claim:

PROPOSITION 3.3.5.

(a) *The functor*

$${}^{L}\mathrm{LKE}_{{}^{\leq n}\mathrm{Sch}^{\mathrm{aff}} \hookrightarrow \mathrm{Sch}^{\mathrm{aff}}} : {}^{\leq n}\mathrm{Stk} \hookrightarrow \mathrm{Stk}$$

sends ${}^{\leq n}\mathrm{Sch}$ *to* Sch.

(b) *If* Z *is an object of* ${}^{\leq n}\mathrm{Sch}$ *with a Zariski atlas* $\underset{i}{\sqcup} S_i \to Z$*, then*

$$\underset{i}{\sqcup} S_i \to {}^{L}\mathrm{LKE}_{{}^{\leq n}\mathrm{Sch}^{\mathrm{aff}} \hookrightarrow \mathrm{Sch}^{\mathrm{aff}}}(Z)$$

is a Zariski atlas.

PROOF. Follows from Proposition 3.2.4.

□

3.3.6. We shall call a scheme 'n-coconnective' if it is n-coconnective as an object of Stk.

We obtain that the functor ${}^L\mathrm{LKE}_{{}^{\leq n}\mathrm{Sch}^{\mathrm{aff}} \hookrightarrow \mathrm{Sch}^{\mathrm{aff}}}$ identifies the category ${}^{\leq n}\mathrm{Sch}$ with that of n-coconnective schemes.

We emphasize that an n-coconnective scheme is *not* necessarily n-coconnective as a prestack, but it is n-coconnective as a stack.

3.3.7. We have the following characterization of n-coconnective schemes:

PROPOSITION 3.3.8. *For $Z \in \mathrm{Sch}$ the following conditions are equivalent:*

(i) *Z is n-coconnective.*

(ii) *For every $Z' \in \mathrm{Sch}$ equipped with an affine open embedding $Z' \to Z$, we have $Z \in {}^{\leq n}\mathrm{Sch}$.*

(iii) *Z admits a Zariski atlas by affine schemes belonging to ${}^{\leq n}\mathrm{Sch}^{\mathrm{aff}}$.*

PROOF. The implication (i) $\Rightarrow$ (iii) is Proposition 3.3.5(b). The implication (ii) $\Rightarrow$ (iii) is tautological. We will now show that (iii) implies both (i) and (ii).

Assume first that Z admits a Zariski atlas consisting of affine schemes in ${}^{\leq n}\mathrm{Sch}^{\mathrm{aff}}$. Then we can write Z as

$$L(\underset{a \in A}{\mathrm{colim}}\, S_a), \tag{3.1}$$

for some diagram of objects $S_a \in {}^{\leq n}\mathrm{Sch}^{\mathrm{aff}}$, see Lemma 3.1.6. Concretely, the colimit in question is the geometric realization of the Čech nerve of the given atlas.

In particular, $\underset{a \in A}{\mathrm{colim}}\, S_a \in {}^{\leq n}\mathrm{PreStk}$. And hence, $Z \in {}^{\leq n}\mathrm{Stk}$.

For any affine open embedding $Z' \to Z$, the pullback of this atlas gives a Zariski atlas for Z' with a similar property. This implies that in this case Z' also belongs to ${}^{\leq n}\mathrm{Sch}$.

□

3.4. Schemes and convergence.

3.4.1. We claim:

PROPOSITION 3.4.2. *A scheme, regarded as an object of* PreStk, *is convergent.*

PROOF. Let Z be a scheme and let us be given a map $f_0 : {}^{cl}T \to Z$, where $T \in \mathrm{Sch}^{\mathrm{aff}}$. We will show that the datum of a lift of f_0 to a map $f : T \to Z$ is equivalent to the datum of a compatible family of lifts $f_n : {}^{\leq n}T \to Z$.

Let $\underset{i}{\sqcup} S_i \to Z$ and $\underset{i \in I}{\sqcup} T_i = T' \to T$ be as in the proof of Proposition 3.3.3.

As in *loc.cit.*, the datum of a map $f : T \to Z$ that restricts to f_0 is equivalent to the datum of a point of

$$\mathrm{Tot}(\mathrm{Maps}(T'^{\bullet}/T, Z) \underset{\mathrm{Maps}({}^{cl}T'^{\bullet}/{}^{cl}T, Z)}{\times} \{f_0\}|_{{}^{cl}T'^{\bullet}/{}^{cl}T}).$$

The datum of a compatible family of maps f_n is equivalent to the datum of a point of

$$\mathrm{Tot}\left(\lim_n \mathrm{Maps}({}^{\leq n}T'^{\bullet}/{}^{\leq n}T, Z) \underset{\mathrm{Maps}({}^{cl}T'^{\bullet}/{}^{cl}T, Z)}{\times} \{f_0\}|_{{}^{cl}T'^{\bullet}/{}^{cl}T}\right).$$

Now, we claim that the restriction map

$$\text{(3.2)}\quad \mathrm{Maps}(T'^\bullet/T, Z) \underset{\mathrm{Maps}({}^{\mathrm{cl}}T'^\bullet/{}^{\mathrm{cl}}T, Z)}{\times} \{f_0\}|_{{}^{\mathrm{cl}}T'^\bullet/{}^{\mathrm{cl}}T} \to$$
$$\to \lim_n \mathrm{Maps}({}^{\leq n}T'^\bullet/{}^{\leq n}T, Z) \underset{\mathrm{Maps}({}^{\mathrm{cl}}T'^\bullet/{}^{\mathrm{cl}}T, Z)}{\times} \{f_0\}|_{{}^{\mathrm{cl}}T'^\bullet/{}^{\mathrm{cl}}T}$$

is an isomorphism simplex-wise.

Indeed, by Corollary 3.1.8(b), for every $m \geq 0$, the spaces of m-simplices in the two sides in (3.2) are products over the set of $(m+1)$-tuples $(i_0, ..., i_m)$ of elements of I of

$$\mathrm{Maps}(T_{i_0} \underset{T}{\times} ... \underset{T}{\times} T_{i_m}, S_{i_0}) \underset{\mathrm{Maps}({}^{\mathrm{cl}}T_{i_0} \underset{{}^{\mathrm{cl}}T}{\times} ... \underset{{}^{\mathrm{cl}}T}{\times} {}^{\mathrm{cl}}T_{i_m}, S_{i_0})}{\times} \{f_0\}|_{{}^{\mathrm{cl}}T_{i_0} \underset{{}^{\mathrm{cl}}T}{\times} ... \underset{{}^{\mathrm{cl}}T}{\times} {}^{\mathrm{cl}}T_{i_m}}$$

and

$$\lim_n \mathrm{Maps}({}^{\leq n}T_{i_0} \underset{{}^{\leq n}T}{\times} ... \underset{{}^{\leq n}T}{\times} {}^{\leq n}T_{i_m}, S_{i_0}) \underset{\mathrm{Maps}({}^{\mathrm{cl}}T_{i_0} \underset{{}^{\mathrm{cl}}T}{\times} ... \underset{{}^{\mathrm{cl}}T}{\times} {}^{\mathrm{cl}}T_{i_m}, S_{i_0})}{\times} \{f_0\}|_{{}^{\mathrm{cl}}T_{i_0} \underset{{}^{\mathrm{cl}}T}{\times} ... \underset{{}^{\mathrm{cl}}T}{\times} {}^{\mathrm{cl}}T_{i_m}},$$

respectively.

Now, the required isomorphism follows from the fact that each

$$\mathrm{Maps}(T_{i_0} \underset{T}{\times} ... \underset{T}{\times} T_{i_m}, S_{i_0}) \to \lim_n \mathrm{Maps}({}^{\leq n}T_{i_0} \underset{{}^{\leq n}T}{\times} ... \underset{{}^{\leq n}T}{\times} {}^{\leq n}T_{i_m}, S_{i_0})$$

is an isomorphism (the convergence of S_{i_0} as a prestack).

□

3.4.3. We have the following partial converse to Proposition 3.4.2:

PROPOSITION 3.4.4. *Let Z be an object of ${}^{\mathrm{conv}}\mathrm{PreStk}$, such that for every n, the corresponding object ${}^{\leq n}Z \in {}^{\leq n}\mathrm{PreStk}$ belongs to ${}^{\leq n}\mathrm{Sch}$. Then $Z \in \mathrm{Sch}$.*

PROOF. Let $\underset{i}{\sqcup} \widetilde{S}_i \to {}^{\mathrm{cl}}Z$ be a Zariski atlas of ${}^{\mathrm{cl}}Z$. By Corollary 3.1.8(b), for every i we have a compatible family of open embeddings

$$S_{i,n} \to {}^{\leq n}Z.$$

Set

$$S_i = \underset{n}{\mathrm{colim}}\, S_{i,n},$$

where the colimit is taken in $\mathrm{Sch}^{\mathrm{aff}}$. By construction, we have $S_{i,n} = {}^{\leq n}S_i$, and the convergence property of Z implies that we have a well-defined map $S_i \to Z$.

We claim now that Z is a scheme with $\underset{i}{\sqcup} S_i \to Z$ providing a Zariski atlas. Indeed, this follows from Proposition 3.2.2.

□

3.5. Schemes locally (almost) of finite type.

3.5.1. We shall denote by ${}^{\leq n}\mathrm{Sch}_{\mathrm{lft}}$ and $\mathrm{Sch}_{\mathrm{laft}}$ the full subcategories of Stk, given by

$$\mathrm{Sch} \cap {}^{\leq n}\mathrm{Stk}_{\mathrm{lft}} \text{ and } \mathrm{Sch} \cap \mathrm{Stk}_{\mathrm{laft}},$$

respectively.

We will denote by

$${}^{\leq n}\mathrm{Sch}_{\mathrm{ft}} \subset {}^{\leq n}\mathrm{Sch}_{\mathrm{lft}} \text{ and } \mathrm{Sch}_{\mathrm{aft}} \subset \mathrm{Sch}_{\mathrm{laft}}$$

the full subcategories corresponding to quasi-compact schemes.

3.5.2. We have:

PROPOSITION 3.5.3. *For $Z \in {}^{\leq n}\mathrm{Sch}$ (resp., $Z \in \mathrm{Sch}$) the following conditions are equivalent:*

(i) *$Z \in {}^{\leq n}\mathrm{Sch}_{\mathrm{lft}}$ (resp., $Z \in \mathrm{Sch}_{\mathrm{laft}}$);*

(ii) *For an affine open embedding $Z' \to Z$ with $Z' \in {}^{\leq n}\mathrm{Sch}$ (resp., $Z' \in \mathrm{Sch}$), we have $Z' \in {}^{\leq n}\mathrm{Sch}_{\mathrm{lft}}$ (resp., $Z' \in \mathrm{Sch}_{\mathrm{laft}}$);*

(iii) *Z admits a Zariski atlas consisting of affine schemes from ${}^{\leq n}\mathrm{Sch}^{\mathrm{aff}}_{\mathrm{ft}}$ (resp., $\mathrm{Sch}^{\mathrm{aff}}_{\mathrm{aft}}$).*

PROOF. Since schemes are convergent (see Proposition 3.4.2), it suffices to treat the case of $Z \in {}^{\leq n}\mathrm{Sch}$.

Assume first that Z admits a Zariski atlas consisting of affine schemes from ${}^{\leq n}\mathrm{Sch}^{\mathrm{aff}}_{\mathrm{ft}}$. Write

$$Z \simeq {}^{\leq n}L(\operatorname*{colim}_{a \in A} S_a),$$

where $S_a \in {}^{\leq n}\mathrm{Sch}^{\mathrm{aff}}_{\mathrm{ft}}$.

Using Corollary 2.7.10, we obtain that Z lies in the image of the functor $\mathrm{LKE}_{{}^{\leq n}\mathrm{Sch}^{\mathrm{aff}}_{\mathrm{ft}} \hookrightarrow {}^{\leq n}\mathrm{Sch}^{\mathrm{aff}}}$, i.e., it belongs to ${}^{\leq n}\mathrm{Stk}_{\mathrm{lft}}$.

Assume now that Z belongs to ${}^{\leq n}\mathrm{Stk}_{\mathrm{lft}}$. We will show that if we have an affine open embedding $Z' \to Z$, then $Z' \in {}^{\leq n}\mathrm{Stk}_{\mathrm{lft}}$.

Let T be an object of ${}^{\leq n}\mathrm{Sch}^{\mathrm{aff}}$. We need to show that the map

$$\operatorname*{colim}_{a} Z'(T_a) \to Z'(T) \tag{3.3}$$

is an isomorphism, where a runs over the (*filtered*) category $({}^{\leq n}\mathrm{Sch}^{\mathrm{aff}}_{\mathrm{ft}})_{T/}$.

The map $Z'(S) \to Z(S)$ is a monomorphism for any $S \in {}^{\leq n}\mathrm{Sch}^{\mathrm{aff}}$. Hence, since

$$\operatorname*{colim}_{a} Z(T_a) \to Z(T)$$

is an isomorphism, we obtain that (3.3) is a monomorphism, by filteredness.

Hence, it remains to show that any map $T \to Z'$ can be factored as

$$T \to T_a \to Z',$$

where $T_a \in {}^{\leq n}\mathrm{Sch}^{\mathrm{aff}}_{\mathrm{ft}}$.

Consider the composite morphism

$$T \to Z' \to Z,$$

and let $T \to T_b \to Z$ be its factorization with $T_b \in {}^{\leq n}\mathrm{Sch}^{\mathrm{aff}}_{\mathrm{ft}}$, which exists because Z is locally of finite type.

Now set $T_a := T_b \underset{Z}{\times} Z'$.

□

3.6. Properties of morphisms.

3.6.1. Let $f : \mathcal{Y}_1 \to \mathcal{Y}_2$ be a morphism in PreStk. We say that f is *schematic* if for any $S \in \mathrm{Sch}^{\mathrm{aff}}$ and $S \to \mathcal{Y}_2$, the Cartesian product

$$S \underset{\mathcal{Y}_2}{\times} \mathcal{Y}_1$$

is representable by an object of Sch.

The class of schematic maps is tautologically stable under base change. In addition, we claim that the composition of schematic maps is schematic. This is equivalent to the next assertion:

PROPOSITION 3.6.2. *Let Z be a scheme and let $Z' \to Z$ be a schematic map. Then Z' is also a scheme.*

PROOF. It is clear that Z' satisfies étale descent.

Let $\underset{i}{\sqcup} S_i \to Z$ be a Zariski atlas of Z. By assumption, each $S_i \underset{Z}{\times} Z'$ is a scheme. Let

$$\underset{j \in J_i}{\sqcup} T_j \to S_i \underset{Z}{\times} Z'$$

be its Zariski atlas. We claim that

$$\underset{i}{\sqcup} \big(\underset{j \in J_i}{\sqcup} T_j \big) \to \underset{i}{\sqcup} S_i \underset{Z}{\times} Z' \to Z'$$

provides a Zariski atlas for Z'.

Indeed, this is true at the classical level. Hence, by Proposition 3.2.2, it suffices to show that each of the maps

$$T_j \to S_i \underset{Z}{\times} Z' \to Z'$$

is an affine open embedding. However, this is evident, since $T_j \to S_i \underset{Z}{\times} Z'$ is such by construction, and $S_i \underset{Z}{\times} Z' \to Z'$ is such being a base change of an open embedding.

□

3.6.3. The next assertion follows from Proposition 3.4.4:

LEMMA 3.6.4. *Let $f : \mathcal{Y}_1 \to \mathcal{Y}_2$ be a map in ${}^{\mathrm{conv}}\mathrm{PreStk}$. To test the property of f of being schematic (resp., schematic flat/ppf/smooth/étale) it is enough to do so on affine schemes S belonging to ${}^{<\infty}\mathrm{Sch}^{\mathrm{aff}}$. If, moreover, $\mathcal{Y}_1, \mathcal{Y}_2 \in \mathrm{PreStk}_{\mathrm{laft}}$, then it is enough to take $S \in {}^{<\infty}\mathrm{Sch}^{\mathrm{aff}}_{\mathrm{ft}}$.*

3.6.5. Since the properties of a morphism in $\mathrm{Sch}^{\mathrm{aff}}$ of being flat/ppf/smooth/étale/Zariski are local in the Zariski topology of the source, they transfer to the corresponding notions for morphisms in Sch:

A morphism $Z' \to Z$ between schemes is flat/ppf/smooth/étale/Zariski if and only if for some (equivalently, any) Zariski atlas $\underset{i}{\sqcup} S'_i \to Z'$, each of the composite maps $S'_i \to Z$ (which is now a schematic affine map of prestacks) has the corresponding property.

Thus, by base change, we obtain the notion of a schematic flat/ppf/smooth/étale/Zariski morphism in PreStk.

3.6.6. The following is obtained by reduction to the affine case:

LEMMA 3.6.7. *Let $Z' \overset{f}{\to} Z' \overset{g}{\to} Z''$ be morphisms between schemes. Assume that f is surjective*[6] *and flat (resp., ppf, smooth, étale, Zariski). If $g \circ f$ is flat (resp., ppf, smooth, étale, Zariski), then so is g.*

4. (Derived) Artin stacks

In this section we introduce the notion of k-Artin stack, $k = 0, 1, \ldots$. As in the case of schemes, k-Artin stacks are prestacks with some particular properties (but no additional structure).

Our definition is a variation of the definition of *k-geometric stacks* ot *geometric k-stacks* in [**TV2**]. Although for an individual k, our definition will be different from both these notions from [**TV2**], the union over all k produces the same class of objects for all three classes of objects.

We also note that from the point of view of (our version of) the hierarchy of k-Artin stacks, schemes (which are, beyond doubt, a natural object of study) are a red herring: the category of schemes properly contains the category of 0-Artin stacks and is properly contained in the category of 1-Artin stacks. As a related phenomenon, we completely bypass the other important notion: that of algebraic space.

As in the previous sections, we will only be interested in only the most formal aspects of the theory: the notions of n-coconnectivity, finite typeness and convergence.

4.1. Setting up Artin stacks.

For $k \geq 0$, we will define a full subcategory of Stk spanned by objects that we refer to as k-Artin stacks.

In setting up Artin stacks the choice of étale topology is no longer arbitrary. It is made in order to make our system of definitions as simple as possible; see, however, Remark 4.1.4 below.

4.1.1. We start with $k = 0$. We shall say that an object $\mathcal{Y} \in \mathrm{Stk}$ is a 0-Artin stack if it is of the form $L(\underset{i}{\sqcup} S_i)$, where $S_i \in \mathrm{Sch}^{\mathrm{aff}}$. In particular,

$$\mathrm{Stk}^{0\text{-Artn}} \subset \mathrm{Sch}.$$

[6] Surjective=surjective at the level of underlying classical schemes.

4.1.2. To define the notion of k-Artin stack for $k \geq 1$ we proceed inductively.

Along with this notion, we will define what it means for a morphism in PreStk to be k-representable, and for a k-representable morphism what it means to be flat (resp., ppf, smooth, étale, surjective). These notions have an obvious meaning in the case of $k = 0$.

We will inductively assume the following properties:

- Any $(k-1)$-Artin stack is a k-Artin stack;
- Any morphism that is $(k-1)$-representable, is k-representable;
- A $(k-1)$-representable morphism is flat (resp., ppf, smooth, étale, surjective) if and only if it is such when viewed as a k-representable morphism;
- The class of k-representable (resp., k-representable + flat/ppf/smooth/étale/surjective) morphisms is stable under compositions and base change.

It will follow inductively from the construction that the class of k-Artin stacks is closed under fiber products.

4.1.3. Suppose the above notions have been defined for $k' < k$.

We say that $\mathcal{Y} \in \text{Stk}$ is a k-Artin stack if the following conditions hold:

(1) The diagonal map $\mathcal{Y} \to \mathcal{Y} \times \mathcal{Y}$ is $(k-1)$-representable.

(2) There exists $\mathcal{Z} \in \text{Stk}^{(k-1)\text{-Artn}}$ and a map $f : \mathcal{Z} \to \mathcal{Y}$ (which is a $(k-1)$-representable by the previous point), which is smooth and surjective.

We shall call the pair $f : \mathcal{Z} \to \mathcal{Y}$ a (smooth) atlas for $\mathcal{Y}$. Note that we can always choose an atlas with $\mathcal{Z} \in \text{Stk}^{0\text{-Artn}}$.

REMARK 4.1.4. Here we quote two fundamental results of Toën ([**To**, Theorem 2.1]). One says that Artin stacks as defined above actually satisfy ppf descent. Another says that if we require ppf descent, but instead of requiring a smooth atlas, we only require a ppf atlas, we still arrive at the same class of objects.

4.1.5. We will say that $\mathcal{Y} \in \text{Stk}$ is an Artin stack if it is a k-Artin stack for some k.

We let $\text{Stk}^{k\text{-Artn}}$ (resp., Stk^{Artn}) denote the full subcategory of Stk spanned by k-Artin (resp., Artin) stacks.

Note that in our definition, schemes are 1-Artin stacks:

$$\text{Sch} \subset \text{Stk}^{1\text{-Artn}}.$$

4.1.6. We say that a morphism $f : \mathcal{Y}_1 \to \mathcal{Y}_2$ in PreStk is k-representable if for every $S \to \mathcal{Y}_2$ with $S \in \text{Sch}^{\text{aff}}$ the fiber product $S \underset{\mathcal{Y}_2}{\times} \mathcal{Y}_1$ is a k-Artin stack in the above sense.

4.1.7. Let $\mathcal{Y}$ be a k-Artin stack mapping to an affine scheme S. We shall say that this map is flat (resp., ppf, smooth, étale, surjective) if for some atlas $\mathcal{Z} \to \mathcal{Y}$, the composite map of $\mathcal{Z} \to S$ (which is $(k-1)$-representable) is flat (resp., ppf, smooth, étale, surjective). Note that Lemma 3.6.7 implies by induction that if this condition holds for one atlas, then it holds for any other atlas.

4.1.8. We shall say that a k-representable morphism $f : \mathcal{Y}_1 \to \mathcal{Y}_2$ is flat (resp., ppf, smooth, étale, surjective) if for every $S \to \mathcal{Y}_2$ with $S \in \mathrm{Sch}^{\mathrm{aff}}$, the map

$$S \underset{\mathcal{Y}_2}{\times} \mathcal{Y}_1 \to S$$

is flat (resp., ppf, smooth, étale, surjective).

4.1.9. *Quasi-compactness and quasi-separatedness.* Let $\mathcal{Y}$ be a k-Artin stack. We say that $\mathcal{Y}$ is *quasi-compact* if there exists a smooth atlas $f : S \to \mathcal{Y}$ with $S \in \mathrm{Sch}^{\mathrm{aff}}$.

For a k-representable morphism $\mathcal{Y}_1 \to \mathcal{Y}_2$ in PreStk, we say that it is quasi-compact, if its base change by an affine scheme yields a quasi-compact k-Artin stack.

For $0 \le k' \le k$, we define the notion of k'-quasi-separatedness of a k-Artin stack or a k-representable morphism inductively on k'.

We say that a k-Artin stack $\mathcal{Y}$ is 0-quasi-separated0-quasi-separated if the diagonal map $\mathcal{Y} \to \mathcal{Y} \times \mathcal{Y}$ is quasi-compact, as a $(k-1)$-representable map. We say that a k-representable map is 0-quasi-separated if its base change by an affine scheme yields a 0-quasi-separated k-Artin stack.

For $k' > 0$, we say that a k-Artin stack $\mathcal{Y}$ is k'-quasi-separated if the diagonal map $\mathcal{Y} \to \mathcal{Y} \times \mathcal{Y}$ is $(k'-1)$-quasi-separated, as a $(k-1)$-representable map. We shall say that a k-representable map is k'-quasi-separated if its base change by an affine scheme yields a k'-quasi-separated k-Artin stack.

We shall say that a k-Artin stack is quasi-separated if it is k'-quasi-separated for all k', $0 \le k' \le k$. We shall say that a k-representable map is quasi-separated if its base change by an affine scheme yields a quasi-separated k-Artin stack.

4.2. Verification of the induction hypothesis.

4.2.1. Tautologically, the class of representable maps is stable under base change. Moreover, diagram chase shows:

LEMMA 4.2.2.

(a) *Let a morphism $f : \mathcal{Y}_1 \to \mathcal{Y}_2$ in* PreStk *be k-representable. Then the diagonal morphism $\mathcal{Y}_1 \to \mathcal{Y}_1 \underset{\mathcal{Y}_2}{\times} \mathcal{Y}_1$ is $(k-1)$-representable.*

(b) *Any map between k-Artin stacks is k-representable.*

4.2.3. We claim that the class of k-representable maps is stable under compositions. This is equivalent to the following assertion:

PROPOSITION 4.2.4. *Let $f : \mathcal{Y}' \to \mathcal{Y}$ be a k-representable map in* PreStk *where $\mathcal{Y}$ is a k-Artin stack. Then so is $\mathcal{Y}'$.*

PROOF. Consider the diagonal $\mathcal{Y}' \to \mathcal{Y}' \times \mathcal{Y}'$, and factor it as

$$\mathcal{Y}' \to \mathcal{Y}' \underset{\mathcal{Y}}{\times} \mathcal{Y}' \to \mathcal{Y}' \times \mathcal{Y}'.$$

Since f is k-representable we obtain that

$$\mathcal{Y}' \to \mathcal{Y}' \underset{\mathcal{Y}}{\times} \mathcal{Y}'$$

is $(k-1)$-representable (by Lemma 4.2.2(a)). Now, $\mathcal{Y}' \underset{\mathcal{Y}}{\times} \mathcal{Y}' \to \mathcal{Y}' \times \mathcal{Y}'$ is $(k-1)$-representable, being a base change of $\mathcal{Y} \to \mathcal{Y} \times \mathcal{Y}$.

We now need to construct a smooth atlas for $\mathcal{Y}'$. Let $\mathcal{Z} \to \mathcal{Y}$ be a smooth atlas for $\mathcal{Y}$ with $\mathcal{Z} \in \mathrm{Stk}^{0\text{-Artn}}$. By assumption, each $\mathcal{Z} \underset{\mathcal{Y}}{\times} \mathcal{Y}'$ is a k-Artin stack. Choose a smooth atlas $\mathcal{Z}' \to \mathcal{Z} \underset{\mathcal{Y}}{\times} \mathcal{Y}'$. We claim that the composite map

$$\mathcal{Z}' \to \mathcal{Z} \underset{\mathcal{Y}}{\times} \mathcal{Y}' \to \mathcal{Y}'$$

provides a smooth atlas for $\mathcal{Y}'$. Indeed, this map is smooth and surjective, being the composition of $\mathcal{Z}' \to \mathcal{Z}$ (which is smooth and surjective by assumption) and $\mathcal{Z} \underset{\mathcal{Y}}{\times} \mathcal{Y}' \to \mathcal{Y}'$ (which is smooth and surjective, being a base change of $\mathcal{Z} \to \mathcal{Y}$).

□

4.2.5. We claim that that the composition of representable flat/ppf/smooth/étale/surjective maps is itself a flat/ppf/smooth/étale/surjective map. This is equivalent to the following:

PROPOSITION 4.2.6. *Let $\mathcal{Y}' \to \mathcal{Y}$ be a k-representable flat (resp., ppf. smooth, étale, surjective) map, where $\mathcal{Y}$ is a k-Artin stack, equipped with a flat (resp., ppf, smooth, étale, surjective) map to $S \in \mathrm{Sch}^{\mathrm{aff}}$. Then the composite map $\mathcal{Y}' \to S$ is flat (resp., ppf. smooth, étale, surjective).*

PROOF. The required property tautologically holds for the atlas constructed in the proof of Proposition 4.2.4.

□

4.3. Descent properties.

4.3.1. The following results from the definitions:

LEMMA 4.3.2.

(a) *If $f : \mathcal{Z} \to \mathcal{Y}$ is an atlas of a k-Artin stack, then it is an étale surjection.*

(b) *If $\mathcal{Y}_1 \to \mathcal{Y}_2$ is a k-representable morphism, which is étale and surjective, then it is an étale surjection.*

COROLLARY 4.3.3. *Let $\mathcal{Y}$ be a k-Artin stack and let $f : \mathcal{Z} \to \mathcal{Y}$ be a smooth atlas. Then the natural map*

$$L(|\mathcal{Z}^\bullet/\mathcal{Y}|_{\mathrm{PreStk}}) \simeq |\mathcal{Z}^\bullet/\mathcal{Y}|_{\mathrm{Stk}} \to \mathcal{Y}$$

is an isomorphism, where the subscript Stk *(resp.,* PreStk*) indicates that the geometric realization is taken in* Stk *(resp.,* PreStk*).*

COROLLARY 4.3.4. *Let $\mathcal{Y}$ be a k-Artin stack. Then for any n, the restriction ${}^{\leq n}\mathcal{Y} \in {}^{\leq n}\mathrm{PreStk}$ is $(n+k)$-truncated.*

PROOF. We prove the assertion by induction. The assertion for $k = 0$ is a particular case of Proposition 3.3.3. Assume now that the assertion is valid for $k' < k$.

Note that the geometric realization of a m-truncated groupoid object in Spc is $(m+1)$-truncated. Combining this with Lemma 2.5.3, we obtain that it suffices to show that the simplicial prestack $\mathcal{Z}^\bullet/\mathcal{Y}$ has the property that for every n its restriction ${}^{\leq n}(\mathcal{Z}^\bullet/\mathcal{Y})$ is $(n+k-1)$-truncated.

However, each simplex of ${}^{\leq n}(\mathcal{Z}^\bullet/\mathcal{Y})$ belongs to $\mathrm{Stk}^{(k-1)\text{-Artn}}$, and the assertion follows from the induction hypothesis. □

4.3.5. We will now prove an (amplified) converse to Corollary 4.3.3. Let $\mathcal{Y}^\bullet$ be a groupoid-object of Stk (see [**Lu1**, Sect. 6.1.2] for what this means).

Set

$$\mathcal{Y} := |\mathcal{Y}^\bullet|_{\mathrm{Stk}} \simeq L(|\mathcal{Y}^\bullet|_{\mathrm{PreStk}})$$

be its geometric realization. We have

$$\mathcal{Y}^1 \simeq \mathcal{Y}^0 \underset{\mathcal{Y}}{\times} \mathcal{Y}^0 \tag{4.1}$$

(indeed, this tautologically holds before sheafification, and then use the fact that the functor L preserves fiber products).

We claim:

PROPOSITION 4.3.6.

(a) *Assume that in the above situation $\mathcal{Y}^1$ and $\mathcal{Y}^0$ are k-Artin stacks, the maps $\mathcal{Y}^1 \rightrightarrows \mathcal{Y}^0$ are smooth and the map $\mathcal{Y}^1 \to \mathcal{Y}^0 \times \mathcal{Y}^0$ is $(k-1)$-representable. Then $\mathcal{Y}$ is a k-Artin stack.*

(b) *Let*

$$\begin{array}{ccc} \mathcal{Y} & \longleftarrow & \mathcal{Y}' \\ \downarrow & & \downarrow \\ S & \xleftarrow{f} & S' \end{array}$$

be a Cartesian square in Stk *with $S, S' \in \mathrm{Sch}^{\mathrm{aff}}$ and the morphism f being smooth and surjective. Then if $\mathcal{Y}'$ is a k-Artin stack, the so is $\mathcal{Y}$.*

(c) *Suppose that a morphism $f : \mathcal{Y}_1 \to \mathcal{Y}_2$ in* PreStk *is k-representable (resp., k-representable and flat/ppf/smooth/étale/surjective). Then so is the morphism $L(f) : L(\mathcal{Y}_1) \to L(\mathcal{Y}_2)$.*

REMARK 4.3.7. By Remark 4.1.4, statement (b) of the above lemma can be strengthened: one can relax the condition that the morphism f be ppf instead of smooth. I.e., Artin stacks satisfy ppf descent, and not just smooth descent. Statement (a) can be strengthened accordingly, by requiring that the maps $\mathcal{Y}^1 \rightrightarrows \mathcal{Y}^0$ be ppf instead of smooth.

REMARK 4.3.8. The above proposition allows to construct the familiar examples of algebraic stacks. For example, if G is a smooth group-scheme acting on a scheme X, we consider $\mathcal{Y}^1 := G \times X$ as a groupoid acting on $Z^0 := X$, and the resulting 1-Artin stack $\mathcal{Y}$ is what we usually refer to as X/G.

PROOF OF PROPOSITION 4.3.6. We prove all three assertions by induction on k. The base case is $k = 1$, which we will establish together with the induction step. We note that statements (b) and (c) make sense for $k = 0$, and hold due to Proposition 2.4.2(b) and Corollary 2.4.4, respectively.

We begin by proving point (a).

Let us show that the diagonal morphism of $\mathcal{Y}$ is $(k-1)$-representable. By point (c) for $k-1$, it suffices to show that the map

$$|\mathcal{Y}^\bullet|_{\mathrm{PreStk}} \to |\mathcal{Y}^\bullet|_{\mathrm{PreStk}} \times |\mathcal{Y}^\bullet|_{\mathrm{PreStk}}$$

is $(k-1)$-representable. Fix a map $S \to |\mathcal{Y}^\bullet|_{\text{PreStk}} \times |\mathcal{Y}^\bullet|_{\text{PreStk}}$ with $S \in \text{Sch}^{\text{aff}}$. Such a map factors through a map $S \to \mathcal{Y}^0 \times \mathcal{Y}^0$. Hence,

$$S \underset{|\mathcal{Y}^\bullet|_{\text{PreStk}} \times |\mathcal{Y}^\bullet|_{\text{PreStk}}}{\times} |\mathcal{Y}^\bullet|_{\text{PreStk}} \simeq S \underset{\mathcal{Y}^0 \times \mathcal{Y}^0}{\times} (\mathcal{Y}^0 \times \mathcal{Y}^0) \underset{|\mathcal{Y}^\bullet|_{\text{PreStk}} \times |\mathcal{Y}^\bullet|_{\text{PreStk}}}{\times} |\mathcal{Y}^\bullet|_{\text{PreStk}} \simeq$$
$$\simeq S \underset{\mathcal{Y}^0 \times \mathcal{Y}^0}{\times} (\mathcal{Y}^0 \underset{|\mathcal{Y}^\bullet|_{\text{PreStk}}}{\times} \mathcal{Y}^0) \simeq S \underset{\mathcal{Y}^0 \times \mathcal{Y}^0}{\times} \mathcal{Y}^1.$$

A similar argument shows that the map $\mathcal{Y}^0 \to \mathcal{Y}$ is smooth and surjective. Hence, if $\mathcal{Z} \to \mathcal{Y}_0$ is a smooth atlas for $\mathcal{Z}^0$, then the composition $\mathcal{Z} \to \mathcal{Y}^0 \to \mathcal{Y}^1$ is a smooth atlas for $\mathcal{Y}$.

Let us now prove point (b).

Let $\mathcal{Y}^\bullet$ be the Čech nerve of the map $\mathcal{Y}' \to \mathcal{Y}$. In particular $\mathcal{Y}^0 = \mathcal{Y}'$ is a k-Artin stack. The maps $\mathcal{Y}^1 \rightrightarrows \mathcal{Y}^0$ are affine schematic and smooth, being base-changed from $S' \to S$. In particular, $\mathcal{Y}^1$ is also a k-Artin stack. The map $\mathcal{Y}^1 \to \mathcal{Y}^0 \times \mathcal{Y}^0$ is $(k-1)$-representable since the diagonal morphism of $\mathcal{Y}'$ is $(k-1)$-representable.

Since $\mathcal{Y}' \to \mathcal{Y}$ is an étale surjection, we have $\mathcal{Y} \simeq L(|\mathcal{Y}^\bullet|_{\text{PreStk}})$, by Lemmas 4.3.2(b) and 2.3.8. Applying point (a) we obtain that $\mathcal{Y}$ is a k-Artin stack, as desired.

Finally, let us prove point (c).

Let us be given a map $S \to L(\mathcal{Y}_2)$. We need to show that the fiber product $S \underset{L(\mathcal{Y}_2)}{\times} L(\mathcal{Y}_1)$ is a k-Artin stack (resp., a k-Artin stack, whose map to S is flat/ppf/smooth/étale/surjective).

Since $\mathcal{Y}_2 \to L(\mathcal{Y}_2)$ is an étale surjection, we can find an étale covering $S' \to S$ so that the composition $S' \to S \to L(\mathcal{Y}_2)$ factors as $S' \to \mathcal{Y}_2 \to L(\mathcal{Y}_2)$. Consider the Cartesian square

$$\begin{array}{ccc}
S \underset{L(\mathcal{Y}_2)}{\times} L(\mathcal{Y}_1) & \longleftarrow & S' \underset{L(\mathcal{Y}_2)}{\times} L(\mathcal{Y}_1) \\
\Big\downarrow & & \Big\downarrow \\
S & \longleftarrow & S'.
\end{array}$$

By point (b), it suffices to show that $S' \underset{L(\mathcal{Y}_2)}{\times} L(\mathcal{Y}_1)$ is a k-Artin stack (the properties of the map $S' \underset{L(\mathcal{Y}_2)}{\times} L(\mathcal{Y}_1) \to S'$ imply the corresponding properties of the map $S \underset{L(\mathcal{Y}_2)}{\times} L(\mathcal{Y}_1) \to S$ by Corollary 2.4.6.)

However, since the functor L commutes with fiber products, we have

$$S' \underset{L(\mathcal{Y}_2)}{\times} L(\mathcal{Y}_1) \simeq L(S' \underset{\mathcal{Y}_2}{\times} \mathcal{Y}_1),$$

where

$$L(S' \underset{\mathcal{Y}_2}{\times} \mathcal{Y}_1) \simeq S' \underset{\mathcal{Y}_2}{\times} \mathcal{Y}_1,$$

since $S' \underset{\mathcal{Y}_2}{\times} \mathcal{Y}_1$ is a k-Artin stack by assumption.

□

COROLLARY 4.3.9. *Let $\mathcal{Y}$ be an object of* Stk, *and let $f : \mathcal{Z} \to \mathcal{Y}$ be a $(k-1)$-representable, smooth and surjective morphism, where $\mathcal{Z}$ is a k-Artin stack. Then $\mathcal{Y}$ is a k-Artin stack.*

PROOF. Apply Proposition 4.3.6(a) to the Čech nerve of the map $\mathcal{Z} \to \mathcal{Y}$. □

4.4. Artin stacks and n-coconnectivity.

4.4.1. Replacing the category Sch by ${}^{\leq n}\mathrm{Sch}$ in the above discussion, we arrive to the definition of the category ${}^{\leq n}\mathrm{Stk}^{k\text{-Artn}}$.

It is clear that the restriction functor under ${}^{\leq n}\mathrm{Sch} \hookrightarrow \mathrm{Sch}$ sends $\mathrm{Stk}^{k\text{-Artn}}$ to ${}^{\leq n}\mathrm{Stk}^{k\text{-Artn}}$.

4.4.2. We claim:

PROPOSITION 4.4.3.

(a) *The functor*

$$^{L}\mathrm{LKE}_{{}^{\leq n}\mathrm{Sch}^{\mathrm{aff}} \hookrightarrow \mathrm{Sch}^{\mathrm{aff}}} : {}^{\leq n}\mathrm{Stk} \hookrightarrow \mathrm{Stk}$$

sends ${}^{\leq n}\mathrm{Stk}^{k\text{-Artn}}$ *to* $\mathrm{Stk}^{k\text{-Artn}}$.

(b) *If* $\mathcal{Z} \to \mathcal{Y}$ *is a smooth atlas for an object* $\mathcal{Y} \in {}^{\leq n}\mathrm{Stk}^{k\text{-Artn}}$, *then*

$$^{L}\mathrm{LKE}_{{}^{\leq n}\mathrm{Sch}^{\mathrm{aff}} \hookrightarrow \mathrm{Sch}^{\mathrm{aff}}}(\mathcal{Z}) \to {}^{L}\mathrm{LKE}_{{}^{\leq n}\mathrm{Sch}^{\mathrm{aff}} \hookrightarrow \mathrm{Sch}^{\mathrm{aff}}}(\mathcal{Y})$$

is a smooth atlas.

PROOF. We will prove the proposition by induction on k, assuming its validity for $k' < k$. We note that the assertion for a given k' implies the following:

(i) If $\mathcal{Y}_1 \to \mathcal{Y}_2$ is a k'-representable (resp., k'-representable and flat/smooth) map in ${}^{\leq n}\mathrm{PreStk}$, then the induced map in PreStk

$$^{L}\mathrm{LKE}_{{}^{\leq n}\mathrm{Sch}^{\mathrm{aff}} \hookrightarrow \mathrm{Sch}^{\mathrm{aff}}}(\mathcal{Y}_1) \to {}^{L}\mathrm{LKE}_{{}^{\leq n}\mathrm{Sch}^{\mathrm{aff}} \hookrightarrow \mathrm{Sch}^{\mathrm{aff}}}(\mathcal{Y}_2)$$

is also k'-representable (resp., k'-representable and flat/smooth).

(ii) If we have a Cartesian diagram in ${}^{\leq n}\mathrm{Stk}^{k'\text{-Artn}}$

$$\begin{array}{ccc} \mathcal{Y}_1' & \longrightarrow & \mathcal{Y}_1 \\ \downarrow & & \downarrow \\ \mathcal{Y}_2' & \longrightarrow & \mathcal{Y}_2 \end{array} \tag{4.2}$$

with the vertical arrows flat, then the diagram

$$\begin{array}{ccc} ^{L}\mathrm{LKE}_{{}^{\leq n}\mathrm{Sch}^{\mathrm{aff}} \hookrightarrow \mathrm{Sch}^{\mathrm{aff}}}(\mathcal{Y}_1') & \longrightarrow & ^{L}\mathrm{LKE}_{{}^{\leq n}\mathrm{Sch}^{\mathrm{aff}} \hookrightarrow \mathrm{Sch}^{\mathrm{aff}}}(\mathcal{Y}_1) \\ \downarrow & & \downarrow \\ ^{L}\mathrm{LKE}_{{}^{\leq n}\mathrm{Sch}^{\mathrm{aff}} \hookrightarrow \mathrm{Sch}^{\mathrm{aff}}}(\mathcal{Y}_2') & \longrightarrow & ^{L}\mathrm{LKE}_{{}^{\leq n}\mathrm{Sch}^{\mathrm{aff}} \hookrightarrow \mathrm{Sch}^{\mathrm{aff}}}(\mathcal{Y}_2) \end{array} \tag{4.3}$$

is Cartesian as well.

Let us now carry out the induction step.

Let $\mathcal{Y}$ be an object of ${}^{\leq n}\mathrm{Stk}^{k\text{-Artn}}$. By Corollary 4.3.3, for a given smooth atlas $\mathcal{Z} \to \mathcal{Y}$, we can write $\mathcal{Y}$ as $|\mathcal{Z}^\bullet|_{{}^{\leq n}\mathrm{Stk}}$, where $\mathcal{Z}^\bullet$ is the Čech nerve of $\mathcal{Z} \to \mathcal{Y}$, In particular, $\mathcal{Z}^\bullet$ is a groupoid object in ${}^{\leq n}\mathrm{Stk}^{(k-1)\text{-Artn}}$.

By (ii) above, the simplicial object of Stk given by

$$^{L}\mathrm{LKE}_{{}^{\leq n}\mathrm{Sch}^{\mathrm{aff}} \hookrightarrow \mathrm{Sch}^{\mathrm{aff}}}(\mathcal{Z}^\bullet)$$

is a groupoid object. Moreover, by (i) above, it satisfies the assumption of Proposition 4.3.6(a). Hence,

$$\mathcal{Y}' := |{}^L\text{LKE}_{{}^{\leq n}\text{Sch}^{\text{aff}} \hookrightarrow \text{Sch}^{\text{aff}}}(\mathcal{Z}^\bullet)|$$

is an object of $\text{Stk}^{k\text{-Artn}}$.

Furthermore, $\mathcal{Y}'$ is n-coconnective as a stack, whose restriction to ${}^{\leq n}\text{Sch}$ identifies with $\mathcal{Y}$. Therefore,

$$\mathcal{Y}' \simeq {}^L\text{LKE}_{{}^{\leq n}\text{Sch}^{\text{aff}} \hookrightarrow \text{Sch}^{\text{aff}}}(\mathcal{Y}).$$

□

4.4.4. We shall say that an object of $\text{Stk}^{k\text{-Artn}}$ is n-coconnective if it is n-coconnective as an object of Stk. From Proposition 4.4.3, we obtain:

COROLLARY 4.4.5. *The functor* ${}^L\text{LKE}_{{}^{\leq n}\text{Sch}^{\text{aff}} \hookrightarrow \text{Sch}^{\text{aff}}}$ *is an equivalence from* ${}^{\leq n}\text{Stk}^{k\text{-Artn}}$ *to the full subcategory of* $\text{Stk}^{k\text{-Artn}}$, *spanned by n-coconnective k-Artin stacks.*

Warning: We emphasize again that being n-coconnective as a stack does *not* imply being n-coconnective as a prestack.

4.4.6. We will now characterize those k-Artin stacks that are n-coconnective:

PROPOSITION 4.4.7. *Let $\mathcal{Y}$ be a k-Artin stack. The following conditions are equivalent:*

(i) *$\mathcal{Y}$ is n-coconnective.*

(ii) *There exists an atlas $f : Z \to \mathcal{Y}$, where $Z \in {}^{\leq n}\text{Stk}^{0\text{-Artn}}$.*

(iii) *If $\mathcal{Y}' \to \mathcal{Y}$ is a k-representable flat map, then $\mathcal{Y}'$ is n-coconnective as a stack.*

PROOF. We argue inductively on k, assuming the validity for $k' < k$.

The implication (i) ⇒ (ii) follows from Proposition 4.4.3(b).

Let us show that (ii) implies (i). By Corollary 4.3.3, it suffices to show that the Čech nerve of the atlas $Z \to \mathcal{Y}$ consists of $(k-1)$-Artin stacks that are n-coconnective. However, this follows from the implication (i) ⇒ (iii) for $k-1$.

The implication (iii) ⇒ (ii) is tautological: the assumption in (iii) implies that for any smooth atlas $Z \to \mathcal{Y}$, the scheme Z is n-coconnective.

Finally, the implication (i),(ii) ⇒ (iii) follows by retracing the construction of the atlas in the proof of Proposition 4.2.4.

□

4.4.8. *Artin stacks and convergence.* We will now prove:

PROPOSITION 4.4.9.

(a) *Any k-Artin stack, viewed as an object of* PreStk, *is convergent.*

(b) *Let* $\mathcal{Y} \in {}^{\mathrm{conv}}\mathrm{PreStk}$ *be such that for any n, we have* ${}^{\leq n}\mathcal{Y} \in {}^{\leq n}\mathrm{Stk}^{k\text{-Artn}}$. *Then* $\mathcal{Y}$ *is a k-Artin stack.*

PROOF. We proceed by induction on k. For $k = 0$, point (a) follows from Proposition 3.4.2, and point (b) follows by repeating the argument of Proposition 3.4.4.

We first prove point (a), assuming the validity of both (a) and (b) for $k' < k$.

Let $f : \mathcal{Z} \to \mathcal{Y}$ be a smooth atlas $\mathcal{Y}$. By Corollary 4.3.3, we have:

$$\mathcal{Y} \simeq |\mathcal{Z}^\bullet/\mathcal{Y}|_{\mathrm{Stk}}.$$

Consider the induced map ${}^{\mathrm{conv}}f : {}^{\mathrm{conv}}\mathcal{Z} \to {}^{\mathrm{conv}}\mathcal{Y}$. We claim that ${}^{\mathrm{conv}}f$ is $(k-1)$-representable, smooth and surjective. Indeed, for $S \to {}^{\mathrm{conv}}\mathcal{Y}$ with $S \in \mathrm{Sch}^{\mathrm{aff}}$, for every n, we have

$${}^{\leq n}(S \underset{{}^{\mathrm{conv}}\mathcal{Y}}{\times} {}^{\mathrm{conv}}\mathcal{Z}) \simeq {}^{\leq n}S \underset{{}^{\leq n}\mathcal{Y}}{\times} {}^{\leq n}\mathcal{Z} \in {}^{\leq n}\mathrm{Stk}^{(k-1)\text{-Artn}}.$$

Hence, $S \underset{{}^{\mathrm{conv}}\mathcal{Y}}{\times} {}^{\mathrm{conv}}\mathcal{Z}$ is a $(k-1)$-Artin stack by the induction hypothesis. Moreover, since each ${}^{\leq n}S \underset{{}^{\leq n}\mathcal{Y}}{\times} {}^{\leq n}\mathcal{Z}$ is smooth and surjective over ${}^{\leq n}S$, by Lemma 2.1.3, we obtain that $S \underset{{}^{\mathrm{conv}}\mathcal{Y}}{\times} {}^{\mathrm{conv}}\mathcal{Z}$ is smooth and surjective over S.

In particular, by Lemma 4.3.2(b), we obtain that ${}^{\mathrm{conv}}\mathcal{Z} \to {}^{\mathrm{conv}}\mathcal{Y}$ is an étale surjection, and hence

$${}^{\mathrm{conv}}\mathcal{Y} \simeq |{}^{\mathrm{conv}}\mathcal{Z}^\bullet/{}^{\mathrm{conv}}\mathcal{Y}|_{\mathrm{Stk}}.$$

However, we claim that the map of the cosimplicial objects

$$\mathcal{Z}^\bullet/\mathcal{Y} \to {}^{\mathrm{conv}}\mathcal{Z}^\bullet/{}^{\mathrm{conv}}\mathcal{Y}$$

is an isomorphism. Indeed, for every m, we have

$${}^{\mathrm{conv}}\mathcal{Z}^m/{}^{\mathrm{conv}}\mathcal{Y} \simeq {}^{\mathrm{conv}}(\mathcal{Z}^m/\mathcal{Y}),$$

where $\mathcal{Z}^m/\mathcal{Y}$ is a $(k-1)$-Artin stack, and hence $\mathcal{Z}^m/\mathcal{Y} \to {}^{\mathrm{conv}}(\mathcal{Z}^m/\mathcal{Y})$ is an isomorphism by the induction hypothesis.

To prove point (b) we will need to appeal to deformation theory. Choose a smooth atlas $Z_0 \to {}^{\mathrm{cl}}\mathcal{Y}$ with $Z_0 \in {}^{\mathrm{cl}}\mathrm{Stk}^{0\text{-Artn}}$. Then deformation theory (see Volume II, Chapter 1, Sect. 7.4) implies that we can we can construct a compatible system of objects $Z_n \in {}^{\leq n}\mathrm{Stk}^{0\text{-Artn}}$ equipped with smooth maps $Z_n \to {}^{\leq n}\mathcal{Y}$.

Set $Z := \underset{n}{\mathrm{colim}}\, Z_n$, where the colimit is taken in ${}^{\mathrm{conv}}\mathrm{PreStk}$. By the case of $k = 0$, we have $Z \in \mathrm{Stk}^{0\text{-Artn}}$, and since $\mathcal{Y}$ is convergent we have a canonically defined map $Z \to \mathcal{Y}$. Set $\mathcal{Y}^\bullet := Z^\bullet/\mathcal{Y}$. By Lemma 2.5.9, we have $\mathcal{Y} \in \mathrm{Stk}$. Hence, the map

$$|\mathcal{Y}^\bullet|_{\mathrm{Stk}} \to \mathcal{Y}$$

is an isomorphism.

Thus, by Proposition 4.3.6(a), it suffices to show that the map

$$\mathcal{Y}^1 = Z \underset{\mathcal{Y}}{\times} Z \to Z \times Z$$

is $(k-1)$-representable and its composition with either of the the projections $Z\times Z\to Z$ is smooth. By the induction hypothesis and Lemma 2.1.3, it suffices to show that the map

$$Z_n \underset{{}^{\leq n}\mathcal{Y}}{\times} Z_n = {}^{\leq n}Z \underset{{}^{\leq n}\mathcal{Y}}{\times} {}^{\leq n}Z \simeq {}^{\leq n}(Z \underset{\mathcal{Y}}{\times} Z) \to {}^{\leq n}(Z\times Z) \simeq {}^{\leq n}Z \times {}^{\leq n}Z = Z_n \times Z_n$$

has the corresponding properties. However, this follows from the fact that the map $Z_n \to {}^{\leq n}\mathcal{Y}$ is $(k-1)$-representable and smooth.

□

4.5. Artin stacks locally almost of finite type.

4.5.1. The goal of this subsection is to establish the following:

PROPOSITION 4.5.2. *Let $\mathcal{Y}$ be an object of* $\mathrm{Stk}^{k\text{-}\mathrm{Artn}}$ *(resp.,* ${}^{\leq n}\mathrm{Stk}^{k\text{-}\mathrm{Artn}}$*). The following conditions are equivalent:*

(i) $\mathcal{Y} \in \mathrm{Stk}_{\mathrm{laft}}$ *(resp.,* $\mathcal{Y} \in {}^{\leq n}\mathrm{Stk}_{\mathrm{lft}}$*);*

(ii) *$\mathcal{Y}$ admits an atlas $f : Z \to \mathcal{Y}$ with* $Z \in \mathrm{Stk}^{0\text{-}\mathrm{Artn}}_{\mathrm{laft}}$ *(resp.,* $Z \in {}^{\leq n}\mathrm{Stk}^{0\text{-}\mathrm{Artn}}_{\mathrm{lft}}$*);*

(iii) *For a k-representable ppf morphism $Z \to \mathcal{Y}$ with* $Z \in \mathrm{Stk}^{0\text{-}\mathrm{Artn}}$ *(resp.,* $Z \in {}^{\leq n}\mathrm{Stk}^{0\text{-}\mathrm{Artn}}$*), we have* $Z \in \mathrm{Stk}^{0\text{-}\mathrm{Artn}}_{\mathrm{laft}}$ *(resp.,* $Z \in {}^{\leq n}\mathrm{Stk}^{0\text{-}\mathrm{Artn}}_{\mathrm{lft}}$*).*

(iv) *For a k-representable ppf morphism $\mathcal{Y}' \to \mathcal{Y}$, we have* $\mathcal{Y}' \in \mathrm{Stk}_{\mathrm{laft}}$ *(resp.,* $\mathcal{Y}' \in {}^{\leq n}\mathrm{Stk}_{\mathrm{lft}}$*)*

We will call k-Artin stacks satisfying the equivalent conditions of the above proposition 'k-Artin stacks locally almost of finite type'.

Since we know that k-Artin stacks are convergent, it is enough to treat the case of $\mathcal{Y} \in {}^{\leq n}\mathrm{Stk}^{k\text{-}\mathrm{Artn}}$.

4.5.3. The proof of the proposition proceeds by induction, so we are assuming that all four conditions are equivalent for $k' < k$.

The implications (iii) ⇒ (ii) and (iv) ⇒ (iii) are tautological. The construction of the atlas in Proposition 4.2.4 shows that (ii) and (iii) imply (iv).

4.5.4. *Implication (ii) ⇒ (i).* By Corollary 4.3.3, we have

$$\mathcal{Y} \simeq {}^{\leq n}L(|Z^\bullet/\mathcal{Y}|_{{}^{\leq n}\mathrm{PreStk}}).$$

First, the implication (i) ⇒ (iv) for $k-1$ implies that the terms of the Čech nerve of the atlas $Z \to \mathcal{Y}$ consist of objects of ${}^{\leq n}\mathrm{Stk}_{\mathrm{lft}}$. Hence, we can rewrite the expression for $\mathcal{Y}$ as

$${}^{\leq n}L \circ \mathrm{LKE}_{{}^{\leq n}\mathrm{Sch}^{\mathrm{aff}}_{\mathrm{ft}} \hookrightarrow {}^{\leq n}\mathrm{Sch}^{\mathrm{aff}}}(\mathcal{Y}|_{{}^{\leq n}\mathrm{Sch}^{\mathrm{aff}}_{\mathrm{ft}}}).$$

However, by Corollary 4.3.4, the restriction $\mathcal{Y}|_{{}^{\leq n}\mathrm{Sch}^{\mathrm{aff}}_{\mathrm{ft}}}$ is $(n+k)$-truncated. Hence, applying Proposition 2.7.7, we obtain that

$${}^{\leq n}L \circ \mathrm{LKE}_{{}^{\leq n}\mathrm{Sch}^{\mathrm{aff}}_{\mathrm{ft}} \hookrightarrow {}^{\leq n}\mathrm{Sch}^{\mathrm{aff}}}(\mathcal{Y}|_{{}^{\leq n}\mathrm{Sch}^{\mathrm{aff}}_{\mathrm{ft}}}) \simeq \mathrm{LKE}_{{}^{\leq n}\mathrm{Sch}^{\mathrm{aff}}_{\mathrm{ft}} \hookrightarrow {}^{\leq n}\mathrm{Sch}^{\mathrm{aff}}}(\mathcal{Y}|_{{}^{\leq n}\mathrm{Sch}^{\mathrm{aff}}_{\mathrm{ft}}})$$

(i.e., no sheafification is necessary).

Thus, $\mathcal{Y}$, viewed as an object of ${}^{\leq n}\mathrm{PreStk}$, lies in the essential image of $\mathrm{LKE}_{{}^{\leq n}\mathrm{Sch}^{\mathrm{aff}}_{\mathrm{ft}} \hookrightarrow {}^{\leq n}\mathrm{Sch}^{\mathrm{aff}}}$, i.e., belongs to ${}^{\leq n}\mathrm{PreStk}_{\mathrm{lft}}$.

4.5.5. *Implication* $(i) \Rightarrow (iii)$. (J.Lurie)

It is easy to see that we can assume that $Z = S$ is an affine scheme. Let us be given a ppf map $f : S \to \mathcal{Y}$. We wish to show that $S \in {}^{\leq n}\mathrm{Sch}^{\mathrm{aff}}_{\mathrm{ft}}$.

Since $\mathcal{Y} \in {}^{\leq n}\mathrm{PreStk}_{\mathrm{lft}}$, there exists $T \in {}^{\leq n}\mathrm{Sch}^{\mathrm{aff}}$, such that f factors as

$$S \xrightarrow{h} T \xrightarrow{g} \mathcal{Y}.$$

Consider the Cartesian square:

$$\begin{array}{ccc} T \underset{\mathcal{Y}}{\times} S & \xrightarrow{g'} & S \\ {\scriptstyle f'}\downarrow & & \downarrow{\scriptstyle f} \\ T & \xrightarrow{g} & \mathcal{Y}. \end{array}$$

Since the map f is ppf, so is f'. Let $Z' \to T \underset{\mathcal{Y}}{\times} S$ be an atlas with $Z' \in {}^{\leq n}\mathrm{Stk}^{0\text{-}\mathrm{Artn}}$. We obtain that Z' is ppf over T. Since T is of finite type, we obtain that $Z' \in {}^{\leq n}\mathrm{Stk}^{0\text{-}\mathrm{Artn}}_{\mathrm{lft}}$.

Since $T \underset{\mathcal{Y}}{\times} S \in {}^{\leq n}\mathrm{Stk}^{(k-1)\text{-}\mathrm{Artn}}$, by the induction hypothesis, we obtain that

$$T \underset{\mathcal{Y}}{\times} S \in {}^{\leq n}\mathrm{Stk}^{(k-1)\text{-}\mathrm{Artn}}_{\mathrm{lft}} \subset {}^{\leq n}\mathrm{PreStk}_{\mathrm{lft}}.$$

Consider now the maps

$$S \xrightarrow{\mathrm{diag}} S \underset{\mathcal{Y}}{\times} S \xrightarrow{h \times \mathrm{id}} T \underset{\mathcal{Y}}{\times} S \to S,$$

where the last map if the projection on the second factor. The composition is the identity map on S. Hence, S is a retract of $T \underset{\mathcal{Y}}{\times} S$ as an object of ${}^{\leq n}\mathrm{PreStk}$. Since the subcategory ${}^{\leq n}\mathrm{PreStk}_{\mathrm{lft}} \subset {}^{\leq n}\mathrm{PreStk}$ is stable under retracts, we obtain that

$$S \in {}^{\leq n}\mathrm{PreStk}_{\mathrm{lft}} \cap {}^{\leq n}\mathrm{Sch}^{\mathrm{aff}}.$$

Now, the assertion that $S \in {}^{\leq n}\mathrm{Sch}^{\mathrm{aff}}_{\mathrm{ft}}$ follows from Lemma 1.6.6.

CHAPTER 3

Quasi-coherent sheaves on prestacks

Introduction

0.1. What is (derived) algebraic geometry about? Arguably, the object of study of (derived) algebraic geometry is not so much the *geometric objects* (i.e., the most general of which we call *prestacks*, see Chapter 2), but quasi-coherent sheaves on these geometric objects.

This Chapter is devoted to the definition and the study of the most basic properties and structures on quasi-coherent sheaves.

0.1.1. Having at our disposal the theory of ∞-categories, the definition of the category of quasi-coherent sheaves on a prestack is very simple.

First, if our prestack is an affine scheme $S = \text{Spec}(A)$, then

$$\text{QCoh}(S) := A\text{-mod},$$

i.e., this is the DG category of A-modules.

For a general prestack $\mathcal{Y}$, we define $\text{QCoh}(\mathcal{Y})$ to be the *limit* of the categories $\text{QCoh}(S)$ over the category of pairs

$$(S \in \text{Sch}^{\text{aff}},\ y : S \to \mathcal{Y}). \tag{0.1}$$

I.e., an object $\mathcal{F} \in \text{QCoh}(\mathcal{Y})$ is a of assignments for every (S, y) as above of $\mathcal{F}_{S,y} \in \text{QCoh}(S)$ and for every $g : S' \to S$ and $y' \sim y \circ g$ we are given an isomorphism

$$\mathcal{F}_{S',y'} \simeq g^*(\mathcal{F}_{S,y}).$$

These isomorphisms must satisfy a homotopy-coherent system of compatibilities for compositions of morphisms between affine schemes.

We note that the above limit takes place in the ∞-category $\text{DGCat}_{\text{cont}}$, so we really need to input the entire machinery of [**Lu1**]. We also note that it is important that we work with DG categories rather than triangulated categories: limits of the latter are known to be ill-behaved.

0.1.2. Assume for a moment that $\mathcal{Y}$ is a (derived) scheme. Then Proposition 1.4.4 shows that in considering the above limit, it is enough to consider those $(S, y : S \to \mathcal{Y})$ for which s is an open embedding. I.e., we glue the category $\text{QCoh}(\mathcal{Y})$ from the corresponding categories on its open affine subschemes.

Note that this is *not* how most textbooks define the category QCoh on a (derived) scheme. The more usual way is to consider all *sheaves of $\mathcal{O}$-modules* in the Zariski topology, and then pass to the subcategory consisting of objects with quasi-coherent cohomologies.

By contrast, our definition avoids any mention of sheaves that are non quasi-coherent. We regard it as an advantage: in a sense non quasi-coherent sheaves do not fully belong to algebraic geometry.

0.1.3. Generalizing from schemes to Artin stacks, we show that if $\mathcal{Y}$ is an Artin stack, when considering the limit over the category of pairs (0.1), we can replace it by its full subcategory where we require that the map y be smooth (note that since $\mathcal{Y}$ is an Artin stack, it makes sense to talk about a map to it from an affine scheme being smooth).

Moreover, we can replace the latter category by its 1-full subcategory where when considering morphisms

$$(g : S' \to S,\ y' \simeq y \circ g),$$

we only allow those g that are themselves smooth.

So, when considering QCoh on an Artin stack, we do not have to consider maps that are non-smooth.

0.1.4. Another possible approach to the definition of QCoh would have been as the derived category of an abelian category.

Although it is true that for any prestack $\mathcal{Y}$, the category $\text{QCoh}(\mathcal{Y})$ carries a canonical t-structure, the derived category of its heart is not at all equivalent to $\text{QCoh}(\mathcal{Y})$. This equivalence fails already for affine schemes that are not classical.

What one can show, however, is that when $\mathcal{Y}$ is a *classical* algebraic stack, then the bounded below part of $\text{QCoh}(\mathcal{Y})$ is equivalent to the bounded below part of $D(\text{QCoh}(\mathcal{Y})^{\heartsuit})$.

0.2. What is done in this Chapter beyond the definition? So far, we only have the functor

$$\text{QCoh}^*_{\text{PreStk}} : \text{PreStk}^{\text{op}} \to \text{DGCat}_{\text{cont}}$$

that sends a prestack $\mathcal{Y}$ to the category $\text{QCoh}(\mathcal{Y})$ and a morphism $f : \mathcal{Y}' \to \mathcal{Y}$ to the pullback functor

$$f^* : \text{QCoh}(\mathcal{Y}) \to \text{QCoh}(\mathcal{Y}').$$

The rest of this chapter is devoted to exploring some very basic properties of QCoh.

0.2.1. In Sect. 2, for a morphism $f : \mathcal{Y}' \to \mathcal{Y}$ between prestacks we study the functor

$$f_* : \text{QCoh}(\mathcal{Y}') \to \text{QCoh}(\mathcal{Y}),$$

right adjoint to f^* (which exists by the Adjoint Functor Theorem since f^* is continuous).

In general, the functor f_* is ill-behaved. For example, it does not have the *base change property* (see Proposition 2.2.2(b) for what this means). In particular, for $\mathcal{F}' \in \text{QCoh}(\mathcal{Y}')$, we cannot explicitly say what is the value of $f_*(\mathcal{F}')$ on $S \xrightarrow{y} \mathcal{Y}$.

However, the situation is much better when f is schematic quasi-compact (i.e., the base change of f by an affine scheme yields a quasi-compact scheme). In this case, the direct image functor does have the base change property.

0.2.2. In Sect. 3 we show that the functor

$$\mathrm{QCoh}^*_{\mathrm{PreStk}} : \mathrm{PreStk}^{\mathrm{op}} \to \mathrm{DGCat}_{\mathrm{cont}}$$

has a natural *right-lax* symmetric monoidal structure, where the symmetric monoidal structure on $\mathrm{PreStk}^{\mathrm{op}}$ is induced by the Cartesian symmetric monoidal structure on PreStk, and on $\mathrm{DGCat}_{\mathrm{cont}}$, it is given by the Lurie tensor product.

Concretely, this means that for $\mathcal{Y}_1, \mathcal{Y}_2 \in \mathrm{PreStk}$ we have a canonically defined functor

$$\mathrm{QCoh}(\mathcal{Y}_1) \otimes \mathrm{QCoh}(\mathcal{Y}_2) \to \mathrm{QCoh}(\mathcal{Y}_1 \times \mathcal{Y}_2), \tag{0.2}$$

given by the *external tensor product* of quasi-coherent sheaves

$$\mathcal{F}_1, \mathcal{F}_2 \mapsto \mathcal{F}_1 \boxtimes \mathcal{F}_2,$$

We give criteria for when the functor (0.2) is an equivalence. For example, a sufficient condition is that the DG category $\mathrm{QCoh}(\mathcal{Y}_1)$ (or $\mathrm{QCoh}(\mathcal{Y}_2)$) be dualizable.

0.2.3. The symmetric monoidal structure on the functor $\mathrm{QCoh}^*_{\mathrm{PreStk}}$ induces a symmetric monoidal structure on $\mathrm{QCoh}(\mathcal{Y})$ for an individual $\mathcal{Y}$.

We study how various conditions on $\mathrm{QCoh}(\mathcal{Y})$ (such as being dualizable, rigid or compactly generated) interact with each other.

Finally, we study the following question: let

$$\begin{array}{ccc} \mathcal{Y}'_1 & \longrightarrow & \mathcal{Y}' \\ \downarrow & & \downarrow \\ \mathcal{Y}_1 & \longrightarrow & \mathcal{Y} \end{array}$$

be a pullback diagram of prestacks. Under what conditions is the tautological functor

$$\mathrm{QCoh}(\mathcal{Y}_1) \underset{\mathrm{QCoh}(\mathcal{Y})}{\otimes} \mathrm{QCoh}(\mathcal{Y}') \to \mathrm{QCoh}(\mathcal{Y}'_1)$$

an equivalence?

1. The category of quasi-coherent sheaves

In this section we define the functor QCoh^* that maps $\mathrm{PreStk}^{\mathrm{op}}$ to $\mathrm{DGCat}_{\mathrm{cont}}$. We study its basic properties: behavior with respect to n-coconnectivity and finite typeness, descent and t-structure.

We then show that in the case of Artin stacks, QCoh agrees with the more familiar definition of (the derived category of) quasi-coherent sheaves.

1.1. Setting up the theory of quasi-coherent sheaves. The basic input we feed into the theory of QCoh is the fact that the assignment $A \mapsto A\text{-mod}$ is a functor from $(\mathrm{AssocAlg}(\mathrm{Vect}))^{\mathrm{op}}$ to 1-Cat.

1.1.1. Recall that according to Chapter 1, Sect. 3.5.5 we have a canonically defined functor

$$(\text{AssocAlg}(\text{Vect}))^{\text{op}} \to \text{DGCat}_{\text{cont}}, \quad A \mapsto A\text{-mod}$$

Composing with the forgetful functors

$$\text{ComAlg}(\text{Vect}^{\leq 0}) \to \text{ComAlg}(\text{Vect}) \to \text{AssocAlg}(\text{Vect})$$

we obtain the functor

$$(\text{ComAlg}(\text{Vect}^{\leq 0}))^{\text{op}} \to \text{DGCat}_{\text{cont}}. \tag{1.1}$$

1.1.2. We use the functor (1.1) as the initial input for QCoh.

Namely, we interpret (1.1) as a functor

$$\begin{aligned} &\text{QCoh}_{\text{Sch}^{\text{aff}}} : \text{Sch}^{\text{aff}} \to \text{DGCat}_{\text{cont}}, \\ &\qquad S \rightsquigarrow \text{QCoh}(S), \quad (S \xrightarrow{f} S') \rightsquigarrow (\text{QCoh}(S) \xrightarrow{f_*} \text{QCoh}(S')). \end{aligned} \tag{1.2}$$

We will now use the fact that the structure of $(\infty,1)$-category on $\text{DGCat}_{\text{cont}}$ can be canonically extended to a structure of $(\infty,2)$-category, denoted $\text{DGCat}^{2\text{-Cat}}_{\text{cont}}$ (see Chapter 1, Sect. 10.3.9).

Note that for an individual morphism $f : S \to S'$ in Sch^{aff}, the functor

$$\text{QCoh}(S) \xrightarrow{f_*} \text{QCoh}(S')$$

admits a left adjoint in $\text{DGCat}^{2\text{-Cat}}_{\text{cont}}$, denoted f^*.

Hence, applying Chapter 12, Corollary 1.3.4, by *passing to left adjoints*, from $\text{QCoh}_{\text{Sch}^{\text{aff}}}$ we obtain a functor

$$\begin{aligned} &\text{QCoh}^*_{\text{Sch}^{\text{aff}}} : (\text{Sch}^{\text{aff}})^{\text{op}} \to \text{DGCat}_{\text{cont}}, \\ &\qquad S \rightsquigarrow \text{QCoh}(S), \quad (S \xrightarrow{f} S') \rightsquigarrow (\text{QCoh}(S') \xrightarrow{f^*} \text{QCoh}(S)). \end{aligned} \tag{1.3}$$

1.1.3. Finally, we define the functor

$$\text{QCoh}^*_{\text{PreStk}} : \text{PreStk}^{\text{op}} \to \text{DGCat}_{\text{cont}}$$

to be the *right Kan extension* of the functor $\text{QCoh}^*_{\text{Sch}^{\text{aff}}}$ of (1.3) along the fully faithful embedding

$$(\text{Sch}^{\text{aff}})^{\text{op}} \hookrightarrow \text{PreStk}^{\text{op}}.$$

For an individual $\mathcal{Y} \in \text{PreStk}$ we denote the value of $\text{QCoh}^*_{\text{PreStk}}$ on it by $\text{QCoh}(\mathcal{Y})$. For a map $f : \mathcal{Y}' \to \mathcal{Y}$ we denote the corresponding 1-morphism in $\text{DGCat}_{\text{cont}}$ by

$$f^* : \text{QCoh}(\mathcal{Y}') \to \text{QCoh}(\mathcal{Y}).$$

1.1.4. By definition, for an individual $\mathcal{Y} \in \text{PreStk}$, we have

$$\text{QCoh}(\mathcal{Y}) \simeq \lim_{(S \xrightarrow{y} \mathcal{Y})} \text{QCoh}(S), \tag{1.4}$$

where the limit is taken over the category opposite to $(\text{Sch}^{\text{aff}})_{/\mathcal{Y}}$.

Thus, we can think of an object $\mathcal{F} \in \text{QCoh}(\mathcal{Y})$ as an assignment

$$(S \xrightarrow{y} \mathcal{Y}) \rightsquigarrow \mathcal{F}_{S,y} \in \text{QCoh}(S),$$

$$(S' \xrightarrow{f} S) \in (\text{Sch}^{\text{aff}})_{/\mathcal{Y}} \rightsquigarrow (\mathcal{F}_{S',y\circ f} \simeq f^*(\mathcal{F}_{S,y})) \in \text{QCoh}(S'),$$

satisfying a homotopy-coherent system of compatibilities for compositions of morphisms in $(\text{Sch}^{\text{aff}})_{/\mathcal{Y}}$.

1.2. Basic properties of QCoh.

1.2.1. *Quasi-coherent sheaves and n-coconnective prestacks.* Assume that $\mathcal{Y}$ is n-coconnective (see Chapter 2, Sect. 1.3.3), i.e., that when we view $\mathcal{Y}$ as a functor $(\text{Sch}^{\text{aff}})^{\text{op}} \to \text{Spc}$, it is a left Kan extension along the embedding

$$^{\leq n}\text{Sch}^{\text{aff}} \hookrightarrow \text{Sch}^{\text{aff}}.$$

We have:

LEMMA 1.2.2. *Under the above circumstances, the natural map*

$$\text{QCoh}(\mathcal{Y}) \to \lim_{(S \xrightarrow{y} \mathcal{Y}) \in ((^{\leq n}\text{Sch}^{\text{aff}})_{/\mathcal{Y}})^{\text{op}}} \text{QCoh}(S)$$

is an equivalence.

PROOF. Follows from (1.4), since the fact that $\mathcal{Y}$ is n-coconnective exactly means that the functor

$$(^{\leq n}\text{Sch}^{\text{aff}})_{/\mathcal{Y}} \to (\text{Sch}^{\text{aff}})_{/\mathcal{Y}}$$

is cofinal.

□

In other words, the above lemma says that if $\mathcal{Y}$ in n-coconnective, in the definition of quasi-coherent sheaves, it is enough to consider only those affine DG schemes mapping to $\mathcal{Y}$ that are themselves n-coconnective.

In particular, if $\mathcal{Y}$ is a classical prestack, it is sufficient to consider only classical affine schemes mapping to $\mathcal{Y}$.

1.2.3. *Quasi-coherent sheaves on stacks locally of finite type.*
Let $\mathcal{Y} \in \text{PreStk}$ be n-coconnective as above, and assume, moreover, that it is locally of finite type (see Chapter 2, Sect. 1.6). I.e., $\mathcal{Y}|_{^{\leq n}\text{Sch}^{\text{aff}}}$ is the left Kan extension along the embedding

$$^{\leq n}\text{Sch}^{\text{aff}}_{\text{ft}} \hookrightarrow {}^{\leq n}\text{Sch}^{\text{aff}}.$$

LEMMA 1.2.4. *Under the above circumstances, the natural map*

$$\text{QCoh}(\mathcal{Y}) \to \lim_{(S \xrightarrow{y} \mathcal{Y}) \in ((^{\leq n}\text{Sch}^{\text{aff}}_{\text{ft}})_{/\mathcal{Y}}))^{\text{op}}} \text{QCoh}(S)$$

is an equivalence.

PROOF. Follows from Lemma 1.2.2, since the fact that $\mathcal{Y}$ being locally of finite type exactly means that the functor

$$({}^{\leq n}\mathrm{Sch}^{\mathrm{aff}}_{\mathrm{ft}})_{/\mathcal{Y}} \to ({}^{\leq n}\mathrm{Sch}^{\mathrm{aff}})_{/\mathcal{Y}}$$

is cofinal.

□

I.e., for n-coconnective prestacks locally of finite type, in the definition of quasi-coherent sheaves, it is enough to consider only those affine DG schemes mapping to $\mathcal{Y}$ that are themselves n-coconnective *and* are of finite type.

1.2.5. *Non-convergence.* We note, however, that for $S \in \mathrm{Sch}^{\mathrm{aff}}$ the functor

$$\mathrm{QCoh}(S) \to \lim_{n} \mathrm{QCoh}({}^{\leq n}S)$$

is *not* necessarily an equivalence. The simplest counterexample is provided by $S = \mathrm{Spec}(k[\eta])$ with $\deg(\eta) = -2$.

This means, in particular, that for $\mathcal{Y} \in \mathrm{PreStk}_{\mathrm{laft}}$ we cannot express $\mathrm{QCoh}(\mathcal{Y})$ in terms of the categories $\mathrm{QCoh}(S)$ with $S \in \mathrm{Sch}^{\mathrm{aff}}_{\mathrm{aft}}$.

1.3. Descent. In this subsection we will discuss a fundamental feature of the functor QCoh^*, namely, that it satisfies flat descent.

1.3.1. Recall what it means for a functor $(\mathrm{Sch}^{\mathrm{aff}})^{\mathrm{op}} \to \mathrm{Spc}$ to satisfy descent with respect to a given topology, see Chapter 2, Sect. 2.3.1.

We note, however, that this notion make sense when we replace the target category Spc by any ∞-category.

We observe, however, that the notion of descent when the target is some ∞-category $\mathbf{C}$ is expressible in terms of descent with values in Spc:

LEMMA 1.3.2. *Let $F : (\mathrm{Sch}^{\mathrm{aff}})^{\mathrm{op}} \to \mathbf{C}$ be a functor. Then it satisfies descent if and and only if for every $\mathbf{c} \in \mathbf{C}$, the functor*

$$\mathrm{Maps}_{\mathbf{C}}(\mathbf{c}, -) \circ F : (\mathrm{Sch}^{\mathrm{aff}})^{\mathrm{op}} \to \mathrm{Spc}$$

satisfies descent.

1.3.3. The following assertion is a version of Grothendieck's flat descent (see [**Lu5**, Proposition 2.7.14]):

THEOREM 1.3.4. *The composite functor*

$$(\mathrm{Sch}^{\mathrm{aff}})^{\mathrm{op}} \overset{\mathrm{QCoh}^*_{\mathrm{Sch}^{\mathrm{aff}}}}{\longrightarrow} \mathrm{DGCat}_{\mathrm{cont}} \to 1\text{-}\mathrm{Cat}$$

satisfies descent with respect to the flat (and hence, ppf, étale, Zariski) topology.

Since the forgetful functor

$$\mathrm{DGCat}_{\mathrm{cont}} \to 1\text{-}\mathrm{Cat}$$

preserves limits (see Chapter 1, Lemma 2.5.2(b)), we obtain:

COROLLARY 1.3.5. *The functor*

$$\mathrm{QCoh}^*_{\mathrm{Sch}^{\mathrm{aff}}} : (\mathrm{Sch}^{\mathrm{aff}})^{\mathrm{op}} \to \mathrm{DGCat}_{\mathrm{cont}}$$

satisfies descent with respect to the flat (and hence, ppf, étale, Zariski) topology.

1.3.6. Combining Corollary 1.3.5 with Chapter 2, Sect. 2.3.3, we obtain:

COROLLARY 1.3.7. *Let* $f : \mathcal{Y}' \to \mathcal{Y}$ *be a map in* PreStk *that is an equivalence for the flat topology. Then*

$$f^* : \mathrm{QCoh}(\mathcal{Y}) \to \mathrm{QCoh}(\mathcal{Y}')$$

is an equivalence.

From this, we obtain, tautologically:

COROLLARY 1.3.8. *For* $\mathcal{Y} \in \mathrm{PreStk}$, *the canonical map* $\mathcal{Y} \to L(\mathcal{Y})$ *induces an equivalence:*

$$\mathrm{QCoh}(L(\mathcal{Y})) \to \mathrm{QCoh}(\mathcal{Y}).$$

1.3.9. The last corollary has a two-fold significance:

First, to specify a stack we may often have to start from a prestack given explicitly, and then apply the functor L. Corollary 1.3.8 implies that in order to calculate the category QCoh of the resulting stack we can work with the initial prestack.

Secondly, we obtain that for the purposes of QCoh, we will lose no information if we work with the subcategory Stk rather than all of PreStk.

1.3.10. From Chapter 2, Lemma 2.3.8, we obtain:

COROLLARY 1.3.11. *Let* $\mathcal{Y}_1 \to \mathcal{Y}_2$ *be a surjection in the flat topology. Then the natural map*

$$\mathrm{QCoh}(\mathcal{Y}_2) \to \mathrm{Tot}\left(\mathrm{QCoh}(\mathcal{Y}_1^\bullet/\mathcal{Y}_2)\right)$$

is an equivalence.

1.3.12. *Quasi-coherent sheaves on n-coconnective stacks.* Recall the notion of n-coconnective stack, see Chapter 2, Sect. 2.6.3.

Note that if $\mathcal{Y}$ is n-coconnective as a stack, then this does *not* mean that its is n-coconnective as a prestack. However, combining Corollary 1.3.8 and Lemma 1.2.2, we obtain:

COROLLARY 1.3.13. *Let* $\mathcal{Y}$ *be an* n*-coconnective stack. Then the natural map*

$$\mathrm{QCoh}(\mathcal{Y}) \to \lim_{(S \xrightarrow{y} Y) \in ((^{\leq n}\mathrm{Sch}^{\mathrm{aff}})_{/\mathcal{Y}})^{\mathrm{op}}} \mathrm{QCoh}(S)$$

is an equivalence.

1.4. Quasi-coherent sheaves on Artin stacks. The point of this subsection is that when $\mathcal{Y}$ is an Artin stack, in order to recover $\mathrm{QCoh}(\mathcal{Y})$, instead of considering all affine schemes mapping to $\mathcal{Y}$, it is enough to consider only ones that are smooth over $\mathcal{Y}$.

1.4.1. Let $\mathcal{Y}$ be an k-Artin stack (see Chapter 2, Sect. 4.1). We claim that in this case, there is a more concise expression for $\mathrm{QCoh}(\mathcal{Y})$.

Let $(\mathrm{Sch}^{\mathrm{aff}})_{/\mathcal{Y},\mathrm{sm}}$ denote the full subcategory of $(\mathrm{Sch}^{\mathrm{aff}})_{/\mathcal{Y}}$ consisting of those $S \overset{y}{\to} \mathcal{Y}$, for which y is smooth (as a $(k-1)$-representable map).

Let $((\mathrm{Sch}^{\mathrm{aff}})_{\mathrm{sm}})_{/\mathcal{Y}}$ be the 1-full subcategory of $(\mathrm{Sch}^{\mathrm{aff}})_{/\mathcal{Y},\mathrm{sm}}$, where we restrict maps $f : S' \to S$ to also be smooth.

We claim:

PROPOSITION 1.4.2.
(a) *The natural map*

$$\mathrm{QCoh}(\mathcal{Y}) \to \lim_{(S \overset{y}{\to} \mathcal{Y}) \in ((\mathrm{Sch}^{\mathrm{aff}})_{/\mathcal{Y},\mathrm{sm}})^{\mathrm{op}}} \mathrm{QCoh}(S)$$

is an equivalence.
(b) *The natural map*

$$\mathrm{QCoh}(\mathcal{Y}) \to \lim_{(S \overset{y}{\to} \mathcal{Y}) \in (((\mathrm{Sch}^{\mathrm{aff}})_{\mathrm{sm}})_{/\mathcal{Y}})^{\mathrm{op}}} \mathrm{QCoh}(S)$$

is an equivalence.

PROOF. Assume by induction that both statements are true for $k' < k$. The base case of $k = 0$ is obvious: in this case our $\mathcal{Y}$ is a disjoint union of affine schemes.

We are going to construct a map

$$\lim_{(S \overset{y}{\to} \mathcal{Y}) \in (((\mathrm{Sch}^{\mathrm{aff}})_{\mathrm{sm}})_{/\mathcal{Y}})^{\mathrm{op}}} \mathrm{QCoh}(S) \to \mathrm{QCoh}(\mathcal{Y}), \tag{1.5}$$

inverse to the composition

$$\mathrm{QCoh}(\mathcal{Y}) \to \lim_{(S \overset{y}{\to} \mathcal{Y}) \in ((\mathrm{Sch}^{\mathrm{aff}})_{/\mathcal{Y},\mathrm{sm}})^{\mathrm{op}}} \mathrm{QCoh}(S) \to \lim_{(S \overset{y}{\to} \mathcal{Y}) \in (((\mathrm{Sch}^{\mathrm{aff}})_{\mathrm{sm}})_{/\mathcal{Y}})^{\mathrm{op}}} \mathrm{QCoh}(S).$$

Let $f : \mathcal{Z} \to \mathcal{Y}$ be a smooth atlas, where $\mathcal{Z}$ is a $(k-1)$-Artin stack. By Corollary 1.3.11, the map

$$\mathrm{QCoh}(\mathcal{Y}) \to \mathrm{Tot}\,(\mathrm{QCoh}(\mathcal{Z}^\bullet/\mathcal{Y}))$$

is an equivalence. Thus, the datum of a map in (1.5) is equivalent to a map

$$\lim_{(S \overset{y}{\to} \mathcal{Y}) \in (((\mathrm{Sch}^{\mathrm{aff}})_{\mathrm{sm}})_{/\mathcal{Y}})^{\mathrm{op}}} \mathrm{QCoh}(S) \to \mathrm{Tot}\,(\mathrm{QCoh}(\mathcal{Z}^\bullet/\mathcal{Y}))\,. \tag{1.6}$$

Note that the expression in the LHS of (1.6) equals the value on $\mathcal{Y}$ of

$$\mathrm{RKE}_{((\mathrm{Sch}^{\mathrm{aff}})_{\mathrm{sm}})^{\mathrm{op}} \to ((\mathrm{Stk}^{k\text{-}\mathrm{Artn}})_{\mathrm{sm}})^{\mathrm{op}}}(\mathrm{QCoh}^*_{\mathrm{Sch}^{\mathrm{aff}}}|_{(\mathrm{Sch}^{\mathrm{aff}})_{\mathrm{sm}}}).$$

In the above formula, $(\mathrm{Sch}^{\mathrm{aff}})_{\mathrm{sm}}$ (resp., $(\mathrm{Stk}^{k\text{-}\mathrm{Artn}})_{\mathrm{sm}}$) denotes the 1-full subcategory of $\mathrm{Sch}^{\mathrm{aff}}$ (resp., $\mathrm{Stk}^{k\text{-}\mathrm{Artn}}$), where we restrict 1-morphisms to be smooth maps.

The validity of point (b) for $k-1$ is equivalent to the fact that the map

$$\mathrm{RKE}_{((\mathrm{Sch}^{\mathrm{aff}})_{\mathrm{sm}})^{\mathrm{op}} \to ((\mathrm{Stk}^{(k-1)\text{-}\mathrm{Artn}})_{\mathrm{sm}})^{\mathrm{op}}}(\mathrm{QCoh}^*_{\mathrm{Sch}^{\mathrm{aff}}}|_{(\mathrm{Sch}^{\mathrm{aff}})_{\mathrm{sm}}}) \to \\ \to \mathrm{QCoh}^*_{\mathrm{Sch}^{\mathrm{aff}}}|_{(\mathrm{Stk}^{(k-1)\text{-}\mathrm{Artn}})_{\mathrm{sm}}}$$

is an isomorphism.

Hence, by the transitivity of the operation of the right Kan extension, we can rewrite the LHS of (1.6) as

$$\lim_{(\mathcal{Z}'\overset{y}{\to}\mathcal{Y})\in(((\mathrm{Stk}^{(k-1)\text{-Artn}})_{\mathrm{sm}})_{/\mathcal{Y}})^{\mathrm{op}}} \mathrm{QCoh}(\mathcal{Z}').$$

Now, the required map

$$\lim_{(\mathcal{Z}'\overset{y}{\to}\mathcal{Y})\in(((\mathrm{Stk}^{(k-1)\text{-Artn}})_{\mathrm{sm}})_{/\mathcal{Y}})^{\mathrm{op}}} \mathrm{QCoh}(\mathcal{Z}') \to \mathrm{Tot}\,(\mathrm{QCoh}(\mathcal{Z}^\bullet/\mathcal{Y}))$$

is given by restriction, since $\mathcal{Z}^\bullet/\mathcal{Y}$ is a simplicial object in $(\mathrm{Stk}^{(k-1)\text{-Artn}})_{\mathrm{sm}}$.

□

1.4.3. Assume now that $\mathcal{Y} = Z \in \mathrm{Sch}$. Let $(\mathrm{Sch}^{\mathrm{aff}})_{Z,\mathrm{open}}$ denote the full subcategory of $(\mathrm{Sch}^{\mathrm{aff}})_{/Z}$, that consists of those $z : S \to Z$ for which z is an open embedding. Note that morphisms in this category automatically consist of open embeddings.

Then as in Proposition 1.4.2 we prove:

PROPOSITION 1.4.4. *The natural map*

$$\mathrm{QCoh}(Z) \to \lim_{(S\overset{z}{\to}Z)\in((\mathrm{Sch}^{\mathrm{aff}})_{/Z,\mathrm{open}})^{\mathrm{op}}} \mathrm{QCoh}(S)$$

is an equivalence.

1.5. The t-structure. For any prestack $\mathcal{Y}$, the category $\mathrm{QCoh}(\mathcal{Y})$ comes equipped with a t-structure. When $\mathcal{Y}$ is an Artin stack, this t-structure is quite explicit.

1.5.1. Let $\mathcal{Y}$ be an arbitrary prestack. We claim that the category $\mathrm{QCoh}(\mathcal{Y})$ carries a canonical t-structure. Namely, we declare that an object $\mathcal{F} \in \mathrm{QCoh}(\mathcal{Y})$ belongs to $\mathrm{QCoh}(\mathcal{Y})^{\leq 0}$ if for any $S \in \mathrm{Sch}^{\mathrm{aff}}$ and $S \overset{y}{\to} \mathcal{Y}$, the corresponding object $\mathcal{F}_{S,y} \in \mathrm{QCoh}(S)$ belongs to $\mathrm{QCoh}(S)^{\leq 0}$.

This indeed defines a t-structure (see [**Lu2**, Proposition 1.2.1.16]):

Since the subcategory $\mathrm{QCoh}(\mathcal{Y})^{\leq 0}$ is stable under colimits, by the Adjoint Functor Theorem, the embedding

$$\mathrm{QCoh}(\mathcal{Y})^{\leq 0} \hookrightarrow \mathrm{QCoh}(\mathcal{Y})$$

admits a right adjoint.

For a general prestack there is not much that one can say about this t-structure.

1.5.2. *An example.* Let $\mathcal{Y}$ be an affine scheme $\mathrm{Spec}(A)$. We have $\mathrm{QCoh}(\mathcal{Y}) = A\text{-mod}$, while

$$(A\text{-mod})^\heartsuit \simeq (H^0(A)\text{-mod})^\heartsuit,$$

so heart of the t-structure depends in this case only on the underlying classical affine scheme.

1.5.3. Assume now that $\mathcal{Y}$ is a k-Artin stack. In this case one can give an explicit description of the t-structure on $\mathrm{QCoh}(\mathcal{Y})$ in terms of an atlas:

PROPOSITION 1.5.4.
(a) *Let $\mathcal{Y}$ be a k-Artin stack and let $f_i : S_i \to \mathcal{Y}$ be a smooth atlas, where $S_i \in \mathrm{Sch}^{\mathrm{aff}}$. Then an object $\mathcal{F} \in \mathrm{QCoh}(\mathcal{Y})$ belongs to $\mathrm{QCoh}(\mathcal{Y})^{\leq 0}$ (resp., $\mathrm{QCoh}(\mathcal{Y})^{>0}$) if and only if each $f_i^*(\mathcal{F})$ belongs to $\mathrm{QCoh}(S_i)^{\leq 0}$ (resp., $\mathrm{QCoh}(S_i)^{>0}$).*
(b) *Let $\pi : \mathcal{Y}_1 \to \mathcal{Y}_2$ be a flat map between k-Artin stacks. Then the functor $\pi^* : \mathrm{QCoh}(\mathcal{Y}_2) \to \mathrm{QCoh}(\mathcal{Y}_1)$ is t-exact.*

REMARK 1.5.5. Since for an atlas $f_i : S_i \to \mathcal{Y}$ and $Z = \underset{i}{\sqcup} S_i$, the functor

$$\mathrm{QCoh}(\mathcal{Y}) \to \mathrm{Tot}(\mathrm{QCoh}(Z^\bullet/\mathcal{Y}))$$

is an equivalence, we obtain that point (a) is a particular case of point (b).

PROOF. We will argue by induction, assuming that both statements are true for $k' < k$. Let us first prove point (a). It is enough to show that the functor f^* is compatible with the truncation functors.

Denote as above $Z = \underset{i}{\sqcup} S_i$. Let $\mathcal{F}$ be an object of $\mathrm{QCoh}(\mathcal{Y})$, and let

$$\mathcal{F}|_{Z^\bullet/\mathcal{Y}} \in \mathrm{QCoh}(Z^\bullet/\mathcal{Y})$$

be the corresponding object. We claim that

$$i \mapsto \tau^{\leq 0}(\mathcal{F}|_{Z^\bullet/\mathcal{Y}}) \text{ and } i \mapsto \tau^{>0}(\mathcal{F}|_{Z^\bullet/\mathcal{Y}})$$

both belong to $\mathrm{QCoh}(Z^\bullet/\mathcal{Y})$. This follows by the induction hypothesis from the fact that the face maps in the simplicial stack $Z^\bullet/\mathcal{Y}$ are flat.

It is clear that the object $\mathcal{F}' \in \mathrm{QCoh}(\mathcal{Y})$ that corresponds to $\tau^{\leq 0}(\mathcal{F}|_{Z^\bullet/\mathcal{Y}})$ belongs to $\mathrm{QCoh}(\mathcal{Y})^{\leq 0}$.

We claim now that the object $\mathcal{F}'' \in \mathrm{QCoh}(\mathcal{Y})$ that corresponds to $\tau^{>0}(\mathcal{F}|_{Z^\bullet/\mathcal{Y}})$ belongs to $\mathrm{QCoh}(\mathcal{Y})^{>0}$. Indeed, for $\mathcal{F}''' \in \mathrm{QCoh}(\mathcal{Y})^{\leq 0}$, we have

$$\mathrm{Hom}_{\mathrm{QCoh}(\mathcal{Y})}(\mathcal{F}''', \mathcal{F}') \simeq \mathrm{Tot}\left(\mathrm{Hom}_{\mathrm{QCoh}(Z^\bullet/\mathcal{Y})}(\mathcal{F}'''|_{Z^\bullet/\mathcal{Y}}, \tau^{>0}(\mathcal{F}|_{Z^\bullet/\mathcal{Y}}))\right),$$

and the right-hand side vanishes, since $\mathcal{F}'''|_{Z^\bullet/\mathcal{Y}} \in \mathrm{QCoh}(Z^\bullet/\mathcal{Y})^{\leq 0}$.

Let us now prove point (b). By point (a), we can assume that $\mathcal{Y}_1$ is an affine scheme T (replace the initial $\mathcal{Y}_1$ by its atlas). So, we are dealing with a flat map π from an affine scheme T to a k-Artin stack $\mathcal{Y} = \mathcal{Y}_2$. Let $f_i : S_i \to \mathcal{Y}$ be an atlas with $S_i \in \mathrm{Sch}^{\mathrm{aff}}$. Consider the Cartesian square:

$$\begin{array}{ccc} T \underset{\mathcal{Y}}{\times} S_i & \xrightarrow{\pi'} & S_i \\ {\scriptstyle f_i'}\downarrow & & \downarrow{\scriptstyle f_i} \\ T & \xrightarrow{\pi} & \mathcal{Y}. \end{array}$$

Again, by point (a), it is sufficient to show that the functor

$$f'^* \circ \pi^* : \mathrm{QCoh}(\mathcal{Y}) \to \mathrm{QCoh}(T \underset{\mathcal{Y}}{\times} S_i)$$

is exact. However, $f_i'^* \circ \pi^* \simeq \pi'^* \circ f_i^*$, and f_i^* is t-exact by point (a), and π'^* is t-exact by the induction hypothesis.

□

1.5.6. Proposition 1.5.4 has the following corollary:

COROLLARY 1.5.7. *Let $\mathcal{Y}$ be an Artin stack.*

(a) *The t-stucture on* $\mathrm{QCoh}(\mathcal{Y})$ *is compatible with filtered colimits, i.e., the truncation functors on* $\mathrm{QCoh}(\mathcal{Y})$ *are compatible with filtered colimits (or, equivalently, the subcategory* $\mathrm{QCoh}(\mathcal{Y})^{>0}$ *is closed under filtered colimits).*

(b) *The t-structure on* $\mathrm{QCoh}(\mathcal{Y})$ *is* left-complete *and* right-complete *, i.e., for* $\mathcal{F} \in \mathrm{QCoh}(\mathcal{Y})$*, the natural maps*

$$\mathcal{F} \to \lim_{n\in\mathbb{N}} \tau^{\geq -n}(\mathcal{F})$$

$$\operatorname*{colim}_{n\in\mathbb{N}} \tau^{\leq n}(\mathcal{F}) \to \mathcal{F}$$

are isomorphisms, where τ *denotes the truncation functor.*

PROOF. Follows from Proposition 1.4.2(b) and the fact that both assertions are true for affine schemes, using the following lemma:

LEMMA 1.5.8. *Let*

$$I \to \mathrm{DGCat}_{\mathrm{cont}}, \quad i \mapsto \mathbf{C}_i$$

be a diagram of DG categories and continuous functors. Assume that each $\mathbf{C}_i$ *is endowed with a t-structure, and all of the transition functors* $F_{i,j} : \mathbf{C}_i \to \mathbf{C}_j$ *are t-exact. Set* $\mathbf{C} = \operatorname*{colim}_{\in I} \mathbf{C}_i$*. Then:*

(a) *The category* $\mathbf{C}$ *acquires a unique t-structure such that the evaluation functors* $\mathrm{ev}_i : \mathbf{C} \to \mathbf{C}_i$ *are t-exact;*

(b) *If the t-structure on each* $\mathbf{C}_i$ *is compatible with filtered colimits, then so is the one on* $\mathbf{C}$.

(c) *If the t-structure on each* $\mathbf{C}_i$ *is right-complete, then so is the one on* $\mathbf{C}$.

(d) *If the t-structure on each* $\mathbf{C}_i$ *is left-complete, then so is the one on* $\mathbf{C}$.

□

PROOF OF LEMMA 1.5.8. Only the last point is potentially non-obvious (because the transition functors $F_{i,j} : \mathbf{C}_i \to \mathbf{C}_i$ are not assumed to preserve limits). However, it follows from Chapter 1, Lemma 2.6.2.

□

2. Direct image for QCoh

So far we only know how the form the pullback of quasi-coherent sheaves for a map between prestacks. However, in order to have a richer theory, we should also develop the operation of *direct image.*

In general, the functor of direct image is quite ill-behaved. But there are exceptions: notably, when our morphism is schematic and quasi-compact. Or when one deals with Artin stacks and restricts oneself to the *eventually coconnective* (=bounded below) subcategory.

2.1. The functor of direct image. Recall that the functor QCoh^* for affine schemes was obtained by passing to left adjoints from $\mathrm{QCoh}_{\mathrm{Sch}^{\mathrm{aff}}}$, the latter being functorial with respect to the operation of direct image.

For prestacks we apply an inverse procedure: to get $\mathrm{QCoh}_{\mathrm{PreStk}}$ we pass to right adjoints in $\mathrm{QCoh}^*_{\mathrm{PreStk}}$.

2.1.1. Let $f : \mathcal{Y}_1 \to \mathcal{Y}_2$ be a morphism in PreStk, and consider the corresponding functor

$$f^* : \text{QCoh}(\mathcal{Y}_2) \to \text{QCoh}(\mathcal{Y}_1).$$

Applying Lurie's Adjoint Functor Theorem (see Chapter 1, Theorem 2.5.4), we obtain that the above functor f^* admits a *discontinuous* right adjoint, denoted

$$f_* : \text{QCoh}(\mathcal{Y}_1) \to \text{QCoh}(\mathcal{Y}_2).$$

REMARK 2.1.2. In fact, using Chapter 12, Corollary 1.3.4, we obtain that the assignment

$$\mathcal{Y} \rightsquigarrow \text{QCoh}(\mathcal{Y}), \quad (\mathcal{Y}_1 \xrightarrow{f} \mathcal{Y}_2) \rightsquigarrow (\text{QCoh}(\mathcal{Y}_1) \xrightarrow{f_*} \text{QCoh}(\mathcal{Y}_2)) \in \text{DGCat}$$

extends to a functor

$$\text{QCoh} : \text{PreStk} \to \text{DGCat},$$

whose restriction to $\text{Sch}^{\text{aff}} \subset \text{PreStk}$ is the composition of the functor $\text{QCoh}_{\text{Sch}^{\text{aff}}}$ of (1.2) with the forgetful functor $\text{DGCat}_{\text{cont}} \to \text{DGCat}$.

2.1.3. Let

$$\begin{array}{ccc} \mathcal{Y}_1' & \xrightarrow{g'} & \mathcal{Y}_1 \\ {\scriptstyle f'}\downarrow & & \downarrow{\scriptstyle f} \\ \mathcal{Y}_2' & \xrightarrow{g} & \mathcal{Y}_2 \end{array} \tag{2.1}$$

be a Cartesian square in PreStk. By adjunction, we obtain a natural transformation, known as the *base change morphism*

$$g^* \circ f_* \to f'_* \circ g'^*. \tag{2.2}$$

However, in general, (2.2) is *not* an isomorphism.

The simplest counter-example is provided by $\mathcal{Y}_2 = \text{pt}$, $\mathcal{Y}_2' = \mathbb{A}^1$ and $\mathcal{Y}_1$ be a countable disjoint union of copies of pt.

REMARK 2.1.4. The failure of the isomorphism (2.2) says that in general, the functor f_* is difficult to calculate. Concretely, for $\mathcal{F} \in \text{QCoh}(\mathcal{Y}_1)$ and $(S \xrightarrow{y} \mathcal{Y}_2) \in (\text{Sch}^{\text{aff}})_{/\mathcal{Y}_2}$ we do *not* have an explicit expression for $(f_*(\mathcal{F}))_{S,y} \in \text{QCoh}(S)$.

2.2. Direct image for schematic morphisms.
2.2.1. Above we saw that the direct image functor for a general morphism between prestacks does not have good properties. However, the situation improves considerably if we consider the class of *schematic quasi-compact morphisms*, see Chapter 2, Sect. 3.6.1 and 4.1.9 for what this means:

PROPOSITION 2.2.2. *Let $f : \mathcal{Y}_1 \to \mathcal{Y}_2$ be a morphism of prestacks. Assume that f is schematic and quasi-compact.*

(a) *The functor f_* is continuous.*

(b) *The base change property is satisfied, i.e., for any diagram* (2.1), *the map* (2.2) *is an isomorphism.*

PROOF. Note that in order to prove point (b), it is enough to consider the case when

$$\mathcal{Y}_2' = Z_2 \in \mathrm{Sch}^{\mathrm{aff}} \subset \mathrm{Sch}_{\mathrm{qc}},$$

where the super-script "qc" means quasi-compact. In this case $\mathcal{Y}_1' =: Z_1$ is also an object of $\mathrm{Sch}_{\mathrm{qc}}$, by Chapter 2, Proposition 3.6.2.

Note that from the transitivity of the procedure of right Kan extension, for a prestack $\mathcal{Y}$, the map

$$\mathrm{QCoh}(\mathcal{Y}) \to \lim_{(Z \to \mathcal{Y}) \in ((\mathrm{Sch}_{\mathrm{qc}})_{/\mathcal{Y}})^{\mathrm{op}}} \mathrm{QCoh}(Z)$$

is an equivalence.

Note also that the functor

$$(Z_2 \to \mathcal{Y}_2) \in ((\mathrm{Sch}_{\mathrm{qc}})_{/\mathcal{Y}_2})^{\mathrm{op}} \quad \rightsquigarrow Z_2 \underset{\mathcal{Y}_2}{\times} \mathcal{Y}_1 \in ((\mathrm{Sch}_{\mathrm{qc}})_{/\mathcal{Y}_1})^{\mathrm{op}}$$

is cofinal. Indeed, it admits a left adjoint given by

$$(Z_1 \to \mathcal{Y}_1) \mapsto (Z_1 \to \mathcal{Y}_1 \to \mathcal{Y}_2).$$

Hence, the functor

$$\mathrm{QCoh}(\mathcal{Y}_1) \to \lim_{(Z_2 \to \mathcal{Y}_2) \in ((\mathrm{Sch}_{\mathrm{qc}})_{/\mathcal{Y}})^{\mathrm{op}}} \mathrm{QCoh}(Z_2 \underset{\mathcal{Y}_2}{\times} \mathcal{Y}_1)$$

is an equivalence.

Hence, applying Chapter 1, Lemma 2.6.2 and 2.6.4, we obtain that it suffices to prove that the functor f_* is continuous for a morphism between quasi-compact schemes

$$W \xrightarrow{f} Z$$

and that the natural transformation (2.2) is an isomorphism when all the prestacks involved are quasi-compact schemes

$$\begin{array}{ccc} W' & \xrightarrow{g'} & W \\ {\scriptstyle f'}\downarrow & & \downarrow{\scriptstyle f} \\ Z' & \xrightarrow{g} & Z \end{array}$$

Applying Chapter 1, Lemma 2.6.2 and 2.6.4 again, and using the fact that the functor QCoh^* satisfies Zariski descent, we can assume that Z and Z' are affine. We will prove the assertion by induction on the number of affines by which we can cover W.

The base of the induction thus is when W (and hence also W') is affine. In this case, the functor f_* is the same functor as in (1.2), and hence is continuous. To check the isomorphism (2.2), it is enough to do so in the generator of $\mathrm{QCoh}(Z)$, i.e., on $\mathcal{O}_Z$, and this is a tautology.

Let now $W = U_1 \cup U_2$; denote $U_{1,2} := U_1 \underset{W}{\times} U_2$. Denote

$$f_1 := f|_{U_1},\ f_2 := f|_{U_2},\ f_{1,2} := f|_{U_{1,2}}.$$

By the induction hypothesis, we can assume that the assertion of the proposition holds for the morphisms $f_1, f_2, f_{1,2}$. However, it is easy to see that the functor f_* can be explicitly described as

$$\mathcal{F} \mapsto (f_1)_*(\mathcal{F}|_{U_1}) \underset{(f_{1,2})_*(\mathcal{F}|_{U_{1,2}})}{\times} (f_2)_*(\mathcal{F}|_{U_2}),$$

and this implies the required assertion for f_*.

□

REMARK 2.2.3. The following strengthening of Proposition 2.2.2 is established in [**DrGa1**, Corollary 1.4.5]:

Instead of requiring that $f : \mathcal{Y}_1 \to \mathcal{Y}_2$ be schematic quasi-compact, it suffices to ask that the base change of f be an affine scheme yields a *QCA algebraic stack*, see [**DrGa1**, Definition 1.1.8] for what this means.

2.2.4. Let us denote by $\mathrm{PreStk}_{\mathrm{sch,qc}}$ the 1-subcategory of PreStk where we restrict 1-morphisms to be schematic and quasi-compact.

Consider the functor

$$\mathrm{QCoh}^*_{\mathrm{PreStk}_{\mathrm{sch,qc\text{-}qs}}} : \mathrm{QCoh}^*_{\mathrm{PreStk}}|_{\mathrm{PreStk}_{\mathrm{sch,qc\text{-}qs}}} : (\mathrm{PreStk}_{\mathrm{sch,qc}})^{\mathrm{op}} \to \mathrm{DGCat}_{\mathrm{cont}}.$$

Combining Proposition 2.2.2(a) and Chapter 12, Corollary 1.3.4 for the target $(\infty, 2)$-category $\mathrm{DGCat}^{2\text{-Cat}}_{\mathrm{cont}}$, we obtain that by passing to right adjoints we can obtain from the functor $\mathrm{QCoh}^*_{\mathrm{PreStk}_{\mathrm{sch,qc\text{-}qs}}}$ a canonically defined functor

$$\mathrm{QCoh}_{\mathrm{PreStk}_{\mathrm{sch,qc\text{-}qs}}} : \mathrm{PreStk}_{\mathrm{sch,qc}} \to \mathrm{DGCat}_{\mathrm{cont}}.$$

By construction, the restriction of $\mathrm{QCoh}_{\mathrm{PreStk}_{\mathrm{sch,qc\text{-}qs}}}$ to $\mathrm{Sch}^{\mathrm{aff}}$ is the functor $\mathrm{QCoh}_{\mathrm{Sch}^{\mathrm{aff}}}$ of (1.2).

2.3. Direct image for a map between Artin stacks. Let $f : \mathcal{Y}_1 \to \mathcal{Y}_2$ be a map between Artin stacks. In general, the functor $f_* : \mathrm{QCoh}(\mathcal{Y}_1) \to \mathrm{QCoh}(\mathcal{Y}_2)$ will still be discontinuous. But the situation improves if one restricts one's attention to the bounded below subcategory.

2.3.1. Let $f : \mathcal{Y}_1 \to \mathcal{Y}_2$ be a map between Artin stacks. Assume that π is quasi-compact and quasi-separated (see Chapter 2, Sect. 4.1.9 for what this means).

We have:

PROPOSITION 2.3.2.

(a) *The restriction $f_*|_{\mathrm{QCoh}(\mathcal{Y}_1)^{\geq 0}}$ maps to $\mathrm{QCoh}(\mathcal{Y}_2)^{\geq 0}$, and commutes with filtered colimits.*

(b) *Let*

$$\begin{array}{ccc} \mathcal{Y}_1' & \xrightarrow{g'} & \mathcal{Y}_1 \\ {\scriptstyle f'}\downarrow & & \downarrow{\scriptstyle f} \\ \mathcal{Y}_2' & \xrightarrow{g} & \mathcal{Y}_2 \end{array}$$

be a Cartesian square, where the morphism g is flat. Then the diagram of functors

$$\begin{array}{ccc} \mathrm{QCoh}(\mathcal{Y}'_1)^+ & \xleftarrow{g'^*} & \mathrm{QCoh}(\mathcal{Y}_1)^+ \\ {\scriptstyle f'_*}\downarrow & & \downarrow{\scriptstyle f_*} \\ \mathrm{QCoh}(\mathcal{Y}'_2)^+ & \xleftarrow{g^*} & \mathrm{QCoh}(\mathcal{Y}_2)^+ \end{array}$$

is commutative.

PROOF. Let f be k-representable. We argue by induction on k, assuming that the assertion is true for $k' < k$. We will prove point (a); point (b) is proved similarly.

The base of the induction (i.e., the case of $k=0$) follows from Prop. 2.2.2(a).

Let

$$Z \to \mathcal{Y}_1$$

be a smooth (or even flat) atlas with $Z \in \mathrm{Sch}$. Let f^i denote the composition

$$Z^i/\mathcal{Y}_1 \to \mathcal{Y}_1 \xrightarrow{f} \mathcal{Y}_2.$$

By Corollary 1.3.11, for $\mathcal{F} \in \mathrm{QCoh}(\mathcal{Y}_1)$, we have:

$$f_*(\mathcal{F}) \simeq \mathrm{Tot}\left(f^\bullet_*(\mathcal{F}|_{Z^\bullet/\mathcal{Y}_1})\right).$$

Note that each f_i is $(k-1)$-representable, quasi-compact and quasi-separated.

By the induction hypothesis, for $\mathcal{F} \in \mathrm{QCoh}(\mathcal{Y}_1)^{\geq 0}$, each term of the co-simplicial object

$$i \mapsto f^i_*(\mathcal{F}|_{Z^i/\mathcal{Y}_1}) \tag{2.3}$$

is in $\mathrm{QCoh}(\mathcal{Y}_2)^{\geq 0}$. Hence, so is $\mathrm{Tot}\left(f^\bullet_*(\mathcal{F}|_{Z^\bullet/\mathcal{Y}_1})\right)$.

Recall that the t-structure on $\mathrm{QCoh}(\mathcal{Y}_2)$ is right-complete (see Cor. 1.5.7(b)). Hence, in order to show that that $f_*|_{\mathrm{QCoh}(\mathcal{Y}_1)^{\geq 0}}$ commutes with filtered colimits, it is enough to do so for its composition with the truncation functor

$$\tau^{\leq m} : \mathrm{QCoh}(\mathcal{Y}_2) \to \mathrm{QCoh}(\mathcal{Y}_2)^{\leq m}$$

for every $m \geq 0$.

Note, however, that since the terms of $f^\bullet_*(\mathcal{F}|_{Z^\bullet/\mathcal{Y}_1})$ belong to $\mathrm{QCoh}(\mathcal{Y}_2)^{\geq 0}$, we have

$$\tau^{\leq m}\left(\mathrm{Tot}\left(f^\bullet_*(\mathcal{F}|_{Z^\bullet/\mathcal{Y}_1})\right)\right) \simeq \tau^{\leq m}\left(\mathrm{Tot}^{\leq m}\left(f^\bullet_*(\mathcal{F}|_{Z^\bullet/\mathcal{Y}_1})\right)\right),$$

where $\mathrm{Tot}^{\leq m}$ denotes the limit over the subcategory $\mathbf{\Delta}^{\leq m} \subset \mathbf{\Delta}$.

Now, $\mathrm{Tot}^{\leq m}$ is a *finite* limit, and hence it preserves filtered colimits.

□

2.4. Classical algebraic stacks. In this subsection for $\mathcal{Y}$ a *classical algebraic stack*, we relate the category $\mathrm{QCoh}(\mathcal{Y})$ to some (potentially) more familiar notion.

2.4.1. According to [**Lu2**, Sect. 1.3.3], for any cocomplete stable ∞-category $\mathbf{C}$, equipped with a left and right complete t-structure, there is a canonically defined functor

$$D(\mathbf{C}^{\heartsuit}) \to \mathbf{C},$$

where $D(-)$ denotes the *derived stable ∞-category*, attached to a given abelian category, see [**Lu2**, Sect. 1.3.2]. In general, this functor is very far from being an equivalence.

In particular, for any Artin stack $\mathcal{Y}$ we obtain a canonical t-exact functor

$$D(\mathrm{QCoh}(\mathcal{Y})^{\heartsuit}) \to \mathrm{QCoh}(\mathcal{Y}).$$

2.4.2. Assume now that $\mathcal{Y}$ is a quasi-compact and quasi-separated algebraic stack (i.e., a 1-Artin stack), and assume that it is classical (see Chapter 2, Sect. 4.4.4) for what this means.

PROPOSITION 2.4.3. *Under the above circumstances, the functor*

$$D(\mathrm{QCoh}(\mathcal{Y})^{\heartsuit})^{+} \to \mathrm{QCoh}(\mathcal{Y})^{+}$$

is an equivalence.

REMARK 2.4.4. The above proposition implies that, under the specified assumptions, the category $\mathrm{QCoh}(\mathcal{Y})$ identifies with the left-completion of

$$D(\mathrm{QCoh}(\mathcal{Y})^{\heartsuit}).$$

We do not know what are the general conditions that guarantee that $D(\mathrm{QCoh}(\mathcal{Y})^{\heartsuit})$ itself is left-complete. For example, this is true for quasi-compact schemes. It is also easy to see that this is true for algebraic stacks of the form Z/G, where Z is a quasi-projective DG scheme and G an algebraic group acting linearly on Z (recall that we are working over a field of characteristic 0).

PROOF OF PROPOSITION 2.4.3. The proof will follow from the following general lemma:

LEMMA 2.4.5. *Let $\mathbf{C}$ be a DG category equipped with a t-structure that commutes with filtered colimits, and which is right-complete. Assume that for every object $\mathbf{c} \in \mathbf{C}^{\heartsuit}$ there exists an injection $\mathbf{c} \to \mathbf{c}_0$, where $\mathbf{c}_0 \in \mathbf{C}^{\heartsuit}$ is such that $\mathrm{Hom}_{\mathbf{C}}(\mathbf{c}', \mathbf{c}_0[n]) = 0$ for $n > 0$ and all $\mathbf{c}' \in \mathbf{C}^{\heartsuit}$. Then the natural functor*

$$D(\mathbf{C}^{\heartsuit})^{+} \to \mathbf{C}^{+}$$

is an equivalence.

We apply this lemma to $\mathbf{C} = \mathrm{QCoh}(\mathcal{Y})$. Let $f : S \to \mathcal{Y}$ be a map, where S is a classical affine scheme. Since the diagonal morphism of $\mathcal{Y}$ is affine, the map f itself is affine. Hence, by Proposition 1.5.4(a), if $\mathcal{F}_S \in \mathrm{QCoh}(S)^{\heartsuit}$, then $f_*(\mathcal{F}_S) \in \mathrm{QCoh}(\mathcal{Y})^{\heartsuit}$. Moreover, by Proposition 1.5.4(b), if f is flat and $\mathcal{F}_S \in \mathrm{QCoh}(S)^{\heartsuit}$ is injective, we have

$$\mathrm{Hom}_{\mathrm{QCoh}(\mathcal{Y})}(\mathcal{F}', f_*(\mathcal{F}_S)[n]) = 0, \ \forall\, \mathcal{F}' \in \mathrm{QCoh}(\mathcal{Y})^{\heartsuit}, \ \forall\, n > 0.$$

If $\mathcal{F}$ is an object of $\mathrm{QCoh}(\mathcal{Y})^{\heartsuit}$, let $f : S \to \mathcal{Y}$ be a flat atlas with $S \in \mathrm{Sch}^{\mathrm{aff}}$. Since $\mathcal{Y}$ was assumed classical, S is classical as well. Choose an injective $f^*(\mathcal{F}) \hookrightarrow \mathcal{F}_S$, and $\mathcal{F}$ embed into $f_*(\mathcal{F}_S)$.

□

REMARK 2.4.6. The same proof shows that (the homotopy category of)

$$D(\mathrm{QCoh}(\mathcal{Y})^{\heartsuit})^+$$

identifies with the eventually coconnective part of the quasi-coherent derived category of $\mathcal{Y}$ as defined in [**LM**].

3. The symmetric monoidal structure

In this section we will study the symmetric monoidal structure on QCoh^* as a functor, and the symmetric monoidal structure on $\mathrm{QCoh}(\mathcal{Y})$ as a category for a given prestack $\mathcal{Y}$.

3.1. The symmetric monoidal structure on QCoh as a functor. In this subsection we return to the setting of Sect. 1.1. We will show that the functor $\mathrm{QCoh}^*_{\mathrm{PreStk}}$ has a natural *right-lax symmetric monoidal structure*.

3.1.1. First, according to Chapter 1, Sect. 8.5.10, the functor

$$A \mapsto A\text{-mod}, \quad (\mathrm{AssocAlg}(\mathrm{Vect}))^{\mathrm{op}} \to \mathrm{DGCat}_{\mathrm{cont}}$$

has a natural symmetric monoidal structure, where AssocAlg(Vect) is viewed as a symmetric monoidal category via the operation of tensor product of algebras, and $\mathrm{DGCat}_{\mathrm{cont}}$ is viewed as a symmetric monoidal category via the Lurie tensor product.

Composing with the forgetful functors

$$\mathrm{ComAlg}(\mathrm{Vect}^{\leq 0}) \to \mathrm{ComAlg}(\mathrm{Vect}) \to \mathrm{AssocAlg}(\mathrm{Vect}),$$

we obtain that the functor

$$(\mathrm{ComAlg}(\mathrm{Vect}^{\leq 0}))^{\mathrm{op}} \to \mathrm{DGCat}_{\mathrm{cont}}, \quad A \mapsto A\text{-mod} \tag{3.1}$$

has a natural symmetric monoidal structure. Note that according to Chapter 1, Sect. 3.6.6, the symmetric monoidal structure on $(\mathrm{ComAlg}(\mathrm{Vect}^{\leq 0}))^{\mathrm{op}}$ is *Cartesian.*

3.1.2. Thus, we obtain that the functor

$$\mathrm{QCoh}_{\mathrm{Sch}^{\mathrm{aff}}} : \mathrm{Sch}^{\mathrm{aff}} \to \mathrm{DGCat}_{\mathrm{cont}}$$

has a naturally defined symmetric monoidal structure, where $\mathrm{Sch}^{\mathrm{aff}}$ is endowed with the Cartesian symmetric monoidal structure.

Applying Chapter 9, Sect. 3.1 (in the simplest case of *vert* = *horiz* = *adm* = all, *co-adm* = isom) we obtain that the functor

$$\mathrm{QCoh}^*_{\mathrm{Sch}^{\mathrm{aff}}} : (\mathrm{Sch}^{\mathrm{aff}})^{\mathrm{op}} \to \mathrm{DGCat}_{\mathrm{cont}}$$

also acquires a symmetric monoidal structure, where the symmetric monoidal structure on $(\mathrm{Sch}^{\mathrm{aff}})^{\mathrm{op}}$ is induced by the Cartesian symmetric monoidal structure on $\mathrm{Sch}^{\mathrm{aff}}$.

3.1.3. Finally, applying Chapter 9, Sect. 3.2 (in the simplest case of *vert* = *adm* = isom) we obtain that the functor

$$\mathrm{QCoh}^*_{\mathrm{PreStk}} : \mathrm{PreStk}^{\mathrm{op}} \to \mathrm{DGCat}_{\mathrm{cont}}$$

acquires a *right-lax* symmetric monoidal structure, where the symmetric monoidal structure on $(\mathrm{Sch}^{\mathrm{aff}})^{\mathrm{op}}$ is induced by the Cartesian symmetric monoidal structure on PreStk.

3.1.4. In concrete terms, the meaning of the above construction is that for $\mathcal{Y}_1, \mathcal{Y}_2 \in$ PreStk we have a canonically defined functor

$$\mathrm{QCoh}(\mathcal{Y}_1) \otimes \mathrm{QCoh}(\mathcal{Y}_2) \to \mathrm{QCoh}(\mathcal{Y}_1 \times \mathcal{Y}_2), \tag{3.2}$$

denoted

$$\mathcal{F}_1, \mathcal{F}_2 \mapsto \mathcal{F}_1 \boxtimes \mathcal{F}_2.$$

By construction, the functor (3.2) is an equivalence if $\mathcal{Y}_1, \mathcal{Y}_2 \in \mathrm{Sch}^{\mathrm{aff}}$.

3.1.5. The functor (3.2) can be explicitly described as follows.

$$\mathrm{QCoh}(\mathcal{Y}_1) \otimes \mathrm{QCoh}(\mathcal{Y}_2) \simeq \left(\underset{S_1 \overset{y_1}{\to} \mathcal{Y}_1}{\underset{\longleftarrow}{\lim}} \mathrm{QCoh}(S_1) \right) \otimes \left(\underset{S_2 \overset{y_2}{\to} \mathcal{Y}_2}{\underset{\longleftarrow}{\lim}} \mathrm{QCoh}(S_2) \right), \tag{3.3}$$

whereas

$$\mathrm{QCoh}(\mathcal{Y}_1 \times \mathcal{Y}_2) \simeq \underset{S \overset{y}{\to} \mathcal{Y}_1 \times \mathcal{Y}_2}{\underset{\longleftarrow}{\lim}} \mathrm{QCoh}(S).$$

However, the functor

$$(\mathrm{Sch}^{\mathrm{aff}})_{/\mathcal{Y}_1} \times (\mathrm{Sch}^{\mathrm{aff}})_{/\mathcal{Y}_2} \to (\mathrm{Sch}^{\mathrm{aff}})_{/\mathcal{Y}}, \quad S_1, S_2 \mapsto S_1 \times S_2$$

is cofinal, so we can rewrite

$$\mathrm{QCoh}(\mathcal{Y}_1 \times \mathcal{Y}_2) \simeq \underset{S_1 \overset{y_1}{\to} \mathcal{Y}_1, S_2 \overset{y_2}{\to} \mathcal{Y}_2}{\lim} \mathrm{QCoh}(S_1) \otimes \mathrm{QCoh}(S_2). \tag{3.4}$$

Now, the map (3.2) is the tautological map from (3.3) to (3.4) (swapping the limit with the tensor product).

3.1.6. We claim:

PROPOSITION 3.1.7. *Assume that $\mathcal{Y}_1$ is such that the category* QCoh($\mathcal{Y}_1$), *viewed as an object of* $\mathrm{DGCat}_{\mathrm{cont}}$, *is dualizable. Then for any $\mathcal{Y}_2$, the functor* (3.2) *is an equivalence.*

PROOF. We need to show that the map from (3.3) to (3.4) is an isomorphism. We can write it as a composition

$$\mathrm{QCoh}(\mathcal{Y}_1) \otimes \left(\underset{S_2 \overset{y_2}{\to} \mathcal{Y}_2}{\lim} \mathrm{QCoh}(S_2) \right) \to \underset{S_2 \overset{y_2}{\to} \mathcal{Y}_2}{\lim} \mathrm{QCoh}(\mathcal{Y}_1) \otimes \mathrm{QCoh}(S_2) \simeq$$
$$\simeq \underset{S_2 \overset{y_2}{\to} \mathcal{Y}_2}{\lim} \left(\left(\underset{S_1 \overset{y_1}{\to} \mathcal{Y}_1}{\lim} \mathrm{QCoh}(S_1) \right) \otimes \mathrm{QCoh}(S_2) \right) \to$$
$$\to \underset{S_2 \overset{y_2}{\to} \mathcal{Y}_2}{\lim} \left(\underset{S_1 \overset{y_1}{\to} \mathcal{Y}_1}{\lim} (\mathrm{QCoh}(S_1) \otimes \mathrm{QCoh}(S_2)) \right).$$

The first arrow is an isomorphism since tensoring with a dualizable category commutes with limits (see Chapter 1, Sect. 4.3.2). The third arrow is an isomorphism for the same reason, as QCoh(S) for an affine scheme S is dualizable.

□

3.2. The symmetric monoidal structure on QCoh **of a prestack.** In this subsection we will use the right-lax symmetric monoidal structure on the functor $\mathrm{QCoh}^*_{\mathrm{PreStk}}$ to construct a symmetric monoidal structure on each $\mathrm{QCoh}(\mathcal{Y})$ for $\mathcal{Y} \in \mathrm{PreStk}$.

3.2.1. Being right-lax symmetric monoidal, the functor $\mathrm{QCoh}^*_{\mathrm{PreStk}}$ sends commutative algebra objects in $\mathrm{PreStk}^{\mathrm{op}}$ to commutative algebra objects on $\mathrm{DGCat}_{\mathrm{cont}}$.

However, since the symmetric monoidal structure on $\mathrm{PreStk}^{\mathrm{op}}$ is coCartesian, the forgetful functor

$$\mathrm{ComAlg}(\mathrm{PreStk}^{\mathrm{op}}) \to \mathrm{PreStk}^{\mathrm{op}}$$

is an equivalence (see [**Lu2**, Corollary 2.4.3.10]).

We obtain that the functor $\mathrm{QCoh}^*_{\mathrm{PreStk}}$ naturally lifts to a functor

$$\mathrm{PreStk}^{\mathrm{op}} \to \mathrm{ComAlg}(\mathrm{DGCat}_{\mathrm{cont}}) =: \mathrm{DGCat}^{\mathrm{SymMon}}_{\mathrm{cont}}.$$

3.2.2. In other words, for any $\mathcal{Y} \in \mathrm{PreStk}$, the DG category $\mathrm{QCoh}(\mathcal{Y})$ acquires a canonical symmetric monoidal structure, explicitly given by

$$\mathcal{F}_1, \mathcal{F}_2 \mapsto \mathcal{F}_1 \otimes \mathcal{F}_2 := (\mathrm{diag}_{\mathcal{Y}})^*(\mathcal{F}_1 \boxtimes \mathcal{F}_2).$$

Furthermore, for a morphism $f : \mathcal{Y}' \to \mathcal{Y}$ in PreStk, the functor

$$f^* : \mathrm{QCoh}(\mathcal{Y}) \to \mathrm{QCoh}(\mathcal{Y}')$$

is naturally symmetric monoidal.

3.2.3. Consider again the direct image functor

$$f_* : \mathrm{QCoh}(\mathcal{Y}') \to \mathrm{QCoh}(\mathcal{Y}).$$

Being a right adjoint to a symmetric monoidal functor, the functor f_* is *right-lax symmetric monoidal* (this the commutative version of Chapter 1, Lemma 3.2.4).

In particular, for $\mathcal{F} \in \mathrm{QCoh}(\mathcal{Y})$, $\mathcal{F}' \in \mathrm{QCoh}(\mathcal{F}')$, we have a canonically defined map

$$\mathcal{F} \otimes f_*(\mathcal{F}') \to f_*(f^*(\mathcal{F}) \otimes \mathcal{F}')), \tag{3.5}$$

called the *projection formula map*.

In general, the map (3.5) is *not* an isomorphism. However, as in Proposition 2.2.2, one shows:

LEMMA 3.2.4. *Assume that f is schematic quasi-compact. Then the map* (3.5) *is an isomorphism.*

3.2.5. Let $f : \mathcal{Y}' \to \mathcal{Y}$ be again a map between prestacks. Since the functor f_* is right-lax (symmetric) monoidal, it maps algebras in $\text{QCoh}(\mathcal{Y}')$ to algebras in $\text{QCoh}(\mathcal{Y})$.

In particular, the object

$$f_*(\mathcal{O}_{\mathcal{Y}'}) = \mathcal{A} \in \text{QCoh}(\mathcal{Y})$$

has a natural structure of commutative algebra.

Moreover, by Chapter 1, Sect. 3.7.3, the functor f_* naturally factors as

$$\text{QCoh}(\mathcal{Y}') \to \mathcal{A}\text{-mod}(\text{QCoh}(\mathcal{Y})) \overset{\mathbf{oblv}_{\mathcal{A}}}{\longrightarrow} \text{QCoh}(\mathcal{Y}).$$

In general, the above functor

$$\text{QCoh}(\mathcal{Y}') \to \mathcal{A}\text{-mod}(\text{QCoh}(\mathcal{Y})) \tag{3.6}$$

is *not* an equivalence.

3.2.6. Let

$$\begin{array}{ccc} \mathcal{Y}'_1 & \longrightarrow & \mathcal{Y}_1 \\ {\scriptstyle f'}\downarrow & & \downarrow{\scriptstyle f} \\ \mathcal{Y}'_2 & \overset{g}{\longrightarrow} & \mathcal{Y}_2 \end{array}$$

be a Cartesian square of prestacks.

Note that we have a canonical map:

$$\text{QCoh}(\mathcal{Y}_1) \underset{\text{QCoh}(\mathcal{Y}_2)}{\otimes} \text{QCoh}(\mathcal{Y}'_2) \to \text{QCoh}(\mathcal{Y}'_1). \tag{3.7}$$

In general, the functor (3.7) is *not* an equivalence. Here is a counter-example:

Take $\mathcal{Y}_1 = \mathcal{Y}_2 = \text{pt}$, and $\mathcal{Y} = \text{pt}/A$, where A is an abelian variety. Then $\mathcal{Y}_1 \underset{\mathcal{Y}}{\times} \mathcal{Y}_2 \simeq A$, while

$$\text{QCoh}(\text{pt}/A) \simeq H\text{-mod},$$

where $H = (\Gamma(A, \mathcal{O}_A))^\vee$ is an algebra with respect to convolution, and is isomorphic to $\text{Sym}(H^1(X, \mathcal{O}_A)^\vee[1])$. So

$$\text{Vect} \underset{H\text{-mod}}{\otimes} \text{Vect} \simeq \text{Sym}(H^1(X, \mathcal{O}_A)^\vee[2])\text{-mod}.$$

3.3. The quasi-affine case.

3.3.1. We shall say that an object $X \in \text{Sch}$ is *quasi-affine* if it quasi-compact and admits an open embedding into an affine scheme.

We shall say that a morphism $f : \mathcal{Y}' \to \mathcal{Y}$ in PreStk is *schematic quasi-affine* if its base change by an affine scheme yields a quasi-affine scheme.

3.3.2. We claim:

PROPOSITION 3.3.3. *Let $f : \mathcal{Y}' \to \mathcal{Y}$ be schematic quasi-affine. Then the functor* (3.6) *is an equivalence.*

PROOF. Let first f be arbitrary. We note that we have a canonical homomorphism of monads acting on $\operatorname{QCoh}(\mathcal{Y})$.

$$\mathcal{A} \otimes - \to f_* \circ f^*.$$

Suppose that f is schematic and quasi-compact. In this case, by Lemma 3.2.4, the above map of monads is an isomorphism. Hence, in order to prove the proposition, it remains to show that the functor f_* satisfies the hypothesis of the Barr-Beck-Lurie theorem, see Chapter 1, Proposition 3.7.7.

Now, since f was assumed schematic quasi-compact, the functor f_* commutes with all colimits, by Proposition 2.2.2(a). Thus, it remains to show that f_* is conservative. By Proposition 2.2.2(b), the latter assertion reduces to the case when $\mathcal{Y} \in \operatorname{Sch}^{\mathrm{aff}}$.

Thus, it remains to show that the functor of *global sections* on QCoh of a quasi-affine scheme X is conservative. Let $j : X \hookrightarrow S$ be an open embedding, where $S \in \operatorname{Sch}^{\mathrm{aff}}$. Since the functor of global sections over S is conservative, it remains to show that the functor j_* is fully faithful.

However, we claim that j_* admits a left inverse, namely, j^*. Indeed, this follows from Proposition 2.2.2(b) for the Cartesian square

$$\begin{array}{ccc} X & \xrightarrow{\mathrm{id}} & X \\ {\scriptstyle \mathrm{id}}\downarrow & & \downarrow{\scriptstyle j} \\ X & \xrightarrow{j} & S. \end{array}$$

□

3.3.4. Here is another favorable feature of quasi-affine maps:

PROPOSITION 3.3.5. *Assume in the situation of Sect. 3.2.6, the map f (and hence f') is quasi-affine. Then then map* (3.7) *is an equivalence.*

PROOF. By Proposition 3.3.3

$$\operatorname{QCoh}(\mathcal{Y}'_1) \simeq f'_*(\mathcal{O}_{\mathcal{Y}'_1})\text{-mod}(\operatorname{QCoh}(\mathcal{Y}'_2)).$$

Now, again by Proposition 3.3.3 and Chapter 1, Corollary 8.5.7,

$$\operatorname{QCoh}(\mathcal{Y}_1) \underset{\operatorname{QCoh}(\mathcal{Y}_2)}{\otimes} \operatorname{QCoh}(\mathcal{Y}'_2) \simeq f_*(\mathcal{O}_{\mathcal{Y}_1})\text{-mod}(\operatorname{QCoh}(\mathcal{Y}_2)) \underset{\operatorname{QCoh}(\mathcal{Y}_2)}{\otimes} \operatorname{QCoh}(\mathcal{Y}'_2) \simeq$$
$$\simeq g^*(f_*(\mathcal{O}_{\mathcal{Y}_1}))\text{-mod}(\operatorname{QCoh}(\mathcal{Y}'_2)).$$

Finally, by Proposition 2.2.2(2),

$$f'_*(\mathcal{O}_{\mathcal{Y}'_1}) \simeq g^*(f_*(\mathcal{O}_{\mathcal{Y}_1}))$$

as algebras in $\operatorname{QCoh}(\mathcal{Y}'_2)$.

□

3.4. When is QCoh rigid?

3.4.1. Recall the notion of *rigid* stable monoidal category, Chapter 1, Sect. 9.1.

The following assertion provides a partial converse to Proposition 3.1.7:

PROPOSITION 3.4.2. *Let $\mathcal{Y}$ be a prestack, such that the diagonal map* $\text{diag}_{\mathcal{Y}}$ *is schematic and quasi-compact, and such that the object* $\mathcal{O}_{\mathcal{Y}} \in \text{QCoh}(\mathcal{Y})$ *is compact. Then the following conditions are equivalent:*

- (i) *The functor* $\text{QCoh}(\mathcal{Y}) \otimes \text{QCoh}(\mathcal{Y}') \to \text{QCoh}(\mathcal{Y} \times \mathcal{Y}')$ *is an equivalence for any* $\mathcal{Y}'$.
- (ii) *The functor* $\text{QCoh}(\mathcal{Y}) \otimes \text{QCoh}(\mathcal{Y}) \to \text{QCoh}(\mathcal{Y} \times \mathcal{Y})$ *is an equivalence.*
- (iii) *The category* $\mathcal{Y}$ *is rigid as a stable monoidal category.*
- (iv) *The category* $\text{QCoh}(\mathcal{Y})$ *is dualizable.*

PROOF. The implicatios (i) $\Rightarrow$ (ii) is tautological, and the implication (iii) $\Rightarrow$ (iv) follows froom Chapter 1, Sect. 9.2.1.

The implication (iv) $\Rightarrow$ (i) is the content of Proposition 3.1.7. It remains to show (ii) $\Rightarrow$ (iii).

Given (ii), we can identify the map $\text{mult}^*_{\text{QCoh}(\mathcal{Y})}$ (we are using the notation of Chapter 1, Sect. 9.1.1) with

$$(\text{diag}_{\mathcal{Y}})^* : \text{QCoh}(\mathcal{Y} \times \mathcal{Y}) \to \text{QCoh}(\mathcal{Y}).$$

The fact that it satisfies the assumptions of *loc. cit.* follows from Prop. 2.2.2(b). □

3.4.3. Let $f : \mathcal{Y}_1 \to \mathcal{Y}_2$ be a map between prestacks, such that both $\text{QCoh}(\mathcal{Y}_1)$ and $\text{QCoh}(\mathcal{Y}_2)$ are rigid. From Chapter 1, Lemma 9.2.6(2) we obtain:

LEMMA 3.4.4. *The functor* $f_* : \text{QCoh}(\mathcal{Y}_1) \to \text{QCoh}(\mathcal{Y}_2)$ *is continuous, and under the identifications*

$$\text{QCoh}(\mathcal{Y}_i)^\vee \simeq \text{QCoh}(\mathcal{Y}_i),$$

we have $f_* \simeq (f^*)^\vee$.

3.5. Passable prestacks.

3.5.1. We shall say that a prestack $\mathcal{Y}$ is *passable* if

(1) The diagonal morphism of $\mathcal{Y}$ is quasi-affine;

(2) $\mathcal{O}_{\mathcal{Y}} \in \text{QCoh}(\mathcal{Y})$ is compact;

(3) The category $\text{QCoh}(\mathcal{Y})$ is dualizable.

For example, any stack which is perfect (see Sect. 3.7.1) below is passable. In particular, any quasi-compact scheme is passable when viewed as a prestack.

By Proposition 3.4.2, we obtain that if $\mathcal{Y}$ is passable, then $\text{QCoh}(\mathcal{Y})$ is rigid as a monoidal category.

3.5.2. We are going to show that passable prestacks are adapted to having the map in (3.7) an equivalence:

Let $\mathcal{Y}$ be a passable prestack. Let $\mathcal{Y}_1$ and $\mathcal{Y}_2$ be prestacks mapping to $\mathcal{Y}$.

PROPOSITION 3.5.3. *If under the above circumstances* $\mathrm{QCoh}(\mathcal{Y}_1)$ *is dualizable as a category, the natural functor*

$$\mathrm{QCoh}(\mathcal{Y}_1) \underset{\mathrm{QCoh}(\mathcal{Y})}{\otimes} \mathrm{QCoh}(\mathcal{Y}_2) \to \mathrm{QCoh}(\mathcal{Y}_1 \underset{\mathcal{Y}}{\times} \mathcal{Y}_2)$$

is an equivalence.

PROOF. By Chapter 1, Propositions 9.4.4 and Sect. 4.3.2, the rigidity of $\mathrm{QCoh}(\mathcal{Y})$ and the fact that $\mathrm{QCoh}(\mathcal{Y}_1)$ is dualizable imply that the operation

$$\mathbf{C} \mapsto \mathrm{QCoh}(\mathcal{Y}_1) \underset{\mathrm{QCoh}(\mathcal{Y})}{\otimes} \mathbf{C}$$

preserves limits.

This allows to replace $\mathcal{Y}_2$ by $S \in \mathrm{Sch}^{\mathrm{aff}}$. But then the map $S \to \mathcal{Y}$ is quasi-compact and quasi-affine, and we find ourselves in the situation of Proposition 3.3.5. □

3.6. The perfect subcategory.

3.6.1. Recall the notion of *dualizable* object in a symmetric monoidal category, see Chapter 1, Sect. 4.1.

For a prestack $\mathcal{Y}$, we let

$$\mathrm{QCoh}(\mathcal{Y})^{\mathrm{perf}} \subset \mathrm{QCoh}(\mathcal{Y})$$

denote the full subcategory consisting of dualizable objects.

For a map $f : \mathcal{Y}' \to \mathcal{Y}$, the functor $f^* : \mathrm{QCoh}(\mathcal{Y}) \to \mathrm{QCoh}(\mathcal{Y}')$ clearly sends $\mathrm{QCoh}(\mathcal{Y})^{\mathrm{perf}}$ to $\mathrm{QCoh}(\mathcal{Y}')^{\mathrm{perf}}$.

Thus, we obtain a well-defined functor

$$\mathcal{Y} \mapsto \mathrm{QCoh}(\mathcal{Y})^{\mathrm{perf}}, \quad \mathrm{PreStk}^{\mathrm{op}} \to 1\text{-}\mathrm{Cat}. \tag{3.8}$$

3.6.2. We have the following basic assertion:

LEMMA 3.6.3. *Let*

$$I \to 1\text{-}\mathrm{Cat}^{\mathrm{Mon}}, \quad i \mapsto \mathbf{A}_i$$

be a functor, and denote $\mathbf{A} := \lim\limits_i \mathbf{A}_i$. *Then an object* $\mathbf{a} \in \mathbf{A}$ *is left dualizable if and only if* $\mathrm{ev}_i(\mathbf{a}) \in \mathbf{A}_i$ *is dualizable for every* i.

As a corollary we obtain:

COROLLARY 3.6.4.

(a) *An object* $\mathcal{F} \in \mathrm{QCoh}(\mathcal{Y})$ *is perfect if and only if for every* $(S \overset{y}{\to} Y) \in (\mathrm{Sch}^{\mathrm{aff}})_{/\mathcal{Y}}$, *the corresponding* $\mathcal{F}_{S,y} \in \mathrm{QCoh}(S)$ *is perfect.*

(b) *The functor* (3.8) *maps isomorphically to the right Kan extension to its restriction to* $\mathrm{Sch}^{\mathrm{aff}}$.

Moreover, combing with Theorem 1.3.4, we obtain:

COROLLARY 3.6.5. *The restriction of the functor* (3.8) *to* $\mathrm{Sch}^{\mathrm{aff}}$ *satisfies flat descent.*

3.6.6. *Perfectness and compactness.* Let S be an affine DG scheme. We recall the following (this is a particular case of Chapter 1, Corollary 9.1.7):

LEMMA 3.6.7. *For $M \in \mathrm{QCoh}(S)$ the following conditions are equivalent:*

(i) *M is compact;*

(ii) *M is dualizable as an object of the symmetric monoidal category $\mathrm{QCoh}(S)$.*

We now claim:

PROPOSITION 3.6.8.

(1) *Suppose that the diagonal morphism of $\mathcal{Y}$ is schematic and quasi-compact. Then any compact object of $\mathrm{QCoh}(\mathcal{Y})$ is perfect.*

(2) *Suppose that $\mathcal{O}_{\mathcal{Y}} \in \mathrm{QCoh}(\mathcal{Y})$ is compact. Then any perfect object of $\mathrm{QCoh}(\mathcal{Y})$ is compact.*

PROOF. Point (2) follows from Chapter 1, Lemma 8.8.4: dualizability implies compactness in any symmetric monoidal stable category in which the unit is compact.

To prove point (1), taking into account Lemma 3.6.7, we have to show that the functor $\mathcal{F} \mapsto f^*(\mathcal{F})$ for $f : S \to \mathcal{Y}$ with S affine, sends compact objects to compact ones. However, this is true, since the right adjoint of f^*, i.e., the functor f_* is continuous, by Proposition 2.2.2(a).

□

3.6.9. Finally, we note:

PROPOSITION 3.6.10. *The functor*

$$S \mapsto \mathrm{QCoh}(S)^{\mathrm{perf}}, \quad (\mathrm{Sch}^{\mathrm{aff}})^{\mathrm{op}} \to 1\text{-}\mathrm{Cat}$$

is convergent.

PROOF. By definition, we need to show that for $S \in \mathrm{Sch}^{\mathrm{aff}}$, the family of functors

$$\mathcal{F} \mapsto \mathcal{F}|_{\leq n_S},$$

given by restriction, defines an equivalence

$$\mathrm{QCoh}(S)^{\mathrm{perf}} \to \lim_n \mathrm{QCoh}({}^{\leq n}S)^{\mathrm{perf}}.$$

However, we claim that more is true: namely, the functor

$$\mathrm{QCoh}(S)^- \to \lim_n \mathrm{QCoh}({}^{\leq n}S)^-$$

is an equivalence[1]. Indeed, its inverse is given by sending a compatible family $\mathcal{F}_n \in \mathrm{QCoh}({}^{\leq n}S)^-$ to

$$\lim_n (i_n)_*(\mathcal{F}_n),$$

where i_n denotes the tautological map ${}^{\leq n}S \to S$.

□

[1]Note, however, that the corresponding fact is *false* for all of QCoh.

3.7. Perfect prestacks.

3.7.1. Following [**BFN**], we shall say that a prestack $\mathcal{Y}$ is perfect if

(1) The diagonal morphism of $\mathcal{Y}$ is affine;
(2) The functor $\text{Ind}(\text{QCoh}(\mathcal{Y})^{\text{perf}}) \to \text{QCoh}(\mathcal{Y})$ is an equivalence.

3.7.2. By Proposition 3.6.8, the above conditions can be reformulated as follows:

(1) The diagonal morphism of $\mathcal{Y}$ is affine,
(2) $\mathcal{O}_{\mathcal{Y}} \in \text{QCoh}(\mathcal{Y})$ is compact,
(3) $\text{QCoh}(\mathcal{Y})$ is compactly generated.

3.7.3. Since every compactly generated category is dualizable, we obtain that every perfect stack is passable, see Sect. 3.5.1 for what this means.

3.7.4. *Examples.* In [**BFN**], following the arguments of [**TT**] and [**Ne**], it is shown that any quasi-compact scheme, considered as a prestack, is perfect.

Moreover, in [**BFN**] it is shown that if $\mathcal{Y}$ is of the form X/G, where G is an algebraic group and X is a scheme endowed with a G-equivariant ample line bundle, then $\mathcal{Y}$ is perfect (under our assumption that the ground field k is of char. 0).

Part II

Ind-coherent Sheaves

Introduction

1. Ind-coherent sheaves vs quasi-coherent sheaves

One of the primary goals of this book is to construct the theory of ind-coherent sheaves as a theory of $\mathcal{O}$-modules on prestacks that exists alongside the theory of quasi-coherent sheaves.

We shall now try to explain what we mean by a 'theory', and highlight the formal features that the two theories have in common and those that set them apart.

1.1. For us QCoh is ultimately a functor

$$\mathrm{QCoh}^*_{\mathrm{PreStk}} : (\mathrm{PreStk})^{\mathrm{op}} \to \mathrm{DGCat}_{\mathrm{cont}}.$$

I.e., it is a functorial assignment

$$(\mathcal{X} \in \mathrm{PreStk}) \rightsquigarrow (\mathrm{QCoh}(\mathcal{X}) \in \mathrm{DGCat}_{\mathrm{cont}})$$

and

$$(\mathcal{X} \xrightarrow{f} \mathcal{Y}) \rightsquigarrow (f^* : \mathrm{QCoh}(\mathcal{Y}) \to \mathrm{QCoh}(\mathcal{X})).$$

Moreover, the functor $\mathrm{QCoh}^*_{\mathrm{PreStk}}$ has a natural right-lax symmetric monoidal structure, where PreStk is a symmetric monoidal category with respect to the Cartesian product, and $\mathrm{DGCat}_{\mathrm{cont}}$ is symmetric monoidal category with respect to the $\otimes$ tensor product of DG categories.

NB: Here it is of crucial importance that we work with $\mathrm{DGCat}_{\mathrm{cont}}$ (and not DGCat): the operation of tensor product of DG categories is only functorial with respect to continuous (i.e., colimit preserving) functors.

Thus, for $\mathcal{X}, \mathcal{Y} \in \mathrm{PreStk}$, we have a well-defined functor

$$\mathrm{QCoh}(\mathcal{X}) \otimes \mathrm{QCoh}(\mathcal{Y}) \to \mathrm{QCoh}(\mathcal{X} \times \mathcal{Y}), \quad \mathcal{F}, \mathcal{G} \mapsto \mathcal{F} \boxtimes \mathcal{G}. \tag{1.1}$$

1.2. The functor (1.1) is an equivalence if $\mathcal{X}$ and $\mathcal{Y}$ are schemes (in fact, it is an equivalence of just one of them is a scheme).

The functor $\mathrm{QCoh}^*_{\mathrm{PreStk}}$ has the following features:

(i) If $\mathcal{X} \xrightarrow{f} \mathcal{Y}$ is a *schematic and quasi-compact* morphism between prestacks, the above functor f^* admits a *continuous* right adjoint

$$f_* : \mathrm{QCoh}(\mathcal{X}) \to \mathrm{QCoh}(\mathcal{Y}).$$

Moreover, if

$$\begin{array}{ccc} \mathcal{X}' & \xrightarrow{g_{\mathcal{X}}} & \mathcal{X} \\ {\scriptstyle f'}\downarrow & & \downarrow{\scriptstyle f} \\ \mathcal{Y}' & \xrightarrow{g_{\mathcal{Y}}} & \mathcal{Y} \end{array}$$

is a Cartesian diagram of prestacks with vertical maps being schematic, the natural transformation of functors

$$g_{\mathcal{Y}}^* \circ f_* \to f'_* \circ g_{\mathcal{X}}^*, \quad \mathrm{QCoh}(\mathcal{X}) \rightrightarrows \mathrm{QCoh}(\mathcal{Y}')$$

that *arises by adjunction* from the isomorphism of functors

$$(f')^* \circ g_{\mathcal{Y}}^* \simeq g_{\mathcal{X}}^* \circ f^*,$$

is an isomorphism.

(ii) If $\mathcal{X} = X \in \mathrm{Sch}$, then the functor

$$\mathrm{QCoh}(X) \otimes \mathrm{QCoh}(X) \simeq \mathrm{QCoh}(X \times X) \xrightarrow{\Delta_X^*} \mathrm{QCoh}(X) \xrightarrow{\Gamma(X,-)} \mathrm{Vect}$$

defines the counit of a duality, thereby giving rise to an equivalence

$$\mathbf{D}_X^{\mathrm{naive}} : \mathrm{QCoh}(X)^\vee \to \mathrm{QCoh}(X).$$

In the above formula $\Gamma(X,-)$ is the functor $(p_X)_* : \mathrm{QCoh}(X) \to \mathrm{QCoh}(\mathrm{pt}) = \mathrm{Vect}$, where p_X is the tautological projection $X \to \mathrm{pt}$.

1.3. Here is what the theory of IndCoh will do. First and foremost it will be a functor

$$\mathrm{IndCoh}^!_{\mathrm{PreStk}_{\mathrm{laft}}} : (\mathrm{PreStk}_{\mathrm{laft}})^{\mathrm{op}} \to \mathrm{DGCat}_{\mathrm{cont}}.$$

I.e., it is a functorial assignment

$$(\mathcal{X} \in \mathrm{PreStk}_{\mathrm{laft}}) \rightsquigarrow (\mathrm{IndCoh}(\mathcal{X}) \in \mathrm{DGCat}_{\mathrm{cont}})$$

and

$$(\mathcal{X} \xrightarrow{f} \mathcal{Y}) \rightsquigarrow (f^! : \mathrm{IndCoh}(\mathcal{Y}) \to \mathrm{IndCoh}(\mathcal{X})).$$

As in the case of QCoh, the functor $\mathrm{IndCoh}^!_{\mathrm{PreStk}_{\mathrm{laft}}}$ has a natural right-lax symmetric monoidal structure.

If we work over the ground field of characteristic 0 (which is our assumption throughout), then the corresponding functor

$$\mathrm{IndCoh}(\mathcal{X}) \otimes \mathrm{IndCoh}(\mathcal{Y}) \to \mathrm{IndCoh}(\mathcal{X} \times \mathcal{Y}), \quad \mathcal{F}, \mathcal{G} \mapsto \mathcal{F} \boxtimes \mathcal{G}. \tag{1.2}$$

is an equivalence if either $\mathcal{X}$ or $\mathcal{Y}$ is a scheme.

Already here, there is one piece of difference from the case of QCoh: the functor (1.2) is guaranteed to be an equivalence on a far larger class of algebro-geometric objects. Namely, it suffices to require that $\mathcal{X}$ (or $\mathcal{Y}$) be an *inf-scheme*. We refer the reader to Volume II, Chapter 2 where it is explained what inf-schemes are. Here we will just say that this is a class of prestacks that includes formal schemes and de Rham prestacks of schemes, and is closed under fiber products.

1.4. Here are some features of the functor $\mathrm{IndCoh}^!_{\mathrm{PreStk}_{\mathrm{laft}}}$:

(i) If $\mathcal{X} \xrightarrow{f} \mathcal{Y}$ is a schematic (more generally, *inf-schematic*) morphism between prestacks, we have a well-defined continuous functor

$$f_*^{\mathrm{IndCoh}} : \mathrm{IndCoh}(\mathcal{X}) \to \mathrm{IndCoh}(\mathcal{Y}),$$

and if if

$$\begin{array}{ccc} \mathcal{X}' & \xrightarrow{g_{\mathcal{X}}} & \mathcal{X} \\ {\scriptstyle f'}\downarrow & & \downarrow{\scriptstyle f} \\ \mathcal{Y}' & \xrightarrow{g_{\mathcal{Y}}} & \mathcal{Y} \end{array}$$

is a Cartesian diagram of laft prestacks with vertical maps being schematic (more generally, *inf-schematic*), then we are *given* an isomorphism of functors

$$g^!_{\mathcal{Y}} \circ f_*^{\mathrm{IndCoh}} \to (f')_*^{\mathrm{IndCoh}} \circ g^!_{\mathcal{X}}, \quad \mathrm{QCoh}(\mathcal{X}) \rightrightarrows \mathrm{QCoh}(\mathcal{Y}'). \tag{1.3}$$

However, unlike the case of QCoh, for a general f, the functor f_*^{IndCoh} is not the adjoint of $f^!$ on either side. In particular, the isomorphism (1.3) does not come by adjunction from some *a priori* defined map. So, (1.3) is really an additional piece of data.

That said, if f is an open embedding, it is stipulated that f_*^{IndCoh} should be the *right* adjoint of $f^!$, and in this case, the map $\to$ in (1.3) should come by adjunction from the isomorphism

$$(f')^! \circ g^!_{\mathcal{Y}} \simeq g^!_{\mathcal{X}} \circ f^!.$$

Also, it is stipulated that if f is proper, then f_*^{IndCoh} should be the *left* adjoint of $f^!$, and in this case, the map $\leftarrow$ in (1.3) should come by adjunction from the isomorphism

$$(f')^! \circ g^!_{\mathcal{Y}} \simeq g^!_{\mathcal{X}} \circ f^!.$$

(ii) If $\mathcal{X} = X \in \mathrm{Sch}$ (more generally, $\mathcal{X}$ can be an inf-scheme), then the functor

$$\mathrm{IndCoh}(X) \otimes \mathrm{IndCoh}(X) \simeq \mathrm{IndCoh}(X \times X) \xrightarrow{\Delta^!_X} \mathrm{IndCoh}(X) \xrightarrow{\Gamma^{\mathrm{IndCoh}}(X,-)} \mathrm{Vect}$$

defines the counit of a duality, thereby giving rise to an equivalence

$$\mathbf{D}_X^{\mathrm{Serre}} : \mathrm{IndCoh}(X)^\vee \to \mathrm{IndCoh}(X).$$

In the above formula $\Gamma^{\mathrm{IndCoh}}(X,-)$ is the functor

$$(p_X)_*^{\mathrm{IndCoh}} : \mathrm{IndCoh}(X) \to \mathrm{IndCoh}(\mathrm{pt}) = \mathrm{Vect},$$

where p_X is the tautological projection $X \to \mathrm{pt}$.

1.5. To summarise, we can say that the category $\mathrm{IndCoh}(\mathcal{X})$ and the functor f_*^{IndCoh} is guaranteed to be better behaved on a larger class of objects and morphisms (than QCoh and f_*).

But the nature of the relationship between pullbacks and push-forwards for IndCoh is quite different from that of QCoh.

Finally, we should say that there will exist a natural transformation

$$\mathrm{QCoh}^*_{\mathrm{PreStk}}|_{\mathrm{PreStk}_{\mathrm{laft}}} =: \mathrm{QCoh}^*_{\mathrm{PreStk}_{\mathrm{laft}}} \overset{\Upsilon_{\mathrm{PreStk}_{\mathrm{laft}}}}{\longrightarrow} \mathrm{IndCoh}^!_{\mathrm{PreStk}_{\mathrm{laft}}}$$

as (symmetric monoidal) functors

$$(\mathrm{PreStk}_{\mathrm{laft}})^{\mathrm{op}} \to \mathrm{DGCat}_{\mathrm{cont}}.$$

The corresponding functor

$$\Upsilon_{\mathcal{X}} : \mathrm{QCoh}(\mathcal{X}) \to \mathrm{IndCoh}(\mathcal{X})$$

will, of course, not be an equivalence in general. However:

(a) If $\mathcal{X} = X \in \mathrm{Sch}_{\mathrm{aft}}$, then Υ_X is an equivalence if and only if X is a smooth classical scheme.

(b) If $\mathcal{X} = X_{\mathrm{dR}}$, for $X \in \mathrm{Sch}_{\mathrm{aft}}$, the functor $\Upsilon_{X_{\mathrm{dR}}}$ is always an equivalence.

2. How to construct IndCoh?

One should say that it is quite a long way to construct IndCoh having the above pieces of structure: it will take us all of Part II and Volume II, Part I of the book to do so. Here we will outline the strategy of how this is done.

2.1. In Chapter 4 we begin by constructing the category $\mathrm{IndCoh}(X)$ for an individual object $X \in \mathrm{Sch}_{\mathrm{aft}}$.

We start with the usual category $\mathrm{QCoh}(X)$ and consider its (non-cocomplete) subcategory $\mathrm{Coh}(X) \subset \mathrm{QCoh}(X)$ consisting of bounded complexes with coherent cohomologies. We let $\mathrm{IndCoh}(X)$ to be the ind-completion of of $\mathrm{Coh}(X)$.

We obtain that $\mathrm{IndCoh}(X)$ is a compactly generated category, equipped with a t-structure and a tautologically defined t-exact functor

$$\Psi_X : \mathrm{IndCoh}(X) \to \mathrm{QCoh}(X)$$

that induces an equivalence on the eventually coconnective subcategories, i.e., the corresponding functors

$$\mathrm{IndCoh}(X)^{\geq -n} \to \mathrm{QCoh}(X)^{\geq -n}$$

are equivalences for any n.

Thus, $\mathrm{IndCoh}(X)$ begins life as a 'small modification' of $\mathrm{QCoh}(X)$–the two categories only differ at $-\infty$. But once we construct IndCoh as a full-fledged theory, it will be quite different from QCoh, as was explained in Sect. 1 above.

2.2. Having defined the category IndCoh(X) for an individual object $X \in \mathrm{Sch}_{\mathrm{aft}}$ we proceed to defining the *-push forward functor

$$f_*^{\mathrm{IndCoh}} : \mathrm{IndCoh}(X) \to \mathrm{IndCoh}(Y)$$

for a morphism $f : X \to Y$ between schemes.

The functor f_*^{IndCoh} is essentially inherited from QCoh: it is uniquely determined by the requirement that it should be left t-exact and make the diagram

$$\begin{array}{ccc} \mathrm{IndCoh}(X) & \xrightarrow{\Psi_X} & \mathrm{QCoh}(X) \\ {\scriptstyle f_*^{\mathrm{IndCoh}}}\downarrow & & \downarrow{\scriptstyle f_*} \\ \mathrm{IndCoh}(Y) & \xrightarrow{\Psi_Y} & \mathrm{QCoh}(Y) \end{array}$$

commute.

Furthermore, we show that the assigment

$$X \rightsquigarrow \mathrm{IndCoh}(X), \quad (X \xrightarrow{f} Y) \rightsquigarrow f_*^{\mathrm{IndCoh}}$$

naturally extends to a functor

$$\mathrm{IndCoh}_{\mathrm{Sch}_{\mathrm{aft}}} : \mathrm{Sch}_{\mathrm{aft}} \to \mathrm{DGCat}_{\mathrm{cont}} . \tag{2.1}$$

2.3. Our subsequent task is to construct the !-pullback functors for IndCoh, equipped with base change isomorphisms (1.3) against *-push forwards.

When a map $X \xrightarrow{f} Y$ is proper, we define $f^!$ to be the *right* adjoint of f_*^{IndCoh}, and when it is an open embedding, we define $f^!$ to be the *left* adjoint of f_*^{IndCoh}.

In each of these cases, base change against *-push forwards is a property and not an additional piece of structure, because the corresponding map in one direction[1] comes by adjunction from a tautological isomorphism.

For a general f, we decompose it as a composition

$$f = f_1 \circ f_2 \tag{2.2}$$

with f_1 an open embedding and f_2 a proper map, and we wish to define $f^!$ to be $f_2^! \circ f_1^!$. The challenge is to show that definition is canonically independent of the decomposition (2.2), and that it is functorial with respect to compositions of maps.

Furthermore, we need to show that $f^!$ thus defined is equipped with base change isomorphisms (1.3), and that these isomorphisms are compatible with compositions etc. However, before proving these compatibilities, we need to formulate them in the ∞-categorical level, and this brings us to the paradigm of the *category of correspondences*.

[1] But the direction of the map is different for proper maps and open embeddings.

2.4. In Chapter 5 we introduce, following a suggestion of J. Lurie, an $(\infty, 2)$-category, denoted $\mathrm{Corr}(\mathrm{Sch}_{\mathrm{aft}})^{\mathrm{proper}}$.

Its objects are $X \in \mathrm{Sch}_{\mathrm{aft}}$. The $(\infty, 1)$-category of morphisms between X_0 and X_1 has as objects diagrams

$$\begin{array}{ccc} X_{0,1} & \xrightarrow{g} & X_0 \\ {\scriptstyle f}\downarrow & & \\ X_1. & & \end{array}$$

and as morphisms (i.e., 2-morphisms in $\mathrm{Corr}(\mathrm{Sch}_{\mathrm{aft}})^{\mathrm{proper}}$) diagrams

$$\begin{array}{ccccc} X^s_{0,1} & & & & \\ & \overset{h}{\searrow} & & \overset{f^s}{\searrow} & \\ {\scriptstyle g^s}\downarrow & & X^t_{0,1} & \xrightarrow{f^t} & X_0 \\ & & \downarrow{\scriptstyle g^t} & & \\ & & X_1, & & \end{array}$$

where h is proper and the superscripts 's' and 't' stand for 'source' and 'target', respectively.

This $(\infty, 2)$-category is equipped with 1-fully faithful functors

$$\mathrm{Sch}_{\mathrm{aft}} \to \mathrm{Corr}(\mathrm{Sch}_{\mathrm{aft}})^{\mathrm{proper}} \leftarrow (\mathrm{Sch}_{\mathrm{aft}})^{\mathrm{op}}. \tag{2.3}$$

2.5. We refer the reader to the introduction to Chapter 5, where it is explained that a proper way to encode IndCoh equipped with both functorialities (!-pullback and *-pushforward) is a functor

$$\mathrm{IndCoh}_{\mathrm{Corr}(\mathrm{Sch}_{\mathrm{aft}})^{\mathrm{proper}}} : \mathrm{Corr}(\mathrm{Sch}_{\mathrm{aft}})^{\mathrm{proper}} \to \mathrm{DGCat}_{\mathrm{cont}}, \tag{2.4}$$

whose restriction to $\mathrm{Sch}_{\mathrm{aft}}$ (under the functor $\to$ in (2.3)) is the functor

$$\mathrm{IndCoh}_{\mathrm{Sch}_{\mathrm{aft}}} : \mathrm{Sch}_{\mathrm{aft}} \to \mathrm{DGCat}_{\mathrm{cont}}$$

of (2.1), and whose restriction to $(\mathrm{Sch}_{\mathrm{aft}})^{\mathrm{op}}$ (under the functor $\leftarrow$ in (2.3)) is the functor

$$\mathrm{IndCoh}^!_{\mathrm{Sch}_{\mathrm{aft}}} : (\mathrm{Sch}_{\mathrm{aft}})^{\mathrm{op}} \to \mathrm{DGCat}_{\mathrm{cont}},$$

encoding the !-pullback.

Thus, in order to construct the theory of IndCoh on schemes, we need to extend the functor (2.1) to a functor (2.4). We prove in Chapter 5, Theorem 2.1.4 that such an extension exists and is unique.

2.6. Having thus constructed the theory of IndCoh on schemes, we need to extend it to prestacks, so that is satisfies (i) from Sect. 1.4.

This is done by the procedure of right Kan extension on the suitable categories of correspondences.

The extension from schemes to inf-schemes (resp., allowing inf-schematic maps between prestacks instead of schematic ones) requires quite a bit more work, and will be the subject of Volume II, Part I of the book.

2.7. Finally, we show that the functor $\mathrm{IndCoh}_{\mathrm{Corr}(\mathrm{Sch}_{\mathrm{aft}})^{\mathrm{proper}}}$ has a natural symmetric monoidal structure.

From here we formally deduce the Serre duality structure on $\mathrm{IndCoh}(X)$ for $X \in \mathrm{Sch}_{\mathrm{aft}}$, mentioned in (ii) from Sect. 1.4.

2.8. By the construction of $\mathrm{IndCoh}(X)$ for a scheme X, it carries an action of the (symmetric) monoidal category $\mathrm{QCoh}(X)$.

In Chapter 6 we formulate and prove how this structure is compatible with the functor $\mathrm{IndCoh}_{\mathrm{Corr}(\mathrm{Sch}_{\mathrm{aft}})^{\mathrm{proper}}}$ of (2.4).

One consequence of this compatibility is the canonically defined natural transformation

$$\Upsilon_{\mathrm{Sch}_{\mathrm{aft}}} : \mathrm{QCoh}^*_{\mathrm{Sch}_{\mathrm{aft}}} \to \mathrm{IndCoh}^!_{\mathrm{Sch}_{\mathrm{aft}}}$$

that *right-Kan-extends* to the natural transformation

$$\Upsilon_{\mathrm{PreStk}_{\mathrm{laft}}} : \mathrm{QCoh}^*_{\mathrm{PreStk}_{\mathrm{laft}}} \to \mathrm{IndCoh}^!_{\mathrm{PreStk}_{\mathrm{laft}}}$$

mentioned in Sect. 1.5.

NB: for a scheme X we have a pair of functors

$$\mathrm{IndCoh}(X) \overset{\Psi_X}{\to} \mathrm{QCoh}(X) \text{ and } \mathrm{QCoh}(X) \overset{\Upsilon_X}{\to} \mathrm{IndCoh}(X).$$

We will show that these functors are *mutually dual*, where we identify

$$\mathrm{QCoh}(X)^\vee \simeq \mathrm{QCoh}(X) \text{ and } \mathrm{IndCoh}(X)^\vee \simeq \mathrm{IndCoh}(X)$$

via the functors $\mathbf{D}_X^{\mathrm{naive}}$ and $\mathbf{D}_X^{\mathrm{Serre}}$, respectively.

We note also that whereas the functor

$$\Upsilon_{\mathcal{X}} : \mathrm{QCoh}(\mathcal{X}) \to \mathrm{IndCoh}(\mathcal{X})$$

is defined for any prestack $\mathcal{X}$, the functor $\Psi_{\mathcal{X}}$ is *not*; the latter is really a feature of schemes (or, more generally, Artin stacks). So, the functor Ψ_X that was so necessary for the initial stages of the development of IndCoh in a sense loses its significance further along the development of the theory.

CHAPTER 4

Ind-coherent sheaves on schemes

Introduction

In this Chapter we initiate the study of ind-coherent sheaves, which is the main subject of this book. Here we will define and study the category $\text{IndCoh}(X)$ for X being a scheme (assumed almost of finite type), and its basic functorality properties for maps of schemes.

In subsequent Chapters will be extend the definition IndCoh to a much wider class of algebro-geometric objects, namely, prestacks locally almost of finite type. The latter will allow us to create a paradigm that contains both D-modules and $\mathcal{O}$-modules.

0.1. Why IndCoh**?** The basic question is: why bother with IndCoh? I.e., why is the usual QCoh not good enough?

0.1.1. There are multiple reasons for why one would like to have the theory of IndCoh. Here are two mutually related reasons that can be spelled out already for schemes.

(i) For a proper morphism $f : X \to Y$ between schemes, the functor of !-pullback, *right adjoint* to the *-direct image, is *not necessarily continuous* when viewed as a functor

$$\text{QCoh}(Y) \to \text{QCoh}(X).$$

But it is continuous, when viewed as a functor $\text{IndCoh}(Y) \to \text{IndCoh}(X)$. Since for various reasons, explained elsewhere in the book, we wanted to stay within the world of cocomplete categories and continuous functors, the above phenomenon was, for us, the main reason to introduce and study IndCoh.

(ii) Many categories that naturally arise in geometric representation theory are $\text{IndCoh}(X)$ (for some scheme X), and not $\text{QCoh}(X)$. A remarkable set of examples of this are the categories appearing on the spectral side of the geometric Langlands theory (see, e.g., [**Bezr**] ot [**AG**]). A baby example of this would be the Koszul duality that says that the category A-mod for

$$A = k[\xi], \quad \deg(\xi) = 2$$

is equivalent to $\text{IndCoh}(X)$, where $X = \text{pt} \underset{\mathbb{A}^1}{\times} \text{pt}$. This is while $\text{QCoh}(X)$ is the subcategory of A-mod consisting of objects on which the generator ξ acts locally nilpotently.

0.1.2. We should emphasize, however, that one should not be tempted to think that IndCoh is a 'better object' than QCoh. In fact, both categories are needed and they interact in interesting ways, see Chapter 6.

0.1.3. We would also like to mention that the category $\mathrm{IndCoh}(X)$ has appeared significantly before the present book (and its predecessor [**Ga1**]). Namely, if X is classical, it was introduced in the work of H. Krause [**Kr**], and it was subsequently studied by him and his collaborators.

Specifically, in *loc.cit.*, $\mathrm{IndCoh}(X)$ appeared as the category of *injective complexes* on X.

0.1.4. In this chapter we start from (i) mentioned above, and develop the theory of IndCoh for schemes so that the !-pullback for a proper morphism is continuous.

In Chapter 5, Sect. 2.1 we will expand the functoriality of IndCoh by showing that it admits !-pullbacks for arbitrary (i.e., not necessarily proper) morphisms, and that these !-pullbacks satisfy *base change* against *-push forwards. The difficulty here is that base change is not a property but an extra piece of structure, and one needs to introduce a new categorical device, the *category of correspondences* to account for it[1].

0.1.5. Having defined !-pullbacks for arbitrary morphisms, we will be able to define $\mathrm{IndCoh}(\mathcal{X})$, where $\mathcal{X}$ is now an object of $\mathrm{PreStk}_{\mathrm{laft}}$, see Chapter 5, Sect. 3.4. We should emphasize that, whereas in the case of schemes IndCoh can be thought of as a small modification of QCoh, for general prestacks the two categories are very different. The former is functorial with respect to the !-pullback, and the latter is functorial with respect to the *-pullback.

For a map $f : \mathcal{X} \to \mathcal{Y}$ between prestacks we will have the !-pullback functor

$$f^! : \mathrm{IndCoh}(\mathcal{Y}) \to \mathrm{IndCoh}(\mathcal{X}).$$

However, for a general f there is no conceivable way to define the *-push forward functor from $\mathrm{IndCoh}(\mathcal{X})$ to $\mathrm{IndCoh}(\mathcal{Y})$ so that it satisfies base change against the !-pullback.

That said, in Volume II, Part I of the book we will single a class of morphsisms, called *inf-schematic*, for which the push-forward functor f_*^{IndCoh} is defined and has the desired base change property.

This will allow to extend the formalism of IndCoh as a functor out of the category of correspondences from schemes to *inf-schemes* (these are algebro-geometric objects that include formal schemes as well as de Rham prestacks of schemes). In this way we will obtain a convenient formalism that allows to treat D-modules and $\mathcal{O}$-modules within the same framework.

0.2. What is done in this chapter?

0.2.1. In Sect. 1 we introduce $\mathrm{IndCoh}(X)$ for a scheme X. We show that it is endowed with a t-structure and a t-exact functor

$$\Psi_X : \mathrm{IndCoh}(X) \to \mathrm{QCoh}(X),$$

which induces an equivalence on the eventually coconnective subcategories, i.e., the induced functor $\mathrm{IndCoh}(X)^+ \to \mathrm{QCoh}(X)^+$ is an equivalence.

Thus, $\mathrm{IndCoh}(X)$ is only different from QCoh 'at $-\infty$'. So, one can say that the whole point here is *convergence*, i.e., convergence of spectral sequences.

[1]The idea of the category of correspondences was suggested to us by J. Lurie.

0.2.2. In Sect. 2 we introduce the direct image functor

$$f_*^{\mathrm{IndCoh}} : \mathrm{IndCoh}(X) \to \mathrm{IndCoh}(Y)$$

for a morphism $f : X \to Y$ between schemes.

This functor is 'inherited' from QCoh via the equivalence

$$\Psi : \mathrm{IndCoh}(-)^+ \to \mathrm{QCoh}(-)^+.$$

We then extend the assignment

$$X \rightsquigarrow \mathrm{IndCoh}(X), \quad f \rightsquigarrow f_*^{\mathrm{IndCoh}}$$

to a functor

$$\mathrm{Sch} \to \mathrm{DGCat}_{\mathrm{cont}}.$$

0.2.3. In Sect. 3 we study the functor of (the usual) *-pullback

$$f^{\mathrm{IndCoh},*} : \mathrm{IndCoh}(Y) \to \mathrm{IndCoh}(X)$$

for a morphism $f : X \to Y$. This functor is supposed to be the *left* adjoint of f_*^{IndCoh}.

However, there is a caveat: the functor $f^{\mathrm{IndCoh},*}$ is only defined for morphisms f that are *of finite Tor amplitude*. A functor that is defined for all morphisms is introduced in Sect. 5: this is the !-pullback.

It is fair to say that QCoh is well-adapted to the *-pullback and IndCoh is well-adapted to the !-pullback. (But if f is of finite Tor amplitude, both functors exist and are continuous for both categories.)

0.2.4. In Sect. 4 we study the behavior of IndCoh under open embeddings. In particular, we show that it satisfies Zariski descent.

0.2.5. Beyond the definition of IndCoh, Sect. 5 is the central in this chapter. In this section we show that if $f : X \to Y$ is a proper morphism, then the functor f_*^{IndCoh} admits a continuous right adjoint, denoted

$$f^! : \mathrm{IndCoh}(Y) \to \mathrm{IndCoh}(X).$$

We show that the !-pullback (so far only defined for proper maps) satisfies base change against the *-push forward (unlike the general base change, this instance of base change is a *property* and *not an extra piece of structure*).

Finally, we establish the following crucial piece of compatibility that will eventually imply that the !-pullback is defined for all maps. Let

$$\begin{array}{ccc} X' & \xrightarrow{g_X} & X \\ {\scriptstyle f'}\downarrow & & \downarrow{\scriptstyle f} \\ Y' & \xrightarrow{g_Y} & Y \end{array}$$

be a Cartesian diagram, where the vertical arrows are proper and the horizontal ones are open embeddings.

In this case, we have a canonically defined natural transformation

$$g_X^{\mathrm{IndCoh},*} \circ f^! \to f'^! \circ g_Y^{\mathrm{IndCoh},*}.$$

We show that this natural transformation is an isomorphism.

0.2.6. In Sect. 6 we study several additional properties of the assignment

$$X \rightsquigarrow \text{IndCoh}(X).$$

We show:

(i) For a closed subscheme $Y \subset X$, the subcategory $\text{IndCoh}_Y(X)$ of $\text{IndCoh}(X)$ consisting of objects that vanish when restricted to $X - Y$, is compactly generated by $\text{Coh}_Y(X)$.

(ii) If $f : X \to Y$ is a proper and point-wise surjective map, then the functor $f^! : \text{IndCoh}(Y) \to \text{IndCoh}(X)$ is conservative.

(iii) For two schemes X_1 and X_2, the external tensor product functor

$$\text{IndCoh}(X_1) \otimes \text{IndCoh}(X_2) \to \text{IndCoh}(X_1 \times X_2)$$

is an equivalence.

(iv) The assignment $X \rightsquigarrow \text{IndCoh}(X)$ is *convergent* in the sense of Chapter 2, Sect. 1.4, i.e., the functor

$$\text{IndCoh}(X) \to \lim_n \text{IndCoh}({}^{\leq n}X)$$

is an equivalence, where ${}^{\leq n}X$ denotes the n-coconnective truncation of X. Note that the corresponding assertion is *false* for QCoh.

0.2.7. Finally, in Sect. 7 we establish the proper descent for IndCoh: if $X \to Y$ is a proper map, which is surjective at the level of k-points, then the functor

$$\text{IndCoh}(Y) \to \text{Tot}(\text{IndCoh}(X^\bullet))$$

is an equivalence, where $X^\bullet$ is the simplicial scheme equal to the Čech nerve of $X \to Y$. In Chapter 5 we will strengthen this, and show that IndCoh satisfies h-descent (and in particular, ppf descent).

1. Ind-coherent sheaves on a scheme

In this section we introduce the category $\text{IndCoh}(X)$ and study its basic properties. The material here repeats [**Ga1**, Sect. 1].

1.1. Definition of the category.

In this subsection we define $\text{IndCoh}(X)$ and the functors that connect it to the usual category $\text{QCoh}(X)$ of quasi-coherent sheaves.

1.1.1. For $X \in \text{Sch}_{\text{aft}}$ we consider the category $\text{QCoh}(X)$ and its full (but not cocomplete) subcategory $\text{Coh}(X)$, consisting of bounded complexes with coherent cohomologies.

We define the category $\text{IndCoh}(X)$ by

$$\text{IndCoh}(X) := \text{Ind}(\text{Coh}(X)).$$

1.1.2. By construction, we have a naturally defined functor

$$\Psi_X : \mathrm{IndCoh}(X) \to \mathrm{QCoh}(X)$$

obtained by ind-extension of the tautological inlcusion $\mathrm{Coh}(X) \hookrightarrow \mathrm{QCoh}(X)$.

We have:

LEMMA 1.1.3. *Assume that X is a smooth classical scheme. Then Ψ_X is an equivalence.*

PROOF. It is known by [**TT**] (see also [**Ne**]) that for X classical, $\mathrm{QCoh}(X) \simeq \mathrm{Ind}(\mathrm{QCoh}(X)^{\mathrm{perf}})$. Now, for X a regular classical scheme, we have

$$\mathrm{QCoh}(X)^{\mathrm{perf}} = \mathrm{Coh}(X),$$

as subcategories of $\mathrm{QCoh}(X)$. □

REMARK 1.1.4. It is shown in [**Ga1**, Proposition 1.5.4] that the assertion of the above lemma is in fact 'if and only if'.

1.1.5. We give the following definition:

DEFINITION 1.1.6. *We shall say that $X \in \mathrm{Sch}_{\mathrm{aft}}$ is* eventually coconnective *if the structure sheaf $\mathcal{O}_X$ belongs to* $\mathrm{Coh}(X)$.

I.e., X is eventually coconnective if, Zariski locally, the structure sheaf had non-zero cohomologies in finitely many degrees.

We have:

LEMMA 1.1.7. *If X is eventually coconnective, the functor Ψ_X admits a left adjoint, to be denoted Ξ_X, and this left adjoint is fully faithful.*

PROOF. If X is eventually coconnective, we have

$$\mathrm{QCoh}(X)^{\mathrm{perf}} \subset \mathrm{Coh}(X),$$

and, using the fact that $\mathrm{QCoh}(X) \simeq \mathrm{Ind}(\mathrm{QCoh}(X)^{\mathrm{perf}})$, the functor Ξ_X is obtained as the ind-extension of the above embedding.

The composition $\Psi_X \circ \Xi_X$ is the ind-extension of the functor

$$\mathrm{QCoh}(X)^{\mathrm{perf}} \hookrightarrow \mathrm{Coh}(X) \hookrightarrow \mathrm{QCoh}(X),$$

and it is manifest that its map to $\mathrm{Id}_{\mathrm{QCoh}(X)}$ is an isomorphism. □

REMARK 1.1.8. In [**Ga1**, Proposition 1.5.2] it is shown that Ψ_X admits a left adjoint *if and only if* X is eventually coconnective.

1.2. t-structure. Some of the most basic operations on the IndCoh category (such as the functor of direct image studied in the next section) are inherited from those on QCoh using the t-structures on both categories. The crucial fact is that the *eventually coconnective* (a.k.a., bounded below) parts of the two categories are equivalent.

The goal of this subsection is to define the t-structure on $\mathrm{IndCoh}(X)$ and establish its basic properties.

1.2.1. We claim:

PROPOSITION 1.2.2. *The category* IndCoh(X) *carries a unique t-structure that satisfies*

$$\mathrm{IndCoh}(X)^{\leq 0} = \{\mathcal{F} \in \mathrm{IndCoh}(X) \,|\, \Psi_X(\mathcal{F}) \in \mathrm{QCoh}(X)^{\leq 0}\}.$$

Moreover:

(a) *The functor* Ψ_X *is t-exact.*

(b) *This t-structure is compatible with filtered colimits (i.e., the subcategory* IndCoh$(X)^{\geq 0}$ *is closed under filtered colimits).*

(c) *The induced functor*

$$\Psi_X : \mathrm{IndCoh}(X)^{\geq n} \to \mathrm{QCoh}(X)^{\geq n}$$

is an equivalence for any n.

As a corollary we obtain:

COROLLARY 1.2.3. *The functor* Ψ_X *defines an equivalence* IndCoh$(X)^+ \to$ QCoh$(X)^+$.

PROOF OF PROPOSITION 1.2.2. It is clear that the condition of the proposition determines the t-structure uniquely. To establish its properties we will use the following general assertion:

LEMMA 1.2.4. *Let* $\mathbf{C}_0$ *be a (non-cocomplete) DG category, endowed with a t-structure. Then* $\mathbf{C} := \mathrm{Ind}(\mathbf{C}_0)$ *carries a unique t-structure, which is compatible with filtered colimits, and for which the tautological inclusion* $\mathbf{C}_0 \hookrightarrow \mathbf{C}$ *is t-exact. Moreover:*

(1) *The subcategory* $\mathbf{C}^{\leq 0}$ *(resp.,* $\mathbf{C}^{\geq 0}$*) is compactly generated under filtered colimits by* $\mathbf{C}_0^{\leq 0}$ *(resp.,* $\mathbf{C}_0^{\geq 0}$*).*

(2) *Let* $\mathbf{D}$ *be another DG category endowed with a t-structure which is compatible with filtered colimits, and let* $F : \mathbf{C} \to \mathbf{D}$ *a continuous functor. Then* F *is t-exact (resp., left t-exact, right t-exact) if and only if* $F|_{\mathbf{C}_0}$ *is.*

We apply Lemma 1.2.4(1) to $\mathbf{C}_0 = \mathrm{Coh}(X)$ and obtain a (a priori different) t-structure on IndCoh(X), which satisfies point (b) of the proposition. It also satisfies point (a) of the proposition, by Lemma 1.2.4(2) applied to $\mathbf{D} = \mathrm{QCoh}(X)$ and $F = \Psi_X$.

To show that this t-structure coincides with the one introduced earlier, it suffices to show that Ψ_X is conservative when restricted to IndCoh$(X)^{\geq 0}$. Hence, it remains to show that the t-structure, given by Lemma 1.2.4, satisfies points (c) of the proposition.

To prove point (c), it is sufficient to consider the case of $n = 0$. Using Lemma 1.2.4(1), the required assertion follows from the next statement:

LEMMA 1.2.5. *The category* QCoh$(X)^{\geq 0}$ *is compactly generated under filtered colimits by* Coh$(X)^{\geq 0}$.

□

1.2.6. Note that Proposition 1.2.2 implies that, as long as the functor Ψ_X is *not* an equivalence (i.e., X is not a classical smooth scheme), the category $\text{IndCoh}(X)$ is *not* left-complete in its t-structure. The latter means that for $\mathcal{F} \in \text{IndCoh}(X)$, the canonical arrow

$$\mathcal{F} \to \lim_n \tau^{\geq -n}(\mathcal{F})$$

is not necessarily an isomorphism.

Furthermore, we see that for any X, the functor Ψ_X realizes $\text{QCoh}(X)$ as the left completion of $\text{IndCoh}(X)$.

1.2.7. From Proposition 1.2.2 we also obtain the following:

COROLLARY 1.2.8. *The inclusion* $\text{Coh}(X) \subset \text{IndCoh}(X)^c$ *is an equality.*

PROOF. Since the category $\text{Coh}(X)$ compactly generates $\text{IndCoh}(X)$, the category $\text{IndCoh}(X)^c$ is the Karoubi-completion of $\text{Coh}(X)$. I.e., every object $\mathcal{F} \in \text{IndCoh}(X)^c$ can be realized as a direct summand of an object $\mathcal{F}' \in \text{Coh}(X)$. In particular, $\mathcal{F} \in \text{IndCoh}(X)^+$.

The object $\Psi_X(\mathcal{F})$ is a direct summand of $\Psi_X(\mathcal{F}')$. Hence, $\Psi_X(\mathcal{F})$, regarded as an object of $\text{QCoh}(X)$, belongs to $\text{Coh}(X)$. Let us denote this object of $\text{Coh}(X)$ by $\widetilde{\mathcal{F}}$.

Thus, we can regard $\mathcal{F}$ and $\widetilde{\mathcal{F}}$ as objects of $\text{IndCoh}(X)^+$ such that

$$\Psi_X(\mathcal{F}) \simeq \Psi_X(\widetilde{\mathcal{F}}).$$

Applying Proposition 1.2.2(c), we obtain that $\mathcal{F} \simeq \widetilde{\mathcal{F}}$ as objects of $\text{IndCoh}(X)$. □

1.2.9. *The monoidal action of* QCoh. We claim:

PROPOSITION 1.2.10. *There exists a uniquely defined monoidal action of* $\text{QCoh}(X)$, *viewed as a monoidal category, on* $\text{IndCoh}(X)$, *such that the functor* Ψ_X *is compatible with the* $\text{QCoh}(X)$*-actions.*

PROOF. The action in question is obtained by ind-extension of the action of the non-cocomplete monoidal category $\text{QCoh}(X)^{\text{perf}}$ on $\text{Coh}(X)$.

To prove uniqueness, by Corollaries 1.2.3 and 1.2.8, it suffices to show that, given an action of $\text{QCoh}(X)$ on $\text{IndCoh}(X)$, the objects of $\text{QCoh}(X)^{\text{perf}} \subset \text{QCoh}(X)$ map compact objects of $\text{IndCoh}(X)$ to compact ones. However, this follows from the fact that objects in $\text{QCoh}(X)^{\text{perf}}$ are dualizable in the monoidal category $\text{QCoh}(X)$. □

2. The direct image functor

The assignment

$$X \rightsquigarrow \text{IndCoh}(X)$$

is 'very functorial'. However, all of this functoriality is born from a single cource: the operation of direct image, defined in this section.

2.1. Direct image for an individual morphism. In this subsection we perform the first step in developing the formalism of direct image for IndCoh: we define the corresponding functor for one given morphism between schemes.

2.1.1. Let $f : X \to Y$ be a morphism in $\mathrm{Sch}_{\mathrm{aft}}$. We claim:

PROPOSITION 2.1.2. *There exists a uniquely defined functor*

$$f_*^{\mathrm{IndCoh}} : \mathrm{IndCoh}(X) \to \mathrm{IndCoh}(Y)$$

that is left t-exact and makes the diagram

$$\begin{array}{ccc} \mathrm{IndCoh}(X) & \xrightarrow{\Psi_X} & \mathrm{QCoh}(X) \\ {\scriptstyle f_*^{\mathrm{IndCoh}}}\downarrow & & \downarrow{\scriptstyle f_*} \\ \mathrm{IndCoh}(Y) & \xrightarrow{\Psi_Y} & \mathrm{QCoh}(Y) \end{array}$$

commute.

PROOF. By continuity, the functor f_*^{IndCoh} is the ind-extension of its retstriction to

$$\mathrm{Coh}(X) \subset \mathrm{IndCoh}(X).$$

The commutative diagram in the proposition implies that

$$\Psi_Y \circ f_*^{\mathrm{IndCoh}}|_{\mathrm{Coh}(X)} = f_*|_{\mathrm{Coh}(X)},$$

as functors $\mathrm{Coh}(X) \to \mathrm{QCoh}(Y)$. Furthermore, $f_*^{\mathrm{IndCoh}}|_{\mathrm{Coh}(X)}$ is a functor that takes values in $\mathrm{QCoh}(Y)^+$.

Note that $\Psi_Y|_{\mathrm{QCoh}(Y)^+}$ is invertible by Proposition 1.2.2(c). Hence,

$$f_*^{\mathrm{IndCoh}}|_{\mathrm{Coh}(X)}$$

is recovered as

$$(\Psi_Y|_{\mathrm{QCoh}(Y)^+})^{-1} \circ (f_*|_{\mathrm{Coh}(X)}).$$

□

2.1.3. Recall (see Chapter 3, Sects. 3.5.1 and 3.7.4) that for $X \in \mathrm{Sch}_{\mathrm{aft}}$, the monoidal category $\mathrm{QCoh}(X)$ is *rigid* (see Chapter 1, Sect. 9.1 for what this means). Hence, by Chapter 1, Lemma 9.3.6, for a morphism $f : X \to Y$, the functor

$$f_* : \mathrm{QCoh}(X) \to \mathrm{QCoh}(Y)$$

has a canonical structure of morphism in $\mathrm{QCoh}(Y)$-**mod**, where $\mathrm{QCoh}(Y)$ acts on $\mathrm{QCoh}(X)$ via f^*.

As in Proposition 2.1.2 and Proposition 1.2.10, one shows:

PROPOSITION 2.1.4. *For a morphism $f : X \to Y$, the functor*

$$f_*^{\mathrm{IndCoh}} : \mathrm{IndCoh}(X) \to \mathrm{IndCoh}(Y)$$

has a unique structure of 1*-morphism in* $\mathrm{QCoh}(Y)$-**mod** *which makes the square*

$$\begin{array}{ccc} \mathrm{IndCoh}(X) & \xrightarrow{\Psi_X} & \mathrm{QCoh}(X) \\ {\scriptstyle f_*^{\mathrm{IndCoh}}}\downarrow & & \downarrow{\scriptstyle f_*} \\ \mathrm{IndCoh}(Y) & \xrightarrow{\Psi_Y} & \mathrm{QCoh}(Y) \end{array}$$

commute in $\mathrm{QCoh}(Y)$-**mod**.

2.2. Upgrading to a functor. We now claim that the assigment

$$X \rightsquigarrow \mathrm{IndCoh}(X), \quad f \rightsquigarrow f_*^{\mathrm{IndCoh}}$$

upgrades to a functor

$$\mathrm{Sch}_{\mathrm{aft}} \to \mathrm{DGCat}_{\mathrm{cont}}, \tag{2.1}$$

to be denoted $\mathrm{IndCoh}_{\mathrm{Sch}_{\mathrm{aft}}}$.

Such an extension is not altogether automatic because we live in the world of higher categories. But constructing it will not be very difficult.

2.2.1. First, we consider the functor

$$\mathrm{QCoh}^*_{\mathrm{Sch}_{\mathrm{aft}}} : (\mathrm{Sch}_{\mathrm{aft}})^{\mathrm{op}} \to \mathrm{DGCat}_{\mathrm{cont}},$$

obtained by restriction from the functor

$$\mathrm{QCoh}^*_{\mathrm{PreStk}} : \mathrm{PreStk}^{\mathrm{op}} \to \mathrm{DGCat}_{\mathrm{cont}}$$

og Chapter 3, Sect. 1.1.3.

Applying Chapter 1, Sect. 8.4.2, we obtain a functor

$$\mathrm{QCoh}_{\mathrm{Sch}_{\mathrm{aft}}} : \mathrm{Sch}_{\mathrm{aft}} \to \mathrm{DGCat}_{\mathrm{cont}},$$

obtained from $\mathrm{QCoh}^*_{\mathrm{Sch}_{\mathrm{aft}}}$ by passage to *right adjoints.*

2.2.2. Now, we claim:

PROPOSITION 2.2.3. *There exists a uniquely defined functor*

$$\mathrm{IndCoh}_{\mathrm{Sch}_{\mathrm{aft}}} : \mathrm{Sch}_{\mathrm{aft}} \to \mathrm{DGCat}_{\mathrm{cont}},$$

equipped with a natural transformation

$$\Psi_{\mathrm{Sch}_{\mathrm{aft}}} : \mathrm{IndCoh}_{\mathrm{Sch}_{\mathrm{aft}}} \to \mathrm{QCoh}_{\mathrm{Sch}_{\mathrm{aft}}},$$

which at the level of objects and 1*-morphisms is given by the assignment*

$$X \rightsquigarrow \mathrm{IndCoh}(X), \quad f \rightsquigarrow f_*^{\mathrm{IndCoh}}.$$

The rest of this subsection is devoted to the proof of this proposition.

2.2.4. Consider the following $(\infty, 1)$-categories:

$$\mathrm{DGCat}_+^{\mathrm{cont}} \text{ and } \mathrm{DGCat}^t_{\mathrm{cont}} :$$

The category $\mathrm{DGCat}_+^{\mathrm{cont}}$ consists of non-cocomplete DG categories $\mathbf{C}$, endowed with a t-structure, such that $\mathbf{C} = \mathbf{C}^+$. We also require that $\mathbf{C}^{\geq 0}$ contains filtered colimits and that the embedding $\mathbf{C}^{\geq 0} \hookrightarrow \mathbf{C}$ commutes with filtered colimits. As 1-morphisms we take those exact functors $F : \mathbf{C}_1 \to \mathbf{C}_2$ that are *left t-exact up to a finite shift*, and such that $F|_{\mathbf{C}_1^{\geq 0}}$ commutes with filtered colimits. The higher categorical structure is uniquely determined by the requirement that the forgetful functor

$$\mathrm{DGCat}_+^{\mathrm{cont}} \to \mathrm{DGCat}$$

be 1-fully faithful (see Chapter 1, Sect. 1.2.4 for what this means).

The category $\mathrm{DGCat}^t_{\mathrm{cont}}$ consists of cocomplete DG categories $\mathbf{C}$, endowed with a t-structure, such that $\mathbf{C}^{\geq 0}$ is closed under filtered colimits, and such that $\mathbf{C}$ is compactly generated by objects from $\mathbf{C}^+$. As 1-morphisms we allow those exact functors $F : \mathbf{C}_1 \to \mathbf{C}_2$ that are continuous and *left t-exact up to a finite shift.*

The higher categorical structure is uniquely determined by the requirement that the forgetful functor

$$\mathrm{DGCat}^t_{\mathrm{cont}} \to \mathrm{DGCat}_{\mathrm{cont}}$$

be 1-fully faithful.

We have a naturally defined functor

$$\mathrm{DGCat}^t_{\mathrm{cont}} \to \mathrm{DGCat}^{\mathrm{cont}}_+, \quad \mathbf{C} \mapsto \mathbf{C}^+. \tag{2.2}$$

LEMMA 2.2.5. *The functor* (2.2) *is 1-fully faihful.*

2.2.6. We will use the following general assertion. Let $T : \mathbf{D}' \to \mathbf{D}$ be a functor between $(\infty, 1)$-categories, which is 1-fully faithful. Let $\mathbf{I}$ be another $(\infty, 1)$-category, and let

$$\mathbf{i} \rightsquigarrow F'(\mathbf{i}), \tag{2.3}$$

be an assignment, such that the assignment

$$\mathbf{i} \mapsto T \circ F'(\mathbf{i})$$

has been extended to a functor $F : \mathbf{I} \to \mathbf{D}$.

LEMMA 2.2.7. *Suppose that for every* $\alpha \in \mathrm{Maps}_{\mathbf{I}}(\mathbf{i}_1, \mathbf{i}_2)$, *the point* $F(\alpha) \in \mathrm{Maps}_{\mathbf{D}}(F(\mathbf{i}_1), F(\mathbf{i}_2))$ *lies in the connected component corresponding to the image of*

$$\mathrm{Maps}_{\mathbf{D}'}(F'(\mathbf{i}_1), F'(\mathbf{i}_2)) \to \mathrm{Maps}_{\mathbf{D}}(F(\mathbf{i}_1), F(\mathbf{i}_2)).$$

Then there exists a unique extension of (2.3) *to a functor* $F' : \mathbf{I} \to \mathbf{D}$ *equipped with an isomorphism* $T \circ F' \simeq F$.

Let now F'_1 and F'_2 be two assignments as in (2.3), satisfying the assumption of Lemma 2.2.7. Let us be given an assignment

$$\mathbf{i} \rightsquigarrow \psi'_{\mathbf{i}} \in \mathrm{Maps}_{\mathbf{D}'}(F'_1(\mathbf{i}), F'_2(\mathbf{i})). \tag{2.4}$$

LEMMA 2.2.8. *Suppose that the assignment*

$$\mathbf{i} \rightsquigarrow T(\psi'_{\mathbf{i}}) \in \mathrm{Maps}_{\mathbf{D}}(F_1(\mathbf{i}), F_2(\mathbf{i}))$$

has been extended to a natural transformation $\psi : F_1 \to F_2$. *Then there exists a unique extension of* (2.4) *to a natural transformation* $\psi : F'_1 \to F'_2$ *equipped with an isomorphism* $T \circ \psi' \simeq \psi$.

2.2.9. We are now ready to prove Proposition 2.2.3:

Step 1. We start with the functor

$$\mathrm{QCoh}_{\mathrm{Sch}_{\mathrm{aft}}} : \mathrm{Sch}_{\mathrm{aft}} \to \mathrm{DGCat}_{\mathrm{cont}},$$

and consider

$$\mathbf{I} = \mathrm{Sch}_{\mathrm{aft}},\ \mathbf{D} = \mathrm{DGCat}_{\mathrm{cont}},\ \mathbf{D}' := \mathrm{DGCat}^t_{\mathrm{cont}},\ F = \mathrm{QCoh}_{\mathrm{Sch}_{\mathrm{aft}}},$$

and the assignment

$$(X \in \mathrm{Sch}_{\mathrm{aft}}) \rightsquigarrow (\mathrm{QCoh}(X) \in \mathrm{DGCat}^t_{\mathrm{cont}}).$$

Apping Lemma 2.2.7, we obtain a functor

$$\mathrm{QCoh}^t_{\mathrm{Sch}_{\mathrm{aft}}} : \mathrm{Sch}_{\mathrm{aft}} \to \mathrm{DGCat}^t_{\mathrm{cont}}. \tag{2.5}$$

Step 2. Note that Proposition 2.1.2 defines a functor

$$\mathrm{IndCoh}^t_{\mathrm{Sch}_{\mathrm{aft}}} : \mathrm{Sch}_{\mathrm{aft}} \to \mathrm{DGCat}^t_{\mathrm{cont}},$$

and the natural transformation

$$\Psi^t_{\mathrm{Sch}_{\mathrm{aft}}} : \mathrm{IndCoh}^t_{\mathrm{Sch}_{\mathrm{aft}}} \to \mathrm{QCoh}^t_{\mathrm{Sch}_{\mathrm{aft}}}$$

at the level of objects and 1*-morphisms.*

Since the functor $\mathrm{DGCat}^t_{\mathrm{cont}} \to \mathrm{DGCat}_{\mathrm{cont}}$ is 1-fully faithful, by Lemmas 2.2.7 and 2.2.8, the existence and uniqueness of the pair $(\mathrm{IndCoh}_{\mathrm{Sch}_{\mathrm{aft}}}, \Psi_{\mathrm{Sch}_{\mathrm{aft}}})$ with a fixed behavior on objects and 1-morphisms, is equivalent to that of

$$(\mathrm{IndCoh}^t_{\mathrm{Sch}_{\mathrm{aft}}}, \Psi^t_{\mathrm{Sch}_{\mathrm{aft}}})$$

.

Step 3. By Lemma 2.2.5, combined with Lemmas 2.2.7 and 2.2.8, we obtain that the existence and uniqueness of the pair $(\mathrm{IndCoh}^t_{\mathrm{Sch}_{\mathrm{aft}}}, \Psi^t_{\mathrm{Sch}_{\mathrm{aft}}})$, with a fixed behavior on objects and 1-morphisms is equivalent to the existence and uniqueness of the pair

$$(\mathrm{IndCoh}^+_{\mathrm{Sch}_{\mathrm{aft}}}, \Psi^+_{\mathrm{Sch}_{\mathrm{aft}}}),$$

obtained by composing with the functor (2.2).

The latter, however, is given by

$$\mathrm{IndCoh}^+_{\mathrm{Sch}_{\mathrm{aft}}} := \mathrm{QCoh}^+_{\mathrm{Sch}_{\mathrm{aft}}} \text{ and } \Psi^+_{\mathrm{Sch}_{\mathrm{aft}}} := \mathrm{Id}.$$

□

3. The functor of 'usual' inverse image

We now construct another piece of functoriality in the assignment

$$X \rightsquigarrow \mathrm{IndCoh}(X),$$

namely, the functor of * pullback.

Unlike the case of QCoh, its role in the theory is rather limited–a 'more important' functor is that of !-pullback. However, the *-pullback is a necessary step in the construction of the !-pullback, which is why we discuss it.

3.1. Inverse image with respect to eventually coconnective morphisms. Unlike the case of QCoh, the functor of *-pullback on IndCoh is not defined for all maps of schemes, but only for *eventually coconnective* ones. In this subsection we give the corresponding construction.

3.1.1. Let $f : X \to Y$ be a morphism in $\mathrm{Sch}_{\mathrm{aft}}$.

DEFINITION 3.1.2. *We shall say that f is* eventually coconnective *if the functor*

$$f^* : \mathrm{QCoh}(Y) \to \mathrm{QCoh}(X)$$

sends $\mathrm{Coh}(Y) \subset \mathrm{QCoh}(Y)$ *to* $\mathrm{QCoh}(X)^+$.

It is easy to see that if f is eventually coconnective, then it sends $\mathrm{Coh}(Y)$ to $\mathrm{Coh}(X)$: indeed, for any morphism f, the functor f^* sends objects of $\mathrm{QCoh}(Y)^-$ *with coherent cohomologies* to objects with a similar property on X.

In addition, we have the following fact, established in [**Ga1**, Lemma 3.4.2]:

LEMMA 3.1.3. *The following conditions are equivalent:*
(a) f *is eventually coconnective;*
(b) f *is of finite Tor amplitude, i.e., is left t-exact up to a finite cohomological shift.*

COROLLARY 3.1.4. *The class of eventually coconnective morphisms is stable under base change.*

3.1.5. Let $f : X \to Y$ be eventually coconnective. Ind-extending the functor

$$f^* : \mathrm{Coh}(Y) \to \mathrm{Coh}(X)$$

we obtain a functor

$$f^{\mathrm{IndCoh},*} : \mathrm{IndCoh}(Y) \to \mathrm{IndCoh}(X),$$

which makes the diagram

$$\begin{array}{ccc} \mathrm{IndCoh}(X) & \xrightarrow{\Psi_X} & \mathrm{QCoh}(X) \\ {\scriptstyle f^{\mathrm{IndCoh},*}}\uparrow & & \uparrow{\scriptstyle f^*} \\ \mathrm{IndCoh}(Y) & \xrightarrow{\Psi_Y} & \mathrm{QCoh}(Y) \end{array}$$

commute.

We have:

PROPOSITION 3.1.6. *The functor* $f^{\mathrm{IndCoh},*}$ *is a left adjoint to* f_*^{IndCoh}.

PROOF. It is sufficient to construct a functorial isomorphism

$$\mathrm{Maps}_{\mathrm{IndCoh}(X)}(f^{\mathrm{IndCoh},*}(\mathcal{F}_Y), \mathcal{F}_X) \simeq \mathrm{Maps}_{\mathrm{IndCoh}(Y)}(\mathcal{F}_Y, f_*^{\mathrm{IndCoh}}(\mathcal{F}_X)) \tag{3.1}$$

for $\mathcal{F}_Y \in \mathrm{Coh}(Y)$ and $\mathcal{F}_X \in \mathrm{Coh}(X)$. However, by construction, the left-hand side in (3.1) is

$$\mathrm{Maps}_{\mathrm{Coh}(X)}(f^*(\mathcal{F}_Y), \mathcal{F}_X)) \simeq \mathrm{Maps}_{\mathrm{QCoh}(X)}(f^*(\mathcal{F}_Y), \mathcal{F}_X)),$$

while the right-hand side maps isomorphically by the functor Ψ_Y to

$$\mathrm{Maps}_{\mathrm{QCoh}(Y)}(\mathcal{F}_Y, f_*(\mathcal{F}_X)).$$

Now, (3.1) follows from the (f^*, f_*)-adjunction on QCoh. □

REMARK 3.1.7. It is shown in [**Ga1**], that the functor f_*^{IndCoh} admits a left adjoint *if and only if* the morphism f is eventually coconnective.

3.1.8. Note that by Chapter 1, Lemma 9.3.6, the functor $f^{\mathrm{IndCoh},*}$ carries a canonical structure of morphism in $\mathrm{QCoh}(Y)$-**mod**. It is easy to see that this is the same structure as obtained by ind-extending the structure of compatibility with the action of $\mathrm{QCoh}(Y)^{\mathrm{perf}}$ on

$$f^* : \mathrm{Coh}(Y) \to \mathrm{Coh}(X).$$

3.1.9. Let $(\mathrm{Sch}_{\mathrm{aft}})_{\text{event-coconn}} \subset \mathrm{Sch}_{\mathrm{aft}}$ be the 1-full subcategory, where we restrict 1-morphisms to maps that are eventually coconnective.

By Chapter 1, Sect. 8.4.2, combining Propositions 2.2.3 and 3.1.6, we obtain:

COROLLARY 3.1.10. *The assignment*

$$X \rightsquigarrow \mathrm{IndCoh}(X), \quad f \rightsquigarrow f^{\mathrm{IndCoh},*}$$

canonically extends to a functor, to be denoted $\mathrm{IndCoh}^*_{(\mathrm{Sch}_{\mathrm{aft}})_{\text{event-coconn}}}$,

$$((\mathrm{Sch}_{\mathrm{aft}})_{\text{event-coconn}})^{\mathrm{op}} \to \mathrm{DGCat}_{\mathrm{cont}},$$

obtained from

$$\mathrm{IndCoh}_{\mathrm{Sch}_{\mathrm{aft}}}|_{(\mathrm{Sch}_{\mathrm{aft}})_{\text{event-coconn}}}$$

by adjunction.

3.2. Base change for eventually coconnective morphisms. An important property of the *-pullback (and one which is crucial for the construction of the !-pullback) is *base change*. It closely mimics the corresponding phenomenon for QCoh.

3.2.1. Let

$$\begin{array}{ccc} X_1 & \xrightarrow{g_X} & X_2 \\ {\scriptstyle f_1}\downarrow & & \downarrow{\scriptstyle f_2} \\ Y_1 & \xrightarrow{g_Y} & Y_2 \end{array}$$

be a Cartesian diagram in $\mathrm{Sch}_{\mathrm{aft}}$.

Suppose that f_2 is eventually coconnective. By Corollary 3.1.4, the morphism f_1 is also eventually coconnective. Then the isomorphism of functors

$$(g_Y)_*^{\mathrm{IndCoh}} \circ (f_1)_*^{\mathrm{IndCoh}} \simeq (f_2)_*^{\mathrm{IndCoh}} \circ (g_X)_*^{\mathrm{IndCoh}}$$

gives rise to a natural transformation.

$$(f_2)^{\mathrm{IndCoh},*} \circ (g_Y)_*^{\mathrm{IndCoh}} \to (g_X)_*^{\mathrm{IndCoh}} \circ (f_1)^{\mathrm{IndCoh},*}. \tag{3.2}$$

PROPOSITION 3.2.2. *The map* (3.2) *is an isomorphism.*

PROOF. It is enough to show that

$$(f_2)^{\mathrm{IndCoh},*} \circ (g_Y)_*^{\mathrm{IndCoh}}(\mathcal{F}) \to (g_X)_*^{\mathrm{IndCoh}} \circ (f_1)^{\mathrm{IndCoh},*}(\mathcal{F}) \tag{3.3}$$

is an isomorphism for $\mathcal{F} \in \mathrm{Coh}(Y_1)$.

In this case both sides of (3.3) belong to $\mathrm{IndCoh}(X_2)^+$. By Proposition 1.2.2, it is therefore sufficient to show that the map

$$\Psi_{X_2} \circ (f_2)^{\mathrm{IndCoh},*} \circ (g_Y)_*^{\mathrm{IndCoh}} \to \Psi_{X_2} \circ (g_X)_*^{\mathrm{IndCoh}} \circ (f_1)^{\mathrm{IndCoh},*} \tag{3.4}$$

is an isomorphism.

We have:

$$\Psi_{X_2} \circ (f_2)^{\mathrm{IndCoh},*} \circ (g_Y)_*^{\mathrm{IndCoh}} \simeq (f_2)^* \circ \Psi_{Y_2} \circ (g_Y)_*^{\mathrm{IndCoh}} \simeq (f_2)^* \circ (g_Y)_* \circ \Psi_{Y_1}$$

and

$$\Psi_{X_2} \circ (g_X)_*^{\mathrm{IndCoh}} \circ (f_1)^{\mathrm{IndCoh},*} \simeq (g_X)_* \circ \Psi_{X_1} \circ (f_1)^{\mathrm{IndCoh},*} \simeq (g_X)_* \circ (f_1)^* \circ \Psi_{Y_1}.$$

Now, it follows from the construction of the $(f^{\mathrm{IndCoh},*}, f_*^{\mathrm{IndCoh}})$-adjunction, that the map in (3.4) corresponds to the map

$$(f_2)^* \circ (g_Y)_* \circ \Psi_{Y_1} \to (g_X)_* \circ (f_1)^* \circ \Psi_{Y_1},$$

obtained from the (f^*, f_*)-adjunction.

Hence, (3.4) is an isomorphism by base change for QCoh.

□

3.3. Tensoring up. In this section we study the following question: given a map $f : X \to Y$, how closely can we approximate $\mathrm{IndCoh}(X)$ from knowing $\mathrm{IndCoh}(Y)$ and the QCoh categories on both schemes.

In the process we will come across several *convergence-type* assertions, that are of significant technical importance: some maps that are isomorphisms in QCoh are much less obviously so in the IndCoh context.

3.3.1. Let $f : X \to Y$ be an eventually coconnective map. Regarding the functor $f^{\mathrm{IndCoh},*}$ as a map in $\mathrm{QCoh}(Y)$-**mod**, we obtain a functor

$$(\mathrm{Id}_{\mathrm{QCoh}(X)} \otimes f^{\mathrm{IndCoh},*}) : \mathrm{QCoh}(X) \underset{\mathrm{QCoh}(Y)}{\otimes} \mathrm{IndCoh}(Y) \to \mathrm{IndCoh}(X). \tag{3.5}$$

We claim:

PROPOSITION 3.3.2. *The functor* (3.5) *is fully faithful.*

REMARK 3.3.3. It is shown in [**Ga1**, Proposition 4.4.9] that the functor (3.5) is an equivalence when f is smooth. In Proposition 4.1.6, we will prove this in the case when f is an open embedding.

3.3.4. Note that for a diagram of $(\infty, 1)$-categories

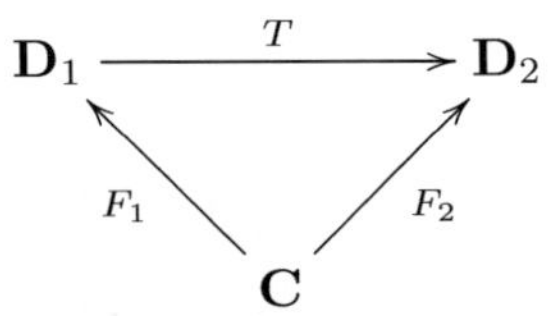

if the functors F_1 and F_2 admit right adjoints, we have a natural transformation of the resulting endo-functors of $\mathbf{C}$:

$$F_1^R \circ F_1 \to F_2^R \circ F_2.$$

Furthermore, if T is fully faithful, then the above natural transformation is an isomorphism.

From here we obtain that the functor $(\mathrm{Id}_{\mathrm{QCoh}(X)} \otimes f^{\mathrm{IndCoh},*})$ gives rise to a map of endo-functors of $\mathrm{IndCoh}(Y)$:

$$(f_*(\mathcal{O}_X) \otimes -) \simeq ((f_* \circ f^*) \otimes \mathrm{Id}_{\mathrm{IndCoh}(Y)}) \to f_*^{\mathrm{IndCoh}} \circ f^{\mathrm{IndCoh},*}, \tag{3.6}$$

where $f_*(\mathcal{O}_X) \otimes -$ denotes the functor of action of $f_*(\mathcal{O}_X) \in \mathrm{QCoh}(Y)$ on $\mathrm{IndCoh}(Y)$.

Thus, from Proposition 3.3.2 we obtain:

COROLLARY 3.3.5. *The map* (3.6) *is an isomorphism.*

3.3.6. The rest of the subsection is devoted to the proof of Proposition 3.3.2. We note that the left-hand side in (3.5) is compactly generated by objects of the form

$$\mathcal{E}_X \otimes \mathcal{F}_Y \in \operatorname{QCoh}(X) \underset{\operatorname{QCoh}(Y)}{\otimes} \operatorname{IndCoh}(Y),$$

where $\mathcal{E}_X \in \operatorname{QCoh}(X)^{\operatorname{perf}}$ and $\mathcal{F}_Y \in \operatorname{Coh}(Y)$. Moreover, the functor

$$(\operatorname{Id}_{\operatorname{QCoh}(X)} \otimes f^{\operatorname{IndCoh},*})$$

sends these objects to compact objects in $\operatorname{IndCoh}(X)$.

Hence, it is enough to show that for $\mathcal{E}^1_X, \mathcal{E}^2_X$ and $\mathcal{F}^1_Y, \mathcal{F}^2_Y$ as above, the map

$$\begin{aligned}(3.7)\quad & \operatorname{Maps}_{\operatorname{QCoh}(X) \underset{\operatorname{QCoh}(Y)}{\otimes} \operatorname{IndCoh}(Y)}(\mathcal{E}^1_X \otimes \mathcal{F}^1_Y, \mathcal{E}^2_X \otimes \mathcal{F}^2_Y) \to \\ & \to \operatorname{Maps}_{\operatorname{IndCoh}(X)}(\mathcal{E}^1_X \otimes f^{\operatorname{IndCoh},*}(\mathcal{F}^1), \mathcal{E}^2_X \otimes f^{\operatorname{IndCoh},*}(\mathcal{F}^2))\end{aligned}$$

is an isomorphism, where in the right-hand side $\otimes$ denotes the action of QCoh on IndCoh.

We can rewrite the map in (3.7) as

$$\begin{aligned}(3.8)\quad & \operatorname{Maps}_{\operatorname{QCoh}(X) \underset{\operatorname{QCoh}(Y)}{\otimes} \operatorname{IndCoh}(Y)}(\mathcal{O}_X \otimes \mathcal{F}^1_Y, \mathcal{E}_X \otimes \mathcal{F}^2_Y) \to \\ & \to \operatorname{Maps}_{\operatorname{IndCoh}(X)}(\mathcal{O}_X \otimes f^{\operatorname{IndCoh},*}(\mathcal{F}^1), \mathcal{E}_X \otimes f^{\operatorname{IndCoh},*}(\mathcal{F}^2)),\end{aligned}$$

where $\mathcal{E}_X \simeq \mathcal{E}^2_X \otimes (\mathcal{E}^1_X)^\vee$.

Furthermore, we rewrite the map in (3.8) as

$$\operatorname{Maps}_{\operatorname{IndCoh}(Y)}(\mathcal{F}^1_Y, f_*(\mathcal{E}) \otimes \mathcal{F}^2_Y) \to \operatorname{Maps}_{\operatorname{IndCoh}(Y)}(\mathcal{F}^1_Y, f^{\operatorname{IndCoh}}_*(\mathcal{E} \otimes f^{\operatorname{IndCoh},*}(\mathcal{F}^2_Y))).$$

I.e., we are reduced to showing that the following version of the projection formula:

PROPOSITION 3.3.7. *For an eventally coconnective map, the natural transformation between the functors*

$$\operatorname{QCoh}(X) \times \operatorname{IndCoh}(Y) \rightrightarrows \operatorname{IndCoh}(Y),$$

that sends $\mathcal{E}_X \in \operatorname{QCoh}(X)$ and $\mathcal{F}_Y \in \operatorname{IndCoh}(\mathcal{Y})$ to the map

$$(3.9)\qquad f_*(\mathcal{E}_X) \otimes \mathcal{F}_Y \to f^{\operatorname{IndCoh}}_*(\mathcal{E}_X \otimes f^{\operatorname{IndCoh},*}(\mathcal{F}_Y)),$$

is an isomorphism.

REMARK 3.3.8. Note that there is another kind of projection formula, that encodes the compatibility of $f^{\operatorname{IndCoh}}_*$ with the monoidal action of $\operatorname{QCoh}(Y)$, and which holds tautologically for any morphism f, see Proposition 2.1.4:

For $\mathcal{E}_Y \in \operatorname{QCoh}(Y)$ and $\mathcal{F}_X \in \operatorname{IndCoh}(X)$ we have:

$$f^{\operatorname{IndCoh}}_*(f^*(\mathcal{E}_Y) \otimes \mathcal{F}_X) \simeq \mathcal{E}_Y \otimes f^{\operatorname{IndCoh}}_*(\mathcal{F}_X).$$

3.3.9. *Proof of Proposition* 3.3.7. It is enough to prove the isomorphism (3.9) holds for $\mathcal{E}_X \in \mathrm{QCoh}(X)^{\mathrm{perf}}$ and $\mathcal{F}_Y \in \mathrm{Coh}(Y)$.

We also note that the map (3.9) becomes an isomorphism after applying the functor Ψ_Y, by the usual projection formula for QCoh. For $\mathcal{E}_X \in \mathrm{QCoh}(X)^{\mathrm{perf}}$ and $\mathcal{F}_Y \in \mathrm{Coh}(Y)$ we have

$$f_*^{\mathrm{IndCoh}}(\mathcal{E}_X \otimes f^{\mathrm{IndCoh},*}(\mathcal{F}_Y)) \in \mathrm{IndCoh}(Y)^+.$$

Hence, by Proposition 1.2.2, it suffices to show that in this case

$$f_*(\mathcal{E}_X) \otimes \mathcal{F}_Y \in \mathrm{IndCoh}(Y)^+.$$

We note that the object $f_*(\mathcal{E}_X) \in \mathrm{QCoh}(Y)^b$ is of bounded Tor dimension. The required fact follows from the next general observation:

LEMMA 3.3.10. *For $X \in \mathrm{Sch}_{\mathrm{aft}}$ and $\mathcal{E} \in \mathrm{QCoh}(X)^b$, whose Tor dimension is bounded on the left by an integer n, the functor*

$$\mathcal{E} \otimes - : \mathrm{IndCoh}(X) \to \mathrm{IndCoh}(X)$$

has a cohomological amplitude bounded on the left by n.

□

3.3.11. *Proof of Lemma* 3.3.10. We need to show that the functor $\mathcal{E} \otimes -$ sends $\mathrm{IndCoh}(X)^{\geq 0}$ to $\mathrm{IndCoh}(X)^{\geq -n}$. By Lemma 1.2.4(1), it is sufficient to show that this functor sends $\mathrm{Coh}(X)^{\geq 0}$ to $\mathrm{IndCoh}(X)^{\geq -n}$. By cohomological devissage, the latter is equivalent to sending $\mathrm{Coh}(X)^{\heartsuit}$ to $\mathrm{IndCoh}(X)^{\geq -n}$.

Let i denote the closed embedding ${}^{\mathrm{cl}}X =: X' \to X$. The functor i_*^{IndCoh} induces an equivalence $\mathrm{Coh}(X')^{\heartsuit} \to \mathrm{Coh}(X)^{\heartsuit}$. So, it is enough to show that for $\mathcal{F}' \in \mathrm{Coh}(X')^{\heartsuit}$, we have

$$\mathcal{E} \otimes i_*^{\mathrm{IndCoh}}(\mathcal{F}') \in \mathrm{IndCoh}(X)^{\geq -n}.$$

We have:

$$\mathcal{E} \otimes i_*^{\mathrm{IndCoh}}(\mathcal{F}') \simeq i_*^{\mathrm{IndCoh}}(i^*(\mathcal{E}) \otimes \mathcal{F}').$$

Note that the functor i_*^{IndCoh} is t-exact (since i_* is), and $i^*(\mathcal{E})$ has Tor dimension bounded by the same integer n.

This reduces the assertion of the lemma to the case when X is classical. Further, by Proposition 4.2.4 (which will be proved independently later), the statement is Zariski local, so we can assume that X is affine.

In the latter case, the assumption on $\mathcal{E}$ implies that it can be represented by a complex of flat $\mathcal{O}_X$-modules that lives in the cohomological degrees $\geq -n$. This reduces the assertion further to the case when $\mathcal{E}$ is a flat $\mathcal{O}_X$-module in degree 0. In this case we claim that the functor

$$\mathcal{E} \otimes - : \mathrm{IndCoh}(X) \to \mathrm{IndCoh}(X)$$

is t-exact.

The latter follows from Lazard's lemma: such an $\mathcal{E}$ is a filtered colimit of locally free $\mathcal{O}_X$-modules $\mathcal{E}'$, while for each such $\mathcal{E}'$, the functor $\mathcal{E}' \otimes - : \mathrm{IndCoh}(X) \to \mathrm{IndCoh}(X)$ is by definition the ind-extension of the functor

$$\mathcal{E}' \otimes - : \mathrm{Coh}(X) \to \mathrm{Coh}(X),$$

and the latter is t-exact.

□

4. Open embeddings

The behavior of direct and inverse image functors for open embeddings is, obviously, an important piece of information about IndCoh.

4.1. Restriction to an open. In this subsection we show that the behavior of IndCoh with respect to open embeddings is 'exactly the same' as that of QCoh.

4.1.1. Let now $j : \overset{\circ}{X} \hookrightarrow X$ be an open embedding. We claim:

PROPOSITION 4.1.2. *The functor* $j_*^{\mathrm{IndCoh}} : \mathrm{IndCoh}(\overset{\circ}{X}) \to \mathrm{IndCoh}(X)$ *is fully faithful.*

PROOF. We need to show that the co-unit of the adjunction

$$j^{\mathrm{IndCoh},*} \circ j_*^{\mathrm{IndCoh}} \to \mathrm{Id}_{\mathrm{IndCoh}(\overset{\circ}{X})}$$

is an isomorphism.

Since the functors in question are continuous, it is enough to check that

$$j^{\mathrm{IndCoh},*} \circ j_*^{\mathrm{IndCoh}}(\mathcal{F}) \to \mathcal{F}$$

is an isomorphism for $\mathcal{F} \in \mathrm{Coh}(X)$. However, in this case both $j^{\mathrm{IndCoh},*} \circ j_*^{\mathrm{IndCoh}}(\mathcal{F})$ and $\mathcal{F}$ belong to $\mathrm{IndCoh}(X)^+$, so by Proposition 1.2.2, it is sufficient to check that

$$\Psi_{\overset{\circ}{X}} \circ j^{\mathrm{IndCoh},*} \circ j_*^{\mathrm{IndCoh}} \to \Psi_{\overset{\circ}{X}}$$

is an isomorphism.

However,

$$\Psi_{\overset{\circ}{X}} \circ j^{\mathrm{IndCoh},*} \circ j_*^{\mathrm{IndCoh}} \simeq j^{\mathrm{IndCoh},*} \circ \Psi_X \circ j_*^{\mathrm{IndCoh}} \simeq j^{\mathrm{IndCoh},*} \circ j_*^{\mathrm{IndCoh}} \circ \Psi_{\overset{\circ}{X}},$$

and it follows from the construction of the $(f^{\mathrm{IndCoh},*}, f_*^{\mathrm{IndCoh}})$-adjunction that the resulting map

$$j^{\mathrm{IndCoh},*} \circ j_*^{\mathrm{IndCoh}} \circ \Psi_{\overset{\circ}{X}} \to \Psi_{\overset{\circ}{X}}$$

comes from the co-unit of the (j^*, j_*)-adjunction. Therefore, it is an isomorphism, as

$$j^* \circ j_* \to \mathrm{Id}_{\mathrm{QCoh}(X)}$$

is an isomorphism.

□

4.1.3. The next assertion follows immediately from Lemma 1.2.4:

LEMMA 4.1.4. *For an open embedding* j*, the functor* $j^{\mathrm{IndCoh},*}$ *is t-exact.*

4.1.5. Finally, let us recall the functor (3.5). We claim:

PROPOSITION 4.1.6. *Assume that f is an open embedding. Then the functor* (3.5) *is an equivalence.*

PROOF. We already know that the functor in question is fully faithful. Hence, it remains to show that its essential image generates the target category. But this follows from Proposition 4.1.2. □

4.2. Zariski descent. In this subsection we will show that IndCoh can be glued locally from a Zariski cover, in a way completely parallel to QCoh.

4.2.1. Let now $f : U \to X$ be a Zariski cover, i.e., U is the disjoint union of open subschemes of X, whose union is all of X. Let $U^\bullet$ be the Čech nerve of f. The functors of (IndCoh, $*$)-pullback define a cosimplicial category

$$\text{IndCoh}(U^\bullet),$$

which is augmented by $\text{IndCoh}(X)$.

We claim:

PROPOSITION 4.2.2. *The functor*

$$\text{IndCoh}(X) \to \text{Tot}(\text{IndCoh}(U^\bullet))$$

is an equivalence.

PROOF. The usual argument reduces the assertion of the proposition to the following. Let $X = U_1 \cup U_2$; $U_{12} = U_1 \cap U_2$. Let

$$U_1 \overset{j_1}{\hookrightarrow} X,\ U_2 \overset{j_2}{\hookrightarrow} X,\ U_{12} \overset{j_{12}}{\hookrightarrow} X,\ U_{12} \overset{j_{12,1}}{\hookrightarrow} U_1,\ U_{12} \overset{j_{12,2}}{\hookrightarrow} U_2$$

denote the corresponding open embeddings.

We need to show that the functor

$$\text{IndCoh}(X) \to \text{IndCoh}(U_1) \underset{\text{IndCoh}(U_{12})}{\times} \text{IndCoh}(U_1)$$

that sends $\mathcal{F} \in \text{IndCoh}(X)$ to the datum of

$$\{j_1^{\text{IndCoh},*}(\mathcal{F}_1), j_2^{\text{IndCoh},*}(\mathcal{F}_2), j_{12,1}^{\text{IndCoh},*}(j_1^{\text{IndCoh},*}(\mathcal{F})) \simeq j_{12}^{\text{IndCoh},*}(\mathcal{F}) \simeq \\ \simeq j_{12,2}^{\text{IndCoh},*}(j_2^{\text{IndCoh},*}(\mathcal{F}))\}$$

is an equivalence.

We construct a right adjoint functor

$$\text{IndCoh}(U_1) \underset{\text{IndCoh}(U_{12})}{\times} \text{IndCoh}(U_1) \to \text{IndCoh}(X)$$

by sending

$$\{\mathcal{F}_1 \in \text{IndCoh}(U_1), \mathcal{F}_2 \in \text{IndCoh}(U_2), \mathcal{F}_{12} \in \text{IndCoh}(U_{12}), j_{12,1}^{\text{IndCoh},*}(\mathcal{F}_1) \simeq \mathcal{F}_{12} \simeq \\ \simeq j_{12,2}^{\text{IndCoh},*}(\mathcal{F}_2)\}$$

to

$$\ker\Big(((j_1)_*^{\text{IndCoh}}(\mathcal{F}_1) \oplus (j_2)_*^{\text{IndCoh}}(\mathcal{F}_1)) \to (j_{12})_*^{\text{IndCoh}}(\mathcal{F}_{12})\Big),$$

where the maps $(j_i)_*^{\mathrm{IndCoh}}(\mathcal{F}_i) \to (j_{12})_*^{\mathrm{IndCoh}}(\mathcal{F}_{12})$ are

$$(j_i)_*^{\mathrm{IndCoh}}(\mathcal{F}_i) \to (j_i)_*^{\mathrm{IndCoh}} \circ (j_{12,i})_*^{\mathrm{IndCoh}} \circ (j_{12,i})^{\mathrm{IndCoh},*}(\mathcal{F}_i) = \\ = (j_{12})_*^{\mathrm{IndCoh}} \circ (j_{12,i})^{\mathrm{IndCoh},*}(\mathcal{F}_i) \simeq (j_{12})_*^{\mathrm{IndCoh}}(\mathcal{F}_{12}).$$

It is straightforward to see from Propositions 4.1.2 and 3.2.2 that the composition

$$(4.1)\quad \mathrm{IndCoh}(U_1) \underset{\mathrm{IndCoh}(U_{12})}{\times} \mathrm{IndCoh}(U_1) \to \mathrm{IndCoh}(X) \to \\ \to \mathrm{IndCoh}(U_1) \underset{\mathrm{IndCoh}(U_{12})}{\times} \mathrm{IndCoh}(U_1)$$

is canonically isomorphic to the identity functor.

To prove that the composition

$$\mathrm{IndCoh}(X) \to \mathrm{IndCoh}(U_1) \underset{\mathrm{IndCoh}(U_{12})}{\times} \mathrm{IndCoh}(U_1) \to \mathrm{IndCoh}(X)$$

is also isomorphic to the identity functor, it is sufficient to show that for $\mathcal{F} \in \mathrm{IndCoh}(X)$, the canonical map from it to

$$(4.2)\quad \ker\Big(\big((j_1)_*^{\mathrm{IndCoh}} \circ (j_1)^{\mathrm{IndCoh},*}(\mathcal{F}) \oplus (j_2)_*^{\mathrm{IndCoh}} \circ (j_2)^{\mathrm{IndCoh},*}(\mathcal{F})\big) \to \\ \to (j_{12})_*^{\mathrm{IndCoh}} \circ j_{12}^{\mathrm{IndCoh},*}(\mathcal{F})\Big)$$

is an isomorphism.

Since all functors in question are continuous, it is sufficient to do so for $\mathcal{F} \in \mathrm{Coh}(X)$. In this case, both sides of (4.2) belong to $\mathrm{IndCoh}(X)^+$. So, it is enough to prove that the map in question becomes an isomorphism after applying the functor Ψ_X. However, in this case we are dealing with the map

$$\Psi_X(\mathcal{F}) \to \ker\Big(\big((j_1)_* \circ (j_1)^*(\Psi_X(\mathcal{F})) \oplus (j_2)_* \circ (j_2)^*(\Psi_X(\mathcal{F}))\big) \to (j_{12})_* \circ j_{12}^*(\Psi_X(\mathcal{F}))\Big),$$

which is known to be an isomorphism.

□

4.2.3. We also have

PROPOSITION 4.2.4. *Let $f : U \to X$ be a Zariski cover. Then $\mathcal{F} \in \mathrm{IndCoh}(X)$ belongs to $\mathrm{IndCoh}(X)^{\leq 0}$ (resp., $\mathrm{IndCoh}(X)^{\geq 0}$) if and only if $f^{\mathrm{IndCoh},*}(\mathcal{F})$ does.*

PROOF. The 'only if' direction for both statetements follows from Lemma 4.1.4.

For the 'if' direction, assuming that $f^*(\mathcal{F}) \in \mathrm{IndCoh}(X)^{\leq 0}$, it is sufficient to show that $\Psi_X(\mathcal{F}) \in \mathrm{QCoh}(X)^{\leq 0}$, and the assertion follows from the corresponding assertion for QCoh.

If $f^{\mathrm{IndCoh},*}(\mathcal{F}) \in \mathrm{IndCoh}(X)^{\geq 0}$, the assertion follows from the construction of the inverse functor

$$\mathrm{Tot}(\mathrm{IndCoh}(U^\bullet)) \to \mathrm{IndCoh}(X).$$

□

5. Proper maps

If until now the theory of IndCoh has run in parallel to (and was inherited from that of) QCoh, in this section we will come across to the main point of difference between the two: the functor of !-pullback, studied in this section.

5.1. The !-pullback. In this subsection we introduce the functor of !-pullback for proper maps. Its extension for arbitrary maps between schemes is the subject of Chapter 5, Sect. 3.

5.1.1. Let $f : X \to Y$ be a map in $\mathrm{Sch}_{\mathrm{aft}}$. We recall the following definition:

DEFINITION 5.1.2. *The map f is said to be proper (resp., closed embedding) if the corresponding map ${}^{\mathrm{cl}}X \to {}^{\mathrm{cl}}Y$ has this property.*

5.1.3. Let $f : X \to Y$ be a proper map. We claim:

LEMMA 5.1.4. *The functor*

$$f_*^{\mathrm{IndCoh}} : \mathrm{IndCoh}(X) \to \mathrm{IndCoh}(Y)$$

sends $\mathrm{Coh}(X) \subset \mathrm{IndCoh}(X)$ *to* $\mathrm{Coh}(Y) \subset \mathrm{IndCoh}(Y)$.

PROOF. By the construction of f_*^{IndCoh}, it is sufficient to show that the functor

$$f_* : \mathrm{QCoh}(X) \to \mathrm{QCoh}(Y)$$

$\mathrm{Coh}(X) \subset \mathrm{QCoh}(X)$ to $\mathrm{Coh}(Y) \subset \mathrm{QCoh}(Y)$.

First, we note that the assertion holds when f is a closed embedding.

In general, by the devissage with respect to the t-structure, it is sufficient to show that for $\mathcal{F} \in \mathrm{Coh}(X)^{\heartsuit}$, we have

$$f_*(\mathcal{F}) \in \mathrm{Coh}(Y).$$

Let i denote the canonical closed embedding ${}^{\mathrm{cl}}X \hookrightarrow X$. The functor i_* is an equivalence $\mathrm{Coh}({}^{\mathrm{cl}}X)^{\heartsuit} \to \mathrm{Coh}(X)^{\heartsuit}$. Hence, $\mathcal{F} = i_*(\mathcal{F}')$ for $\mathcal{F}' \in \mathrm{Coh}({}^{\mathrm{cl}}X)$. This reduces the assertion to the case when X is classical.

We factor the map $f : X \to Y$ as $X \to {}^{\mathrm{cl}}Y \overset{i}{\hookrightarrow} Y$. Since i is a closed embedding, we have reduced the assertion to the case when Y is classical as well. In the latter case, the assertion is well-known.

□

5.1.5. The above lemma implies that the functor f_*^{IndCoh} sends $\mathrm{IndCoh}(X)^c$ to $\mathrm{IndCoh}(Y)^c$.

Hence, f_*^{IndCoh} admits a continuous right adjoint, to be denoted

$$f^! : \mathrm{IndCoh}(X) \to \mathrm{IndCoh}(Y).$$

REMARK 5.1.6. The continuity of the functor $f^!$ is the *raison d'etre* of the category IndCoh, and its main difference from QCoh.

5.1.7. By Chapter 1, Lemma 9.3.6, we obtain that the functor $f^!$ has a natural structure of 1-morphism in $\mathrm{QCoh}(Y)$-**mod**.

5.1.8. Note that the functor f_*^{IndCoh} is *right t-exact, up to a finite shift.* Hence, the functor $f^!$ is left t-exact up to a finite shift. In particular, $f^!$ maps $\mathrm{IndCoh}(Y)^+$ to $\mathrm{IndCoh}(X)^+$.

Let $f^{\mathrm{QCoh},!}$ denote the *not necessarily continuous* right adjoint to f_*. It also has the property that it maps $\mathrm{QCoh}(Y)^+$ to $\mathrm{QCoh}(X)^+$.

LEMMA 5.1.9. *The diagram*

$$\begin{array}{ccc} \mathrm{IndCoh}(X)^+ & \xleftarrow{f^!} & \mathrm{IndCoh}(Y)^+ \\ {\scriptstyle\Psi_X}\downarrow & & \downarrow{\scriptstyle\Psi_Y} \\ \mathrm{QCoh}(X)^+ & \xleftarrow{f^{\mathrm{QCoh},!}} & \mathrm{QCoh}(Y)^+ \end{array}$$

obtained by passing to right adjoints along the horizontal arrows in

$$\begin{array}{ccc} \mathrm{IndCoh}(X)^+ & \xrightarrow{f_*^{\mathrm{IndCoh}}} & \mathrm{IndCoh}(Y)^+ \\ {\scriptstyle\Psi_X}\downarrow & & \downarrow{\scriptstyle\Psi_Y} \\ \mathrm{QCoh}(X)^+ & \xrightarrow{f_*} & \mathrm{QCoh}(Y)^+, \end{array}$$

commutes.

PROOF. Follows from the fact that the vertical arrows are equivalences, by Proposition 1.2.2. □

REMARK 5.1.10. It is not in general true that the diagram

$$\begin{array}{ccc} \mathrm{IndCoh}(X) & \xleftarrow{f^!} & \mathrm{IndCoh}(Y) \\ {\scriptstyle\Psi_X}\downarrow & & \downarrow{\scriptstyle\Psi_Y} \\ \mathrm{QCoh}(X) & \xleftarrow{f^{\mathrm{QCoh},!}} & \mathrm{QCoh}(Y) \end{array}$$

obtained by passing to right adjoints along the horizontal arrows in

$$\begin{array}{ccc} \mathrm{IndCoh}(X) & \xrightarrow{f_*^{\mathrm{IndCoh}}} & \mathrm{IndCoh}(Y) \\ {\scriptstyle\Psi_X}\downarrow & & \downarrow{\scriptstyle\Psi_Y} \\ \mathrm{QCoh}(X) & \xrightarrow{f_*} & \mathrm{QCoh}(Y), \end{array}$$

commutes.

For example, take $X = \mathrm{pt} = \mathrm{Spec}(k)$, $Y = \mathrm{Spec}(k[t]/t^2)$ and $0 \neq \mathcal{F} \in \mathrm{IndCoh}(Y)$ be in the kernel of the functor Ψ_Y. Then $\Psi_X \circ f^!(\mathcal{F}) \neq 0$. Indeed, Ψ_X is an equivalence, and $f^!$ is conservative, see Corollary 6.1.5.

5.1.11. Let $(\mathrm{Sch}_{\mathrm{aft}})_{\mathrm{proper}}$ be a 1-full subcategory of $\mathrm{Sch}_{\mathrm{aft}}$ when we restrict 1-morphisms to be proper maps. By Chapter 1, Sect. 8.4.2, we obtain:

COROLLARY 5.1.12. *There exists a canonically defined functor*

$$\mathrm{IndCoh}^!_{(\mathrm{Sch}_{\mathrm{aft}})_{\mathrm{proper}}} : ((\mathrm{Sch}_{\mathrm{aft}})_{\mathrm{proper}})^{\mathrm{op}} \to \mathrm{DGCat}_{\mathrm{cont}},$$

obtained from

$$\mathrm{IndCoh}_{(\mathrm{Sch}_{\mathrm{aft}})_{\mathrm{proper}}} := \mathrm{IndCoh}_{(\mathrm{Sch}_{\mathrm{aft}})}|_{((\mathrm{Sch}_{\mathrm{aft}})_{\mathrm{proper}})^{\mathrm{op}}}$$

by passing to right adjoints.

5.2. Base change for proper maps. A crucial property of the !-pullback is *base change* against the *-direct image. We establish it in this subsection.

5.2.1. Let

$$\begin{array}{ccc} X_1 & \xrightarrow{g_X} & X_2 \\ {\scriptstyle f_1}\downarrow & & \downarrow{\scriptstyle f_2} \\ Y_1 & \xrightarrow{g_Y} & Y_2 \end{array}$$

be a Cartesian diagram in $\mathrm{Sch}_{\mathrm{aft}}$, with the vertical maps being proper.

The isomorphism of functors

$$(f_2)^{\mathrm{IndCoh}}_* \circ (g_X)^{\mathrm{IndCoh}}_* \simeq (g_Y)^{\mathrm{IndCoh}}_* \circ (f_1)^{\mathrm{IndCoh}}_*$$

gives rise to a natural transformation:

$$(g_X)^{\mathrm{IndCoh}}_* \circ f_1^! \to f_2^! \circ (g_Y)^{\mathrm{IndCoh}}_*. \tag{5.1}$$

We will prove:

PROPOSITION 5.2.2. *The map* (5.1) *is an isomorphism.*

PROOF. Since all functors involved are continuous, it is enough to show that the map

$$(g_X)^{\mathrm{IndCoh}}_* \circ f_1^!(\mathcal{F}) \to f_2^! \circ (g_Y)^{\mathrm{IndCoh}}_*(\mathcal{F})$$

is an isomorphism for $\mathcal{F} \in \mathrm{Coh}(Y_1)$. Hence, it is enough to show that (5.1) is an isomorphism when restricted to $\mathrm{IndCoh}(Y_1)^+$.

By Lemma 5.1.9 and Proposition 1.2.2, this reduces the assertion to showing that the natural transformation

$$(g_X)_* \circ f_1^{\mathrm{QCoh},!} \to f_2^{\mathrm{QCoh},!} \circ (g_Y)_* \tag{5.2}$$

is an isomorphism for the functors

$$\begin{array}{ccc} \mathrm{QCoh}(X_1)^+ & \xrightarrow{(g_X)_*} & \mathrm{QCoh}(X_2)^+ \\ {\scriptstyle f_1^{\mathrm{QCoh},!}}\uparrow & & \uparrow{\scriptstyle f_2^{\mathrm{QCoh},!}} \\ \mathrm{QCoh}(Y_1)^+ & \xrightarrow{(g_Y)_*} & \mathrm{QCoh}(Y_2)^+, \end{array}$$

where the natural transformation comes from the commutative diagram

$$\begin{array}{ccc} \mathrm{QCoh}(X_1)^+ & \xrightarrow{(g_X)_*} & \mathrm{QCoh}(X_2)^+ \\ {\scriptstyle (f_1)_*}\downarrow & & \downarrow{\scriptstyle (f_2)_*} \\ \mathrm{QCoh}(Y_1)^+ & \xrightarrow{(g_Y)_*} & \mathrm{QCoh}(Y_2)^+ \end{array}$$

by passing to right adjoint along the vertical arrows.

We consider the commutative diagram

$$\begin{array}{ccc} \mathrm{QCoh}(X_1) & \xrightarrow{(g_X)_*} & \mathrm{QCoh}(X_2) \\ {\scriptstyle (f_1)_*}\downarrow & & \downarrow{\scriptstyle (f_2)_*} \\ \mathrm{QCoh}(Y_1) & \xrightarrow{(g_Y)_*} & \mathrm{QCoh}(Y_2), \end{array}$$

and the the diagram

$$\begin{array}{ccc} \mathrm{QCoh}(X_1) & \xrightarrow{(g_X)_*} & \mathrm{QCoh}(X_2) \\ {\scriptstyle f_1^{\mathrm{QCoh},!}}\uparrow & & \uparrow{\scriptstyle f_2^{\mathrm{QCoh},!}} \\ \mathrm{QCoh}(Y_1) & \xrightarrow{(g_Y)_*} & \mathrm{QCoh}(Y_2), \end{array}$$

obtained by passing to right adjoints along the vertical arrows. (Note, however, that the functors involved are no longer continuous).

We claim that the resulting natural transformation

$$(g_X)_* \circ f_1^{\mathrm{QCoh},!} \to f_2^{\mathrm{QCoh},!} \circ (g_Y)_* \tag{5.3}$$

between the functors

$$\mathrm{QCoh}(Y_1) \rightrightarrows \mathrm{QCoh}(X_2)$$

is an isomorphism. This would imply that (5.2) is an isomorphism by restricting to the eventually coconnective subcategory.

To prove that (5.3) is an isomorphism, we note that this map is obtained by passing to right adjoints in the natural transformation

$$(g_Y)^* \circ (f_2)_* \to (f_1)_* \circ (g_X)^* \tag{5.4}$$

as functors

$$\mathrm{QCoh}(X_2) \rightrightarrows \mathrm{QCoh}(Y_1)$$

in the commutative diagram

$$\begin{array}{ccc} \mathrm{QCoh}(X_1) & \xleftarrow{(g_X)^*} & \mathrm{QCoh}(X_2) \\ {\scriptstyle (f_1)_*}\downarrow & & \downarrow{\scriptstyle (f_2)_*} \\ \mathrm{QCoh}(Y_1) & \xleftarrow{(g_Y)^*} & \mathrm{QCoh}(Y_2). \end{array}$$

Now, (5.4) is an isomorphism by the usual base change for QCoh. Hence, (5.3) is an isomorphism as well.

□

5.3. Pullback compatibilty. The !-pullback for *arbitrary* maps between schemes will be defined in such a way that it is the !-pullback for proper morphisms, and the *-pullback for open embeddings.

Hence, if we want that !-pullback to be well-defined, a certain compatibility must take place, when we decompose a morphism in two different ways as a composition of a proper morphism and an open embedding. A basic case of such compatibility is established in this subsection.

5.3.1. Let

$$\begin{array}{ccc} X_1 & \xrightarrow{g_X} & X_2 \\ {\scriptstyle f_1}\downarrow & & \downarrow{\scriptstyle f_2} \\ Y_1 & \xrightarrow{g_Y} & Y_2 \end{array}$$

be a Cartesian diagram in $\mathrm{Sch}_{\mathrm{aft}}$, with the vertical maps being proper, and horizontal maps being eventually coconnective.

We start with the base change isomorphism

$$(f_1)_*^{\mathrm{IndCoh}} \circ g_X^{\mathrm{IndCoh},*} \simeq g_Y^{\mathrm{IndCoh},*} \circ (f_2)_*^{\mathrm{IndCoh}}$$

of Proposition 3.2.2, and by the $(f_*^{\mathrm{IndCoh}}, f^!)$-adjunction obtain a map

$$g_X^{\mathrm{IndCoh},*} \circ f_2^! \to f_1^! \circ g_Y^{\mathrm{IndCoh},*}. \tag{5.5}$$

REMARK 5.3.2. Note that one can get another map

$$g_X^{\mathrm{IndCoh},*} \circ f_2^! \to f_1^! \circ g_Y^{\mathrm{IndCoh},*}, \tag{5.6}$$

namely, via the $(g^{\mathrm{IndCoh},*}, g_*^{\mathrm{IndCoh}})$-adjunction from the isomorphism

$$f_2^! \circ (g_Y)_*^{\mathrm{IndCoh}} \simeq (g_X)_*^{\mathrm{IndCoh}} \circ f_1^!$$

of Proposition 5.2.2. A diagram chase shows that the map (5.6) is canonically the same as (5.5).

5.3.3. We are going to prove:

PROPOSITION 5.3.4. *Suppose that g_Y (and hence g_X) are open embeddings. Then the map* (5.5) *is an isomorphism.*

REMARK 5.3.5. It is shown in [**Ga1**, Proposition 7.1.6] that the map (5.5) is an isomorphism for any eventually coconnective g_Y.

PROOF. By Proposition 4.1.2, it suffices to show that the induced map

$$(g_X)_*^{\mathrm{IndCoh}} \circ g_X^{\mathrm{IndCoh},*} \circ f_2^! \to (g_X)_*^{\mathrm{IndCoh}} \circ f_1^! \circ g_Y^{\mathrm{IndCoh},*}$$

is an isomorphism.

Using Proposition 5.2.2, we have

$$(g_X)_*^{\mathrm{IndCoh}} \circ f_1^! \circ g_Y^{\mathrm{IndCoh},*} \simeq f_2^! \circ (g_Y)_*^{\mathrm{IndCoh}} \circ g_Y^{\mathrm{IndCoh},*}.$$

Hence, we need to show that the map

$$(g_X)_*^{\mathrm{IndCoh}} \circ g_X^{\mathrm{IndCoh},*} \circ f_2^! \to f_2^! \circ (g_Y)_*^{\mathrm{IndCoh}} \circ g_Y^{\mathrm{IndCoh},*} \tag{5.7}$$

is an isomorphism.

By Corollary 3.3.2, for $\mathcal{F} \in \mathrm{IndCoh}(Y_2)$, we have canonical isomorphisms

$$(g_X)_*^{\mathrm{IndCoh}} \circ g_X^{\mathrm{IndCoh},*} \circ f_2^!(\mathcal{F}) \simeq (g_X)_*(\mathcal{O}_{X_1}) \otimes f_2^!(\mathcal{F})$$

and

$$(g_Y)_*^{\mathrm{IndCoh}} \circ g_Y^{\mathrm{IndCoh},*}(\mathcal{F}) \simeq (g_Y)_*(\mathcal{O}_{Y_1}) \otimes \mathcal{F},$$

where $\otimes$ denotes the action of QCoh on IndCoh.

By the compatibilty of the action of QCoh with the !-pullback, we have

$$f_2^! \circ (g_Y)_*^{\mathrm{IndCoh}} \circ g_Y^{\mathrm{IndCoh},*}(\mathcal{F}) \simeq f_2^*((g_Y)_*(\mathcal{O}_{Y_1})) \otimes f_2^!(\mathcal{F}) \simeq (g_X)_*(\mathcal{O}_{X_1}) \otimes f_2^!(\mathcal{F}).$$

Now, diagram chase shows that, under the above identifications, the map (5.7) is the identify map endomorphism of $(g_X)_*(\mathcal{O}_{X_1}) \otimes f_2^!(\mathcal{F})$.

□

6. Closed embeddings

The behavior of IndCoh with respect to closed embedding is 'better' than that of QCoh. The main point of difference is that the direct image functor under a closed embedding for IndCoh preserves compactness, which is not the case for QCoh.

6.1. Category with support. In this subsection we study the full subcategory of $\text{IndCoh}(X)$ corresponding with objects 'with support' on a given closed subscheme.

6.1.1. Let X be an object of Sch_{aft}, and let $i : Z \to X$ be a closed embedding. Let $j : U \to X$ be the complementary open.

We let $\text{IndCoh}(X)_Z$ denote the full subcategory of $\text{IndCoh}(X)$ equal to

$$\ker\Big(j^{\text{IndCoh},*} : \text{IndCoh}(X) \to \text{IndCoh}(U)\Big).$$

Note that the embedding

$$\text{IndCoh}(X)_Z \hookrightarrow \text{IndCoh}(Z)$$

admits a right adjoint given by sending $\mathcal{F}$ to

$$\ker(\mathcal{F} \to j_*^{\text{IndCoh}} \circ j^{\text{IndCoh},*}(\mathcal{F})),$$

which by Corollary 3.3.5 is the same as

$$\ker(\mathcal{O}_X \to j_*(\mathcal{O}_U)) \otimes \mathcal{F}.$$

6.1.2. We claim:

PROPOSITION 6.1.3.

(a) *The subcategory* $\text{IndCoh}(X)_Z \subset \text{IndCoh}(X)$ *is compatible with the t-structure (i.e., is preserved by the truncation functors).*

(b) *The subcategory* $\text{IndCoh}(X)_Z \subset \text{IndCoh}(X)$ *is generated by the essential image of the functor* $i_*^{\text{IndCoh}} : \text{IndCoh}(Z) \to \text{IndCoh}(X)$.

(c) *The functor* $i^! : \text{IndCoh}(X) \to \text{IndCoh}(Z)$ *is conservative, when restricted to* $\text{IndCoh}(X)_Z$.

(d) *The category* $\text{IndCoh}(X)_Z$ *identifies with the ind-completion of*

$$\text{Coh}(X)_Z := \ker\Big(j^* : \text{Coh}(X) \to \text{Coh}(U)\Big).$$

PROOF. Point (a) follows from the fact that the functor $j^{\text{IndCoh},*}$ is t-exact.

Let us prove point (b). The category $\text{IndCoh}(X)$ is generated by $\text{IndCoh}(X)^+$. The description of the right adjoint to $\text{IndCoh}(X)_Z \hookrightarrow \text{IndCoh}(Z)$ implies that the same is true for $\text{IndCoh}(X)_Z$.

For every $\mathcal{F} \in \text{IndCoh}(X)^+$, the map

$$\underset{n}{\text{colim}}\, \tau^{\leq n}(\mathcal{F}) \to \mathcal{F},$$

is an isomorphism, by Proposition 1.2.2 and because the corresponding fact is true in QCoh^+.

Hence, by point (a), every $\mathcal{F} \in \text{IndCoh}(X)^+_Z$ is a colimit of cohomologically bounded objects from $\text{IndCoh}(X)_Z$. This implies that $\text{IndCoh}(X)_Z$ is generated by

$$\text{IndCoh}(X)^{\heartsuit}_Z = \text{Coh}(X)^{\heartsuit}_Z.$$

Now, it is easy to see that every object $\mathcal{F} \in \text{Coh}(X)^{\heartsuit}_Z$ has a filtration

$$\mathcal{F} = \bigcup_n \mathcal{F}_n$$

with $\mathcal{F}_n/\mathcal{F}_{n-1} \in i_*(\text{Coh}(Z)^{\heartsuit})$. This proves point (b).

Point (c) follows from point (b) by adjunction.

To prove point (d), it suffices to note that the objects from $\text{Coh}(X)_Z$ are compact in $\text{IndCoh}(X)_Z$ (because they are compact in $\text{IndCoh}(X)$) and that they generate $\text{IndCoh}(X)_Z$, by point (b).

□

6.1.4. In what follows we will use the following terminology: for a (derived) scheme X, we will denote by ${}^{\text{red}}X$ the classical *reduced* scheme, underlying reduced scheme of the classical scheme ${}^{\text{cl}}X$.

We shall say that a map of (derived) schemes $f : X' \to X$ is a *nil-isomorphism*, i.e., a map such that ${}^{\text{red}}X' \to {}^{\text{red}}X$ is an isomorphism.

As a corollary of Proposition 6.1.3 we obtain:

COROLLARY 6.1.5. *Let $f : X' \to X$ be a nil-isomorphism, i.e., a map such that ${}^{\text{red}}X' \to {}^{\text{red}}X$ is an isomorphism. Then the functor $f^!$ is conservative. Equivalently, the essential image of* $\text{IndCoh}(X')$ *under f^{IndCoh}_* generates* $\text{IndCoh}(X)$.

PROOF. We can assume that $X' = {}^{\text{red}}X$, so f is also a closed embedding. In this case the assertion of the corollary follows from Proposition 6.1.3(b). □

6.2. A conservativeness result for proper maps. The main result established in this subsection, Proposition 6.2.2, is of technical significance.

6.2.1. We shall now use Proposition 6.1.3 to prove the following:

PROPOSITION 6.2.2. *Let $f : X \to Y$ be a proper map, which is surjective at the level of geometric points. Then the functor $f^! : \text{IndCoh}(Y) \to \text{IndCoh}(X)$ is conservative.*

The rest of this subsection is devoted to the proof of the proposition.

6.2.3. By Corollary 6.1.5 we can assume that both X and Y are classical and reduced. We argue by Noetherian induction, assuming that the statement is true for all proper closed subschemes of Y. We need to show that the essential image of $\text{IndCoh}(X)$ under f^{IndCoh}_* generates $\text{IndCoh}(Y)$.

By Proposition 6.1.3(a), it is sufficient to show that Y contains an open subscheme $\overset{\circ}{Y} \subset Y$ such that for $\overset{\circ}{X} := f^{-1}(\overset{\circ}{Y})$, the essential image of $\text{IndCoh}(\overset{\circ}{X})$ under $(f|_{\overset{\circ}{X}})^{\text{IndCoh}}_*$ generates $\text{IndCoh}(\overset{\circ}{Y})$.

Since Y is classical and reduced, it contains a non-empty open smooth subscheme, which we take to be $\overset{\circ}{Y}$. By Lemma 1.1.3 and Lemma 1.1.7, we are reduced to showing the following.

LEMMA 6.2.4. *Let $f : X \to Y$ be a proper surjective morphism of classical schemes with Y smooth. Then the essential image of* $\mathrm{QCoh}(X)$ *under f_* generates* $\mathrm{QCoh}(Y)$.

□

PROOF OF LEMMA 6.2.4. Let $\mathbf{C} \subset \mathrm{QCoh}(Y)$ be the full subcategory generated by the essential image of $\mathrm{QCoh}(X)$ under f_*. Note that $\mathbf{C}$ is a monoidal ideal, since the functor f_* respects the action of $\mathrm{QCoh}(Y)$. Consider $\mathcal{E} := f_*(\mathcal{O}_X) \in \mathbf{C}$. This is an object of $\mathrm{QCoh}(Y)$, whose fiber at every geometric point of Y is non-zero.

However, it is easy to see that for any Noetherian classical scheme Y and $\mathcal{E} \in \mathbf{C} \subset \mathrm{QCoh}(Y)$ with the above properties, we have

$$\mathbf{C} = \mathrm{QCoh}(Y).$$

□

6.3. Products. It is known that for a pair of quasi-compact schemes X_1 and X_2, tensor product defines an equivalence of DG categories

$$\mathrm{QCoh}(X_1) \otimes \mathrm{QCoh}(X_2) \to \mathrm{QCoh}(X_1 \times X_2).$$

In this subsection we will establish a similar assertion for IndCoh. This is not altogether tautological; for example the validity of this fact replies on the assumption that the ground field k be perfect.

6.3.1. Let X_1 and X_2 be two objects of $\mathrm{Sch}_{\mathrm{aft}}$. We claim:

LEMMA 6.3.2. *There exists a uniquely defined functor*

$$\boxtimes : \mathrm{IndCoh}(X_1) \otimes \mathrm{IndCoh}(X_2) \to \mathrm{IndCoh}(X_1 \times X_2) \tag{6.1}$$

that preserves compactness and makes the diagram

$$\begin{array}{ccc} \mathrm{IndCoh}(X_1) \otimes \mathrm{IndCoh}(X_2) & \xrightarrow{\boxtimes} & \mathrm{IndCoh}(X_1 \times X_2) \\ {\scriptstyle \Psi_{X_1} \times \Psi_{X_2}} \downarrow & & \downarrow {\scriptstyle \Psi_{X_1 \times X_2}} \\ \mathrm{QCoh}(X_1) \otimes \mathrm{QCoh}(X_2) & \xrightarrow{\boxtimes} & \mathrm{QCoh}(X_1 \times X_2) \end{array}$$

commute.

PROOF. The anticlock-wise composition sends the compact generators of the category

$$\mathrm{IndCoh}(X_1) \otimes \mathrm{IndCoh}(X_2)$$

(i.e., objects of the form $\mathcal{F}_1 \otimes \mathcal{F}_2$ for $\mathcal{F}_i \in \mathrm{Coh}(X_i)$) to $\mathrm{QCoh}(X_1 \times X_2)^+$. Hence, it sends all of $(\mathrm{IndCoh}(X_1) \otimes \mathrm{IndCoh}(X_2))^c$ to $\mathrm{QCoh}(X_1 \times X_2)^+$.

The sought-for functor is the ind-extension of

$$(\mathrm{IndCoh}(X_1) \otimes \mathrm{IndCoh}(X_2))^c \to \mathrm{QCoh}(X_1 \times X_2)^+ \overset{\Psi_{X_1 \times X_2}}{\simeq} \mathrm{IndCoh}(X_1 \times X_2)^+.$$

This functor preserves compactness since the objects

$$\mathcal{F}_1 \boxtimes \mathcal{F}_2 \in \mathrm{QCoh}(X_1 \times X_2)^+, \quad \mathcal{F}_i \in \mathrm{Coh}(X_i)$$

belong to $\mathrm{Coh}(X_1 \times X_2)$.

□

6.3.3. We now claim:

PROPOSITION 6.3.4.

(a) *The functor* (6.1) *is fully faithful.*

(b) *If the ground field k is perfect, then* (6.1) *is an equivalence.*

PROOF. Since the functor (6.1) preserves compactness, for point (a) it is sufficient to show that for $\mathcal{F}'_i, \mathcal{F}''_i \in \mathrm{Coh}(X_i)$, $i = 1, 2$, the map

$$\mathcal{M}aps_{\mathrm{IndCoh}(X_1)\otimes\mathrm{IndCoh}(X_2)}(\mathcal{F}'_1 \otimes \mathcal{F}'_2, \mathcal{F}''_1 \otimes \mathcal{F}''_2) \to \mathcal{M}aps_{\mathrm{IndCoh}(X_1\times X_2)}(\mathcal{F}'_1 \boxtimes \mathcal{F}'_2, \mathcal{F}''_1 \boxtimes \mathcal{F}''_2)$$

is an isomorphism.

We have a commutative diagram

$$\begin{array}{ccc}
\mathcal{M}aps_{\mathrm{IndCoh}(X_1)\otimes\mathrm{IndCoh}(X_2)}(\mathcal{F}'_1 \otimes \mathcal{F}'_2, \mathcal{F}''_1 \otimes \mathcal{F}''_2) & \longrightarrow & \mathcal{M}aps_{\mathrm{IndCoh}(X_1\times X_2)}(\mathcal{F}'_1 \boxtimes \mathcal{F}'_2, \mathcal{F}''_1 \boxtimes \mathcal{F}''_2) \\
\sim\uparrow & & \\
\mathcal{M}aps_{\mathrm{Coh}(X_1)}(\mathcal{F}'_1, \mathcal{F}''_1) \otimes \mathcal{M}aps_{\mathrm{Coh}(X_2)}(\mathcal{F}'_2, \mathcal{F}''_2) & & \downarrow\sim \\
\sim\downarrow & & \\
\mathcal{M}aps_{\mathrm{QCoh}(X_1)}(\mathcal{F}'_1, \mathcal{F}''_1) \otimes \mathcal{M}aps_{\mathrm{QCoh}(X_2)}(\mathcal{F}'_2, \mathcal{F}''_2) & \longrightarrow & \mathcal{M}aps_{\mathrm{QCoh}(X_1\times X_2)}(\mathcal{F}'_1 \boxtimes \mathcal{F}'_2, \mathcal{F}''_1 \boxtimes \mathcal{F}''_2).
\end{array}$$

Hence, it remains to show that

$$\mathcal{M}aps_{\mathrm{QCoh}(X_1)}(\mathcal{F}'_1, \mathcal{F}''_1) \otimes \mathcal{M}aps_{\mathrm{QCoh}(X_2)}(\mathcal{F}'_2, \mathcal{F}''_2) \to \to \mathcal{M}aps_{\mathrm{QCoh}(X_1\times X_2)}(\mathcal{F}'_1 \boxtimes \mathcal{F}'_2, \mathcal{F}''_1 \boxtimes \mathcal{F}''_2)$$

is an isomorphism. This is not immediate since the objects $\mathcal{F}'_i \in \mathrm{QCoh}(S_i)$ are not compact. To circumvent this, we proceed as follows.

It is enough to show that

$$\tau^{\leq n}\left(\mathcal{M}aps_{\mathrm{QCoh}(X_1)}(\mathcal{F}'_1, \mathcal{F}''_1) \otimes \mathcal{M}aps_{\mathrm{QCoh}(X_2)}(\mathcal{F}'_2, \mathcal{F}''_2)\right) \to \to \tau^{\leq n}\left(\mathcal{M}aps_{\mathrm{QCoh}(X_1\times X_2)}(\mathcal{F}'_1 \boxtimes \mathcal{F}'_2, \mathcal{F}''_1 \boxtimes \mathcal{F}''_2)\right)$$

is an isomorphism for any fixed n.

Choose $\alpha_1 : \widetilde{\mathcal{F}}'_1 \to \mathcal{F}'_1$ (resp., $\alpha_2 : \widetilde{\mathcal{F}}'_2 \to \mathcal{F}'_2$) with $\widetilde{\mathcal{F}}'_1$ (resp., $\widetilde{\mathcal{F}}'_2$) in $\mathrm{QCoh}(X_1)^{\mathrm{perf}}$ (resp., $\mathrm{QCoh}(X_2)^{\mathrm{perf}}$), such that

$$\mathrm{Cone}(\alpha_1) \in \mathrm{QCoh}(X_1)^{\leq -N} \text{ and } \mathrm{Cone}(\alpha_2) \in \mathrm{QCoh}(X_2)^{\leq -N}$$

for $N \gg 0$.

By choosing N large enough, we can ensure that

$$\tau^{\leq m}\left(\mathcal{M}aps_{\mathrm{QCoh}(S_i)}(\mathcal{F}'_i, \mathcal{F}''_i)\right) \to \tau^{\leq m}\left(\mathcal{M}aps_{\mathrm{QCoh}(S_i)}(\widetilde{\mathcal{F}}'_i, \mathcal{F}''_i)\right)$$

is an isomorphism for a given integer m. This implies that for $N \gg 0$ and our fixed n, the maps

$$\tau^{\leq n}\left(\mathcal{M}aps_{\mathrm{QCoh}(X_1)}(\mathcal{F}'_1, \mathcal{F}''_1) \otimes \mathcal{M}aps_{\mathrm{QCoh}(X_2)}(\mathcal{F}'_2, \mathcal{F}''_2)\right) \to \to \tau^{\leq n}\left(\mathcal{M}aps_{\mathrm{QCoh}(X_1)}(\widetilde{\mathcal{F}}'_1, \mathcal{F}''_1) \otimes \mathcal{M}aps_{\mathrm{QCoh}(X_2)}(\widetilde{\mathcal{F}}'_2, \mathcal{F}''_2)\right)$$

and

$$\tau^{\leq n}\left(\mathcal{M}aps_{\mathrm{QCoh}(X_1\times X_2)}(\mathcal{F}'_1\boxtimes\mathcal{F}'_2,\mathcal{F}''_1\boxtimes\mathcal{F}''_2)\right)\to$$
$$\to\tau^{\leq n}\left(\mathcal{M}aps_{\mathrm{QCoh}(X_1\times X_2)}(\widetilde{\mathcal{F}}'_1\boxtimes\widetilde{\mathcal{F}}'_2,\mathcal{F}''_1\boxtimes\mathcal{F}''_2)\right)$$

are isomorphisms.

Hence, it is enough to show that

$$\mathcal{M}aps_{\mathrm{QCoh}(X_1)}(\widetilde{\mathcal{F}}'_1,\mathcal{F}''_1)\otimes\mathcal{M}aps_{\mathrm{QCoh}(X_2)}(\widetilde{\mathcal{F}}'_2,\mathcal{F}''_2)\to$$
$$\to\mathcal{M}aps_{\mathrm{QCoh}(X_1\times X_2)}(\widetilde{\mathcal{F}}'_1\boxtimes\widetilde{\mathcal{F}}'_2,\mathcal{F}''_1\boxtimes\mathcal{F}''_2)$$

is an isomorphism. But this follows from the fact that the fuctor

$$\mathrm{QCoh}(X_1)\boxtimes\mathrm{QCoh}(X_2)\to\mathrm{QCoh}(X_1\times X_2) \tag{6.2}$$

is an equivalence.

This finishes the proof of point (a).

To prove point (b), we have to show that the essential image of the functor (6.1) generates $\mathrm{IndCoh}(X_1\times X_2)$. By Corollary 6.1.5 we can assume that both X_1 and X_2 are classical and reduced.

We argue by Noetherian induction, assuming that the statement is true for all proper closed subschemes $X'_i\subset X_i$.

By Proposition 6.1.3(b), it is sufficient to show that X_1 and X_2 contain non-empty open subschemes $\overset{\circ}{X}_i\subset X_i$, for which the statement of the proposition is true.

Since X_i are classical and reduced, we can take $\overset{\circ}{X}_i$ to be a non-empty open smooth subscheme of X_i. Note that the assumption that k be perfect implies that $\overset{\circ}{X}_1\times\overset{\circ}{X}_2$ is also smooth. Now, the assertion follows from Lemma 1.1.3 and the fact that (6.2) is an equivalence.

□

6.3.5. *Upgrading to a functor.* Consider the category $\mathrm{Sch}_{\mathrm{aft}}$ as endowed with a symmetric monoidal structure given by Cartesian product. We consider the category $\mathrm{DGCat}_{\mathrm{cont}}$ also as a symmetric monoidal ∞-category with respect to the operation of tensor product.

First we recall that the functor

$$\mathrm{QCoh}^*_{\mathrm{Sch}_{\mathrm{aft}}}:(\mathrm{Sch}_{\mathrm{aft}})^{\mathrm{op}}\to\mathrm{DGCat}_{\mathrm{cont}}$$

has a natural symmetric monoidal structure.

Indeed, this follows from Chapter 3, Sect. 3.1.3 and Proposition 3.1.7.

6.3.6. Passing to adjoints, by Chapter 9, Sect. 3.1.3, we obtain that the functor

$$\mathrm{QCoh}_{\mathrm{Sch}_{\mathrm{aft}}} : \mathrm{Sch}_{\mathrm{aft}} \to \mathrm{DGCat}_{\mathrm{cont}}$$

also has a natural symmetric monoidal structure.

Now, as in Proposition 2.2.3 one shows:

PROPOSITION 6.3.7. *There exists a unique symmetric monoidal structure on the functor*

$$\mathrm{IndCoh}_{\mathrm{Sch}_{\mathrm{aft}}} : \mathrm{Sch}_{\mathrm{aft}} \to \mathrm{DGCat}_{\mathrm{cont}}$$

and the natural transformation $\Psi_{\mathrm{Sch}_{\mathrm{aft}}}$ *that at the level of objects is given by Lemma* 6.3.2.

6.4. Convergence. In this subsection we will establish another crucial property of IndCoh that distinguishes it from QCoh. Namely, we will show that for a given scheme, its category IndCoh can be recovered from IndCoh on the n-coconnective truncations of this scheme.

6.4.1. Let X be an object of $\mathrm{Sch}_{\mathrm{aft}}$. For each n let i_n denote the closed embedding ${}^{\leq n}X \to X$, and for $n_1 \leq n_2$, let i_{n_1,n_2} denote the closed embedding

$${}^{\leq n_1}X \to {}^{\leq n_2}X.$$

By Chapter 1, Proposition 2.5.7, we have a canonical equivalence

$$\operatorname*{colim}_n \mathrm{IndCoh}({}^{\leq n}X) \simeq \lim_n \mathrm{IndCoh}({}^{\leq n}X),$$

where in the left-hand side the transition functors are $(i_{n_1,n_2})_*^{\mathrm{IndCoh}}$, and in the right-hand side $(i_{n_1,n_2})^!$.

The functors $i_n^!$ define a functor

$$\mathrm{IndCoh}(X) \to \lim_n \mathrm{IndCoh}({}^{\leq n}X), \tag{6.3}$$

whose left adjoint

$$\operatorname*{colim}_n \mathrm{IndCoh}({}^{\leq n}X) \to \mathrm{IndCoh}(X) \tag{6.4}$$

is given by the compatible family of functors $(i_n)_*^{\mathrm{IndCoh}}$.

6.4.2. We are going to establish the following property of the category IndCoh:

PROPOSITION 6.4.3. *The functors* (6.3) *and* (6.4) *are mutually inverse equivalences.*

PROOF. First, we note that Corollary 6.1.5 shows that the functor (6.3) is conservative. It is clear that the functor (6.4) sends compact objects to compact ones.

Hence, it remains to prove the following: for

$$\mathcal{F}_1, \mathcal{F}_2 \in \mathrm{Coh}({}^{\leq 0}X)^{\heartsuit} \simeq \mathrm{Coh}(X)^{\heartsuit}$$

and $k \in \mathbb{N}$, the map

$$\operatorname*{colim}_n \mathrm{Maps}_{\mathrm{Coh}({}^{\leq n}X)}(\mathcal{F}_1, \mathcal{F}_2[k]) \to \mathrm{Maps}_{\mathrm{Coh}(X)}(\mathcal{F}_1, \mathcal{F}_2[k])$$

is an isomorphism.

We will prove more generally the following:

LEMMA 6.4.4. *For*

$$\mathcal{F}_1 \in (\mathrm{QCoh}({}^{\leq n}X))^{\leq 0},\ \mathcal{F}_2 \in (\mathrm{QCoh}({}^{\leq n}X))^{\geq -k},$$

the map

$$\mathrm{Maps}_{\mathrm{QCoh}({}^{\leq n}X)}(\mathcal{F}_1, \mathcal{F}_2) \to \mathrm{Maps}_{\mathrm{QCoh}(X)}((i_n)_*(\mathcal{F}_1), (i_n)_*(\mathcal{F}_2)) \tag{6.5}$$

is an isomorphism for $n \geq k$.

□

6.4.5. *Proof of Lemma* 6.4.4. We rewrite the right-hand side as

$$\mathrm{Maps}_{\mathrm{QCoh}({}^{\leq n}X)}((i_n)^* \circ (i_n)_*(\mathcal{F}_1), \mathcal{F}_2),$$

and we claim that

$$\mathrm{Cone}((i_n)^* \circ (i_n)_*(\mathcal{F}_1) \to \mathcal{F}_1) \in (\mathrm{QCoh}({}^{\leq n}X))^{\leq -n-1},$$

which is equivalent to

$$\mathrm{Cone}((i_n)_* \circ (i_n)^* \circ (i_n)_*(\mathcal{F}_1) \to (i_n)_*(\mathcal{F}_1)) \in \mathrm{QCoh}(X)^{\leq -n-1},$$

and further equivalent to

$$\mathrm{Cone}((i_n)_*(\mathcal{F}_1) \to (i_n)_* \circ (i_n)^* \circ (i_n)_*(\mathcal{F}_1)) \in \mathrm{QCoh}(X)^{\leq -n}.$$

In fact, we claim that for $\mathcal{F} \in \mathrm{QCoh}(X)^{\leq 0}$,

$$\mathrm{Cone}(\mathcal{F} \to (i_n)_* \circ (i_n)^*(\mathcal{F})) \in \mathrm{QCoh}(X)^{\leq -n}.$$

Indeed,

$$\mathrm{Cone}(\mathcal{F} \to (i_n)_* \circ (i_n)^*(\mathcal{F}))) \simeq \mathrm{Cone}(\mathcal{O}_X \to (i_n)_*(\mathcal{O}_{{}^{\leq n}X})) \otimes \mathcal{F},$$

and the assertion follows.

□

7. Groupoids and descent

In this section we will show that the category IndCoh satisfies descent with respect to proper surjective maps. We will later strengthen this to show that IndCoh satisfies h-descent.

7.1. The Beck-Chevalley condition. The Beck-Chevalley condition gives a sufficient condition for when the totalization of a given co-simplicial category can be described as co-modules over a co-monad acting on the category of 0-simplices.

7.1.1. Let us recall the following general framework.

Let $\mathbf{C}^\bullet$ be a co-simplicial ∞-category. Consider the corresponding category $\mathbf{C}^{\bullet+1}$, and the co-simplicial functor

$$\mathbf{C}^{\bullet+1} \leftarrow \mathbf{C}^\bullet : \mathfrak{s}^\bullet.$$

Note that the co-simplicial category $\mathbf{C}^{\bullet+1}$ is augmented and split by $\mathbf{C}^0$. Hence, we have a canonical equivalence

$$\mathbf{C}^0 \simeq \operatorname{Tot}(\mathbf{C}^{\bullet+1}),$$

so that the composed functor

$$\mathbf{C}^0 \simeq \operatorname{Tot}(\mathbf{C}^{\bullet+1}) \overset{\operatorname{Tot}(\mathfrak{s}^\bullet)}{\longleftarrow} \operatorname{Tot}(\mathbf{C}^\bullet)$$

identifies with $\mathrm{ev}^0_{\mathbf{C}^\bullet}$.

Furthermore, the functor

$$\mathbf{C}^0 \simeq \operatorname{Tot}(\mathbf{C}^{\bullet+1}) \overset{\mathrm{ev}^0_{\mathbf{C}^{\bullet+1}}}{\longrightarrow} \mathbf{C}^1$$

identifies with p_s, where p_s, p_t are the two functors

$$\mathbf{C}^0 \rightrightarrows \mathbf{C}^1.$$

7.1.2. Recall the following definition:

DEFINITION 7.1.3. *We shall say that $\mathbf{C}^\bullet$ satisfies the Beck-Chevalley condition if for each n the functor*

$$\mathbf{C}^{n+1} \leftarrow \mathbf{C}^n : \mathfrak{s}^n$$

admits a left adjoint (to be denoted by $\mathfrak{t}^n$), and for every map $[m] \to [n]$ in $\mathbf{\Delta}$, the diagram

$$\begin{array}{ccc} \mathbf{C}^{n+1} & \xrightarrow{\mathfrak{t}^n} & \mathbf{C}^n \\ \uparrow & & \uparrow \\ \mathbf{C}^{m+1} & \xrightarrow{\mathfrak{t}^m} & \mathbf{C}^m, \end{array}$$

that a priori commutes up to a natural transformation, actually commutes.

We have:

LEMMA 7.1.4. *Suppose that $\mathbf{C}^\bullet$ satisfies the Beck-Chevalley condition. Then:*

(a) *The functor*

$$\mathbf{C}^0 \leftarrow \operatorname{Tot}(\mathbf{C}^\bullet) : \mathrm{ev}^0_{\mathbf{C}}$$

admits a left adjoint.

(b) *The monad*

$$\mathrm{ev}^0_{\mathbf{C}} \circ (\mathrm{ev}^0_{\mathbf{C}})^L,$$

viewed as an endo-functor of $\mathbf{C}^0$, identifies with $(p_t)^L \circ p_s$, where $(p^t)^L$ is the left adjoint of p^t.

(c) *The adjoint pair*

$$(\mathrm{ev}^0_{\mathbf{C}^\bullet})^L : \mathbf{C}^0 \rightleftarrows \operatorname{Tot}(\mathbf{C}^\bullet) : \mathrm{ev}^0_{\mathbf{C}^\bullet}$$

is monadic.

PROOF. The Beck-Chevalley condition implies that the simplex-wise left adjoints $\mathfrak{t}^n$ form a co-simplicial functor

$$\mathfrak{t}^\bullet : \mathbf{C}^{\bullet+1} \leftarrow \mathbf{C}^\bullet.$$

In particular, we obtain a pair of adjoint functors

$$\mathrm{Tot}(\mathfrak{t}^\bullet) : \mathrm{Tot}(\mathbf{C}^{\bullet+1}) \rightleftarrows \mathrm{Tot}(\mathbf{C}^\bullet) : \mathrm{Tot}(\mathfrak{s}^\bullet),$$

that commute with evaluation on n-simplicies for every n.

Note also that $\mathfrak{s}^0 \simeq p^t$ and so $\mathfrak{t}^0 \simeq (p^t)^L$. Now, the required assertion concering

$$\mathrm{ev}^0_{\mathbf{C}^\bullet} \circ (\mathrm{ev}^0_{\mathbf{C}^\bullet})^L$$

follows from the commutative diagram

$$\begin{array}{ccc} \mathbf{C}^1 & \xrightarrow{\mathfrak{t}^0} & \mathbf{C}^0 \\ {\scriptstyle \mathrm{ev}^0_{\mathbf{C}^{\bullet+1}}}\big\uparrow & & \big\uparrow{\scriptstyle \mathrm{ev}^0_{\mathbf{C}^\bullet}} \\ \mathbf{C}^0 \simeq \mathrm{Tot}(\mathbf{C}^{\bullet+1}) & \xrightarrow{\mathrm{Tot}(\mathfrak{t}^\bullet)} & \mathrm{Tot}(\mathbf{C}^\bullet). \end{array}$$

Finally, it is easy to see that the functor $\mathrm{ev}^0_{\mathbf{C}^\bullet}$ is conservative and commutes with $\mathrm{ev}^0_{\mathbf{C}^\bullet}$-split geometric realizations. Hence, it satisfies the conditions of the Barr-Beck-Lurie theorem, and therefore is monadic.

□

7.2. Proper descent. We will now prove proper descent for IndCoh.

7.2.1. Let $X^\bullet$ be a groupoid simplicial object in $\mathrm{Sch}_{\mathrm{aft}}$ (see [**Lu1**], Definition 6.1.2.7 for the notion of groupoid in the context of ∞-categories).

Denote by

$$p_s, p_t : X^1 \rightrightarrows X^0 \tag{7.1}$$

the corresponding maps. Let us assume that the map p_s (and hence also p_t) is proper.

We form a co-simplicial category $\mathrm{IndCoh}(X^\bullet)^!$ using the !-pullback functors, and consider its totalization $\mathrm{Tot}(\mathrm{IndCoh}(X^\bullet)^!)$. Consider the functor of evaluation on 0-simplices:

$$\mathrm{ev}^0 : \mathrm{Tot}(\mathrm{IndCoh}(X^\bullet)) \to \mathrm{IndCoh}(X^0).$$

PROPOSITION 7.2.2.

(a) *Then functor* ev^0 *admits a left adjoint. The resulting monad on* $\mathrm{IndCoh}(X^0)$, *viewed as an endo-functor, is canonically isomorpic to* $(p_t)^{\mathrm{IndCoh}}_* \circ (p_s)^!$. *The adjoint pair*

$$\mathrm{IndCoh}(X^0) \rightleftarrows \mathrm{Tot}(\mathrm{IndCoh}(X^\bullet)^!)$$

is monadic.

(b) *Suppose that* $X^\bullet$ *is the Čech nerve of a map* $f : X^0 \to Y$, *where* f *proper. Assume also that* f *is surjective at the level of geometric points. Then the resulting map*

$$\mathrm{IndCoh}(Y) \to \mathrm{Tot}(\mathrm{IndCoh}(X^\bullet)^!)$$

is an equivalence.

REMARK 7.2.3. Note that the fact that ev^0 admits a left adjoint follows from Chapter 1, Proposition 2.5.7.

Indeed, the maps in $\mathrm{IndCoh}(X^\bullet)^!$ admit left adjoints, and we can interpret $\mathrm{Tot}(\mathrm{IndCoh}(X^\bullet)^!)$ as the geometric realization of the corresponding simplicial category $\mathrm{IndCoh}(X^\bullet)$, with the left adjoint to ev^0 being the corresponding tautological functor

$$\mathrm{IndCoh}(X^0) \to |\mathrm{IndCoh}(X^\bullet)|.$$

7.2.4. *Proof of Proposition* 7.2.2*(a).* By Lemma 7.1.4 we need to show that the co-simplicial category $\mathrm{IndCoh}(X^{\bullet+1})^!$ satisfies the Beck-Chevalley condition. However, this follows immediately from Proposition 5.2.2.

□

7.2.5. *Proof of Proposition* 7.2.2*(b).* Consider the adjoint pair

$$f_*^{\mathrm{IndCoh}} : \mathrm{IndCoh}(X^0) \rightleftarrows \mathrm{IndCoh}(Y) : f^!.$$

By Proposition 6.2.2, the functor $f^!$ is conservative, and continuous. Hence, the above pair is monadic.

The composition

$$\mathrm{IndCoh}(Y) \to \mathrm{Tot}(\mathrm{IndCoh}(X^\bullet)) \to \mathrm{IndCoh}(X^0)$$

identifies with the functor $f^!$. Hence, it remains to show that the map of the corresponding monads

$$\mathrm{ev}^0 \circ (\mathrm{ev}^0)^L \to f^! \circ f_*^{\mathrm{IndCoh}}$$

induces an isomorphism at the level of the underlying endo-functors of $\mathrm{IndCoh}(X^0)$.

By Proposition 7.2.2(a), the left-hand side identifies with

$$(p_t)_*^{\mathrm{IndCoh}} \circ p_s^!.$$

Furthermore, it follows from the construction that the resulting map

$$(p_t)_*^{\mathrm{IndCoh}} \circ p_s^! \to f^! \circ f_*^{\mathrm{IndCoh}}$$

is the base change morphism of Proposition 5.2.2 for the Cartesian diagram

$$\begin{array}{ccc} X^1 & \xrightarrow{p_s} & X^0 \\ {\scriptstyle p^t}\downarrow & & \downarrow{\scriptstyle f} \\ X^0 & \xrightarrow{f} & Y. \end{array}$$

Hence, the required isomorphism follows from Proposition 5.2.2.

□

CHAPTER 5

Ind-coherent sheaves as a functor out of the category of correspondences

Introduction

0.1. The !-pullback and base change.

0.1.1. In Chapter 4 we constructed the functor

(0.1)
$$\mathrm{IndCoh}_{\mathrm{Sch_{aft}}} : \mathrm{Sch_{aft}} \to \mathrm{DGCat_{cont}}, \quad X \rightsquigarrow \mathrm{IndCoh}(X), \quad (X \xrightarrow{f} Y) \rightsquigarrow f_*^{\mathrm{IndCoh}}.$$

In addition, we constructed the functors

(0.2)
$$\mathrm{IndCoh}^!_{(\mathrm{Sch_{aft}})_{\mathrm{proper}}} : ((\mathrm{Sch_{aft}})_{\mathrm{proper}})^{\mathrm{op}} \to \mathrm{DGCat_{cont}},$$
$$X \rightsquigarrow \mathrm{IndCoh}(X), \quad (X \xrightarrow{f} Y) \rightsquigarrow f^!$$

and

(0.3)
$$\mathrm{IndCoh}^*_{(\mathrm{Sch_{aft}})_{\mathrm{open}}} : ((\mathrm{Sch_{aft}})_{\mathrm{open}})^{\mathrm{op}} \to \mathrm{DGCat_{cont}},$$
$$X \rightsquigarrow \mathrm{IndCoh}(X), \quad (X \xrightarrow{f} Y) \rightsquigarrow f^{\mathrm{IndCoh},*},$$

where (0.2) is obtained from (0.1) by passing to *right* adjoints along proper maps, and (0.3) is obtained from (0.1) by passing to *left* adjoints along open embeddings.

The goal of the present chapter is to combine the above functors to a single piece of structure.

0.1.2. It is easy to phrase (but not to prove!) what it means to combine the functors (0.2) and (0.3): we will have a single functor

(0.4)
$$\mathrm{IndCoh}^!_{\mathrm{Sch_{aft}}} : (\mathrm{Sch_{aft}})^{\mathrm{op}} \to \mathrm{DGCat_{cont}}, \quad X \rightsquigarrow \mathrm{IndCoh}(X), \quad (X \xrightarrow{f} Y) \rightsquigarrow f^!.$$

It is trickier to say what kind of structure encodes both (0.1) and (0.4). The idea that we want to express is that these two functors are compatible via base change. I.e., for a Cartesian diagram in Sch

(0.5)
$$\begin{array}{ccc} X' & \xrightarrow{g_X} & X \\ {\scriptstyle f'}\downarrow & & \downarrow{\scriptstyle f} \\ Y' & \xrightarrow{g_Y} & Y \end{array}$$

we want to be *given* an isomorphism of functors

(0.6)
$$g_Y^! \circ f_*^{\mathrm{IndCoh}} \simeq (f')_*^{\mathrm{IndCoh}} \circ g_X^!.$$

The problem is that for a general diagram (0.5), there is no adjunction that gives rise to a map in (0.6) in either direction. Namely, if g_Y is proper, the natural map points $\leftarrow$, and when g_Y is an open embedding, the natural map points $\rightarrow$.

So, in general, the isomorphism (0.6) is really an additional piece of data, and once we want to say that these isomorphisms are compatible with the compositions of f's a and g's, we need to specify what we mean by that, i.e., what a *homotopy-compatible* system of isomorphisms means in this case.

0.1.3. Here enters the idea of the *category of correspondences*, suggested to us by J. Lurie, and developed in Chapter 7. This is the category, denoted $\text{Corr}(\text{Sch}_{\text{aft}})$, whose objects are $X \in \text{Sch}_{\text{aft}}$, and whose 1-morphisms are diagrams

$$\begin{array}{ccc} X_{0,1} & \xrightarrow{g} & X_0 \\ {\scriptstyle f}\downarrow & & \\ X_1. & & \end{array} \tag{0.7}$$

Compositions of 1-morphisms are given by fiber products: the composition of (0.7) with the 1-morphism

$$\begin{array}{ccc} X_{1,2} & \longrightarrow & X_1 \\ \downarrow & & \\ X_2 & & \end{array}$$

is given by the diagram

$$\begin{array}{ccc} X_{0,2} & \longrightarrow & X_0 \\ \downarrow & & \\ X_2, & & \end{array}$$

where $X_{0,2} := X_{1,2} \underset{X_1}{\times} X_{0,1}$. We refer the reader to Chapter 7, where it is explained how to define $\text{Corr}(\text{Sch}_{\text{aft}})$ as an ∞-category.

0.1.4. The main goal of this chapter is to define IndCoh as a functor

$$\text{IndCoh}_{\text{Corr}(\text{Sch}_{\text{aft}})} : \text{Corr}(\text{Sch}_{\text{aft}}) \to \text{DGCat}_{\text{cont}} \tag{0.8}$$

that, at the level of objects sends $X \rightsquigarrow \text{IndCoh}(X)$, and at the level of 1-morphisms sends the diagram (0.7) to $f_*^{\text{IndCoh}} \circ g^!$.

The functor in (0.8) will encode the initial functor (0.1) by restricting to the 1-full subcategory of $\text{Corr}(\text{Sch}_{\text{aft}})$, where we only allow 1-morphisms (0.7) with α being an isomorphism (this subcategory is tautologically equivalent to Sch_{aft}).

The functor in (0.8) will encode the initial functor (0.4) by restricting to the 1-full subcategory of $\text{Corr}(\text{Sch}_{\text{aft}})$, where we only allow 1-morphisms (0.7) with β being an isomorphism (this subcategory is tautologically equivalent to $(\text{Sch}_{\text{aft}})^{\text{op}}$).

0.1.5. In order to construct the functor $\mathrm{IndCoh}_{\mathrm{Corr}(\mathrm{Sch}_{\mathrm{aft}})}$ we will apply the machinery developed in Chapter 7. It turns out that the data of $\mathrm{IndCoh}_{\mathrm{Corr}(\mathrm{Sch}_{\mathrm{aft}})}$ is *uniquely recovered* from the data of the functor $\mathrm{IndCoh}_{\mathrm{Sch}_{\mathrm{aft}}}$ of (0.1).

However, there is one caveat: in order for the uniqueness statement mentioned above to hold, and in order to perform the construction of $\mathrm{IndCoh}_{\mathrm{Corr}(\mathrm{Sch}_{\mathrm{aft}})}$, one needs to work not with $(\infty,1)$-category $\mathrm{Corr}(\mathrm{Sch}_{\mathrm{aft}})$, but with the $(\infty,2)$-category $\mathrm{Corr}(\mathrm{Sch}_{\mathrm{aft}})^{\mathrm{proper}}$.

The latter $(\infty,2)$-category is one where we allow non-invertible 2-morphisms of the following kind: a 2-morphism from the 1-morphism (0.7) to the 1-morphism

$$\begin{array}{ccc} X'_{0,1} & \xrightarrow{g'} & X'_0 \\ {\scriptstyle f'}\big\downarrow & & \\ X'_1 & & \end{array}$$

is a commutative diagram

$$\begin{array}{ccccc} X_{0,1} & & & & \\ & \overset{h}{\searrow} & & \overset{f}{\searrow\searrow} & \\ & & X'_{0,1} & \xrightarrow[f']{} & X_0 \\ \quad\ \overset{g}{\searrow} & & \big\downarrow{\scriptstyle g'} & & \\ & & X_1, & & \end{array} \tag{0.9}$$

where h is proper.

What we will actually construct is the functor

$$\mathrm{IndCoh}_{\mathrm{Corr}(\mathrm{Sch}_{\mathrm{aft}})^{\mathrm{proper}}} : \mathrm{Corr}(\mathrm{Sch}_{\mathrm{aft}})^{\mathrm{proper}} \to \mathrm{DGCat}^{2\text{-Cat}}_{\mathrm{cont}}. \tag{0.10}$$

The additional piece of data that is contained in $\mathrm{IndCoh}_{\mathrm{Corr}(\mathrm{Sch}_{\mathrm{aft}})^{\mathrm{proper}}}$ as compared to $\mathrm{IndCoh}_{\mathrm{Corr}(\mathrm{Sch}_{\mathrm{aft}})}$ is that of *adjunction* between $f^!$ and f^{IndCoh}_* for a proper morphism f.

0.2. The !-pullback and IndCoh **on prestacks.** Having constructed the functor

$$\mathrm{IndCoh}_{\mathrm{Corr}(\mathrm{Sch}_{\mathrm{aft}})^{\mathrm{proper}}} : \mathrm{Corr}(\mathrm{Sch}_{\mathrm{aft}})^{\mathrm{proper}} \to \mathrm{DGCat}_{\mathrm{cont}},$$

we restrict it to $(\mathrm{Sch}_{\mathrm{aft}})^{\mathrm{op}} \subset \mathrm{Corr}(\mathrm{Sch}_{\mathrm{aft}})^{\mathrm{proper}}$ and obtain the functor

$$\mathrm{IndCoh}^!_{\mathrm{Sch}_{\mathrm{aft}}} : (\mathrm{Sch}_{\mathrm{aft}})^{\mathrm{op}} \to \mathrm{DGCat}_{\mathrm{cont}}$$

of (0.4).

REMARK 0.2.1. My emphasize that even if one is only interested in the functor (0.4), one has to employ the machinery of $(\infty,2)$-categories of correspondences in order to construct it.

0.2.2. Starting with the functor $\text{IndCoh}^!_{\text{Sch}_{\text{aft}}}$, we will *right-Kan-extend* it to a functor

$$\text{IndCoh}^!_{\text{PreStk}_{\text{aft}}} : (\text{PreStk}_{\text{aft}})^{\text{op}} \to \text{DGCat}_{\text{cont}}.$$

I.e., we now have a well-defined category $\text{IndCoh}(\mathcal{X})$ for $\mathcal{X} \in \text{PreStk}_{\text{aft}}$.

0.2.3. The assignment

$$\mathcal{X} \rightsquigarrow \text{IndCoh}(\mathcal{X})$$

provides a theory of $\mathcal{O}$-modules on prestacks, that exists alongside of QCoh; the former is functorial with respect to the !-pullback, while the assignment

$$\mathcal{X} \rightsquigarrow \text{QCoh}(\mathcal{X})$$

is functorial with respect to the *-pullback.

In Chapter 6, Sect. 3.3 we will see that the categories $\text{QCoh}(\mathcal{X})$ and $\text{QCoh}(\mathcal{X})$ are related by a functor

$$\Upsilon_{\mathcal{X}} : \text{QCoh}(\mathcal{X}) \to \text{IndCoh}(\mathcal{X}), \quad \mathcal{F} \mapsto \mathcal{F} \otimes \omega_{\mathcal{X}}, \tag{0.11}$$

where $\omega_{\mathcal{X}} \in \text{IndCoh}(\mathcal{X})$ is the dualizing object, and $\otimes$ is the action of $\text{QCoh}(\mathcal{X})$ on $\text{IndCoh}(\mathcal{X})$, defined in Chapter 6, Sect. 3.3.

However, in general, the functor $\Upsilon_{\mathcal{X}}$ is far from being an equivalence.

REMARK 0.2.4. Recall that when $\mathcal{X} = X$ is a scheme, we have a different functor relating $\text{IndCoh}(X)$ and $\text{QCoh}(X)$, namely

$$\Psi_X : \text{IndCoh}(X) \to \text{QCoh}(X)$$

(this functor was instrumental of getting the theory of IndCoh off the ground; we used it in order to defined the *-push forward functors for IndCoh).

In Chapter 6, Sect. 4.4 we will see that the functors

$$\Psi_X : \text{IndCoh}(X) \to \text{QCoh}(X) \text{ and } \Upsilon_X : \text{IndCoh}(X) \to \text{QCoh}(X)$$

are naturally *duals* of one another.

However, for a general prestack $\mathcal{X}$, only the functor $\Upsilon_{\mathcal{X}}$ makes sense; the functor $\Psi_{\mathcal{X}}$ is a feature of schemes (or, more generally, Artin stacks).

0.2.5. Here is a typical manifestation of the usefulness of the category $\text{IndCoh}(\mathcal{X})$ for a prestack $\mathcal{X}$.

Let us take $\mathcal{X}$ to be an *ind-scheme*. In this case, $\text{IndCoh}(\mathcal{X})$ is compactly generated by the direct images of $\text{Coh}(X)$ for closed subschemes $X \to \mathcal{X}$.

This is while it is not clear (and probably not true) that $\text{QCoh}(\mathcal{X})$ is compactly generated.

Note, however, that in [**GaRo1**, Theorem 10.1.1] it is proved that if $\mathcal{X}$ is a *formally smooth*, ind-scheme then the functor $\Upsilon_{\mathcal{X}}$ of (0.11) is an equivalence.

0.2.6. Let us now take $\mathcal{X} = X_{dR}$, where $X \in \text{Sch}_{\text{aft}}$. It is shown in [**GaRo2**, Proposition 2.4.4] that in this case the functor

$$\Upsilon_{X_{dR}} : \text{QCoh}(X_{dR}) \to \text{IndCoh}(X_{dR})$$

is an equivalence.

We can view

$$\text{QCoh}(X_{dR}) =: \text{D-mod}(X) := \text{IndCoh}(X_{dR})$$

as the two incarnations of the category of D-modules on X: as 'left D-modules' and as 'right D-modules'. Correspondingly, we have the two forgetful functors

$$\text{D-mod}(X) = \text{QCoh}(X_{dR}) \to \text{QCoh}(X)$$

and

$$\text{D-mod}(X) = \text{IndCoh}(X_{dR}) \to \text{IndCoh}(X),$$

corresponding to the $*$- and !-pullback, respectively, with respect to the map $X \to X_{dR}$.

In [**GaRo2**, Sects. 2-4] is it shown that the above 'right' forgetful functor $\text{D-mod}(X) \to \text{IndCoh}(X)$ has much better properties than the the 'left' forgetful functor $\text{D-mod}(X) \to \text{IndCoh}(X)$.

This is closely related to the fact that the category $\text{IndCoh}(X^\wedge)$ is better behaved (see Sect. 0.2.5 above) than the category $\text{QCoh}(X^\wedge)$, where $X^\wedge$ is the ind-scheme

$$X^\wedge := X \underset{X_{dR}}{\times} X,$$

i.e., the formal completion of X in $X \times X$.

0.3. What is done in this chapter?

0.3.1. In Sect. 1 we collect some geometric preliminaries needed for the proof of the main theorem (Theorem 2.1.4) in Sect. 2.

Namely, we show that the operation of the *closure of the image of a morphism* is well-behaved in the context of derived algebraic geometry. Specifically, for a morphism $X \xrightarrow{f} Y$, its closure is the initial object in the category of factorizations of f as

$$X \to X' \xrightarrow{f'} Y,$$

where f' is a closed embedding (i.e., the corresponding map of classical schemes ${}^{cl}X' \to {}^{cl}Y$ is a closed embedding).

The main result of this section is Proposition 1.3.2, which establishes the *transitivity property* of the operation of closure of the image of a morphism. Namely, it says that for a composition of morphisms

$$X \xrightarrow{f} Y \xrightarrow{g} Z,$$

if Y' denotes the closure of the image of f, and $g' := g|_{Y'}$, then the canonical map from the closure of the image of $g \circ f$ maps isomorphically to the closure of the image of g'.

0.3.2. The central section of this Chapter is Sect. 2, where we construct the functor

$$\mathrm{IndCoh}_{\mathrm{Corr}(\mathrm{Sch}_{\mathrm{aft}})^{\mathrm{proper}}}$$

of (0.10), starting from the functor $\mathrm{IndCoh}_{\mathrm{Sch}_{\mathrm{aft}}}$ of (0.1). This is done by applying Chapter 7, Theorem 5.2.4.

In order to apply this theorem, we need to check one condition of geometric nature. Namely, we need to show that for a given morphism between schemes $X \overset{f}{\to} Y$, the *category of its factorizations* as

$$X \overset{j}{\to} Z \overset{g}{\to} Y$$

with j an open embedding and g is proper, is *contractible.*

We prove the required contractibility assertion by appealing to the classical Nagata theorem, and using Proposition 1.3.2, about the operation of closure of the image of a morphism, mentioned above.

0.3.3. In Sect. 3 we study the functor

$$\mathrm{IndCoh}^!_{\mathrm{Sch}_{\mathrm{aft}}} : (\mathrm{Sch}_{\mathrm{aft}})^{\mathrm{op}} \to \mathrm{DGCat}_{\mathrm{cont}}$$

of (0.4) that is obtained from $\mathrm{IndCoh}_{\mathrm{Corr}(\mathrm{Sch}_{\mathrm{aft}})^{\mathrm{proper}}}$ by restriction to the 1-full subcategory

$$(\mathrm{Sch}_{\mathrm{aft}})^{\mathrm{op}} \to \mathrm{Corr}(\mathrm{Sch}_{\mathrm{aft}})^{\mathrm{proper}},$$

see Sect. 0.1.4.

The main point of this section is that, having the functor $\mathrm{IndCoh}^!_{\mathrm{Sch}_{\mathrm{aft}}}$ at our disposal, we can extend it to a functor

$$\mathrm{IndCoh}^!_{\mathrm{PreStk}_{\mathrm{laft}}} : (\mathrm{PreStk}_{\mathrm{laft}})^{\mathrm{op}} \to \mathrm{DGCat}_{\mathrm{cont}}.$$

The latter extension procedure is simply the right Kan extension along the embedding

$$(\mathrm{Sch}_{\mathrm{aft}})^{\mathrm{op}} \hookrightarrow (\mathrm{PreStk}_{\mathrm{laft}})^{\mathrm{op}}.$$

In other words, for a prestack $\mathcal{Y}$, an object $\mathcal{F} \in \mathrm{IndCoh}(\mathcal{Y})$ is a compatible family of objects

$$\mathcal{F}_{X,y} \in \mathrm{IndCoh}(X), \quad (X \overset{y}{\to} \mathcal{Y}) \in \mathrm{Sch}_{/\mathcal{Y}},$$

where the compatibility is understood in the sense of the !-pullback functor.

Furthermore, we can canonically extend the functor $\mathrm{IndCoh}_{\mathrm{Corr}(\mathrm{Sch}_{\mathrm{aft}})^{\mathrm{proper}}}$ to a functor

$$\mathrm{Corr}(\mathrm{PreStk}_{\mathrm{laft}})^{\mathrm{sch}\,\&\,\mathrm{proper}}_{\mathrm{sch\text{-}qc;all}} \to \mathrm{DGCat}^{2\text{-}\mathrm{Cat}}_{\mathrm{cont}},$$

where $\mathrm{Corr}(\mathrm{PreStk}_{\mathrm{laft}})^{\mathrm{sch}\,\&\,\mathrm{proper}}_{\mathrm{sch\text{-}qc;all}}$ is an $(\infty, 2)$-category, whose objects are

$$\mathcal{X} \in \mathrm{PreStk}_{\mathrm{laft}},$$

1-morphisms are diagrams

$$\begin{array}{ccc} \mathcal{X}_{0,1} & \xrightarrow{g} & \mathcal{X}_0 \\ {\scriptstyle f}\downarrow & & \\ \mathcal{X}_1, & & \end{array} \tag{0.12}$$

with g arbitrary and f *schematic quasi-compact*, and 2-morphisms are diagrams

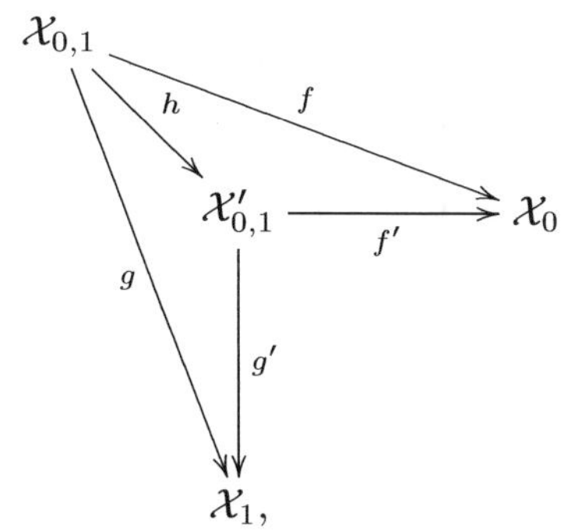

with h *schematic and proper*.

0.3.4. In Sect. 4 we show that the functor $\text{IndCoh}_{\text{Corr}(\text{Sch}_{\text{aft}})^{\text{proper}}}$ of (0.10) has a natural symmetric monoidal structure, where $\text{Corr}(\text{Sch}_{\text{aft}})^{\text{proper}}$ acquires a structure of symmetric monoidal $(\infty, 2)$-category from the operation of Cartesian product on Sch_{aft}.

We show that the symmetric monoidal structure on $\text{IndCoh}_{\text{Corr}(\text{Sch}_{\text{aft}})}$ gives rise to the *Serre duality equivalence*

$$\mathbf{D}_X^{\text{Serre}} : \text{IndCoh}(X)^\vee \simeq \text{IndCoh}(X), \quad X \in \text{Sch}_{\text{aft}}. \tag{0.13}$$

At the level of the subcategories of compact objects,

$$\text{Coh}(X) \simeq \text{IndCoh}(X)^c,$$

the functor $\text{Coh}(X)^{\text{op}} \to \text{Coh}(X)$, corresponding to $\mathbf{D}_X^{\text{Serre}}$, is the usual (contravariant) Serre duality auto-equivalence

$$\mathbb{D}_X^{\text{Serre}} : \text{Coh}(X)^{\text{op}} \simeq \text{Coh}(X).$$

Under the equivalence (0.10), for a morphism $X \xrightarrow{f} Y$, the functor $f^!$ identifies with the *dual* of the functor f_*^{IndCoh}.

0.3.5. In Sect. 5 we apply the theory developed in the preceding sections show that if

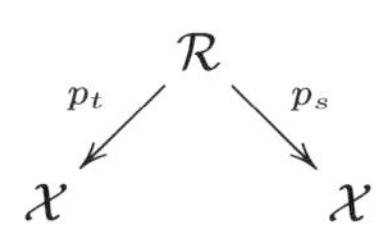

is a groupoid-object in $\text{PreStk}_{\text{laft}}$, where the maps p_s and p_t are schematic, then the category $\text{IndCoh}(\mathcal{R})$ has a natural monoidal structure, and as such it acts on $\text{IndCoh}(\mathcal{X})$.

We show, moreover, that if p_s and p_t are proper, then the dualizing object

$$\omega_{\mathcal{R}} \in \text{IndCoh}(\mathcal{R})$$

acquires a natural structure of associative algebra in $\text{IndCoh}(\mathcal{R})$.

1. Factorizations of morphisms of DG schemes

In this section we will study what happens to the notion of the closure of the image of a morphism between schemes in derived algebraic geometry. The upshot is that there is essentially 'nothing new' as compared to the classical case.

1.1. Colimits of closed embeddings. In this subsection we will show that colimits exist and are well-behaved in the category of closed subschemes of a given ambient scheme.

1.1.1. Recall that a map $X \to Y$ in Sch is called a closed embedding if the map

$$ {}^{\mathrm{cl}}X \to {}^{\mathrm{cl}}Y $$

is a closed embedding of classical schemes.

1.1.2. Let $f : X \to Y$ be a morphism in Sch. We let

$$ \mathrm{Sch}_{X/,\,\text{closed in } Y} $$

denote the full subcategory of $\mathrm{Sch}_{X//Y}$ consisting of diagrams

$$ X \to X' \overset{f'}{\to} Y, $$

where the map f' is a closed embedding.

We claim:

PROPOSITION 1.1.3.

(a) *The category* $\mathrm{Sch}_{X/,\,\text{closed in } Y}$ *has finite colimits (including the initial object).*

(b) *The formation of colimits commutes with Zariski localization on* Y.

PROOF.

Step 1. Assume first that Y is affine, $Y = \mathrm{Spec}(A)$. Let

$$ i \rightsquigarrow (X \to X'_i \overset{f'_i}{\to} Y), \tag{1.1} $$

be a finite diagram in $\mathrm{Sch}_{X/,\,\text{closed in } Y}$.

Set $B := \Gamma(X, \mathcal{O}_X)$. This is a (not necessarily connective) commutative k-algebra. Set also $X'_i = \mathrm{Spec}(B'_i)$. Consider the corresponding diagram

$$ i \rightsquigarrow (A \to B'_i \to B) \tag{1.2} $$

in $\mathrm{ComAlg}_{A//B}$.

Set

$$ (\widetilde{B}' \to B) := \lim_i \, (B'_i \to B), $$

where the limit taken in $\mathrm{ComAlg}_{/B}$. Note that we have a canonical map $A \to \widetilde{B}'$, and

$$ (A \to \widetilde{B}' \to B) \in \mathrm{ComAlg}_{A//B} $$

maps isomorphically to the limit of (1.2) taken in, $\mathrm{ComAlg}_{A//B}$.

Set

$$ B' := \tau^{\leq 0}(\widetilde{B}') \underset{H^0(\widetilde{B}')}{\times} \mathrm{Im}\left(H^0(A) \to H^0(\widetilde{B}')\right), $$

where the fiber product is taken in the category of *connective* commutative algebras (i.e., it is $\tau^{\leq 0}$ of the fiber product taken in the category of all commutative algebras).

We still have the canonical maps

$$A \to B' \to B,$$

and it is easy to see that for $X' := \operatorname{Spec}(B')$, the object

$$(X \to X' \to Y) \in \operatorname{Sch}_{X/,\,\text{closed in } Y}$$

is the colimit of (1.1).

Step 2. To treat the general case it it suffices to show that the formation of colimits in the affine case commutes with Zariski localization. I.e., we need to show that if Y is affine, $\overset{\circ}{Y} \subset Y$ is a basic open, then for $\overset{\circ}{X} := f^{-1}(\overset{\circ}{Y})$, $\overset{\circ}{X}{}'_i := (\phi'_i)^{-1}(\overset{\circ}{Y})$, $\overset{\circ}{X}{}' := (f')^{-1}(\overset{\circ}{Y})$, then the map

$$\operatorname*{colim}_i \overset{\circ}{X}{}'_i \to \overset{\circ}{X}{}',$$

is an isomorphism, where the colimit is taken in $\operatorname{Sch}_{\overset{\circ}{X}/,\,\text{closed } \overset{\circ}{Y}}$.

However, the required isomorphism follows from the description of the colimit in Step 1.

□

1.1.4. We note the following property of colimits in the situation of Proposition 1.1.3.

Let $g : Y \to \widetilde{Y}$ be a closed embedding. Set

$$(X \to X' \to Y) = \operatorname*{colim}_i (X \to X'_i \to Y) \text{ and } (X \to \widetilde{X}' \to \widetilde{Y}) = \operatorname*{colim}_i (X \to X'_i \to \widetilde{Y}),$$

where the colimits are taken in $\operatorname{Sch}_{X/,\,\text{closed in } Y}$ and $\operatorname{Sch}_{X/,\,\text{closed } \widetilde{Y}}$, respectively.

Consider the composition

$$X \to X' \to Y \to \widetilde{Y},$$

and the corresponding object

$$(X \to X' \to \widetilde{Y}) \in \operatorname{Sch}_{X/,\,\text{closed } \widetilde{Y}}.$$

It is endowed with a compatible family of maps

$$(X \to X'_i \to \widetilde{Y}) \to (X \to X' \to \widetilde{Y}).$$

Hence, by the universal property of $(X \to \widetilde{X}' \to \widetilde{Y})$, we obtain a canonically defined map

$$\widetilde{X}' \to X'. \tag{1.3}$$

We claim:

LEMMA 1.1.5. *The map* (1.3) *is an isomorphism.*

PROOF. We construct the inverse map as follows. We note that by the universal property of $(X \to \widetilde{X}' \to \widetilde{Y})$, we have a canonical map

$$(X \to \widetilde{X}' \to \widetilde{Y}) \to (X \to Y \to \widetilde{Y}).$$

This produces a compatible family of maps

$$(X \to X'_i \to Y) \to (X \to \widetilde{X}' \to Y),$$

and hence the desired map

$$X' \to \widetilde{X}'.$$

□

1.1.6. In the situation of Proposition 1.1.3 let us consider the case of $X = \emptyset$. We shall denote the resulting category by $\mathrm{Sch}_{\text{closed in } Y}$. Thus, Proposition 1.1.3 guarantees the existence and compatibility with Zariski localization of finite colimits in $\mathrm{Sch}_{\text{closed in } Y}$.

Explicitly, if $Y = \mathrm{Spec}(A)$ is affine and

$$i \rightsquigarrow Y'_i \subset Y$$

is a diagram of closed subschemes, $Y'_i = \mathrm{Spec}(A'_i)$, then

$$\underset{i}{\mathrm{colim}}\, Y'_i = Y',$$

where

$$Y' = \mathrm{Spec}(A'), \quad A' := \tau^{\leq 0}(\widetilde{A}') \underset{H^0(\widetilde{A}')}{\times} \mathrm{Im}\left(H^0(A) \to H^0(\widetilde{A}')\right), \quad \widetilde{A}' := \underset{i}{\lim}\, A'_i.$$

1.2. The closure. In this subsection we will define the notion of closure of the image of a morphism of schemes.

1.2.1. In what follows, in the situation of Proposition 1.1.3, we shall refer to the initial object in the category $\mathrm{Sch}_{X/,\text{closed in } Y}$ as the *closure* of X and Y, and denote it by $\overline{f(X)}$.

Explicitly, if $Y = \mathrm{Spec}(A)$ is affine, we have:

(1.4)

$$\overline{f(X)} = \mathrm{Spec}(A'), \quad A' = \tau^{\leq 0}(\Gamma(X, \mathcal{O}_X)) \underset{H^0(\Gamma(X,\mathcal{O}_X))}{\times} \mathrm{Im}\left(H^0(A) \to H^0(\Gamma(X, \mathcal{O}_X))\right).$$

A particular case of Lemma 1.1.5 says:

COROLLARY 1.2.2. *If $f : X \to Y$ is a closed embedding, then $X \to \overline{f(X)}$ is an isomorphism.*

1.2.3. The following property of the operation of taking the closure will be used in the sequel. Let us be in the situation of Proposition 1.1.3,

$$X = X_1 \cup X_2,$$

where $X_i \subset X$ are open and set $X_{12} = X_1 \cap X_2$. Denote $f_i := f|_{X_i}$.

We have a canonical map

$$\overline{f_1(X_1)} \underset{\overline{f_{12}(X_{12})}}{\sqcup} \overline{f_2(X_2)} \to \overline{f(X)}, \tag{1.5}$$

where the colimit is taken in $\mathrm{Sch}_{\text{closed in } Y}$.

LEMMA 1.2.4. *The map (1.5) is an isomorphism.*

PROOF. Follows by reducing to the case when Y is affine, and in the latter case by (1.4). □

1.2.5. We give the following definition:

DEFINITION 1.2.6. *A map $f: X \to Y$ is said to be a* locally closed embedding *if Y contains an open $\overset{\circ}{Y} \subset Y$, such that f defines a closed embedding $X \to \overset{\circ}{Y}$.*

We have:

LEMMA 1.2.7. *Suppose that f is a locally closed embedding. Then f defines an open embedding of X into $\overline{f(X)}$.*

PROOF. Follows by combining Corollary 1.2.2 and Proposition 1.1.3(b). □

1.3. Transitivity of closure. The basic fact established in this subsection, Proposition 1.3.2, will be of crucial technical importance for the proof of Theorem 2.1.4.

1.3.1. Consider a diagram

$$X \overset{f}{\to} Y \overset{g}{\to} Z.$$

Set $Y' := \overline{f(X)}$ and $g' := g|_{Y'}$. By the universal property of closure, we have a canonical map

$$\overline{g \circ f(X)} \to \overline{g'(Y')}. \tag{1.6}$$

We claim:

PROPOSITION 1.3.2. *The map* (1.6) *is an isomorphism.*

The rest of this subsection is devoted to the proof of this proposition.

1.3.3. *Step 1.* As in the proof of Proposition 1.1.3, the assertion reduces to the case when $Z = \operatorname{Spec}(A)$ is affine. Assume first that Y is affine as well, $Y = \operatorname{Spec}(B)$. Then we have the following descriptions of the two sides in (1.6).

Set $C := \Gamma(X, \mathcal{O}_X)$. We have $Y' := \operatorname{Spec}(B')$, where

$$B' = \tau^{\leq 0}(C) \underset{H^0(C)}{\times} \operatorname{Im}\left(H^0(B) \to H^0(C)\right),$$

where here and below the fiber product is taken in the category of connective commutative algebras.

Furthermore, $\overline{g \circ f(X)} = \operatorname{Spec}(A')$, where

$$A' = \tau^{\leq 0}(C) \underset{H^0(C)}{\times} \operatorname{Im}\left(H^0(A) \to H^0(C)\right).$$

Finally, $\overline{g'(Y')} = \operatorname{Spec}(A'')$, where

$$A'' = B' \underset{H^0(B')}{\times} \operatorname{Im}\left(H^0(A) \to H^0(B')\right).$$

Note that

$$H^0(B') = \operatorname{Im}\left(H^0(B) \to H^0(C)\right) \text{ and } \operatorname{Im}\left(H^0(A) \to H^0(B')\right) = \operatorname{Im}(H^0(A) \to H^0(C)).$$

The map (1.6) corresponds to the homomorphism

$$A'' = B' \underset{H^0(B')}{\times} \operatorname{Im}(H^0(A) \to H^0(B')) =$$

$$= \left(\tau^{\leq 0}(C) \underset{H^0(C)}{\times} \operatorname{Im}(H^0(B) \to H^0(C))\right) \underset{\operatorname{Im}(H^0(B)\to H^0(C))}{\times} \operatorname{Im}(H^0(A) \to H^0(C)) \simeq$$
$$\simeq \tau^{\leq 0}(C) \underset{H^0(C)}{\times} \operatorname{Im}(H^0(A) \to H^0(C)) \simeq A',$$

which is an isomorphism, as required.

1.3.4. *Step 2.* Let Y be arbitrary. Choose an open affine cover $Y = \underset{i}{\cup} Y_i$ and set $X_i = f^{-1}(Y_i)$. Then the assertion of the proposition follows from Step 1 using Lemma 1.2.4.

□

2. IndCoh as a functor from the category of correspondences

This section realizes one of the main goal of our book, namely, the construction of IndCoh as a functor out of the category of correspondences.

It will turn out that IndCoh, equipped with the operation of direct image, and left and right adjoints, corresponding to open embeddings and proper morphisms, respectively will uniquely extend to the sought-for formalism of correspondences.

2.1. The category of correspondences. In this subsection we introduce the category of correspondences on schemes and state our main theorem.

2.1.1. We consider the category $\mathrm{Sch}_{\mathrm{aft}}$ equipped with the following classes of morphisms:

$$vert = \text{all}, \; horiz = \text{all}, \; adm = \text{proper},$$

and consider the corresponding category

$$\operatorname{Corr}(\mathrm{Sch}_{\mathrm{aft}})^{\mathrm{proper}}_{\mathrm{all;all}},$$

see Chapter 7, Sect. 1.

Our goal in this section is to extend the functor

$$\operatorname{IndCoh}_{\mathrm{Sch}_{\mathrm{aft}}} : \mathrm{Sch}_{\mathrm{aft}} \to \mathrm{DGCat}_{\mathrm{cont}}$$

of Chapter 4, Sect. 2.2 to a functor

$$\operatorname{IndCoh}_{\operatorname{Corr}(\mathrm{Sch}_{\mathrm{aft}})^{\mathrm{proper}}_{\mathrm{all;all}}} : \operatorname{Corr}(\mathrm{Sch}_{\mathrm{aft}})^{\mathrm{proper}}_{\mathrm{all;all}} \to \mathrm{DGCat}^{2\text{-Cat}}_{\mathrm{cont}}.$$

We shall do so in several stages.

2.1.2. We start with the functor

$$\mathrm{IndCoh}_{\mathrm{Sch}_{\mathrm{aft}}} : \mathrm{Sch}_{\mathrm{aft}} \to \mathrm{DGCat}_{\mathrm{cont}}$$

and consider the class of morphisms

$$\mathrm{open} \subset \mathrm{all}.$$

By Chapter 4, Proposition 3.2.2, the functor $\mathrm{IndCoh}_{\mathrm{Sch}_{\mathrm{aft}}}$, viewed as a functor

$$\mathrm{Sch}_{\mathrm{aft}} \to \left(\mathrm{DGCat}^{2\text{-Cat}}_{\mathrm{cont}}\right)^{2\text{-op}},$$

satisfies the *left Beck-Chevalley condition* with respect to the class[1] open ⊂ all.

Applying Chapter 7, Theorem 3.2.2(a), we extend $\mathrm{IndCoh}_{\mathrm{Sch}_{\mathrm{aft}}}$ to a functor

$$\mathrm{IndCoh}_{\mathrm{Corr}(\mathrm{Sch}_{\mathrm{aft}})^{\mathrm{open}}_{\mathrm{all};\mathrm{open}}} : \mathrm{Corr}(\mathrm{Sch}_{\mathrm{aft}})^{\mathrm{open}}_{\mathrm{all};\mathrm{open}} \to \left(\mathrm{DGCat}^{2\text{-Cat}}_{\mathrm{cont}}\right)^{2\text{-op}}.$$

We restrict the latter functor to

$$\mathrm{Corr}(\mathrm{Sch}_{\mathrm{aft}})_{\mathrm{all};\mathrm{open}} \subset \mathrm{Corr}(\mathrm{Sch}_{\mathrm{aft}})^{\mathrm{open}}_{\mathrm{all};\mathrm{open}},$$

and denote the resulting functor by $\mathrm{IndCoh}_{\mathrm{Corr}(\mathrm{Sch}_{\mathrm{aft}})_{\mathrm{all};\mathrm{open}}}$, viewed as a functor

$$\mathrm{Corr}(\mathrm{Sch}_{\mathrm{aft}})_{\mathrm{all};\mathrm{open}} \to \mathrm{DGCat}_{\mathrm{cont}}.$$

We note that due to the uniqueness assertion in Chapter 7, Theorem 4.1.3, the restriction procedure

$$\mathrm{IndCoh}_{\mathrm{Corr}(\mathrm{Sch}_{\mathrm{aft}})^{\mathrm{open}}_{\mathrm{all};\mathrm{open}}} \rightsquigarrow \mathrm{IndCoh}_{\mathrm{Corr}(\mathrm{Sch}_{\mathrm{aft}})_{\mathrm{all};\mathrm{open}}}$$

loses no information. I.e., the datum of the functor $\mathrm{IndCoh}_{\mathrm{Corr}(\mathrm{Sch}_{\mathrm{aft}})_{\mathrm{all};\mathrm{open}}}$ is *equivalent* to that of $\mathrm{IndCoh}_{\mathrm{Corr}(\mathrm{Sch}_{\mathrm{aft}})^{\mathrm{open}}_{\mathrm{all};\mathrm{open}}}$.

2.1.3. The main result of this section reads:

THEOREM 2.1.4. *There exists a unque extension of the functor*

$$\mathrm{IndCoh}_{\mathrm{Corr}(\mathrm{Sch}_{\mathrm{aft}})_{\mathrm{all};\mathrm{open}}}$$

to a functor

$$\mathrm{IndCoh}_{\mathrm{Corr}(\mathrm{Sch}_{\mathrm{aft}})^{\mathrm{proper}}_{\mathrm{all};\mathrm{all}}} : \mathrm{Corr}(\mathrm{Sch}_{\mathrm{aft}})^{\mathrm{proper}}_{\mathrm{all};\mathrm{all}} \to \mathrm{DGCat}^{2\text{-Cat}}_{\mathrm{cont}}.$$

We will deduce this theorem from Chapter 7, Theorem 5.2.4. We refer the reader to Chapter 7, Sects. 5.1 and 5.2 where the notations involved in this theorem are introduced.

[1] We note that the left Beck-Chevalley condition is intrinsic to the target $(\infty, 2)$-category, in our case $\left(\mathrm{DGCat}^{2\text{-Cat}}_{\mathrm{cont}}\right)^{2\text{-op}}$.

2.1.5. *Proof of Theorem* 2.1.4. We start with the following three classes of morphisms

$$vert = \text{all}, \; horiz = \text{all}, \; co\text{-}adm = \text{open}, \; adm = \text{proper}.$$

We note that the class *open* ∩ *proper* is that of embeddings of a connected component. This implies that the condition of Chapter 7, Sect. 5.1.2 holds.

The fact that

$$\text{IndCoh}_{\text{Corr}(\text{Sch}_{\text{aft}})_{\text{all;open}}}|_{\text{Sch}_{\text{aft}}} = \text{IndCoh}_{\text{Sch}_{\text{aft}}}|_{\text{Sch}_{\text{aft}}}$$

satisfies the *left Beck-Chevalley condition* with respect to the class of proper maps is the content of Chapter 4, Proposition 5.2.1.

Finally, the fact that the condition of Chapter 7, Sect. 5.2.2 holds is the content of Chapter 4, Proposition 5.3.4.

Hence, in order to deduce Theorem 2.1.4 from Chapter 7, Theorem 5.2.4, it remains to verify that the factorization condition of Chapter 7, Sect. 5.1.3. I.e. we need to prove the following:

PROPOSITION 2.1.6. *For a morphism* $f : X \to Y$ *in* Sch_{aft}, *the category* $\text{Factor}(f)$ *of factorizations of* f *as*

$$X \overset{j}{\to} Z \overset{g}{\to} Y, \tag{2.1}$$

where j *is an open embedding, and* g *is proper, is contractible.*

□

2.2. Proof of Proposition 2.1.6. Modulo the classical Nagata theorem, the proof will be a simple manipulation with the notion of closure, developed in the previous section.

2.2.1. *Step 1.* Recall the notation ${}^{\text{red}}X$ for $X \in \text{Sch}$, see Chapter 4, Sect. 6.1.4.

First we show that $\text{Factor}(f)$ is non-empty. By Nagata's theorem, we can factor the morphism

$${}^{\text{red}}X \to {}^{\text{red}}Y$$

as

$${}^{\text{red}}X \to Z_{\text{red}} \to {}^{\text{red}}Y,$$

where ${}^{\text{red}}X \to Z_{\text{red}}$ is an open embedding and $Z_{\text{red}} \to {}^{\text{red}}Y$ is proper.

We define an object of $\text{Factor}(f)$ by setting

$$Z := X \underset{{}^{\text{red}}X}{\sqcup} Z_{\text{red}},$$

where we use Volume II, Chapter 1, Corollary 1.3.5(a) for the existence and the properties of push-out in this situation.

2.2.2. *Step 2.* Let $\text{Factor}(f)_{\text{dense}} \subset \text{Factor}(f)$ be the full subcategory consisting of those objects

$$X \overset{j}{\to} Z \overset{g}{\to} Y,$$

for which the map

$$\overline{j(X)} \to Z$$

is an isomorphism.

We claim that the tautological embedding

$$\text{Factor}(f)_{\text{dense}} \hookrightarrow \text{Factor}(f)$$

admits a right adjoint, which sends a given object (2.1) to

$$X \to \overline{j(X)} \to Y.$$

Indeed, the fact that the map $X \to \overline{j(X)}$ is an open embedding follows from Proposition 1.1.3(b). The fact that the above operation indeed produces a right adoint follows from Proposition 1.3.2.

Hence, it suffices to show that the category $\text{Factor}(f)_{\text{dense}}$ is contractible. Since it is non-empty (by Step 1), it suffices to show that it contains products.

2.2.3. *Step 3.* Given two objects

$$(X \to Z_1 \to Y) \text{ and } (X \to Z_2 \to Y)$$

of $\text{Factor}(f)_{\text{dense}}$ consider

$$W := Z_1 \underset{Y}{\times} Z_2,$$

and let h denote the resulting map $X \to W$.

Set $Z := \overline{h(X)}$. We claim that the map $X \to Z$ is an open embedding. Indeed, consider the open subscheme of $\overset{\circ}{W} \subset W$ equal to $X \underset{Y}{\times} X$. By Proposition 1.1.3(b),

$$\overset{\circ}{Z} := Z \cap \overset{\circ}{W}$$

is the closure of the map

$$\Delta_{X/Y} : X \to X \underset{Y}{\times} X.$$

However, $X \to \overline{\Delta_{X/Y}(X)}$ is an isomorphism by Corollary 1.2.2.

2.2.4. *Step 4.* Finally, we claim that the resulting object

$$X \to Z \to Y$$

is the product of $X \to Z_1 \to Y$ and $X \to Z_2 \to Y$ in $\text{Factor}(f)_{\text{dense}}$.

Indeed, let

$$X \to Z' \to Y$$

be another object of $\text{Factor}(f)_{\text{dense}}$, endowed with maps to $X \to Z_1 \to Y$ and to $X \to Z_2 \to Y$. Let i denote the resulting morphism

$$Z' \to Z_1 \underset{Y}{\times} Z_2 = W.$$

We have a canonical map in $\text{Factor}(f)$

$$(X \to Z' \to Y) \to (X \to \overline{i(Z')} \to Y).$$

However, from Proposition 1.3.2 we obtain that the natural map

$$Z \to \overline{i(Z')}$$

is an isomorphism. This gives rise to the desired map

$$(X \to Z' \to Y) \to (X \to Z \to Y).$$

□

3. The functor of !-pullback

Having defined IndCoh as a functor out of the category of correspondences, restricting to 'horizontal morphisms', we in particular obtain the functor of !-pullback, which is now defined on all morphisms.

In this section we study the basic properties of this functor.

3.1. Definition of the functor. In this subsection we summarize the basic properties of the !-pullback that follow formally from Theorem 2.1.4.

3.1.1. We let $\text{IndCoh}^!_{\text{Sch}_{\text{aft}}}$ denote the restriction of the functor

$$\text{IndCoh}_{\text{Corr}(\text{Sch}_{\text{aft}})^{\text{proper}}_{\text{all;all}}}$$

to

$$(\text{Sch}_{\text{aft}})^{\text{op}} \to \text{Corr}(\text{Sch}_{\text{aft}})^{\text{proper}}_{\text{all;all}}.$$

In particular, for a morphism $f : X \to Y$, we let $f^! : \text{IndCoh}(Y) \to \text{IndCoh}(X)$ the resulting morphism.

The functor $\text{IndCoh}^!_{\text{Sch}_{\text{aft}}}$ is essentially defined by the following two properties:

- The restriction $\text{IndCoh}^!_{\text{Sch}_{\text{aft}}}|_{((\text{Sch}_{\text{aft}})_{\text{proper}})^{\text{op}}}$ identifies with
$$\text{IndCoh}^!_{(\text{Sch}_{\text{aft}})_{\text{proper}}}.$$
- The restriction $\text{IndCoh}^!_{\text{Sch}_{\text{aft}}}|_{((\text{Sch}_{\text{aft}})_{\text{open}})^{\text{op}}}$ identifies with
$$\text{IndCoh}^*_{(\text{Sch}_{\text{aft}})_{\text{open}}}.$$

In the above formula,

$$\text{IndCoh}^*_{(\text{Sch}_{\text{aft}})_{\text{open}}} := \text{IndCoh}^*_{(\text{Sch}_{\text{aft}})_{\text{event-coconn}}}|_{((\text{Sch}_{\text{aft}})_{\text{open}})^{\text{op}}},$$

see Chapter 4, Corollary 3.1.10, where the functor

$$\text{IndCoh}^*_{(\text{Sch}_{\text{aft}})_{\text{event-coconn}}} : ((\text{Sch}_{\text{aft}})_{\text{event-coconn}})^{\text{op}} \to \text{DGCat}_{\text{cont}}$$

is introduced.

3.1.2. In what follows we shall denote by $\omega_X \in \text{IndCoh}(X)$ the canonical object equal to

$$p^!_X(k),$$

where $p_X : X \to \text{pt}$.

3.1.3. *Base change.* Let

$$\begin{array}{ccc} X_1 & \xrightarrow{g_X} & X_2 \\ {\scriptstyle f_1}\downarrow & & \downarrow{\scriptstyle f_2} \\ Y_1 & \xrightarrow{g_Y} & Y_2 \end{array}$$

be a Cartesian diagram in $\mathrm{Sch}_{\mathrm{aft}}$. As the main corollary of Theorem 2.1.4 we obtain:

COROLLARY 3.1.4. *There exists a canonical isomorphism of functors*

$$g_Y^! \circ (f_2)_*^{\mathrm{IndCoh}} \simeq (f_1)_*^{\mathrm{IndCoh}} \circ g_X^!, \tag{3.1}$$

compatible with compositions of vertical and horizontal morphisms in the natural sense. Furthermore:

(a) *Suppose that g_Y (and hence g_X) is proper. Then the morphism $\leftarrow$ in* (3.1) *arises from the $(g_*^{\mathrm{IndCoh}}, g^!)$-adjunction from the isomorphism*

$$(f_2)_*^{\mathrm{IndCoh}} \circ (g_X)_*^{\mathrm{IndCoh}} \simeq (g_Y)_*^{\mathrm{IndCoh}} \circ (f_1)_*^{\mathrm{IndCoh}}.$$

(b) *Suppose that f_2 (and hence f_1) is proper. Then the morphism $\leftarrow$ in* (3.1) *arises from the $(f_*^{\mathrm{IndCoh}}, f^!)$-adjunction from the isomorphism*

$$f_1^! \circ g_Y^! \simeq g_X^! \circ f_2^!.$$

(c) *Suppose that g_Y (and hence g_X) is an open embedding. Then the morphism $\to$ in* (3.1) *arises from the $(g^!, g_*^{\mathrm{IndCoh}})$-adjunction from the isomorphism*

$$(f_2)_*^{\mathrm{IndCoh}} \circ (g_X)_*^{\mathrm{IndCoh}} \simeq (g_Y)_*^{\mathrm{IndCoh}} \circ (f_1)_*^{\mathrm{IndCoh}}.$$

(d) *Suppose that f_2 (and hence f_1) is an open embedding. Then the morphism $\to$ in* (3.1) *arises from the $(f^!, f_*^{\mathrm{IndCoh}})$-adjunction from the isomorphism*

$$f_1^! \circ g_Y^! \simeq g_X^! \circ f_2^!.$$

REMARK 3.1.5. The real content of Theorem 2.1.4 is that there exists a uniquely defined family of functors $f^!$, that satisfies the properties listed in Corollary 3.1.4 and those of Sect. 3.1.1.

3.2. Some properties.

3.2.1. Let $\mathrm{IndCoh}^!_{\mathrm{Sch}^{\mathrm{aff}}_{\mathrm{aft}}}$ denote the restriction

$$\mathrm{IndCoh}^!_{\mathrm{Sch}_{\mathrm{aft}}}|_{(\mathrm{Sch}^{\mathrm{aff}}_{\mathrm{aft}})^{\mathrm{op}}}.$$

We claim:

LEMMA 3.2.2. *The functor*

$$\mathrm{IndCoh}^!_{\mathrm{Sch}_{\mathrm{aft}}} \to \mathrm{RKE}_{(\mathrm{Sch}^{\mathrm{aff}}_{\mathrm{aft}})^{\mathrm{op}} \to (\mathrm{Sch}_{\mathrm{aft}})^{\mathrm{op}}}(\mathrm{IndCoh}^!_{\mathrm{Sch}^{\mathrm{aff}}_{\mathrm{aft}}}) \to \mathrm{IndCoh}^!_{\mathrm{Sch}_{\mathrm{aft}}}$$

is an isomorphism.

PROOF. Follows from the fact that $\mathrm{IndCoh}^!_{\mathrm{Sch}_{\mathrm{aft}}}$ satisfies Zariski descent (Chapter 4, Proposition 4.2.2), combined with the fact that affine schemes form a basis for the Zariski topology:

For a given $X \in \mathrm{Sch}_{\mathrm{aft}}$, we need to show that the functor

$$\mathrm{IndCoh}(X) \to \lim_{S \to X} \mathrm{IndCoh}(S) \tag{3.2}$$

is an equivalence, where the limit is taken over the index category $((\mathrm{Sch}^{\mathrm{aff}}_{\mathrm{aft}})_{/X})^{\mathrm{op}}$.

Choose a Zariski cover $U \to X$ with $U \in \mathrm{Sch}^{\mathrm{aff}}_{\mathrm{aft}}$, and let $U^\bullet$ be its Čech nerve. We extend (3.2) to a string of functors

$$\mathrm{IndCoh}(X) \to \lim_{S\to X} \mathrm{IndCoh}(S) \to \mathrm{Tot}(\mathrm{IndCoh}(U^\bullet)) \to \lim_{S\to X} \Big(\mathrm{Tot}(\mathrm{IndCoh}(S \underset{X}{\times} U^\bullet)\Big).$$

Now, Zariski descent for IndCoh implies that the two composites

$$\mathrm{IndCoh}(X) \to \lim_{S\to X} \mathrm{IndCoh}(S) \to \mathrm{Tot}(\mathrm{IndCoh}(U^\bullet))$$

and

$$\lim_{S\to X} \mathrm{IndCoh}(S) \to \mathrm{Tot}(\mathrm{IndCoh}(U^\bullet)) \to \lim_{S\to X} \Big(\mathrm{Tot}(\mathrm{IndCoh}(S \underset{X}{\times} U^\bullet)\Big)$$

are both equivalences.

□

3.2.3. *Convergence.* Let

$$\mathrm{IndCoh}^!_{<\infty\mathrm{Sch}_{\mathrm{ft}}} \text{ and } \mathrm{IndCoh}^!_{<\infty\mathrm{Sch}^{\mathrm{aff}}_{\mathrm{ft}}}$$

denote the restrictions of $\mathrm{IndCoh}^!_{\mathrm{Sch}_{\mathrm{aft}}}$ to the corresponding subcategories.

We claim:

LEMMA 3.2.4. *The functors*

$$\mathrm{IndCoh}^!_{\mathrm{Sch}_{\mathrm{aft}}} \to \mathrm{RKE}_{({}^{<\infty}\mathrm{Sch}_{\mathrm{ft}})^{\mathrm{op}} \to (\mathrm{Sch}_{\mathrm{aft}})^{\mathrm{op}}}(\mathrm{IndCoh}^!_{<\infty\mathrm{Sch}_{\mathrm{ft}}})$$

and

$$\mathrm{IndCoh}^!_{\mathrm{Sch}^{\mathrm{aff}}_{\mathrm{aft}}} \to \mathrm{RKE}_{({}^{<\infty}\mathrm{Sch}^{\mathrm{aff}}_{\mathrm{ft}})^{\mathrm{op}} \to (\mathrm{Sch}^{\mathrm{aff}}_{\mathrm{aft}})^{\mathrm{op}}}(\mathrm{IndCoh}^!_{<\infty\mathrm{Sch}^{\mathrm{aff}}_{\mathrm{aft}}})$$

are isomorphisms.

PROOF. Both statements are equivalent to the assertion that for $X \in \mathrm{Sch}_{\mathrm{aft}}$, the functor

$$\mathrm{IndCoh}(X) \to \lim_{n} \mathrm{IndCoh}({}^{\leq n}X)$$

is an equivalence.

The latter assertion is the content of Chapter 4, Proposition 6.4.3.

□

3.3. h-descent. We will now use proper descent for IndCoh to show that it in fact has h-descent.

3.3.1. Let $\mathbf{C}$ be a category with Cartesian products, and let α be an isomorphism class of 1-morphisms, closed under base change.

We define the Grothendieck topology generated by α to be the minimal Grothendieck topology that contains all morphisms from α and has the following "2-out-of-3" property:

If $X \overset{f}{\to} Y \overset{g}{\to} Z$ are maps in $\mathbf{C}$ such that f and $g \circ f$ are coverings, then so is g.

The following is well-known:

LEMMA 3.3.2. *Let $\mathcal{F}$ be a presheaf on $\mathbf{C}$ that satisfies descent with respect to morphisms from the class α. Then $\mathcal{F}$ is a sheaf with respect to the Grothendieck topology generated by α.*

3.3.3. We recall that the h-topology on $\mathrm{Sch}_{\mathrm{aft}}$ is the one generated by the class of proper surjective maps and Zariski covers.

From Lemma 3.3.2, combined with Chapter 4, Propositions 4.2.2 and 7.2.2 we obtain:

COROLLARY 3.3.4. *The functor*

$$\mathrm{IndCoh}^!_{\mathrm{Sch}_{\mathrm{aft}}} : (\mathrm{Sch}_{\mathrm{aft}})^{\mathrm{op}} \to \mathrm{DGCat}_{\mathrm{op}}$$

satisfies h-descent.

3.3.5. We have:

LEMMA 3.3.6. *Any ppf covering is an h-covering.*

PROOF. Let $f : X \to Y$ be an ppf covering. Consider the Cartesian square

$$\begin{array}{ccc} {}^{\mathrm{cl}}Y \underset{Y}{\times} X & \longrightarrow & X \\ \downarrow & & \downarrow \\ {}^{\mathrm{cl}}Y & \longrightarrow & Y. \end{array}$$

It suffices to show that ${}^{\mathrm{cl}}Y \underset{Y}{\times} X \to {}^{\mathrm{cl}}Y$ is an h-covering. By flatness, ${}^{\mathrm{cl}}Y \underset{Y}{\times} X$ is classical. Hence, we are reduced to an assertion at the classical level, in which case it is well-known.

□

Hence, combining, we obtain:

COROLLARY 3.3.7. *The functor*

$$\mathrm{IndCoh}^!_{\mathrm{Sch}_{\mathrm{aft}}} : (\mathrm{Sch}_{\mathrm{aft}})^{\mathrm{op}} \to \mathrm{DGCat}_{\mathrm{op}}$$

satisfies ppf-descent.

3.4. Extension to prestacks. The functor of !-pullback for arbitrary morphisms of schemes allows to define the category IndCoh on arbitrary prestacks (locally almost of finite type).

3.4.1. We consider the category $\mathrm{PreStk}_{\mathrm{laft}}$ and define the functor

$$\mathrm{IndCoh}^!_{\mathrm{PreStk}_{\mathrm{laft}}} : (\mathrm{PreStk}_{\mathrm{laft}})^{\mathrm{op}} \to \mathrm{DGCat}_{\mathrm{cont}}$$

as the right Kan extension of $\mathrm{IndCoh}^!_{\mathrm{Sch}^{\mathrm{aff}}_{\mathrm{aft}}}$ along the Yoneda functor

$$(\mathrm{Sch}^{\mathrm{aff}}_{\mathrm{aft}})^{\mathrm{op}} \hookrightarrow (\mathrm{PreStk}_{\mathrm{laft}})^{\mathrm{op}}.$$

According to Lemmas 3.2.2 and 3.2.4, we can equivalently define $\mathrm{IndCoh}^!_{\mathrm{PreStk}_{\mathrm{laft}}}$ as the right Kan extension of

$$\mathrm{IndCoh}^!_{\mathrm{Sch}_{\mathrm{aft}}},\ \mathrm{IndCoh}^!_{<\infty\mathrm{Sch}_{\mathrm{ft}}} \text{ or } \mathrm{IndCoh}^!_{<\infty\mathrm{Sch}^{\mathrm{aff}}_{\mathrm{ft}}}$$

from the corresponding subcategories.

For $\mathcal{X} \in \mathrm{PreStk}_{\mathrm{laft}}$ we let $\mathrm{IndCoh}(\mathcal{X})$ denote the value of $\mathrm{IndCoh}^!_{\mathrm{PreStk}_{\mathrm{laft}}}$ on it. For a morphism $f : \mathcal{X}_1 \to \mathcal{X}_2$ we let

$$f^! : \mathrm{IndCoh}(\mathcal{X}_2) \to \mathrm{IndCoh}(\mathcal{X}_1)$$

denote the corresponding functor.

For $\mathcal{X} \in \mathrm{PreStk}_{\mathrm{laft}}$, we let $\omega_X \in \mathrm{IndCoh}(X)$ denote the canonical object equal to $p^!_{\mathcal{X}}(k)$, where $p_{\mathcal{X}} : \mathcal{X} \to \mathrm{pt}$.

3.4.2. We now consider the category

$$\mathrm{Corr}(\mathrm{PreStk}_{\mathrm{laft}})^{\mathrm{sch}\,\&\,\mathrm{proper}}_{\mathrm{sch\text{-}qc;all}},$$

where

$$\mathrm{sch} \text{ and } \mathrm{sch}\,\&\,\mathrm{proper}$$

signify the classes of schematic and quasi-compact (resp., schematic and proper) morphisms between prestacks.

We claim:

THEOREM 3.4.3. *There exists a uniquely defined functor*

$$\mathrm{IndCoh}_{\mathrm{Corr}(\mathrm{PreStk}_{\mathrm{laft}})^{\mathrm{sch}\,\&\,\mathrm{proper}}_{\mathrm{sch\text{-}qc;all}}} : \mathrm{Corr}(\mathrm{PreStk}_{\mathrm{laft}})^{\mathrm{sch}\,\&\,\mathrm{proper}}_{\mathrm{sch\text{-}qc;all}} \to \mathrm{DGCat}^{2\text{-Cat}}_{\mathrm{cont}},$$

equipped with isomorphisms

$$\mathrm{IndCoh}_{\mathrm{Corr}(\mathrm{PreStk}_{\mathrm{laft}})^{\mathrm{sch}\,\&\,\mathrm{proper}}_{\mathrm{sch\text{-}qc;all}}}|_{(\mathrm{PreStk}_{\mathrm{laft}})^{\mathrm{op}}} \simeq \mathrm{IndCoh}^!_{\mathrm{PreStk}_{\mathrm{laft}}}$$

and

$$\mathrm{IndCoh}_{\mathrm{Corr}(\mathrm{PreStk}_{\mathrm{laft}})^{\mathrm{sch}\,\&\,\mathrm{proper}}_{\mathrm{sch\text{-}qc;all}}}|_{(\mathrm{Sch}_{\mathrm{aft}})^{\mathrm{proper}}_{\mathrm{corr:all;all}}} \simeq \mathrm{IndCoh}_{(\mathrm{Sch}_{\mathrm{aft}})^{\mathrm{proper}}_{\mathrm{corr:all;all}}},$$

where the latter two isomorphisms are compatible in a natural sense.

PROOF. Follows from Chapter 8, Theorem 6.1.5. □

3.4.4. The actual content of Theorem 3.4.3 can be summarized as follows:

First, for any *schematic quasi-compact* morphism $f : \mathcal{X} \to \mathcal{Y}$ we have a well-defined functor

$$f^{\mathrm{IndCoh}}_* : \mathrm{IndCoh}(\mathcal{X}) \to \mathrm{IndCoh}(\mathcal{Y}).$$

Furthermore, if $f : \mathcal{X} \to \mathcal{Y}$ is schematic and proper, the functor f^{IndCoh}_* is the left adjoint of $f^!$.

When $\mathcal{Y}$ is a scheme (and hence $\mathcal{X}$ is one as well), the above functor f^{IndCoh}_* is the usual f^{IndCoh}_* defined in this case, and similarly for the $(f^{\mathrm{IndCoh}}_*, f^!)$-adjunction.

Second, let

$$\text{(3.3)} \qquad \begin{array}{ccc} \mathcal{X}_1 & \xrightarrow{g_X} & \mathcal{X}_2 \\ {\scriptstyle f_1}\downarrow & & \downarrow{\scriptstyle f_2} \\ \mathcal{Y}_1 & \xrightarrow{g_Y} & \mathcal{Y}_2 \end{array}$$

be a Cartesian diagram in $\mathrm{PreStk}_{\mathrm{laft}}$, where the vertical maps are schematic. Then we have a canonical isomorphism of functors

$$\text{(3.4)} \qquad g^!_Y \circ (f_2)^{\mathrm{IndCoh}}_* \simeq (f_1)^{\mathrm{IndCoh}}_* \circ g^!_X,$$

compatible with compositions. Furthermore, if the vertical (resp., horizontal) morphisms are proper (resp., schematic and proper), the map $\leftarrow$ in (3.4) comes by adjunction in a way similar to Corollary 3.1.4(a) (resp., Corollary 3.1.4(b)).

3.4.5. In Volume II, Chapter 3, Proposition 5.3.6, we will show that for a morphism $f : \mathcal{X} \to \mathcal{Y}$, which is an open embedding, the functor f_*^{IndCoh} is the *right* adjoint of $f^!$.

Furthermore, if in the Cartesian diagram (3.3) the vertical (resp., horizontal) morphisms are open embeddings, the map $\to$ in (3.4) comes by adjunction in a way similar to Corollary 3.1.4(c) (resp., Corollary 3.1.4(d)).

3.4.6. For future use, we note that the statement and proof of Chapter 4, Proposition 7.2.2 remains valid for groupoid objects in $(\mathrm{PreStk}_{\mathrm{laft}})_{\mathrm{sch}\,\&\,\mathrm{proper}}$.

4. Multiplicative structure and duality

In this section we will show that the functor IndCoh, when viewed as a functor out of the category of correspondences, and equipped with a natural symmetric monoidal structure, encodes Serre duality.

4.1. IndCoh **as a symmetric monoidal functor.** In this subsection we show that the functor IndCoh possesses a natural symmetric monoidal structure.

4.1.1. We recall that by Chapter 4, Proposition 6.3.6, the functor

$$\mathrm{IndCoh}_{\mathrm{Sch}_{\mathrm{aft}}} : \mathrm{Sch}_{\mathrm{aft}} \to \mathrm{DGCat}_{\mathrm{cont}}$$

carries a natural symmetric monoidal structure.

Applying Chapter 9, Proposition 3.1.5, we obtain:

THEOREM 4.1.2. *The functor*

$$\mathrm{IndCoh}_{\mathrm{Corr}(\mathrm{Sch}_{\mathrm{aft}})^{\mathrm{proper}}_{\mathrm{all;all}}} : \mathrm{Corr}(\mathrm{Sch}_{\mathrm{aft}})^{\mathrm{proper}}_{\mathrm{all;all}} \to \mathrm{DGCat}^{2\text{-}\mathrm{Cat}}_{\mathrm{cont}}$$

carries a canonical symmetric monoidal structure that extends one on $\mathrm{IndCoh}_{\mathrm{Sch}_{\mathrm{aft}}}$.

In particular, we obtain that the functors

$$\mathrm{IndCoh}_{\mathrm{Corr}(\mathrm{Sch}_{\mathrm{aft}})_{\mathrm{all;all}}} : \mathrm{Corr}(\mathrm{Sch}_{\mathrm{aft}})_{\mathrm{all;all}} \to \mathrm{DGCat}_{\mathrm{cont}}$$

and

$$\mathrm{IndCoh}^!_{\mathrm{Sch}_{\mathrm{aft}}} : (\mathrm{Sch}_{\mathrm{aft}})^{\mathrm{op}} \to \mathrm{DGCat}_{\mathrm{cont}}$$

both carry natural symmetric monoidal structures.

4.1.3. Note that the symmetric monoidal structure on $\mathrm{IndCoh}^!_{\mathrm{Sch}_{\mathrm{aft}}}$ automatically upgrades the functor $\mathrm{IndCoh}^!_{\mathrm{Sch}_{\mathrm{aft}}}$ to a functor

$$(\mathrm{Sch}_{\mathrm{aft}})^{\mathrm{op}} \to \mathrm{DGCat}^{\mathrm{SymMon}}_{\mathrm{cont}}, \tag{4.1}$$

due to the fact that the identity functor on $(\mathrm{Sch}_{\mathrm{aft}})^{\mathrm{op}}$ naturally lifts to a functor

$$(\mathrm{Sch}_{\mathrm{aft}})^{\mathrm{op}} \to \mathrm{ComAlg}\left((\mathrm{Sch}_{\mathrm{aft}})^{\mathrm{op}}\right)$$

via the diagonal map.

Explicitly, the monoidal operation on $\mathrm{IndCoh}(X)$ is given by

$$\mathrm{IndCoh}(X) \otimes \mathrm{IndCoh}(X) \overset{\boxtimes}{\to} \mathrm{IndCoh}(X \times X) \xrightarrow{\Delta_X^!} \mathrm{IndCoh}(X).$$

We shall denote the above monoidal operation by $\overset{!}{\otimes}$:

$$\mathcal{F}_1, \mathcal{F}_2 \in \mathrm{IndCoh}(X) \mapsto \mathcal{F}_1 \overset{!}{\otimes} \mathcal{F}_2 \in \mathrm{IndCoh}(X).$$

The unit for this symmetric monoidal structure is given by $\omega_X \in \text{IndCoh}(X)$.

4.1.4. Applying the functor of right Kan extension along

$$(\text{Sch}_{\text{aft}})^{\text{op}} \to (\text{PreStk}_{\text{laft}})^{\text{op}}$$

of the functor (4.1), we obtain that the functor

$$\text{IndCoh}^!_{\text{PreStk}_{\text{laft}}} : (\text{PreStk}_{\text{laft}})^{\text{op}} \to \text{DGCat}_{\text{cont}} \tag{4.2}$$

naturally upgrades to a functor

$$(\text{PreStk}_{\text{laft}})^{\text{op}} \to \text{DGCat}^{\text{SymMon}}_{\text{cont}}. \tag{4.3}$$

The functor (4.3) is tautologically right-lax symmetric monoidal with respect to the *coCartesian* symmetric monoidal structures on the source and the target. Since the forgetful functor

$$\text{DGCat}^{\text{SymMon}}_{\text{cont}} \to \text{DGCat}_{\text{cont}}$$

is symmetric monoidal when viewed with respect to the coCartesian symmetric monoidal structures on the source and the Lurie tensor product on the target (see Chapter 1, Sect. 3.3.6), we obtain that the functor $\text{IndCoh}^!_{\text{PreStk}_{\text{laft}}}$ of (4.2) acquires a natural right-lax symmetric monoidal structure.

4.1.5. The above right-lax symmetric monoidal structure on $\text{IndCoh}^!_{\text{PreStk}_{\text{laft}}}$ can be enhanced:

Indeed, applying Chapter 9, Proposition 3.2.4, we obtain that the functor

$$\text{IndCoh}_{\text{Corr}(\text{PreStk}_{\text{laft}})^{\text{sch \& proper}}_{\text{sch-qc;all}}} : \text{Corr}(\text{PreStk}_{\text{laft}})^{\text{sch \& proper}}_{\text{sch-qc;all}} \to \text{DGCat}^{2\text{-Cat}}_{\text{cont}}$$

carries a canonical right-lax symmetric monoidal structure.

4.2. Duality. In this subsection we will formally deduce Serre duality for schemes from the symmetric monoidal structure on IndCoh.

4.2.1. Let $\mathbf{O}$ be a symmetric monoidal category, and let $\mathbf{O}^{\text{dualizable}} \subset \mathbf{O}$ be the full subcategory spanned by dualizable objects. This subcategory carries a canonical symmetric monoidal anti-involution

$$\left(\mathbf{O}^{\text{dualizable}}\right)^{\text{op}} \overset{\text{dualization}}{\longrightarrow} \mathbf{O}^{\text{dualizable}},$$

given by the passage to the dual object, see Chapter 1, Sect. 4.1.4:

$$\mathbf{o} \mapsto \mathbf{o}^\vee.$$

Note that that if

$$F : \mathbf{O}_1 \to \mathbf{O}_2$$

is a symmetric monoidal functor between symmetric monoidal categories, then it maps

$$\mathbf{O}_1^{\text{dualizable}} \to \mathbf{O}_2^{\text{dualizable}},$$

and the following diagram commutes

$$\begin{array}{ccc} (\mathbf{O}_1^{\text{dualizable}})^{\text{op}} & \xrightarrow{F^{\text{op}}} & (\mathbf{O}_2^{\text{dualizable}})^{\text{op}} \\ {\scriptstyle\text{dualization}}\downarrow & & \downarrow{\scriptstyle\text{dualization}} \\ \mathbf{O}_1^{\text{dualizable}} & \xrightarrow{F} & \mathbf{O}_2^{\text{dualizable}}. \end{array}$$

4.2.2. Recall now that by Chapter 9, Sect. 2.2 the category $\mathrm{Corr}(\mathrm{Sch}_{\mathrm{aft}})_{\mathrm{all;all}}$ carries a canonical anti-involution ϖ, which is the identity on objects, and at the level of 1-morphisms is maps a 1-morphism

$$\begin{array}{ccc} X_{12} & \xrightarrow{f} & X_1 \\ {\scriptstyle g}\downarrow & & \\ X_2 & & \end{array}$$

to

$$\begin{array}{ccc} X_{12} & \xrightarrow{g} & X_2 \\ {\scriptstyle f}\downarrow & & \\ X_1. & & \end{array}$$

Moreover, by Chapter 9, Proposition 2.3.4, we have:

THEOREM 4.2.3. *The inclusion*

$$\left(\mathrm{Corr}(\mathrm{Sch}_{\mathrm{aft}})_{\mathrm{all;all}}\right)^{\mathrm{dualizable}} \subset \mathrm{Corr}(\mathrm{Sch}_{\mathrm{aft}})_{\mathrm{all;all}}$$

is an isomorphism. The anti-involution ϖ identifies canonically with the dualization functor

$$\left(\left(\mathrm{Corr}(\mathrm{Sch}_{\mathrm{aft}})_{\mathrm{all;all}}\right)^{\mathrm{dualizable}}\right)^{\mathrm{op}} \to \left(\mathrm{Corr}(\mathrm{Sch}_{\mathrm{aft}})_{\mathrm{all;all}}\right)^{\mathrm{dualizable}}.$$

4.2.4. Combining Theorem 4.2.3 with Theorem 4.1.2 we obtain:

THEOREM 4.2.5. *We have the following commutative diagram of functors*

$$\begin{array}{ccc} \left(\mathrm{Corr}(\mathrm{Sch}_{\mathrm{aft}})_{\mathrm{all;all}}\right)^{\mathrm{op}} & \xrightarrow{\left(\mathrm{IndCoh}_{\mathrm{Corr}(\mathrm{Sch}_{\mathrm{aft}})_{\mathrm{all;all}}}\right)^{\mathrm{op}}} & \left(\mathrm{DGCat}^{\mathrm{dualizable}}_{\mathrm{cont}}\right)^{\mathrm{op}} \\ {\scriptstyle \varpi}\downarrow & & \downarrow{\scriptstyle \mathrm{dualization}} \\ \mathrm{Corr}(\mathrm{Sch}_{\mathrm{aft}})_{\mathrm{all;all}} & \xrightarrow{\mathrm{IndCoh}_{\mathrm{Corr}(\mathrm{Sch}_{\mathrm{aft}})_{\mathrm{all;all}}}} & \mathrm{DGCat}^{\mathrm{dualizable}}_{\mathrm{cont}}. \end{array}$$

4.2.6. Let us explain the concrete meaning of this theorem. It says that for $X \in \mathrm{Sch}_{\mathrm{aft}}$ there is a canonical equivalence

$$\mathbf{D}_X^{\mathrm{Serre}} : \mathrm{IndCoh}(X)^\vee \simeq \mathrm{IndCoh}(X),$$

and for a map $f : X \to Y$ an isomorphism

$$(f^!)^\vee \simeq f_*^{\mathrm{IndCoh}},$$

where $(f^!)^\vee$ is viewed as a functor

$$\mathrm{IndCoh}(X) \overset{(\mathbf{D}_X^{\mathrm{Serre}})^{-1}}{\longrightarrow} \mathrm{IndCoh}(X)^\vee \overset{(f^!)^\vee}{\longrightarrow} \mathrm{IndCoh}(Y)^\vee \overset{\mathbf{D}_Y^{\mathrm{Serre}}}{\longrightarrow} \mathrm{IndCoh}(Y).$$

4.2.7. Let us write down explicitly the unit and co-unit for the identification $\mathbf{D}_X^{\mathrm{Serre}}$:

The co-unit, denoted ϵ_X is given by

$$\mathrm{IndCoh}(X)\otimes\mathrm{IndCoh}(X)\overset{\boxtimes}{\simeq}\mathrm{IndCoh}(X\times X)\overset{\Delta_X^!}{\longrightarrow}$$
$$\mathrm{IndCoh}(X)\overset{(p_X)_*^{\mathrm{IndCoh}}}{\longrightarrow}\mathrm{IndCoh}(\mathrm{pt})=\mathrm{Vect},$$

where $p_X : X \to \mathrm{pt}$.

The unit, denoted μ_X is given by

$$\mathrm{Vect}=\mathrm{IndCoh}(\mathrm{pt})\overset{p_X^!}{\longrightarrow}\mathrm{IndCoh}(X)\overset{(\Delta_X)_*^{\mathrm{IndCoh}}}{\longrightarrow}$$
$$\mathrm{IndCoh}(X\times X)\overset{\boxtimes}{\simeq}\mathrm{IndCoh}(X)\otimes\mathrm{IndCoh}(X).$$

REMARK 4.2.8. One does not need to rely on Theorems 4.2.3 and 4.1.2 in order to show that the maps μ_X and ϵ_X, defined above, give rise to an identification

$$\mathrm{IndCoh}(X)^\vee\simeq\mathrm{IndCoh}(X).$$

Indeed, the fact that the composition

$$\mathrm{IndCoh}(X)\overset{\mathrm{Id}_{\mathrm{IndCoh}(X)}\otimes\mu_X}{\longrightarrow}\mathrm{IndCoh}(X)\otimes\mathrm{IndCoh}(X)$$
$$\otimes\,\mathrm{IndCoh}(X)\overset{\epsilon_X\otimes\mathrm{Id}_{\mathrm{IndCoh}(X)}}{\longrightarrow}\mathrm{IndCoh}(X)$$

is isomorphic to the identity functor follows by base change from the diagram

$$\begin{array}{ccccc}
X & \xrightarrow{\Delta_X} & X\times X & \xrightarrow{\mathrm{id}_X\times p_X} & X\\
{\scriptstyle\Delta_X}\downarrow & & \downarrow{\scriptstyle\mathrm{id}_X\times\Delta_X} & & \\
X\times X & \xrightarrow{\Delta_X\times\mathrm{id}_X} & X\times X\times X & & \\
{\scriptstyle p_X\times\mathrm{id}_X}\downarrow & & & & \\
X, & & & &
\end{array}$$

and similarly for the other composition. A similar diagram chase implies the isomorphism

$$(f^!)^\vee\simeq f_*^{\mathrm{IndCoh}}.$$

REMARK 4.2.9. Let us also note that one does not need the (difficult) Theorem 2.1.4 either in order to construct the pairing ϵ_X:

Indeed, both functors involved in ϵ_X, namely, $\Delta_X^!$ and $(p_X)_*^{\mathrm{IndCoh}}$ are 'elementary'.

If one believes that the functor ϵ_X defined in the above way is the co-unit of a duality (which is a property, and not an extra structure), then one can recover the object $\omega_X\in\mathrm{IndCoh}(X)$. Namely,

$$\omega_X:=(p_X\times\mathrm{id}_X)_*^{\mathrm{IndCoh}}(\mu_X(k)).$$

4.2.10. *Relation to the usual Serre duality.* By passage to compact objects, the equivalence

$$\mathbf{D}_X^{\mathrm{Serre}} : \mathrm{IndCoh}(X)^\vee \simeq \mathrm{IndCoh}(X)$$

gives rise to an equivalence

$$\mathbb{D}_X^{\mathrm{Serre}} : (\mathrm{Coh}(X))^{\mathrm{op}} \simeq \mathrm{Coh}(X).$$

It is shown in [**Ga1**, Proposition 8.3.5] that $\mathbb{D}_X^{\mathrm{Serre}}$ is the usual Serre duality anti-equivalence of $\mathrm{Coh}(X)$, given by internal Hom into ω_X.

4.3. An alternative construction of the !-pullback. In this subsection we show how one can avoid using the formalism of correspondences if one only wants to construct the functor

$$\mathrm{IndCoh}^!_{\mathrm{Sch}_{\mathrm{aft}}} : (\mathrm{Sch}_{\mathrm{aft}})^{\mathrm{op}} \to \mathrm{DGCat}_{\mathrm{cont}}.$$

4.3.1. Note that even without having the formalism of the !-pullback, we know that for $X \in \mathrm{Sch}_{\mathrm{aft}}$, the functor

$$\epsilon_X : \mathrm{IndCoh}(X) \otimes \mathrm{IndCoh}(X) \to \mathrm{Vect},$$

defined as

$$\mathrm{IndCoh}(X) \otimes \mathrm{IndCoh}(X) \overset{\sim}{\to} \mathrm{IndCoh}(X \boxtimes X) \xrightarrow{\Delta_X^!} \mathrm{IndCoh}(X) \xrightarrow{(p_X)_*^{\mathrm{IndCoh}}} \mathrm{Vect}$$

gives[2] rise to the co-unit of an adjunction.

Indeed, the corresponding unit of the adjunction μ_X can be defined as follows. Choose an open embedding $X \overset{j}{\hookrightarrow} \overline{X}$, where $\overline{X}$ is proper and set

$$\widetilde{\omega}_X := j^{\mathrm{IndCoh},*} \circ (p_{\overline{X}})^!(k).$$

Then one readily checks that the object

$$(\Delta_X)_*^{\mathrm{IndCoh}}(\widetilde{\omega}_X) \simeq \mathrm{IndCoh}(X \boxtimes X) \simeq \mathrm{IndCoh}(X) \otimes \mathrm{IndCoh}(X),$$

viewed as a functor $\mathrm{Vect} \to \mathrm{IndCoh}(X) \otimes \mathrm{IndCoh}(X)$ provides the unit of the adjunction.

4.3.2. Thus, if we start with the functor

(4.4)

$$\mathrm{IndCoh}_{\mathrm{Sch}_{\mathrm{aft}}} : \mathrm{Sch}_{\mathrm{aft}} \to \mathrm{DGCat}_{\mathrm{cont}}, \quad X \mapsto \mathrm{IndCoh}(X), \quad (X \overset{f}{\to} Y) \mapsto f_*^{\mathrm{IndCoh}},$$

we obtain that it takes values in the full subcategory

$$(\mathrm{DGCat}_{\mathrm{cont}})^{\mathrm{dualizable}} \subset \mathrm{DGCat}_{\mathrm{cont}}.$$

Applying the dualization functor

$$((\mathrm{DGCat}_{\mathrm{cont}})^{\mathrm{dualizable}})^{\mathrm{op}} \to (\mathrm{DGCat}_{\mathrm{cont}})^{\mathrm{dualizable}},$$

from (4.4), we obtain the desired functor

$$\mathrm{IndCoh}^!_{\mathrm{Sch}_{\mathrm{aft}}} : (\mathrm{Sch}_{\mathrm{aft}})^{\mathrm{op}} \to \mathrm{DGCat}_{\mathrm{cont}}, \quad X \mapsto \mathrm{IndCoh}(X), \quad (X \overset{f}{\to} Y) \mapsto f^!. \tag{4.5}$$

In other words, for a morphism $f : X \to Y$, the functor

$$f^! : \mathrm{IndCoh}(Y) \to \mathrm{IndCoh}(X)$$

[2]Here we use the assumption that our schemes are assumed separated, so the morphism Δ_X is a closed embedding, and thus $\Delta_X^!$ is a priori defined as the right adjoint of $(\Delta_X)_*^{\mathrm{IndCoh}}$.

is *defined* as the dual of f_*^{IndCoh} under the self-dualities given by ϵ_X and ϵ_Y, respectively.

4.3.3. Let $\omega_X \in \mathrm{IndCoh}(X)$ denote the object $(p_X)^!(k)$. Unwinding the definitions we obtain that ω_X identifies with

$$(p_X \times \mathrm{id})_*^{\mathrm{IndCoh}} \circ \mu_X(k).$$

One can also give an explicit construction of the functor

$$f^! : \mathrm{IndCoh}(Y) \to \mathrm{IndCoh}(X)$$

for a morphism $f : X \to Y$. Namely,

$$f^!(\mathcal{F}) \simeq (\mathrm{Graph}_f)^!(\omega_X \boxtimes \mathcal{F}),$$

where $\mathrm{Graph}_f : X \to X \times Y$ is a closed embedding because Y is separated, and so $(\mathrm{Graph}_f)^!$ is defined as the right adjoint of $(\mathrm{Graph}_f)_*^{\mathrm{IndCoh}}$.

4.3.4. Since (4.4) is symmetric monoidal, the functor (4.5) also acquires a natural symmetric monoidal structure.

As in Sect. 4.1.3, the symmetric monoidal structure on $\mathrm{IndCoh}^!_{\mathrm{Sch}_{\mathrm{aft}}}$ makes $\mathrm{IndCoh}(X)$ into a symmetric monoidal category under the operation of $\overset{!}{\otimes}$-tensor product. By construction, $\omega_X \in \mathrm{IndCoh}(X)$ is the unit of this symmetric monoidal structure.

Note, however, that the construction of the *non-unital* symmetric monoidal structure on $\mathrm{IndCoh}(X)$ only uses the !-pullback functor for diagonal morphisms, which are closed embeddings.

Thus, the object $\omega_X \in \mathrm{IndCoh}(X)$ can be uniquely characterized as being *the* unit in the above non-unital symmetric monoidal category[3].

REMARK 4.3.5. The idea that the isomorphism

$$\omega_X \simeq \omega_X \overset{!}{\otimes} \omega_X$$

characterizes ω_X uniquely is borrowed from [**YZ**, Theorem 5.11 and Proposition 6.1].

4.3.6. Finally, let us see that for $f : X \to Y$, the functor $f^! : \mathrm{IndCoh}(Y) \to \mathrm{IndCoh}(Y)$ constructed above identifies with the functor that we had initially denoted by $f^!$, i.e., the right adjoint of f_*^{IndCoh}. To distinguish the two, let us keep the notation $f^!$ for the latter functor.

We need to construct an identification between $(f^!)^\vee$ and f_*^{IndCoh}. Unwinding the definitions, the functor

$$(f^!)^\vee : \mathrm{IndCoh}(X) \to \mathrm{IndCoh}(Y)$$

is given by

$$\mathcal{F} \mapsto (\mathrm{id}_Y \times p_X)_*^{\mathrm{IndCoh}} \circ (\mathrm{id}_Y \times \Delta_X)^! \circ (\mathrm{id}_Y \times f \times \mathrm{id}_X)(\mu_Y(k) \boxtimes \mathcal{F}).$$

[3]The essential uniqueness of a unit is established in [**Lu2**, Corollary 5.4.4.7].

I.e., this is pull-push of $\mu_Y(k) \boxtimes \mathcal{F} \in \text{IndCoh}(Y \times Y \times X)$ along the clockwise circuit of the following diagram

$$\begin{array}{ccc} Y \times Y \times X & \xleftarrow{\text{id}_Y \times \text{Graph}_f} & Y \times X \\ \Big\downarrow{\scriptstyle \text{id}_Y \times p_Y \times f} & & \Big\downarrow{\scriptstyle \text{id}_Y \times p_X} \\ Y \times Y & \xleftarrow{\Delta_Y} & Y, \end{array}$$

in which the horizontal arrows are closed embeddings.

Applying base change, we replace the above functor by push-pull along the counterclockwise circuit, and we obtain

$$\begin{aligned} (\Delta_Y)^! \circ (\text{id}_Y \times p_Y \times f)_*^{\text{IndCoh}}(\mu_Y(k) \boxtimes \mathcal{F}) & \\ \simeq (\Delta_Y)^! \left((\text{id}_Y \times p_Y)_*^{\text{IndCoh}}(\mu_Y(k)) \boxtimes f_*^{\text{IndCoh}}(\mathcal{F})\right) & \\ \simeq (\Delta_Y)^!(\omega_Y \boxtimes f^!(\mathcal{F})) & \\ \simeq f_*^{\text{IndCoh}}(\mathcal{F}), & \end{aligned}$$

as required.

5. Convolution monoidal categories and algebras

In this section[4] we will apply the the formalism of IndCoh as a functor out of the category of correspondences to carry out the following construction and its generalizations:

Let $\mathcal{R} \rightrightarrows X$ be a Segal object in the category of schemes (see below for what this means). Then the category $\text{IndCoh}(\mathcal{R})$ has a natural monoidal structure, and $\omega_{\mathcal{R}} \in \text{IndCoh}(\mathcal{R})$ defines a monad acting on $\text{IndCoh}(X)$.

5.1. Convolution algebras.

In this subsection we will show that monoid-objects give rise to convolution algebras.

5.1.1. Let $\mathcal{R}^\bullet$ be a *Segal object* in $\text{PreStk}_{\text{laft}}$ acting on a given $\mathcal{X} \in \text{PreStk}_{\text{laft}}$.

I.e., $\mathcal{R}^\bullet$ is a simplicial object in $\text{PreStk}_{\text{laft}}$, equipped with an identification $\mathcal{R}^0 = \mathcal{X}$, and such that for any $n \geq 2$, the map

$$\mathcal{R}^n \to \mathcal{R}^1 \underset{\mathcal{X}}{\times} \ldots \underset{\mathcal{X}}{\times} \mathcal{R}^1,$$

given by the product of the maps

$$[1] \to [n], \quad 0 \mapsto i, 1 \mapsto i+1, \quad i = 0, ..., n-1,$$

is an isomorphism.

REMARK 5.1.2. An alternative terminology for such $\mathcal{R}^\bullet$ is *category-object.* Indeed, the above condition is equivalent to requiring that for any $\mathcal{Y} \in \text{PreStk}_{\text{laft}}$, the simplicial space

$$\text{Maps}(\mathcal{Y}, \mathcal{R}^\bullet)$$

be a Segal space. Note we *do not* require it to be a *complete* Segal space.

[4]The contents of this section were suggested to us by S. Raskin.

5.1.3. In what follows we will denote $\mathcal{R} = \mathcal{R}^1$. We will informally think of a Segal object $\mathcal{R}^\bullet$ as the prestack $\mathcal{R}$, equipped with the source and target maps

$$p_s, p_t : \mathcal{R} \rightrightarrows \mathcal{X},$$

and the multiplication map

$$\text{mult} : \mathcal{R} \underset{t,\mathcal{X},s}{\times} \mathcal{R} \to \mathcal{R}.$$

For the duration of this subsection we will assume:

- The target map $p_t : \mathcal{R} \rightrightarrows \mathcal{X}$ is schematic;
- The multiplication map $\text{mult} : \mathcal{R} \underset{t,\mathcal{X},s}{\times} \mathcal{R} \to \mathcal{R}$ is proper.

5.1.4. Applying Chapter 9, Proposition 4.1.4 and Variant 4.1.6, we obtain that $\mathcal{R}^\bullet$ defines a monad $\mathsf{M}_\mathcal{R}$, acting on $\mathcal{X}$, i.e., an associative algebra object in the monoidal $(\infty, 1)$-category

$$\mathbf{Maps}_{\text{Corr}(\text{PreStk}_{\text{laft}})_{\text{sch;all}}}(\mathcal{X}, \mathcal{X}).$$

Concretely, the 1-morphism $\mathcal{X} \to \mathcal{X}$, corresponding to $\mathsf{M}_\mathcal{R}$ is given by the diagram

$$\begin{array}{ccc} \mathcal{R} & \xrightarrow{p_s} & \mathcal{X} \\ {\scriptstyle p_t}\downarrow & & \\ \mathcal{X}, & & \end{array}$$

and the multiplication on $\mathsf{M}_\mathcal{R}$ is given by the diagram

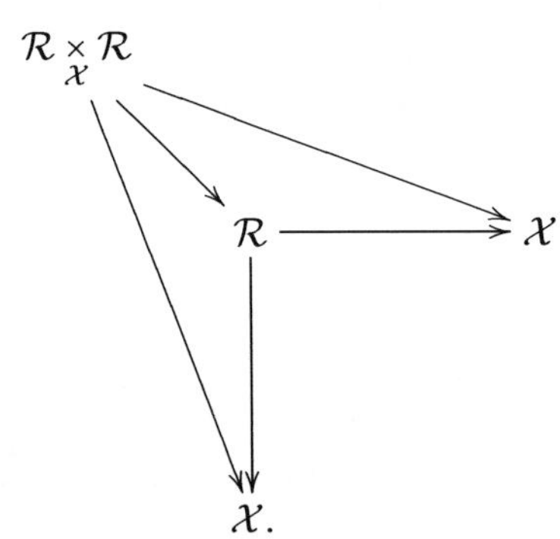

(5.1)

5.1.5. Applying the functor

$$\text{IndCoh}_{\text{Corr}(\text{PreStk}_{\text{laft}})^{\text{sch \& proper}}_{\text{sch-qc;all}}} : \text{Corr}(\text{PreStk}_{\text{laft}})^{\text{sch \& proper}}_{\text{sch-qc;all}} \to \text{DGCat}^{2\text{-Cat}}_{\text{cont}},$$

we obtain that to $\mathsf{M}_\mathcal{R}$ there corresponds a monad $\text{IndCoh}(\mathsf{M}_\mathcal{R})$ acting on $\text{IndCoh}(\mathcal{X})$.

It follows from the definitions that as an endo-functor of $\text{IndCoh}(\mathcal{X})$, the monad $\text{IndCoh}(\mathsf{M}_\mathcal{R})$ is given by

$$(p_t)^{\text{IndCoh}}_* \circ (p_s)^!.$$

I.e., the above construction formalizes the idea of a pull-push monad, corresponding to a Segal object $\mathcal{R}^\bullet$.

5.1.6. Assume now that $\mathcal{R}^\bullet$ is a groupoid object of $\mathrm{PreStk}_{\mathrm{laft}}$, equal to the Čech nerve of a proper schematic map $g : \mathcal{X} \to \mathcal{Y}$.

In this case, it follows from Chapter 9, Sect. 4.3.4 and Variant 4.3.5 that the monad $\mathsf{M}_{\mathcal{R}}$ is canonically isomorphic to one corresponding to the composite of g (viewed as a 1-morphism in the $(\infty, 2)$-category $\mathrm{Corr}(\mathrm{PreStk}_{\mathrm{laft}})^{\mathrm{sch}\,\&\,\mathrm{proper}}_{\mathrm{sch\text{-}qc;all}}$) with its right adjoint.

5.1.7. Assume now that $\mathcal{R}^\bullet$ is a groupoid object of $\mathrm{PreStk}_{\mathrm{laft}}$, with the maps $p_s, p_t : \mathcal{R} \to \mathcal{X}$ being proper.

In then according to Sect. 3.4.6, the endo-functor $(p_t)^{\mathrm{IndCoh}}_* \circ (p_s)^!$ acquires an (a priori different) structure of monad.

We claim, however that the above two ways of introducing a structure of monad on the endo-functor $(p_t)^{\mathrm{IndCoh}}_* \circ (p_s)^!$ coincide. Indeed, this follows from Sect. 5.1.6, applied to

$$\mathcal{Y} := |\mathcal{R}^\bullet|.$$

5.2. Convolution monoidal categories. In this subsection we will show that IndCoh of a Segal object in $\mathrm{PreStk}_{\mathrm{laft}}$ carries a natural monoidal structure.

5.2.1. Let $\mathcal{R}^\bullet$ be a Segal object in $\mathrm{PreStk}_{\mathrm{laft}}$ acting on $\mathcal{X}$. We impose the following conditions:

The maps $p_t : \mathcal{R} \to \mathcal{X}$ *and* $\mathrm{mult} : \mathcal{R} \underset{t,\mathcal{X},s}{\times} \mathcal{R} \to \mathcal{R}$ *are both schematic.*

5.2.2. By Chapter 9, Theorem 4.4.2 and Variant 4.4.7, the object $\mathcal{R}$ acquires a natural structure of algebra in the (symmetric) monoidal category

$$\mathrm{Corr}(\mathrm{PreStk}_{\mathrm{laft}})_{\mathrm{sch;all}}.$$

Moreover, according to Chapter 9, Sect. 4.5.2 and Variant 4.5.5, the object

$$\mathcal{X} \in \mathrm{Corr}(\mathrm{PreStk}_{\mathrm{laft}})_{\mathrm{sch;all}}$$

is naturally a module for $\mathcal{R}$.

Applying the right-lax (symmetric) monoidal functor

$$\mathrm{IndCoh}_{\mathrm{Corr}(\mathrm{PreStk}_{\mathrm{laft}})_{\mathrm{sch;all}}} : \mathrm{Corr}(\mathrm{PreStk}_{\mathrm{laft}})_{\mathrm{sch;all}} \to \mathrm{DGCat}_{\mathrm{cont}},$$

we obtain that the DG category $\mathrm{IndCoh}(\mathcal{R})$ acquires a structure of monoidal DG category (i.e., a structure of associative algebra in $\mathrm{DGCat}_{\mathrm{cont}}$), and $\mathrm{IndCoh}(\mathcal{X})$ acquires a structure of $\mathrm{IndCoh}(\mathcal{R})$-module.

Unwinding the definitions, we obtain that the binary operation on $\mathrm{IndCoh}(\mathcal{R})$ is given by the *convolution* product, i.e., pull-push along the diagram

$$\begin{array}{ccc} \mathcal{R} \underset{\mathcal{X}}{\times} \mathcal{R} & \longrightarrow & \mathcal{R} \times \mathcal{R} \\ \downarrow & & \\ \mathcal{R}, & & \end{array}$$

and the action of IndCoh($\mathcal{R}$) on IndCoh($\mathcal{X}$) is given by pull-push along the diagram

$$\begin{array}{ccc} \mathcal{R} & \xrightarrow{p_s \times \mathrm{id}} & \mathcal{X} \times \mathcal{R} \\ {\scriptstyle p_t}\downarrow & & \\ \mathcal{X}. & & \end{array}$$

5.2.3. Consider a particular case when $\mathcal{X} = X \in \mathrm{Sch}_{\mathrm{aft}}$, and $\mathcal{X}^n = X^{\times(n+1)}$. So

$$\mathcal{R} = X \times X.$$

We obtain that $\mathrm{IndCoh}(X \times X)$ acquires a structure of monoidal category, equipped with an action on $\mathrm{IndCoh}(X)$.

I.e., we obtain a monoidal functor

$$\mathrm{IndCoh}(X \times X) \to \mathrm{Funct}_{\mathrm{cont}}(\mathrm{IndCoh}(X), \mathrm{IndCoh}(X)). \tag{5.2}$$

By construction, as a functor of plain categories, (5.2) identifies with

$$\mathrm{IndCoh}(X \times X) \simeq \mathrm{IndCoh}(X) \otimes \mathrm{IndCoh}(X) \overset{(\mathbf{D}_X^{\mathrm{Serre}})^{-1} \otimes \mathrm{Id}}{\longrightarrow} \mathrm{IndCoh}(X)^{\vee} \otimes$$
$$\otimes \mathrm{IndCoh}(X) \simeq \mathrm{Funct}_{\mathrm{cont}}(\mathrm{IndCoh}(X), \mathrm{IndCoh}(X)).$$

In particular, the functor (5.2) is an equivalence of monoidal categories.

5.3. The case of QCoh. In this subsection we will explain the variant of the constructions in this subsection for QCoh instead of IndCoh.

5.3.1. First, starting from the functor

$$\mathrm{Sch} \overset{\mathrm{QCoh}_{\mathrm{Sch}}}{\longrightarrow} \mathrm{DGCat}_{\mathrm{cont}} \to \mathrm{DGCat}^{2\text{-}\mathrm{Cat}}_{\mathrm{cont}},$$

and using the usual base change property for QCoh, we apply Chapter 7, Theorem 3.2.2(a) and we obtain a functor

$$\mathrm{QCoh}_{\mathrm{Corr}(\mathrm{Sch})^{\mathrm{all}}_{\mathrm{all};\mathrm{all}}} : \mathrm{Corr}(\mathrm{Sch})^{\mathrm{all}}_{\mathrm{all};\mathrm{all}} \to (\mathrm{DGCat}^{2\text{-}\mathrm{Cat}}_{\mathrm{cont}})^{2\text{-}\mathrm{op}}.$$

Moreover, by Chapter 9, Proposition 3.1.5, the above functor carries a natural (symmetric) monoidal structure.

5.3.2. Further, applying Chapter 8, Theorem 6.1.5, from the functor

$$\mathrm{QCoh}_{\mathrm{Corr}(\mathrm{Sch})^{\mathrm{all}}_{\mathrm{all};\mathrm{all}}}$$

constructed above, we obtain the functor

$$\mathrm{QCoh}_{\mathrm{Corr}(\mathrm{PreStk})^{\mathrm{sch}}_{\mathrm{sch};\mathrm{all}}} : \mathrm{Corr}(\mathrm{PreStk})^{\mathrm{sch}}_{\mathrm{sch};\mathrm{all}} \to (\mathrm{DGCat}^{2\text{-}\mathrm{Cat}}_{\mathrm{cont}})^{2\text{-}\mathrm{op}}.$$

By Chapter 9, Proposition 3.2.4, the latter functor carries a right-lax (symmetric) monoidal structure.

5.3.3. Hence, the above discussion of convolution categories and algebras applies almost verbatim, when we replace IndCoh by QCoh, with the only difference that in whatever applies to 2-categorical phenomena, the direction of 2-morphisms gets reversed.

In particular, the geometric constructions that gave rise to *algebras* in the monoidal categories IndCoh($\mathcal{R}$) will produce *co-algebras* in the monoidal categories QCoh($\mathcal{R}$).

CHAPTER 6

Interaction of QCoh and IndCoh

Introduction

One of the first things one notices about the category $\text{IndCoh}(X)$ (for a scheme X) is that it is equipped with an action of the (symmetric) monoidal category $\text{QCoh}(X)$, see Chapter 4, Sect. 1.2.9.

In this chapter we will study how (or, rather, in what sense) this action extends, when we want to consider IndCoh as a functor out of the category of correspondences.

We should say that the contents of this chapter are rather technical (and are largely included for completeness), and thus can be skipped on the first pass.

0.1. Why does this chapter exist?

0.1.1. The first question to ask is, indeed, why bother? The true answer is that if we really care about IndCoh as a functor out of category of correspondences and about the action of $\text{QCoh}(X)$ on $\text{IndCoh}(X)$, then we must understand how they interact.

However, in addition to that, the material of this chapter will have some practical consequences.

0.1.2. We recall that, say, for an individual scheme X, both categories $\text{QCoh}(X)$ and $\text{IndCoh}(X)$ have a canonical symmetric monoidal structure. We will show that the action of $\text{QCoh}(X)$ on $\text{IndCoh}(X)$ comes from a symmetric monoidal functor

$$\Upsilon_X : \text{QCoh}(X) \to \text{IndCoh}(X),$$

which as a plain functor looks like

$$\mathcal{F} \mapsto \mathcal{F} \otimes \omega_X.$$

The functoriality properties established in this chapter will allow us to extend the assignment $X \rightsquigarrow \Upsilon_X$ from schemes to prestacks.

0.1.3. We will see that the functor *dual* to Υ_X with respect to the Serre auto-duality of $\text{IndCoh}(X)$ and the naive auto-duality of $\text{QCoh}(X)$, is the natural transformation

$$\Psi_X : \text{IndCoh}(X) \to \text{QCoh}(X),$$

again in a way functorial with respect to X.

0.2. The action of $\text{QCoh}(X)$ on $\text{IndCoh}(X)$ and correspondences. Let us explain how we encode the action of $\text{QCoh}(X)$ on $\text{IndCoh}(X)$ in the framework of the $(\infty, 2)$-categories of correspondences.

0.2.1. Recall that in Chapter 5 we extended the assignment $X \rightsquigarrow \text{IndCoh}(X)$ into a functor

$$\text{IndCoh}_{\text{Corr}(\text{Sch}_{\text{aft}})^{\text{proper}}_{\text{all;all}}} : \text{Corr}(\text{Sch}_{\text{aft}})^{\text{proper}}_{\text{all;all}} \to \text{DGCat}^{2\text{-Cat}}_{\text{cont}}.$$

We now consider the assignment

$$X \rightsquigarrow (\text{QCoh}(X), \text{IndCoh}(X)),$$

where we regard $\text{QCoh}(X)$ as a monoidal DG category (i.e., an algebra object in $\text{DGCat}_{\text{cont}}$), and $\text{IndCoh}(X)$ as a $\text{QCoh}(X)$-module category, i.e., an object of $\text{QCoh}(X)$-**mod**.

We want to extend this assignment to a functor out of $\text{Corr}(\text{Sch}_{\text{aft}})^{\text{proper}}_{\text{all;all}}$. The challenge is to identify the target $(\infty, 2)$-category, so that it will account for the pieces of structure that we need, also one for which such a construction will be possible.

0.2.2. The sought-for $(\infty, 2)$-category, denoted $\text{DGCat}^{\text{Mon+Mod,ext}}_{\text{cont}}$, is introduced in Sect. 1. As expected, its objects are pairs $(\mathbf{O}, \mathbf{C})$, where $\mathbf{O}$ is a monoidal DG category and $\mathbf{C}$ is a module $\mathbf{O}$-category.

But 1-morphisms are less obvious. We refer the reader to Sect. 1.1 for the definition. It is designed so that there is a natural forgetful functor

$$\text{DGCat}^{\text{Mon+Mod,ext}}_{\text{cont}} \to \text{DGCat}^{2\text{-Cat}}_{\text{cont}} \tag{0.1}$$

that at the level of objects sends $(\mathbf{O}, \mathbf{C})$ to $\mathbf{C}$.

Here is how the desired functor

$$(\text{QCoh}, \text{IndCoh})_{\text{Corr}(\text{Sch}_{\text{aft}})^{\text{proper}}_{\text{all;all}}} : \text{Corr}(\text{Sch}_{\text{aft}})^{\text{proper}}_{\text{all;all}} \to \text{DGCat}^{\text{Mon+Mod,ext}}_{\text{cont}} \tag{0.2}$$

is constructed[1].

0.2.3. Consider the $(\infty, 1)$-category $\text{DGCat}^{\text{Mon}^{\text{op}}+\text{Mod}}_{\text{cont}}$, whose objects are pairs $(\mathbf{O}, \mathbf{C})$, but where the space of morphisms from $(\mathbf{O}_1, \mathbf{C}_1)$ to $(\mathbf{O}_2, \mathbf{C}_2)$ is that of pairs

$$(F_{\mathbf{O}} : \mathbf{O}_2 \to \mathbf{O}_1, F_{\mathbf{C}} : \mathbf{C}_1 \to \mathbf{C}_2),$$

where $F_{\mathbf{O}}$ is a monoidal functor (note the direction of the arrow), and $F_{\mathbf{C}}$ is a map of $\mathbf{O}_2$-module categories.

As our initial input we start with the functor

$$(\text{QCoh}^*, \text{IndCoh}_*)_{\text{Sch}_{\text{aft}}} : \text{Sch}_{\text{aft}} \to \text{DGCat}^{\text{Mon}^{\text{op}}+\text{Mod}}_{\text{cont}} \tag{0.3}$$

that sends a scheme X to $(\text{QCoh}(X), \text{IndCoh}(X))$ and a morphism $X \xrightarrow{f} Y$ to the pair $(f^*, f^{\text{IndCoh}}_*)$.

0.2.4. Next, from the definition of $\text{DGCat}^{\text{Mon+Mod,ext}}_{\text{cont}}$ it follows that there exists a canonically defined functor

$$\text{DGCat}^{\text{Mon}^{\text{op}}+\text{Mod}}_{\text{cont}} \to \text{DGCat}^{\text{Mon+Mod,ext}}_{\text{cont}}.$$

Precomposing with (0.3) we obtain a functor

$$(\text{QCoh}, \text{IndCoh})_{\text{Sch}_{\text{aft}}} : \text{Sch}_{\text{aft}} \to \text{DGCat}^{\text{Mon+Mod,ext}}_{\text{cont}}. \tag{0.4}$$

Now, to get from (0.4) to (0.2) we repeat the procedure of Chapter 5, Sect. 2.1.

[1] This essentially mimics the construction of $\text{IndCoh}_{\text{Corr}(\text{Sch}_{\text{aft}})^{\text{proper}}_{\text{all;all}}}$ in Chapter 5, Sect. 2.1.

0.3. The natural transformation Υ. We shall now explain how the existence of the functor $(\mathrm{QCoh}, \mathrm{IndCoh})_{\mathrm{Corr}(\mathrm{Sch}_{\mathrm{aft}})^{\mathrm{proper}}_{\mathrm{all;all}}}$ in (0.2) leads to the natural transformation from Sect. 0.1.1.

0.3.1. First, we note that $(\mathrm{QCoh}, \mathrm{IndCoh})_{\mathrm{Corr}(\mathrm{Sch}_{\mathrm{aft}})^{\mathrm{proper}}_{\mathrm{all;all}}}$ comes equipped with a canonically defined symmetric monoidal structure, where $\mathrm{Corr}(\mathrm{Sch}_{\mathrm{aft}})^{\mathrm{proper}}_{\mathrm{all;all}}$ acquires a symmetric monoidal structure from the operation of Cartesian product on $\mathrm{Sch}_{\mathrm{aft}}$, and the symmetric monoidal structure on $\mathrm{DGCat}^{\mathrm{Mon+Mod,ext}}_{\mathrm{cont}}$ is given by component-wise tensor product.

0.3.2. Let $\mathrm{DGCat}^{\mathrm{Mon+Mod}}_{\mathrm{cont}}$ denote the $(\infty,1)$-category, whose objects are pairs $(\mathbf{O}, \mathbf{C})$, but where the space of morphisms from $(\mathbf{O}_1, \mathbf{C}_1)$ to $(\mathbf{O}_2, \mathbf{C}_2)$ is that of pairs

$$(F_{\mathbf{O}} : \mathbf{O}_1 \to \mathbf{O}_2, F_{\mathbf{C}} : \mathbf{C}_1 \to \mathbf{C}_2),$$

where $F_{\mathbf{O}}$ is a monoidal functor and $F_{\mathbf{C}}$ is a map of $\mathbf{O}_1$-module categories. (Note that the difference between $\mathrm{DGCat}^{\mathrm{Mon+Mod}}_{\mathrm{cont}}$ and $\mathrm{DGCat}^{\mathrm{Mon^{op}+Mod}}_{\mathrm{cont}}$ is in the direction of the arrow $F_{\mathbf{O}}$.)

From the definition of $\mathrm{DGCat}^{\mathrm{Mon+Mod,ext}}_{\mathrm{cont}}$ it follows that there exists a canonically defined (symmetric monoidal) functor

$$\mathrm{DGCat}^{\mathrm{Mon+Mod}}_{\mathrm{cont}} \to \mathrm{DGCat}^{\mathrm{Mon+Mod,ext}}_{\mathrm{cont}}.$$

0.3.3. Restricting $(\mathrm{QCoh}, \mathrm{IndCoh})_{\mathrm{Corr}(\mathrm{Sch}_{\mathrm{aft}})^{\mathrm{proper}}_{\mathrm{all;all}}}$ to $(\mathrm{Sch}_{\mathrm{aft}})^{\mathrm{op}} \subset \mathrm{Corr}(\mathrm{Sch}_{\mathrm{aft}})^{\mathrm{proper}}_{\mathrm{all;all}}$ we obtain a functor

$$(\mathrm{Sch}_{\mathrm{aft}})^{\mathrm{op}} \to \mathrm{DGCat}^{\mathrm{Mon+Mod,ext}}_{\mathrm{cont}},$$

and one shows that it factors through a canonically defined (symmetric monoidal) functor

$$(\mathrm{QCoh}^*, \mathrm{IndCoh}^!)_{\mathrm{Sch}_{\mathrm{aft}}} : (\mathrm{Sch}_{\mathrm{aft}})^{\mathrm{op}} \to \mathrm{DGCat}^{\mathrm{Mon+Mod}}_{\mathrm{cont}}. \tag{0.5}$$

Since every object in $\mathrm{Sch}_{\mathrm{aft}}$ has a canonical structure of co-commutative co-algebra (via the diagonal map), the functor (0.5) gives rise to a functor

$$(\mathrm{Sch}_{\mathrm{aft}})^{\mathrm{op}} \to \mathrm{ComAlg}(\mathrm{DGCat}^{\mathrm{Mon+Mod}}_{\mathrm{cont}}). \tag{0.6}$$

0.3.4. We note that the category $\mathrm{ComAlg}(\mathrm{DGCat}^{\mathrm{Mon+Mod}}_{\mathrm{cont}})$ identifies with that of triples

$$(\mathbf{O}, \mathbf{O}', \alpha : \mathbf{O} \to \mathbf{O}'),$$

where $\mathbf{O}$ and $\mathbf{O}'$ are symmetric monoidal categories, and α is a symmetric monoidal functor.

Hence, the data of the functor (0.6) is equivalent to that of a natural transformation

$$\Upsilon : \mathrm{QCoh}^*_{\mathrm{Sch}_{\mathrm{aft}}} \to \mathrm{IndCoh}^!_{\mathrm{Sch}_{\mathrm{aft}}}, \tag{0.7}$$

where $\mathrm{QCoh}^*_{\mathrm{Sch}_{\mathrm{aft}}}$ and $\mathrm{IndCoh}^!_{\mathrm{Sch}_{\mathrm{aft}}}$ are viewed as functors

$$(\mathrm{Sch}_{\mathrm{aft}})^{\mathrm{op}} \to \mathrm{ComAlg}(\mathrm{DGCat}_{\mathrm{cont}}) = \mathrm{DGCat}^{\mathrm{SymMon}}_{\mathrm{cont}}.$$

0.3.5. For an individual scheme X we thus obtain a functor

$$\Upsilon_X : \mathrm{QCoh}(X) \to \mathrm{IndCoh}(X),$$

which is obtained by the action of $\mathrm{QCoh}(X)$ on the monoidal unit in $\mathrm{IndCoh}(X)$, i.e., $\omega_X \in \mathrm{IndCoh}(X)$.

For a morphism $X \xrightarrow{f} Y$, we have the following commutative diagram

$$\begin{array}{ccc} \mathrm{QCoh}(X) & \xrightarrow{\Upsilon_X} & \mathrm{IndCoh}(X) \\ {\scriptstyle f^*}\uparrow & & \uparrow{\scriptstyle f^!} \\ \mathrm{QCoh}(Y) & \xrightarrow{\Upsilon_Y} & \mathrm{IndCoh}(Y). \end{array}$$

The above observation allows to extend the assignment $X \rightsquigarrow \Upsilon_X$ from schemes to prestacks. I.e., for every $\mathcal{Y} \in \mathrm{PreStk}$ we also have a canonically defined (symmetric monoidal) functor

$$\Upsilon_{\mathcal{Y}} : \mathrm{QCoh}(\mathcal{Y}) \to \mathrm{IndCoh}(\mathcal{Y}).$$

0.4. Relationship with Ψ. Let us finally explain the relationship between the functors

$$\Upsilon_X : \mathrm{QCoh}(X) \to \mathrm{IndCoh}(X) \text{ and } \Psi_X : \mathrm{IndCoh}(X) \to \mathrm{QCoh}(X).$$

REMARK 0.4.1. Recall that the functor Ψ_X played a fundamental role in the initial steps of setting up the theory of ind-coherent sheaves. Specifically, it was used in the definition of the direct image functor

$$X \xrightarrow{f} Y \quad \rightsquigarrow \quad \mathrm{IndCoh}(X) \xrightarrow{f_*^{\mathrm{IndCoh}}} \mathrm{IndCoh}(Y).$$

However, Ψ is really a feature of schemes. In particular, it does not have an intrinsic meaning for prestacks.

0.4.2. Recall the categories $\mathrm{DGCat}_{\mathrm{cont}}^{\mathrm{Mon+Mod}}$ and $\mathrm{DGCat}_{\mathrm{cont}}^{\mathrm{Mon}^{\mathrm{op}}+\mathrm{Mod}}$, and note that they contain full subcategories

$$(\mathrm{DGCat}_{\mathrm{cont}}^{\mathrm{Mon+Mod}})^{\mathrm{dualizable}} \subset \mathrm{DGCat}_{\mathrm{cont}}^{\mathrm{Mon+Mod}}$$

and

$$(\mathrm{DGCat}_{\mathrm{cont}}^{\mathrm{Mon}^{\mathrm{op}}+\mathrm{Mod}})^{\mathrm{dualizable}} \subset \mathrm{DGCat}_{\mathrm{cont}}^{\mathrm{Mon}^{\mathrm{op}}+\mathrm{Mod}},$$

respectively that consist of pairs $(\mathbf{O}, \mathbf{C})$ where $\mathbf{C}$ is dualizable as a plain DG category.

The operation of dualization $(\mathbf{O}, \mathbf{C}) \mapsto (\mathbf{O}, \mathbf{C}^\vee)$ defines an equivalence

$$((\mathrm{DGCat}_{\mathrm{cont}}^{\mathrm{Mon}^{\mathrm{op}}+\mathrm{Mod}})^{\mathrm{dualizable}})^{\mathrm{op}} \to (\mathrm{DGCat}_{\mathrm{cont}}^{\mathrm{Mon+Mod}})^{\mathrm{dualizable}}. \tag{0.8}$$

0.4.3. Recall (see Chapter 5, Sect. 4.2) that Serre duality for IndCoh was a formal consequence of the existence of the functor $\text{IndCoh}_{\text{Corr}(\text{Sch}_{\text{aft}})^{\text{proper}}_{\text{all;all}}}$, equipped with its symmetric monoidal structure.

In the same way, we use the functor $(\text{QCoh}, \text{IndCoh})_{\text{Corr}(\text{Sch}_{\text{aft}})^{\text{proper}}_{\text{all;all}}}$, equipped with its symmetric monoidal structure, to show that the composition of the functor $(\text{QCoh}^*, \text{IndCoh}_*)_{\text{Sch}_{\text{aft}}}$ of (0.3) with (0.8) identifies with the functor

$$(\text{QCoh}^*, \text{IndCoh}^!)_{\text{Sch}_{\text{aft}}}$$

of (0.5).

0.4.4. Next, by construction, the functor $(\text{QCoh}^*, \text{IndCoh}_*)_{\text{Sch}_{\text{aft}}}$ comes equipped with the natural transformation

$$(\text{Id}, \Psi)_{\text{Sch}_{\text{aft}}} : (\text{QCoh}^*, \text{IndCoh}_*)_{\text{Sch}_{\text{aft}}} \to (\text{QCoh}^*, \text{QCoh}_*)_{\text{Sch}_{\text{aft}}}$$

as functors $\text{Sch}_{\text{aft}} \to \text{DGCat}^{\text{Mon+Mod}}_{\text{cont}}$. Applying (0.8), we obtain a natural transformation

$$(0.9) \qquad (\text{Id}, \Psi^\vee)_{\text{Sch}_{\text{aft}}} : (\text{QCoh}^*, \text{QCoh}^*)_{\text{Sch}_{\text{aft}}} \to (\text{QCoh}^*, \text{IndCoh}^!)_{\text{Sch}_{\text{aft}}}$$

as functors $(\text{Sch}_{\text{aft}})^{\text{op}} \to \text{DGCat}^{\text{Mon}^{\text{op}}+\text{Mod}}_{\text{cont}}$.

0.4.5. What we show is that that the above natural transformation (0.9) is canonically isomorphic to the natural transformation

$$(0.10) \qquad (\text{Id}, \Upsilon)_{\text{Sch}_{\text{aft}}} : (\text{QCoh}^*, \text{QCoh}^*)_{\text{Sch}_{\text{aft}}} \to (\text{QCoh}^*, \text{IndCoh}^!)_{\text{Sch}_{\text{aft}}},$$

the latter being part of the data of the functor (0.7).

0.4.6. For an individual scheme X this means that the functors

$$\Psi_X : \text{IndCoh}(X) \to \text{QCoh}(X) \text{ and } \Upsilon_X : \text{QCoh}(X) \to \text{IndCoh}(X)$$

are canonically duals of each other.

Here $\text{IndCoh}(X)$ is identified with its own dual via the Serre duality functor $\mathbf{D}^{\text{Serre}}_X$ (see Chapter 5, Sect. 4.2.6). The category $\text{QCoh}(X)$ is identified with its own dual via the "naive" duality functor

$$\mathbf{D}^{\text{naive}}_X : \text{QCoh}(X)^\vee \simeq \text{QCoh}(X),$$

whose evaluation map $\text{QCoh}(X) \otimes \text{QCoh}(X) \to \text{Vect}$ is

$$\text{QCoh}(X) \otimes \text{QCoh}(X) \overset{\otimes}{\to} \text{QCoh}(X) \overset{\Gamma(X,-)}{\longrightarrow} \text{Vect}.$$

0.5. What is done in this chapter?

0.5.1. In Sect. 1 we define the $(\infty,2)$-category $\text{DGCat}^{\text{Mon+Mod,ext}}_{\text{cont}}$ that will be the recipient of the functor

$$(0.11) \quad (\text{QCoh}, \text{IndCoh})_{\text{Corr}(\text{Sch}_{\text{aft}})^{\text{proper}}_{\text{all;all}}} : \text{Corr}(\text{Sch}_{\text{aft}})^{\text{proper}}_{\text{all;all}} \to \text{DGCat}^{\text{Mon+Mod,ext}}_{\text{cont}}.$$

In doing so we allow ourselves a certain sloppiness: we say what the space of objects of $\text{DGCat}^{\text{Mon+Mod,ext}}_{\text{cont}}$ is, and what is the $(\infty,1)$-category of morphisms between any two objects.

We leave it to the reader to complete this to an actual definition of a $(\infty,2)$-category (as those are defined in Chapter 10, Sect. 2.1).

0.5.2. In Sect. 2 we carry out the construction of the functor (0.11) along the lines indicated in Sect. 0.2.

0.5.3. In Sect. 3 we discuss the symmetric monoidal structure on the functor (0.11), and how it gives rise to the natural transformation Υ, as described in Sect. 0.3.

0.5.4. In Sect. 4 we discuss the self-duality feature of the assignment

$$X \quad \rightsquigarrow \quad \mathrm{QCoh}(X) \in \mathrm{DGCat}^{\mathrm{Mon}}_{\mathrm{cont}}, \mathrm{IndCoh}(X) \in \mathrm{QCoh}(X)\text{-}\mathbf{mod},$$

and the relationship between the natural transformations

$$\Upsilon_X : \mathrm{QCoh}(X) \to \mathrm{IndCoh}(X)$$

and

$$\Psi_X : \mathrm{IndCoh}(X) \to \mathrm{QCoh}(X).$$

1. The $(\infty, 2)$-category of pairs

In this section we introduce a general framework that encodes the $(\infty, 2)$-category of pairs $(\mathbf{O}, \mathbf{C})$, where $\mathbf{O}$ us a monoidal DG category, and $\mathbf{C}$ is an $\mathbf{O}$-module category.

This $(\infty, 2)$-category will be the recipient of the functor from the $(\infty, 2)$-category of correspondences that assigns to a scheme X the pair

$$(\mathrm{QCoh}(X), \mathrm{IndCoh}(X).$$

The reason that we need this rather involved $(\infty, 2)$-category instead of the more easily defined underlying 1-category is that $(\infty, 2)$-category are necessary for the construction of the assignment

$$X \rightsquigarrow (\mathrm{QCoh}(X), \mathrm{IndCoh}(X)$$

as a functor, see Chapter 5, Sect. 2.1.

1.1. The category $\mathrm{DGCat}^{\mathrm{Mon+Mod,ext}}_{\mathrm{cont}}$. In this subsection we introduce our two category of pairs.

1.1.1. We introduce the $(\infty, 2)$-category $\mathrm{DGCat}^{\mathrm{Mon+Mod,ext}}_{\mathrm{cont}}$ as follows. Its objects are pairs $(\mathbf{O}, \mathbf{C})$, where $\mathbf{O} \in \mathrm{DGCat}^{\mathrm{Mon}}_{\mathrm{cont}}$, and $\mathbf{C} \in \mathbf{O}\text{-}\mathbf{mod}$.

Given two objects $(\mathbf{O}_1, \mathbf{C}_1)$ and $(\mathbf{O}_2, \mathbf{C}_2)$ of $\mathrm{DGCat}^{\mathrm{Mon+Mod,ext}}_{\mathrm{cont}}$, the objects of the $(\infty, 1)$-category of 1-morphisms $(\mathbf{O}_1, \mathbf{C}_1) \to (\mathbf{O}_2, \mathbf{C}_2)$ are the data of:

- An $(\mathbf{O}_2, \mathbf{O}_1)$-bimodule category $\mathbf{M}$;
- A map $F : \mathbf{M} \underset{\mathbf{O}_1}{\otimes} \mathbf{C}_1 \to \mathbf{C}_2$ in $\mathbf{O}_2\text{-}\mathbf{mod}$;
- A distinguished object $\mathbf{1}_{\mathbf{M}} \in \mathbf{M}$.

1.1.2. Given two objects $(\mathbf{M}^s, F^s, \mathbf{1}_{\mathbf{M}^s})$ and $(\mathbf{M}^t, F^t, \mathbf{1}_{\mathbf{M}^t})$ as above, the space of 2-morphisms

$$(\mathbf{M}^s, F^s, \mathbf{1}_{\mathbf{M}^s}) \to (\mathbf{M}^t, F^t, \mathbf{1}_{\mathbf{M}^t})$$

is that of the following data:

- A map of bimodules $\Phi : \mathbf{M}^t \to \mathbf{M}^s$ (note the direction of the arrow!);
- A natural transformation of maps of $\mathbf{O}_2$-bimodules
$$T : F^s \circ (\Phi \otimes \mathrm{Id}_{\mathbf{C}_1}) \Rightarrow F^t;$$
- A map in $\mathbf{M}^s$ as a plain DG category
$$\psi : \mathbf{1}_{\mathbf{M}^s} \to \Phi(\mathbf{1}_{\mathbf{M}^t}).$$

1.1.3. Compositions of 1-morphisms are defined naturally: for

$$(\mathbf{M}_{1,2}, F_{1,2}, \mathbf{1}_{\mathbf{M}_{1,2}}) : (\mathbf{O}_1, \mathbf{C}_1) \to (\mathbf{O}_2, \mathbf{C}_2)$$

and

$$(\mathbf{M}_{2,3}, F_{2,3}, \mathbf{1}_{\mathbf{M}_{2,3}}) : (\mathbf{O}_2, \mathbf{C}_2) \to (\mathbf{O}_3, \mathbf{C}_3),$$

their composition is defined by means of

$$\mathbf{M}_{1,3} := \mathbf{M}_{2,3} \underset{\mathbf{O}_2}{\otimes} \mathbf{M}_{1,2},$$

the data of $F_{1,3}$ equal to the composition

$$\mathbf{M}_{2,3} \underset{\mathbf{O}_2}{\otimes} \mathbf{M}_{1,2} \underset{\mathbf{O}_1}{\otimes} \mathbf{C}_1 \xrightarrow{F_{1,2}} \mathbf{M}_{2,3} \underset{\mathbf{O}_2}{\otimes} \mathbf{C}_2 \xrightarrow{F_{2,3}} \mathbf{C}_3,$$

and the data of $\mathbf{1}_{\mathbf{M}_{1,3}}$ being

$$\mathbf{1}_{\mathbf{M}_{2,3}} \otimes \mathbf{1}_{\mathbf{M}_{1,2}} \in \mathbf{M}_{2,3} \underset{\mathbf{O}_2}{\otimes} \mathbf{M}_{1,2}.$$

Compositions of 2-morphisms are also defined naturally.

The higher-categorical structure on $\mathrm{DGCat}_{\mathrm{cont}}^{\mathrm{Mon+Mod,ext}}$ is defined in a standard fashion.

1.2. Some forgetful functors. In this subsection we discuss two (obvious) forgetful functors from $\mathrm{DGCat}_{\mathrm{cont}}^{\mathrm{Mon+Mod,ext}}$ to some more familiar 2-categories.

1.2.1. First, we observe that $\mathrm{DGCat}_{\mathrm{cont}}^{\mathrm{Mon+Mod,ext}}$ comes equipped with a forgetful functor to $\mathrm{DGCat}_{\mathrm{cont}}^{2\text{-Cat}}$.

At the level of objects we send $(\mathbf{O}, \mathbf{C})$ to $\mathbf{C}$. At the level of 1-morphisms, given

$$(\mathbf{M}, F, \mathbf{1}_{\mathbf{M}}) : (\mathbf{O}_1, \mathbf{C}_1) \to (\mathbf{O}_2, \mathbf{C}_2)$$

we define the corresponding functor between plain DG categories $\overline{F} : \mathbf{C}_1 \to \mathbf{C}_2$ as the composition

$$\mathbf{C}_1 \xrightarrow{\mathbf{1}_{\mathbf{M}} \otimes \mathrm{Id}_{\mathbf{C}_1}} \mathbf{M} \underset{\mathbf{O}_1}{\otimes} \mathbf{C}_1 \xrightarrow{F} \mathbf{C}_2.$$

Given a 2-morphism

$$(\Phi, T, \psi) : (\mathbf{M}^s, F^s, \mathbf{1}_{\mathbf{M}^s}) \to (\mathbf{M}^t, F^t, \mathbf{1}_{\mathbf{M}^t}),$$

the corresponding natural transformation $\overline{F}^s \to \overline{F}^t$ is the composition

$$\overline{F}^s := F^s \circ (\mathbf{1}_{\mathbf{M}^s} \otimes \mathrm{Id}_{\mathbf{C}_1}) \overset{\psi}{\Rightarrow} F^s \circ (\Phi \otimes \mathrm{Id}_{\mathbf{C}_1}) \circ (\mathbf{1}_{\mathbf{M}^t} \otimes \mathrm{Id}_{\mathbf{C}_1}) \overset{T}{\Rightarrow} F^t \circ (\mathbf{1}_{\mathbf{M}^t} \otimes \mathrm{Id}_{\mathbf{C}_1}) = \overline{F}^t.$$

1.2.2. Let $\mathrm{DGCat}^{\mathrm{Mon,ext}}$ denote the $(\infty,2)$-category, where:

- The objects are $\mathbf{O} \in \mathrm{DGCat}^{\mathrm{Mon}}_{\mathrm{cont}}$;
- Given $\mathbf{O}_1, \mathbf{O}_2 \in \mathrm{DGCat}^{\mathrm{Mon}}_{\mathrm{cont}}$, the objects of $(\infty,1)$-category of 1-morphisms from $\mathbf{O}_1$ to $\mathbf{O}_2$ are $(\mathbf{O}_2, \mathbf{O}_1)$-bimodule categories;
- For a pair of 1-morphims from $\mathbf{O}_1$ to $\mathbf{O}_2$, given by bimodule categories $\mathbf{M}^s$ and $\mathbf{M}^t$ respectively, the space of 2-morphisms from $\mathbf{M}^s$ to $\mathbf{M}^t$ is that of maps of bimodules $\mathbf{M}^t \to \mathbf{M}^s$ (note the direction of the arrow)[2].

We have a naturally defined forgetful functor

$$\mathrm{DGCat}^{\mathrm{Mon+Mod,ext}}_{\mathrm{cont}} \to \mathrm{DGCat}^{\mathrm{Mon,ext}}$$

that sends $(\mathbf{O}, \mathbf{C})$ to $\mathbf{O}$.

1.3. Two 2-full subcategories. In this subsection we single out two 1-subcategories of the $(\infty,2)$-category $\mathrm{DGCat}^{\mathrm{Mon+Mod,ext}}_{\mathrm{cont}}$ that we will ultimately be interested in.

1.3.1. Let $\mathrm{DGCat}^{\mathrm{Mon+Mod}}_{\mathrm{cont}}$ be the $(\infty,1)$-category, where

- The objects are the same as those of $\mathrm{DGCat}^{\mathrm{Mon+Mod,ext}}_{\mathrm{cont}}$;
- 1-morphisms between $(\mathbf{O}_1, \mathbf{C}_1) \to (\mathbf{O}_2, \mathbf{C}_2)$ are pairs $(R_{\mathbf{O}}, R_{\mathbf{C}})$, where
  - $R_{\mathbf{O}} : \mathbf{O}_1 \to \mathbf{O}_2$ is a 1-morphism in $\mathrm{DGCat}^{\mathrm{Mon}}_{\mathrm{cont}}$;
  - $R_{\mathbf{C}} : \mathbf{C}_1 \to \mathbf{C}_2$ is a map of $\mathbf{O}_1$-module categories.

1.3.2. We claim that there is a canonically defined 1-fully faithful functor [3]

$$\mathrm{DGCat}^{\mathrm{Mon+Mod}}_{\mathrm{cont}} \to (\mathrm{DGCat}^{\mathrm{Mon+Mod,ext}}_{\mathrm{cont}})^{1\text{-}\mathrm{Cat}}.$$

Indeed, the functor in question is the identity on objects. At the level of 1-morphisms its essential image corresponds to those pairs $(\mathbf{M}, \mathbf{1}_{\mathbf{M}})$, for which the functor

$$\mathbf{O}_2 \to \mathbf{M},$$

defined by $\mathbf{1}_{\mathbf{M}}$, is an equivalence.

1.3.3. Let $\mathrm{DGCat}^{\mathrm{Mon}^{\mathrm{op}}+\mathrm{Mod}}_{\mathrm{cont}}$ be the $(\infty,1)$-category, where

- The objects are the same as those of $\mathrm{DGCat}^{\mathrm{Mon+Mod,ext}}_{\mathrm{cont}}$;
- 1-morphisms between $(\mathbf{O}_1, \mathbf{C}_1) \to (\mathbf{O}_2, \mathbf{C}_2)$ are pairs $(R_{\mathbf{O}}, R_{\mathbf{C}})$, where
  - $R_{\mathbf{O}} : \mathbf{O}_2 \to \mathbf{O}_1$ (note the direction of the arrow!) is a 1-morphism in $\mathrm{DGCat}^{\mathrm{Mon}}_{\mathrm{cont}}$;
  - $R_{\mathbf{C}} : \mathbf{C}_1 \to \mathbf{C}_2$ is a map of $\mathbf{O}_2$-module categories.

[2] We emphasize that the latter is considered as a space and not as an $(\infty,1)$-category.

[3] We remind that a functor between $(\infty,2)$-categories is said to be 1-fully faithful if it defines a fully faithful functor on $(\infty,1)$-categories of 1-morphisms.

1.3.4. We claim that there is a canonically defined 2-fully faithful functor

$$\mathrm{DGCat}_{\mathrm{cont}}^{\mathrm{Mon}^{\mathrm{op}}+\mathrm{Mod}} \to (\mathrm{DGCat}_{\mathrm{cont}}^{\mathrm{Mon+Mod,ext}})^{1\text{-Cat}}.$$

Indeed, the functor in question is the identity on objects. At the level of 1-morphisms its essential image corresponds to those pairs $(\mathbf{M}, \mathbf{1}_{\mathbf{M}})$, for which the functor

$$\mathbf{O}_1 \to \mathbf{M},$$

defined by $\mathbf{1}_{\mathbf{M}}$, is an equivalence.

2. The functor of IndCoh, equipped with the action of QCoh

In this section we will construct the assignment

$$X \rightsquigarrow (\mathrm{QCoh}(X), \mathrm{IndCoh}(X))$$

as a functor from $\mathrm{IndCoh}_{\mathrm{Corr}(\mathrm{Sch}_{\mathrm{aft}})_{\mathrm{all;all}}^{\mathrm{proper}}}$ to the $(\infty,2)$-category $\mathrm{DGCat}_{\mathrm{cont}}^{\mathrm{Mon+Mod,ext}}$ defined in the previous section.

2.1. The goal. In this subsection we explain the idea behind the assignment

$$X \rightsquigarrow (\mathrm{QCoh}(X), \mathrm{IndCoh}(X)).$$

2.1.1. In Chapter 5, Sect. 2, we constructed the functor

$$\mathrm{IndCoh}_{\mathrm{Corr}(\mathrm{Sch}_{\mathrm{aft}})_{\mathrm{all;all}}^{\mathrm{proper}}} : \mathrm{Corr}(\mathrm{Sch}_{\mathrm{aft}})_{\mathrm{all;all}}^{\mathrm{proper}} \to \mathrm{DGCat}_{\mathrm{cont}}^{2\text{-Cat}}.$$

In this section we will extend this functor to a functor

$$(\mathrm{QCoh}, \mathrm{IndCoh})_{\mathrm{Corr}(\mathrm{Sch}_{\mathrm{aft}})_{\mathrm{all;all}}^{\mathrm{proper}}} : \mathrm{Corr}(\mathrm{Sch}_{\mathrm{aft}})_{\mathrm{all;all}}^{\mathrm{proper}} \to \mathrm{DGCat}_{\mathrm{cont}}^{\mathrm{Mon+Mod,ext}}.$$

The original functor $\mathrm{IndCoh}_{\mathrm{Corr}(\mathrm{Sch}_{\mathrm{aft}})_{\mathrm{all;all}}^{\mathrm{proper}}}$ is recovered from

$$(\mathrm{QCoh}, \mathrm{IndCoh})_{\mathrm{Corr}(\mathrm{Sch}_{\mathrm{aft}})_{\mathrm{all;all}}^{\mathrm{proper}}}$$

by composing with the forgetful functor of Sect. 1.2.1.

2.1.2. The meaning of the functor $(\mathrm{QCoh}, \mathrm{IndCoh})_{\mathrm{Corr}(\mathrm{Sch}_{\mathrm{aft}})_{\mathrm{all;all}}^{\mathrm{proper}}}$ is that it 'remembers' the category $\mathrm{IndCoh}(-)$ together with the action og $\mathrm{QCoh}(-)$.

Namely, we recall that for $X \in \mathrm{Sch}_{\mathrm{aft}}$, the category $\mathrm{IndCoh}(X)$ is naturally a module over the monoidal category $\mathrm{QCoh}(X)$, see Chapter 4, Sect. 1.2.9.

Now, the functor $(\mathrm{QCoh}, \mathrm{IndCoh})_{\mathrm{Corr}(\mathrm{Sch}_{\mathrm{aft}})_{\mathrm{all;all}}^{\mathrm{proper}}}$ encodes the fact that for $f : X \to Y$, the functors

$$f^! : \mathrm{IndCoh}(Y) \to \mathrm{IndCoh}(X) \text{ and } f_*^{\mathrm{IndCoh}} : \mathrm{IndCoh}(X) \to \mathrm{IndCoh}(Y)$$

each has a natural structure of morphim in $\mathrm{QCoh}(Y)$-**mod**, where $\mathrm{QCoh}(Y)$ acts on $\mathrm{IndCoh}(X)$ via the monoidal functor $f^* : \mathrm{QCoh}(Y) \to \mathrm{QCoh}(X)$.

Moreover, if f is proper, the $(f_*^{\mathrm{IndCoh}}, f^!)$-adjunction also upgrades to one in the $(\infty,2)$-category $\mathrm{QCoh}(Y)\text{-}\mathbf{mod}^{2\text{-Cat}}$.

2.2. The input. As in the case of IndCoh, the only input for the construction of the functor $(\mathrm{QCoh}, \mathrm{IndCoh})_{\mathrm{Corr}(\mathrm{Sch}_{\mathrm{aft}})^{\mathrm{proper}}_{\mathrm{all;all}}}$ is the ability to take direct images. However, this time we need to do it for IndCoh and QCoh simultaneously, in a compatible way.

In this subsection we construct the required direct image procedure.

2.2.1. Recall the categories $\mathrm{DGCat}^{\mathrm{Mon+Mod}}_{\mathrm{cont}}$ and $\mathrm{DGCat}^{\mathrm{Mon^{op}+Mod}}_{\mathrm{cont}}$.

Let

$$(\mathrm{DGCat}^{\mathrm{Mon+Mod}}_{\mathrm{cont}})_{\mathrm{adjtble}} \subset \mathrm{DGCat}^{\mathrm{Mon+Mod}}_{\mathrm{cont}}$$

and

$$(\mathrm{DGCat}^{\mathrm{Mon^{op}+Mod}}_{\mathrm{cont}})_{\mathrm{adjtble}} \subset \mathrm{DGCat}^{\mathrm{Mon^{op}+Mod}}_{\mathrm{cont}}$$

be 1-full subcategories, where we restrict 1-morphisms to those pairs $(R_{\mathbf{O}}, R_{\mathbf{C}})$, where we require that $R_{\mathbf{C}}$ admit a right (resp., left) adjoint in the $(\infty, 2)$-category of $\mathbf{O}_1\text{-}\mathbf{mod}^{2\text{-Cat}}$ (resp., $\mathbf{O}_2\text{-}\mathbf{mod}^{2\text{-Cat}}$).

The operation of passing to the right/left adjoint (see Chapter 12, Corollary 1.3.4) defines an equivalence

$$((\mathrm{DGCat}^{\mathrm{Mon+Mod}}_{\mathrm{cont}})_{\mathrm{adjtble}})^{\mathrm{op}} \simeq (\mathrm{DGCat}^{\mathrm{Mon^{op}+Mod}}_{\mathrm{cont}})_{\mathrm{adjtble}}. \tag{2.1}$$

2.2.2. Note that we have a tautologically defined functor of $(\infty, 1)$-categories

$$\mathrm{DGCat}^{\mathrm{Mon}}_{\mathrm{cont}} \to \mathrm{DGCat}^{\mathrm{Mon+Mod}}_{\mathrm{cont}}, \quad \mathbf{O} \mapsto (\mathbf{O}, \mathbf{O}). \tag{2.2}$$

We start with the functor $\mathrm{QCoh}^*_{\mathrm{Sch}_{\mathrm{aft}}}$, considered as a functor

$$(\mathrm{Sch}_{\mathrm{aft}})^{\mathrm{op}} \to \mathrm{DGCat}^{\mathrm{Mon}}_{\mathrm{cont}},$$

and consider its composition with (2.2). We obtain a functor

$$(\mathrm{QCoh}^*, \mathrm{QCoh}^*)_{\mathrm{Sch}_{\mathrm{aft}}} : (\mathrm{Sch}_{\mathrm{aft}})^{\mathrm{op}} \to \mathrm{DGCat}^{\mathrm{Mon+Mod}}_{\mathrm{cont}}. \tag{2.3}$$

2.2.3. It is easy to see that the functor (2.3) factors (automatically, canonically) via the subcategory

$$(\mathrm{DGCat}^{\mathrm{Mon+Mod}}_{\mathrm{cont}})_{\mathrm{adjtble}} \subset \mathrm{DGCat}^{\mathrm{Mon+Mod}}_{\mathrm{cont}}.$$

Thus, we obtain a functor

$$(\mathrm{Sch}_{\mathrm{aft}})^{\mathrm{op}} \to (\mathrm{DGCat}^{\mathrm{Mon+Mod}}_{\mathrm{cont}})_{\mathrm{adjtble}}. \tag{2.4}$$

Composing (2.5) with the equivalence (2.1) we obtain a functor

$$\mathrm{Sch}_{\mathrm{aft}} \to (\mathrm{DGCat}^{\mathrm{Mon^{op}+Mod}}_{\mathrm{cont}})_{\mathrm{adjtble}}. \tag{2.5}$$

We follow (2.5) by the forgetful functor

$$(\mathrm{DGCat}^{\mathrm{Mon^{op}+Mod}}_{\mathrm{cont}})_{\mathrm{adjtble}} \subset \mathrm{DGCat}^{\mathrm{Mon^{op}+Mod}}_{\mathrm{cont}}$$

and obtain a functor

$$(\mathrm{QCoh}^*, \mathrm{QCoh}_*)_{\mathrm{Sch}_{\mathrm{aft}}} : \mathrm{Sch}_{\mathrm{aft}} \to \mathrm{DGCat}^{\mathrm{Mon^{op}+Mod}}_{\mathrm{cont}}. \tag{2.6}$$

Explicitly, the functor (2.6) sends $X \in \mathrm{Sch}_{\mathrm{aft}}$ to the pair $(\mathrm{QCoh}(X), \mathrm{QCoh}(X))$, and a morphism $f : X \to Y$ to the pair

$$f^* : \mathrm{QCoh}(Y) \to \mathrm{QCoh}(X), \quad f_* : \mathrm{QCoh}(X) \to \mathrm{QCoh}(Y).$$

2.2.4. Recall that for an individual object $X \in \mathrm{Sch}_{\mathrm{aft}}$, the DG category $\mathrm{IndCoh}(X)$ carries a canonical action of the monoidal category $\mathrm{QCoh}(X)$, see Chapter 4, Sect. 1.2.9.

By construction, the functor

$$\Psi_X : \mathrm{IndCoh}(X) \to \mathrm{QCoh}(X)$$

carries a unique structure of map of $\mathrm{QCoh}(X)$-module categories.

Furthermore, the following results from the construction in Chapter 4, Proposition 2.2.1:

LEMMA 2.2.5. *For a map $f : X \to Y$, the functor*

$$f_*^{\mathrm{IndCoh}} : \mathrm{IndCoh}(X) \to \mathrm{IndCoh}(Y)$$

can be equipped with a unique structure of map of $\mathrm{QCoh}(Y)$-module categories, in such a way that

$$\begin{array}{ccc} \mathrm{IndCoh}(X) & \xrightarrow{\Psi_X} & \mathrm{QCoh}(X) \\ {\scriptstyle f_*^{\mathrm{IndCoh}}}\downarrow & & \downarrow{\scriptstyle f_*} \\ \mathrm{IndCoh}(Y) & \xrightarrow{\Psi_Y} & \mathrm{QCoh}(Y) \end{array}$$

is a commutative diagram in $\mathrm{QCoh}(Y)$- **mod**.

From here we obtain:

COROLLARY 2.2.6. *There exists a uniquely defined functor*

$$(\mathrm{QCoh}^*, \mathrm{IndCoh}_*)_{\mathrm{Sch}_{\mathrm{aft}}} : \mathrm{Sch}_{\mathrm{aft}} \to \mathrm{DGCat}_{\mathrm{cont}}^{\mathrm{Mon}^{\mathrm{op}} + \mathrm{Mod}},$$

equipped with a natural transformation

$$(\mathrm{Id}, \Psi)_{\mathrm{Sch}_{\mathrm{aft}}} : (\mathrm{QCoh}^*, \mathrm{IndCoh}_*)_{\mathrm{Sch}_{\mathrm{aft}}} \Rightarrow (\mathrm{QCoh}^*, \mathrm{QCoh}_*)_{\mathrm{Sch}_{\mathrm{aft}}},$$

such that

- *The composition of $(\mathrm{QCoh}^*, \mathrm{IndCoh}_*)_{\mathrm{Sch}_{\mathrm{aft}}}$ and $(\mathrm{Id}, \Psi)_{\mathrm{Sch}_{\mathrm{aft}}}$ with the forgetful functor*

  $$\mathrm{DGCat}_{\mathrm{cont}}^{\mathrm{Mon}^{\mathrm{op}} + \mathrm{Mod}} \to (\mathrm{DGCat}_{\mathrm{cont}}^{\mathrm{Mon}})^{\mathrm{op}}$$

  is the identity on $\mathrm{QCoh}^*_{\mathrm{Sch}_{\mathrm{aft}}}$;
- *At the level of objects and 1-morphisms, $(\mathrm{Id}, \Psi)_{\mathrm{Sch}_{\mathrm{aft}}}$ is given by the structure specified in Lemma* 2.2.5.
- *The composition of $(\mathrm{QCoh}^*, \mathrm{IndCoh}_*)_{\mathrm{Sch}_{\mathrm{aft}}}$ and $(\mathrm{Id}, \Psi)_{\mathrm{Sch}_{\mathrm{aft}}}$ with the forgetful functor*

  $$\mathrm{DGCat}_{\mathrm{cont}}^{\mathrm{Mon}^{\mathrm{op}} + \mathrm{Mod}} \to \mathrm{DGCat}_{\mathrm{cont}}$$

  is the pair $(\mathrm{IndCoh}_{\mathrm{Sch}_{\mathrm{aft}}}, \Psi_{\mathrm{Sch}_{\mathrm{aft}}})$ of Chapter 4, Proposition 2.2.3.

2.3. The construction. In this subsection we will finally construct the sought-for functor $(\mathrm{QCoh}, \mathrm{IndCoh})_{\mathrm{Corr}(\mathrm{Sch}_{\mathrm{aft}})^{\mathrm{proper}}_{\mathrm{all;all}}}$. The method will be analogous to that by which we constructed the functor $\mathrm{IndCoh}_{\mathrm{Corr}(\mathrm{Sch}_{\mathrm{aft}})^{\mathrm{proper}}_{\mathrm{all;all}}}$ in Chapter 5, Sect. 2.1.

2.3.1. As in the case of $\mathrm{IndCoh}_{\mathrm{Corr}(\mathrm{Sch}_{\mathrm{aft}})^{\mathrm{proper}}_{\mathrm{all;all}}}$, the point of departure for the sought-for functor $(\mathrm{QCoh},\mathrm{IndCoh})_{\mathrm{Corr}(\mathrm{Sch}_{\mathrm{aft}})^{\mathrm{proper}}_{\mathrm{all;all}}}$ is a functor

$$(\mathrm{QCoh},\mathrm{IndCoh})_{\mathrm{Sch}_{\mathrm{aft}}}:\mathrm{Sch}_{\mathrm{aft}}\to\mathrm{DGCat}^{\mathrm{Mon+Mod,ext}}_{\mathrm{cont}}. \tag{2.7}$$

To construct the functor (2.7) we proceed as follows. We start with the functor

$$(\mathrm{QCoh}^*,\mathrm{IndCoh}_*)_{\mathrm{Sch}_{\mathrm{aft}}}:\mathrm{Sch}_{\mathrm{aft}}\to\mathrm{DGCat}^{\mathrm{Mon}^{\mathrm{op}}+\mathrm{Mod}}_{\mathrm{cont}}, \tag{2.8}$$

of Corollary 2.2.6, and compose it with the functor of Sect. 1.3.4 to obtain the desired functor $(\mathrm{QCoh},\mathrm{IndCoh})_{\mathrm{Sch}_{\mathrm{aft}}}$ in (2.7).

2.3.2. We shall now extend the functor (2.7) to a functor

$$(\mathrm{QCoh},\mathrm{IndCoh})_{\mathrm{Corr}(\mathrm{Sch}_{\mathrm{aft}})_{\mathrm{all;open}}}:\mathrm{Corr}(\mathrm{Sch}_{\mathrm{aft}})_{\mathrm{all;open}}\to\mathrm{DGCat}^{\mathrm{Mon+Mod,ext}}_{\mathrm{cont}}. \tag{2.9}$$

As in Chapter 5, Sect. 2.1.2, in order to do so, it suffices to prove:

PROPOSITION 2.3.3. *The functor* $(\mathrm{QCoh},\mathrm{IndCoh})_{\mathrm{Sch}_{\mathrm{aft}}}$, *viewed as a functor*

$$\mathrm{Sch}_{\mathrm{aft}}\to\left(\mathrm{DGCat}^{\mathrm{Mon+Mod,ext}}_{\mathrm{cont}}\right)^{2\text{-op}},$$

satisfies the left Beck-Chevalley condition *with respect to open embeddings.*

This proposition will be proved in Sect. 2.4. We proceed with the construction of the functor $(\mathrm{QCoh},\mathrm{IndCoh})_{\mathrm{Corr}(\mathrm{Sch}_{\mathrm{aft}})^{\mathrm{proper}}_{\mathrm{all;all}}}$.

2.3.4. We will now show that the functor (2.9) admits a unique extension to a functor

$$(\mathrm{QCoh},\mathrm{IndCoh})_{\mathrm{Corr}(\mathrm{Sch}_{\mathrm{aft}})^{\mathrm{proper}}_{\mathrm{all;all}}}:\mathrm{Corr}(\mathrm{Sch}_{\mathrm{aft}})^{\mathrm{proper}}_{\mathrm{all;all}}\to\mathrm{DGCat}^{\mathrm{Mon+Mod,ext}}_{\mathrm{cont}}. \tag{2.10}$$

As in Chapter 5, Sect. 2.1.5, we need to verify the following two statements. One is the next proposition, proved in Sect. 2.5:

PROPOSITION 2.3.5. *The functor* $(\mathrm{QCoh},\mathrm{IndCoh})_{\mathrm{Sch}_{\mathrm{aft}}}$ *satisfies the* left Beck-Chevalley condition *with respect to the class of proper maps.*

Another is that the condition of Chapter 7, Sect. 5.2.2 is satisfied. This will be proved in Sect. 2.6.

2.4. Open adjunction. In this subsection we will prove Proposition 2.3.3.

2.4.1. Recall that the value of $(\mathrm{QCoh},\mathrm{IndCoh})_{\mathrm{Sch}_{\mathrm{aft}}}$ on $X\in\mathrm{Sch}_{\mathrm{aft}}$ is

$$(\mathrm{QCoh}(X),\mathrm{IndCoh}(X)),$$

where $\mathrm{QCoh}(X)$ acts on $\mathrm{IndCoh}(X)$ as in Chapter 4, Sect. 1.1.5.

For a morphism $f:X\to Y$, the 1-morphism

$$(\mathrm{QCoh}(X),\mathrm{IndCoh}(X))\to(\mathrm{QCoh}(Y),\mathrm{IndCoh}(Y))\in\mathrm{DGCat}^{\mathrm{Mon+Mod,ext}}_{\mathrm{cont}}. \tag{2.11}$$

is given by the pair $(\mathbf{M},F,\mathbf{1}_{\mathbf{M}})$, where

- $$\mathbf{M}:=\mathrm{QCoh}(X),$$
  regarded as an $\mathrm{QCoh}(X)$-module tautologically and as a $\mathrm{QCoh}(Y)$-module via the functor $f^*:\mathrm{QCoh}(Y)\to\mathrm{QCoh}(X)$.

•
$$F : \mathrm{IndCoh}(X) \simeq \mathrm{QCoh}(X) \underset{\mathrm{QCoh}(X)}{\otimes} \mathrm{IndCoh}(X) \to \mathrm{IndCoh}(Y)$$
is the functor f_*^{IndCoh};
• $\mathbf{1}_{\mathbf{M}} = \mathcal{O}_X$.

2.4.2. Let f be an open embedding. We need to show that in this case the corresponding 1-morphism (2.11) admits a *left* adjoint, and that the corresponding base change property holds.

We construct the left adjoint
$$(\mathrm{QCoh}(Y), \mathrm{IndCoh}(Y)) \to (\mathrm{QCoh}(X), \mathrm{IndCoh}(X)) \in \mathrm{DGCat}_{\mathrm{cont}}^{\mathrm{Mon+Mod,ext}}$$
as follows. It is given by the pair $(\mathbf{N}, G, \mathbf{1}_{\mathbf{N}})$, where

•
$$\mathbf{N} := \mathrm{QCoh}(X),$$
regarded as an $\mathrm{QCoh}(X)$-module tautologically and as a $\mathrm{QCoh}(Y)$-module via the functor $f^* : \mathrm{QCoh}(Y) \to \mathrm{QCoh}(X)$.

•
$$G : \mathrm{QCoh}(X) \underset{\mathrm{QCoh}(Y)}{\otimes} \mathrm{IndCoh}(Y) \to \mathrm{IndCoh}(X)$$
is obtained by tensoring up from the functor $f^{\mathrm{IndCoh},*} : \mathrm{QCoh}(Y) \to \mathrm{QCoh}(Y)$;
• $\mathbf{1}_{\mathbf{N}} = \mathcal{O}_X$.

Let us construct the adjunction data.

2.4.3. The composition
$$\begin{gathered}(\mathbf{M}, F, \mathbf{1}_{\mathbf{M}}) \circ (\mathbf{N}, G, \mathbf{1}_{\mathbf{N}}) : \\ (\mathrm{QCoh}(Y), \mathrm{IndCoh}(Y)) \to (\mathrm{QCoh}(X), \mathrm{IndCoh}(X)) \to (\mathrm{QCoh}(Y), \mathrm{IndCoh}(Y))\end{gathered}$$
is given by:

• The $(\mathrm{QCoh}(Y), \mathrm{QCoh}(Y))$-bimodule $\mathrm{QCoh}(X)$;
• The functor
$$\mathrm{QCoh}(X) \underset{\mathrm{QCoh}(Y)}{\otimes} \mathrm{IndCoh}(Y) \to \mathrm{IndCoh}(Y),$$
which under the identification
$$\mathrm{IndCoh}(X) \simeq \mathrm{QCoh}(X) \underset{\mathrm{QCoh}(Y)}{\otimes} \mathrm{IndCoh}(Y)$$
of Chapter 4, Proposition 4.1.6 goes over to
$$f_*^{\mathrm{IndCoh}} : \mathrm{IndCoh}(X) \to \mathrm{IndCoh}(Y);$$
• The object $\mathcal{O}_X \in \mathrm{QCoh}(X)$.

The unit of the adjunction is a 2-morphism (Φ, T, ψ)
$$\mathrm{Id}_{(\mathrm{QCoh}(Y), \mathrm{IndCoh}(Y))} \to (\mathbf{M}, F, \mathbf{1}_{\mathbf{M}}) \circ (\mathbf{N}, G, \mathbf{1}_{\mathbf{N}}),$$

where

- Φ is the functor $f_* : \mathrm{QCoh}(X) \to \mathrm{QCoh}(Y)$;
- T is the identity natural transformation on f_*^{IndCoh};
- ψ is the canonical map from $\mathcal{O}_X$ to $f_*(\mathcal{O}_Y)$.

2.4.4. The composition

$$(\mathbf{N}, G, \mathbf{1_N}) \circ (\mathbf{M}, F, \mathbf{1_M}) :$$
$$(\mathrm{QCoh}(X), \mathrm{IndCoh}(X)) \to (\mathrm{QCoh}(Y), \mathrm{IndCoh}(Y)) \to (\mathrm{QCoh}(X), \mathrm{IndCoh}(X))$$

is given by:

- The bimodule $\mathrm{QCoh}(X) \underset{\mathrm{QCoh}(Y)}{\otimes} \mathrm{QCoh}(X) \simeq \mathrm{QCoh}(X)$;
- The identity functor on $\mathrm{IndCoh}(X)$;
- The object $\mathcal{O}_X \in \mathrm{QCoh}(X)$.

The co-unit for the adjunction is a 2-morphism (Φ, T, ψ)

$$(\mathbf{N}, G, \mathbf{1_N}) \circ (\mathbf{M}, F, \mathbf{1_M}) \to \mathrm{Id}_{(\mathrm{QCoh}(X), \mathrm{IndCoh}(X))},$$

where

- Φ is the identity functor;
- T is the identity functor;
- ψ is the identity map.

The fact that the unit and co-unit specified above indeed satisfy the adjunction axioms is a straightforward verification.

2.4.5. We will now verify the base change property for the open adjunction. Let

$$\begin{array}{ccc} U_1 & \xrightarrow{j_1} & X_1 \\ {\scriptstyle f_U}\downarrow & & \downarrow{\scriptstyle f_X} \\ U_2 & \xrightarrow{j_2} & X_2 \end{array}$$

be a Cartesian diagram in $\mathrm{Sch}_{\mathrm{aft}}$, in which the horizontal arrows are open. Consider the commutative diagram of 1-morphisms in $\mathrm{DGCat}^{\mathrm{Mon+Mod,ext}}_{\mathrm{cont}}$

$$\begin{array}{ccc} (\mathrm{QCoh}(U_1), \mathrm{IndCoh}(U_1)) & \longrightarrow & (\mathrm{QCoh}(X_1), \mathrm{IndCoh}(X_1)) \\ \downarrow & & \downarrow \\ (\mathrm{QCoh}(U_2), \mathrm{IndCoh}(U_2)) & \longrightarrow & (\mathrm{QCoh}(X_2), \mathrm{IndCoh}(X_2)). \end{array}$$

By passing to left adjoints along the horizontal arrows, we obtain a diagram that commutes up to a 2-morphism as indicated:

(2.12)
$$\begin{array}{ccc} (\mathrm{QCoh}(U_1), \mathrm{IndCoh}(U_1)) & \longleftarrow & (\mathrm{QCoh}(X_1), \mathrm{IndCoh}(X_1)) \\ \downarrow & \nwarrow\!\!\nwarrow & \downarrow \\ (\mathrm{QCoh}(U_2), \mathrm{IndCoh}(U_2)) & \longleftarrow & (\mathrm{QCoh}(X_2), \mathrm{IndCoh}(X_2)). \end{array}$$

We need to show that the 2-morphism in question is an isomorphism.

For the clockwise circuit in (2.12), the corresponding $(\mathrm{QCoh}(X_1), \mathrm{QCoh}(U_2))$-bimodule is $\mathrm{QCoh}(U_1)$, equipped with the distinuished object $\mathcal{O}_{U_1} \in \mathrm{QCoh}(U_1)$.

For the anti-clockwise circuit, the corresponding $(\mathrm{QCoh}(X_1), \mathrm{QCoh}(U_2))$-bimodule is

$$\mathrm{QCoh}(U_2) \underset{\mathrm{QCoh}(X_2)}{\otimes} \mathrm{QCoh}(X_1),$$

equippped with the distinguished object $\mathcal{O}_{U_2} \underset{X_2}{\boxtimes} \mathcal{O}_{X_1}$.

The datum Φ of the 2-morphism in (2.12) is the canonical equivalence

$$\mathrm{QCoh}(U_1) \simeq \mathrm{QCoh}(U_2) \underset{\mathrm{QCoh}(X_2)}{\otimes} \mathrm{QCoh}(X_1),$$

and the datum ψ is the identity map on $\mathrm{QCoh}(U_1)$.

Under the above identification

$$\mathrm{QCoh}(U_2) \underset{\mathrm{QCoh}(X_2)}{\otimes} \mathrm{QCoh}(X_1) \simeq \mathrm{QCoh}(U_1),$$

the datum of T of the 2-morphism in (2.12) is the identity map on the functor

$$(f_U)_*^{\mathrm{IndCoh}} : \mathrm{IndCoh}(U_1) \simeq \mathrm{QCoh}(U_1) \underset{\mathrm{QCoh}(X_1)}{\otimes} \mathrm{IndCoh}(X_1) \to \mathrm{IndCoh}(U_2).$$

2.5. Proper adjunction. In this subsection we will prove Proposition 2.3.5.

2.5.1. Let now f be proper. We need to show that in this case the corresponding 1-morphism (2.11) admits a *right* adjoint.

We construct the right adjoint

$$(\mathrm{QCoh}(Y), \mathrm{IndCoh}(Y)) \to (\mathrm{QCoh}(X), \mathrm{IndCoh}(X)) \in \mathrm{DGCat}_{\mathrm{cont}}^{\mathrm{Mon+Mod,ext}}$$

as follows. It is given by the pair $(\mathbf{N}, G, \mathbf{1_N})$, where

- $\mathbf{N} := \mathrm{QCoh}(X)$;
- The functor

$$G : \mathrm{QCoh}(X) \underset{\mathrm{QCoh}(Y)}{\otimes} \mathrm{IndCoh}(Y) \to \mathrm{IndCoh}(X)$$

  is obtained by tensoring up from the functor $f^! : \mathrm{IndCoh}(Y) \to \mathrm{IndCoh}(X)$ (here we are using Chapter 4, Sect. 5.1.7).
- The object $\mathbf{1_N}$ is $\mathcal{O}_X \in \mathrm{QCoh}(X)$.

Let us construct the adjunction data.

2.5.2. The composition

$$(\mathbf{M}, F, \mathbf{1_M}) \circ (\mathbf{N}, G, \mathbf{1_N}) :$$
$$(\mathrm{QCoh}(Y), \mathrm{IndCoh}(Y)) \to (\mathrm{QCoh}(X), \mathrm{IndCoh}(X)) \to (\mathrm{QCoh}(Y), \mathrm{IndCoh}(Y))$$

is given by:

- The $(\mathrm{QCoh}(Y), \mathrm{QCoh}(Y))$-bimodule $\mathrm{QCoh}(X)$;
- The functor $\mathrm{QCoh}(X) \underset{\mathrm{QCoh}(Y)}{\otimes} \mathrm{IndCoh}(Y) \to \mathrm{IndCoh}(Y)$ is

$$(\mathcal{E}, \mathcal{F}) \mapsto f_*^{\mathrm{IndCoh}}(\mathcal{E} \otimes f^!(\mathcal{F}));$$

- The object $\mathcal{O}_X \in \mathrm{QCoh}(X)$.

The co-unit of the adjunction is a 2-morphism (Φ, T, ψ)

$$(\mathbf{M}, F, \mathbf{1_M}) \circ (\mathbf{N}, G, \mathbf{1_N}) \to \mathrm{Id}_{(\mathrm{QCoh}(Y), \mathrm{IndCoh}(Y))},$$

where

- Φ is the functor f^*;
- T is the natural transformation between $f_*^{\mathrm{IndCoh}} \circ f^!$ and $\mathrm{Id}_{\mathrm{IndCoh}(Y)}$ equal to the co-unit of the $(f_*^{\mathrm{IndCoh}}, f^!)$-adjunction
- ψ is the natural isomorphism.

2.5.3. The composition

$$\begin{gathered}(\mathbf{N}, G, \mathbf{1_N}) \circ (\mathbf{M}, F, \mathbf{1_M}) : \\ (\mathrm{QCoh}(X), \mathrm{IndCoh}(X)) \to (\mathrm{QCoh}(Y), \mathrm{IndCoh}(Y)) \to (\mathrm{QCoh}(X), \mathrm{IndCoh}(X))\end{gathered}$$

is given by:

- The $(\mathrm{QCoh}(X), \mathrm{QCoh}(X))$-bimodule
$$\mathrm{QCoh}(X) \underset{\mathrm{QCoh}(Y)}{\otimes} \mathrm{QCoh}(X) \simeq \mathrm{QCoh}(X \underset{Y}{\times} X);$$
- The functor $\mathrm{QCoh}(X \underset{Y}{\times} X) \underset{\mathrm{QCoh}(X)}{\otimes} \mathrm{IndCoh}(X) \to \mathrm{IndCoh}(X)$ is
$$(\mathcal{E}, \mathcal{F}) \mapsto (p_2)_*^{\mathrm{IndCoh}}(\mathcal{E} \otimes p_1^!(\mathcal{F}));$$
- The object $\mathcal{O}_{X \underset{Y}{\times} X} \in \mathrm{QCoh}(X \underset{Y}{\times} X)$;

The unit of the adjunction is a 2-morphism (Φ, T, ψ)

$$\mathrm{Id}_{(\mathrm{QCoh}(X), \mathrm{IndCoh}(X))} \to (\mathbf{N}, G, \mathbf{1_N}) \circ (\mathbf{M}, F, \mathbf{1_M}),$$

where

- Φ is the functor $\Delta_{X/Y}^*$, where $\Delta_{X/Y}$ is the diagonal map
$$X \to X \underset{Y}{\times} X;$$
- T is the natural transformation $\Delta_{X/Y}^*(\mathcal{E}) \otimes \mathcal{F} \to (p_2)_*^{\mathrm{IndCoh}}(\mathcal{E} \otimes p_1^!(\mathcal{F}))$ equal to
$$\begin{aligned}\Delta_{X/Y}^*(\mathcal{E}) \otimes \mathcal{F} \simeq (p_2)_*^{\mathrm{IndCoh}} \circ (\Delta_{X/Y})_*^{\mathrm{IndCoh}} \left(\Delta_{X/Y}^*(\mathcal{E}) \otimes \mathcal{F}\right) \simeq \\ \simeq (p_2)_*^{\mathrm{IndCoh}} \left(\mathcal{E} \otimes (\Delta_{X/Y})_*^{\mathrm{IndCoh}}(\mathcal{F})\right) \\ \simeq (p_2)_*^{\mathrm{IndCoh}} \left(\mathcal{E} \otimes (\Delta_{X/Y})_*^{\mathrm{IndCoh}} \circ \Delta_{X/Y}^! \circ p_1^!(\mathcal{F})\right) \to\to (p_2)_*^{\mathrm{IndCoh}} \left(\mathcal{E} \otimes p_1^!(\mathcal{F})\right),\end{aligned}$$
where the last arrow coves from the co-unit for the $((\Delta_{X/Y})_*^{\mathrm{IndCoh}}, \Delta_{X/Y}^!)$-adjunction.
- ψ is the natural isomorphism.

Again, the fact that the unit and co-unit specified above indeed satisfy the adjunction axioms is a straightforward verification.

2.5.4. We will now verify the base change property for the proper adjunction. Let

$$\begin{array}{ccc} Y_1 & \xrightarrow{g_1} & X_1 \\ {\scriptstyle f_Y}\downarrow & & \downarrow{\scriptstyle f_X} \\ Y_2 & \xrightarrow{g_2} & X_2 \end{array}$$

be a Cartesian diagram in $\mathrm{Sch}_{\mathrm{aft}}$, in which the vertical arrows are proper. Consider the commutative diagram of 1-morphisms in $\mathrm{DGCat}^{\mathrm{Mon+Mod,ext}}_{\mathrm{cont}}$

$$\begin{array}{ccc} (\mathrm{QCoh}(Y_1), \mathrm{IndCoh}(Y_1)) & \longrightarrow & (\mathrm{QCoh}(X_1), \mathrm{IndCoh}(X_1)) \\ \downarrow & & \downarrow \\ (\mathrm{QCoh}(Y_2), \mathrm{IndCoh}(Y_2)) & \longrightarrow & (\mathrm{QCoh}(X_2), \mathrm{IndCoh}(X_2)). \end{array}$$

By passing to right adjoints along the vertical arrows, we obtain a diagram that commutes up to a 2-morphism as indicated:

$$\begin{array}{ccc} (\mathrm{QCoh}(Y_1), \mathrm{IndCoh}(Y_1)) & \longrightarrow & (\mathrm{QCoh}(X_1), \mathrm{IndCoh}(X_1)) \\ \uparrow & \searrow\!\!\!\searrow & \uparrow \\ (\mathrm{QCoh}(Y_2), \mathrm{IndCoh}(Y_2)) & \longrightarrow & (\mathrm{QCoh}(X_2), \mathrm{IndCoh}(X_2)). \end{array} \tag{2.13}$$

We need to show that the 2-morphism in question is an isomorphism.

For the clockwise circuit in (2.13), the corresponding $(\mathrm{QCoh}(X_1), \mathrm{QCoh}(Y_2))$-module is by definition $\mathrm{QCoh}(Y_1)$, equipped with the distinguished object $\mathcal{O}_{Y_1} \in \mathrm{QCoh}(Y_1)$.

For the anti-clockwise circuit in diagram (2.13), the corresponding $(\mathrm{QCoh}(X_1), \mathrm{QCoh}(Y_2))$-module is

$$\mathrm{QCoh}(X_1) \underset{\mathrm{QCoh}(X_2)}{\otimes} \mathrm{QCoh}(Y_2),$$

equipped with the distinguished object $\mathcal{O}_{X_1} \underset{X_1}{\boxtimes} \mathcal{O}_{Y_2}$.

The datum of Φ of the 2-morphism in (2.13) is the canonical functor

$$\mathrm{QCoh}(X_1) \underset{\mathrm{QCoh}(X_2)}{\otimes} \mathrm{QCoh}(Y_2) \to \mathrm{QCoh}(Y_1),$$

which is known to be an equivalence (see, e.g., Chapter 3, Proposition 3.5.3). Under this identification, the datum of ψ of the 2-morphism in (2.13) is the identity isomorphism on $\mathcal{O}_{Y_1}$.

It remains to show that the natural transformation T is an isomorphism.

For a triple

$$\mathcal{E}_1 \in \mathrm{QCoh}(X_1), \quad \mathcal{E}_2 \in \mathrm{QCoh}(Y_2), \quad \mathcal{F} \in \mathrm{IndCoh}(Y_2)$$

and the corresponding object

$$(\mathcal{E}_1 \otimes \mathcal{E}_2) \otimes \mathcal{F} \in \left(\mathrm{QCoh}(X_1) \underset{\mathrm{QCoh}(X_2)}{\otimes} \mathrm{QCoh}(Y_2)\right) \underset{\mathrm{QCoh}(Y_2)}{\otimes} \mathrm{IndCoh}(Y_2),$$

the functor F^t sends it to

$$\mathcal{E}_1 \otimes \left(f_X^! \circ (g_2)_*^{\mathrm{IndCoh}}(\mathcal{E}_2 \otimes \mathcal{F})\right),$$

and the functor $F^s \circ (\Phi \otimes \mathrm{Id}_{\mathbf{C}_1})$ sends it to

$$(g_1)_*^{\mathrm{IndCoh}}\left((g_1^*(\mathcal{E}_1) \otimes f_Y^*(\mathcal{E}_2)) \otimes f_Y^!(\mathcal{F})\right).$$

Under the above identifications, the natutal transformation T acts as follows:

$$(g_1)_*^{\mathrm{IndCoh}}\left((g_1^*(\mathcal{E}_1) \otimes f_Y^*(\mathcal{E}_2)) \otimes f_Y^!(\mathcal{F})\right) \simeq (g_1)_*^{\mathrm{IndCoh}}\left((g_1^*(\mathcal{E}_1) \otimes f_Y^!(\mathcal{E}_2 \otimes \mathcal{F})\right) \simeq$$
$$\simeq \mathcal{E}_1 \otimes (g_1)_*^{\mathrm{IndCoh}}\left(f_Y^!(\mathcal{E}_2 \otimes \mathcal{F})\right) \simeq \mathcal{E}_1 \otimes f_X^!\left((g_2)_*^{\mathrm{IndCoh}}(\mathcal{E}_2 \otimes \mathcal{F})\right),$$

which is an isomorphism, as required.

2.6. Verification of compatibility. In this subsection we will show that the condition of Chapter 7, Sect. 5.2.2 is satisfied for the functor $(\mathrm{QCoh}, \mathrm{IndCoh})_{\mathrm{Sch}_{\mathrm{aft}}}$.

2.6.1. Let

$$\begin{array}{ccc} U_1 & \xrightarrow{j_1} & X_1 \\ {\scriptstyle f_U}\downarrow & & \downarrow{\scriptstyle f_X} \\ U_2 & \xrightarrow{j_2} & X_2 \end{array}$$

be a Cartesian diagram in $\mathrm{Sch}_{\mathrm{aft}}$, with the vertical arrows being proper and the horizontal arrows being open.

According to Chapter 7, Sect. 5.2.2, from the base change isomorphism of Sect. 2.4.5, we obtain a 2-morphism

$$\begin{array}{ccc} (\mathrm{QCoh}(U_1), \mathrm{IndCoh}(U_1)) & \longleftarrow & (\mathrm{QCoh}(X_1), \mathrm{IndCoh}(X_1)) \\ \uparrow & \swarrow\!\!\!\!\swarrow & \uparrow \\ (\mathrm{QCoh}(U_2), \mathrm{IndCoh}(U_2)) & \longleftarrow & (\mathrm{QCoh}(X_2), \mathrm{IndCoh}(X_2)). \end{array} \tag{2.14}$$

We need to show that this morphism is an isomorphism.

2.6.2. By the description of the left and right adjoint functors in Sects. 2.4 and 2.5, the $(\mathrm{QCoh}(U_1), \mathrm{QCoh}(X_2))$-bimodule corresponding to both circuits in the diagram (2.14) is the category $\mathrm{QCoh}(U_1)$, equipped with the distinguished object $\mathcal{O}_{U_1} \in \mathrm{QCoh}(U_1)$.

Under this identification, the data of Φ and γ in the 2-morphism in (2.14) are the identity maps. Hence, it remains to show that the natural transformation T is an isomorphism.

2.6.3. The natural transformation T is a 2-morphism in $\operatorname{QCoh}(U_1)\text{-}\mathbf{mod}^{2\text{-Cat}}$ between two functors

$$\operatorname{QCoh}(U_1) \underset{\operatorname{QCoh}(X_2)}{\otimes} \operatorname{IndCoh}(X_2) \rightrightarrows \operatorname{IndCoh}(U_1).$$

Such functors and natural transformations are in bijection with those in $\operatorname{QCoh}(X_2)\text{-}\mathbf{mod}^{2\text{-Cat}}$

$$\operatorname{IndCoh}(X_2) \rightrightarrows \operatorname{IndCoh}(U_1).$$

Now, the assertion follows from the fact that in the diagram

$$
\begin{array}{ccc}
\operatorname{IndCoh}(U_1) & \xleftarrow{\;j_1^{\operatorname{IndCoh},*}\;} & \operatorname{IndCoh}(X_1) \\
{\scriptstyle f_U^!}\Big\uparrow & \swarrow\!\!\swarrow & \Big\uparrow{\scriptstyle f_X^!} \\
\operatorname{IndCoh}(U_2) & \xleftarrow[\;j_2^{\operatorname{IndCoh},*}\;]{} & \operatorname{IndCoh}(X_2)
\end{array}
$$

the 2-morphism is an isomorphism (which is Chapter 4, Proposition 5.3.4).

3. The multiplicative structure

In this section we will further amplify the functor

$$(\operatorname{QCoh}, \operatorname{IndCoh})_{\operatorname{Corr}(\operatorname{Sch}_{\operatorname{aft}})^{\operatorname{proper}}_{\operatorname{all};\operatorname{all}}} : \operatorname{Corr}(\operatorname{Sch}_{\operatorname{aft}})^{\operatorname{proper}}_{\operatorname{all};\operatorname{all}} \to \operatorname{DGCat}^{\operatorname{Mon+Mod},\operatorname{ext}}_{\operatorname{cont}}.$$

Namely, we will show that it has a natural symmetric monoidal structure.

This will imply certain expected properties of IndCoh, regarded as equipped with an action of QCoh.

3.1. Upgrading to a symmetric monoidal functor.

In this subsection we state the existence of $(\operatorname{QCoh}, \operatorname{IndCoh})_{\operatorname{Corr}(\operatorname{Sch}_{\operatorname{aft}})^{\operatorname{proper}}_{\operatorname{all};\operatorname{all}}}$ as a symmetric monoidal functor.

3.1.1. Recall that the $(\infty, 2)$-category $\operatorname{Corr}(\operatorname{Sch}_{\operatorname{aft}})^{\operatorname{proper}}_{\operatorname{all};\operatorname{all}}$ carries a natural symmetric monoidal functor, which at the level of objects is given by Cartesian product.

Note that the $(\infty, 2)$-category $\operatorname{DGCat}^{\operatorname{Mon+Mod},\operatorname{ext}}_{\operatorname{cont}}$ also carries a symmetric monoidal structure, given by term-wise tensor product.

3.1.2. As in Chapter 5, Theorem 4.1.2, we have:

THEOREM 3.1.3. *The functor*

$$(\operatorname{QCoh}, \operatorname{IndCoh})_{\operatorname{Corr}(\operatorname{Sch}_{\operatorname{aft}})^{\operatorname{proper}}_{\operatorname{all};\operatorname{all}}} : \operatorname{Corr}(\operatorname{Sch}_{\operatorname{aft}})^{\operatorname{proper}}_{\operatorname{all};\operatorname{all}} \to \operatorname{DGCat}^{\operatorname{Mon+Mod},\operatorname{ext}}_{\operatorname{cont}}$$

carries a canonical symmetric monoidal structure.

3.2. Consequences for the !-pullback. In this subsection we will show that the action of QCoh on IndCoh comes from a symmetric monoidal functor

$$\mathrm{QCoh}(-) \to \mathrm{IndCoh}(-),$$

whose formation is compatible with pullbacks (the *-pullback for QCoh and the !-pullback for IndCoh).

3.2.1. Let us restrict the functor $(\mathrm{QCoh}, \mathrm{IndCoh})_{\mathrm{Corr}(\mathrm{Sch}_{\mathrm{aft}})^{\mathrm{proper}}_{\mathrm{all;all}}}$ to

$$(\mathrm{Sch}_{\mathrm{aft}})^{\mathrm{op}} \subset \mathrm{Corr}(\mathrm{Sch}_{\mathrm{aft}})^{\mathrm{proper}}_{\mathrm{all;all}}.$$

We obtain a functor

$$(\mathrm{Sch}_{\mathrm{aft}})^{\mathrm{op}} \to \mathrm{DGCat}^{\mathrm{Mon+Mod,ext}}_{\mathrm{cont}}. \tag{3.1}$$

However, the explicit description of the !-pullback functors in Sects. 2.4 and 2.5 imply that the functor (3.1) factors (automatically canonically) through the 1-fullly faithful functor

$$\mathrm{DGCat}^{\mathrm{Mon+Mod}}_{\mathrm{cont}} \to \mathrm{DGCat}^{\mathrm{Mon+Mod,ext}}_{\mathrm{cont}}.$$

We denote the resulting functor

$$(\mathrm{Sch}_{\mathrm{aft}})^{\mathrm{op}} \to \mathrm{DGCat}^{\mathrm{Mon+Mod}}_{\mathrm{cont}}$$

by $(\mathrm{QCoh}^*, \mathrm{IndCoh}^!)_{\mathrm{Sch}_{\mathrm{aft}}}$.

REMARK 3.2.2. The meaning of the functor $(\mathrm{QCoh}^*, \mathrm{IndCoh}^!)_{\mathrm{Sch}_{\mathrm{aft}}}$ is that it encodes that for a map $f : X \to Y$, the functor $f^! : \mathrm{IndCoh}(Y) \to \mathrm{IndCoh}(X)$ has a natural structure of map of $\mathrm{QCoh}(Y)$-module categories. By contrast with $(\mathrm{QCoh}, \mathrm{IndCoh})_{\mathrm{Corr}(\mathrm{Sch}_{\mathrm{aft}})^{\mathrm{proper}}_{\mathrm{all;all}}}$, we discard the information pertaining to the functor f^{IndCoh}_*.

3.2.3. Taking into account Theorem 3.1.3, we obtain that the functor

$$(\mathrm{QCoh}^*, \mathrm{IndCoh}^!)_{\mathrm{Sch}_{\mathrm{aft}}}$$

has a natural symmetric monoidal structure with respect to the symmetric monoidal structure on $(\mathrm{Sch}_{\mathrm{aft}})^{\mathrm{op}}$, induced by the Cartesian product on $\mathrm{Sch}_{\mathrm{aft}}$ and the symmetric monoidal structure on $\mathrm{DGCat}^{\mathrm{Mon+Mod}}_{\mathrm{cont}}$, given by term-wise tensor product.

Recall now (see Chapter 5, Sect. 4.1.3) that any symmetric monoidal functor from $(\mathrm{Sch}_{\mathrm{aft}})^{\mathrm{op}}$ with values in a symmetric monoidal category naturally upgrades to a functor from $(\mathrm{Sch}_{\mathrm{aft}})^{\mathrm{op}}$ with values in the category of commutative algebras in that symmetric monoidal category.

In particular, we obtain that the functor $(\mathrm{QCoh}^*, \mathrm{IndCoh}^!)_{\mathrm{Sch}_{\mathrm{aft}}}$ naturally extends to a functor

$$(\mathrm{Sch}_{\mathrm{aft}})^{\mathrm{op}} \to \mathrm{ComAlg}\left(\mathrm{DGCat}^{\mathrm{Mon+Mod}}_{\mathrm{cont}}\right). \tag{3.2}$$

The composition of the functor (3.2) with the forgetful functor

$$\mathrm{ComAlg}\left(\mathrm{DGCat}^{\mathrm{Mon+Mod}}_{\mathrm{cont}}\right) \to \mathrm{ComAlg}\left(\mathrm{DGCat}_{\mathrm{cont}}\right) = \mathrm{DGCat}^{\mathrm{SymMon}}_{\mathrm{cont}}$$

is the functor of Chapter 5, Formula (4.1).

3.2.4. Note that the category $\mathrm{ComAlg}\left(\mathrm{DGCat}_{\mathrm{cont}}^{\mathrm{Mon+Mod}}\right)$ can be identified with the category

$$\mathrm{Funct}([1], \mathrm{DGCat}_{\mathrm{cont}}^{\mathrm{SymMon}}),$$

i.e., the category of triples

$$(\mathbf{O}, \mathbf{O}', \alpha : \mathbf{O} \to \mathbf{O}'),$$

where $\mathbf{O}$ and $\mathbf{O}'$ are symmetric monoidal categories, and α is a symmetric monoidal functor.

Hence, the content of the functor (3.2) is that the assignment

$$X \rightsquigarrow (\mathrm{QCoh}(X), \mathrm{IndCoh}(X))$$

is the functor from the category (opposite to that) of schemes to the category of pairs of symmetric monoidal DG categories, where:

- $\mathrm{QCoh}(X)$ is a symmetric monoidal DG category via the usual $*$-tensor product operation;
- $\mathrm{IndCoh}(X)$ is a symmetric monoidal DG category via the usual !-tensor product operation (see Chapter 5, Sect. 4.1.3)
- The symmetric monoidal functor $\mathrm{QCoh}(X) \to \mathrm{IndCoh}(X)$ is given by the action of $\mathrm{QCoh}(X)$ on the unit object in $\mathrm{IndCoh}(X)$, when $\mathrm{IndCoh}(X)$ is regarded as a $\mathrm{QCoh}(X)$-module category.

3.2.5. As a consequence, we obtain a natural transformation between the functors

$$(\mathrm{QCoh}^*, \mathrm{QCoh}^*)_{\mathrm{Sch}_{\mathrm{aft}}} \Rightarrow (\mathrm{QCoh}^*, \mathrm{IndCoh}^!)_{\mathrm{Sch}_{\mathrm{aft}}}, \quad (\mathrm{Sch}_{\mathrm{aft}})^{\mathrm{op}} \to \mathrm{DGCat}_{\mathrm{cont}}^{\mathrm{Mon+Mod}}.$$

In particular, we obtain a natural transformation between the functors

$$\mathrm{QCoh}^*_{\mathrm{Sch}_{\mathrm{aft}}} \Rightarrow \mathrm{IndCoh}^!_{\mathrm{Sch}_{\mathrm{aft}}}, \quad (\mathrm{Sch}_{\mathrm{aft}})^{\mathrm{op}} \to \mathrm{DGCat}_{\mathrm{cont}}.$$

We denote the latter transformation by $\Upsilon_{\mathrm{Sch}_{\mathrm{aft}}}$ and the former by

$$(\mathrm{Id}, \Upsilon)_{\mathrm{Sch}_{\mathrm{aft}}}.$$

For an individual scheme X, we will denote the corresponding fonctor

$$\mathrm{QCoh}(X) \to \mathrm{IndCoh}(X)$$

by Υ_X.

By construction, this functor is given by

$$\mathcal{E} \mapsto \mathcal{E} \otimes \omega_X, \quad \mathcal{E} \in \mathrm{QCoh}(X),$$

where $\otimes$ is the action of $\mathrm{QCoh}(X)$ on IndCoh_X, and ω_X is the dualizing object of $\mathrm{IndCoh}(X)$ (the !-pullback of k under $X \to \mathrm{pt}$, or equivalently, the unit for the !-symmetric monoidal structure on $\mathrm{IndCoh}(X)$).

3.3. Extension to prestacks. We will now use the theory developed above, to show that for a prestack $\mathcal{X}$, the category $\mathrm{IndCoh}(\mathcal{X})$ acquires a natural action of the monoidal category $\mathrm{QCoh}(\mathcal{X})$.

3.3.1. We consider again the functor

$$(\mathrm{QCoh}^*, \mathrm{IndCoh}^!)_{\mathrm{Sch}_{\mathrm{aft}}} : (\mathrm{Sch}_{\mathrm{aft}})^{\mathrm{op}} \to \mathrm{ComAlg}\left(\mathrm{DGCat}_{\mathrm{cont}}^{\mathrm{Mon+Mod}}\right)$$

of (3.2) and apply the right Kan extension along

$$(\mathrm{Sch}_{\mathrm{aft}})^{\mathrm{op}} \to (\mathrm{PreStk}_{\mathrm{laft}})^{\mathrm{op}}.$$

Denote the resulting functor by

$$(\widetilde{\mathrm{QCoh}}^*, \mathrm{IndCoh}^!)_{\mathrm{Sch}_{\mathrm{aft}}} : (\mathrm{PreStk}_{\mathrm{laft}})^{\mathrm{op}} \to \mathrm{ComAlg}\left(\mathrm{DGCat}_{\mathrm{cont}}^{\mathrm{Mon+Mod}}\right).$$

The value of this funcor on a given prestack $\mathcal{Y}$ is $(\widetilde{\mathrm{QCoh}}(\mathcal{Y}), \mathrm{IndCoh}(\mathcal{Y}))$, where

$$\widetilde{\mathrm{QCoh}}^* : (\mathrm{PreStk}_{\mathrm{laft}})^{\mathrm{op}} \to \mathrm{DGCat}_{\mathrm{cont}}^{\mathrm{SymMon}}$$

is right Kan extension of $\mathrm{QCoh}^*_{\mathrm{Sch}_{\mathrm{aft}}}$ along $(\mathrm{Sch}_{\mathrm{aft}})^{\mathrm{op}} \to (\mathrm{PreStk}_{\mathrm{laft}})^{\mathrm{op}}$, i.e.,

$$\widetilde{\mathrm{QCoh}}(\mathcal{Y}) := \lim_{X \in ((\mathrm{Sch}_{\mathrm{aft}})_{/\mathcal{Y}})^{\mathrm{op}}} \mathrm{QCoh}(X).$$

REMARK 3.3.2. The difference between $\widetilde{\mathrm{QCoh}}(\mathcal{Y})$ and $\mathrm{QCoh}(\mathcal{Y})$ is that in the latter we take the limit of $\mathrm{QCoh}(X)$ over all schemes X mapping to $\mathcal{Y}$, and in the former only $X \in \mathrm{Sch}_{\mathrm{aft}}$. Under some (mild) conditions on $\mathcal{Y}$, the restriction functor $\mathrm{QCoh}(\mathcal{Y}) \to \widetilde{\mathrm{QCoh}}(\mathcal{Y})$ is an equivalence. For example, this happens if $\mathcal{Y}$ admits *deformation theory*, see Volume II, Chapter 1, Theorem 9.1.4.

3.3.3. Note that there is a canonically defined natural transformation of functors

$$\mathrm{QCoh}_{\mathrm{PreStk}_{\mathrm{laft}}} \Rightarrow \widetilde{\mathrm{QCoh}}_{\mathrm{PreStk}_{\mathrm{laft}}}, (\mathrm{PreStk}_{\mathrm{laft}})^{\mathrm{op}} \to \mathrm{DGCat}_{\mathrm{cont}}^{\mathrm{SymMon}}.$$

Composing with the functor $(\widetilde{\mathrm{QCoh}}^*, \mathrm{IndCoh}^!)_{\mathrm{PreStk}_{\mathrm{laft}}}$, we obtain a functor

$$(\mathrm{QCoh}^*, \mathrm{IndCoh}^!)_{\mathrm{PreStk}_{\mathrm{laft}}} : (\mathrm{PreStk}_{\mathrm{laft}})^{\mathrm{op}} \to \mathrm{ComAlg}\left(\mathrm{DGCat}_{\mathrm{cont}}^{\mathrm{Mon+Mod}}\right).$$

The value of the latter functor on $\mathcal{Y} \in \mathrm{PreStk}_{\mathrm{laft}}$ is now

$$(\mathrm{QCoh}(\mathcal{Y}), \mathrm{IndCoh}(\mathcal{Y})).$$

3.3.4. The content of the functor $(\mathrm{QCoh}^*, \mathrm{IndCoh}^!)_{\mathrm{PreStk}_{\mathrm{laft}}}$ is the natural transformation

$$\Upsilon_{\mathrm{PreStk}_{\mathrm{laft}}} : \mathrm{QCoh}^*_{\mathrm{PreStk}_{\mathrm{laft}}} \Rightarrow \mathrm{IndCoh}^!_{\mathrm{PreStk}_{\mathrm{laft}}},$$

as functors

$$\mathrm{PreStk}_{\mathrm{laft}} \rightrightarrows \mathrm{DGCat}_{\mathrm{cont}}^{\mathrm{SymMon}}.$$

For an individual $\mathcal{Y} \in \mathrm{PreStk}_{\mathrm{laft}}$ we shall denote the resulting functor

$$\mathrm{QCoh}(\mathcal{Y}) \to \mathrm{IndCoh}(\mathcal{Y})$$

by $\Upsilon_{\mathcal{Y}}$.

3.3.5. Applying the forgetful functor $\mathrm{ComAlg}\left(\mathrm{DGCat}_{\mathrm{cont}}^{\mathrm{Mon+Mod}}\right) \to \mathrm{DGCat}_{\mathrm{cont}}^{\mathrm{Mon+Mod}}$, we can view $(\mathrm{QCoh}^*, \mathrm{IndCoh}^!)_{\mathrm{PreStk_{laft}}}$ as a functor

$$(\mathrm{PreStk_{laft}})^{\mathrm{op}} \to \mathrm{DGCat}_{\mathrm{cont}}^{\mathrm{Mon+Mod}}.$$

I.e., for $\mathcal{Y} \in \mathrm{PreStk_{laft}}$, the DG category $\mathrm{IndCoh}(\mathcal{Y})$ acquires a structure of $\mathrm{QCoh}(\mathcal{Y})$-module, functorially with respect to the !-pullback on IndCoh and *-pullback on QCoh.

The functor $\Upsilon_{\mathcal{Y}}$ is given by the monoidal action of $\mathrm{QCoh}(\mathcal{Y})$ on the object $\omega_{\mathcal{Y}} \in \mathrm{IndCoh}(\mathcal{Y})$.

3.3.6. The following assertion is often useful:

LEMMA 3.3.7. *For any $\mathcal{Y} \in \mathrm{PreStk_{laft}}$, the functor*

$$\Upsilon_{\mathcal{Y}} : \mathrm{QCoh}(\mathcal{Y})^{\mathrm{perf}} \to \mathrm{IndCoh}(\mathcal{Y})$$

is fully faithful and the essential image consists of objects dualizable with respect to the $\overset{!}{\otimes}$ symmetric monoidal structure.

PROOF. Since both functors $\mathrm{QCoh}(-)^{\mathrm{perf}}$ and $\mathrm{IndCoh}(-)$ are convergent (by Chapter 3, Proposition 3.6.10 and Chapter 4, Proposition 6.4.3, respectively), the assertion reduces to the case when $\mathcal{Z} = S \in {}^{<\infty}\mathrm{Sch}_{\mathrm{ft}}^{\mathrm{aff}}$.

In this case, the functor Υ_S is fully faithful on all of $\mathrm{QCoh}(S)$.

Let $\mathcal{F} \in \mathrm{IndCoh}(S)$ be a dualizable object. Since the unit object $\omega_S \in \mathrm{IndCoh}(S)$ is compact, we obtain that $\mathcal{F}$ is compact, i.e., it belongs to $\mathrm{Coh}(S)$.

Consider $\mathcal{E} := \mathbb{D}_S^{\mathrm{Serre}}(\mathcal{F}) \in \mathrm{Coh}(S)$. It suffices to show that $\mathcal{E} \in \mathrm{QCoh}(S)^{\mathrm{perf}} \subset \mathrm{Coh}(S)$. For that it suffices to show that all the *-fibers of $\mathcal{E}$ are finite-dimensional.

But the *-fibers of $\mathcal{E}$ are the duals of the !-fibers of $\mathcal{F}$, and the latter are finite-dimensional by the dualizability hypothesis: indeed, taking the !-fiber is a symmetric monoidal functor from $(\mathrm{IndCoh}(Z), \overset{!}{\otimes})$ to Vect.

□

4. Duality

In this section we will study the interaction of the Serre duality on IndCoh with the naive (i.e., usual) duality on QCoh.

4.1. Duality on the category $\mathrm{DGCat}_{\mathrm{cont}}^{\mathrm{Mon+Mod,ext}}$. In this subsection we will explicitly describe dualizable objects in the symmetric monoidal category

$$\mathrm{DGCat}_{\mathrm{cont}}^{\mathrm{Mon+Mod,ext}},$$

and how the duality involution on dualizable objects looks like.

4.1.1. Let $(\mathbf{O}, \mathbf{C})$ be an object of $\mathrm{DGCat}_{\mathrm{cont}}^{\mathrm{Mon+Mod,ext}}$.

The forgetful symmetric monoidal functor

$$\mathrm{DGCat}_{\mathrm{cont}}^{\mathrm{Mon+Mod,ext}} \to \mathrm{DGCat}_{\mathrm{cont}}$$

implies that if $(\mathbf{O}, \mathbf{C})$ is dualizable with respect to the symmetric monoidal structure on $\mathrm{DGCat}_{\mathrm{cont}}^{\mathrm{Mon+Mod,ext}}$, then $\mathbf{C}$ is dualizable as a plain DG category.

4.1.2. Vice versa, we claim that if $\mathbf{C}$ is dualizable as a DG category, then $(\mathbf{O},\mathbf{C})$ is dualizable in $\mathrm{DGCat}_{\mathrm{cont}}^{\mathrm{Mon+Mod,ext}}$, with the dual being $(\mathbf{O}^{\mathrm{rev\text{-}mult}},\mathbf{C}^\vee)$, where $\mathbf{O}^{\mathrm{rev\text{-}mult}}$ is the monoidal category, obtained from $\mathbf{O}$ by reveresing the multiplication.

Namely, the unit

$$(\mathrm{Vect},\mathrm{Vect}) \to (\mathbf{O},\mathbf{C}) \otimes (\mathbf{O}^{\mathrm{rev\text{-}mult}},\mathbf{C}^\vee) = (\mathbf{O}\otimes\mathbf{O}^{\mathrm{rev\text{-}mult}},\mathbf{C}\otimes\mathbf{C}^\vee)$$

is given by the data $(\mathbf{M},F,\mathbf{1}_{\mathbf{M}})$, where:

- $\mathbf{M}=\mathbf{O}$, regarded as a $\mathbf{O}\otimes\mathbf{O}^{\mathrm{rev\text{-}mult}}$-module category;
- F is the functor of $\mathbf{O}\to\mathbf{C}\otimes\mathbf{C}^\vee$, given by the action of $\mathbf{O}$ on $\mathbf{C}$;
- $\mathbf{1}_{\mathbf{M}}$ is the unit object of $\mathbf{O}$.

The co-unit

$$(\mathbf{O}^{\mathrm{rev\text{-}mult}}\otimes\mathbf{O},\mathbf{C}^\vee\otimes\mathbf{C}) = (\mathbf{O}^{\mathrm{rev\text{-}mult}},\mathbf{C}^\vee)\otimes(\mathbf{O},\mathbf{C}) \to (\mathrm{Vect},\mathrm{Vect})$$

is given by the data $(\mathbf{M},F,\mathbf{1}_{\mathbf{M}})$, where:

- $\mathbf{M}=\mathbf{O}$, regarded as a *right* $\mathbf{O}^{\mathrm{rev\text{-}mult}}\otimes\mathbf{O}$-module category;
- F is the functor
$$\mathbf{O}\underset{\mathbf{O}^{\mathrm{rev\text{-}mult}}\otimes\mathbf{O}}{\otimes}(\mathbf{C}^\vee\otimes\mathbf{C}) \simeq \mathbf{C}^\vee\underset{\mathbf{O}}{\otimes}\mathbf{C}\to\mathrm{Vect},$$
where the last arrow is the canonical pairing;
- $\mathbf{1}_{\mathbf{M}}$ is the unit object of $\mathbf{O}$.

4.1.3. Let

$$(\mathrm{DGCat}_{\mathrm{cont}}^{\mathrm{Mon+Mod}})^{\mathrm{dualizable}}\subset\mathrm{DGCat}_{\mathrm{cont}}^{\mathrm{Mon+Mod}}$$

and

$$(\mathrm{DGCat}_{\mathrm{cont}}^{\mathrm{Mon^{op}+Mod}})^{\mathrm{dualizable}}\subset\mathrm{DGCat}_{\mathrm{cont}}^{\mathrm{Mon^{op}+Mod}}$$

be the full subcategories corresponding to the pairs $(\mathbf{O},\mathbf{C})$ in which $\mathbf{C}$ is dualizable as a plain DG category.

Note that we have a canonically defined functor

$$((\mathrm{DGCat}_{\mathrm{cont}}^{\mathrm{Mon+Mod}})^{\mathrm{dualizable}})^{\mathrm{op}}\to(\mathrm{DGCat}_{\mathrm{cont}}^{\mathrm{Mon^{op}+Mod}})^{\mathrm{dualizable}},\quad (\mathbf{O},\mathbf{C})\mapsto(\mathbf{O}^{\mathrm{rev\text{-}mult}},\mathbf{C}^\vee). \tag{4.1}$$

By Sect. 4.1.2, the functors

$$\mathrm{DGCat}_{\mathrm{cont}}^{\mathrm{Mon+Mod}}\to\mathrm{DGCat}_{\mathrm{cont}}^{\mathrm{Mon+Mod,ext}}\leftarrow\mathrm{DGCat}_{\mathrm{cont}}^{\mathrm{Mon^{op}+Mod}}$$

send the subcategories

$$(\mathrm{DGCat}_{\mathrm{cont}}^{\mathrm{Mon+Mod}})^{\mathrm{dualizable}}\text{ and }(\mathrm{DGCat}_{\mathrm{cont}}^{\mathrm{Mon^{op}+Mod}})^{\mathrm{dualizable}}$$

to $(\mathrm{DGCat}_{\mathrm{cont}}^{\mathrm{Mon+Mod,ext}})^{\mathrm{dualizable}}$.

Moreover the construction of Sect. 4.1.2 can be upgraded to the following statement:

LEMMA 4.1.4. *The following square*

$$\begin{array}{ccc} ((\mathrm{DGCat}^{\mathrm{Mon+Mod}}_{\mathrm{cont}})^{\mathrm{dualizable}})^{\mathrm{op}} & \longrightarrow & ((\mathrm{DGCat}^{\mathrm{Mon+Mod,ext}}_{\mathrm{cont}})^{\mathrm{dualizable}})^{\mathrm{op}} \\ {\scriptstyle (4.1)}\downarrow & & \downarrow{\scriptstyle \text{dualization in } \mathrm{DGCat}^{\mathrm{Mon+Mod,ext}}_{\mathrm{cont}}} \\ (\mathrm{DGCat}^{\mathrm{Mon}^{\mathrm{op}}+\mathrm{Mod}}_{\mathrm{cont}})^{\mathrm{dualizable}} & \longrightarrow & (\mathrm{DGCat}^{\mathrm{Mon+Mod,ext}}_{\mathrm{cont}})^{\mathrm{dualizable}} \end{array}$$

canonically commutes.

4.2. The linearity structure on Serre duality. In this subsection we study how Serre duality on IndCoh is compatible with the action of QCoh.

4.2.1. Consider again the symmetric monoidal functor

$$(\mathrm{QCoh}, \mathrm{IndCoh})_{\mathrm{Corr}(\mathrm{Sch}_{\mathrm{aft}})^{\mathrm{proper}}_{\mathrm{all;all}}} : \mathrm{Corr}(\mathrm{Sch}_{\mathrm{aft}})^{\mathrm{proper}}_{\mathrm{all;all}} \to \mathrm{DGCat}^{\mathrm{Mon+Mod,ext}}_{\mathrm{cont}}$$

of Theorem 3.1.3. Consider its restriction

$$(\mathrm{QCoh}, \mathrm{IndCoh})_{\mathrm{Corr}(\mathrm{Sch}_{\mathrm{aft}})_{\mathrm{all;all}}} : \mathrm{Corr}(\mathrm{Sch}_{\mathrm{aft}})_{\mathrm{all;all}} \to \mathrm{DGCat}^{\mathrm{Mon+Mod,ext}}_{\mathrm{cont}}.$$

As in Chapter 5, Theorem 4.2.3, we obtain:

COROLLARY 4.2.2. *The following diagram of functors canonically commutes:*

$$\begin{array}{ccc} (\mathrm{Corr}(\mathrm{Sch}_{\mathrm{aft}})_{\mathrm{all;all}})^{\mathrm{op}} & \xrightarrow{((\mathrm{QCoh},\mathrm{IndCoh})_{\mathrm{Corr}(\mathrm{Sch}_{\mathrm{aft}})_{\mathrm{all;all}}})^{\mathrm{op}}} & ((\mathrm{DGCat}^{\mathrm{Mon+Mod,ext}}_{\mathrm{cont}})^{\mathrm{dualizable}})^{\mathrm{op}} \\ {\scriptstyle \varpi}\downarrow & & \downarrow{\scriptstyle \text{dualization in } \mathrm{DGCat}^{\mathrm{Mon+Mod,ext}}_{\mathrm{cont}}} \\ \mathrm{Corr}(\mathrm{Sch}_{\mathrm{aft}})_{\mathrm{all;all}} & \xrightarrow{(\mathrm{QCoh},\mathrm{IndCoh})_{\mathrm{Corr}(\mathrm{Sch}_{\mathrm{aft}})_{\mathrm{all;all}}}} & (\mathrm{DGCat}^{\mathrm{Mon+Mod,ext}}_{\mathrm{cont}})^{\mathrm{dualizable}}, \end{array}$$

where ϖ is as Chapter 5, Sect. 4.2.2.

4.2.3. Combining Corollary 4.2.2 with Lemma 4.1.4, we obtain:

COROLLARY 4.2.4. *The following diagram canonically commutes*

$$\begin{array}{ccc} \mathrm{Sch}_{\mathrm{aft}} & \xrightarrow{(\mathrm{QCoh}^*,\mathrm{IndCoh}_*)_{\mathrm{Sch}_{\mathrm{aft}}}} & (\mathrm{DGCat}^{\mathrm{Mon}^{\mathrm{op}}+\mathrm{Mod}}_{\mathrm{cont}})^{\mathrm{dualizable}} \\ {\scriptstyle \mathrm{Id}}\downarrow & & \downarrow{\scriptstyle (4.1)} \\ \mathrm{Sch}_{\mathrm{aft}} & \xrightarrow{((\mathrm{QCoh}^*,\mathrm{IndCoh}^!)_{\mathrm{Sch}_{\mathrm{aft}}})^{\mathrm{op}}} & ((\mathrm{DGCat}^{\mathrm{Mon+Mod}}_{\mathrm{cont}})^{\mathrm{dualizable}})^{\mathrm{op}}. \end{array}$$

The content of Corollary 4.2.4 is that for an individual $X \in \mathrm{Sch}_{\mathrm{aft}}$, the Serre duality equivalence

$$\mathbf{D}^{\mathrm{Serre}}_X : \mathrm{IndCoh}(X)^\vee \simeq \mathrm{IndCoh}(X)$$

is compatible with the action of $\mathrm{QCoh}(X)$.

Furthermore, for a morphism $f : X \to Y$, the identification

$$f^! \simeq (f^{\mathrm{IndCoh}}_*)^\vee$$

is also compatible with the action of $\mathrm{QCoh}(Y)$.

4.3. Digression: QCoh **as a functor out of the category of correspondences.** The contents of this subsection are more or less tautological: they encode that the operation of direct image on QCoh is compatible with the action of QCoh on itself by tensor products.

4.3.1. The goal of this subsection is to prove the following analog of Corollary 4.2.4 for the pair of functors $(\text{QCoh}^*, \text{QCoh}^*)_{\text{Sch}_{\text{aft}}}$ and $(\text{QCoh}^*, \text{QCoh}_*)_{\text{Sch}_{\text{aft}}}$ of (2.3) and (2.6), respectively:

PROPOSITION 4.3.2. *The following diagram canonically commutes*

$$\begin{array}{ccc} \text{Sch}_{\text{aft}} & \xrightarrow{(\text{QCoh}^*, \text{QCoh}_*)_{\text{Sch}_{\text{aft}}}} & (\text{DGCat}_{\text{cont}}^{\text{Mon}^{\text{op}}+\text{Mod}})^{\text{dualizable}} \\ {\scriptstyle \text{Id}}\downarrow & & \downarrow{\scriptstyle (4.1)} \\ \text{Sch}_{\text{aft}} & \xrightarrow{((\text{QCoh}^*, \text{QCoh}^*)_{\text{Sch}_{\text{aft}}})^{\text{op}}} & ((\text{DGCat}_{\text{cont}}^{\text{Mon}+\text{Mod}})^{\text{dualizable}})^{\text{op}}. \end{array}$$

The content of Proposition 4.3.2 is that for an individual $X \in \text{Sch}_{\text{aft}}$, there is a canonical duality equivalence

$$\mathbf{D}_X^{\text{naive}} : \text{QCoh}(X)^\vee \simeq \text{QCoh}(X), \tag{4.2}$$

which is compatible with the action of $\text{QCoh}(X)$.

Furthermore, for a morphism $f : X \to Y$, we have an identification

$$f^* \simeq (f_*)^\vee,$$

which is also compatible with the action of $\text{QCoh}(Y)$.

4.3.3. It will follow from the construction below that the duality (4.2) is that given by the unit

$$\text{Vect} \overset{k \to \mathcal{O}_X}{\longrightarrow} \text{QCoh}(X) \overset{(\Delta_X)_*}{\longrightarrow} \text{QCoh}(X \times X) \simeq \text{QCoh}(X) \otimes \text{QCoh}(X),$$

and the co-unit

$$\text{QCoh}(X) \otimes \text{QCoh}(X) \simeq \text{QCoh}(X \times X) \overset{\Delta_X^*}{\longrightarrow} \text{QCoh}(X) \overset{\Gamma(X,-)}{\longrightarrow} k.$$

Equivalently, the duality (4.2) is induced by the anti-self equivalence of

$$\text{QCoh}(X)^c \simeq \text{QCoh}(X)^{\text{perf}},$$

given by

$$\mathcal{E} \mapsto \mathcal{E}^\vee := \underline{\text{Hom}}_{\mathcal{O}_X}(\mathcal{E}, \mathcal{O}_X).$$

4.3.4. As in the case of Corollary 4.2.4, in order to prove Proposition 4.3.2, it is sufficient to construct a symmetric monoidal functor

$$(\text{QCoh}, \text{QCoh})_{\text{Corr}(\text{Sch}_{\text{aft}})_{\text{all;all}}} : \text{Corr}(\text{Sch}_{\text{aft}})_{\text{all;all}} \to \text{DGCat}_{\text{cont}}^{\text{Mon}+\text{Mod},\text{ext}}. \tag{4.3}$$

We will obtain $(\text{QCoh}, \text{QCoh})_{\text{Corr}(\text{Sch}_{\text{aft}})_{\text{all;all}}}$ by restriction from a symmetric monoidal functor

$$(\text{QCoh}, \text{QCoh})_{\text{Corr}(\text{Sch}_{\text{aft}})^{\text{all}}_{\text{all;all}}} : \text{Corr}(\text{Sch}_{\text{aft}})^{\text{all}}_{\text{all;all}} \to (\text{DGCat}_{\text{cont}}^{\text{Mon}+\text{Mod},\text{ext}})^{2\text{-op}}. \tag{4.4}$$

4.3.5. To construct (4.4), we start with the functor

$$(\mathrm{QCoh}^*, \mathrm{QCoh}^*)_{\mathrm{Sch}_{\mathrm{aft}}} : (\mathrm{Sch}_{\mathrm{aft}})^{\mathrm{op}} \to \mathrm{DGCat}_{\mathrm{cont}}^{\mathrm{Mon+Mod}},$$

of (2.3) and follow it by the functor

$$\mathrm{DGCat}_{\mathrm{cont}}^{\mathrm{Mon+Mod}} \to \mathrm{DGCat}_{\mathrm{cont}}^{\mathrm{Mon+Mod,ext}}$$

of Sect. 1.3.4.

We obtain a functor

$$(\mathrm{Sch}_{\mathrm{aft}})^{\mathrm{op}} \to \mathrm{DGCat}_{\mathrm{cont}}^{\mathrm{Mon+Mod,ext}}. \tag{4.5}$$

4.3.6. We now claim that the functor (4.5), viewed as a functor

$$(\mathrm{Sch}_{\mathrm{aft}})^{\mathrm{op}} \to (\mathrm{DGCat}_{\mathrm{cont}}^{\mathrm{Mon+Mod,ext}})^{2\text{-op}}$$

satisfies the right Beck-Chevalley condition.

This is proved in the same was as in Sect. 2.4.

Now, applying Chapter 7, Theorem 3.2.2(b), we obtain that the functor (4.5) uniquely gives rise to the sought-for functor (4.4).

4.4. Compatibility with the functor Ψ. In this subsection we will use the theory developed above to show that the dual of the functor Υ identifies with the functor Ψ of Chapter 4, Sect. 1.1.2.

4.4.1. Recall (see Corollary 2.2.6) that the functor

$$(\mathrm{QCoh}^*, \mathrm{IndCoh}_*)_{\mathrm{Sch}_{\mathrm{aft}}} : \mathrm{Sch}_{\mathrm{aft}} \to \mathrm{DGCat}_{\mathrm{cont}}^{\mathrm{Mon}^{\mathrm{op}}+\mathrm{Mod}}$$

was constructed in such a way that it was equipped with a natural transformation

$$(\mathrm{Id}, \Psi)_{\mathrm{Sch}_{\mathrm{aft}}} : (\mathrm{QCoh}^*, \mathrm{IndCoh}_*)_{\mathrm{Sch}_{\mathrm{aft}}} \Rightarrow (\mathrm{QCoh}^*, \mathrm{QCoh}_*)_{\mathrm{Sch}_{\mathrm{aft}}}.$$

Applying the functor (4.1), and taking into account Corollary 4.2.4 and Proposition 4.3.2, from the natural transformation $(\mathrm{Id}, \Psi)_{\mathrm{Sch}_{\mathrm{aft}}}$, we obtain a natural transformation

$$(\mathrm{Id}, \Psi^\vee)_{\mathrm{Sch}_{\mathrm{aft}}} : (\mathrm{QCoh}^*, \mathrm{QCoh}^*)_{\mathrm{Sch}_{\mathrm{aft}}} \to (\mathrm{QCoh}^*, \mathrm{IndCoh}^!)_{\mathrm{Sch}_{\mathrm{aft}}}$$

as functors

$$(\mathrm{Sch}_{\mathrm{aft}})^{\mathrm{op}} \to \mathrm{DGCat}_{\mathrm{cont}}^{\mathrm{Mon+Mod}}.$$

For an individual $X \in \mathrm{Sch}_{\mathrm{aft}}$, we let the resulting functor

$$\mathrm{QCoh}(X) \to \mathrm{IndCoh}(X)$$

be denoted by $\Psi_X^\vee$.

4.4.2. The goal of this subsection is to prove the following:

THEOREM 4.4.3. *There is a canonical isomomorphism of natural transformations*

$$(\mathrm{Id}, \Psi^\vee)_{\mathrm{Sch_{aft}}} \simeq (\mathrm{Id}, \Upsilon)_{\mathrm{Sch_{aft}}},$$

where $(\mathrm{Id}, \Upsilon)_{\mathrm{Sch_{aft}}}$ *is as in Sect.* 3.2.5.

The content of this theorem is that for an individual $X \in \mathrm{Sch_{aft}}$, with respect to the identifications

$$\mathbf{D}_X^{\mathrm{Serre}} : \mathrm{IndCoh}(X)^\vee \simeq \mathrm{IndCoh}(X) \text{ and } \mathbf{D}_X^{\mathrm{naive}} : \mathrm{QCoh}(X)^\vee \simeq \mathrm{QCoh}(X),$$

the dual of the functor

$$\Psi_X : \mathrm{IndCoh}(X) \to \mathrm{QCoh}(X)$$

is the functor

$$\Upsilon_X : \mathrm{QCoh}(X) \to \mathrm{IndCoh}(X), \quad \mathcal{E} \mapsto \mathcal{E} \otimes \omega_X.$$

Furthermore, these identifications of functors are compatible with respect to maps $f : X \to Y$.

4.4.4. *Proof of Theorem* 4.4.3. Since the functor $(\mathrm{QCoh}^*, \mathrm{QCoh}^*)_{\mathrm{Sch_{aft}}}$ corresponds to the 'free module on one generator', a datum of a natural transformation out of it to some other functor

$$\mathsf{F} : (\mathrm{Sch_{aft}})^{\mathrm{op}} \to \mathrm{DGCat}_{\mathrm{cont}}^{\mathrm{Mon+Mod}},$$

whose composition with $\mathrm{DGCat}_{\mathrm{cont}}^{\mathrm{Mon+Mod}} \to \mathrm{DGCat}_{\mathrm{cont}}^{\mathrm{Mon}}$ is the functor $\mathrm{QCoh}^*_{\mathrm{Sch_{aft}}}$, is equivalent to the datum of a 1-morphism

$$(\mathrm{QCoh}^*, \mathrm{QCoh}^*)_{\mathrm{Sch_{aft}}}(\mathrm{pt}) \to \mathsf{F}(\mathrm{pt}).$$

I.e., in order to prove the theorem, it is sufficient to perform the identification

$$\Psi_X^\vee \simeq \Upsilon_X$$

for $X = \mathrm{pt}$. However, the latter is evident. □

Part III

Categories of Correspondences

Introduction

1. Why correspondences?

This part introduces one of the two main innovations in this book–the $(\infty, 2)$-category of correspondences as a way to encode bi-variant functors and the six functor formalism. This idea was suggested to us by J. Lurie.

1.1. Let us start with a category $\mathbf{C}$ (with finite limits), equipped with two classes of morphisms *vert* and *horiz* (both closed under composition). The category of correspondences is designed to perform the following function.

Suppose we want to encode a *bi-variant* functor Φ from $\mathbf{C}$ to some target $(\infty, 1)$-category $\mathbb{S}$. I.e., to $\mathbf{c} \in \mathbf{C}$ we assign $\Phi(\mathbf{c}) \in \mathbb{S}$, and to a 1-morphism $\mathbf{c}_1 \overset{\gamma}{\to} \mathbf{c}_2$ in $\mathbf{C}$, we assign a 1-morphism

$$\Phi(\gamma) : \Phi(\mathbf{c}_1) \to \Phi(\mathbf{c}_2) \text{ if } \gamma \in \mathit{vert}$$

and a 1-morphism

$$\Phi^!(\gamma) : \Phi(\mathbf{c}_2) \to \Phi(\mathbf{c}_1) \text{ if } \gamma \in \mathit{horiz},$$

equipped with the following pieces of structure:

(1) Compatibility of both $\Phi(-)$ and $\Phi^!(-)$ with compositions of 1-morphisms in $\mathbf{C}$;

(2) For a Cartesian square

$$\begin{array}{ccc} \mathbf{c}_{0,1} & \xrightarrow{\alpha_0} & \mathbf{c}_{0,0} \\ {\scriptstyle \beta_1}\downarrow & & \downarrow{\scriptstyle \beta_0} \\ \mathbf{c}_{1,1} & \xrightarrow{\alpha_1} & \mathbf{c}_{1,0} \end{array}$$

with vertical arrows in *vert* and horizontal arrows in *horiz*, we are supposed to be *given* an identification (called *base change isomorhism*)

$$\Phi(\beta_1) \circ \Phi^!(\alpha_0) \simeq \Phi^!(\alpha_1) \circ \Phi(\beta_0).$$

The above pieces of data must satisfy a *homotopy-coherent system of compatibilities*. The partial list consists of the following:

- The data making Φ into a functor $\mathbf{C}_{vert} \to \mathbb{S}$, and the data making $\Phi^!$ into a functor $(\mathbf{C}_{horiz})^{\mathrm{op}} \to \mathbb{S}$.
- The compatibility of base-change isomorphisms with compositions;

However, the above is really only the beginning of an infinite tail of compatibilities, as it always happens in higher category theory.

So, if we want a workable theory, we need to find a convenient way to package this information, preferably in terms of one of the existing packages, such as the notion of functor between two given $(\infty,1)$-categories.

The category $\operatorname{Corr}(\mathbf{C})_{vert;horiz}$ allows us to do just that. Namely, we show (Chapter 7, Theorem 2.1.3) that the datum of a functor as above is equivalent to the datum of a functor

$$\operatorname{Corr}(\mathbf{C})_{vert;horiz} \to \mathbb{S}.$$

1.2. The idea of $\operatorname{Corr}(\mathbf{C})_{vert;horiz}$ is very simple. Its objects are the same as objects of $\mathbf{C}$. But its 1-morphisms are diagrams

$$\begin{array}{ccc} \mathbf{c}_{0,1} & \longrightarrow & \mathbf{c}_0 \\ \downarrow & & \\ \mathbf{c}_1 & & \end{array} \tag{1.1}$$

with the vertical arrow in *vert* and the horizontal arrow in *horiz*.

The composition of the 1-morphism (1.1) with a 1-morphism

$$\begin{array}{ccc} \mathbf{c}_{1,2} & \longrightarrow & \mathbf{c}_1 \\ \downarrow & & \\ \mathbf{c}_2 & & \end{array}$$

is the 1-morphism

$$\begin{array}{ccc} \mathbf{c}_{1,2} \underset{\mathbf{c}_1}{\times} \mathbf{c}_{0,1} & \longrightarrow & \mathbf{c}_0 \\ \downarrow & & \\ \mathbf{c}_2. & & \end{array}$$

What may be a little less obvious is to how to give the definition of $\operatorname{Corr}(\mathbf{C})_{vert;horiz}$ in the ∞-context (and without appealing to a particular model of $(\infty,1)$-categories, i.e., we do not want to talk about simplicial sets).

The definition of $\operatorname{Corr}(\mathbf{C})_{vert;horiz}$ is the subject of Chapter 7, Sect. 1. In fact, the construction is not difficult and quite natural: it is formulated in terms of the interpretation of $(\infty,1)$-categories as *complete Segal spaces*.

1.3. At this point let us comment on the relationship between our approach and (our interpretation of) [**LZ1, LZ2**].

Consider the following bi-simplicial space $\operatorname{Grid}_{\bullet,\bullet}(\mathbf{C})$: its space $\operatorname{Grid}_{m,n}(\mathbf{C})$ of $[m]\times[n]$-simplicies are $m\times n$-grids of objects of $\mathbf{C}$, in which every square is Cartesian, all vertical arrows are in *vert* and all horizontal arrow are in *horiz*.

Then Chapter 7, Theorem 2.1.3 says that the datum of a functor

$$\operatorname{Corr}(\mathbf{C})_{vert;horiz} \to \mathbb{S}$$

is equivalent to that of a map of bi-simplicial spaces

$$\operatorname{Grid}_{\bullet,\bullet}(\mathbf{C}) \to \operatorname{Maps}([\bullet]\times[\bullet],\mathbb{S}).$$

The authors of [**LZ1, LZ2**] construct their datum in terms of the latter map of bi-simplicial spaces.

2. The six functor formalism

Let us now explain how the $(\infty, 2)$-category of correspondences encodes the six functor formalism.

2.1. The setup. The general setup for the six functor formalism is the following. Suppose that we have a category $\mathbf{C}$ of 'geometric objects', e.g., the category of topological spaces, schemes, prestacks, etc. To each object $X \in \mathbf{C}$, we associate a category

$$X \rightsquigarrow \mathrm{Sh}(X) \in \mathrm{DGCat}_{\mathrm{cont}},$$

of 'sheaves on X', e.g., $\mathrm{IndCoh}(X)$ or $\mathrm{Dmod}(X)$. This assignment comes with the following additional data, in particular making it natural in $X \in \mathbf{C}$:

(1) (functoriality) For every map $f : X \to Y \in \mathbf{C}$, there are two pairs of adjoint functors

$$f_! : \mathrm{Sh}(X) \rightleftarrows \mathrm{Sh}(Y) : f^!, \text{ and } f^* : \mathrm{Sh}(Y) \rightleftarrows \mathrm{Sh}(X) : f_*$$

which are natural in f, i.e. each of them is given by a functor $\mathbf{C} \to \mathrm{DGCat}_{\mathrm{cont}}$ (or $\mathbf{C}^{\mathrm{op}} \to \mathrm{DGCat}_{\mathrm{cont}}$). These are four of the six functors in the six functor formalism.

Note the data of an adjoint pair is uniquely determined by one of the functors. In the case of IndCoh and D-modules, we only have the right adjoint functors $f^!$ and f_* exist in general. For this reason, we will describe the formalism in terms of these functors without explicit reference to their adjoints.

(2) (proper adjunction) Given $f : X \to Y \in \mathbf{C}$, there is a natural transformation

$$f_! \to f_*,$$

which is natural in f and is an isomorphism when f is proper.

Equivalently, there is a natural transformation

$$\mathrm{id} \to f^! \circ f_*,$$

which is the unit of an adjunction when f is proper.

(3) (open adjunction) If $f : X \to Y \in \mathbf{C}$ is an open immersion, there is a natural isomorphism

$$f^! \circ f_* \simeq \mathrm{id},$$

which is the counit of an adjunction. In particular, in this case, we have an isomorphism

$$f^! \simeq f^*.$$

(4) (proper base change) For a Cartesian square

$$\begin{array}{ccc} X' & \xrightarrow{g'} & X \\ {\scriptstyle f'}\downarrow & & \downarrow{\scriptstyle f} \\ Y' & \xrightarrow{g} & Y \end{array}$$

in $\mathbf{C}$, there is a natural base change isomorphism

$$f'_* \circ g'^! \simeq g^! \circ f_*.$$

In the case that f is proper (resp. open), this isomorphism is given by the natural transformation arising from proper (resp. open) adjunction above.

(5) (duality) For each $X \in \mathbf{C}$, the DG category $\mathrm{Sh}(X)$ is self-dual; see Chapter 5, Sect. 4.6 for an explanation of how this recovers the usual Verdier or Serre duality on sheaves. Moreover, for each morphism $f : X \to Y \in \mathbf{C}$, the functors $f^!$ and $f_!$ are dual to f_* and f^*, respectively.

(6) (tensor structure) For each $X, Y \in \mathbf{C}$ we have a functor

$$\boxtimes : \mathrm{Sh}(X) \otimes \mathrm{Sh}(Y) \to \mathrm{Sh}(X \times Y),$$

natural in X and Y.

Moreover, $*$-pullback along the diagonal $X \to X \times X$ defines a *closed* symmetric monoidal structure on $\mathrm{Sh}(X)$ for every $X \in \mathbf{C}$, i.e., each $\mathrm{Sh}(X)$ comes with a tensor product $\otimes$ and an inner hom $\underline{\mathrm{Hom}}_{\mathrm{Sh}(X)}$ functor – these are the remaining two of the six functors. Furthermore, for every map $f : X \to Y \in \mathbf{C}$, the functor

$$f^* : \mathrm{Sh}(Y) \to \mathrm{Sh}(X)$$

is equipped with a symmetric monoidal structure.

In the case of IndCoh and D-modules, the functor $\boxtimes$ above is an isomorphism in the case of (ind-inf-)schemes X and Y. However, we only have the !-pullback functor and so we can only define the dual !-tensor structure $\overset{!}{\otimes}$ on $\mathrm{Sh}(X)$ given by !-pullback along the diagonal. In this case, the functor

$$f^! : \mathrm{Sh}(Y) \to \mathrm{Sh}(X)$$

is equipped with a symmetric monoidal structure with respect to the $\overset{!}{\otimes}$ tensor product.

(7) (projection formula) Let $f : X \to Y$ be a morphism in $\mathbf{C}$. Since f^* is a tensor functor by the above, we have that $\mathrm{Sh}(X)$ is a module category over the tensor category $\mathrm{Sh}(Y)$. We further require that the functor $f_!$ be equipped with the structure of a functor of module categories over $\mathrm{Sh}(Y)$. In particular, from this we obtain the familiar natural isomorphisms:

$$f_!(M \otimes f^*(N)) \simeq f_!(M) \otimes N,$$
$$\underline{\mathrm{Hom}}(f_!(M), N) \simeq f_*(\underline{\mathrm{Hom}}(M, f^!(N))), \text{ and}$$
$$f^!(\underline{\mathrm{Hom}}(M, N)) \simeq \underline{\mathrm{Hom}}(f^*(M), f^!(N)),$$

for $M \in \mathrm{Sh}(X)$ and $N \in \mathrm{Sh}(Y)$.

In the case of IndCoh and D-modules, where we have the dual tensor product $\overset{!}{\otimes}$, we require that the functor f_* be equipped with the structure of a functor of module categories over $\mathrm{Sh}(Y)$ with respect to the $\overset{!}{\otimes}$ tensor product. In particular, we obtain the projection formula

$$f_*(M \overset{!}{\otimes} f^!(N)) \simeq f_*(M) \overset{!}{\otimes} N,$$

dual to the one above.

2.2. As explained in Sect. 1, the data of functoriality and proper base change above is equivalent to the data of a functor of $(\infty, 1)$-categories

$$\mathrm{Sh} : \mathrm{Corr}(\mathbf{C})_{\mathrm{all;all}} \to \mathrm{DGCat}_{\mathrm{cont}};$$

namely, an object $X \in \mathrm{Corr}(\mathbf{C})_{\mathrm{all;all}}$ maps to $\mathrm{Sh}(X)$ and a morphism

$$\begin{array}{ccc} Z & \xrightarrow{f} & X \\ {\scriptstyle g}\downarrow & & \\ Y & & \end{array}$$

maps to $g_* \circ f^! : \mathrm{Sh}(X) \to \mathrm{Sh}(Y)$.

The idea is that we can enlarge $\mathrm{Corr}(\mathbf{C})_{\mathrm{all;all}}$ to a symmetric monoidal $(\infty, 2)$-category $\mathrm{Corr}(\mathbf{C})^{\mathrm{proper}}_{\mathrm{all;all}}$ so that *all* of the above data will be encoded by a symmetric monoidal functor of $(\infty, 2)$-categories

$$\mathrm{Sh} : \mathrm{Corr}(\mathbf{C})^{\mathrm{proper}}_{\mathrm{all;all}} \to \mathrm{DGCat}^{\mathrm{2\text{-}Cat}}_{\mathrm{cont}}. \tag{2.1}$$

Suppose that we are in the situation of Sect. 1.1, and in addition to *vert* and *horiz*, we are given a third class of 1-morphisms

$$adm \subset vert \cap horiz.$$

We define the $(\infty, 2)$-category

$$\mathrm{Corr}(\mathbf{C})^{adm}_{vert;horiz}$$

so that its underlying $(\infty, 1)$-category is the $(\infty, 1)$-category $\mathrm{Corr}(\mathbf{C})_{vert;horiz}$ discussed above, but we now allow non-invertible 2-morphisms. Namely, a 2-morphism from the 1-morphism (1.1) to the 1-morphism

$$\begin{array}{ccc} \mathbf{c}'_{0,1} & \longrightarrow & \mathbf{c}_0 \\ \downarrow & & \\ \mathbf{c}_1 & & \end{array}$$

is a commutative diagram

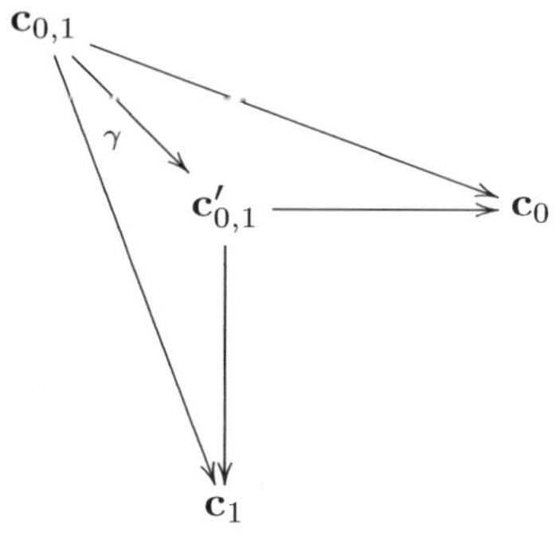

with $\gamma \in adm$.

If $\mathbf{C}$ has a Cartesian symmetric monoidal structure that preserves each of the subcategories $adm, vert$ and $horiz$, then it induces a symmetric monoidal structure on the $(\infty,2)$-category $\mathrm{Corr}(\mathbf{C})^{adm}_{vert;horiz}$. In particular, in the situation of Sect. 2.1, we obtain that the $(\infty,2)$-category

$$\mathrm{Corr}(\mathbf{C})^{\mathrm{proper}}_{\mathrm{all;all}}$$

has a canonical symmetric monoidal structure such that the functor

$$\mathbf{C} \to \mathrm{Corr}(\mathbf{C})^{\mathrm{proper}}_{\mathrm{all;all}}$$

given by "vertical morphisms" is symmetric monoidal with respect to the Cartesian symmetric monoidal structure on $\mathbf{C}$.

2.3. We will now explain how to recover all of the data in Sect. 2.1 from the data of the functor (2.1). We have already seen how functoriality and proper base change is encoded as a functor out of correspondences.

Proper adjunction.

Let $f : X \to Y \in \mathbf{C}$. In this case, we have that the functor $f_* \circ f^! : \mathrm{Sh}(Y) \to \mathrm{Sh}(Y)$ is the image under Sh of the morphism

$$\begin{array}{ccc} X & \longrightarrow & Y \\ \downarrow & & \\ Y & & \end{array}$$

in $\mathrm{Corr}(\mathbf{C})_{\mathrm{all;all}}$. Similarly, $f^! \circ f_* : \mathrm{Sh}(X) \to \mathrm{Sh}(X)$ is given by the image of the composite

$$\begin{array}{ccccc} X \underset{Y}{\times} X & \longrightarrow & X & = & X \\ \downarrow & & \downarrow & & \\ X & \longrightarrow & Y & & \\ \| & & & & \\ X & & & & \end{array}$$

If the diagonal morphism $X \to X \underset{Y}{\times} X$ is proper (as is the case with a separated morphism of schemes), we obtain the desired natural transformation

$$\mathrm{id} \to f^! \circ f_*.$$

Furthermore, if the map $f : X \to Y$ is proper, we also obtain a natural transformation $f_* \circ f^! \to \mathrm{id}$ and it is easy to see that the two natural transformations give the unit and counit of an adjunction.

2.3.1. *Open adjunction.* Similarly, if $f : X \to Y$ is an open embedding, we have that

$$X \underset{Y}{\times} X \simeq X$$

and therefore we obtain the desired isomorphism

$$f^! \circ f_* \simeq \mathrm{id}\,.$$

The assertion that this isomorphism gives a counit of an adjunction is an additional *condition.*

Duality.

A key feature of the symmetric monoidal category $\mathrm{Corr}(\mathbf{C})^{\mathrm{proper}}_{\mathrm{all;all}}$ is that every object X is self-dual. In particular, it is easy to see that the morphisms

$$\begin{array}{ccc} X & \longrightarrow & * \\ {\scriptstyle\Delta}\downarrow & & \\ X \times X & & \end{array} \quad \text{and} \quad \begin{array}{ccc} X & \xrightarrow{\Delta} & X \times X \\ \downarrow & & \\ * & & \end{array}$$

give the unit and counit maps, respectively. Applying the symmetric monoidal functor Sh, we obtain that the DG category $\mathrm{Sh}(X)$ is self-dual.

Moreover, it is straightforward to check that given a map $f : X \to Y \in \mathbf{C}$, the morphisms

$$\begin{array}{ccc} X & \xrightarrow{f} & Y \\ \| & & \\ X & & \end{array} \quad \text{and} \quad \begin{array}{ccc} X & = & X \\ {\scriptstyle f}\downarrow & & \\ Y & & \end{array}$$

are dual in $\mathrm{Corr}(\mathbf{C})^{\mathrm{proper}}_{\mathrm{all;all}}$. Hence the functors $f^!$ and f_* are dual to each other.

Tensor structure.

The symmetric monoidal structure on the functor Sh gives a natural isomorphism

$$\boxtimes : \mathrm{Sh}(X) \otimes \mathrm{Sh}(Y) \xrightarrow{\sim} \mathrm{Sh}(X \times Y);$$

in the case that Sh is only right-lax symmetric monoidal, we would only have a functor.

Moreover, by construction of the symmetric monoidal structure on

$$\mathrm{Corr}(\mathbf{C})^{\mathrm{proper}}_{\mathrm{all;all}},$$

we have that the functor

$$\mathbf{C}^{\mathrm{op}} \to \mathrm{Corr}(\mathbf{C})^{\mathrm{proper}}_{\mathrm{all;all}}$$

given by 'horizontal morphisms' is symmetric monoidal, where the symmetric monoidal structure on $\mathbf{C}^{\mathrm{op}}$ is given by coproduct. In particular, every object $X \in \mathbf{C}^{\mathrm{op}}$ has a canonical structure of a commutative algebra with multiplication given by the opposite of the diagonal map

$$(X \to X \times X)^{\mathrm{op}}$$

(see Chapter 1, Sect. 5.1.8). Thus, $\mathrm{Sh}(X)$ carries a symmetric monoidal structure $\overset{!}{\otimes}$ given by !-restriction along the diagonal $\Delta : X \to X \times X$.

Projection formula.

Suppose that we have an object $Y \in \mathbf{C}$. By the above, we have that Y has a canonical structure of a commutative algebra object in $\mathrm{Corr}(\mathbf{C})^{\mathrm{proper}}_{\mathrm{all;all}}$. Furthermore, if $f : X \to Y$ is a morphism in Y, we have that X has a canonical structure of a module over Y. It is straightforward to see that in this case the morphism

$$\begin{array}{ccc} X & = & X \\ {\scriptstyle f}\downarrow & & \\ Y & & \end{array}$$

has the structure of a morphism of Y-modules in $\mathrm{Corr}(\mathbf{C})^{\mathrm{proper}}_{\mathrm{all;all}}$. In particular, applying the symmetric monoidal functor Sh, we obtain that the functor

$$f_* : \mathrm{Sh}(X) \to \mathrm{Sh}(Y)$$

has the structure of a functor of $\mathrm{Sh}(Y)$-modules, as desired.

3. Constructing functors

Having constructed the categories $\mathrm{Corr}(\mathbf{C})_{vert;horiz}$ and $\mathrm{Corr}(\mathbf{C})^{adm}_{vert;horiz}$, our next problem is how to construct functors

$$\mathrm{Corr}(\mathbf{C})_{vert;horiz} \to 1\text{-Cat}.$$

In our main application, $\mathbf{C} = \mathrm{Sch}_{\mathrm{aft}}$, $\mathbb{S} = 1\text{-Cat}$, and Φ is supposed to send a scheme X to the category $\mathrm{IndCoh}(X)$. We take *vert* and *horiz* to be all morphisms in $\mathrm{Sch}_{\mathrm{aft}}$.

3.1. It turns out, however, that in order to construct functors out of

$$\mathrm{Corr}(\mathbf{C})_{vert;horiz},$$

it is convenient (and necessary, if one wants to retain canonicity) to enlarge it to an $(\infty, 2)$-category $\mathrm{Corr}(\mathbf{C})^{adm}_{vert;horiz}$.

Suppose that $\mathbb{S}$ is an $(\infty, 2)$-category. A functor

$$\Phi : \mathrm{Corr}(\mathbf{C})^{adm}_{vert;horiz} \to \mathbb{S}$$

encodes the following data (in addition to that of its restriction to $\mathrm{Corr}(\mathbf{C})_{vert;horiz}$):

For a 1-morphism $(\mathbf{c} \overset{\gamma}{\to} \mathbf{c}') \in adm$, the 1-morphism

$$\Phi^!(\gamma) : \Phi(\mathbf{c}') \to \Phi(\mathbf{c})$$

in $\mathbb{S}$ identifies with the *right adjoint* of

$$\Phi(\gamma) : \Phi(\mathbf{c}) \to \Phi(\mathbf{c}').$$

We recall that the notion of adjoint morphisms makes sense in an arbitrary $(\infty, 2)$-category.

3.2. The above 2-categorical enhancement plays a crucial role for the following reason.

Suppose that $horiz \subset vert$, and consider the $(\infty, 2)$-category $\mathrm{Corr}(\mathbf{C})^{horiz}_{vert;horiz}$. We have a tautological functor

$$\mathbf{C}_{vert} \simeq \mathrm{Corr}(\mathbf{C})_{vert;\mathrm{isom}} \to \mathrm{Corr}(\mathbf{C})_{vert;horiz} \to \mathrm{Corr}(\mathbf{C})^{horiz}_{vert;horiz}.$$

Then the basic result (Chapter 7, Theorem 3.2.2) is that for any target $\mathbb{S}$, restriction along the above functor identifies the space of functors

$$\mathrm{Corr}(\mathbf{C})^{horiz}_{vert;horiz} \to \mathbb{S}$$

with the full subspace of functors

$$\mathbf{C}_{vert} \to \mathbb{S},$$

consisting of those functions for which for every $\alpha \in horiz$, the corresponding 1-morphism $\Phi(\alpha)$ in $\mathbb{S}$ admits a right adjoint, and the *Beck-Chevalley* conditions are satisfied (see Chapter 7, Sect. 3.1 for what this means).

The above theorem is the initial input for any functor out of any category of correspondences considered in this book.

3.3. For example, let us take $\mathbf{C} = \mathrm{Sch}_{\mathrm{aft}}$ with *vert* = all and *horiz* = proper. Then starting from IndCoh, viewed as a functor

$$\mathrm{Sch}_{\mathrm{aft}} \to \mathrm{DGCat}_{\mathrm{cont}}, \tag{3.1}$$

(with respect to the operation of direct image), we canonically extend it to a functor

$$\mathrm{Corr}(\mathrm{Sch}_{\mathrm{aft}})^{\mathrm{proper}}_{\mathrm{all;proper}} \to \mathrm{DGCat}^{2\text{-Cat}}_{\mathrm{cont}}.$$

Similarly, taking *horiz* = open, and inverting the direction of 2-morphisms, we canonically extend (3.1) to a functor

$$\mathrm{Corr}(\mathrm{Sch}_{\mathrm{aft}})^{\mathrm{open}}_{\mathrm{all;open}} \to (\mathrm{DGCat}^{2\text{-Cat}}_{\mathrm{cont}})^{2\text{-op}}.$$

3.4. In Chapter 7 we prove two fundamental theorems that allow to (uniquely) extend functors defined on one category of correspondences to a larger one. Rather than giving the abstract formulation, we will consider the example of $\mathbf{C} = \mathrm{Sch}_{\mathrm{aft}}$.

Together, these theorems allow to start with IndCoh, viewed as a functor as in (3.1), and extend it to a functor

$$\mathrm{Corr}(\mathrm{Sch}_{\mathrm{aft}})^{\mathrm{proper}}_{\mathrm{all;all}} \to \mathrm{DGCat}^{2\text{-Cat}}_{\mathrm{cont}}.$$

3.5. The first of these theorems, Chapter 7, Theorem 4.1.3, allows to treat the following situation:

Let us be given a functor

$$\Phi : \mathrm{Corr}(\mathrm{Sch}_{\mathrm{aft}})^{\mathrm{closed}}_{\mathrm{all;all}} \to \mathbb{S},$$

and we want to extend to a functor

$$\mathrm{Corr}(\mathrm{Sch}_{\mathrm{aft}})^{\mathrm{proper}}_{\mathrm{all;all}} \to \mathbb{S}.$$

I.e., the initial functor was only defined on 2-morphisms given by closed embeddings, and we want to extend it to 2-morphisms given by proper maps.

The assertion of Chapter 7, Theorem 4.1.3 is that if such an extension exists, it is unique, and one can give explicit conditions for the existence.

The idea here is that for a separated map $f : S \to S'$, the diagram

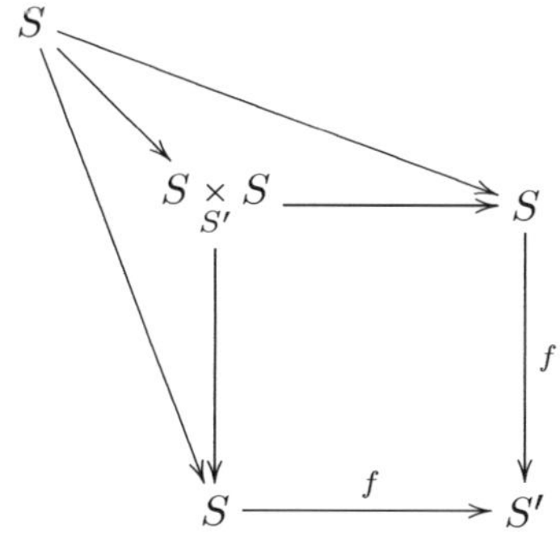

provides a 2-morphism

$$\mathrm{id} \to \Phi^!(f) \circ \Phi(f),$$

which will be the unit of an adjunction if f is proper.

3.6. The second theorem, Chapter 7, Theorem 5.2.4, is designed to treat the following situation. Let us be given a functor

$$\Phi : \mathrm{Corr}(\mathrm{Sch}_{\mathrm{aft}})_{\mathrm{all;open}} \to \mathbb{S} \tag{3.2}$$

(see Sect. 3.3 for the example that we have in mind) and we want to extend to a functor

$$\mathrm{Corr}(\mathrm{Sch}_{\mathrm{aft}})^{\mathrm{proper}}_{\mathrm{all;all}} \to \mathbb{S}. \tag{3.3}$$

Again, the claim is that if such an extension exists, it is unique, and one can give explicit conditions for the existence.

The idea here is the following: as a first step we restrict Φ to

$$\mathrm{Sch}_{\mathrm{aft}} \simeq \mathrm{Corr}(\mathrm{Sch}_{\mathrm{aft}})_{\mathrm{all;isom}} \subset \mathrm{Corr}(\mathrm{Sch}_{\mathrm{aft}})_{\mathrm{all;open}},$$

and then extend it to $\mathrm{Corr}(\mathrm{Sch}_{\mathrm{aft}})^{\mathrm{proper}}_{\mathrm{all;proper}}$, using Chapter 7, Theorem 4.1.3 (see Sect. 3.3 for the example we have in mind).

Thus, we now have $\Phi^!(f)$ defined separately for f open and proper. One now uses Nagata's theorem that any morphism can be factored into a composition of an open morphism, followed by a proper one.

The bulk of the proof consists of showing how the existence of such factorizations leads to the existence and uniqueness of the functor (3.3).

We emphasize that in this theorem the 2-categorical structure on the category of correspondences is essential. I.e., even if we are only interested in the functor

$$\Phi^! : (\mathrm{Sch}_{\mathrm{aft}})^{\mathrm{op}} \to \mathbb{S},$$

we need to pass by 2-categories in order to obtain it from the initial functor (3.2).

4. Extension theorems

The Chapter V.2 contains two results (Theorems 1.1.9 and 6.1.5) that allow to (uniquely) extend a given functor

$$\Phi : \mathrm{Corr}(\mathbf{C})^{adm}_{vert;horiz} \to \mathbb{S}$$

to a functor

$$\Psi : \mathrm{Corr}(\mathbf{D})^{adm}_{vert;horiz} \to \mathbb{S},$$

along the functor $\mathrm{Corr}(\mathbf{C})^{adm}_{vert;horiz} \to \mathrm{Corr}(\mathbf{D})^{adm}_{vert;horiz}$, corresponding to a functor between $(\infty,1)$-categories $\mathbf{C} \to \mathbf{D}$.

4.1. Let us explain the typical situation that Chapter 8, Theorem 1.1.9 is applied to. We start with IndCoh, viewed as a functor

$$\mathrm{Corr}(\mathrm{Sch}_{\mathrm{aft}})^{\text{nil-closed}}_{\mathrm{all;all}} \to \mathrm{DGCat}_{\mathrm{cont}},$$

and we want to (canonically) extend it to a functor[1]

$$\mathrm{Corr}(\mathrm{infSch}_{\mathrm{laft}})^{\text{nil-closed}}_{\mathrm{all;all}} \to \mathrm{DGCat}_{\mathrm{cont}}.$$

[1]Here we really have to work with the class of nil-closed morphisms rather than proper ones, because Chapter 8, Theorem 1.1.9 only applies in this situation. The further extension to proper morphisms is obtained by the procedure described in Sect. 3.5.

We do not really know what is the general 2-categorical paradigm in which such an extension fits (it has features of both the left and right Kan extension).

Again, the 2-categorical structure on the category of correspondences here is essential.

4.2. Let us now explain what Chapter 8, Theorem 6.1.5 says. We start with the functor

$$\mathrm{Corr}(\mathrm{Sch}_{\mathrm{aft}})^{\mathrm{proper}}_{\mathrm{all;all}} \to \mathbb{S}, \tag{4.1}$$

and we wish to (canonically) extend it to a functor

$$\mathrm{Corr}(\mathrm{PreStk}_{\mathrm{laft}})^{\mathrm{sch\ \&\ proper}}_{\mathrm{sch;all}} \to \mathbb{S}. \tag{4.2}$$

Here the subscript sch stands for the class of schematic maps, and the superscript sch & proper stands for the class of schematic and proper maps.

The required extension is the 2-categorical right Kan extension. However, the particular properties of the functor

$$\mathrm{Corr}(\mathrm{Sch}_{\mathrm{aft}})^{\mathrm{proper}}_{\mathrm{all;all}} \to \mathrm{Corr}(\mathrm{PreStk}_{\mathrm{laft}})^{\mathrm{sch\ \&\ proper}}_{\mathrm{sch;all}}$$

make this extension procedure very manageable.

Namely, it turns out that the restriction of (4.2) along

$$\mathrm{Corr}(\mathrm{PreStk}_{\mathrm{laft}})_{\mathrm{sch;all}} \to \mathrm{Corr}(\mathrm{PreStk}_{\mathrm{laft}})^{\mathrm{sch\ \&\ proper}}_{\mathrm{sch;all}}$$

equals the right Kan extension along the functor of $(\infty, 1)$-categories

$$\mathrm{Corr}(\mathrm{Sch}_{\mathrm{aft}})_{\mathrm{all;all}} \to \mathrm{Corr}(\mathrm{PreStk}_{\mathrm{laft}})_{\mathrm{sch;all}}$$

of the restriction of (4.1) along

$$\mathrm{Corr}(\mathrm{Sch}_{\mathrm{aft}})_{\mathrm{all;all}} \to \mathrm{Corr}(\mathrm{Sch}_{\mathrm{aft}})^{\mathrm{proper}}_{\mathrm{all;all}}.$$

I.e., the above 2-categorical right Kan extension is essentially 1-categorical.

Moreover, the further restriction of (4.2) along

$$(\mathrm{PreStk}_{\mathrm{laft}})^{\mathrm{op}} \simeq \mathrm{Corr}(\mathrm{PreStk}_{\mathrm{laft}})_{\mathrm{isom;all}} \to \mathrm{Corr}(\mathrm{PreStk}_{\mathrm{laft}})_{\mathrm{sch;all}}$$

equals the right Kan extension along

$$(\mathrm{Sch}_{\mathrm{aft}})^{\mathrm{op}} \to (\mathrm{PreStk}_{\mathrm{laft}})^{\mathrm{op}}$$

of the further restriction of (4.1) along

$$(\mathrm{Sch}_{\mathrm{aft}})^{\mathrm{op}} \simeq \mathrm{Corr}(\mathrm{Sch}_{\mathrm{aft}})_{\mathrm{isom;all}} \to \mathrm{Corr}(\mathrm{Sch}_{\mathrm{aft}})_{\mathrm{all;all}}.$$

I.e., this extension procedure 'does the right thing' on objects and pullbacks.

A similar discussion applies when we replace $\mathrm{Corr}(\mathrm{Sch}_{\mathrm{aft}})^{\mathrm{proper}}_{\mathrm{all;all}}$ by

$$\mathrm{Corr}(\mathrm{infSch}_{\mathrm{aft}})^{\mathrm{proper}}_{\mathrm{all;all}}$$

and $\mathrm{Corr}(\mathrm{PreStk}_{\mathrm{laft}})^{\mathrm{sch\ \&\ proper}}_{\mathrm{sch;all}}$ by $\mathrm{Corr}(\mathrm{PreStk}_{\mathrm{laft}})^{\mathrm{infsch\ \&\ proper}}_{\mathrm{infsch;all}}$.

5. (Symmetric) monoidal structures

In Chapter 9 we study the symmetric monoidal structure that arises on the $(\infty,2)$-category $\mathrm{Corr}(\mathbf{C})^{adm}_{vert;horiz}$, induced by the *Cartesian* symmetric monoidal structure on $\mathbf{C}$. But in fact, our primary focus will be on the $(\infty,1)$-category $\mathrm{Corr}(\mathbf{C})_{vert;horiz}$.

The essence of Chapter 9 is the following two observations. Assume for simplicity that $vert = horiz =$ all, and consider the $(\infty,1)$-category $\mathrm{Corr}(\mathbf{C}) = \mathrm{Corr}(\mathbf{C})_{\mathrm{all;all}}$.

5.1. The first observation is the following. We note that the category $\mathrm{Corr}(\mathbf{C})$ carries a canonical anti-involution, given by swapping the roles of vertical and horizontal arrows.

We show that this involution canonically identifies with the dualization functor on $\mathrm{Corr}(\mathbf{C})$ for the symmetric monoidal structure on the latter.

As a corollary, we obtain that whenever

$$\Phi : \mathrm{Corr}(\mathbf{C}) \to \mathbf{O}$$

is a symmetric monoidal functor, where $\mathbf{O}$ is a target symmetric monoidal category, for every $\mathbf{c} \in \mathbf{C}$, the corresponding object $\Phi(\mathbf{c}) \in \mathbf{O}$ is canonically self-dual.

This fact is responsible for the Serre duality on IndCoh on schemes: apply the above observation to the functor

$$\mathrm{IndCoh}_{\mathrm{Corr}(\mathrm{Sch}_{\mathrm{aft}})_{\mathrm{all;all}}} : \mathrm{Corr}(\mathrm{Sch}_{\mathrm{aft}})_{\mathrm{all;all}} \to \mathrm{DGCat}_{\mathrm{cont}}.$$

5.2. The second observation has to do with the construction of *convolution categories*.

Let $\mathbf{c}^\bullet$ be a *Segal object* of $\mathbf{C}$. I.e., this is a simplicial object such that for every $n \geq 2$, the map

$$\mathbf{c}^1 \underset{\mathbf{c}^0}{\times} \dots \underset{\mathbf{c}^0}{\times} \mathbf{c}^1,$$

given by the product of the maps

$$[1] \to [n], \quad 0 \mapsto i, 1 \mapsto i+1, \quad i = 0, \dots, n-1,$$

is an isomorphism[2].

We show that $\mathbf{c}^1$, regarded as as object of $\mathrm{Corr}(\mathbf{C})$, carries a canonical structure of associative algebra, (with respect to the symmetric monoidal structure on $\mathrm{Corr}(\mathbf{C})$). For example, the binary operation on $\mathbf{c}^1$ is given by the diagram

$$\begin{array}{ccc} \mathbf{c}^2 & \longrightarrow & \mathbf{c}^1 \times \mathbf{c}^1 \\ \downarrow & & \\ \mathbf{c}^1, & & \end{array}$$

in which the vertical map is given by the active map $[1] \to [2]$, and the horizontal map is given by the product of the two inert maps $[1] \to [2]$.

As a corollary, we obtain that whenever we are given a monoidal functor

$$\Phi : \mathrm{Corr}(\mathbf{C}) \to \mathbf{O},$$

[2] An alternative terminology is *category object*.

where $\mathbf{O}$ is a monoidal category, the object $\Phi(\mathbf{c}^1) \in \mathbf{O}$ acquires a structure of associative algebra.

In particular, taking $\mathbf{C} = \mathrm{Sch}_{\mathrm{aft}}$, $\mathbf{O} = \mathrm{DGCat}_{\mathrm{cont}}$, and Φ to be the functor IndCoh, we obtain that for a Segal object $X^\bullet$ in the category of schemes, the category $\mathrm{IndCoh}(X^1)$ is endowed with a monoidal structure, given by *convolution*. I.e., it is given by pull-push along the diagram

$$\begin{array}{ccc} X_1 \underset{X_0}{\times} X_1 & \longrightarrow & X^1 \times X^1 \\ \big\downarrow & & \\ X^1. & & \end{array}$$

CHAPTER 7

The $(\infty, 2)$-category of correspondences

Introduction

This chapter contains one of the two main innovations in this book: functors out of the $(\infty, 2)$-category of correspondences (the other one being the notion of inf-scheme).

The idea of the $(\infty, 2)$–category of correspondences, as a way to encode bi-variant functors that satisfy base change, was explained to us by J. Lurie. So, in a sense we realize his suggestion, even though our approach to the definition of the $(\infty, 2)$-category of correspondences is different from what he had originally envisaged[1].

0.1. Why do we need it?

0.1.1. Suppose we have an $(\infty, 1)$-category $\mathbf{C}$, and let $\mathbf{S}$ be a target $(\infty, 1)$-category.

We want to express, in a functorial way, what it means to have a *bi-variant* assignment

$$\mathbf{c} \in \mathbf{C} \rightsquigarrow \Phi(\mathbf{c}) \in \mathbf{S}$$

that satisfies base change.

In other words, we want to have a functor

$$\Phi : \mathbf{C} \to \mathbf{S}$$

and also a functor

$$\Phi^! : \mathbf{C}^{\mathrm{op}} \to \mathbf{S}$$

that interact as follows:

(1) At the level of objects, for any $\mathbf{c} \in \mathbf{C}$ we are given an isomorphism $\Phi(\mathbf{c}) \simeq \Phi^!(\mathbf{c})$.
(2) Whenever we have a Cartesian square

$$\begin{array}{ccc} \mathbf{c}_{0,1} & \xrightarrow{\alpha_0} & \mathbf{c}_{0,0} \\ {\scriptstyle\beta_1}\downarrow & & \downarrow{\scriptstyle\beta_0} \\ \mathbf{c}_{1,1} & \xrightarrow{\alpha_1} & \mathbf{c}_{1,0}, \end{array} \tag{0.1}$$

we want to be *given* a base change isomorphism

$$\Phi(\beta_1) \circ \Phi^!(\alpha_0) \simeq \Phi^!(\alpha_1) \circ \Phi(\beta_0). \tag{0.2}$$

[1] Lurie's idea was to construct it combinatorially in terms simplicial *sets* starting from quasi-categories, whereas our approach is independent of a particular model for $(\infty, 1)$-categories.

We emphasize that, in general, for a morphism $\gamma : \mathbf{c} \to \mathbf{c}'$ in $\mathbf{C}$ and $\mathbf{S} = 1\text{-Cat}$, the 1-morphisms

$$\Phi(\gamma) : \Phi(\mathbf{c}) \to \Phi(\mathbf{c}') \text{ and } \Phi^!(\gamma) : \Phi(\mathbf{c}') \to \Phi(\mathbf{c})$$

are not adjoint on either side; therefore, in (0.2) there is no a priori defined map in either direction.

An example to keep in mind is when $\mathbf{C} = \mathrm{Sch}_{\mathrm{aft}}$ and $\mathbf{S} = 1\text{-Cat}$, and we take $\Phi(S) = \mathrm{IndCoh}(S)$ with the morphism $(S_1 \xrightarrow{f} S_2)$ being sent to

$$f_*^{\mathrm{IndCoh}} : \mathrm{IndCoh}(S_1) \to \mathrm{IndCoh}(S_2) \text{ and } f^! : \mathrm{IndCoh}(S_2) \to \mathrm{IndCoh}(S_1),$$

respectively.

0.1.2. A challenge is to even express what it means for the data (1) and (2) above to be functorial. Let us give a typical example of why one would want that.

Say, we want to extend the functor

$$\mathrm{IndCoh} : \mathrm{Sch}_{\mathrm{aft}} \to 1\text{-Cat}$$

to a functor

$$\mathrm{IndCoh} : (\mathrm{PreStk}_{\mathrm{laft}})_{\mathrm{sch}} \to 1\text{-Cat},$$

where $(\mathrm{PreStk}_{\mathrm{laft}})_{\mathrm{sch}}$ is the 1-full subcategory of $\mathrm{PreStk}_{\mathrm{laft}}$, where we restrict morphisms to maps that are schematic.

In other words, we want to assign to a schematic map $\mathcal{Y}_1 \xrightarrow{f} \mathcal{Y}_2$ between prestacks a functor

$$f_*^{\mathrm{IndCoh}} : \mathrm{IndCoh}(\mathcal{Y}_1) \to \mathrm{IndCoh}(\mathcal{Y}_2),$$

and we want this assignment to be functorial in the ∞-categorical sense.

By definition, for $\mathcal{Y} \in \mathrm{PreStk}_{\mathrm{laft}}$, we have

$$\mathrm{IndCoh}(\mathcal{Y}) := \lim_{Y \in (\mathrm{Sch}_{\mathrm{aft}})_{/\mathcal{Y}}} \mathrm{IndCoh}(Y),$$

where for a map $Y' \xrightarrow{g} Y''$ over $\mathcal{Y}$, the corresponding functor $\mathrm{IndCoh}(Y_2) \to \mathrm{IndCoh}(Y_1)$ is $g^!$.

For each $Y_1 \to \mathcal{Y}_1$, set $Y_2 := Y_1 \underset{\mathcal{Y}_2}{\times} \mathcal{Y}_1$, which is a scheme since f was assumed to be schematic, and let $\widetilde{f}$ denote the resulting map $Y_1 \to Y_2$.

The sought-for functor f_*^{IndCoh} is given by the *compatible* family of functors

$$\widetilde{f}_*^{\mathrm{IndCoh}} : \mathrm{IndCoh}(Y_1) \to \mathrm{IndCoh}(Y_2),$$

where the compatibility is exactly encoded by the base change isomorphisms.

We will carry out this construction in detail in Chapter 8, Sect. 6.

0.1.3. Of course, if $\mathbf{S}$ is an ordinary category, one can express the required compatibility conditions on the data (1) and (2) in Sect. 0.1.1 by hand: one specifies the natural transformations (0.2) for each Cartesian square (0.1) requiring that for a Cartesian diagram

$$\begin{array}{ccccc} \mathbf{c}_{0,2} & \xrightarrow{\alpha_0'} & \mathbf{c}_{0,1} & \xrightarrow{\alpha_0''} & \mathbf{c}_{0,0} \\ {\scriptstyle \beta_2}\downarrow & & {\scriptstyle \beta_1}\downarrow & & \downarrow{\scriptstyle \beta_0} \\ \mathbf{c}_{1,2} & \xrightarrow{\alpha_1'} & \mathbf{c}_{1,1} & \xrightarrow{\alpha_1''} & \mathbf{c}_{1,0}, \end{array}$$

the resulting two natural isomorphisms

$$\Phi(\beta_2)\circ\Phi^!(\alpha_0')\circ\Phi^!(\alpha_0'')\rightrightarrows\Phi^!(\alpha_1')\circ\Phi^!(\alpha_1'')\circ\Phi(\beta_0)$$

coincide, and similarly for every Cartesian diagram

$$\begin{array}{ccc} \mathbf{c}_{0,1} & \xrightarrow{\alpha_0} & \mathbf{c}_{0,0} \\ {\scriptstyle \beta_1'}\downarrow & & \downarrow{\scriptstyle \beta_0'} \\ \mathbf{c}_{1,1} & \xrightarrow{\alpha_1} & \mathbf{c}_{1,0} \\ {\scriptstyle \beta_1''}\downarrow & & \downarrow{\scriptstyle \beta_0''} \\ \mathbf{c}_{2,1} & \xrightarrow{\alpha_2} & \mathbf{c}_{2,0}. \end{array}$$

But if $\mathbf{S}$ is an ∞-category, the word 'coincides' must be replaced by 'a specified homotopy', and thus we need to specify the data of infinitely many homotopies for $(m \times n)$-diagrams for every m and n.

0.1.4. That said, a way to formulate the above data at the level of ∞-categories readily presents itself. Here is one possibility: we specify a map between *bi-simplicial spaces* (i.e., a map in the category $\mathrm{Spc}^{\mathbf{\Delta}^{\mathrm{op}}\times\mathbf{\Delta}^{\mathrm{op}}}$) between the following two objects:

The source is the object of $\mathrm{Spc}^{\mathbf{\Delta}^{\mathrm{op}}\times\mathbf{\Delta}^{\mathrm{op}}}$ that assigns to $([m],[n]) \in \mathbf{\Delta}^{\mathrm{op}}\times\mathbf{\Delta}^{\mathrm{op}}$ the full subspace in $\mathrm{Maps}([m]\times[n],\mathbf{C})$ that consists of *Cartesian diagrams.*

The target is the object of $\mathrm{Spc}^{\mathbf{\Delta}^{\mathrm{op}}\times\mathbf{\Delta}^{\mathrm{op}}}$ that assigns to $([m],[n]) \in \mathbf{\Delta}^{\mathrm{op}}\times\mathbf{\Delta}^{\mathrm{op}}$ simply $\mathrm{Maps}([m]\times[n],\mathbf{S})$.

This is a valid formulation, and we will prove (see Theorem 2.1.3) that it is equivalent to the one we will 'officially' take. Its disadvantage is that a datum of a map in $\mathrm{Spc}^{\mathbf{\Delta}^{\mathrm{op}}\times\mathbf{\Delta}^{\mathrm{op}}}$ does not look like a datum of a functor between ∞-categories (however, Theorem 2.1.3, mentioned above, says that it is actually equivalent to one).

0.1.5. Here is an alternative approach, which is the one we will adopt in this book as the definition. Namely, we will introduce an $(\infty,1)$-category that we will denote $\mathrm{Corr}(\mathbf{C})$, and our datum will be simply a functor from $\mathrm{Corr}(\mathbf{C})$ to $\mathbf{S}$.

In order to define $\mathrm{Corr}(\mathbf{C})$, recall following Chapter 10, Sect. 1.2, that the datum of an $(\infty,1)$-category $\mathbf{D}$ is equivalent to one of the *complete Segal space*

$$\mathrm{Seq}_\bullet(\mathbf{D}) \in \mathrm{Spc}^{\mathbf{\Delta}^{\mathrm{op}}},\quad \mathrm{Seq}_n(\mathbf{D}) = \mathrm{Maps}([n],\mathbf{D}).$$

For the desired $(\infty,1)$-category $\operatorname{Corr}(\mathbf{C})$, we take the corresponding object $\operatorname{Seq}_\bullet(\operatorname{Corr}(\mathbf{C}))$ to be the complete Segal space, denoted $\operatorname{Grid}_\bullet^{\geq \operatorname{dgnl}}(\mathbf{C})$, defined as follows. We let $\operatorname{Grid}_n^{\geq \operatorname{dgnl}}(\mathbf{C})$ be the space of functors

$$([n]\times[n]^{\operatorname{op}})^{\geq \operatorname{dgnl}} \to \mathbf{C}$$

for which every inner square is Cartesian.

Here $([n]\times[n]^{\operatorname{op}})^{\geq \operatorname{dgnl}}$ is the ordinary category, equal to the full subcategory of $[n]\times[n]^{\operatorname{op}}$ spanned by the objects $\{i\},\{j\}$ with $i\leq j$.

Note that $\operatorname{Corr}(\mathbf{C})$ receives a pair of functors

$$\mathbf{C}\to \operatorname{Corr}(\mathbf{C}) \leftarrow \mathbf{C}^{\operatorname{op}}$$

corresponding to the projections

$$[n] \leftarrow ([n]\times[n]^{\operatorname{op}})^{\geq \operatorname{dgnl}} \to [n]^{\operatorname{op}},$$

respectively.

0.1.6. Thus, for example, the space of objects of $\operatorname{Corr}(\mathbf{C})$ equals that of $\mathbf{C}$. The space of 1-morphisms in $\operatorname{Corr}(\mathbf{C})$ is that of diagrams

(0.3)
$$\begin{array}{ccc} \mathbf{c}_{0,1} & \longrightarrow & \mathbf{c}_{0,0} \\ \downarrow & & \\ \mathbf{c}_{1,1}. & & \end{array}$$

The space of two-fold compositions is that of diagrams

$$\begin{array}{ccccc} \mathbf{c}_{0,2} & \longrightarrow & \mathbf{c}_{0,1} & \longrightarrow & \mathbf{c}_{0,0} \\ \downarrow & & \downarrow & & \\ \mathbf{c}_{1,2} & \longrightarrow & \mathbf{c}_{1,1} & & \\ \downarrow & & & & \\ \mathbf{c}_{2,2}, & & & & \end{array}$$

in which the square

$$\begin{array}{ccc} \mathbf{c}_{0,2} & \longrightarrow & \mathbf{c}_{0,1} \\ \downarrow & & \downarrow \\ \mathbf{c}_{1,2} & \longrightarrow & \mathbf{c}_{1,1} \end{array}$$

is Cartesian.

0.1.7. If the target $\mathbf{S}$ is an ordinary category, it is an easy exercise to see that the datum of a functor

$$\operatorname{Corr}(\mathbf{C})\to \mathbf{S}$$

is equivalent to that of a bi-simplicial map as in Sect. 0.1.4.

If $\mathbf{S}$ is a general $(\infty,1)$-category such an equivalence is the content of (a particular case of) Theorem 2.1.3, mentioned above.

0.2. The actual reason we need it.

0.2.1. The above was meant to explain why we need something like the category of correspondences if we want to have a bi-variant functor with base change. However, we were led to consider the category $\mathrm{Corr}(\mathbf{C})$ for a different reason.

Namely, we simply wanted to define $\mathrm{IndCoh}^!$ as a functor

$$(\mathrm{Sch}_{\mathrm{aft}})^{\mathrm{op}} \to 1\text{-Cat} \tag{0.4}$$

that assigns to $S \in \mathrm{IndCoh}$ the category $\mathrm{IndCoh}(S)$ and to a morphism $S_1 \xrightarrow{f} S_2$ the functor

$$f^! : \mathrm{IndCoh}(S_2) \to \mathrm{IndCoh}(S_1).$$

The problem, known since at least Hartshorne's 'Residues and duality', is that for an arbitrary morphism f, the functor $f^!$ is not adjoint to anything.

Namely, when f is proper, $f^!$ is the *right* adjoint to f_*^{IndCoh}, and when f is an open embedding, $f^!$ is the *left* adjoint to f_*^{IndCoh}. In general, one decomposes f as a composition $f_1 \circ f_2$ with f_2 an open embedding and f_1 proper, and defines

$$f^! := ((f_2)_*^{\mathrm{IndCoh}})^L \circ ((f_1)_*^{\mathrm{IndCoh}})^R.$$

0.2.2. This gives a valid definition of $f^!$ for a single morphism f, and it is not difficult to show that it is independent of the decomposition of f as $f_1 \circ f_2$. However, to make this construction functorial (i.e., to have IndCoh as a functor as in (0.4)) becomes a challenge.

For example, let us see what happens with a simple composition $f' \circ f''$. We write

$$f' = f_1' \circ f_2' \text{ and } f'' = f_1'' \circ f_2''.$$

Then to show that

$$(f' \circ f'')^! \simeq (f'')^! \circ (f')^!$$

we will need to show that

$$(f_1'')^! \circ (f_2')^! \simeq g^! \circ h^!,$$

where $f_2' \circ f_1'' = h \circ g$ with g an open embedding and h proper (while f_1'' was proper and f_2' was open, so we need to perform a swap).

One can imagine that this becomes quite combinatorially involved in the ∞-categorical setting, where one needs to consider n-fold compositions for any n.

0.2.3. Now, it turns out that the notion of a functor out of the category of correspondences provides a convenient framework to deal with the above issues.

However, there is a caveat: the $(\infty,1)$-category $\mathrm{Corr}(\mathbf{C})$ as defined above is not sufficient. We need to enlarge it to an $(\infty,2)$-category by allowing non-invertible 2-morphisms.

This is not surprising: in the construction of $f^!$ we appeal to the notion of *adjoint functor*, and the latter is a 2-categorical notion; i.e., it is intrinsic not to the $(\infty,1)$-category 1-Cat, but to the $(\infty,2)$-category $\mathbf{1}\text{-}\mathbf{Cat}$.

0.2.4. Given a class of 1-morphisms in $\mathbf{C}$, denoted *adm*, satisfying some reasonable conditions (see Sect. 1.1.1), we will attach to it an $(\infty, 2)$-category $\mathrm{Corr}(\mathbf{C})^{adm}$, such that the underlying $(\infty, 1)$-category is $\mathrm{Corr}(\mathbf{C})$ as was defined above.

In the example of $\mathbf{C} = \mathrm{Sch}_{\mathrm{aft}}$, we take *adm* to be the class of proper morphisms.

Let us explain the idea of $\mathrm{Corr}(\mathbf{C})^{adm}$. As was said above, 1-morphisms in $\mathrm{Corr}(\mathbf{C})^{adm}$ are diagrams (0.3). Given such a 1-morphism and another 1-morphism between the same two objects

$$\begin{array}{ccc} \mathbf{c}'_{0,1} & \longrightarrow & \mathbf{c}_0 \\ \downarrow & & \\ \mathbf{c}_1, & & \end{array}$$

a 2-morphism from the former to the latter is a commutative diagram

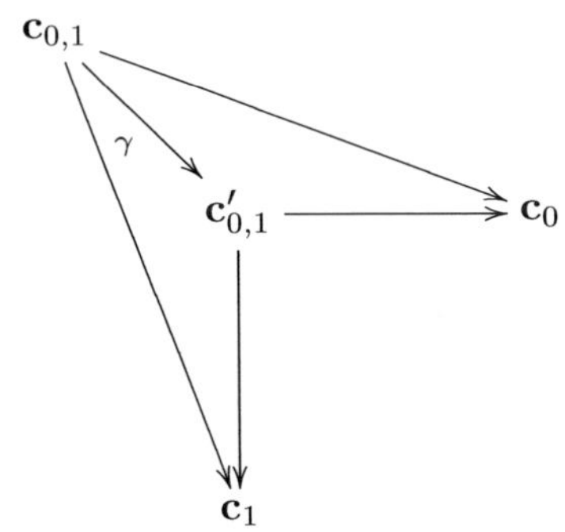

with $\gamma \in adm$.

0.2.5. In order to spell out this definition in the ∞-categorical world, we use the approach to $(\infty, 2)$-categories explained in Chapter 10, Sect. 2. Namely, we will think of a datum of an $(\infty, 2)$-category $\mathbb{T}$ in terms of the corresponding object

$$\mathrm{Seq}_\bullet(\mathbb{T}) \in 1\text{-}\mathrm{Cat}^{\mathbf{\Delta}^{\mathrm{op}}}.$$

For the desired $(\infty, 2)$-category $\mathrm{Corr}(\mathbf{C})^{adm}$, we take the corresponding object of $1\text{-}\mathrm{Cat}^{\mathbf{\Delta}^{\mathrm{op}}}$ to be the simplicial $(\infty, 1)$-category ${}''\mathbf{Grid}_\bullet^{\geq \mathrm{dgnl}}(\mathbf{C})^{adm}$, defined as follows.

For any n, the space underlying ${}''\mathbf{Grid}_n^{\geq \mathrm{dgnl}}(\mathbf{C})^{adm}$ equals $\mathrm{Grid}_n^{\geq \mathrm{dgnl}}(\mathbf{C})$ (as it should be, since we want $(\mathrm{Corr}(\mathbf{C})^{adm})^{1\text{-}\mathrm{Cat}} = \mathrm{Corr}(\mathbf{C})$). Now, ${}''\mathbf{Grid}_n^{\geq \mathrm{dgnl}}(\mathbf{C})^{adm}$ is defined as a 1-full subcategory in

$$\mathrm{Funct}(([n] \times [n]^{\mathrm{op}})^{\geq \mathrm{dgnl}}, \mathbf{C}),$$

where as 1-morphisms we allow natural transformations $\underline{\mathbf{c}} \to \underline{\mathbf{c}}'$ such that for every i, j, the corresponding map

$$\mathbf{c}_{i,j} \to \mathbf{c}'_{i,j}$$

belongs to the class *adm* and is an isomorphism for $i = j$.

0.2.6. Here is the theorem that one can prove regarding the functor IndCoh (this is the combination of Theorems 3.2.2, 4.1.3 and 5.2.4), see Sect. 0.3.7:

THEOREM 0.2.7. *There exists a* uniquely *defined functor*

$$\mathrm{IndCoh}^{\mathrm{proper}}_{\mathrm{all;all}} : \mathrm{Corr}(\mathrm{Sch}_{\mathrm{aft}})^{\mathrm{proper}} \to \mathbf{1\text{-}Cat} \tag{0.5}$$

whose restriction along

$$\mathrm{Sch}_{\mathrm{aft}} \to \mathrm{Corr}(\mathrm{Sch}_{\mathrm{aft}}) \to \mathrm{Corr}(\mathrm{Sch}_{\mathrm{aft}})^{\mathrm{proper}}$$

is identified with the functor

$$\mathrm{IndCoh} : \mathrm{Sch}_{\mathrm{aft}} \to 1\text{-}\mathrm{Cat}\,.$$

The sought-for functor

$$\mathrm{IndCoh}^{!} : (\mathrm{Sch}_{\mathrm{aft}})^{\mathrm{op}} \to 1\text{-}\mathrm{Cat}$$

is obtained from the functor $\mathrm{IndCoh}^{\mathrm{proper}}_{\mathrm{all;all}}$ of (0.5) by restriction along

$$(\mathrm{Sch}_{\mathrm{aft}})^{\mathrm{op}} \to \mathrm{Corr}(\mathrm{Sch}_{\mathrm{aft}}) \to \mathrm{Corr}(\mathrm{Sch}_{\mathrm{aft}})^{\mathrm{proper}}.$$

0.3. What is done in this chapter?

0.3.1. In Sect. 1 we define the $(\infty,2)$-category of correspondences.

The setting here is slightly more general than the one described in Sects. 0.1 and 0.2 (and this generalization is necessary for what we will develop in subsequent sections). Namely, in addition to *adm* we choose two more classes of morphisms in $\mathbf{C}$, denoted *vert* and *horiz*, respectively, so that $adm \subset vert \cap horiz$.

We restrict the class of one 1-morphisms in $\mathrm{Corr}(\mathbf{C})^{adm}$ by only allowing diagrams

$$\begin{array}{ccc} \mathbf{c}_{0,1} & \xrightarrow{\alpha} & \mathbf{c}_0 \\ \Big\downarrow{\scriptstyle\beta} & & \\ \mathbf{c}_1 & & \end{array}$$

where $\alpha \in horiz$ and $\beta \in vert$.

We denote the resulting $(\infty,2)$-category by $\mathrm{Corr}(\mathbf{C})^{adm}_{vert,horiz}$. It is endowed with a pair of functors

$$\mathbf{C}_{vert} \to \mathrm{Corr}(\mathbf{C})^{adm}_{vert,horiz} \leftarrow (\mathbf{C}_{horiz})^{\mathrm{op}}.$$

We describe explicitly the simplicial categories

$${}''\mathbf{Grid}^{\geq \mathrm{dgnl}}_{\bullet}(\mathbf{C})^{adm}_{vert,horiz} = \mathrm{Seq}_{\bullet}(\mathrm{Corr}(\mathbf{C})^{adm}_{vert,horiz}),$$

$${}'\mathbf{Grid}^{\geq \mathrm{dgnl}}_{\bullet}(\mathbf{C})^{adm}_{vert,horiz} = \mathrm{Seq}^{\mathrm{Pair}}_{\bullet}(\mathrm{Corr}(\mathbf{C})^{adm}_{vert,horiz}, \mathbf{C}_{adm})$$

and

$$\mathbf{Grid}^{\geq \mathrm{dgnl}}_{\bullet}(\mathbf{C})^{adm}_{vert,horiz} = \mathrm{Seq}^{\mathrm{Pair}}_{\bullet}(\mathrm{Corr}(\mathbf{C})^{adm}_{vert,horiz}, \mathbf{C}_{vert}).$$

For every $[n] \in \Delta$, each of these categories is a 1-full subcategory inside

$$\mathrm{Funct}(([n] \times [n]^{\mathrm{op}})^{\geq \mathrm{dgnl}}, \mathbf{C}).$$

All three have the same underlying space, denoted $\mathrm{Grid}^{\geq \mathrm{dgnl}}_{n}(\mathbf{C})^{adm}_{vert,horiz}$. Its objects are *half-grids* of objects of $\mathbf{C}$, where each internal square is Cartesian, all vertical arrows belong to *vert* and all horizontal arrows belong to *horiz*.

0.3.2. To specify a functor

$$\mathrm{Corr}(\mathbf{C})^{adm}_{vert,horiz} \to \mathbb{S}, \tag{0.6}$$

where $\mathbb{S}$ is another $(\infty, 2)$-category is equivalent to specifying a map of bi-simplicial spaces in any of the following versions:

$$\mathrm{Seq}_\bullet(''\mathbf{Grid}^{\geq \mathrm{dgnl}}_\bullet(\mathbf{C})^{adm}_{vert,horiz}) \to \mathrm{Sq}_{\bullet,\bullet}(\mathbb{S}),$$

$$\mathrm{Seq}_\bullet('\mathbf{Grid}^{\geq \mathrm{dgnl}}_\bullet(\mathbf{C})^{adm}_{vert,horiz}) \to \mathrm{Sq}_{\bullet,\bullet}(\mathbb{S})$$

or

$$\mathrm{Seq}_\bullet(\mathbf{Grid}^{\geq \mathrm{dgnl}}_\bullet(\mathbf{C})^{adm}_{vert,horiz}) \to \mathrm{Sq}_{\bullet,\bullet}(\mathbb{S}).$$

In Sect. 2 we give two additional formulations of what it takes to specify a functor as in (0.6). These formulations are used in the proofs of the various results establishing the existence and uniqueness of a functor out of a given category of correspondences with specified properties, discussed in the subsequent sections.

The first of these two formulations is given in terms of maps of bi-simplicial spaces

$$\mathrm{defGrid}_{\bullet,\bullet}(\mathbf{C})^{adm}_{vert;horiz} \to \mathrm{Sq}_{\bullet,\bullet}(\mathbb{S}).$$

The main difference between $\mathrm{defGrid}_{m,n}(\mathbf{C})^{adm}_{vert;horiz}$ and the space

$$\mathrm{Grid}^{\geq \mathrm{dgnl}}_n(\mathbf{C})^{adm}_{vert,horiz}$$

is that objects of $\mathrm{defGrid}_{m,n}(\mathbf{C})^{adm}_{vert;horiz}$ are functors

$$[m] \times [n]^{\mathrm{op}} \to \mathbf{C}.$$

I.e., here we are dealing with full $(m \times n)$-grids rather than half-grids. This is in the spirit of the initial idea in Sect. 0.1.4 of how to functorially account for base change.

The second description involves a bi-simplicial *category*, denoted

$$\mathbf{Grid}_{\bullet,\bullet}(\mathbf{C})^{adm}_{vert;horiz};$$

we refer the reader to Sect. 2.2 for the precise statement.

0.3.3. We now ask the following question: how does one ever construct a functor out of a given category of correspondences $\mathrm{Corr}(\mathbf{C})^{adm}_{vert,horiz}$?

In Sect. 3 we state and prove Theorem 3.2.2 that gives an answer to this question (and, quite probably, any functor out of a category of correspondences ultimately comes down to the paradigm described in this theorem).

Namely, suppose we have a category $\mathbf{C}$, with *horiz* a class of 1-morphisms (with some reasonable properties) as well as a a functor

$$\Phi : \mathbf{C} \to \mathbb{S},$$

where $\mathbb{S}$ is some $(\infty, 2)$-category. Assume that this functor has the following *property*[2]: for every arrow $\mathbf{c} \overset{\alpha}{\to} \mathbf{c}'$ in *horiz*, the 1-morphism

$$\Phi(\mathbf{c}) \overset{\Phi(\alpha)}{\to} \Phi(\mathbf{c}')$$

[2]We emphasize that this is a property and *not* an additional piece of data.

admits a right adjoint. Moreover, assume that these right adjoints satisfy a base change *property* against the 1-morphisms

$$\Phi(\mathbf{d}) \overset{\Phi(\beta)}{\to} \Phi(\mathbf{d}'), \quad (\mathbf{d} \overset{\beta}{\to} \mathbf{d}') \in \mathbf{C}.$$

Consider the $(\infty, 2)$-category $\mathrm{Corr}(\mathbf{C})^{horiz}_{\mathrm{all},horiz}$. The statement of Theorem 3.2.2 is that there exists a uniquely defined functor

$$\Phi^{horiz}_{\mathrm{all},horiz} : \mathrm{Corr}(\mathbf{C})^{horiz}_{\mathrm{all},horiz} \to \mathbb{S},$$

whose composition with $\mathbf{C} \to \mathrm{Corr}(\mathbf{C})^{horiz}_{\mathrm{all},horiz}$ is identified with the initial functor Φ.

For example, if we start with the functor

$$\mathrm{IndCoh} : \mathrm{Sch}_{\mathrm{aft}} \to 1\text{-}\mathrm{Cat},$$

the above theorem allows us to (uniquely) extend it to a functor

$$\mathrm{IndCoh}^{\mathrm{proper}}_{\mathrm{all},\mathrm{proper}} : \mathrm{Corr}(\mathrm{Sch}_{\mathrm{aft}})^{\mathrm{proper}}_{\mathrm{all},\mathrm{proper}} \to \mathbf{1}\text{-}\mathbf{Cat}$$

and to a functor

$$\mathrm{IndCoh}^{\mathrm{open}}_{\mathrm{all},\mathrm{open}} : \mathrm{Corr}(\mathrm{Sch}_{\mathrm{aft}})^{\mathrm{open}}_{\mathrm{all},\mathrm{open}} \to (\mathbf{1}\text{-}\mathbf{Cat})^{2\text{-}\mathrm{op}}.$$

0.3.4. In Sect. 4 we prove Theorem 4.1.3, which is the first out of the two basic theorems of this chapter[3] that say that starting from a functor from a given $(\infty, 2)$-category of correspondences, there is a canonical way to extend it to a larger one.

Let $adm' \supset adm$ be a larger class of morphisms, satisfying the following assumption: for any $\gamma : \mathbf{c} \to \mathbf{c}'$ from adm', the diagonal morphism

$$\mathbf{c} \to \mathbf{c} \underset{\mathbf{c}'}{\times} \mathbf{c}$$

belongs to adm.

In this case, Theorem 4.1.3 says that restriction under

$$\mathrm{Corr}(\mathbf{C})^{adm}_{vert,horiz} \to \mathrm{Corr}(\mathbf{C})^{adm'}_{vert,horiz}$$

defines a fully faithful embedding from the space of functors

$$\mathrm{Corr}(\mathbf{C})^{adm'}_{vert,horiz} \to \mathbb{S}$$

to that of functors

$$\mathrm{Corr}(\mathbf{C})^{adm}_{vert,horiz} \to \mathbb{S}.$$

Moreover, we give an explicit description of the essential image of this fully faithful embedding.

[3] A few more theorems of this kind will be given in Chapter 8.

0.3.5. Here are some typical applications of Theorem 4.1.3:

(i) Take $\mathbf{C} = \mathrm{Sch}_{\mathrm{aft}}$ with $adm' = \mathrm{open}$, and let $adm = \mathrm{isom}$, so that

$$\mathrm{Corr}(\mathrm{Sch}_{\mathrm{aft}})^{\mathrm{isom}}_{\mathrm{all,all}} = \mathrm{Corr}(\mathrm{Sch}_{\mathrm{aft}})_{\mathrm{all,all}} = \mathrm{Corr}(\mathrm{Sch}_{\mathrm{aft}})$$

is the $(\infty,1)$-category of correspondences from Sect. 0.1.5. We obtain that the datum of the functor

$$\mathrm{IndCoh}^{\mathrm{open}}_{\mathrm{all,all}} : \mathrm{Corr}(\mathrm{Sch}_{\mathrm{aft}})^{\mathrm{open}}_{\mathrm{all,all}} \to (\mathbf{1}\text{-}\mathbf{Cat})^{2\text{-op}}$$

is equivalent to that of its restriction under $\mathrm{Corr}(\mathrm{Sch}_{\mathrm{aft}}) \to \mathrm{Corr}(\mathrm{Sch}_{\mathrm{aft}})^{\mathrm{open}}_{\mathrm{all,all}}$,

$$\mathrm{IndCoh}_{\mathrm{all,all}} : \mathrm{Corr}(\mathrm{Sch}_{\mathrm{aft}})_{\mathrm{all,all}} \to 1\text{-}\mathrm{Cat}\,.$$

(i') Same as above, but we consider the pair

$$\mathrm{Corr}(\mathrm{Sch}_{\mathrm{aft}})_{\mathrm{all,open}} := \mathrm{Corr}(\mathrm{Sch}_{\mathrm{aft}})^{\mathrm{isom}}_{\mathrm{all,open}} \subset \mathrm{Corr}(\mathrm{Sch}_{\mathrm{aft}})^{\mathrm{open}}_{\mathrm{all,open}}$$

and the functor

$$\mathrm{IndCoh}^{\mathrm{open}}_{\mathrm{all,open}} : \mathrm{Corr}(\mathrm{Sch}_{\mathrm{aft}})^{\mathrm{open}}_{\mathrm{all,open}} \to (\mathbf{1}\text{-}\mathbf{Cat})^{2\text{-op}}.$$

(ii) Take $\mathbf{C} = \mathrm{Sch}_{\mathrm{aft}}$ with $adm' = \mathrm{proper}$, and let $adm = \mathrm{closed}$ be the class of closed embeddings. Then Theorem 4.1.3 implies that the datum of a functor

$$\mathrm{IndCoh}^{\mathrm{proper}}_{\mathrm{all,all}} : \mathrm{Corr}(\mathrm{Sch}_{\mathrm{aft}})^{\mathrm{proper}}_{\mathrm{all,all}} \to \mathbf{1}\text{-}\mathbf{Cat},$$

can be uniquely recovered from its restriction under

$$\mathrm{Corr}(\mathrm{Sch}_{\mathrm{aft}})^{\mathrm{closed}}_{\mathrm{all,all}} \to \mathrm{Corr}(\mathrm{Sch}_{\mathrm{aft}})^{\mathrm{proper}}_{\mathrm{all,all}},$$

$$\mathrm{IndCoh}^{\mathrm{closed}}_{\mathrm{all,all}} : \mathrm{Corr}(\mathrm{Sch}_{\mathrm{aft}})^{\mathrm{closed}}_{\mathrm{all,all}} \to \mathbf{1}\text{-}\mathbf{Cat}.$$

(ii') The same applies to the case when $\mathbf{C} = {}^{\mathrm{cl}}\mathrm{Sch}$ and the functor in question is

$$\mathrm{Dmod}^{\mathrm{proper}}_{\mathrm{all,all}} : \mathrm{Corr}({}^{\mathrm{cl}}\mathrm{Sch}_{\mathrm{aft}})^{\mathrm{proper}}_{\mathrm{all,all}} \to \mathbf{1}\text{-}\mathbf{Cat}.$$

(iii) We take $\mathbf{C} = {}^{\mathrm{cl}}\mathrm{Sch}$ with $adm' = \mathrm{closed}$, while $adm = \mathrm{isom}$, so that

$$\mathrm{Corr}({}^{\mathrm{cl}}\mathrm{Sch})^{\mathrm{isom}}_{\mathrm{all,all}} = \mathrm{Corr}({}^{\mathrm{cl}}\mathrm{Sch})_{\mathrm{all,all}} = \mathrm{Corr}({}^{\mathrm{cl}}\mathrm{Sch})$$

is the $(\infty,1)$-category of correspondences. We start with a functor

$$\mathrm{Dmod}^{\mathrm{closed}}_{\mathrm{all,all}} : \mathrm{Corr}({}^{\mathrm{cl}}\mathrm{Sch})^{\mathrm{closed}}_{\mathrm{all,all}} \to \mathbf{1}\text{-}\mathbf{Cat},$$

and we conclude that it can be uniquely recovered from its restriction under the functor $\mathrm{Corr}({}^{\mathrm{cl}}\mathrm{Sch})_{\mathrm{all,all}} \to \mathrm{Corr}({}^{\mathrm{cl}}\mathrm{Sch})^{\mathrm{closed}}_{\mathrm{all,all}}$,

$$\mathrm{Dmod}_{\mathrm{all,all}} : \mathrm{Corr}({}^{\mathrm{cl}}\mathrm{Sch})_{\mathrm{all,all}} \to 1\text{-}\mathrm{Cat}\,.$$

Note that combining with (ii') and (iii), we obtain that the datum of the functor

$$\mathrm{Dmod}^{\mathrm{proper}}_{\mathrm{all,all}} : \mathrm{Corr}({}^{\mathrm{cl}}\mathrm{Sch})^{\mathrm{proper}}_{\mathrm{all,all}} \to \mathbf{1}\text{-}\mathbf{Cat}$$

can be uniquely recovered from its restriction under

$$\mathrm{Corr}({}^{\mathrm{cl}}\mathrm{Sch})_{\mathrm{all,all}} \to \mathrm{Corr}(\mathrm{Sch}_{\mathrm{aft}})^{\mathrm{proper}}_{\mathrm{all,all}},$$

$$\mathrm{Dmod}_{\mathrm{all,all}} : \mathrm{Corr}({}^{\mathrm{cl}}\mathrm{Sch})_{\mathrm{all,all}} \to 1\text{-}\mathrm{Cat}\,.$$

REMARK 0.3.6. The point is that in (iii) we take fiber products in the category ${}^{\mathrm{cl}}\mathrm{Sch}$, and in this category, for a closed embedding $S \to S'$, the map

$$S \to S \underset{S'}{\times} S$$

is an isomorphism (which is of course completely false in Sch). On the other hand, if we tried to consider IndCoh out of the category ${}^{\mathrm{cl}}\mathrm{Sch}$, it would fail to satisfy base change.

0.3.7. In Sect. 5 we prove the second of the two extension results, Theorem 5.2.4. It is this theorem that allows us to construct the functor

$$\mathrm{IndCoh}^{\mathrm{proper}}_{\mathrm{all,all}} : \mathrm{Corr}(\mathrm{Sch}_{\mathrm{aft}})^{\mathrm{proper}}_{\mathrm{all,all}} \to \mathbf{1}\text{-}\mathbf{Cat},$$

starting from just

$$\mathrm{IndCoh} : \mathrm{Sch}_{\mathrm{aft}} \to 1\text{-}\mathrm{Cat}.$$

In this theorem, we start with four classes of morphisms *vert*, *horiz*, as well as *adm* and *co-adm* with some reasonable assumptions. The example one should keep in mind is when $\mathbf{C} = \mathrm{Sch}_{\mathrm{aft}}$ with $vert = horiz =$ all, $adm =$ proper and $co\text{-}adm =$ open.

We start with a functor

$$\Phi_{vert,co\text{-}adm} : \mathrm{Corr}(\mathbf{C})_{vert,co\text{-}adm} \to \mathbb{S}^{1\text{-}\mathrm{Cat}}, \tag{0.7}$$

and we wish to extend it to a functor

$$\Phi^{adm}_{vert,horiz} : \mathrm{Corr}(\mathbf{C})^{adm}_{vert,horiz} \to \mathbb{S}. \tag{0.8}$$

Theorem 5.2.4 says that under a certain assumption on *horiz*, *adm* and *co-adm*, restriction along $\mathrm{Corr}(\mathbf{C})^{adm}_{vert,horiz} \to \mathrm{Corr}(\mathbf{C})_{vert,co\text{-}adm}$ defines a fully faithful map from the space of functors (0.8) to the space of functors (0.7), whose essential image is explicitly described.

The assumption on our classes of morphisms is that for a given 1-morphism $\mathbf{c}_0 \overset{\alpha}{\to} \mathbf{c}_1$, the category of its factorizations as

$$\mathbf{c}_0 \overset{\epsilon}{\to} \overline{\mathbf{c}}_0 \overset{\gamma}{\to} \mathbf{c}_1, \quad \epsilon \in co\text{-}adm,\ \gamma \in adm$$

is *contractible.*

In our main application, we start with the functor

$$\mathrm{IndCoh}^{\mathrm{open}}_{\mathrm{all,open}} : \mathrm{Corr}(\mathrm{Sch}_{\mathrm{aft}})^{\mathrm{open}}_{\mathrm{all,open}} \to (\mathbf{1}\text{-}\mathbf{Cat})^{2\text{-}\mathrm{op}}$$

(which is uniquely constructed starting from $\mathrm{IndCoh} : \mathrm{Sch}_{\mathrm{aft}} \to 1\text{-}\mathrm{Cat}$, see Sect. 0.3.3); we restrict it to a functor

$$\mathrm{IndCoh}_{\mathrm{all,open}} : \mathrm{Corr}(\mathrm{Sch}_{\mathrm{aft}})_{\mathrm{all,open}} \to 1\text{-}\mathrm{Cat}$$

(this restriction does not lose information, see Example (i') in Sect. 0.3.5 above); and finally apply Theorem 5.2.4 to extend $\mathrm{IndCoh}_{\mathrm{all,open}}$ to the desired functor

$$\mathrm{IndCoh}^{\mathrm{proper}}_{\mathrm{all,all}} : \mathrm{Corr}(\mathrm{Sch}_{\mathrm{aft}})^{\mathrm{proper}}_{\mathrm{all,all}} \to \mathbf{1}\text{-}\mathbf{Cat}.$$

0.3.8. We should remark that if one is only interested in constructing the functor

$$\mathrm{IndCoh}^! : (\mathrm{Sch}_{\mathrm{aft}})^{\mathrm{op}} \to 1\text{-Cat},$$

then one can make do with a vastly simplified version of Theorem 5.2.4.

Namely, starting from the functor (0.7) we can first restrict it to

$$\mathrm{Corr}(\mathbf{C})_{adm,co\text{-}adm},$$

and then apply Theorem 4.1.3 to obtain a functor

$$\mathrm{Corr}(\mathbf{C})^{adm \cap co\text{-}adm}_{adm,co\text{-}adm} \to \mathbb{S}.$$

Then a much simplified version of Steps B and C of the proof will produce from the latter functor the desired functor

$$\Phi^! : (\mathbf{C}_{horiz})^{\mathrm{op}} \to \mathbb{S}.$$

We note, however, that both steps of the procedure just described lose information; whereas the one in Theorem 5.2.4 does not.

1. The 2-category of correspondences

In this section, given an $(\infty,1)$-category $\mathbf{C}$ with three distinguished classes of 1-morphisms *vert*, *horiz* and *adm*, we will construct the corresponding $(\infty,2)$-category of correspondences $\mathrm{Corr}(\mathbf{C})^{adm}_{vert;horiz}$.

1.1. The set-up. In this subsection we will list the requirements on the classes of morphisms *vert*, *horiz* and *adm*, and explain what the desired $(\infty,2)$-category $\mathrm{Corr}(\mathbf{C})^{adm}_{vert;horiz}$ is when $\mathbf{C}$ is an *ordinary* category. In this case, $\mathrm{Corr}(\mathbf{C})^{adm}_{vert;horiz}$ will be an ordinary 2-category, which can specified by saying what are its objects, 1-morphisms and 2-morphisms.

1.1.1. Let $\mathbf{C}$ be an $(\infty,1)$-category. Let $vert, horiz$ be two classes of 1-morphisms in $\mathbf{C}$, and $adm \subset vert \cap horiz$ a third class, such that:

(1) The identity maps of objects of $\mathbf{C}$ belong to all three classes;

(2) If a 1-morphism belongs to a given class, then so do all isomorphic 1-morphisms;

(3) All three classes are closed under compositions.

(4) Given a morphism $\alpha_1 : \mathbf{c}_{1,1} \to \mathbf{c}_{1,0}$ in *horiz* and a morphism $\beta_0 : \mathbf{c}_{0,0} \to \mathbf{c}_{1,0}$ *vert*, the Cartesian square

$$\begin{array}{ccc} \mathbf{c}_{0,1} & \xrightarrow{\alpha_0} & \mathbf{c}_{0,0} \\ {\scriptstyle\beta_1}\downarrow & & \downarrow{\scriptstyle\beta_0} \\ \mathbf{c}_{1,1} & \xrightarrow{\alpha_1} & \mathbf{c}_{1,0} \end{array} \tag{1.1}$$

exists, and $\alpha_0 \in horiz$ and $\beta_1 \in vert$. Moreover, if α_1 (resp., β_0) belongs to *adm*, then so does $\alpha_0 \in horiz$ (resp., $\beta_1 \in vert$).

(5) The class *adm* satisfies the '2 out of 3' property: if

$$\mathbf{c}_1 \overset{\alpha}{\to} \mathbf{c}_2 \overset{\beta}{\to} \mathbf{c}_3$$

are maps with β and $\beta \circ \alpha$ in *adm*, then α is also in *adm*.

For example, if $\mathbf{C}$ contains fiber products for all morphisms, we can take $vert = horiz = adm$ to be the class of all 1-morphisms.

We let

$$\mathbf{C}_{vert}, \ \mathbf{C}_{horiz} \text{ and } \mathbf{C}_{adm}$$

denote the corresponding 1-full subcategories of $\mathbf{C}$.

1.1.2. Note the following consequences of the above conditions.

First, we claim that if $\mathbf{c} \to \mathbf{c}'$ is a 1-morphism that belongs to adm, then so does the diagonal morphism $\mathbf{c} \to \mathbf{c} \underset{\mathbf{c}'}{\times} \mathbf{c}$.

Second, the above observation implies that for a diagram

$$\begin{array}{ccccc} & & \mathbf{c}_1 & & \\ & & \downarrow & \searrow & \\ \mathbf{c}_2 & \longrightarrow & \mathbf{c}_3 & & \mathbf{c}'_1 \\ & \searrow & & \searrow & \downarrow \\ & & \mathbf{c}'_2 & \longrightarrow & \mathbf{c}'_3 \end{array}$$

with the slanted maps in adm, vertical arrows in $vert$ and horizontal arrows in $horiz$, the resulting map

$$\mathbf{c}_1 \underset{\mathbf{c}_3}{\times} \mathbf{c}_2 \to \mathbf{c}'_1 \underset{\mathbf{c}'_3}{\times} \mathbf{c}'_2$$

is in adm.

1.1.3. We wish to define a $(\infty, 2)$-category, denoted $\operatorname{Corr}(\mathbf{C})^{adm}_{vert;horiz}$. We want its objects be the same as objects of $\mathbf{C}$.

For $\mathbf{c}_0, \mathbf{c}_1 \in \operatorname{Corr}(\mathbf{C})$, we want the $(\infty, 1)$-category

$$\operatorname{Maps}_{\operatorname{Corr}(\mathbf{C})^{adm}_{vert;horiz}}(\mathbf{c}_0, \mathbf{c}_1)$$

to have as objects *correspondences*

$$\begin{array}{ccc} \mathbf{c}_{0,1} & \xrightarrow{\alpha} & \mathbf{c}_0 \\ \Big\downarrow{\scriptstyle \beta} & & \\ \mathbf{c}_1 & & \end{array} \tag{1.2}$$

where $\alpha \in horiz$ and $\beta \in vert$.

For a pair of correspondences $(\mathbf{c}_{0,1}, \alpha, \beta)$ and $(\mathbf{c}'_{0,1}, \alpha', \beta')$, we want the space of maps between them to be that of commutative diagrams

(1.3) 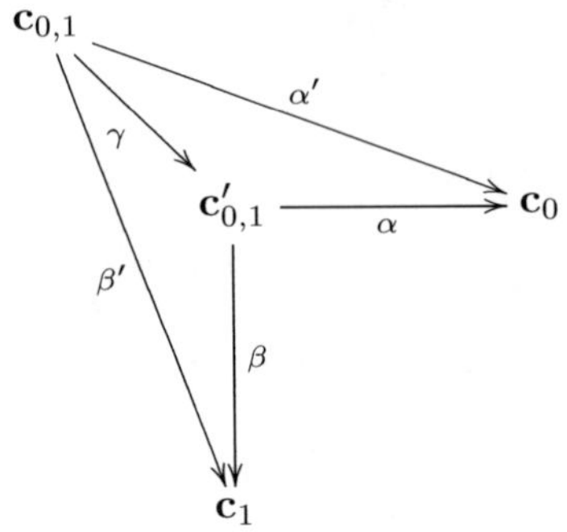

with $\gamma \in adm$.

Composition of 1-morphisms should be given by forming Cartesian products:

$$\begin{array}{ccccc} \mathbf{c}_{0,2} & \longrightarrow & \mathbf{c}_{0,1} & \longrightarrow & \mathbf{c}_0 \\ \downarrow & & \downarrow & & \\ \mathbf{c}_{1,2} & \longrightarrow & \mathbf{c}_1 & & \\ \downarrow & & & & \\ \mathbf{c}_2. & & & & \end{array}$$

1.1.4. It is easy to check that when $\mathbf{C}$ is an ordinary category, the above construction indeed gives rise to an (ordinary) 2-category.

To give the actual definition in the ∞-categorical framework, we shall use the formalism of Chapter 10, Sect. 2.1

1.2. The Segal category of correspondences. In this subsection we will carry out the construction of the sought-for category $\operatorname{Corr}(\mathbf{C})^{adm}_{vert;horiz}$.

The construction will be very intuitive: when we think of the datum of an $(\infty, 2)$-category in terms of its image under the functor

$$\operatorname{Seq}_\bullet : 2\text{-Cat} \to 1\text{-Cat}^{\mathbf{\Delta}^{\mathrm{op}}},$$

it is quite clear what the simplicial $(\infty, 1)$-category corresponding to $\operatorname{Corr}(\mathbf{C})^{adm}_{vert;horiz}$ should be. Namely, for each n, the corresponding $(\infty, 1)$-category, denoted $''\mathbf{Grid}_n^{\geq \mathrm{dgnl}}(\mathbf{C})^{adm}_{vert;horiz}$, is one whose objects are half-grids of size n of objects of $\mathbf{C}$, in which every square is Cartesian.

This category $''\mathbf{Grid}_n^{\geq \mathrm{dgnl}}(\mathbf{C})^{adm}_{vert;horiz}$ of half-grids is easy to make sense of in the ∞-categorical setting, because it is obtained from a category of functors from a (very simple) ordinary category to $\mathbf{C}$.

1.2.1. Let $(\mathbf{C}, vert, horiz, adm)$ be as in Sect. 1.1.1. We will now define the desired object

$$''\mathbf{Grid}_{\bullet}^{\geq \mathrm{dgnl}}(\mathbf{C})_{vert;horiz}^{adm} \in 1\text{-Cat}^{\mathbf{\Delta}^{\mathrm{op}}}.$$

Consider the following co-simplicial object of 1-Cat, which in fact takes values in ordinary categories. For each $n = 0, 1, 2, ...$ we consider the full subcategory

$$([n] \times [n]^{\mathrm{op}})^{\geq \mathrm{dgnl}} \subset [n] \times [n]^{\mathrm{op}},$$

spanned by objects (i, j) with $i \leq j$.

Consider the $(\infty, 1)$-category

$$\mathbf{Maps}\big(([n] \times [n]^{\mathrm{op}})^{\geq \mathrm{dgnl}}, \mathbf{C}\big).$$

I.e., this is the category of commutative diagrams $\underline{\mathbf{c}}$

$$\begin{array}{ccccccccc}
\mathbf{c}_{0,n} & \longrightarrow & \mathbf{c}_{0,n-1} & \longrightarrow & \cdots & \longrightarrow & \mathbf{c}_{0,1} & \longrightarrow & \mathbf{c}_{0,0} \\
\downarrow & & \downarrow & & & & \downarrow & & \\
\mathbf{c}_{1,n} & \longrightarrow & \mathbf{c}_{1,n-1} & \longrightarrow & \cdots & \longrightarrow & \mathbf{c}_{1,1} & & \\
\downarrow & & \downarrow & & \downarrow & & & & \\
\cdots & \longrightarrow & \cdots & \longrightarrow & \cdots & & & & \\
\downarrow & & \downarrow & & & & & & \\
\mathbf{c}_{n-1,n} & \longrightarrow & \mathbf{c}_{n-1,n-1} & & & & & & \\
\downarrow & & & & & & & & \\
\mathbf{c}_{n,n}. & & & & & & & &
\end{array} \tag{1.4}$$

1.2.2. For $n = 0, 1, 2, ...$ we define an $(\infty, 1)$-category $''\mathbf{Grid}_n^{\geq \mathrm{dgnl}}(\mathbf{C})_{vert;horiz}^{adm}$ to be the following 1-full subcategory of $\mathbf{Maps}\big(([n] \times [n]^{\mathrm{op}})^{\geq \mathrm{dgnl}}, \mathbf{C}\big)$.

The objects of $''\mathbf{Grid}_n^{\geq \mathrm{dgnl}}(\mathbf{C})_{vert;horiz}^{adm}$ consist of those diagrams (1.4) that:

(1) All the vertical maps belong to *vert*;

(2) All the horizontal maps belong to *horiz*;

(3) All squares are Cartesian.

We restrict 1-morphisms to those maps $\underline{\mathbf{c}} \to \underline{\mathbf{c}}'$ between diagrams (1.4), for which the corresponding map

$$\mathbf{c}_{i,j} \to \mathbf{c}'_{i,j}$$

belongs to *adm* for all $0 \leq i \leq j \leq n$ and is an *isomorphism* when $i = j$.

1.2.3. It is clear that for a map $[m]\to[n]$ in $\mathbf{\Delta}$, the restriction functor

$$\mathbf{Maps}\big(([n]\times[n]^{\mathrm{op}})^{\geq\,\mathrm{dgnl}},\mathbf{C}\big)\to\mathbf{Maps}\big(([m]\times[m]^{\mathrm{op}})^{\geq\,\mathrm{dgnl}},\mathbf{C}\big)$$

sends

$$''\mathbf{Grid}_n^{\geq\,\mathrm{dgnl}}(\mathbf{C})^{adm}_{vert;horiz}\to{}''\mathbf{Grid}_m^{\geq\,\mathrm{dgnl}}(\mathbf{C})^{adm}_{vert;horiz}.$$

Hence, we obtain a well-defined object

$$''\mathbf{Grid}_\bullet^{\geq\,\mathrm{dgnl}}(\mathbf{C})^{adm}_{vert;horiz}\in 1\text{-}\mathrm{Cat}^{\mathbf{\Delta}^{\mathrm{op}}}.$$

PROPOSITION 1.2.4. *The object* $''\mathbf{Grid}_\bullet^{\geq\,\mathrm{dgnl}}(\mathbf{C})^{adm}_{vert;horiz}\in 1\text{-}\mathrm{Cat}^{\mathbf{\Delta}^{\mathrm{op}}}$ *lies in the essential image of the functor* $\mathrm{Seq}_\bullet$.

PROOF. The fact that 0-simplices of $''\mathbf{Grid}_\bullet^{\geq\,\mathrm{dgnl}}(\mathbf{C})^{adm}_{vert;horiz}$ are a space follows from the definition. The fact that $''\mathbf{Grid}_\bullet^{\geq\,\mathrm{dgnl}}(\mathbf{C})^{adm}_{vert;horiz}$ is a Segal category is also clear. Thus, we only have to check that every invertible 1-simplex is degenerate.

A 1-simplex in $''\mathbf{Grid}_\bullet^{\geq\,\mathrm{dgnl}}(\mathbf{C})^{adm}_{vert;horiz}$ is given by a diagram

$$\begin{array}{ccc}\mathbf{d} & \longrightarrow & \mathbf{c}_0\\ \big\downarrow & & \\ \mathbf{c}_1, & & \end{array}$$

and the fact that it is invertible means that this diagram can be completed to a diagram

$$\begin{array}{ccccccc}\mathbf{d}' & \longrightarrow & \mathbf{c}_1 & \longrightarrow & \mathbf{d}' & \xrightarrow{\beta} & \mathbf{c}_1\\ \big\downarrow & & \big\downarrow & & \big\downarrow{\scriptstyle\alpha} & & \\ \mathbf{c}_0 & \longrightarrow & \mathbf{d} & \longrightarrow & \mathbf{c}_0 & & \\ \big\downarrow & & \big\downarrow & & & & \\ \mathbf{d}' & \xrightarrow{\beta} & \mathbf{c}_1 & & & & \\ \big\downarrow{\scriptstyle\alpha} & & & & & & \\ \mathbf{c}_0, & & & & & & \end{array}$$

with all the squares being Cartesian, and both composite maps $\mathbf{c}_0\to\mathbf{c}_0$ (resp., $\mathbf{c}_1\to\mathbf{c}_1$) are the identity maps. However, this implies that all the arrows in this diagram are isomorphisms.

□

1.2.5. We define the $(\infty,2)$-category

$$\mathrm{Corr}(\mathbf{C})^{adm}_{vert;horiz}\in 2\text{-}\mathrm{Cat},$$

to be such that

$$\mathrm{Seq}_\bullet(\mathrm{Corr}(\mathbf{C})^{adm}_{vert;horiz})={}''\mathbf{Grid}_\bullet^{\geq\,\mathrm{dgnl}}(\mathbf{C})^{adm}_{vert;horiz}.$$

The $(\infty,2)$-category $\mathrm{Corr}(\mathbf{C})^{adm}_{vert;horiz}$ can thus be recovered as

$$\mathrm{Corr}(\mathbf{C})^{adm}_{vert;horiz}\simeq\mathfrak{L}(''\mathbf{Grid}_\bullet^{\geq\,\mathrm{dgnl}}(\mathbf{C})^{adm}_{vert;horiz}),$$

see Chapter 10, Sect. 4.4.1 where the functor $\mathfrak{L}$ is introduced.

Note that we have a canonical identification

$$(\mathrm{Corr}(\mathbf{C})^{adm}_{vert;horiz})^{1\text{-op}} \simeq \mathrm{Corr}(\mathbf{C})^{adm}_{horiz;vert}.$$

1.3. Changing the class of 2-morphisms. In this subsection we show that if we replace the class *adm* (which gives rise to 2-morphisms) by a smaller one, things work as they should.

In particular, we will see that the $(\infty,2)$-category $\mathrm{Corr}(\mathbf{C})^{adm}_{vert;horiz}$ contains $\mathbf{C}_{vert}$ and $\mathbf{C}^{\mathrm{op}}_{horiz}$ as 1-full subcategories.

1.3.1. Let us now be given two classes adm' and adm as in Sect. 1.1.1 with $adm' \subset adm$. On the one hand, we can consider the $(\infty,2)$-category $\mathrm{Corr}(\mathbf{C})^{adm'}_{vert;horiz}$. On the other hand, we can consider the 2-full subcategory

$$\mathrm{Corr}(\mathbf{C})^{adm' \subset adm}_{vert;horiz} \subset \mathrm{Corr}(\mathbf{C})^{adm}_{vert;horiz},$$

obtained by leaving by keeping objects and 1-morphisms the same, but restricting 2-morphisms to those diagrams (1.3), where $\gamma \in adm'$.

The tautological functor

$$\mathrm{Corr}(\mathbf{C})^{adm'}_{vert;horiz} \to \mathrm{Corr}(\mathbf{C})^{adm}_{vert;horiz}$$

factors through a canonical functor

$$\mathrm{Corr}(\mathbf{C})^{adm'}_{vert;horiz} \to \mathrm{Corr}(\mathbf{C})^{adm' \subset adm}_{vert;horiz}. \tag{1.5}$$

The following is tautological:

LEMMA 1.3.2. *The functor* (1.5) *is an equivalence.*

1.3.3. Let us consider a particular case of the above construction when $adm = \mathrm{isom}$, i.e., is the class of all isomorphisms.

In this case, $''\mathbf{Grid}^{\geq \mathrm{dgnl}}_{\bullet}(\mathbf{C})^{\mathrm{isom}}_{vert;horiz}$ belongs to $\mathrm{Spc}^{\mathbf{\Delta}^{\mathrm{op}}}$. Therefore,

$$\mathrm{Corr}(\mathbf{C})^{\mathrm{isom}}_{vert;horiz}$$

is an $(\infty,1)$-category. We shall denote it simply by $\mathrm{Corr}(\mathbf{C})_{vert;horiz}$.

By Lemma 1.3.2, for any *adm*, we have

$$(\mathrm{Corr}(\mathbf{C})^{adm}_{vert;horiz})^{1\text{-Cat}} \simeq \mathrm{Corr}(\mathbf{C})_{vert;horiz}.$$

1.3.4. Let us now take both classes *horiz* and *adm* to be isom. In this case the corresponding $(\infty,1)$-category

$$\mathrm{Corr}(\mathbf{C})_{vert;\mathrm{isom}} = \mathrm{Corr}(\mathbf{C})^{\mathrm{isom}}_{vert;\mathrm{isom}}$$

is canonically equivalent to $\mathbf{C}_{vert}$. Indeed, it is easy to see that for every n, the corresponding category $''\mathbf{Grid}^{\geq \mathrm{dgnl}}_{n}(\mathbf{C})^{\mathrm{isom}}_{vert;\mathrm{isom}}$ is canonically equivalent to

$$\mathbf{Maps}([n], \mathbf{C}_{vert})^{\mathrm{Spc}}.$$

Similarly, the $(\infty,1)$-category

$$\mathrm{Corr}(\mathbf{C})_{\mathrm{isom};horiz} = \mathrm{Corr}(\mathbf{C})^{\mathrm{isom}}_{\mathrm{isom},vert}$$

is canonically equivalent to $(\mathbf{C}_{horiz})^{\mathrm{op}}$.

In particular, we obtain the functors:

$$\mathbf{C}_{vert} \simeq \mathrm{Corr}(\mathbf{C})^{\mathrm{isom}}_{vert;\mathrm{isom}} \to \mathrm{Corr}(\mathbf{C})^{adm}_{vert;horiz}$$

and

$$(\mathbf{C}_{horiz})^{\mathrm{op}} \simeq \mathrm{Corr}(\mathbf{C})^{\mathrm{isom}}_{\mathrm{isom},vert} \to \mathrm{Corr}(\mathbf{C})^{adm}_{vert;horiz}.$$

1.4. Distinguishing a class of 1-morphisms. Recall from Chapter 10, Sect. 4.3.7 that in addition to the functor

$$\mathrm{Seq}_\bullet : 2\text{-Cat} \to 1\text{-Cat}^{\mathbf{\Delta}^{\mathrm{op}}}$$

there is also a functor $\mathrm{Seq}^{\mathrm{Pair}}_\bullet$ that takes as an input an $(\infty, 2)$-category and a class of its 1-morphisms.

In this subsection we will describe the result applying $\mathrm{Seq}^{\mathrm{Pair}}_\bullet$ to

$$\mathrm{Corr}(\mathbf{C})^{adm}_{vert;horiz},$$

with respect to the following two 1-full subcategories

$$\mathbf{C}_{adm} \subset \mathbf{C}_{vert} \subset (\mathrm{Corr}(\mathbf{C})^{adm}_{vert;horiz})^{1\text{-Cat}}.$$

As a result we will see two versions of the simplicial $(\infty, 1)$-category

$$''\mathbf{Grid}^{\geq \mathrm{dgnl}}_n(\mathbf{C})^{adm}_{vert;horiz},$$

denoted $'\mathbf{Grid}^{\geq \mathrm{dgnl}}_n(\mathbf{C})^{adm}_{vert;horiz}$ and $\mathbf{Grid}^{\geq \mathrm{dgnl}}_n(\mathbf{C})^{adm}_{vert;horiz}$, respectively.

Both these versions are useful in the applications.

1.4.1. For a natural number n, we define the $(\infty, 1)$-categories

$$\mathbf{Grid}^{\geq \mathrm{dgnl}}_n(\mathbf{C})^{adm}_{vert;horiz} \text{ and } '\mathbf{Grid}^{\geq \mathrm{dgnl}}_n(\mathbf{C})^{adm}_{vert;horiz}$$

to be 1-full subcategories $\mathbf{Maps}\big(([n] \times [n]^{\mathrm{op}})^{\geq \mathrm{dgnl}}, \mathbf{C}\big)$ having the same objects as the category $''\mathbf{Grid}^{\geq \mathrm{dgnl}}_n(\mathbf{C})^{adm}_{vert;horiz}$, but a larger class of 1-morphisms:

For $\mathbf{Grid}^{\geq \mathrm{dgnl}}_\bullet(\mathbf{C})^{adm}_{vert;horiz}$ we allow those maps $\underline{\mathbf{c}} \to \underline{\mathbf{c}}'$ between diagrams (1.4), such that the corresponding maps

$$\mathbf{c}_{i,j} \to \mathbf{c}'_{i,j}$$

belong to $vert$ for all $0 \leq i \leq j \leq n$, and such that for every $0 \leq j-1 < j \leq n$, the map

$$\mathbf{c}_{i,j} \to \mathbf{c}'_{i,j} \underset{\mathbf{c}'_{i,j-1}}{\times} \mathbf{c}_{i,j-1}$$

belongs to adm (i.e., the defect of the corresponding square to be Cartesian is a 1-morphism from adm).

For $'\mathbf{Grid}^{\geq \mathrm{dgnl}}_\bullet(\mathbf{C})^{adm}_{vert;horiz}$ we allow those maps $\underline{\mathbf{c}} \to \underline{\mathbf{c}}'$ between diagrams (1.4), such that the corresponding maps

$$\mathbf{c}_{i,j} \to \mathbf{c}'_{i,j}$$

belong to adm for all $0 \leq i \leq j \leq n$.

Denote also

$$\mathrm{Grid}^{\geq \mathrm{dgnl}}_n(\mathbf{C})^{adm}_{vert;horiz} := \mathbf{Grid}^{\geq \mathrm{dgnl}}_n(\mathbf{C})^{\mathrm{isom}}_{vert;horiz} \simeq (\mathbf{Grid}^{\geq \mathrm{dgnl}}_n(\mathbf{C})^{adm}_{vert;horiz})^{\mathrm{Spc}} \simeq$$
$$\simeq ('\mathbf{Grid}^{\geq \mathrm{dgnl}}_n(\mathbf{C})^{adm}_{vert;horiz})^{\mathrm{Spc}} \simeq (''\mathbf{Grid}^{\geq \mathrm{dgnl}}_n(\mathbf{C})^{adm}_{vert;horiz})^{\mathrm{Spc}}.$$

1.4.2. As in the case of $''\mathbf{Grid}_n^{\geq \mathrm{dgnl}}(\mathbf{C})^{adm}_{vert;horiz}$, the restriction functor

$$\mathbf{Maps}\left(([n]\times[n]^{\mathrm{op}})^{\geq \mathrm{dgnl}}, \mathbf{C}\right) \to \mathbf{Maps}\left(([m]\times[m]^{\mathrm{op}})^{\geq \mathrm{dgnl}}, \mathbf{C}\right)$$

sends

$$\mathbf{Grid}_n^{\geq \mathrm{dgnl}}(\mathbf{C})^{adm}_{vert;horiz} \to \mathbf{Grid}_m^{\geq \mathrm{dgnl}}(\mathbf{C})^{adm}_{vert;horiz}$$

and

$$'\mathbf{Grid}_n^{\geq \mathrm{dgnl}}(\mathbf{C})^{adm}_{vert;horiz} \to {}'\mathbf{Grid}_m^{\geq \mathrm{dgnl}}(\mathbf{C})^{adm}_{vert;horiz}$$

Thus, we obtain well-defined objects

$$\mathbf{Grid}_\bullet^{\geq \mathrm{dgnl}}(\mathbf{C})^{adm}_{vert;horiz} \text{ and } {}'\mathbf{Grid}_\bullet^{\geq \mathrm{dgnl}}(\mathbf{C})^{adm}_{vert;horiz}$$

in $1\text{-}\mathrm{Cat}^{\mathbf{\Delta}^{\mathrm{op}}}$.

Similarly, we obtain an object

$$\mathrm{Grid}^{\geq \mathrm{dgnl}}(\mathbf{C})^{adm}_{vert;horiz} \in \mathrm{Spc}^{\mathbf{\Delta}^{\mathrm{op}}}.$$

The following results from the definitions:

LEMMA 1.4.3. *The objects*

$$'\mathbf{Grid}_\bullet^{\geq \mathrm{dgnl}}(\mathbf{C})^{adm}_{vert;horiz} \text{ and } \mathbf{Grid}_\bullet^{\geq \mathrm{dgnl}}(\mathbf{C})^{adm}_{vert;horiz}$$

of $1\text{-}\mathrm{Cat}^{\mathbf{\Delta}^{\mathrm{op}}}$ *are both Segal categories.*

1.4.4. We now claim:

PROPOSITION 1.4.5. *We have:*

(a) $\mathbf{Grid}_\bullet^{\geq \mathrm{dgnl}}(\mathbf{C})^{adm}_{vert;horiz} = \mathrm{Seq}_\bullet^{\mathrm{Pair}}(\mathrm{Corr}(\mathbf{C})^{adm}_{vert;horiz}, \mathbf{C}_{vert})$.

(b) $'\mathbf{Grid}_\bullet^{\geq \mathrm{dgnl}}(\mathbf{C})^{adm}_{vert;horiz} = \mathrm{Seq}_\bullet^{\mathrm{Pair}}(\mathrm{Corr}(\mathbf{C})^{adm}_{vert;horiz}, \mathbf{C}_{adm})$.

PROOF. We will give the proof for $\mathbf{Grid}_\bullet^{\geq \mathrm{dgnl}}(\mathbf{C})^{adm}_{vert;horiz}$; the case of

$$'\mathbf{Grid}_\bullet^{\geq \mathrm{dgnl}}(\mathbf{C})^{adm}_{vert;horiz}$$

is similar.

First, we claim that $\mathbf{Grid}_\bullet^{\geq \mathrm{dgnl}}(\mathbf{C})^{adm}_{vert;horiz}$ lies in the essential image of the functor $\mathrm{Seq}_\bullet^{\mathrm{Pair}}$, i.e., that it is a half-symmetric Segal category. We can check this at the level of ordinary Segal categories. Thus, we can replace

$$\mathbf{Grid}_n^{\geq \mathrm{dgnl}}(\mathbf{C})^{adm}_{vert;horiz} \text{ by } (\mathbf{Grid}_n^{\geq \mathrm{dgnl}}(\mathbf{C})^{adm}_{vert;horiz})^{\mathrm{ordn}},$$

while

$$(\mathbf{Grid}_n^{\geq \mathrm{dgnl}}(\mathbf{C})^{adm}_{vert;horiz})^{\mathrm{ordn}} \simeq \mathbf{Grid}_n^{\geq \mathrm{dgnl}}(\mathbf{C}^{\mathrm{ordn}})^{adm}_{vert;horiz}.$$

Now, for $\mathbf{C}^{\mathrm{ordn}}$, its is easy to see that $'\mathbf{Grid}_\bullet^{\geq \mathrm{dgnl}}(\mathbf{C}^{\mathrm{ordn}})^{adm}_{vert;horiz}$ is equivalent to

$$\mathrm{Seq}_\bullet^{\mathrm{Pair}}(\mathrm{Corr}(\mathbf{C}^{\mathrm{ordn}})^{adm}_{vert;horiz}, (\mathbf{C}^{\mathrm{ordn}})_{adm}).$$

Since

$$''\mathbf{Grid}_\bullet^{\geq \mathrm{dgnl}}(\mathbf{C})^{adm}_{vert;horiz} \simeq \mathfrak{r}(\mathbf{Grid}_\bullet^{\geq \mathrm{dgnl}}(\mathbf{C})^{adm}_{vert;horiz}),$$

we conclude that

$$\mathbf{Grid}_\bullet^{\geq \mathrm{dgnl}}(\mathbf{C})^{adm}_{vert;horiz} = \mathrm{Seq}_\bullet^{\mathrm{Pair}}(\mathrm{Corr}(\mathbf{C})^{adm}_{vert;horiz}, \mathbf{D}),$$

where $\mathbf{D}$ is a 1-full subcategory of $(\mathrm{Corr}(\mathbf{C})^{adm}_{vert;horiz})^{1\text{-}\mathrm{Cat}}$, with the same class of objects.

It remains to show that $\mathbf{D} = \mathbf{C}_{vert}$. However, this also follows from the corresponding fact for the underlying ordinary categories.

□

COROLLARY 1.4.6. *We have canonical identifications*

$$\mathfrak{L}^{\mathrm{ext}}(\mathbf{Grid}_{\bullet}^{\geq \mathrm{dgnl}}(\mathbf{C})^{adm}_{vert;horiz}) \simeq \mathrm{Corr}(\mathbf{C})^{adm}_{vert;horiz},$$

and

$$\mathfrak{L}^{\mathrm{ext}}('\mathbf{Grid}_{\bullet}^{\geq \mathrm{dgnl}}(\mathbf{C})^{adm}_{vert;horiz}) \simeq \mathrm{Corr}(\mathbf{C})^{adm}_{vert;horiz},$$

where $\mathfrak{L}^{\mathrm{ext}}$ is as in Chapter 10, Sect. 4.4.1.

2. The category of correspondences via grids

According to the previous section, for an $(\infty,2)$-category $\mathbb{S}$ the data of a functor

$$\mathrm{Corr}(\mathbf{C})^{adm}_{vert;horiz} \to \mathbb{S} \tag{2.1}$$

amounts to any of the following:

(i) A map of bi-simplicial spaces $\mathrm{Seq}_{\bullet}(''\mathbf{Grid}_{\bullet}^{\geq \mathrm{dgnl}}(\mathbf{C})^{adm}_{vert;horiz}) \to \mathrm{Sq}_{\bullet,\bullet}(\mathbb{S})$;

(ii) A map of bi-simplicial spaces $\mathrm{Seq}_{\bullet}('\mathbf{Grid}_{\bullet}^{\geq \mathrm{dgnl}}(\mathbf{C})^{adm}_{vert;horiz}) \to \mathrm{Sq}_{\bullet,\bullet}(\mathbb{S})$;

(iii) A map of bi-simplicial spaces $\mathrm{Seq}_{\bullet}(\mathbf{Grid}_{\bullet}^{\geq \mathrm{dgnl}}(\mathbf{C})^{adm}_{vert;horiz}) \to \mathrm{Sq}_{\bullet,\bullet}(\mathbb{S})$.

In this section we will give yet two more interpretations of what it takes to define a functor (2.1).

One (given by Theorem 2.1.3) will still have the form of maps of from a certain bi-simplicial space to $\mathrm{Sq}_{\bullet,\bullet}(\mathbb{S})$, but the flavor of this new bi-simplicial space will be different: instead of half-grids we will have $m \times n$-grids of objects of $\mathbf{C}$. The other one (given by Theorem 2.2.7) involves three-dimensional grids.

We should say right away that the contents of this section are pure combinatorics, i.e., manipulating diagrams. The reader may do well by absorbing the statements of the two main results, Theorems 2.1.3 and 2.2.7, and skipping the proofs on the first pass.

2.1. The bi-simplicial space of grids with defect.

In this subsection we will introduce the bi-simplicial space of grids with defect and state Theorem 2.1.3.

2.1.1. Consider the following object

$$\mathrm{defGrid}_{\bullet,\bullet}(\mathbf{C})^{adm}_{vert;horiz} \in \mathrm{Spc}^{\mathbf{\Delta}^{\mathrm{op}} \times \mathbf{\Delta}^{\mathrm{op}}}.$$

Namely, the space $\mathrm{defGrid}_{m,n}(\mathbf{C})^{adm}_{vert;horiz}$ is the full subspace in

$$\mathrm{Maps}([m] \times [n]^{\mathrm{op}}, \mathbf{C})$$

that consists of objects $\underline{\mathbf{c}}$ with the following properties:

(1) For any $0 \leq i < i+1 \leq m$, the map $\mathbf{c}_{i,j} \to \mathbf{c}_{i+1,j}$ belongs to *vert*;
(2) For any $0 \leq j-1 < j \leq n$, the map $\mathbf{c}_{i,j} \to \mathbf{c}_{i,j-1}$ belongs to *horiz*;

(3) For any $0 \le i < i+1 \le m$ and $0 \le j-1 < j \le n$ in the commutative square

$$\begin{array}{ccc} \mathbf{c}_{i,j} & \longrightarrow & \mathbf{c}_{i,j-1} \\ \downarrow & & \downarrow \\ \mathbf{c}_{i+1,j} & \longrightarrow & \mathbf{c}_{i+1,j-1}, \end{array} \tag{2.2}$$

its *defect of Cartesianness*, i.e., the map

$$\mathbf{c}_{i,j} \to \mathbf{c}_{i+1,j} \underset{\mathbf{c}_{i+1,j-1}}{\times} \mathbf{c}_{i,j-1},$$

belongs to *adm*.

So, the objects of $\text{defGrid}_{m,n}(\mathbf{C})^{adm}_{vert;horiz}$ are grids of objects of $\mathbf{C}$ of height m and width n, with vertical arrows in *vert*, horizontal arrows in *horiz*, and the defect of Cartesianness of each square in *adm*.

2.1.2. There exists a canonical map in $\text{Spc}^{\mathbf{\Delta}^{\text{op}} \times \mathbf{\Delta}^{\text{op}}}$

$$\text{defGrid}_{\bullet,\bullet}(\mathbf{C})^{adm}_{vert;horiz} \to \text{Sq}_{\bullet,\bullet}(\text{Corr}(\mathbf{C})^{adm}_{vert;horiz}), \tag{2.3}$$

explained in Sect. 2.3.

We will prove:

THEOREM 2.1.3. *For* $\mathbb{S} \in 2\text{-Cat}$, *the map* (2.3) *defines an isomorphism between the space of functors* $\text{Maps}(\text{Corr}(\mathbf{C})^{adm}_{vert;horiz}, \mathbb{S})$ *and the subspace of maps in* $\text{Spc}^{\mathbf{\Delta}^{\text{op}} \times \mathbf{\Delta}^{\text{op}}}$

$$\text{defGrid}_{\bullet,\bullet}(\mathbf{C})^{adm}_{vert;horiz} \to \text{Sq}_{\bullet,\bullet}(\mathbb{S})$$

with the following property:
For every object in $\underline{\mathbf{c}} \in \text{defGrid}_{1,1}(\mathbf{C})^{adm}_{vert;horiz}$, *for which the diagram*

$$\begin{array}{ccc} \mathbf{c}_{0,1} & \longrightarrow & \mathbf{c}_{0,0} \\ \downarrow & & \downarrow \\ \mathbf{c}_{1,1} & \longrightarrow & \mathbf{c}_{1,0} \end{array} \tag{2.4}$$

is Cartesian, *the corresponding object*

$$\begin{array}{ccc} s_{0,1} & \longleftarrow & s_{0,0} \\ \downarrow & \searrow\!\!\!\searrow & \downarrow \\ s_{1,1} & \longleftarrow & s_{1,0} \end{array} \tag{2.5}$$

in $\text{Sq}_{1,1}(\mathbb{S})$ *should represent an* invertible *2-morphism.*

REMARK 2.1.4. When $\mathbb{S}$ is an ordinary category, the statement of Theorem 2.1.3 is easy to verify directly (and we recommend to anyone who wants to study its proof to do this exercise).

In general, the intuition behind Theorem 2.1.3 should also be rather clear: both sides describe the data of a pair of functors

$$\mathbf{C}_{vert} \to \mathbb{S} \text{ and } (\mathbf{C}_{horiz})^{\text{op}} \to \mathbb{S}$$

and an assignment to a every commutative square (2.4) in $\mathbf{C}$ (with vertical arrows in *vert*, horizontal arrows in *horiz* and the defect of Cartesianness in *adm*) of a datum of a natural transformation (2.5) in $\mathbb{S}$, such that if (2.4) is Cartesian, then the natural transformation in (2.5) is an isomorphism.

2.2. The bi-simplicial category of grids. In this subsection we will formulate Theorem 2.2.7 which gives yet another description of what it takes to define a functor out of the $(\infty,2)$-category $\mathrm{Corr}(\mathbf{C})^{adm}_{vert;horiz}$.

This description uses bi-simplicial $(\infty,1)$-*categories* (rather than spaces), and as a result also *tri-simplicial* spaces.

Having to deal with tri-simplicial spaces may appear as a grueling task, but unfortunately it seems that one has no choice: we will need Theorem 2.2.7 in order to prove Theorem 4.1.3, which is one of the key results of this chapter[4].

2.2.1. We consider the following object of $1\text{-}\mathrm{Cat}^{\mathbf{\Delta}^{\mathrm{op}}\times\mathbf{\Delta}^{\mathrm{op}}}$, denoted

$$\mathbf{Grid}_{\bullet,\bullet}(\mathbf{C})^{adm}_{vert;horiz}.$$

For $m,n=0,1,...$ we let

$$\mathbf{Grid}_{m,n}(\mathbf{C})^{adm}_{vert;horiz}$$

be the 1-full subcategory of

$$\mathbf{Maps}([m]\times[n]^{\mathrm{op}},\mathbf{C}),$$

where we restrict objects to diagrams $\underline{\mathbf{c}}$ satisfying:

(1) For every $0\le i<i+1\le m$, the map $\mathbf{c}_{i,j}\to\mathbf{c}_{i+1,j}$ belongs to *vert*;
(2) For every $0\le j-1<j\le n$, the map $\mathbf{c}_{i,j}\to\mathbf{c}_{i,j-1}$ belongs to *horiz*;
(3) For every $0\le i<i+1\le m$ and $0\le j-1<j\le n$, the square

$$\begin{array}{ccc}\mathbf{c}_{i,j} & \longrightarrow & \mathbf{c}_{i,j-1}\\ \downarrow & & \downarrow\\ \mathbf{c}_{i+1,j} & \longrightarrow & \mathbf{c}_{i+1,j-1},\end{array}$$

is Cartesian.

We restrict 1-morphisms to those maps of diagrams $\underline{\mathbf{c}}\to\underline{\mathbf{c}}'$ such that for every $0\le i\le m$ and $0\le j\le n$, the map

$$\mathbf{c}_{i,j}\to\mathbf{c}'_{i,j}$$

belongs to *vert*, and for every $0\le i\le m$ and $0\le j-1<j\le n$, the defect of Cartesianness of the square

$$\begin{array}{ccc}\mathbf{c}_{i,j} & \longrightarrow & \mathbf{c}_{i,j-1}\\ \downarrow & & \downarrow\\ \mathbf{c}'_{i,j} & \longrightarrow & \mathbf{c}'_{i,j-1}\end{array}$$

belongs to *adm*.

[4]As a dubious comfort to the reader, let us point out that the proof of Theorem 4.1.3 uses quadri-simplicial spaces, so tri-simplicial ones are not yet the worst.

REMARK 2.2.2. Note that we impose no condition on the defect of the commutative diagrams

$$\begin{array}{ccc} \mathbf{c}_{i,j} & \longrightarrow & \mathbf{c}_{i+1,j} \\ \downarrow & & \downarrow \\ \mathbf{c}'_{i,j} & \longrightarrow & \mathbf{c}'_{i+1,j}. \end{array}$$

2.2.3. Consider also the following object of $1\text{-Cat}^{\mathbf{\Delta}^{\mathrm{op}}\times\mathbf{\Delta}^{\mathrm{op}}}$, denoted

$$'\mathbf{Grid}_{\bullet,\bullet}(\mathbf{C})^{adm}_{vert;horiz}.$$

For $m,n=0,1,...$ we let

$$'\mathbf{Grid}_{m,n}(\mathbf{C})^{adm}_{vert;horiz}$$

be the 1-full subcategory of $\mathbf{Grid}_{m,n}(\mathbf{C})^{adm}_{vert;horiz}$, which has the same objects, but where we restrict 1-morphisms to those maps of diagrams $\underline{\mathbf{c}}\to\underline{\mathbf{c}}'$ such that for every $0\le i\le m$ and $0\le j\le n$, the map $\mathbf{c}_{i,j}\to\mathbf{c}'_{i,j}$ is in adm.

Denote also

$$\mathrm{Grid}_{\bullet,\bullet}(\mathbf{C})^{adm}_{vert;horiz}\in\mathrm{Spc}^{\mathbf{\Delta}^{\mathrm{op}}\times\mathbf{\Delta}^{\mathrm{op}}},$$

where

$$\mathrm{Grid}_{m,n}(\mathbf{C})^{adm}_{vert;horiz}:=(\mathbf{Grid}_{m,n}(\mathbf{C})^{adm}_{vert;horiz})^{\mathrm{Spc}}.$$

REMARK 2.2.4. The difference between $\mathbf{Grid}_{m,n}(\mathbf{C})^{adm}_{vert;horiz}$ (respectively $'\mathbf{Grid}_{m,n}(\mathbf{C})^{adm}_{vert;horiz}$) and $\mathrm{defGrid}_{m,n}(\mathbf{C})^{adm}_{vert;horiz}$ is that the former has fewer objects, but we allow non-invertible morphisms.

2.2.5. Applying the functor $\mathrm{Seq}_\bullet:1\text{-Cat}\to\mathrm{Spc}^{\mathbf{\Delta}^{\mathrm{op}}}$ term-wise to

$$\mathbf{Grid}_{\bullet,\bullet}(\mathbf{C})^{adm}_{vert;horiz}\text{ and }'\mathbf{Grid}_{\bullet,\bullet}(\mathbf{C})^{adm}_{vert;horiz}$$

we obtain objects

$$\mathrm{Seq}_\bullet(\mathbf{Grid}_{\bullet,\bullet}(\mathbf{C})^{adm}_{vert;horiz})\text{ and }\mathrm{Seq}_\bullet('\mathbf{Grid}_{\bullet,\bullet}(\mathbf{C})^{adm}_{vert;horiz})$$

in $\mathrm{Spc}^{\mathbf{\Delta}^{\mathrm{op}}\times\mathbf{\Delta}^{\mathrm{op}}\times\mathbf{\Delta}^{\mathrm{op}}}$, so that the corresponding spaces of (l,m,n)-simplices are

$$\mathrm{Seq}_l(\mathbf{Grid}_{m,n}(\mathbf{C})^{adm}_{vert;horiz})\text{ and }\mathrm{Seq}_l('\mathbf{Grid}_{m,n}(\mathbf{C})^{adm}_{vert;horiz}),$$

respectively.

2.2.6. Recall now (see Chapter 10, Sect. 4.6.1) that to an $(\infty,2)$-category $\mathbb{S}$ we can canonically attach an object

$$\mathrm{Cu}_{\bullet,\bullet,\bullet}(\mathbb{S})\in\mathrm{Spc}^{\mathbf{\Delta}^{\mathrm{op}}\times\mathbf{\Delta}^{\mathrm{op}}\times\mathbf{\Delta}^{\mathrm{op}}}.$$

There are canonically defined maps

$$\mathrm{Seq}_\bullet(\mathbf{Grid}_{\bullet,\bullet}(\mathbf{C})^{adm}_{vert;horiz})\to\mathrm{Cu}_{\bullet,\bullet,\bullet}(\mathrm{Corr}(\mathbf{C})^{adm}_{vert;horiz}),\tag{2.6}$$

$$\mathrm{Seq}_\bullet('\mathbf{Grid}_{\bullet,\bullet}(\mathbf{C})^{adm}_{vert;horiz})\to\mathrm{Cu}_{\bullet,\bullet,\bullet}(\mathrm{Corr}(\mathbf{C})^{adm}_{vert;horiz}),\tag{2.7}$$

explained in Sect. 2.4 below.

We will prove:

THEOREM 2.2.7.

(a) *For an $(\infty,2)$-category $\mathbb{S}$, the map (2.6) induces an isomorphism from*

$$\mathrm{Maps}(\mathrm{Corr}(\mathbf{C})^{adm}_{vert;horiz}, \mathbb{S})$$

to the subspace of maps in $\mathrm{Spc}^{\mathbf{\Delta}^{\mathrm{op}}\times\mathbf{\Delta}^{\mathrm{op}}\times\mathbf{\Delta}^{\mathrm{op}}}$

$$\mathrm{Maps}(\mathrm{Seq}_\bullet(\mathbf{Grid}_{\bullet,\bullet}(\mathbf{C})^{adm}_{vert;horiz}), \mathrm{Cu}_{\bullet,\bullet,\bullet}(\mathbb{S})),$$

consisting of those maps, for which for every $(0,1,1)$-simplex in

$$\mathrm{Seq}_\bullet(\mathbf{Grid}_{\bullet,\bullet}(\mathbf{C})^{adm}_{vert;horiz}),$$

the corresponding $(0,1,1)$-simplex in $\mathrm{Cu}_{\bullet,\bullet,\bullet}(\mathbb{S})$, thought of as a $(1,1)$-simplex in $\mathrm{Sq}_{\bullet,\bullet}(\mathbb{S})$, represents an invertible *2-morphism in $\mathbb{S}$.*

(b) *Ditto for $'\mathbf{Grid}_{\bullet,\bullet}(\mathbf{C})^{adm}_{vert;horiz}$ instead of $\mathbf{Grid}_{\bullet,\bullet}(\mathbf{C})^{adm}_{vert;horiz}$.*

REMARK 2.2.8. The content of Remark 2.1.4 applies also to Theorem 2.2.7: it is rather clear why this kind of statement should be true (convince yourself for $\mathbb{S}$ ordinary).

2.3. Construction of the map-I. In this subsection we will construct the map (2.3).

The reader who prefers to take Theorem 2.1.3 on faith, may choose to skip this subsection (or better tinker with the relevant objects and invent what the map (2.3) should be).

2.3.1. The map (2.3) is constructed as a composition of a map

$$\mathrm{defGrid}_{\bullet,\bullet}(\mathbf{C})^{adm}_{vert;horiz} \to \mathrm{Seq}_\bullet(\mathbf{Grid}^{\geq \mathrm{dgnl}}_\bullet(\mathbf{C})^{adm}_{vert;horiz}), \tag{2.8}$$

followed by the map

$$\mathrm{Seq}_\bullet(\mathbf{Grid}^{\geq \mathrm{dgnl}}_\bullet(\mathbf{C})^{adm}_{vert;horiz}) \simeq$$
$$\simeq \mathrm{Sq}^{\mathrm{Pair}}_{\bullet,\bullet}(\mathrm{Corr}(\mathbf{C})^{adm}_{vert;horiz}, \mathbf{C}_{vert}) \to \mathrm{Sq}_{\bullet,\bullet}(\mathrm{Corr}(\mathbf{C})^{adm}_{vert;horiz}),$$

where the first arrow is the isomorphism of Proposition 1.4.5(a).

2.3.2. In its turn, the map (2.8) is constructed as follows. For every n, let

$$\mathbf{Maps}([n]^{\mathrm{op}}, \mathbf{C})^{adm}_{vert;horiz}$$

be the 1-full subcategory of $\mathbf{Maps}([n]^{\mathrm{op}}, \mathbf{C})$, where we restrict objects to those n-strings $\underline{\mathbf{c}}$, for which for every $0 \leq j-1 < j \leq n$, the corresponding map $\mathbf{c}_j \to \mathbf{c}_{j-1}$ belongs to *horiz*, and where we restrict 1-morphisms to those maps $\underline{\mathbf{c}} \to \underline{\mathbf{c}}'$ that satisfy:

(1) For every $0 \leq j \leq n$, the map $\mathbf{c}_j \to \mathbf{c}'_j$ belongs to *vert*;
(2) For every $0 \leq j-1 < j \leq n$, the defect of Cartesianness of the square

$$\begin{array}{ccc} \mathbf{c}_j & \longrightarrow & \mathbf{c}_{j-1} \\ \downarrow & & \downarrow \\ \mathbf{c}'_j & \longrightarrow & \mathbf{c}'_{j-1} \end{array}$$

belongs to *adm*.

It is easy to see that

$$\text{defGrid}_{\bullet,n}(\mathbf{C})^{adm}_{vert;horiz} \simeq \text{Seq}_{\bullet}(\mathbf{Maps}([n]^{\text{op}}, \mathbf{C})^{adm}_{vert;horiz}),$$

as simplicial spaces.

2.3.3. Now, the sought-for map (2.8) comes from a canonically defined functor

$$\mathbf{Maps}([n]^{\text{op}}, \mathbf{C})^{adm}_{vert;horiz} \to \mathbf{Grid}^{\geq \text{dgnl}}_n(\mathbf{C})^{adm}_{vert;horiz}. \tag{2.9}$$

In fact, the above functor is a fully faithful embedding, whose image consists of those half-grids, in which the vertical maps are isomorphisms.

2.3.4. In what follows, we will use the following notation. Let $\mathbf{I}$ and $\mathbf{J}$ be $(\infty,1)$-categories. Let

$$\text{Maps}(\mathbf{I} \times \mathbf{J}^{\text{op}}, \mathbf{C})^{adm}_{vert;horiz} \subset \text{Maps}(\mathbf{I} \times \mathbf{J}^{\text{op}}, \mathbf{C})$$

be the subspace consisting of those functors such that:

(1) for every morphism $(i_0 \to i_1) \in \mathbf{I}$ and every object $j \in \mathbf{J}$, the image of the morphism $(i_0, j) \to (i_1, j)$ lies in *vert*;
(2) for every object $i \in \mathbf{I}$ and every morphism $(j_0 \to j_1) \in \mathbf{J}$, the image of the morphism $(i, j_0) \to (i, j_1)$ lies in *horiz*; and
(3) for every pair of morphisms $(i_0 \to i_1) \in \mathbf{I}$ and $(j_0 \to j_1) \in \mathbf{J}$, the defect of Cartesianness of the resulting diagram in $\mathbf{C}$

$$\begin{array}{ccc} \mathbf{c}_{i_0,j_1} & \longrightarrow & \mathbf{c}_{i_0,j_0} \\ \downarrow & & \downarrow \\ \mathbf{c}_{i_1,j_1} & \longrightarrow & \mathbf{c}_{i_1,j_0} \end{array} \tag{2.10}$$

lies in *adm*.

2.3.5. We claim that the datum of a map of bi-simplicial spaces

$$\text{defGrid}_{\bullet,\bullet}(\mathbf{C})^{adm}_{vert;horiz} \to \text{Sq}_{\bullet,\bullet}(\mathbb{S}) \tag{2.11}$$

gives rise to a map

$$\text{Maps}(\mathbf{I} \times \mathbf{J}^{\text{op}}, \mathbf{C})^{adm}_{vert;horiz} \to \text{Maps}(\mathbf{I} \circledast \mathbf{J}, \mathbb{S}) \tag{2.12}$$

that behaves functorially in $\mathbf{I}$ and $\mathbf{J}$.

In the above formula, $\circledast$ is the Gray tensor product, see Chapter 10, Sect. 3.2.

Moreover, if the map (2.11) satisfies the additional condition of Theorem 2.1.3, then the map (2.12) has the property that it sends every square (2.10) that is Cartesian to a diagram in $\mathbb{S}$ representing an invertible 2-morphism.

Indeed, each side in (2.12) is the limit over the index category

$$([m] \to \mathbf{I}) \times ([n] \to \mathbf{J})$$

of terms equal to

$$\text{defGrid}_{m,n}(\mathbf{C})^{adm}_{vert;horiz} \text{ and } \text{Sq}_{m,n}(\mathbb{S}),$$

respectively.

2.4. Construction of the map-II. In this subsection we will carry out the construction of the map (2.6). The case of (2.7) is similar.

The reader who prefers to take Theorem 2.2.7 on faith may choose to skip this subsection.

2.4.1. Recall (see Chapter 10, Proposition 3.2.9) that for an $(\infty,2)$-category $\mathbb{S}$ and n, there exists a canonical monomorphism of bi-simplicial spaces

$$\mathrm{Sq}_{\bullet,\bullet}(\mathrm{Seq}_n^{\mathrm{ext}}(\mathbb{S})) \to \mathrm{Cu}_{\bullet,\bullet,n}(\mathbb{S}), \tag{2.13}$$

where $\mathrm{Seq}_n^{\mathrm{ext}}(\mathbb{S}) \in 1\text{-Cat}$ is regarded as an $(\infty,2)$-category. Explicitly,

$$\mathrm{Sq}_{l,m}(\mathrm{Seq}_n^{\mathrm{ext}}(\mathbb{S})) = \mathrm{Maps}(([l]\times[m])\circledast[n],\mathbb{S})$$

and

$$\mathrm{Cu}_{\bullet,\bullet,n}(\mathbb{S}) = \mathrm{Maps}([l]\circledast[m]\circledast[n],\mathbb{S}),$$

and the map (2.13) comes from

$$[l]\circledast[m]\circledast[n] \simeq ([l]\circledast[m])\circledast[n] \to ([l]\times[m])\circledast[n].$$

2.4.2. Recall the category

$$\mathbf{Maps}([n]^{\mathrm{op}},\mathbf{C})^{adm}_{vert;horiz}$$

(see Sect. 2.3.2).

The functor (2.6) will be defined as the composition of a map

$$\mathrm{Seq}_\bullet(\mathbf{Grid}_{\bullet,n}(\mathbf{C})^{adm}_{vert;horiz}) \to \mathrm{Sq}_{\bullet,\bullet}(\mathbf{Maps}([n]^{\mathrm{op}},\mathbf{C})^{adm}_{vert;horiz}) \tag{2.14}$$

(where $\mathbf{Maps}([n]^{\mathrm{op}},\mathbf{C}) \in 1\text{-Cat}$ is regarded as an $(\infty,2)$-category), followed by

$$\begin{aligned}\mathrm{Sq}_{\bullet,\bullet}(\mathbf{Maps}([n]^{\mathrm{op}},\mathbf{C})^{adm}_{vert;horiz}) &\to \mathrm{Sq}_{\bullet,\bullet}(\mathbf{Grid}_n^{\geq\mathrm{dgnl}}(\mathbf{C})^{adm}_{vert;horiz}) \overset{\text{Proposition 1.4.5}}{\simeq}\\ \simeq \mathrm{Sq}_{\bullet,\bullet}(\mathrm{Seq}_n^{\mathrm{Pair}}(\mathrm{Corr}(\mathbf{C})^{adm}_{vert;horiz},\mathbf{C}_{vert})) &\to \mathrm{Sq}_{\bullet,\bullet}(\mathrm{Seq}_n^{\mathrm{ext}}(\mathrm{Corr}(\mathbf{C})^{adm}_{vert;horiz})) \to\\ &\to \mathrm{Cu}_{\bullet,\bullet,n}(\mathrm{Corr}(\mathbf{C})^{adm}_{vert;horiz}),\end{aligned}$$

where the first arrow is induced by the functor (2.9) and the last arrow from (2.13).

2.4.3. To define (2.14), let us write both sides out explicitly. The space of (l,m)-simplices in the right-hand side is the full subspace of

$$\mathrm{Maps}([l]\times[m]\times[n]^{\mathrm{op}},\mathbf{C}),$$

consisting of diagrams $\underline{\mathbf{c}}$ satisfying the following conditions:

(1) For every i,j and k the map $\mathbf{c}_{i,j,k}\to\mathbf{c}_{i,j,k-1}$ belongs to *horiz*;
(2) For every i,j and k the map $\mathbf{c}_{i,j,k}\to\mathbf{c}_{i,j+1,k}$ belongs to *vert*;
(3) For every i,j and k the map $\mathbf{c}_{i,j,k}\to\mathbf{c}_{i+1,j,k}$ belongs to *vert*;
(4) For every i,j and k, the defect of Cartesianness of the diagram

$$\begin{array}{ccc}\mathbf{c}_{i,j,k} & \longrightarrow & \mathbf{c}_{i,j,k-1}\\ \big\downarrow & & \big\downarrow\\ \mathbf{c}_{i+1,j,k} & \longrightarrow & \mathbf{c}_{i+1,j,k-1}\end{array}$$

belongs to *adm*.

(5) For every i, j and k, the defect of Cartesianness of the diagram

$$\begin{array}{ccc} \mathbf{c}_{i,j,k} & \longrightarrow & \mathbf{c}_{i,j,k-1} \\ \downarrow & & \downarrow \\ \mathbf{c}_{i,j+1,k} & \longrightarrow & \mathbf{c}_{i,j+1,k-1} \end{array}$$

belongs to *adm*.

Now, the left-hand side in (2.14) is the subspace of the space of diagrams as above, where we strengthen condition (5) as follows:

We require that the square

$$\begin{array}{ccc} \mathbf{c}_{i,j,k} & \longrightarrow & \mathbf{c}_{i,j,k-1} \\ \downarrow & & \downarrow \\ \mathbf{c}_{i,j+1,k} & \longrightarrow & \mathbf{c}_{i,j+1,k-1} \end{array}$$

be Cartesian.

2.4.4. One can view the resulting map

$$\mathrm{Maps}(\mathrm{Corr}(\mathbf{C})^{adm}_{vert;horiz}, \mathbb{S}) \to \mathrm{Maps}(\mathrm{Seq}_\bullet(\mathbf{Grid}_{\bullet,\bullet}(\mathbf{C})^{adm}_{vert;horiz}), \mathrm{Cu}_{\bullet,\bullet,\bullet}(\mathbb{S}))$$

in Theorem 2.2.7 (which we have just defined by completing the construction of (2.6)) also as follows.

It is the composition of the map

$$\mathrm{Maps}(\mathrm{Corr}(\mathbf{C})^{adm}_{vert;horiz}, \mathbb{S}) \to \mathrm{Maps}(\mathrm{defGrid}_{\bullet,\bullet}(\mathbf{C})^{adm}_{vert;horiz}, \mathrm{Sq}_{\bullet,\bullet}(\mathbb{S}))$$

in Theorem 2.1.3, followed by a map

$$\begin{aligned} (2.15)\quad & \mathrm{Maps}_{\mathrm{Spc}^{\mathbf{\Delta}^{\mathrm{op}}\times\mathbf{\Delta}^{\mathrm{op}}}}(\mathrm{defGrid}_{\bullet,\bullet}(\mathbf{C})^{adm}_{vert;horiz}, \mathrm{Sq}_{\bullet,\bullet}(\mathbb{S})) \to \\ & \to \mathrm{Maps}_{\mathrm{Spc}^{\mathbf{\Delta}^{\mathrm{op}}\times\mathbf{\Delta}^{\mathrm{op}}\times\mathbf{\Delta}^{\mathrm{op}}}}(\mathrm{Seq}_\bullet(\mathbf{Grid}_{\bullet,\bullet}(\mathbf{C})^{adm}_{vert;horiz}), \mathrm{Cu}_{\bullet,\bullet,\bullet}(\mathbb{S})), \end{aligned}$$

constructed as follows.

2.4.5. Given a map

$$\mathrm{defGrid}_{\bullet,\bullet}(\mathbf{C})^{adm}_{vert;horiz} \to \mathrm{Sq}_{\bullet,\bullet}(\mathbb{S}), \tag{2.16}$$

we need to construct maps

$$\mathrm{Seq}_l(\mathbf{Grid}_{m,n}(\mathbf{C})^{adm}_{vert;horiz}) \to \mathrm{Cu}_{l,m,n}(\mathbb{S}) \tag{2.17}$$

that depend functorially on $[l],[m],[n] \in \mathbf{\Delta}^{\mathrm{op}} \times \mathbf{\Delta}^{\mathrm{op}} \times \mathbf{\Delta}^{\mathrm{op}}$.

By Sect. 2.3.5, a map in (2.16) gives rise to a map

$$\mathrm{Maps}(([l]\times[m])\times[n]^{\mathrm{op}}, \mathbf{C})^{adm}_{vert;horiz} \to \mathrm{Maps}(([l]\times[m])\circledast[n]), \mathbb{S}),$$

while

$$\mathrm{Seq}_l(\mathbf{Grid}_{m,n}(\mathbf{C})^{adm}_{vert;horiz}) \subset \mathrm{Maps}(([l]\times[m])\times[n]^{\mathrm{op}}, \mathbf{C})^{adm}_{vert;horiz}$$

and

$$\mathrm{Maps}(([l]\times[m])\circledast[n]), \mathbb{S}) \subset \mathrm{Cu}_{l,m,n}(\mathbb{S}).$$

Now, we obtain the desired map in (2.17) as the composite:

$$\operatorname{Seq}_l(\mathbf{Grid}_{m,n}(\mathbf{C})^{adm}_{vert;horiz}) \to \operatorname{Maps}(([l]\times[m])\times[n]^{\mathrm{op}}, \mathbf{C})^{adm}_{vert;horiz} \to$$
$$\to \operatorname{Maps}(([l]\times[m])\circledast[n]), \mathbb{S}) \to \operatorname{Cu}_{l,m,n}(\mathbb{S}).$$

2.5. Proof of Theorems 2.1.3 and 2.2.7: initial remarks. We shall only consider point (a) of Theorem 2.2.7, point (b) being similar.

The proof will consist of constructing the inverse maps. Its key idea (which is completely straightforward once you understand what should map where) is given in Sect. 2.5.4.

2.5.1. It it is easy to see that the essential image of the map in Theorem 2.1.3 belongs to

$$\operatorname{Maps}^0(\operatorname{defGrid}_{\bullet,\bullet}(\mathbf{C})^{adm}_{vert;horiz}, \operatorname{Sq}_{\bullet,\bullet}(\mathbb{S})) \subset$$
$$\subset \operatorname{Maps}_{\operatorname{Spc}^{\mathbf{\Delta}^{\mathrm{op}}\times\mathbf{\Delta}^{\mathrm{op}}}}(\operatorname{defGrid}_{\bullet,\bullet}(\mathbf{C})^{adm}_{vert;horiz}, \operatorname{Sq}_{\bullet,\bullet}(\mathbb{S})),$$

where $\operatorname{Maps}^0(\operatorname{defGrid}_{\bullet,\bullet}(\mathbf{C})^{adm}_{vert;horiz}, \operatorname{Sq}_{\bullet,\bullet}(\mathbb{S}))$ is the subspace, singled out by the condition in Theorem 2.1.3.

Furthermore, the map in Theorem 2.2.7 is the composition of the above map

$$\operatorname{Maps}(\operatorname{Corr}(\mathbf{C})^{adm}_{vert;horiz}, \mathbb{S}) \to \operatorname{Maps}^0(\operatorname{defGrid}_{\bullet,\bullet}(\mathbf{C})^{adm}_{vert;horiz}, \operatorname{Sq}_{\bullet,\bullet}(\mathbb{S}))$$

and a map

$$\operatorname{Maps}^0(\operatorname{defGrid}_{\bullet,\bullet}(\mathbf{C})^{adm}_{vert;horiz}, \operatorname{Sq}_{\bullet,\bullet}(\mathbb{S})) \to \tag{2.18}$$
$$\to \operatorname{Maps}^0(\operatorname{Seq}_\bullet(\mathbf{Grid}_{\bullet,\bullet}(\mathbf{C})^{adm}_{vert;horiz}), \operatorname{Cu}_{\bullet,\bullet,\bullet}(\mathbb{S})),$$

where

$$\operatorname{Maps}^0(\operatorname{Seq}_\bullet(\mathbf{Grid}_{\bullet,\bullet}(\mathbf{C})^{adm}_{vert;horiz}), \operatorname{Cu}_{\bullet,\bullet,\bullet}(\mathbb{S})) \subset$$
$$\subset \operatorname{Maps}_{\operatorname{Spc}^{\mathbf{\Delta}^{\mathrm{op}}\times\mathbf{\Delta}^{\mathrm{op}}\times\mathbf{\Delta}^{\mathrm{op}}}}(\operatorname{Seq}_\bullet(\mathbf{Grid}_{\bullet,\bullet}(\mathbf{C})^{adm}_{vert;horiz}), \operatorname{Cu}_{\bullet,\bullet,\bullet}(\mathbb{S}))$$

is the subspace singled out by the condition in Theorem 2.2.7, and the map (2.18) is induced by (2.15).

2.5.2. We will construct a map

$$\operatorname{Maps}^0(\operatorname{Seq}_\bullet(\mathbf{Grid}_{\bullet,\bullet}(\mathbf{C})^{adm}_{vert;horiz}), \operatorname{Cu}_{\bullet,\bullet,\bullet}(\mathbb{S})) \to \tag{2.19}$$
$$\to \operatorname{Maps}_{\operatorname{Spc}^{\mathbf{\Delta}^{\mathrm{op}}\times\mathbf{\Delta}^{\mathrm{op}}}}(\mathbf{Grid}^{\geq \mathrm{dgnl}}_\bullet(\mathbf{C})^{adm}_{vert;horiz}, \operatorname{Seq}^{\mathrm{ext}}_\bullet(\mathbb{S})),$$

where

$$\operatorname{Maps}_{\operatorname{Spc}^{\mathbf{\Delta}^{\mathrm{op}}\times\mathbf{\Delta}^{\mathrm{op}}}}(\mathbf{Grid}^{\geq \mathrm{dgnl}}_\bullet(\mathbf{C})^{adm}_{vert;horiz}, \operatorname{Seq}^{\mathrm{ext}}_\bullet(\mathbb{S})) \simeq \operatorname{Maps}(\operatorname{Corr}(\mathbf{C})^{adm}_{vert;horiz}, \mathbb{S})$$

by Corollary 1.4.6.

It will be a tedious but straightforward verification that each of the three compositions

$$\operatorname{Maps}(\operatorname{Corr}(\mathbf{C})^{adm}_{vert;horiz}, \mathbb{S}) \to \operatorname{Maps}(\operatorname{Corr}(\mathbf{C})^{adm}_{vert;horiz}, \mathbb{S}),$$

$$\operatorname{Maps}^0(\operatorname{defGrid}_{\bullet,\bullet}(\mathbf{C})^{adm}_{vert;horiz}, \operatorname{Sq}_{\bullet,\bullet}(\mathbb{S})) \to \operatorname{Maps}^0(\operatorname{defGrid}_{\bullet,\bullet}(\mathbf{C})^{adm}_{vert;horiz}, \operatorname{Sq}_{\bullet,\bullet}(\mathbb{S}))$$

and

$$\begin{aligned}\mathrm{Maps}^0(\mathrm{Seq}_\bullet(\mathbf{Grid}_{\bullet,\bullet}(\mathbf{C})^{adm}_{vert;horiz}),\mathrm{Cu}_{\bullet,\bullet,\bullet}(\mathbb{S}))\to\\ \to\mathrm{Maps}^0(\mathrm{Seq}_\bullet(\mathbf{Grid}_{\bullet,\bullet}(\mathbf{C})^{adm}_{vert;horiz}),\mathrm{Cu}_{\bullet,\bullet,\bullet}(\mathbb{S}))\end{aligned}$$

is canonically isomorphic to the identity map. We will leave this verification to the reader.

2.5.3. Note that the left-hand side in (2.19) is a subspace in the space of maps of tri-simplicial spaces, from $\mathrm{Seq}_\bullet(\mathbf{Grid}_{\bullet,\bullet}(\mathbf{C})^{adm}_{vert;horiz})$ to $\mathrm{Maps}([\bullet]\circledast([\bullet]\times[\bullet]),\mathbb{S})$, where the latter assigns to i,j,k the space

$$\mathrm{Maps}([i]\circledast([j]\times[k]),\mathbb{S}).$$

By definition, the right-hand side in (2.19) is the space of maps of bi-simplicial spaces, from $\mathrm{Seq}_\bullet(\mathbf{Grid}^{\geq\,\mathrm{dgnl}}_\bullet(\mathbf{C})^{adm}_{vert;horiz})$ to $\mathrm{Maps}([\bullet]\circledast[\bullet],\mathbb{S})$, where the latter assigns to m,n the space

$$\mathrm{Maps}([m]\circledast[n],\mathbb{S}).$$

Note that the diagonal map $[n]\to([n]\times[n])^{\geq\mathrm{dgnl}}$ defines a map of bi-simplicial spaces

$$\mathrm{Maps}([\bullet]\circledast([\bullet]\overset{=}{\times}[\bullet])^{\geq\mathrm{dgnl}},\mathbb{S})\to\mathrm{Maps}([\bullet]\circledast[\bullet],\mathbb{S}),$$

where the former bi-simplicial space assigns to m,n the space

$$\mathrm{Maps}([m]\circledast([n]\times[n])^{\geq\mathrm{dgnl}},\mathbb{S}).$$

We will construct a map

(2.20)
$$\begin{aligned}\mathrm{Maps}_{\mathrm{Spc}^{\mathbf{\Delta}^{\mathrm{op}}\times\mathbf{\Delta}^{\mathrm{op}}\times\mathbf{\Delta}^{\mathrm{op}}}}(\mathrm{Seq}_\bullet(\mathbf{Grid}_{\bullet,\bullet}(\mathbf{C})^{adm}_{vert;horiz}),\mathrm{Maps}([\bullet]\circledast([\bullet]\times[\bullet]),\mathbb{S}))\to\\ \to\mathrm{Maps}_{\mathrm{Spc}^{\mathbf{\Delta}^{\mathrm{op}}\times\mathbf{\Delta}^{\mathrm{op}}}}(\mathrm{Seq}_\bullet(\mathbf{Grid}^{\geq\,\mathrm{dgnl}}_\bullet(\mathbf{C})^{adm}_{vert;horiz}),\mathrm{Maps}([\bullet]\circledast([\bullet]\overset{=}{\times}[\bullet])^{\geq\mathrm{dgnl}},\mathbb{S})).\end{aligned}$$

Then, by composing from (2.20) we obtain a map

$$\begin{aligned}&\mathrm{Maps}^0(\mathrm{Seq}_\bullet(\mathbf{Grid}_{\bullet,\bullet}(\mathbf{C})^{adm}_{vert;horiz}),\mathrm{Cu}_{\bullet,\bullet,\bullet}(\mathbb{S}))\hookrightarrow\\ &\to\mathrm{Maps}_{\mathrm{Spc}^{\mathbf{\Delta}^{\mathrm{op}}\times\mathbf{\Delta}^{\mathrm{op}}\times\mathbf{\Delta}^{\mathrm{op}}}}(\mathrm{Seq}_\bullet(\mathbf{Grid}_{\bullet,\bullet}(\mathbf{C})^{adm}_{vert;horiz}),\mathrm{Maps}([\bullet]\circledast([\bullet]\times[\bullet]),\mathbb{S}))\overset{(2.20)}{\to}\\ &\to\mathrm{Maps}_{\mathrm{Spc}^{\mathbf{\Delta}^{\mathrm{op}}\times\mathbf{\Delta}^{\mathrm{op}}}}(\mathrm{Seq}_\bullet(\mathbf{Grid}^{\geq\,\mathrm{dgnl}}_\bullet(\mathbf{C})^{adm}_{vert;horiz}),\mathrm{Maps}([\bullet]\circledast([\bullet]\overset{=}{\times}[\bullet])^{\geq\mathrm{dgnl}},\mathbb{S}))\to\\ &\to\mathrm{Maps}_{\mathrm{Spc}^{\mathbf{\Delta}^{\mathrm{op}}\times\mathbf{\Delta}^{\mathrm{op}}}}(\mathrm{Seq}_\bullet(\mathbf{Grid}^{\geq\,\mathrm{dgnl}}_\bullet(\mathbf{C})^{adm}_{vert;horiz}),\mathrm{Maps}([\bullet]\circledast[\bullet],\mathbb{S}))=\\ &=\mathrm{Maps}_{\mathrm{Spc}^{\mathbf{\Delta}^{\mathrm{op}}\times\mathbf{\Delta}^{\mathrm{op}}}}(\mathbf{Grid}^{\geq\,\mathrm{dgnl}}_\bullet(\mathbf{C})^{adm}_{vert;horiz},\mathrm{Seq}^{\mathrm{ext}}_\bullet(\mathbb{S})).\end{aligned}$$

giving rise to the desired map (2.19).

2.5.4. *The key idea.* Let us explain where the map (2.20) will come from. For simplicity, let us consider the corresponding map

$$\begin{aligned}(2.21)\quad&\mathrm{Maps}_{\mathrm{Spc}^{\mathbf{\Delta}^{\mathrm{op}}\times\mathbf{\Delta}^{\mathrm{op}}}}(\mathrm{Grid}_{\bullet,\bullet}(\mathbf{C})^{adm}_{vert;horiz},\mathrm{Maps}([\bullet]\times[\bullet],\mathbb{S}))\to\\ &\to\mathrm{Maps}_{\mathrm{Spc}^{\mathbf{\Delta}^{\mathrm{op}}}}(\mathrm{Grid}^{\geq\,\mathrm{dgnl}}_\bullet(\mathbf{C})^{adm}_{vert;horiz},\mathrm{Maps}(([\bullet]\overset{=}{\times}[\bullet])^{\geq\mathrm{dgnl}},\mathbb{S})),\end{aligned}$$

where we recall that

$$\mathrm{Grid}_{m,n}(\mathbf{C})^{adm}_{vert;horiz}:=(\mathbf{Grid}_{m,n}(\mathbf{C})^{adm}_{vert;horiz})^{\mathrm{Spc}}$$

and

$$\mathrm{Grid}_n^{\geq \mathrm{dgnl}}(\mathbf{C})^{adm}_{vert;horiz} := (\mathbf{Grid}_n^{\geq \mathrm{dgnl}}(\mathbf{C})^{adm}_{vert;horiz})^{\mathrm{Spc}}.$$

The idea behind the existence of the map (2.21) is the following motto 'if we know how to map grids of objects in $\mathbf{C}$ to diagrams on $\mathbb{S}$, then we can extend this to half-grids'.

We will turn this motto into an actual construction in the next few subsections.

2.6. Digression: clusters. By a *cluster* we mean a category that 'looks like' $[m] \times [n]$ or $([n] \times [n])^{\geq \mathrm{dgnl}}$. This class of categories will come handy for the construction of the map (2.20).

2.6.1. Let $\mathbf{Q}$ be an $(\infty,1)$-category, equipped with a pair of 1-full subcategories $\mathbf{Q}_{vert}$ and $\mathbf{Q}_{horiz}$. To it we associate a bi-simplicial space $\mathrm{Sq}_{\bullet,\bullet}(\mathbf{Q}_{vert,horiz})$ as follows.

For every m,n, the space $\mathrm{Sq}_{m,n}(\mathbf{Q}_{vert,horiz})$ is a subspace of $\mathrm{Sq}_{m,n}(\mathbf{Q})$ consisting of objects such that for every $[1] \times [0] \to [m] \times [n]$ (resp., $[0] \times [1] \to [m] \times [n]$) the resulting object of $\mathrm{Sq}_{0,1}(\mathbf{Q})$ (resp., $\mathrm{Sq}_{1,0}(\mathbf{Q})$) belongs to *vert* (resp., *horiz*).

For any $(\infty,1)$-category $\mathbf{D}$ we have a tautologically defined map

$$\mathrm{Maps}(\mathbf{Q},\mathbf{D}) \to \mathrm{Maps}(\mathrm{Sq}_{\bullet,\bullet}(\mathbf{Q}_{vert,horiz}), \mathrm{Sq}_{\bullet,\bullet}(\mathbf{D})) \tag{2.22}$$

The map (2.22) does not have to be an isomorphism in general. Note, however, that if (2.22) is an isomorphism for any $\mathbf{D}$, then the triple $(\mathbf{Q}, \mathbf{Q}_{vert}, \mathbf{Q}_{horiz})$ is uniquely determined by the bi-simplicial space $\mathrm{Sq}_{\bullet,\bullet}(\mathbf{Q}_{vert,horiz})$.

Below we will describe a class of triples $(\mathbf{Q}, \mathbf{Q}_{vert}, \mathbf{Q}_{horiz})$ for which (2.22) is an isomorphism.

2.6.2. Let Q be a convex subset of $\{0,...,m\} \times \{0,...,n\}$ for some m and n. To such Q we attach a triple $(\mathbf{Q}, \mathbf{Q}_{vert}, \mathbf{Q}_{horiz})$ by letting $\mathbf{Q}$ be the full subcategory of $[m] \times [n]$ spanned by Q, and $\mathbf{Q}_{vert}$ (resp., $\mathbf{Q}_{horiz}$) be given by vertical (resp., horizontal) arrows (where we are thinking of the first coordinate as the vertical direction, and the second coordinate as the horizontal diction).

By a *cluster* we shall mean a triple $(\mathbf{Q}, \mathbf{Q}_{vert}, \mathbf{Q}_{horiz})$ which is equivalent to one coming from a convex subset $Q \subset \{0,...,m\} \times \{0,...,n\}$ as above.

In Sect. 2.8 we will prove:

PROPOSITION 2.6.3. *If* $(\mathbf{Q}, \mathbf{Q}_{vert}, \mathbf{Q}_{horiz})$ *is a cluster, then the map* (2.22) *is an isomorphism for any* $\mathbf{D}$.

2.6.4. For a bi-simplicial space B, let $B^{\text{horiz-op}}$ be the bi-simplicial space obtained by applying the involution $\mathrm{rev} : \mathbf{\Delta} \to \mathbf{\Delta}$ along the second copy of $\mathbf{\Delta}^{\mathrm{op}}$ in $\mathbf{\Delta}^{\mathrm{op}} \times \mathbf{\Delta}^{\mathrm{op}}$.

Note that if $(\mathbf{Q}, \mathbf{Q}_{vert}, \mathbf{Q}_{horiz})$ is a cluster, there exists a canonically defined cluster

$$(\mathbf{Q}, \mathbf{Q}_{vert}, \mathbf{Q}_{horiz})^{\text{horiz-op}},$$

characterized by the property that

$$\mathrm{Sq}_{\bullet,\bullet}((\mathbf{Q}, \mathbf{Q}_{vert}, \mathbf{Q}_{horiz})^{\text{horiz-op}}) = (\mathrm{Sq}_{\bullet,\bullet}(\mathbf{Q}, \mathbf{Q}_{vert}, \mathbf{Q}_{horiz}))^{\text{horiz-op}}.$$

We let $Q(n)$ denote the cluster given by the subset

$$(\{0,...n\} \times \{0,...,n\})^{\geq \mathrm{dgnl}} \subset \{0,...n\} \times \{0,...,n\}.$$

Note that the underlying category is $([n] \times [n])^{\geq \mathrm{dgnl}}$.

Consider the cluster $Q(n)^{\text{horiz-op}}$. Note that its underlying category is $([n] \times [n]^{\mathrm{op}})^{\geq \mathrm{dgnl}}$.

2.7. Proof of Theorems 2.1.3 and 2.2.7: continuation. We recall that we need to construct the map (2.20). The idea is that we can construct something a lot more general: namely, a map from the left-hand side in (2.20) to

$$\mathrm{Maps}_{\mathrm{Spc}^{\boldsymbol{\Delta}^{\mathrm{op}}}}(\mathrm{Seq}_\bullet(\mathbf{Q}^{\text{horiz-op}}(\mathbf{C})^{adm}_{vert;horiz}), \mathrm{Maps}([\bullet] \circledast \mathbf{Q}, \mathbb{S}))$$

for any cluster Q (see below for the notation $\mathbf{Q}^{\text{horiz-op}}(\mathbf{C})^{adm}_{vert;horiz}$).

Such a map is more or less tautological, modulo Proposition 2.6.3.

2.7.1. Let Q be a cluster with the underlying category $\mathbf{Q}$ and the attached bi-simplicial space $\mathrm{Sq}_{\bullet,\bullet}(\mathbf{Q}_{vert,horiz})$. We will also consider the cluster $Q^{\text{horiz-op}}$ and the corresponding category, denoted $\mathbf{Q}^{\text{horiz-op}}$.

Analogous to the definition of the category

$$\mathbf{Grid}^{\geq \mathrm{dgnl}}_n(\mathbf{C})^{adm}_{vert;horiz},$$

we have the category $\mathbf{Q}^{\text{horiz-op}}(\mathbf{C})^{adm}_{vert;horiz}$, which is a 1-full subcategory in

$$\mathbf{Maps}(\mathbf{Q}^{\text{horiz-op}}, \mathbf{C}).$$

We will construct maps

(2.23)
$$\mathrm{Maps}_{\mathrm{Spc}^{\boldsymbol{\Delta}^{\mathrm{op}} \times \boldsymbol{\Delta}^{\mathrm{op}} \times \boldsymbol{\Delta}^{\mathrm{op}}}}(\mathrm{Seq}_\bullet(\mathbf{Grid}_{\bullet,\bullet}(\mathbf{C})^{adm}_{vert;horiz}), \mathrm{Maps}([\bullet] \circledast ([\bullet] \times [\bullet]), \mathbb{S})) \to$$
$$\to \mathrm{Maps}(\mathrm{Seq}_m(\mathbf{Q}^{\text{horiz-op}}(\mathbf{C})^{adm}_{vert;horiz}), \mathrm{Maps}([m] \circledast \mathbf{Q}, \mathbb{S})),$$

functorial in Q and $[m] \in \boldsymbol{\Delta}^{\mathrm{op}}$. By taking $Q = Q(n)$, we will obtain the desired map (2.20).

2.7.2. Consider the tri-simplicial space

$$\mathrm{Seq}_\bullet([m]) \boxtimes \mathrm{Sq}_{\bullet,\bullet}(\mathbf{Q}_{vert,horiz}),$$

where $\boxtimes$ denotes the product operation

$$\mathrm{Spc}^{\boldsymbol{\Delta}^{\mathrm{op}}} \times \mathrm{Spc}^{\boldsymbol{\Delta}^{\mathrm{op}} \times \boldsymbol{\Delta}^{\mathrm{op}}} \to \mathrm{Spc}^{\boldsymbol{\Delta}^{\mathrm{op}} \times \boldsymbol{\Delta}^{\mathrm{op}} \times \boldsymbol{\Delta}^{\mathrm{op}}}.$$

We have a naturally defined map from the space

$$\mathrm{Maps}_{\mathrm{Spc}^{\boldsymbol{\Delta}^{\mathrm{op}} \times \boldsymbol{\Delta}^{\mathrm{op}} \times \boldsymbol{\Delta}^{\mathrm{op}}}}(\mathrm{Seq}_\bullet(\mathbf{Grid}_{\bullet,\bullet}(\mathbf{C})^{adm}_{vert;horiz}), \mathrm{Maps}([\bullet] \circledast ([\bullet] \times [\bullet]), \mathbb{S}))$$

to the space of maps

(2.24)
$$\mathrm{Maps}_{\mathrm{Spc}^{\boldsymbol{\Delta}^{\mathrm{op}} \times \boldsymbol{\Delta}^{\mathrm{op}} \times \boldsymbol{\Delta}^{\mathrm{op}}}}(\mathrm{Seq}_\bullet([m]) \boxtimes \mathrm{Sq}_{\bullet,\bullet}(\mathbf{Q}_{vert,horiz}), \mathrm{Seq}_\bullet(\mathbf{Grid}_{\bullet,\bullet}(\mathbf{C})^{adm}_{vert;horiz})) \to$$
$$\mathrm{Maps}_{\mathrm{Spc}^{\boldsymbol{\Delta}^{\mathrm{op}} \times \boldsymbol{\Delta}^{\mathrm{op}} \times \boldsymbol{\Delta}^{\mathrm{op}}}}(\mathrm{Seq}_\bullet([m]) \boxtimes \mathrm{Sq}_{\bullet,\bullet}(\mathbf{Q}_{vert,horiz}), \mathrm{Maps}([\bullet] \circledast ([\bullet] \times [\bullet]), \mathbb{S})).$$

We will how that the left-hand side in (2.24) identifies canonically with

$$\mathrm{Seq}_m(\mathbf{Q}^{\text{horiz-op}}(\mathbf{C})^{adm}_{vert;horiz})$$

and the right-hand side with

$$\mathrm{Maps}([m]\circledast \mathbf{Q},\mathbb{S}),$$

as required.

2.7.3. We shall first analyze the right-hand side in (2.24). We have:

$$\begin{aligned}&\mathrm{Maps}_{\mathrm{Spc}^{\mathbf{\Delta}^{\mathrm{op}}\times\mathbf{\Delta}^{\mathrm{op}}\times\mathbf{\Delta}^{\mathrm{op}}}}(\mathrm{Seq}_\bullet([m])\boxtimes \mathrm{Sq}_{\bullet,\bullet}(\mathbf{Q}_{vert,horiz}),\mathrm{Maps}([\bullet]\circledast([\bullet]\times[\bullet]),\mathbb{S}))\simeq\\ &\simeq \mathrm{Maps}_{\mathrm{Spc}^{\mathbf{\Delta}^{\mathrm{op}}\times\mathbf{\Delta}^{\mathrm{op}}}}(\mathrm{Sq}_{\bullet,\bullet}(\mathbf{Q}_{vert,horiz}),\mathrm{Maps}([m]\circledast([\bullet]\times[\bullet]),\mathbb{S}))\simeq\\ &\simeq \mathrm{Maps}_{\mathrm{Spc}^{\mathbf{\Delta}^{\mathrm{op}}\times\mathbf{\Delta}^{\mathrm{op}}}}(\mathrm{Sq}_{\bullet,\bullet}(\mathbf{Q}_{vert,horiz}),\mathrm{Maps}([\bullet]\times[\bullet],\mathrm{Funct}([m],\mathbb{S})_{\text{lef-lax}})\simeq\\ &\simeq \mathrm{Maps}_{\mathrm{Spc}^{\mathbf{\Delta}^{\mathrm{op}}\times\mathbf{\Delta}^{\mathrm{op}}}}(\mathrm{Sq}_{\bullet,\bullet}(\mathbf{Q}_{vert,horiz}),\mathrm{Sq}_{\bullet,\bullet}(\mathrm{Funct}([m],\mathbb{S})_{\text{left-lax}})),\end{aligned}$$

which, using Proposition 2.6.3, we identify with

$$\mathrm{Maps}(\mathbf{Q},\mathrm{Sq}_{\bullet,\bullet}(\mathrm{Funct}([m],\mathbb{S})_{\text{left-lax}}))\simeq \mathrm{Maps}([m]\circledast \mathbf{Q},\mathbb{S}),$$

as required.

In the above formula, the notation $\mathrm{Funct}(-,-)_{\text{left-lax}}$ is stands for the $(\infty,2)$-category, whose objects are functors, but whose 1-morphisms are *left-lax* natural transformations, see Chapter 10, Sect. 3.2.7.

2.7.4. Let us now analyze the left-hand side in (2.24). We have:

$$\begin{aligned}&\mathrm{Maps}_{\mathrm{Spc}^{\mathbf{\Delta}^{\mathrm{op}}\times\mathbf{\Delta}^{\mathrm{op}}\times\mathbf{\Delta}^{\mathrm{op}}}}(\mathrm{Seq}_\bullet([m])\boxtimes \mathrm{Sq}_{\bullet,\bullet}(\mathbf{Q}_{vert,horiz}),\mathrm{Seq}_\bullet(\mathbf{Grid}_{\bullet,\bullet}(\mathbf{C})^{adm}_{vert;horiz}))\simeq\\ &\simeq \mathrm{Maps}_{\mathrm{Spc}^{\mathbf{\Delta}^{\mathrm{op}}\times\mathbf{\Delta}^{\mathrm{op}}}}(\mathrm{Sq}_{\bullet,\bullet}(\mathbf{Q}_{vert,horiz}),\mathrm{Seq}_m(\mathbf{Grid}_{\bullet,\bullet}(\mathbf{C})^{adm}_{vert;horiz})).\end{aligned}$$

Note that the latter expression is a subspace in

$$\begin{aligned}&\mathrm{Maps}_{\mathrm{Spc}^{\mathbf{\Delta}^{\mathrm{op}}\times\mathbf{\Delta}^{\mathrm{op}}}}(\mathrm{Sq}_{\bullet,\bullet}(\mathbf{Q}_{vert,horiz}),\mathrm{Maps}([m]\times([\bullet]\times[\bullet]^{\mathrm{op}}),\mathbf{C}))\simeq\\ &\simeq \mathrm{Maps}_{\mathrm{Spc}^{\mathbf{\Delta}^{\mathrm{op}}\times\mathbf{\Delta}^{\mathrm{op}}}}(\mathrm{Sq}_{\bullet,\bullet}(\mathbf{Q}_{vert,horiz}),\mathrm{Maps}([\bullet]\times[\bullet]^{\mathrm{op}},\mathbf{Maps}([m],\mathbf{C})))\simeq\\ &\simeq \mathrm{Maps}_{\mathrm{Spc}^{\mathbf{\Delta}^{\mathrm{op}}\times\mathbf{\Delta}^{\mathrm{op}}}}(\mathrm{Sq}_{\bullet,\bullet}(\mathbf{Q}_{vert,horiz})^{\text{horiz-op}},\mathrm{Maps}([\bullet]\times[\bullet],\mathbf{Maps}([m],\mathbf{C})))\simeq\\ &\simeq \mathrm{Maps}(\mathrm{Sq}_{\bullet,\bullet}(\mathbf{Q}_{vert,horiz})^{\text{horiz-op}},\mathrm{Sq}_{\bullet,\bullet}(\mathbf{Maps}([m],\mathbf{C}))),\end{aligned}$$

which, using Proposition 2.6.3, we identify

$$\mathrm{Maps}(\mathbf{Q}^{\text{horiz-op}},\mathbf{Maps}([m],\mathbf{C}))\simeq \mathrm{Maps}([m]\times\mathbf{Q}^{\text{horiz-op}},\mathbf{C}).$$

Similarly, $\mathrm{Seq}_m(\mathbf{Q}^{\text{horiz-op}}(\mathbf{C})^{adm}_{vert;horiz})$ is a subspace in

$$\mathrm{Maps}([m]\times\mathbf{Q}^{\text{horiz-op}},\mathbf{C}).$$

Now, under the identification obtained in this way, we have:

$$\begin{aligned}&\mathrm{Maps}_{\mathrm{Spc}^{\mathbf{\Delta}^{\mathrm{op}}\times\mathbf{\Delta}^{\mathrm{op}}}}(\mathrm{Sq}_{\bullet,\bullet}(\mathbf{Q}_{vert,horiz}),\mathrm{Maps}([m]\times([\bullet]\times[\bullet]^{\mathrm{op}}),\mathbf{C}))\simeq\\ &\simeq \mathrm{Maps}([m]\times\mathbf{Q}^{\text{horiz-op}},\mathbf{C}),\end{aligned}$$

the subspaces

$$\begin{aligned}&\mathrm{Maps}_{\mathrm{Spc}^{\mathbf{\Delta}^{\mathrm{op}}\times\mathbf{\Delta}^{\mathrm{op}}}}(\mathrm{Sq}_{\bullet,\bullet}(\mathbf{Q}_{vert,horiz}),\mathrm{Seq}_m(\mathbf{Grid}_{\bullet,\bullet}(\mathbf{C})^{adm}_{vert;horiz}))\subset\\ &\mathrm{Maps}_{\mathrm{Spc}^{\mathbf{\Delta}^{\mathrm{op}}\times\mathbf{\Delta}^{\mathrm{op}}}}(\mathrm{Sq}_{\bullet,\bullet}(\mathbf{Q}_{vert,horiz}),\mathrm{Maps}([m]\times([\bullet]\times[\bullet]^{\mathrm{op}}),\mathbf{C}))\end{aligned}$$

and

$$\mathrm{Seq}_m(\mathbf{Q}^{\text{horiz-op}}(\mathbf{C})^{adm}_{vert;horiz})\subset \mathrm{Maps}([m]\times\mathbf{Q}^{\text{horiz-op}},\mathbf{C})$$

correspond to one another, as required.

2.8. Double Segal spaces and the proof of Proposition 2.6.3. The idea of the proof is that any cluster can be assembled from pieces for which the assertion is manifestly true.

2.8.1. By a double Segal space we shall mean a bi-simplicial space B such that for every fixed m (resp., n), the simplicial space $B_{m,\bullet}$ (resp., $B_{\bullet,n}$) is a Segal space.

By Chapter 10, Sect. 4.1.7, the essential image of the functor

$$\mathrm{Sq}_{\bullet,\bullet} : 2\text{-}\mathrm{Cat} \to \mathrm{Spc}^{\mathbf{\Delta}^{\mathrm{op}} \times \mathbf{\Delta}^{\mathrm{op}}}$$

is a double Segal space.

2.8.2. Let Q be a cluster, realized as a convex subset of some $\{0, ..., m\} \times \{0, ..., n\}$. For a horizontal line

$$m_1 \times \{0, ..., n\} \subset \{0, ..., m\} \times \{0, ..., n\}$$

let $Q^{\leq m_1}$, $Q^{\geq m_1}$ and $Q^{=m_1}$ be the parts of Q that lie below, above or on that line, respectively.

Consider the corresponding bi-simplicial spaces

$$\mathrm{Sq}_{\bullet,\bullet}(\mathbf{Q}^{\leq m_1}_{vert,horiz})\ \mathrm{Sq}_{\bullet,\bullet}(\mathbf{Q}^{\geq m_1}_{vert,horiz}),\ \mathrm{Sq}_{\bullet,\bullet}(\mathbf{Q}^{= m_1}_{vert,horiz}) \text{ and } \mathrm{Sq}_{\bullet,\bullet}(\mathbf{Q}_{vert,horiz})$$

and the categories

$$\mathbf{Q}^{\leq m_1},\ \mathbf{Q}^{\geq m_1},\ \mathbf{Q}^{=m_1} \text{ and } \mathbf{Q}.$$

We have the natural maps

$$\mathrm{Sq}_{\bullet,\bullet}(\mathbf{Q}^{\leq m_1}_{vert,horiz}) \underset{\mathrm{Sq}_{\bullet,\bullet}(\mathbf{Q}^{\leq m_1}_{vert,horiz})}{\sqcup} \mathrm{Sq}_{\bullet,\bullet}(\mathbf{Q}^{\leq m_1}_{vert,horiz}) \to \mathrm{Sq}_{\bullet,\bullet}(\mathbf{Q}_{vert,horiz}) \tag{2.25}$$

and

$$\mathbf{Q}^{\leq m_1} \underset{\mathbf{Q}^{=m_1}}{\sqcup} \mathbf{Q}^{\geq m_1} \to \mathbf{Q} \tag{2.26}$$

We will prove:

LEMMA 2.8.3.

(a) *The map*

$$\mathrm{Maps}(\mathrm{Sq}_{\bullet,\bullet}(\mathbf{Q}_{vert,horiz}), B) \to$$
$$\to \mathrm{Maps}(\mathrm{Sq}_{\bullet,\bullet}(\mathbf{Q}^{\leq m_1}_{vert,horiz}), B) \underset{\mathrm{Maps}(\mathrm{Sq}_{\bullet,\bullet}(\mathbf{Q}^{=m_1}_{vert,horiz}), B)}{\times} \mathrm{Maps}(\mathrm{Sq}_{\bullet,\bullet}(\mathbf{Q}^{\geq m_1}_{vert,horiz}), B),$$

induced by (2.25), *is an isomorphism whenever* B *is a double Segal space.*

(b) *The map* (2.26) *is an isomorphism in* 1-Cat.

2.8.4. Let us show how this lemma implies Proposition 2.6.3.

First, by symmetry, we have an analog of Lemma 2.8.3 when instead of horizontal lines we consider vertical ones. This reduces the verification of Proposition 2.6.3 to the following four cases: (i) $Q = \{0\} \times \{0\}$, (ii) $Q = \{0,1\} \times \{0\}$, (iii) $Q = \{0\} \times \{0,1\}$ and (iv) $Q = \{0,1\} \times \{0,1\}$.

In each of these cases, the assertion of Proposition 2.6.3 is manifest.

2.8.5. *Proof of Lemma* 2.8.3. We will prove point (a) of the lemma; point (b) is similar but simpler.

We have

$$\mathrm{Sq}_{\bullet,\bullet}(\mathbf{Q}_{vert,horiz}) = \underset{[i]\to[m],[j]\to[n],\mathrm{Im}([i]\times[j])\subset Q}{\mathrm{colim}} \mathrm{Seq}_\bullet([i]) \boxtimes \mathrm{Seq}_\bullet([j]).$$

Cofinal in the above index category is the full subcategory, denoted E, that consists of those maps for which one of the following three scenarios happens:

(1) The image of $\{0,...,i\}$ in $\{0,...,m\}$ is $< m_1$;
(2) The image of $\{0,...,i\}$ in $\{0,...,m\}$ is $> m_1$;
(3) The element $m_1 \in \{0,...,m\}$ has a *unique* preimage in $\{0,...,i\}$.

For an object

$$([i]\to[m],[j]\to[n],\mathrm{Im}([i]\times[j])\subset Q) =: e \in E$$

consider the fiber product

$$B_e := (\mathrm{Sq}_{\bullet,\bullet}(\mathbf{Q}^{\leq m_1}_{vert,horiz}) \underset{\mathrm{Sq}_{\bullet,\bullet}(\mathbf{Q}^{=m_1}_{vert,horiz})}{\sqcup} \mathrm{Sq}_{\bullet,\bullet}(\mathbf{Q}^{\geq m_1}_{vert,horiz})) \underset{\mathrm{Sq}_{\bullet,\bullet}(\mathbf{Q}_{vert,horiz})}{\times} (\mathrm{Seq}_\bullet([i]) \boxtimes \mathrm{Seq}_\bullet([j])),$$

taken in the category $\mathrm{Spc}^{\mathbf{\Delta}^{\mathrm{op}}\times\mathbf{\Delta}^{\mathrm{op}}}$.

We have:

$$\mathrm{Sq}_{\bullet,\bullet}(\mathbf{Q}_{vert,horiz}) \simeq \underset{e\in E}{\mathrm{colim}}\, \mathrm{Seq}_\bullet([i]) \boxtimes \mathrm{Seq}_\bullet([j]),$$

and since fiber products in $\mathrm{Spc}^{\mathbf{\Delta}^{\mathrm{op}}\times\mathbf{\Delta}^{\mathrm{op}}}$ commute with colimits, we also have

$$\mathrm{Sq}_{\bullet,\bullet}(\mathbf{Q}^{\leq m_1}_{vert,horiz}) \underset{\mathrm{Sq}_{\bullet,\bullet}(\mathbf{Q}^{=m_1}_{vert,horiz})}{\sqcup} \mathrm{Sq}_{\bullet,\bullet}(\mathbf{Q}^{\geq m_1}_{vert,horiz}) \simeq \underset{e\in E}{\mathrm{colim}}\, B_e.$$

It remains to show that for every $e \in E$ and a double Segal space B, the map

$$B_e \to \mathrm{Seq}_\bullet([i]) \boxtimes \mathrm{Seq}_\bullet([j])$$

induces an isomorphism

$$B_{i,j} = \mathrm{Maps}(\mathrm{Seq}_\bullet([i]) \boxtimes \mathrm{Seq}_\bullet([j]), B) \to \mathrm{Maps}(B_e, B). \tag{2.27}$$

However, this follows from the fact that B_e has the form

(1) $\mathrm{Seq}_\bullet([i]) \boxtimes \mathrm{Seq}_\bullet([j])$;
(2) $\mathrm{Seq}_\bullet([i]) \boxtimes \mathrm{Seq}_\bullet([j])$;
(3) $(\mathrm{Seq}_\bullet([i_1])\boxtimes\mathrm{Seq}_\bullet([j])) \underset{\mathrm{Seq}_\bullet([0])\boxtimes\mathrm{Seq}_\bullet([j])}{\sqcup} (\mathrm{Seq}_\bullet([i_2]))\boxtimes\mathrm{Seq}_\bullet([j]))$ for $[i_1] \underset{[0]}{\sqcup} [i_2] \simeq [i]$

in each of the three scenarios above. In the first two cases, the map (2.27) is an isomorphism for any bi-simplicial category B. In the third case, it follows from the definition of double Segal spaces:

$$B_{i,j} \to B_{i_1,j} \underset{B_{0,j}}{\times} B_{i_2,j}$$

is an isomorphism. □

3. The universal property of the category of correspondences

One of the main themes of this book is the construction of functors out of given $(\infty, 2)$-category of correspondences. How does one construct such a functor?

In turns out that there is one case, where a datum of a functor

$$\mathrm{Corr}(\mathbf{C})^{adm}_{vert;horiz} \to \mathbb{S}$$

is equivalent to a datum of a functor $\Phi : \mathbf{C}_{vert} \to (\mathbb{S})^{1\text{-Cat}}$, having a particular property.

We emphasize that this a property, and not an additional piece of data. Moreover, this property essentially occurs at the level of the underlying ordinary 2-categories. It is called the *left Beck-Chevalley condition*, and it says that for every 1-morphism α in $\mathbf{C}_{horiz}$, the corresponding 1-morphism $\Phi(\alpha)$ in $\mathbb{S}$ *admits a right adjoint*, and that these right adjoints satisfy *base change* against $\Phi(\beta)$ for $\beta \in \mathbf{C}_{vert}$.

All other instances of a functor out of a $(\infty, 2)$-category of correspondences, considered in this book, will be obtained from this case, by various extension procedures, considered in the subsequent sections in this chapter and the next.

3.1. The Beck-Chevalley conditions. In this subsection we will give the definition of the left and right Beck-Chevalley conditions.

3.1.1. Assume that in the context of Sect. 1.1.1, we have $horiz \subset vert$, and $adm = horiz$. Let $\mathbb{S}$ be an $(\infty, 2)$-category, and let

$$\Phi : \mathbf{C}_{vert} \to \mathbb{S}$$

be a functor.

DEFINITION 3.1.2. *We shall say that Φ satisfies the* left *Beck-Chevalley condition with respect to horiz, if for every 1-morphism $\alpha : \mathbf{c} \to \mathbf{c}'$ with $\alpha \in horiz$, the corresponding 1-morphism*

$$\Phi(\alpha) : \Phi(\mathbf{c}) \to \Phi(\mathbf{c}')$$

admits a right adjoint, to be denoted $\Phi^!(\alpha)$, such that for every Cartesian diagram

$$\begin{array}{ccc} \mathbf{c}_{0,1} & \xrightarrow{\alpha_0} & \mathbf{c}_{0,0} \\ {\scriptstyle \beta_1}\downarrow & & \downarrow{\scriptstyle \beta_0} \\ \mathbf{c}_{1,1} & \xrightarrow{\alpha_1} & \mathbf{c}_{1,0} \end{array} \tag{3.1}$$

with $\alpha_0, \alpha_1 \in horiz$ and $\beta_0, \beta_1 \in vert$, the 2-morphism

$$\Phi(\beta_1) \circ \Phi^!(\alpha_0) \to \Phi^!(\alpha_1) \circ \Phi(\beta_0),$$

arising by adjunction from the isomorphism

$$\Phi(\alpha_1) \circ \Phi(\beta_1) \simeq \Phi(\beta_0) \circ \Phi(\alpha_0),$$

is an isomorphism.

In particular, from the existence of the right adjoints $\Phi^!(\alpha)$, we obtain a well-defined functor

$$\Phi^! : (\mathbf{C}_{horiz})^{\mathrm{op}} \to \mathbb{S},$$

see Chapter 12, Sect. 1.3.

REMARK 3.1.3. Note that a functor $\Phi : \mathbf{C}_{vert} \to \mathbb{S}$ satisfies the (left) Beck-Chevalley condition if and only if its composition with $\mathbb{S} \to \mathbb{S}^{\mathrm{ordn}}$ does. So, the Beck-Chevalley condition is something that can be checked at the level of ordinary 2-categories.

3.1.4. Let us now assume that $vert \subset horiz$ and $adm = vert$. Let $\mathbb{S}$ be a 2-category, and let

$$\Phi^! : (\mathbf{C}_{horiz})^{\mathrm{op}} \to \mathbb{S}$$

be a functor.

DEFINITION 3.1.5. *We shall say that $\Phi^!$ satisfies the* right *Beck-Chevalley condition with respect to vert, if for every 1-morphism $\beta : \mathbf{c} \to \mathbf{c}'$ with $\beta \in vert$, the corresponding 1-morphism*

$$\Phi^!(\beta) : \Phi^!(\mathbf{c}') \to \Phi^!(\mathbf{c})$$

admits a left adjoint, to be denoted $\Phi(\beta)$, such that for every Cartesian diagram

$$\begin{array}{ccc} \mathbf{c}_{0,1} & \xrightarrow{\alpha_0} & \mathbf{c}_{0,0} \\ {\scriptstyle\beta_1}\downarrow & & \downarrow{\scriptstyle\beta_0} \\ \mathbf{c}_{1,1} & \xrightarrow{\alpha_1} & \mathbf{c}_{1,0} \end{array} \tag{3.2}$$

with $\alpha_0, \alpha_1 \in horiz$ and $\beta_0, \beta_1 \in vert$, the 2-morphism

$$\Phi(\beta_1) \circ \Phi^!(\alpha_0) \to \Phi^!(\alpha_1) \circ \Phi(\beta_0), \tag{3.3}$$

arising by adjunction from the isomorphism

$$\Phi^!(\alpha_0) \circ \Phi^!(\beta_0) \simeq \Phi^!(\beta_1) \circ \Phi^!(\alpha_1),$$

is an isomorphism.

In particular, if $\Phi^!$ satisfies the right Beck-Chevalley condition with respect to *vert*, we obtain a well-defined functor

$$\Phi : \mathbf{C}_{vert} \to \mathbb{S},$$

see Chapter 12, Sect. 1.3.

3.2. Statement of the universal property. In this subsection we state the main result of this section: it describes functors out of a $(\infty, 2)$-category of correspondences in terms of a *1-categorical datum.*

3.2.1. The goal of this section is to prove the following theorem:

THEOREM 3.2.2.

(a) *Suppose that $horiz \subset vert$ and $adm = horiz$ satisfies condition (5) from Sect.* 1.1.1. *Then restriction along $\mathbf{C}_{vert} \to \mathrm{Corr}(\mathbf{C})^{horiz}_{vert;horiz}$ defines an equivalence between the space of functors*

$$\Phi^{horiz}_{vert;horiz} : \mathrm{Corr}(\mathbf{C})^{horiz}_{vert;horiz} \to \mathbb{S}$$

and the subspace of functors

$$\Phi : \mathbf{C}_{vert} \to \mathbb{S}$$

that satisfy the left Beck-Chevalley condition with respect to horiz. For $\Phi^{horiz}_{vert;horiz}$ as above, the resulting functor $\Phi^! := \Phi^{horiz}_{vert;horiz}|_{(\mathbf{C}_{horiz})^{\mathrm{op}}}$ is obtained from $\Phi|_{\mathbf{C}_{horiz}}$ by passing to right adjoints.

(b) *Suppose that vert ⊂ horiz and adm = vert satisfies condition (5) from Sect.* 1.1.1. *Then restriction along* $(\mathbf{C}_{horiz})^{\mathrm{op}} \to \mathrm{Corr}(\mathbf{C})^{vert}_{vert;horiz}$ *defines an equivalence between the space of functors*

$$\Phi^{vert}_{vert;horiz} : \mathrm{Corr}(\mathbf{C})^{vert}_{vert;horiz} \to \mathbb{S}$$

and the subspace of functors

$$\Phi^! : (\mathbf{C}_{horiz})^{\mathrm{op}} \to \mathbb{S}$$

that satisfy the right Beck-Chevalley condition with respect to vert. For $\Phi^{vert}_{vert;horiz}$ *as above, the resulting functor* $\Phi := \Phi^{vert}_{vert;horiz}|_{\mathbf{C}_{vert}}$ *is obtained from* $\Phi^!|_{(\mathbf{C}_{vert})^{\mathrm{op}}}$ *by passing to left adjoints.*

The rest of this section is devoted to the proof of Theorem 3.2.2. By symmetry, it suffices to treat case (b) of the theorem.

3.2.3. We will first establish the easy direction. Namely, we will start with a functor

$$\Phi^{vert}_{vert;horiz} : \mathrm{Corr}(\mathbf{C})^{vert}_{vert;horiz} \to \mathbb{S},$$

and we will show that its restriction

$$\Phi^! := \Phi^{vert}_{vert;horiz}|_{(\mathbf{C}_{horiz})^{\mathrm{op}}} : (\mathbf{C}_{horiz})^{\mathrm{op}} \to \mathbb{S}$$

satisfies the left Beck-Chevalley condition and that $\Phi := \Phi^{vert}_{vert;horiz}|_{\mathbf{C}_{vert}}$ is obtained from $\Phi^!|_{(\mathbf{C}_{vert})^{\mathrm{op}}}$ by passing to left adjoints.

REMARK 3.2.4. We note, however, that this step is logically unnecessary: in Sect. 3.3 we will be able to establish the desired isomorphism directly.

3.2.5. Note that according to Remark 3.1.3, we can assume that $\mathbb{S}$ is an ordinary 2-category. Since

$$(\mathrm{Corr}(\mathbf{C})^{vert}_{vert;horiz})^{\mathrm{ordn}} \simeq \mathrm{Corr}(\mathbf{C}^{\mathrm{ordn}})^{vert}_{vert;horiz},$$

we can assume that $\mathbf{C}$ is an ordinary 1-category. Hence, we can use a hands-on description of the 2-category $\mathrm{Corr}(\mathbf{C})^{vert}_{vert;horiz}$ given in Sect. 1.1.3.

Furthermore, it suffices to consider the universal case, namely when

$$\mathbb{S} = \mathrm{Corr}(\mathbf{C})^{vert}_{vert;horiz}$$

and $\Phi^!$ is the tautological inclusion

$$(\mathbf{C}_{\mathrm{horiz}})^{\mathrm{op}} \to \mathrm{Corr}(\mathbf{C})^{vert}_{vert;horiz}.$$

3.2.6. Given $\beta : \mathbf{c} \to \mathbf{c}'$ in *vert*, we need to show that the corresponding 1-morphism $\mathbf{c}' \to \mathbf{c}$ in $\mathrm{Corr}(\mathbf{C})^{vert}_{vert;horiz}$, i.e.,

$$\begin{array}{ccc} \mathbf{c} & \xrightarrow{\beta} & \mathbf{c}' \\ {\scriptstyle \mathrm{id}_{\mathbf{c}}}\downarrow & & \\ \mathbf{c}, & & \end{array} \tag{3.4}$$

admits a left adjoint. We claim that the left adjoint in question is given by the diagram

$$
\begin{array}{ccc}
\mathbf{c} & \xrightarrow{\mathrm{id}_{\mathbf{c}}} & \mathbf{c} \\
{\scriptstyle \beta}\big\downarrow & & \\
\mathbf{c}'. & &
\end{array}
\tag{3.5}
$$

Let us construct the corresponding unit and co-unit of the adjunction.

The composition (3.4)∘(3.5) is given by the diagram

$$
\begin{array}{ccc}
\mathbf{c} \underset{\mathbf{c}'}{\times} \mathbf{c} & \longrightarrow & \mathbf{c} \\
\big\downarrow & & \\
\mathbf{c} & &
\end{array}
$$

The unit of the adjunction is given by the diagram:

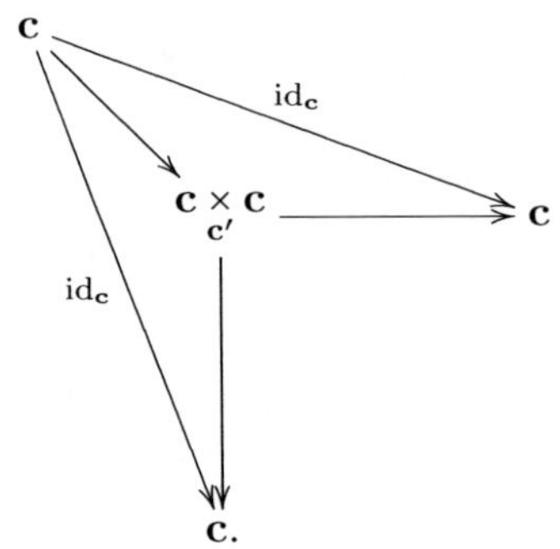

The composition (3.5)∘(3.4) is given by the diagram

$$
\begin{array}{ccc}
\mathbf{c} & \xrightarrow{\beta} & \mathbf{c}' \\
{\scriptstyle \beta}\big\downarrow & & \\
\mathbf{c}'. & &
\end{array}
$$

The co-unit of the adjunction is given by the diagram:

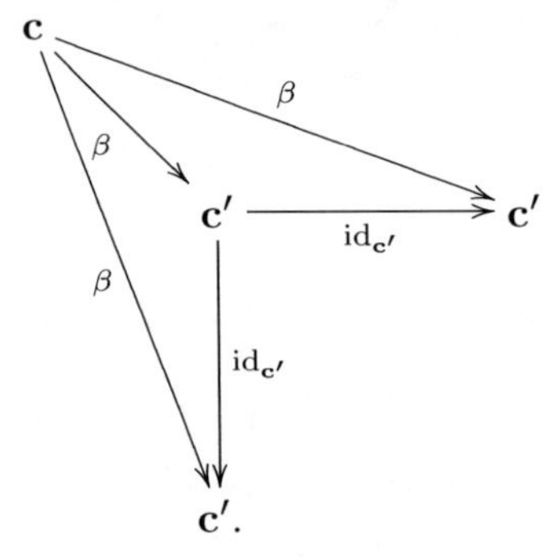

The fact that unit and co-unit maps thus constructed satisfy the adjunction identities is a straightforward verification.

3.3. Construction of a functor out of the category of correspondences. In this subsection we will prove Theorem 3.2.2.

If $\mathbb{S}$ is an ordinary 2-category, the description of the adjoints given in Sect. 3.2.6 above gives an elementary proof.

However, to give a proof in the ∞-categorical setting, we need a more explicit description of the notion of adjoint functor. Such a description is furnished by Chapter 12, Theorem 1.2.4.

3.3.1. Let $\Phi^! : (\mathbf{C}_{horiz})^{op} \to \mathbb{S}$ be a functor, such that for every 1-morphism $\beta : \mathbf{c} \to \mathbf{c}'$ with $\beta \in vert$, the corresponding 1-morphism

$$\Phi^!(\beta) : \Phi^!(\mathbf{c}') \to \Phi^!(\mathbf{c})$$

admits a left adjoint.

According to Chapter 12, Theorem 1.2.4, the datum of such $\Phi^!$ is equivalent to the datum of a map of bi-simplicial spaces

$$\mathrm{Sq}^{\mathrm{Pair}}_{\bullet,\bullet}((\mathbf{C}_{horiz})^{\mathrm{op}}, (\mathbf{C}_{vert})^{\mathrm{op}})^{\text{vert-op}} \to \mathrm{Sq}_{\bullet,\bullet}(\mathbb{S}).$$

By construction, for a commutative square

$$\begin{array}{ccc} \mathbf{c}_{0,1} & \xrightarrow{\alpha_0} & \mathbf{c}_{0,0} \\ {\scriptstyle \beta_1}\downarrow & & \downarrow{\scriptstyle \beta_0} \\ \mathbf{c}_{1,1} & \xrightarrow{\alpha_1} & \mathbf{c}_{1,0}, \end{array} \tag{3.6}$$

the resulting 2-morphism

$$\Phi(\beta_1) \circ \Phi^!(\alpha_0) \to \Phi^!(\alpha_1) \circ \Phi(\beta_0)$$

is obtained by adjunction from the tautological isomorphism

$$\Phi^!(\alpha_0) \circ \Phi^!(\beta_0) \simeq \Phi^!(\beta_1) \circ \Phi^!(\alpha_1).$$

3.3.2. According to Theorem 2.1.3, the datum of a functor

$$\Phi^{vert}_{vert;horiz} : \mathrm{Corr}(\mathbf{C})^{vert}_{vert;horiz} \to \mathbb{S}$$

is equivalent to the datum of a map of bi-simplicial spaces

$$\mathrm{defGrid}_{\bullet,\bullet}(\mathbf{C})^{vert}_{vert;horiz} \to \mathrm{Sq}_{\bullet,\bullet}(\mathbb{S})$$

such that for every object in $\underline{\mathbf{c}} \in \mathrm{defGrid}_{1,1}(\mathbf{C})^{vert}_{vert;horiz}$ corresponding to a diagram (3.6) that is Cartesian, the 2-morphism in the corresponding object in $\mathrm{Sq}_{1,1}(\mathbb{S})$ is an isomorphism.

3.3.3. Note, however, that we have a natural monomorphism of bi-simplicial spaces

$$\mathrm{defGrid}_{\bullet,\bullet}(\mathbf{C})^{vert}_{vert;horiz} \to \mathrm{Sq}^{\mathrm{Pair}}_{\bullet,\bullet}((\mathbf{C}_{horiz})^{\mathrm{op}}, (\mathbf{C}_{vert})^{\mathrm{op}})^{\text{vert-op}} \tag{3.7}$$

Therefore, the assertion of Theorem 3.2.2(b) manifestly follows from:

PROPOSITION 3.3.4. *The map* (3.7) *is an isomorphism of bi-simplicial spaces.*

PROOF. We need to show that for any commutative diagram in $\mathbf{C}$

$$\begin{array}{ccc} x_1 & \longrightarrow & x_2 \\ \downarrow & & \downarrow \\ y_1 & \longrightarrow & y_2 \end{array}$$

with vertical maps in *vert* and horizontal maps in *horiz*, the corresponding map

$$x_1 \to x_2 \underset{y_2}{\times} y_1$$

lies in *vert*. However, this map is given by the composite

$$x_1 \to x_2 \underset{y_2}{\times} x_1 \to x_2 \underset{y_2}{\times} y_1$$

where the first map is a base change of $x_2 \to x_2 \underset{y_2}{\times} x_2$ and the second is a base change of $x_1 \to y_1$, both of which lie in *vert*, by assumption. □

4. Enlarging the class of 2-morphisms at no cost

In this section we will prove the first of the two results of the type that given a functor from one $(\infty, 2)$-category of correspondences, we can canonically extend it to a functor from another $(\infty, 2)$-category of correspondences that has a larger class of 2-morphisms.

4.1. The setting. In this subsection we explain the setting for the main result of this section, Theorem 4.1.3.

4.1.1. Let $(\mathbf{C}, vert, horiz, adm)$ be as in Sect. 1.1.1. Let $(\mathbf{C}, vert, horiz, adm')$ be another such data with

$$adm \subset adm'.$$

We shall also assume that the following condition holds:

For a 1-morphism

$$\gamma' : \mathbf{c} \to \mathbf{c}'$$

with $\gamma \in adm'$, the diagonal map

$$\mathbf{c} \to \mathbf{c} \underset{\mathbf{c}'}{\times} \mathbf{c}$$

belongs to *adm*.

4.1.2. Let $\Phi^{adm}_{vert;horiz}$ be a functor

$$\Phi^{adm}_{vert;horiz} : \mathrm{Corr}(\mathbf{C})^{adm}_{vert;horiz} \to \mathbb{S}.$$

For a morphism $\gamma : \mathbf{c}_0 \to \mathbf{c}_1$ in adm', consider the commutative (but not necessarily) Cartesian) square

$$\begin{array}{ccc} \mathbf{c} & \xrightarrow{\mathrm{id}} & \mathbf{c} \\ {\scriptstyle \mathrm{id}}\downarrow & & \downarrow{\scriptstyle \gamma} \\ \mathbf{c} & \xrightarrow{\gamma} & \mathbf{c}' \end{array}$$

which gives rise to a (not necessarily invertible) 2-morphism

$$\mathrm{id} \to \Phi^!(\gamma) \circ \Phi(\gamma).$$

We impose the condition that the above 2-morphism define the unit of an adjunction.

Under the above circumstances, we claim:

THEOREM 4.1.3. *The functor* $\Phi^{adm}_{vert;horiz}$ *admits a unique extension to a functor*

$$\Phi^{adm'}_{vert;horiz} : \mathrm{Corr}(\mathbf{C})^{adm'}_{vert;horiz} \to \mathbb{S}.$$

The rest of this section is devoted to the proof of this theorem.

REMARK 4.1.4. The proof of Theorem 4.1.3 is essentially combinatorics, i.e., playing with various diagrams. The reader may find it useful to first prove Theorem 4.1.3 in the case when $\mathbb{S}$ is an ordinary 2-category; in this case, this is simple exercise.

4.1.5. Starting from $\Phi^{adm}_{vert;horiz}$, we shall construct the data of the functor $\Phi^{adm'}_{vert;horiz}$ as a map of tri-simplicial spaces

$$\mathrm{Seq}_\bullet({}'\mathbf{Grid}_{\bullet,\bullet}(\mathbf{C})^{adm'}_{vert;horiz}) \to \mathrm{Cu}_{\bullet,\bullet,\bullet}(\mathbb{S}), \tag{4.1}$$

satisfying the additional condition of Theorem 2.2.7. I.e., we need to construct a map

$$\mathrm{Seq}_l({}'\mathbf{Grid}_{m,n}(\mathbf{C})^{adm'}_{vert;horiz}) \to \mathrm{Maps}([l] \circledast [m] \circledast [n], \mathbb{S}), \tag{4.2}$$

functorial in

$$[l] \times [m] \times [n] \in \mathbf{\Delta}^{\mathrm{op}} \times \mathbf{\Delta}^{\mathrm{op}} \times \mathbf{\Delta}^{\mathrm{op}}.$$

The fact that this extension is uniquely defined will follow from the construction.

4.2. Idea of the construction.

4.2.1. As was said above, starting from the data of $\Phi^{adm}_{vert;horiz}$, we need to construct the map (4.2). The problem that we will have to confront already occurs when $l = 1$, $m = 0$ and $n = 1$. I.e., given a square

$$\begin{array}{ccc} \mathbf{c}^0_1 & \xrightarrow{\alpha_0} & \mathbf{c}^0_0 \\ {\scriptstyle \gamma_1}\downarrow & & \downarrow{\scriptstyle \gamma_0} \\ \mathbf{c}^1_1 & \xrightarrow{\alpha_1} & \mathbf{c}^1_0 \end{array}$$

and $\alpha_0, \alpha_1 \in horiz$ and $\gamma_0, \gamma_1 \in adm'$, whose defect of Cartesianness belongs to adm', we want to construct a diagram

(4.3)

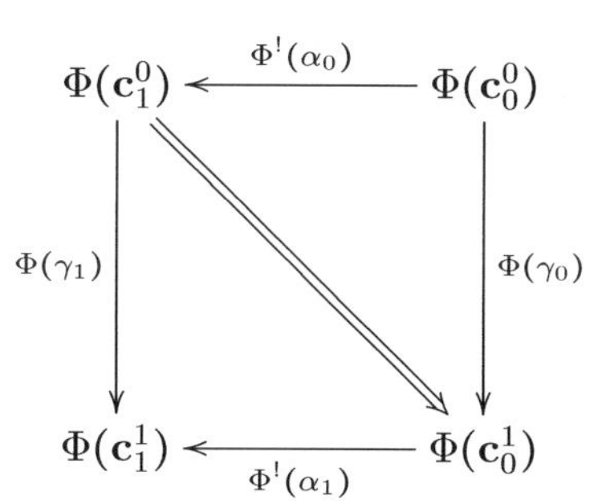

The problem is that the data of $\Phi^{adm}_{vert;horiz}$ does not produce such diagrams: according to Theorem 2.1.3, we can a priori only construct such diagrams for squares in which the defect of Cartesianness belongs to *adm*.

4.2.2. The trick is the following. Consider the 3-dimensional diagram

(4.4)

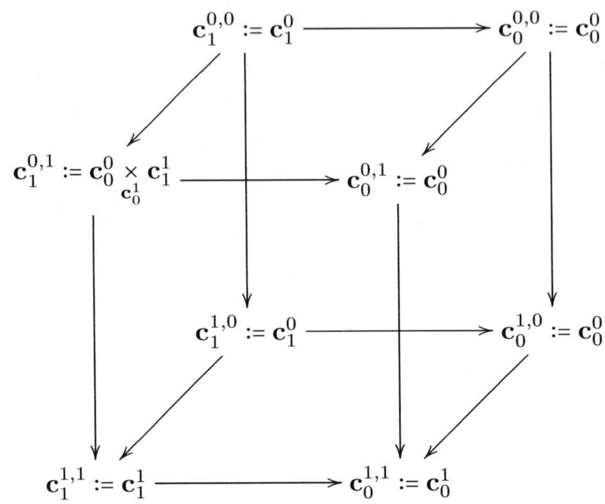

We will construct the diagram (4.3) by first constructing the diagram

(4.5)

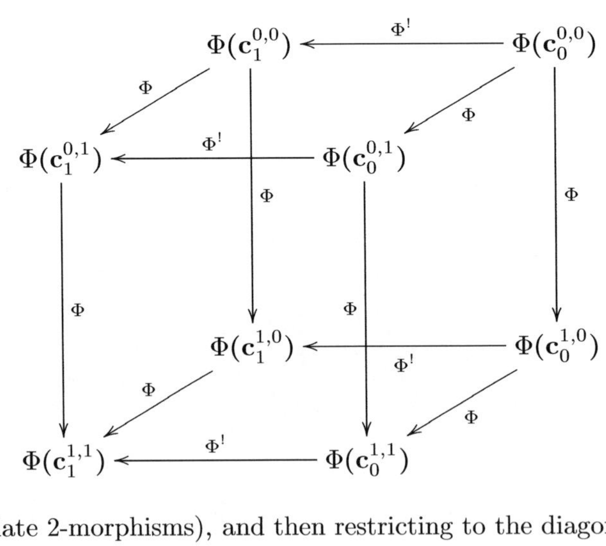

(with appropriate 2-morphisms), and then restricting to the diagonal

$$\begin{array}{ccc} \Phi(\mathbf{c}_1^{0,0}) & \longleftarrow & \Phi(\mathbf{c}_0^{0,0}) \\ \downarrow & & \downarrow \\ \Phi(\mathbf{c}_1^{1,1}) & \longleftarrow & \Phi(\mathbf{c}_0^{1,1}). \end{array}$$

4.2.3. We will obtain the diagram (4.5) from the diagram

(4.6)

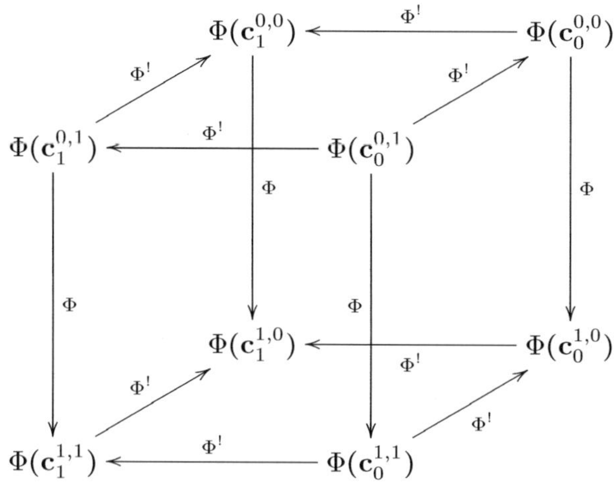

by passing to right adjoints along the slanted arrows.

4.2.4. Now, the point is that the data of the latter diagram, i.e., diagram (4.6), *is* contained in the datum of $\Phi^{adm}_{vert;horiz}$ in its guise as a map of bi-simplicial spaces

$$\text{defGrid}_{\bullet,\bullet}(\mathbf{C})^{adm}_{vert;horiz} \to \text{Sq}_{\bullet,\bullet}(\mathbb{S}). \tag{4.7}$$

Namely, let us recall from Sect. 2.3.5, that the data of a map (4.7) assigns to a functor $\mathbf{I} \times \mathbf{J}^{\text{op}} \to \mathbf{C}$ (that sends arrows along $\mathbf{I}$ to *vert*, arrows along $\mathbf{J}^{\text{op}}$ to *horiz* and where defect of Cartesianness of squares belongs to *adm*) a functor

$$\mathbf{I} \circledast \mathbf{J} \to \mathbb{S}.$$

We take $\mathbf{I} = [1]$ and $\mathbf{J} = [1] \times [1]$, and we take the functor

$$\mathbf{I} \times \mathbf{J}^{\text{op}} \to \mathbf{C}$$

to be given by the diagram (4.4), where $\mathbf{I}$ corresponds to the first upper index (i.e., the "k" in $\mathbf{c}_j^{k,l}$). In other words, $\mathbf{I}$ is the direction depicted as vertical in the diagram (4.4).

The key observation is that that the condition on $adm \subset adm'$ from Sect. 4.1.1 implies that the defect of the Cartesianness of the relevant squares in (4.4) (i.e., the squares where one side is vertical), does belong to *adm*.

The resulting functor

$$[1] \circledast ([1] \times [1]) \to \mathbb{S}$$

exactly produces the desired diagram (4.6).

4.3. Proof of Theorem 4.1.3, the key construction. In this subsection we will formally implement the idea explained above.

The map (4.2) will be constructed as a composition of several maps.

4.3.1. Recall the notation

$$\mathrm{Maps}(\mathbf{I} \times \mathbf{J}^{op}, \mathbf{C})^{adm}_{vert;horiz} \subset \mathrm{Maps}(\mathbf{I} \times \mathbf{J}^{\mathrm{op}}, \mathbf{C})$$

from Sect. 2.3.4.

As a first step in constructing the map (4.2), we will produce a functor

$$\mathrm{Seq}_l(\text{'}\mathbf{Grid}_{m,n}(\mathbf{C})^{adm'}_{vert;horiz}) \to \mathrm{Maps}(([l]\times[m])\times([n]\times[l]^{\mathrm{op}})^{\mathrm{op}}, \mathbf{C})^{adm}_{vert;horiz}. \tag{4.8}$$

The functor (4.8) is given by the following explicit procedure. To an object of the space $\mathrm{Seq}_l(\text{'}\mathbf{Grid}_{m,n}(\mathbf{C})^{adm'}_{vert;horiz})$, given by

$$\mathbf{c}^{l'}_{m',n'}, \quad 0 \le l' \le l,\ 0 \le m' \le m,\ 0 \le n' \le n$$

we assign a map

$$[l] \times [m] \times [n]^{\mathrm{op}} \times [l] \to \mathbf{C} \tag{4.9}$$

that sends

$$(k', m', n', l') \mapsto \mathbf{c}^{k',l'}_{m',n'} := \begin{cases} \mathbf{c}^{k'}_{m',0} \underset{\mathbf{c}^{l'}_{m,0}}{\times} \mathbf{c}^{l'}_{m,n'} & \text{for } k' \le l'; \\ \mathbf{c}^{k'}_{m',0} \underset{\mathbf{c}^{k'}_{m,0}}{\times} \mathbf{c}^{l'}_{m,n'} & \text{for } k' \ge l'. \end{cases}$$

It is easy to see that the map (4.9) thus constructed above has the following properties:

(1) For fixed k', l', each square

$$\begin{array}{ccc} \mathbf{c}^{k',l'}_{m',n'} & \longrightarrow & \mathbf{c}^{k',l'}_{m',n'-1} \\ \downarrow & & \downarrow \\ \mathbf{c}^{k',l'}_{m'+1,n'} & \longrightarrow & \mathbf{c}^{k',l'}_{m'+1,n'-1} \end{array}$$

is Cartesian.

(2) For fixed n' and k', the square

$$\begin{array}{ccc} \mathbf{c}^{k',l'}_{m',n'} & \longrightarrow & \mathbf{c}^{k',l'}_{m'+1,n'} \\ \downarrow & & \downarrow \\ \mathbf{c}^{k',l'+1}_{m',n'} & \longrightarrow & \mathbf{c}^{k',l'+1}_{m'+1,n'} \end{array}$$

is Cartesian.

(3) For fixed m' and l', the square

$$\begin{array}{ccc} \mathbf{c}^{k',l'}_{m',n'} & \longrightarrow & \mathbf{c}^{k',l'}_{m',n'-1} \\ \downarrow & & \downarrow \\ \mathbf{c}^{k'+1,l'}_{m',n'} & \longrightarrow & \mathbf{c}^{k'+1,l'}_{m',n'-1} \end{array}$$

is Cartesian.

(4) For fixed m' and n', the defect of Cartesianness of the square

$$\begin{array}{ccc} \mathbf{c}^{k',l'}_{m',n'} & \longrightarrow & \mathbf{c}^{k',l'+1}_{m',n'} \\ \downarrow & & \downarrow \\ \mathbf{c}^{k'+1,l'}_{m',n'} & \longrightarrow & \mathbf{c}^{k'+1,l'+1}_{m',n'} \end{array}$$

belongs to *adm*.

(5) For fixed k', m', n', the map $\mathbf{c}^{k',l'}_{m',n'} \to \mathbf{c}^{k',l'+1}_{m',n'}$ belongs to adm'.

(6) For fixed l', m', n', the map $\mathbf{c}^{k',l'}_{m',n'} \to \mathbf{c}^{k+1',l'}_{m',n'}$ belongs to adm'.

In particular, we obtain that the map (4.9) indeed belongs to

$$\mathrm{Maps}(([l]\times[m])\times([n]\times[l]^{\mathrm{op}})^{\mathrm{op}},\mathbf{C})^{adm}_{vert;horiz}.$$

REMARK 4.3.2. For an object $\underline{\mathbf{c}} \in \mathrm{Seq}_l(\text{'}\mathbf{Grid}_{m,n}(\mathbf{C})^{adm'}_{vert;horiz})$, given by

$$[l]\times[m]\times[n]^{op}\to\mathbf{C},$$

the corresponding map (4.9) is uniquely characterized by properties (2) and (3) above and the following.

(1) The composite map with the diagonal

$$[l]\times[m]\times[n]^{op}\to[l]\times[m]\times[n]^{op}\times[l]\to\mathbf{C}$$

is isomorphic to $\underline{\mathbf{c}}$;

(2) For $k'\leq l'$ and for all m', the map

$$\mathbf{c}^{k'}_{m',0}\simeq\mathbf{c}^{k',k'}_{m',0}\to\mathbf{c}^{k',l'}_{m',0}$$

is an isomorphism;

(3) For $l'\leq k'$ and for all n', the map

$$\mathbf{c}^{l'}_{m,n'}\simeq\mathbf{c}^{l',l'}_{m,n'}\to\mathbf{c}^{k',l'}_{m,n'}$$

is an isomorphism.

4.3.3. By Theorem 2.1.3 and Sect. 2.3.5, the functor $\Phi^{adm}_{vert;horiz}$ gives rise to a map

$$\mathrm{Maps}(([l]\times[m])\times([n]\times[l]^{\mathrm{op}})^{\mathrm{op}},\mathbf{C})^{adm}_{vert;horiz}\to\mathrm{Maps}(([l]\times[m])\circledast([n]\times[l]^{\mathrm{op}}),\mathbb{S}).\tag{4.10}$$

Let

$$\mathrm{Maps}^0(([l]\times[m])\times([n]\times[l]^{\mathrm{op}})^{\mathrm{op}},\mathbf{C})^{adm}_{vert;horiz}\subset \\ \subset\mathrm{Maps}(([l]\times[m])\times([n]\times[l]^{\mathrm{op}})^{\mathrm{op}},\mathbf{C})^{adm}_{vert;horiz}$$

be the subspace consisting of maps satisfying properties (1) and (3) in Sect. 4.3.1.

Then the map (4.10) has the property that the image of the composition

$$\begin{aligned}&\mathrm{Maps}^0(([l]\times[m])\times([n]\times[l]^{\mathrm{op}})^{\mathrm{op}},\mathbf{C})^{adm}_{vert;horiz}\\ &\quad\to\mathrm{Maps}(([l]\times[m])\times([n]\times[l]^{\mathrm{op}})^{\mathrm{op}},\mathbf{C})^{adm}_{vert;horiz}\\ &\quad\to\mathrm{Maps}(([l]\times[m])\circledast([n]\times[l]^{\mathrm{op}}),\mathbb{S})\\ &\quad\to\mathrm{Maps}(([l]\times[m])\circledast[n]\circledast[l]^{\mathrm{op}},\mathbb{S})\end{aligned}$$

belongs to

$$\mathrm{Maps}(([l] \times [m] \times [n]) \circledast [l]^{\mathrm{op}}, \mathbb{S}) \subset \mathrm{Maps}(([l] \times [m]) \circledast [n] \circledast [l]^{\mathrm{op}}, \mathbb{S}).$$

I.e., we have a well-defined map
(4.11)

$$\mathrm{Maps}^0(([l] \times [m]) \times ([n] \times [l]^{\mathrm{op}})^{\mathrm{op}}, \mathbf{C})^{adm}_{vert;horiz} \to \mathrm{Maps}(([l] \times [m] \times [n]) \circledast [l]^{\mathrm{op}}, \mathbb{S}).$$

Composing with (4.8), we obtain a map

(4.12)

$$\mathrm{Seq}_l({}'\mathbf{Grid}_{m,n}(\mathbf{C})^{adm'}_{vert;horiz}) \to \mathrm{Maps}^0(([l] \times [m]) \times ([n] \times [l]^{\mathrm{op}})^{\mathrm{op}}, \mathbf{C})^{adm}_{vert;horiz} \to$$
$$\to \mathrm{Maps}(([l] \times [m] \times [n]) \circledast [l]^{\mathrm{op}}, \mathbb{S}).$$

4.3.4. Let

$$\mathrm{Maps}^0(([l] \times [m] \times [n]) \circledast [l]^{\mathrm{op}}, \mathbb{S}) \subset \mathrm{Maps}(([l] \times [m] \times [n]) \circledast [l]^{\mathrm{op}}, \mathbb{S})$$

be the subspace of maps that for every fixed k', m', n', the resulting 1-morphism

$$s^{k',l'+1}_{m',n'} \to s^{k',l'}_{m',n'}$$

admits a left adjoint for every $0 \le l' < l' + 1 \le l$.

We shall now use the fact that for a map γ in $\mathbf{C}$ that belongs to adm', the 1-morphsim $\Phi^!(\gamma)$ in $\mathbb{S}$ admits a left adjoint (see the assumption on $\Phi^{adm}_{vert;horiz}$ in Sect. 4.1.2).

Using property (5) in Sect. 4.3.1, this implies that the image of the map (4.12) belongs to $\mathrm{Maps}^0(([l] \times [m] \times [n]) \circledast [l]^{\mathrm{op}}, \mathbb{S})$.

By Chapter 12, Corollary 3.1.7, we have a canonically defined map

(4.13) $$\mathrm{Maps}^0(([l] \times [m] \times [n]) \circledast [l]^{\mathrm{op}}, \mathbb{S}) \to \mathrm{Maps}([l] \circledast ([l] \times [m] \times [n]), \mathbb{S}),$$

giving by passing to left adjoints along the $[l]^{\mathrm{op}}$-direction.

Thus, composing (4.12) and (4.13), we obtain a map

(4.14) $$\mathrm{Seq}_l({}'\mathbf{Grid}_{m,n}(\mathbf{C})^{adm'}_{vert;horiz}) \to \mathrm{Maps}([l] \circledast ([l] \times [m] \times [n]), \mathbb{S}).$$

4.3.5. Consider the composition of (4.14) with the embedding

$$\mathrm{Maps}([l] \circledast ([l] \times [m] \times [n]), \mathbb{S}) \to \mathrm{Maps}([l] \circledast ([l] \times [m]) \circledast [n], \mathbb{S}).$$

We thus obtain a map

(4.15) $$\mathrm{Seq}_l({}'\mathbf{Grid}_{m,n}(\mathbf{C})^{adm'}_{vert;horiz}) \to \mathrm{Maps}([l] \circledast ([l] \times [m]) \circledast [n], \mathbb{S}).$$

We claim that its image belongs to the subspace

$$\mathrm{Maps}(([l] \times [l] \times [m]) \circledast [n], \mathbb{S}) \subset \mathrm{Maps}([l] \circledast ([l] \times [m]) \circledast [n], \mathbb{S}).$$

This follows from properties (2) and (4) in Sect. 4.3.1, and the next lemma:

LEMMA 4.3.6. *Let*

$$\begin{array}{ccc} \mathbf{c}_0 & \xrightarrow{\alpha} & \mathbf{c}_1 \\ {\scriptstyle\gamma_0}\downarrow & & \downarrow{\scriptstyle\gamma_1} \\ \mathbf{c}'_0 & \xrightarrow{\alpha'} & \mathbf{c}'_1 \end{array}$$

be a commutative diagram in $\mathbf{C}$ *with* $\gamma_0, \gamma_1 \in adm'$ *and* $\alpha, \alpha' \in horiz$, *and whose defect of Cartesianness belongs to adm. Then the 2-morphism*

$$(\Phi^!(\gamma_0))^L \circ \Phi^!(\alpha) \to \Phi^!(\alpha') \circ (\Phi^!(\gamma_1))^L,$$

arising by adjunction from the isomorphism

$$\Phi^!(\alpha) \circ \Phi^!(\gamma_1) \simeq \Phi^!(\gamma_0) \circ \Phi^!(\alpha)$$

identifies with the 2-morphism

$$\Phi(\gamma_0) \circ \Phi^!(\alpha) \to \Phi^!(\alpha') \circ \Phi^!(\gamma_1),$$

obtained from the functor $\Phi^{adm}_{vert;horiz}$.

PROOF. Follows from the assumption in Sect. 4.1.2 by diagram chase. □

REMARK 4.3.7. It is easy to see the the conclusion of Lemma 4.3.6 is in fact equivalent to the assumption in Sect. 4.1.2.

4.3.8. Thus, we obtain a map

$$\mathrm{Seq}_l({}'\mathbf{Grid}_{m,n}(\mathbf{C})^{adm'}_{vert;horiz}) \to \mathrm{Maps}(([l] \times [l] \times [m]) \circledast [n], \mathbb{S}). \tag{4.16}$$

Finally, using the diagonal map

$$[l] \to [l] \times [l],$$

we obtain a map

$$\mathrm{Seq}_l({}'\mathbf{Grid}_{m,n}(\mathbf{C})^{adm'}_{vert;horiz}) \to \mathrm{Maps}(([l] \times [m]) \circledast [n], \mathbb{S}). \tag{4.17}$$

The composition of (4.17) with the embedding

$$\mathrm{Maps}(([l] \times [m]) \circledast [n], \mathbb{S}) \hookrightarrow \mathrm{Maps}([l] \circledast [m] \circledast [n], \mathbb{S})$$

is a map

$$\mathrm{Seq}_l({}'\mathbf{Grid}_{m,n}(\mathbf{C})^{adm'}_{vert;horiz}) \to \mathrm{Maps}([l] \circledast [m] \circledast [n], \mathbb{S}). \tag{4.18}$$

This is the desired map of (4.2).

By constriction, its image belongs to

$$\mathrm{Maps}([l] \circledast ([m] \times [n]), \mathbb{S}) \subset \mathrm{Maps}([l] \circledast [m] \circledast [n], \mathbb{S}).$$

4.4. Verification of the tri-simplicial functoriality. In order to be more explicit, we will describe the situation for an individual map in $\mathbf{\Delta} \times \mathbf{\Delta} \times \mathbf{\Delta}$

$$[l_1] \to [l_2],\ [m_1] \to [m_2],\ [n_1] \to [n_2]. \tag{4.19}$$

4.4.1. Let $\underline{\mathbf{c}}_2$ be an object of $\operatorname{Seq}_{l_2}({}'\mathbf{Grid}_{m_2,n_2}(\mathbf{C})^{adm_2}_{vert;horiz})$. We let $\underline{\mathbf{c}}_1$ denote the object of $\operatorname{Seq}_{l_1}({}'\mathbf{Grid}_{m_1,n_1}(\mathbf{C})^{adm'}_{vert;horiz})$, obtained from $\underline{\mathbf{c}}_2$ by restricting along (4.19).

Let $\underline{s}_2$ be the point of $\operatorname{Maps}([l_2] \circledast [m_2] \circledast [n_2], \mathbb{S})$ corresponding to $\underline{\mathbf{c}}_2$ via the map (4.2). Let $\underline{s}_1$ be the point of $\operatorname{Maps}([l_1] \circledast [m_1] \circledast [n_1], \mathbb{S})$ corresponding to $\underline{\mathbf{c}}_1$ via the map (4.2).

Let $\widetilde{\underline{s}}_1$ be the point of $\operatorname{Maps}([l_1] \circledast [m_1] \circledast [n_1], \mathbb{S})$ obtained from $\underline{s}_2$ by restricting along (4.19).

We need to establish a canonical isomorphism

$$\underline{s}_1 \simeq \widetilde{\underline{s}}_1. \tag{4.20}$$

4.4.2. Let $\overline{\mathbf{c}}_2$ denote the object of

$$\operatorname{Maps}^0(([l_2] \times [m_2]) \times ([n_2] \times [l_2]^{\mathrm{op}})^{\mathrm{op}}, \mathbf{C})^{adm}_{vert;horiz}$$

obtained from $\underline{\mathbf{c}}_2$ by the map (4.8), see Sect. 4.3.3 for the notation

$$\operatorname{Maps}^0(-, \mathbf{C})^{adm}_{vert;horiz}.$$

Let $\overline{\mathbf{c}}_1$ denote the object of

$$\operatorname{Maps}^0(([l_1] \times [m_1]) \times ([n_1] \times [l_1]^{\mathrm{op}})^{\mathrm{op}}, \mathbf{C})^{adm}_{vert;horiz}$$

obtained from $\underline{\mathbf{c}}_1$ by the map (4.8).

Let $\widetilde{\overline{\mathbf{c}}}_1$ denote the object of

$$\operatorname{Maps}^0(([l_1] \times [m_1]) \times ([n_1] \times [l_1]^{\mathrm{op}})^{\mathrm{op}}, \mathbf{C})^{adm}_{vert;horiz}$$

obtained from $\overline{\mathbf{c}}_2$ by restricting along (4.19).

Let $\overline{s}_2$ be the object of

$$\operatorname{Maps}(([l_2] \times [l_2]) \circledast [m_2] \circledast [n_2], \mathbb{S}),$$

obtained from $\overline{\mathbf{c}}_2$ via the map (4.11) and passing to left adjoints along the last variable.

Let $\overline{s}_1$ and $\widetilde{\overline{s}}_1$ be the objects of

$$\operatorname{Maps}(([l_1] \times [l_1]) \circledast [m_1] \circledast [n_1], \mathbb{S}),$$

obtained from $\overline{\mathbf{c}}_1$ and $\widetilde{\overline{\mathbf{c}}}_1$, respectively, by the same procedure.

We shall now construct a natural transformation

$$\overline{s} \in \operatorname{Maps}([1] \times (([l_1] \times [l_1]) \circledast [m_1] \circledast [n_1]), \mathbb{S}), \tag{4.21}$$

whose restriction to $\{0\} \in [1]$ and $\{1\} \in [1]$ identifies with $\overline{s}_1$ and $\widetilde{\overline{s}}_1$, respectively.

REMARK 4.4.3. Note, however, that this natural transformation will *not* be an isomorphism.

4.4.4. We note that there is a canonically defined map

$$\overline{\mathbf{c}}_1 \to \widetilde{\overline{\mathbf{c}}}_1,$$

which can be regarded as an object $\overline{\mathbf{c}}$ of

$$\mathrm{Maps}([1] \times ([l_1] \times [m_1]) \times ([n_1] \times [l_1]^{\mathrm{op}})^{\mathrm{op}}, \mathbf{C}),$$

and that this object in fact belongs to

$$\mathrm{Maps}^0(([1] \times [l_1] \times [m_1]) \times ([n_1] \times [l_1]^{\mathrm{op}})^{\mathrm{op}}, \mathbf{C})^{adm}_{vert;horiz}.$$

Consider the object of

$$\mathrm{Maps}(([1] \times [l_1] \times [m_1]) \circledast ([n_1] \times [l_1]^{\mathrm{op}}), \mathbb{S})$$

attached to $\overline{\mathbf{c}}$ by means of the functor $\Phi^{adm}_{vert;horiz}$.

The image of the above object under

$$\mathrm{Maps}(([1]\times[l_1]\times[m_1])\circledast([n_1]\times[l_1]^{\mathrm{op}}), \mathbb{S}) \to \mathrm{Maps}(([1]\times[l_1]\times[m_1])\circledast[n_1]\circledast[l_1]^{\mathrm{op}}, \mathbb{S})$$

belongs

$$\mathrm{Maps}(([1]\times[l_1]\times[m_1]\times[n_1])\circledast[l_1]^{\mathrm{op}}, \mathbb{S}) \subset \mathrm{Maps}(([1]\times[l_1]\times[m_1])\circledast[n_1]\circledast[l_1]^{\mathrm{op}}, \mathbb{S}),$$

and furthermore to

$$\mathrm{Maps}^0(([1]\times[l_1]\times[m_1]\times[n_1])\circledast[l_1]^{\mathrm{op}}, \mathbb{S}) \subset \mathrm{Maps}(([1]\times[l_1]\times[m_1]\times[n_1])\circledast[l_1]^{\mathrm{op}}, \mathbb{S})$$

(see Sect. 4.3.4 for the notation $\mathrm{Maps}^0(-, \mathbb{S})$).

4.4.5. Passing to left adjoints along the last variable, we obtain an object of

$$\mathrm{Maps}([l_1] \circledast ([1] \times [l_1] \times [m_1] \times [n_1]), \mathbb{S}).$$

The image of the latter object under

$$\mathrm{Maps}([l_1] \circledast ([1] \times [l_1] \times [m_1] \times [n_1]), \mathbb{S}) \to \mathrm{Maps}([l_1] \circledast ([1] \times [l_1] \times [m_1]) \circledast [n_1], \mathbb{S})$$

belongs to

$$\mathrm{Maps}(([l_1] \times [1] \times [l_1] \times [m_1]) \circledast [n_1], \mathbb{S}) \subset \mathrm{Maps}([l_1] \circledast ([1] \times [l_1] \times [m_1]) \circledast [n_1], \mathbb{S}).$$

Further, the image of the latter object under

$$\mathrm{Maps}(([l_1]\times[1]\times[l_1]\times[m_1])\circledast[n_1], \mathbb{S}) = \mathrm{Maps}(([1]\times[l_1]\times[l_1]\times[m_1])\circledast[n_1], \mathbb{S}) \to$$
$$\to \mathrm{Maps}([1] \circledast ([l_1] \times [l_1] \times [m_1]) \circledast [n_1], \mathbb{S})$$

belongs to

$$\mathrm{Maps}([1] \times (([l_1] \times [l_1] \times [m_1]) \circledast [n_1]), \mathbb{S}) \subset \mathrm{Maps}([1] \circledast ([l_1] \times [l_1] \times [m_1]) \circledast [n_1], \mathbb{S}).$$

Let us map

$$\mathrm{Maps}([1]\times(([l_1]\times[l_1]\times[m_1])\circledast[n_1]), \mathbb{S}) \to \mathrm{Maps}([1]\times(([l_1]\times[l_1])\circledast[m_1]\circledast[n_1]), \mathbb{S}),$$

and denote the resulting object of

$$\mathrm{Maps}([1] \times (([l_1] \times [l_1]) \circledast [m_1] \circledast [n_1]), \mathbb{S})$$

by $\overline{s}$. respectively.

This $\overline{s}$ is the desired natural transformation in (4.21).

4.4.6. Restricting $\overline{s}$ under $[l_1] \to [l_1] \times [l_1]$, we obtain an object of

$$\mathrm{Maps}([1] \times ([l_1] \circledast [m_1] \circledast [n_1]), \mathbb{S}),$$

denoted $\underline{s}$.

By construction, the restrictions of $\underline{s}$ to $\{0\} \in [1]$ and $\{1\} \in [1]$ identify with $\underline{s}_1$ and $\widetilde{\underline{s}}_1$, respectively.

This provides the natural transformation from $\underline{s}_1$ to $\widetilde{\underline{s}}_1$. However, we claim that this natural transformation *is* an isomorphism, thereby providing the isomorphism in (4.20).

Indeed, this follows from the fact that the object $\overline{\mathbf{c}}$, regarded as a natural transformation from $\overline{\mathbf{c}}_1$ to $\widetilde{\overline{\mathbf{c}}}_1$, viewed as functors

$$[l_1] \times [m_1] \times [n_1]^{\mathrm{op}} \times [l_1] \to \mathbf{C},$$

becomes an isomorphism when restricted to the diagonal copy of $[l_1] \times [m_1] \times [n_1]^{\mathrm{op}}$.

5. Functors constructed by factorization

In the section we will describe the second result that allows to start from a functor out of an $(\infty, 2)$-category of correspondences and (canonically) extend it to a larger $(\infty, 2)$-category of correspondences.

This setting arises in practice when we want to construct the !-pullback as a functor

$$(\mathrm{Sch}_{\mathrm{aft}})^{\mathrm{op}} \to \mathrm{DGCat}_{\mathrm{cont}}, \tag{5.1}$$

starting from the functor

$$\mathrm{Sch}_{\mathrm{aft}} \to \mathrm{DGCat}_{\mathrm{cont}}, \quad X \rightsquigarrow \mathrm{IndCoh}(X), \quad (X \xrightarrow{f} Y) \rightsquigarrow f_*^{\mathrm{IndCoh}}.$$

The idea is that $f^!$ should be the *right* adjoint of f_*^{IndCoh} if f is proper and the *left* adjoint of f_*^{IndCoh} if f is an open embedding. For a general f we want to decompose it as $f_1 \circ f_2$, where f_2 is an open embedding and f_1 is proper. The challenge is to establish the independence of $f^!$ of such a factorization in a functorial way (i.e., in the context of ∞-categories).

It appears, however, that even if the original problem of defining functor (5.1) does not involve the 2-category of correspondences, the construction does necessarily use one (in particular, at a crucial stage we will invoke the already established Theorem 4.1.3).

5.1. Set-up for the source. In this subsection we will describe what categories of correspondences are involved in the extension procedure that is that goal of this section.

5.1.1. We start with a datum of an $(\infty,1)$-category $\mathbf{C}$ equipped with three classes of 1-morphisms $(vert, horiz, adm)$ as in Sect. 1.1.1.

We fix yet one more class $co\text{-}adm \subset horiz$, such that the triple

$$(vert, co\text{-}adm, \text{isom})$$

also satisfies the conditions of Sect. 1.1.1.

We shall assume that *horiz* and *co-adm* also satisfy the '2 out of 3' property. I.e., for a pair of composable morphisms β_1, β_2, if both β_1 and $\beta_1 \circ \beta_2$ belong to *horiz* (resp., *co-adm*), then so does β_2.

5.1.2. We now impose the following additional condition on the pair $(co\text{-}adm, adm)$. Namely, we require that for a Cartesian diagram

$$\begin{array}{ccc} \mathbf{c} \underset{\mathbf{c}'}{\times} \mathbf{c} & \xrightarrow{\delta_1} & \mathbf{c} \\ {\scriptstyle\delta_2}\downarrow & & \downarrow{\scriptstyle\delta} \\ \mathbf{c} & \xrightarrow{\delta} & \mathbf{c}' \end{array} \tag{5.2}$$

with $\delta \in adm \cap co\text{-}adm$, the maps δ_1 and δ_2 both be *isomorphisms*. In other words, we require that the diagram

$$\begin{array}{ccc} \mathbf{c} & \xrightarrow{\text{id}} & \mathbf{c} \\ {\scriptstyle\text{id}}\downarrow & & \downarrow{\scriptstyle\delta} \\ \mathbf{c} & \xrightarrow{\delta} & \mathbf{c}' \end{array}$$

be Cartesian. Equivalently, we require that every map $\delta \in adm \cap co\text{-}adm$ be a *monomorphism*.

5.1.3. The example to keep in mind is that of $\mathbf{C} = \text{Sch}_{\text{aft}}$, with $vert = horiz$ being all morphisms, *adm* being proper morphisms and *co-adm* being open embeddings.

In this case, the class $adm \cap co\text{-}adm$ consists of embeddings of unions of connected components.

5.1.4. We now impose the following crucial condition on the relationship between the classes *adm*, *co-adm* and *horiz*.

For a 1-morphism $\alpha : \mathbf{c}_0 \to \mathbf{c}_1$ in *horiz*, consider the $(\infty,1)$-category $\text{Factor}(\alpha)$, whose objects are

$$\mathbf{c}_0 \xrightarrow{\epsilon} \overline{\mathbf{c}}_0 \xrightarrow{\gamma} \mathbf{c}_1,$$

where $\epsilon \in co\text{-}adm$ and $\gamma \in adm$, and whose morphisms are commutative diagrams

$$\begin{array}{ccccc} & & \overline{\mathbf{c}}_0' & & \\ & {\scriptstyle\epsilon'}\nearrow & & \searrow{\scriptstyle\gamma'} & \\ \mathbf{c}_0 & & \downarrow{\scriptstyle\beta} & & \mathbf{c}_1. \\ & {\scriptstyle\epsilon''}\searrow & & \nearrow{\scriptstyle\gamma''} & \\ & & \overline{\mathbf{c}}_0'' & & \end{array}$$

Note that by the '2-out-of 3' property, the morphism β automatically belongs to *adm*.

We impose the condition that for any $\alpha : \mathbf{c}_0 \to \mathbf{c}_1$ in *horiz*, the category $\text{Factor}(\alpha)$ be *contractible*.

5.2. Set-up for the functor. In this subsection we will describe what kind of functors out of our categories of correspondences we will consider, and formulate Theorem 5.2.4.

5.2.1. Let $\mathbb{S}$ be an $(\infty,2)$-category. We start with functors

$$\Phi^{adm}_{vert;adm} : \mathrm{Corr}(\mathbf{C})^{adm}_{vert;adm} \to \mathbb{S}$$

and

$$\Phi^{\mathrm{isom}}_{vert;co\text{-}adm} : \mathrm{Corr}(\mathbf{C})^{\mathrm{isom}}_{vert;co\text{-}adm} \to \mathbb{S}$$

together with an identification of the corresponding functors

$$\Phi^{adm}_{vert;adm}|_{\mathbf{C}_{vert}} \simeq \Phi^{\mathrm{isom}}_{vert;co\text{-}adm}|_{\mathbf{C}_{vert}}.$$

Denote

$$\Phi := \Phi^{adm}_{vert;adm}|_{\mathbf{C}_{vert}}.$$

Note that by Theorem 3.2.2, the functor $\Phi^{adm}_{vert;adm}$ is uniquely reconstructed from that of Φ, and it exists if and only if Φ satisfies the left Beck-Chevalley condition with respect to $adm \subset vert$. So, the above data is uniquely recovered from that of $\Phi^{\mathrm{isom}}_{vert;co\text{-}adm}$.

In what follows, we shall denote by $\Phi^*_{co\text{-}adm}$ the restriction

$$\Phi_{vert;co\text{-}adm}|_{(\mathbf{C}_{co\text{-}adm})^{\mathrm{op}}},$$

and by $\Phi^!_{adm}$ the restriction

$$\Phi^{adm}_{vert;adm}|_{(\mathbf{C}_{adm})^{\mathrm{op}}}.$$

5.2.2. We now impose the following additional condition on $\Phi^{\mathrm{isom}}_{vert;co\text{-}adm}$:

Let

$$\begin{array}{ccc} \mathbf{c}_{0,1} & \xrightarrow{\epsilon_0} & \mathbf{c}_{0,0} \\ {\scriptstyle\gamma_1}\downarrow & & \downarrow{\scriptstyle\gamma_0} \\ \mathbf{c}_{1,1} & \xrightarrow{\epsilon_1} & \mathbf{c}_{1,0} \end{array} \tag{5.3}$$

be a Cartesian diagram with $\epsilon_i \in co\text{-}adm$ and $\gamma_i \in adm$. Consider the 2-morphism

$$\Phi^*_{co\text{-}adm}(\epsilon_0) \circ \Phi^!_{adm}(\gamma_0) \to \Phi^!_{adm}(\gamma_1) \circ \Phi^*_{co\text{-}adm}(\epsilon_1) \tag{5.4}$$

arising by adjunction from the isomorphism

$$\Phi(\gamma_1) \circ \Phi^*_{co\text{-}adm}(\epsilon_0) \simeq \Phi^*_{co\text{-}adm}(\epsilon_1) \circ \Phi(\gamma_0),$$

the latter being a part of the data of $\Phi^{\mathrm{isom}}_{vert;co\text{-}adm}$.

We need that (5.4) be an isomorphism for all Cartesian diagrams as above.

5.2.3. We claim:

THEOREM 5.2.4. *Restriction along*

$$\mathrm{Corr}(\mathbf{C})^{\mathrm{isom}}_{vert;co\text{-}adm} \to \mathrm{Corr}(\mathbf{C})^{adm}_{vert;horiz}$$

defines an isomorphism between the space of functors

$$\Phi^{adm}_{vert;horiz} : \mathrm{Corr}(\mathbf{C})^{adm}_{vert;horiz} \to \mathbb{S}$$

and that of functors

$$\Phi^{\mathrm{isom}}_{vert;co\text{-}adm} : \mathrm{Corr}(\mathbf{C})^{\mathrm{isom}}_{vert;co\text{-}adm} \to \mathbb{S},$$

for which

$$\Phi := \Phi^{\mathrm{isom}}_{vert;co\text{-}adm}|_{\mathbf{C}_{vert}}$$

satisfies the left Beck-Chevalley condition with respect to $adm \subset vert$, and such that the condition from Sect. 5.2.2 *holds.*

5.3. Proof of Theorem 5.2.4, initial remarks. In this subsection we will explain the strategy of the proof of Theorem 5.2.4.

5.3.1. First, we shall carry out the easy direction. Namely, we will show that if we start with a functor

$$\Phi^{adm}_{vert;horiz} : \mathrm{Corr}(\mathbf{C})^{adm}_{vert;horiz} \to \mathbb{S},$$

then the functor

$$\Phi^{\mathrm{isom}}_{vert;co\text{-}adm} : \mathrm{Corr}(\mathbf{C})^{\mathrm{isom}}_{vert;co\text{-}adm} \to \mathbb{S},$$

obtained by restriction, satisfies left Beck-Chevalley condition with respect to $adm \subset vert$, and such that the condition from Sect. 5.2.2 holds.

First, the fact that Φ satisfies the left Beck-Chevalley condition with respect to $adm \subset vert$ follows from (the easy direction of) Theorem 3.2.2.

For $\Phi^{adm}_{vert;horiz}$ as above, let $\Phi^!$ denote the restriction

$$\Phi^{adm}_{vert;horiz}|_{(\mathbf{C}_{horiz})^{\mathrm{op}}}.$$

By assumption,

$$\Phi^!|_{(\mathbf{C}_{co\text{-}adm})^{\mathrm{op}}} \simeq \Phi^*_{co\text{-}adm} \text{ and } \Phi^!|_{(\mathbf{C}_{adm})^{\mathrm{op}}} \simeq \Phi^!_{adm}.$$

Furthermore, for a Cartesian diagram (5.3), the (iso)morphism

$$\Phi(\gamma_1) \circ \Phi^!(\epsilon_0) \to \Phi^!(\epsilon_1) \circ \Phi(\gamma_0),$$

is one arising by adjunction from the isomorphism

$$\Phi^!(\epsilon_0) \circ \Phi^!(\gamma_0) \simeq \Phi^!(\gamma_0 \circ \epsilon_0) \simeq \Phi^!(\epsilon_1 \circ \gamma_1) \simeq \Phi^!(\gamma_1) \circ \Phi^!(\epsilon_1).$$

In particular, the latter equals the 2-morphism (5.4), which is therefore also an isomorphism.

5.3.2. We are now going to tackle the difficult direction in Theorem 5.2.4. By Theorem 2.1.3, the datum of a functor $\Phi^{adm}_{vert;horiz}$ is equivalent to that of a map of bi-simplicial spaces

$$\text{defGrid}_{\bullet,\bullet}(\mathbf{C})^{adm}_{vert;horiz} \to \text{Sq}_{\bullet,\bullet}(\mathbb{S}), \tag{5.5}$$

satisfying the additional condition from Theorem 2.1.3.

Given $\Phi^{\text{isom}}_{vert;co\text{-}adm}$, we shall produce (5.5) in three steps:

Step A. We first extend the initial functor

$$\Phi^{\text{isom}}_{vert;co\text{-}adm} : \text{Corr}(\mathbf{C})^{\text{isom}}_{vert;co\text{-}adm} \to \mathbb{S}$$

to a functor

$$\Phi^{adm\cap co\text{-}adm}_{vert;co\text{-}adm} : \text{Corr}(\mathbf{C})^{adm\cap co\text{-}adm}_{vert;co\text{-}adm} \to \mathbb{S}. \tag{5.6}$$

The existence and uniqueness of the functor $\Phi^{adm\cap co\text{-}adm}_{vert;co\text{-}adm}$ in (5.6) follows immediately from the assumptions of Theorem 5.2.4 and Theorem 4.1.3.

Step B. We will introduce a bi-simplicial category $\text{Factor}_{\bullet,\bullet}(\mathbf{C})$, equipped with a bi-simplicial functor

$$\text{Factor}_{\bullet,\bullet}(\mathbf{C}) \to \text{defGrid}_{\bullet,\bullet}(\mathbf{C})^{adm}_{vert;horiz}. \tag{5.7}$$

We will use $\Phi^{adm\cap co\text{-}adm}_{vert;co\text{-}adm}$ to construct a bi-simplicial functor $\Phi_{\text{Factor}_{\bullet,\bullet}}$

$$\Phi_{\text{Factor}_{m,n}} : \text{Factor}_{m,n}(\mathbf{C}) \to \text{Sq}_{m,n}(\mathbb{S}). \tag{5.8}$$

Step C. Finally, we shall use the 'contractibility of the space of factorizations' condition from Sect. 5.1.4, to show that each of the functors

$$\text{Factor}_{m,n}(\mathbf{C}) \to \text{defGrid}_{m,n}(\mathbf{C})^{adm}_{vert;horiz} \tag{5.9}$$

has contractible fibers[5]. This will imply that the bi-simplicial functor $\Phi_{\text{Factor}_{\bullet,\bullet}}$ uniquely factors through the sought-for bi-simplicial map (5.5) via the projection (5.7).

5.4. Step B: introduction.

5.4.1. We define the category $\text{Factor}_{m,n}(\mathbf{C})$ as a 1-full subcategory of

$$\mathbf{Maps}\left([m] \times ([n]^{\text{op}} \times [n]^{\text{op}})^{\geq \text{dgnl}}, \mathbf{C}\right),$$

as follows.

At the level of objects we take those diagrams $\underline{\mathbf{c}}$ that satisfy:

(1) The maps $\mathbf{c}_{i,j,k} \to \mathbf{c}_{i+1,j,k}$ belong to *vert*;
(2) The maps $\mathbf{c}_{i,j,k} \to \mathbf{c}_{i,j-1,k}$ belong to *co-adm*;
(3) The maps $\mathbf{c}_{i,j,k} \to \mathbf{c}_{i,j,k-1}$ belong to *adm*;
(4) The defect of Cartesianness of the squares

$$\begin{array}{ccc} \mathbf{c}_{i,j,k} & \longrightarrow & \mathbf{c}_{i,j-1,k} \\ \downarrow & & \downarrow \\ \mathbf{c}_{i+1,j,k} & \longrightarrow & \mathbf{c}_{i+1,j-1,k} \end{array}$$

belongs to $adm \cap co\text{-}adm$.

[5] In particular, $|\text{Factor}_{m,n}(\mathbf{C})| \simeq \text{defGrid}_{m,n}(\mathbf{C})^{adm}_{vert;horiz}$, where the notation $|-|$ is as in Chapter 1, Sect. 2.1.5.

As 1-morphisms we allow those maps between diagrams $\underline{\mathbf{c}} \to \underline{\mathbf{c}}'$ for which the maps

$$\mathbf{c}_{i,j,k} \to \mathbf{c}'_{i,j,k}$$

belong to *adm* and are isomorphisms for $j = k$.

5.4.2. For example, when $m = 1$ and $n = 2$, objects of $\mathrm{Factor}_{m,n}(\mathbf{C})$ are the diagrams

(5.10)

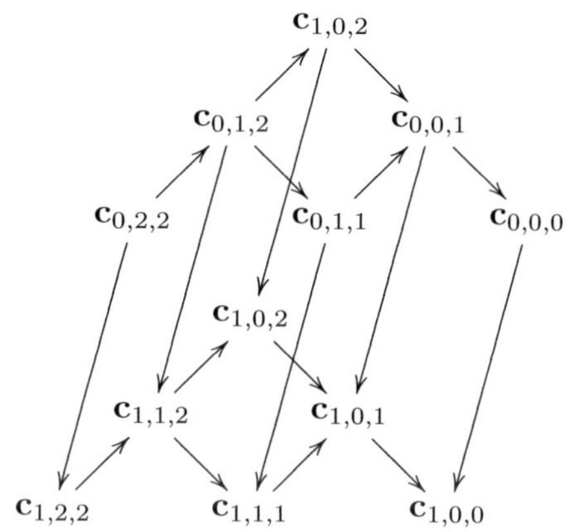

with the long slanted arrows in *vert*, northeast pointing arrows in *co-adm*, and southeast pointing arrows in *adm*.

5.4.3. The functor

$$\mathrm{Factor}_{m,n}(\mathbf{C}) \to \mathbf{Grid}_{m,n}(\mathbf{C})^{adm}_{vert;horiz}$$

is obtained from the diagonal embedding

$$[n]^{\mathrm{op}} \to [n]^{\mathrm{op}} \times [n]^{\mathrm{op}}.$$

5.4.4. In order to perform Step B we need to construct the functor

$$\Phi_{\mathrm{Factor}_{m,n}} : \mathrm{Factor}_{m,n}(\mathbf{C}) \to \mathrm{Sq}_{m,n}(\mathbb{S}),$$

functorially in $([n],[m]) \in \mathbf{\Delta}^{\mathrm{op}} \times \mathbf{\Delta}^{\mathrm{op}}$.

5.4.5. Let explain the idea of this construction for $m = 2$ and $n = 1$. Namely, to a diagram as in (5.10) we want to attach a diagram

(5.11)

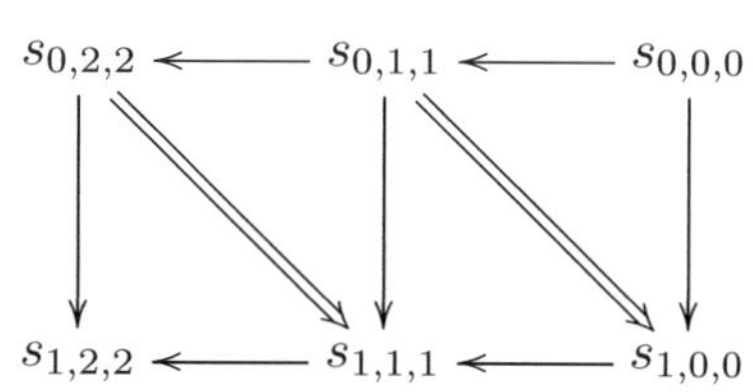

We will do it in several steps. First, starting from (5.10) we will use the functor $\Phi^{adm \cap co\text{-}adm}_{vert;co\text{-}adm}$ to produce a diagram

(5.12)

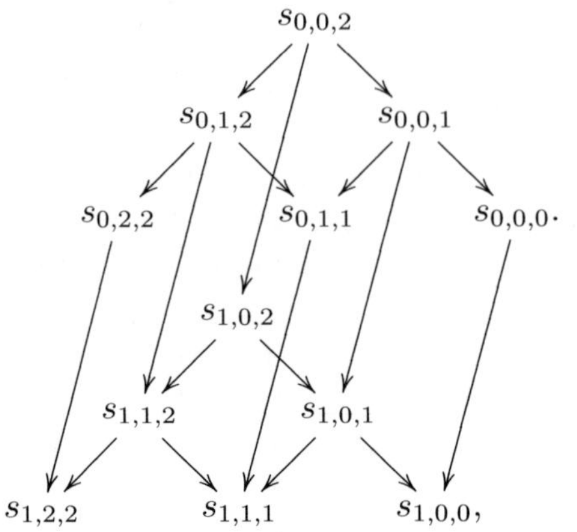

(here the squares are not necessarily commutative, but have 2-morphisms along appropriate faces), where along the long slanted arrows we take the 1-morphisms Φ, along the southwest pointing arrows we take the 1-morphisms $\Phi^*_{co\text{-}adm}$, and along the southeast pointing arrows we take the 1-morphisms Φ.

From the diagram (5.12), by taking right adjoints along the southeast pointing arrows, we obtain the diagram

(5.13)

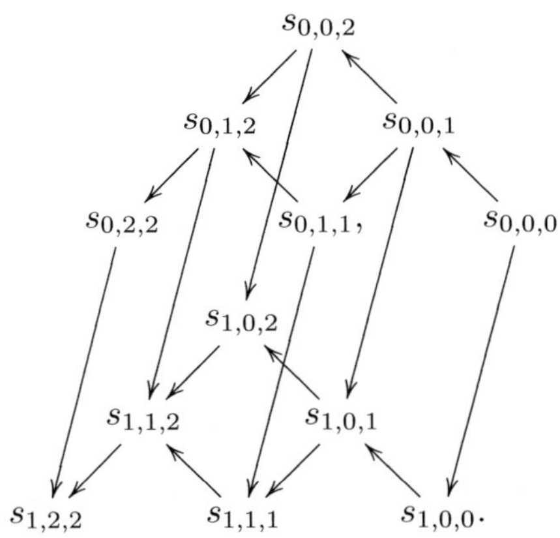

Finally, the desired diagram (5.11) is obtained from the diagram (5.13) by passing to the diagonal (by letting the 2nd and the 3rd indices be equal).

5.5. Step B: preparations. In order to prepare for Step B, we need to perform a certain manipulation with the functor $\Phi^{adm \cap co\text{-}adm}_{vert;co\text{-}adm}$ constructed in Step A.

5.5.1. Let us explain what we want to construct.

Let us be given a functor

$$\underline{\mathbf{c}} : [l]^{\mathrm{op}} \times [m] \times [n]^{\mathrm{op}} \to \mathbf{C},$$

such that:

(1) For every l', m', the map $\mathbf{c}_{l',m',n'} \to \mathbf{c}_{l',m',n'-1}$ belongs to *co-adm*;
(2) For every l', n', the map $\mathbf{c}_{l',m',n'} \to \mathbf{c}_{l',m'+1,n'}$ belongs to *vert*;

(3) For every m', n', the map $\mathbf{c}_{l',m',n'} \to \mathbf{c}_{l'-1,m',n'}$ belongs to *adm*;
(4) For every n', the square

$$\begin{array}{ccc} \mathbf{c}_{l',m',n'} & \longrightarrow & \mathbf{c}_{l'-1,m',n'} \\ \downarrow & & \downarrow \\ \mathbf{c}_{l',m'+1,n'} & \longrightarrow & \mathbf{c}_{l'-1,m'+1,n'} \end{array}$$

is Cartesian;
(5) For every m', the defect of Cartesianness of the square

$$\begin{array}{ccc} \mathbf{c}_{l',m',n'} & \longrightarrow & \mathbf{c}_{l'-1,m',n'} \\ \downarrow & & \downarrow \\ \mathbf{c}_{l',m',n'-1} & \longrightarrow & \mathbf{c}_{l'-1,m',n'-1} \end{array}$$

belongs to $adm \cap co\text{-}adm$;
(6) For every l', the defect of Cartesianness of the square

$$\begin{array}{ccc} \mathbf{c}_{l',m',n'} & \longrightarrow & \mathbf{c}_{l',m',n'-1} \\ \downarrow & & \downarrow \\ \mathbf{c}_{l',m'+1,n'} & \longrightarrow & \mathbf{c}_{l',m'+1,n'-1} \end{array}$$

belongs to $adm \cap co\text{-}adm$.

We claim that to any such $\underline{\mathbf{c}}$ we can attach a functor

$$\underline{s} : [m] \circledast [n] \circledast [l] \to \mathbb{S},$$

such that:

(1) For every l', m', n', we have $s_{m',n',l'} = \Phi(\mathbf{c}_{l',m',n'})$;
(2) For every l', m', the 1-morphism $s_{m',n'-1,l'} \to s_{m',n',l'}$ is obtained by applying $\Phi^*_{co\text{-}adm}$ to the arrow $\mathbf{c}_{l',m',n'} \to \mathbf{c}_{l',m',n'-1}$;
(3) For every m', n', the morphism $s_{m',n',l'-1} \to s_{m',n',l'}$ is obtained by applying $\Phi^!_{adm}$ to the arrow $\mathbf{c}_{l',m',n'} \to \mathbf{c}_{l'-1,m',n'}$;
(4) For every l', n', the morphism $s_{m',n',l'} \to s_{m'+1,n',l'}$ is obtained by applying Φ to the arrow $\mathbf{c}_{l',m',n'} \to \mathbf{c}_{l',m'+1,n'}$;
(5) For every m', the 2-morphism in the diagram

$$\begin{array}{ccc} s_{m',n',l'} & \longleftarrow & s_{m',n'-1,l'} \\ \uparrow & \swarrow & \uparrow \\ s_{m',n',l'-1} & \longleftarrow & s_{m',n'-1,l'-1} \end{array}$$

is an isomorphism.

We now explain how this is done.

5.5.2. First, by Theorem 2.1.3 and Sect. 2.3.5, the functor $\Phi^{adm\cap co\text{-}adm}_{vert;co\text{-}adm}$ gives rise to a map in $\mathrm{Spc}^{\mathbf{\Delta}^{\mathrm{op}}\times\mathbf{\Delta}^{\mathrm{op}}\times\mathbf{\Delta}^{\mathrm{op}}}$

(5.14)
$$\mathrm{Maps}(([l]^{\mathrm{op}}\times[m])\times[n]^{\mathrm{op}},\mathbf{C})^{adm\cap co\text{-}adm}_{vert;co\text{-}adm}\to \mathrm{Maps}(([l]^{\mathrm{op}}\times[m])\circledast[n],\mathbb{S})\to$$
$$\to \mathrm{Maps}([l]^{\mathrm{op}}\circledast[m]\circledast[n],\mathbb{S}).$$

Let

$$\mathrm{Maps}^{adm}(([l]^{\mathrm{op}}\times[m])\times[n]^{\mathrm{op}},\mathbf{C})^{adm\cap co\text{-}adm}_{vert;co\text{-}adm}\subset$$
$$\subset \mathrm{Maps}(([l]^{\mathrm{op}}\times[m])\times[n]^{\mathrm{op}},\mathbf{C})^{adm\cap co\text{-}adm}_{vert;co\text{-}adm}$$

be the subspace consisting of those diagrams such that for every fixed $0\le m'\le m$ and $0\le n'\le n$, the map

$$\mathbf{c}_{l',m',n'}\to\mathbf{c}_{l'-1,m',n'}$$

belongs to adm, for $0\le l'-1<l'\le l$.

5.5.3. By the assumption on Φ, the image of

$$\mathrm{Maps}^{adm}(([l]^{\mathrm{op}}\times[m])\times[n]^{\mathrm{op}},\mathbf{C})^{adm\cap co\text{-}adm}_{vert;co\text{-}adm}$$

under the map (5.14) belongs to the subspace

$$\mathrm{Maps}^{0}([l]^{\mathrm{op}}\circledast[m]\circledast[n],\mathbb{S})\subset\mathrm{Maps}([l]^{\mathrm{op}}\circledast[m]\circledast[n],\mathbb{S}),$$

consisting of functors such that for every fixed $0\le m'\le m$ and $0\le n'\le n$, the 1-morphism

$$s_{l',m',n'}\to s_{l'-1,m',n'}$$

admits a right adjoint, for $0\le l'-1<l'\le l$.

By Chapter 12, Corollary 3.1.7, we have a canonically defined map

$$\mathrm{Maps}^{0}([l]^{\mathrm{op}}\circledast[m]\circledast[n],\mathbb{S})\to\mathrm{Maps}([m]\circledast[n]\circledast[l],\mathbb{S}).$$

Thus, we obtain a map in $\mathrm{Spc}^{\mathbf{\Delta}^{\mathrm{op}}\times\mathbf{\Delta}^{\mathrm{op}}\times\mathbf{\Delta}^{\mathrm{op}}}$

(5.15) $$\mathrm{Maps}^{adm}(([l]^{\mathrm{op}}\times[m])\times[n]^{\mathrm{op}},\mathbf{C})^{adm\cap co\text{-}adm}_{vert;co\text{-}adm}\to\mathrm{Maps}([m]\circledast[n]\circledast[l],\mathbb{S}),$$

functorial in $[l],[m],[n]\in\mathrm{Spc}^{\mathbf{\Delta}^{\mathrm{op}}\times\mathbf{\Delta}^{\mathrm{op}}\times\mathbf{\Delta}^{\mathrm{op}}}$.

We claim:

PROPOSITION 5.5.4. *The image of the map* (5.15) *belongs to* $\mathrm{Maps}([m]\circledast([n]\times[l]),\mathbb{S})$.

PROOF. The statement of the proposition is equivalent to the following one: let

$$\begin{array}{ccc}\mathbf{c}_0 & \xrightarrow{\epsilon} & \mathbf{c}_1\\ {\scriptstyle\gamma_0}\downarrow & & \downarrow{\scriptstyle\gamma_1}\\ \mathbf{c}'_0 & \xrightarrow{\epsilon'} & \mathbf{c}'_1,\end{array}$$

be a commutative diagram with the vertical maps belong to adm and the horizontal ones to $co\text{-}adm$ (in which case, its defect of Cartesianness automatically belongs to $adm\cap co\text{-}adm$, see Lemma 5.6.2 below). Consider the 2-morphism

(5.16) $$\Phi^{*}_{co\text{-}adm}(\epsilon)\circ\Phi^{!}_{adm}(\gamma_1)\to\Phi^{!}_{adm}(\gamma_0)\circ\Phi^{*}_{co\text{-}adm}(\epsilon')$$

arising by adjunction from the map

$$\Phi(\gamma_0) \circ \Phi^*_{co\text{-}adm}(\epsilon) \to \Phi^*_{co\text{-}adm}(\epsilon') \circ \Phi(\gamma_1),$$

the latter being part of the data supplied by $\Phi^{adm \cap co\text{-}adm}_{vert;co\text{-}adm}$. Then the claim is that the 2-morphism (5.16) is an isomorphism.

The above statement can be split into two cases. One is when the above diagram is Cartesian, in which case, the assertion coincides with the assumption on $\Phi^{\text{isom}}_{vert;co\text{-}adm}$ from Sect. 5.2.2.

The second case is when we are dealing with the diagram of the form

$$\begin{array}{ccc} \mathbf{c} & \xrightarrow{\text{id}} & \mathbf{c} \\ {\scriptstyle \text{id}}\downarrow & & \downarrow{\scriptstyle \delta} \\ \mathbf{c} & \xrightarrow{\delta} & \mathbf{c}', \end{array} \tag{5.17}$$

where $\delta \in adm \cap co\text{-}adm$. But such a diagram is automatically Cartesian by the condition of Sect. 5.1.2.

□

REMARK 5.5.5. Note that from diagram (5.17) it follows that for $\delta \in adm \cap co\text{-}adm$ we have a canonical isomorphism

$$\Phi^!_{adm}(\delta) \simeq \Phi^*_{co\text{-}adm}(\delta), \tag{5.18}$$

characterized uniquely by the property that the isomorphism

$$\text{id} \simeq \Phi^*_{co\text{-}adm}(\delta) \circ \Phi(\delta)$$

defines the unit of an adjunction.

5.6. Step B: the construction. We will now turn the idea described in Sect. 5.4.5 into a formal construction. I.e., we will define the functor

$$\Phi_{\text{Factor}_{\bullet,\bullet}} : \text{Factor}_{\bullet,\bullet}(\mathbf{C}) \to \text{Sq}_{\bullet,\bullet}(\mathbb{S}).$$

This will combine the construction from Sect. 5.5 and the manipulation that was employed in the proof of Theorem 2.1.3, namely, we will use *clusters*.

5.6.1. Let $\mathbf{defGrid}^{vert}(\mathbf{C})^{adm \cap co\text{-}adm}_{co\text{-}adm,adm}$ denote the following bi-simplicial $(\infty,1)$-*category*. For each m,n, the $(\infty,1)$-category $\mathbf{defGrid}^{vert}_{m,n}(\mathbf{C})^{adm \cap co\text{-}adm}_{co\text{-}adm,adm}$ is a 1-full subcategory in

$$\mathbf{Maps}([m]^{\text{op}} \times [n]^{\text{op}}, \mathbf{C}).$$

Its objects are commutative diagrams $\underline{\mathbf{c}}$ that satisfy:

(1) For every i, the map $\mathbf{c}_{i,j} \to \mathbf{c}_{i,j-1}$ belongs to *adm*.
(2) For every j, the map $\mathbf{c}_{i,j} \to \mathbf{c}_{i-1,j}$ belongs to *co-adm*.

As 1-morphisms we allow those maps between diagrams $\underline{\mathbf{c}} \to \underline{\mathbf{c}}'$ that satisfy:

(1) For every i and j, the map $\mathbf{c}_{i,j} \to \mathbf{c}'_{i,j}$ belongs to *vert*.

(2) For a fixed j, the defect of Cartesianness of the square

$$\begin{array}{ccc} \mathbf{c}_{i,j} & \longrightarrow & \mathbf{c}_{i-1,j} \\ \downarrow & & \downarrow \\ \mathbf{c}'_{i,j} & \longrightarrow & \mathbf{c}'_{i-1,j} \end{array}$$

belongs to $adm \cap co\text{-}adm$.

We note:

LEMMA 5.6.2. *For a commutative square*

$$\begin{array}{ccc} \mathbf{c}_0 & \longrightarrow & \mathbf{c}_1 \\ \downarrow & & \downarrow \\ \mathbf{c}'_0 & \longrightarrow & \mathbf{c}'_1, \end{array}$$

in which the vertical maps belong to adm and the horizontal to vert or horiz (resp., the horizontal ones to co-adm and vertical ones to vert), its defect of Cartesianness belongs to adm (resp., co-adm).

PROOF. Follows from the '2 out of 3' property of the classes $co\text{-}adm$ and adm. □

Hence, we obtain that there is canonical isomorphism

$$\mathrm{Seq}_l(\mathbf{defGrid}^{vert}_{m,n}(\mathbf{C})^{adm\cap co\text{-}adm}_{co\text{-}adm,adm}) \simeq \mathrm{Maps}^{adm}(([n]^{\mathrm{op}} \times [l]) \times [m]^{\mathrm{op}}, \mathbf{C})^{adm\cap co\text{-}adm}_{vert;co\text{-}adm},$$

see Sect. 5.5.2 for the notation $\mathrm{Maps}^{adm}(-,-)^{adm\cap co\text{-}adm}_{vert;co\text{-}adm}$.

Therefore, from Proposition 5.5.4, we obtain a map

$$\mathrm{Seq}_l(\mathbf{defGrid}^{vert}_{m,n}(\mathbf{C})^{adm\cap co\text{-}adm}_{co\text{-}adm,adm}) \to \mathrm{Maps}([l] \circledast ([m] \times [n]), \mathbb{S}), \tag{5.19}$$

functorial in $[l],[m],[n] \in \mathrm{Spc}^{\mathbf{\Delta}^{\mathrm{op}}\times\mathbf{\Delta}^{\mathrm{op}}\times\mathbf{\Delta}^{\mathrm{op}}}$.

5.6.3. Let Q be a cluster, see Sect. 2.6.2 for what this means. Let $\mathbf{Q}$ be the category underlying Q.

We define the $(\infty,1)$-category

$$\mathbf{defQ}^{vert}(\mathbf{C})^{adm\cap co\text{-}adm}_{co\text{-}adm,adm}$$

analogously to $\mathbf{defGrid}^{vert}_{m,n}(\mathbf{C})^{adm\cap co\text{-}adm}_{co\text{-}adm,adm}$, so that we recover the latter when

$$Q = (\{0,...,m\} \times \{0,...,n\})^{\mathrm{op}}.$$

As in Sect. 2.7, the map (5.19), gives maps

$$\mathrm{Seq}_l(\mathbf{defQ}^{vert}(\mathbf{C})^{adm\cap co\text{-}adm}_{co\text{-}adm,adm}) \to \mathrm{Maps}([l] \circledast \mathbf{Q}^{\mathrm{op}}, \mathbb{S}), \tag{5.20}$$

functorial in $[l] \in \mathbf{\Delta}^{\mathrm{op}}$ and the cluster Q.

In other words, we obtain canonically defined functors

$$\mathbf{defQ}^{vert}(\mathbf{C})^{adm\cap co\text{-}adm}_{co\text{-}adm,adm} \to \mathrm{Funct}(\mathbf{Q}^{\mathrm{op}}, \mathbb{S})_{\text{right-lax}} \tag{5.21}$$

that depend functorially on Q.

5.6.4. Taking $Q = ((\{0,...,n\} \times \{0,...,n\})^{\geq \mathrm{dgnl}})^{\mathrm{op}}$, from (5.21) we obtain a map

$$\mathbf{defGrid}^{vert}_{n,n}(\mathbf{C})^{adm \cap co\text{-}adm}_{co\text{-}adm,adm} \to \mathrm{Funct}(([n] \times [n])^{\geq \mathrm{dgnl}}, \mathbb{S})_{\text{right-lax}},$$

and composing with the diagonal embedding $[n] \to ([n] \times [n])^{\geq \mathrm{dgnl}}$, we obtain a map

$$\mathbf{defGrid}^{vert}_{n,n}(\mathbf{C})^{adm \cap co\text{-}adm}_{co\text{-}adm,adm} \to \mathrm{Funct}([n], \mathbb{S})_{\text{right-lax}}. \tag{5.22}$$

Note that we have a tautologically defined functor

$$\mathrm{Seq}_l(\mathrm{Factor}_{m,n}(\mathbf{C})) \to \mathrm{Maps}([l] \times [m], \mathbf{defGrid}^{vert}_{n,n}(\mathbf{C})^{adm \cap co\text{-}adm}_{co\text{-}adm,adm}).$$

Composing with (5.22), we obtain a functor

$$\begin{aligned}\mathrm{Seq}_l(\mathrm{Factor}_{m,n}(\mathbf{C})) &\to \mathrm{Maps}([l] \times [m], \mathrm{Funct}([n], \mathbb{S})_{\text{right-lax}}) \\ &= \mathrm{Maps}(([l] \times [m]) \circledast [n], \mathbb{S}) \to \mathrm{Maps}([l] \circledast [m] \circledast [n], \mathbb{S}) \\ &= \mathrm{Seq}_l(\mathrm{Funct}([m] \circledast [n], \mathbb{S})_{\text{right-lax}}),\end{aligned}$$

i.e., a functor

$$\mathrm{Factor}_{m,n}(\mathbf{C}) \to \mathrm{Funct}([m] \circledast [n], \mathbb{S})_{\text{right-lax}}. \tag{5.23}$$

We claim:

LEMMA 5.6.5. *The functor* (5.23) *sends every arrow in* $\mathrm{Factor}_{m,n}(\mathbf{C})$ *to an isomorphism in* $\mathrm{Funct}([m] \circledast [n], \mathbb{S})_{\text{right-lax}}$.

PROOF. Follows from the condition that for a map of objects in $\mathrm{Factor}_{m,n}(\mathbf{C})$, i.e., a natural transformation between functors

$$[m] \times ([n]^{\mathrm{op}} \times [n]^{\mathrm{op}})^{\geq \mathrm{dgnl}} \to \mathbf{C},$$

the induced natural transformation of the functors

$$[m] \times [n]^{\mathrm{op}} \to [m] \times ([n]^{\mathrm{op}} \times [n]^{\mathrm{op}})^{\geq \mathrm{dgnl}} \to \mathbf{C}$$

is an isomorphism. □

5.6.6. The above lemma implies that the functor (5.23) factors though

$$(\mathrm{Funct}([m] \circledast [n], \mathbb{S})_{\text{right-lax}})^{\mathrm{Spc}} \simeq \mathrm{Maps}([m] \circledast [n], \mathbb{S}) =: \mathrm{Sq}_{m,n}(\mathbb{S}).$$

The resulting functor

$$\mathrm{Factor}_{m,n}(\mathbf{C}) \to \mathrm{Sq}_{m,n}(\mathbb{S})$$

is the sought-for functor $\Phi_{\mathrm{Factor}_{m,n}}$ of (5.8).

5.7. Step C.

5.7.1. Let us first explain the idea of the proof when $m = 0$ and $n = 1$. In this case, the category $\text{Factor}_{0,1}(\mathbf{C})$ has as objects diagrams

$$\begin{array}{ccc} & \mathbf{c}_{0,1} & \\ {}^{\epsilon}\nearrow & & \searrow^{\gamma} \\ \mathbf{c}_{1,1} & & \mathbf{c}_{0,0}. \end{array}$$

and morphisms are given by diagrams

$$\begin{array}{ccccc} & & \mathbf{c}_{0,1} & & \\ & {}^{\epsilon}\nearrow & \big\downarrow & \searrow^{\gamma} & \\ \mathbf{c}_{1,1} & & {\scriptstyle \beta_{0,1}} & & \mathbf{c}_{0,0} \\ {\scriptstyle \beta_{0,0}}\big\downarrow & & \mathbf{c}'_{0,1} & & \big\downarrow{\scriptstyle \beta_{1,1}} \\ & {}^{\epsilon'}\nearrow & & \searrow^{\gamma'} & \\ \mathbf{c}'_{1,1} & & & & \mathbf{c}'_{0,0}, \end{array}$$

where $\beta_{0,0}$ and $\beta_{1,1}$ are isomorphisms and $\beta_{0,1} \in adm$.

The category $\text{defGrid}_{0,1}(\mathbf{C})^{adm}_{vert,horiz}$ is space of arrows $\mathbf{c}_0 \overset{\alpha}{\to} \mathbf{c}_1$. Hence, we see that the fiber of the functor

$$\text{Factor}_{0,1}(\mathbf{C}) \to \text{defGrid}_{0,1}(\mathbf{C})^{adm}_{vert,horiz}$$

over a given $(\mathbf{c}_0 \overset{\alpha}{\to} \mathbf{c}_1) \in \text{defGrid}_{0,1}(\mathbf{C})^{adm}_{vert,horiz}$ is the category $\text{Factor}(\alpha)$ from Sect. 5.1.4.

Hence, the contractibility of the fiber follows from the assumption in Sect. 5.1.4. The proof in the general case will be a rather straightforward combinatorial game.

5.7.2. The proof of Step C will proceed by induction on m. We shall perform the induction step, because the base of the induction (i.e., the case of $m = 0$) is similar, but simpler. So, we assume that the assertion is valid for $m - 1$, and we will now pass to the case of m.

Consider the map

$$\text{defGrid}_{m,n}(\mathbf{C})^{adm}_{vert,horiz} \to \text{defGrid}_{m-1,n}(\mathbf{C})^{adm}_{vert,horiz},$$

given by restriction along

$$[m-1] \times [n] \to [m] \times [n],$$

where $[m-1] \to [m]$ is given by $i \mapsto i + 1$.

Note that the resulting map

$$\text{Factor}_{m,n}(\mathbf{C}) \to \text{defGrid}_{m,n}(\mathbf{C})^{adm}_{vert,horiz} \underset{\text{defGrid}_{m-1,n}(\mathbf{C})^{adm}_{vert,horiz}}{\times} \text{Factor}_{m-1,n}(\mathbf{C})$$

is a co-Cartesian fibration. Hence, by induction, it suffices to show that it has contractible fibers.

Fix an object of

$$\mathbf{d} \in \text{defGrid}_{m,n}(\mathbf{C})^{adm}_{vert,horiz} \underset{\text{defGrid}_{m-1,n}(\mathbf{C})^{adm}_{vert,horiz}}{\times} \text{Factor}_{m-1,n}(\mathbf{C}),$$

and we will analyze the fiber of $\text{Factor}_{m,n}(\mathbf{C})$ over this object, denote this fiber by $\overline{\mathbf{C}}$.

5.7.3. Denote by $\underline{\mathbf{c}}$ the object of $\text{Maps}([n]^{\text{op}}, \mathbf{C})$ obtained from $\mathbf{d}$ by restriction along

$$\{1\} \times [n]^{\text{op}} \to [m] \times [n]^{\text{op}}.$$

Denote by $\overline{\mathbf{c}}'$ the object of $\text{Maps}(([n]^{\text{op}} \times [n]^{\text{op}})^{\geq \text{dgnl}}, \mathbf{C})$, obtained from $\mathbf{d}$ by restriction along

$$\{0\} \times ([n]^{\text{op}} \times [n]^{\text{op}})^{\geq \text{dgnl}} \to [m-1] \times ([n]^{\text{op}} \times [n]^{\text{op}})^{\geq \text{dgnl}}.$$

Denote by $\underline{\mathbf{c}}'$ the object of $\text{Maps}([n]^{\text{op}}, \mathbf{C})$ obtained from $\overline{\mathbf{c}}'$ be further restriction along

$$[n]^{\text{op}} \to ([n]^{\text{op}} \times [n]^{\text{op}})^{\geq \text{dgnl}}.$$

Note that the data of $\mathbf{d}$ gives a map

$$\underline{\mathbf{c}} \to \underline{\mathbf{c}}'. \tag{5.24}$$

The category $\overline{\mathbf{C}}$ is a 1-full subcategory in the category of functors, denoted $\overline{\mathbf{c}}$,

$$([n]^{\text{op}} \times [n]^{\text{op}})^{\geq \text{dgnl}} \to \mathbf{C},$$

equipped with an identification of their restriction along $[n]^{\text{op}} \to ([n]^{\text{op}} \times [n]^{\text{op}})^{\geq \text{dgnl}}$ with $\underline{\mathbf{c}}$, and with a natural transformation to $\overline{\mathbf{c}}'$, compatible with the identification $\overline{\mathbf{c}}'|_{[n]^{\text{op}}} \simeq \underline{\mathbf{c}}'$ via the natural transformation (5.24).

The category $\overline{\mathbf{C}}$ is obtained by imposing the following conditions. At the level of objects we require:

(1) All maps $\mathbf{c}_{i,j} \to \mathbf{c}_{i-1,j}$ belong to *co-adm*;
(2) All maps $\mathbf{c}_{i,j-1} \to \mathbf{c}_{i,j-1}$ belong to *adm*;
(3) All maps $\mathbf{c}_{i,j} \to \mathbf{c}'_{i,j}$ belong to *vert*;
(4) The defect of Cartesianness of the squares

$$\begin{array}{ccc} \mathbf{c}_{i,j} & \longrightarrow & \mathbf{c}_{i-1,j} \\ \downarrow & & \downarrow \\ \mathbf{c}'_{i,j} & \longrightarrow & \mathbf{c}'_{i-1,j} \end{array}$$

belongs to $adm \cap co\text{-}adm$.

At the level of morphisms we allow maps $\overline{\mathbf{c}} \to \widetilde{\overline{\mathbf{c}}}$ such that for all i, j, the map

$$\mathbf{c}_{i,j} \to \widetilde{\mathbf{c}}_{i,j}$$

belongs to *adm*.

5.7.4. For $0 \leq n' \leq n$ let $\overline{\mathbf{C}}^{\leq n'}$ denote the following variant of $\overline{\mathbf{C}}$: instead of considering functors from $([n]^{\text{op}} \times [n]^{\text{op}})^{\geq \text{dgnl}}$ to $\mathbf{C}$, we consider functors defined on the full subcategory

$$([n]^{\text{op}} \times [n]^{\text{op}})^{\geq \text{dgnl}, \leq \text{dgnl}+n'} \subset ([n]^{\text{op}} \times [n]^{\text{op}})^{\geq \text{dgnl}},$$

spanned by (i,j) with $j \leq i + n'$.

For example, for $n' = n$, we have $\overline{\mathbf{C}}^{\leq n} = \overline{\mathbf{C}}$, and for $n' = 0$, we have $\overline{\mathbf{C}}^{\leq n} = \{*\}$.

Restriction defines functors $\overline{\mathbf{C}}^{\leq n'} \to \overline{\mathbf{C}}^{\leq n'-1}$. We will prove by induction that the categories $\overline{\mathbf{C}}^{\leq n'}$ are contractible. We will use the following lemma:

LEMMA 5.7.5. *Let $F : \mathbf{D}_1 \to \mathbf{D}_2$ be a functor between $(\infty,1)$-categories. Assume that:*

(a) *For every $\mathbf{d}_2 \in \mathbf{D}_2$, the category $\mathbf{D}_1 \underset{\mathbf{D}_2}{\times} \{\mathbf{d}_2\}$ is contractible;*

(b) *For every $\mathbf{d}_1 \in \mathbf{D}_1$ and a morphism $\beta : \mathbf{d}'_2 \to F(\mathbf{d}_1)$, the category of*

$$(\mathbf{d}'_1 \in \mathbf{D}_1, \alpha : \mathbf{d}'_1 \to \mathbf{d}_1, F(\alpha) \simeq \beta)$$

is contractible.

Then F induces an isomorphism between homotopy types.

5.7.6. We will show that the functors

$$\overline{\mathbf{C}}^{\leq n'} \to \overline{\mathbf{C}}^{\leq n'-1}$$

satisfy the conditions of Lemma 5.7.5. We will check condition (a), condition (b) being similar.

5.7.7. Fix an object $\overline{\mathbf{c}}^{\leq n'-1} \in \overline{\mathbf{C}}^{\leq n'-1}$. The fiber of $\overline{\mathbf{C}}^{\leq n'}$ is the product of the following categories, denoted $\mathbf{C}_i$, over the index $0 \leq i \leq n - n' - 1$:

For each i, the category $\mathbf{C}_i$ is that of factorizations of the morphism

$$\mathbf{c}_{i+1,i+n'} \to \mathbf{c}_{i,i+n'-1}$$

(which is part of the data of $\overline{\mathbf{c}}^{\leq n'-1}$) as

$$\mathbf{c}_{i+1,i+n'} \overset{\epsilon}{\to} \mathbf{c}_{i,i+n'} \overset{\gamma}{\to} \mathbf{c}_{i,i+n'-1},$$

equipped with a datum of commutative diagram

$$\begin{array}{ccccc} \mathbf{c}_{i+1,i+n'} & \longrightarrow & \mathbf{c}_{i,i+n'} & \longrightarrow & \mathbf{c}_{i,i+n'-1} \\ \downarrow & & \downarrow{\scriptstyle\beta} & & \downarrow \\ \mathbf{c}'_{i+1,i+n'} & \longrightarrow & \mathbf{c}'_{i,i+n'} & \longrightarrow & \mathbf{c}'_{i,i+n'-1} \end{array}$$

such that

(1) $\epsilon \in$ *co-adm*;
(2) $\gamma \in$ *adm*;
(3) $\beta \in$ *vert*;
(4) The datum of commutation of the outer square

$$\begin{array}{ccc} \mathbf{c}_{i+1,i+n'} & \longrightarrow & \mathbf{c}_{i,i+n'-1} \\ \downarrow & & \downarrow \\ \mathbf{c}'_{i+1,i+n'} & \longrightarrow & \mathbf{c}'_{i,i+n'-1} \end{array}$$

is that coming from $\overline{\mathbf{c}}^{\leq n-1}$;

(5) The defect of Cartesianness of the left square

$$\begin{array}{ccc} \mathbf{c}_{i+1,i+n'} & \longrightarrow & \mathbf{c}_{i,i+n'} \\ \downarrow & & \downarrow{\scriptstyle\beta} \\ \mathbf{c}'_{i+1,i+n'} & \longrightarrow & \mathbf{c}'_{i,i+n'} \end{array}$$

belongs to $adm \cap co\text{-}adm$.

We claim that each of the category $\mathbf{C}_i$ is contractible.

5.7.8. Denote

$$\widetilde{\mathbf{c}}_{i,i+n'} := \mathbf{c}'_{i,i+n'} \underset{\mathbf{c}'_{i,i+n'-1}}{\times} \mathbf{c}_{i,i+n'-1}.$$

Then $\mathbf{C}_i$ is the category $\text{Factor}(\alpha)$ of factorizations of the map

$$\alpha : \mathbf{c}_{i+1,i+n'} \to \widetilde{\mathbf{c}}_{i,i+n'}$$

as

$$\mathbf{c}_{i+1,i+n'} \to \mathbf{c}_{i,i+n'} \to \widetilde{\mathbf{c}}_{i,i+n'}$$

with the map $\mathbf{c}_{i+1,i+n'} \to \mathbf{c}_{i,i+n'}$ being in $co\text{-}adm$ and $\mathbf{c}_{i,i+n'} \to \widetilde{\mathbf{c}}_{i,i+n'}$ being in adm.

We note, however, that the morphisms $\mathbf{c}_{i+1,i+n'} \to \mathbf{c}_{i,i+n'-1}$ and $\mathbf{c}'_{i,i+n'} \to \mathbf{c}'_{i,i+n'-1}$ both belong to $horiz$. Hence, so does the morphism α, by the '2 out of 3' property.

Hence, the category $\text{Factor}(\alpha)$ is contractible by the assumption in Sect. 5.1.4.

CHAPTER 8

Extension theorems for the category of correspondences

Introduction

This Chapter should be regarded as a complement to Chapter 7. Here we prove two more extension theorems that allow, starting from a functor

$$\Phi^{adm}_{vert;horiz} : \mathrm{Corr}(\mathbf{C})^{adm}_{vert;horiz} \to \mathbb{S},$$

and a functor $F : \mathbf{C} \to \mathbf{D}$, to canonically produce a functor

$$\Psi^{adm}_{vert;horiz} : \mathrm{Corr}(\mathbf{D})^{adm}_{vert;horiz} \to \mathbb{S}.$$

The nature of the extensions in this chapter will be very different from that in Chapter 7, Theorems 4.1.3 and 5.2.4: in *loc. cit.* we enlarged our 2-category of correspondences by allowing more 1-morphisms and 2-morphisms. In the present section, we will enlarge the class of objects.

0.1. The bivariant extension procedure. The first of our two extension results, Theorem 1.1.9, is a general framework designed to treat the following situation. We start with the functor

$$\mathrm{IndCoh}_{\mathrm{Corr}(\mathrm{Sch}_{\mathrm{aft}})^{\text{nil-closed}}_{\mathrm{all;all}}} : \mathrm{Corr}(\mathrm{Sch}_{\mathrm{aft}})^{\text{nil-closed}}_{\mathrm{all;all}} \to \mathrm{DGCat}_{\mathrm{cont}}$$

and we want to extend it to a functor

$$\mathrm{IndCoh}_{\mathrm{Corr}(\mathrm{indinfSch}_{\mathrm{aft}})^{\text{nil-closed}}_{\mathrm{all;all}}} : \mathrm{Corr}(\mathrm{indinfSch}_{\mathrm{aft}})^{\text{nil-closed}}_{\mathrm{all;all}} \to \mathrm{DGCat}_{\mathrm{cont}}.$$

0.1.1. Here is how this extension is supposed to behave.

Let us first restrict our attention to the 1-full subcategories

$$\mathrm{Sch}_{\mathrm{aft}} \subset \mathrm{Corr}(\mathrm{Sch}_{\mathrm{aft}})^{\text{nil-closed}}_{\mathrm{all;all}} \text{ and } \mathrm{indinfSch}_{\mathrm{aft}} \subset \mathrm{Corr}(\mathrm{indinfSch}_{\mathrm{aft}})^{\text{nil-closed}}_{\mathrm{all;all}},$$

and the fully faithful embedding

$$\mathrm{Sch}_{\mathrm{aft}} \to \mathrm{indinfSch}_{\mathrm{aft}}. \tag{0.1}$$

We want the restriction $\mathrm{IndCoh}_{\mathrm{indinfSch}_{\mathrm{aft}}}$ of $\mathrm{IndCoh}_{\mathrm{Corr}(\mathrm{indinfSch}_{\mathrm{aft}})^{\text{nil-closed}}_{\mathrm{all;all}}}$ to $\mathrm{indinfSch}_{\mathrm{aft}}$ to be given by *left* Kan extension of

$$\mathrm{IndCoh}_{\mathrm{Corr}(\mathrm{Sch}_{\mathrm{aft}})^{\text{nil-closed}}_{\mathrm{all;all}}}|_{\mathrm{Sch}_{\mathrm{aft}}} := \mathrm{IndCoh}_{\mathrm{Sch}_{\mathrm{aft}}}$$

along (0.1).

0.1.2. Consider now the 1-full subcategories

$(\mathrm{Sch}_{\mathrm{aft}})^{\mathrm{op}} \subset \mathrm{Corr}(\mathrm{Sch}_{\mathrm{aft}})^{\text{nil-closed}}_{\mathrm{all;all}}$ and $(\mathrm{indinfSch}_{\mathrm{aft}})^{\mathrm{op}} \subset \mathrm{Corr}(\mathrm{indinfSch}_{\mathrm{aft}})^{\text{nil-closed}}_{\mathrm{all;all}}$,

and the fully faithful embedding

$$(\mathrm{Sch}_{\mathrm{aft}})^{\mathrm{op}} \to (\mathrm{indinfSch}_{\mathrm{aft}})^{\mathrm{op}}. \tag{0.2}$$

We want the restriction $\mathrm{IndCoh}^!_{\mathrm{indinfSch}_{\mathrm{aft}}}$ of $\mathrm{IndCoh}_{\mathrm{Corr}(\mathrm{indinfSch}_{\mathrm{aft}})^{\text{nil-closed}}_{\mathrm{all;all}}}$ to $(\mathrm{indinfSch}_{\mathrm{aft}})^{\mathrm{op}}$ to be given by *right* Kan extension of

$$\mathrm{IndCoh}_{\mathrm{Corr}(\mathrm{Sch}_{\mathrm{aft}})^{\text{nil-closed}}_{\mathrm{all;all}}}|_{(\mathrm{Sch}_{\mathrm{aft}})^{\mathrm{op}}} =: \mathrm{IndCoh}^!_{\mathrm{Sch}_{\mathrm{aft}}}$$

along (0.2).

0.1.3. So, we see that our extension behaves as a left Kan extension along vertical directions and as a right Kan extension along the horizontal directions. But these two patterns of behavior are closely linked via the 2-categorical structure.

Namely, the left Kan extension behavior along the vertical direction and the right Kan extension behavior along the vertical direction, once restricted to 1-full subcategories corresponding to nil-closed maps, are formal consequences of each other (this is obtained by combining Volume II, Chapter 3, Corollary 4.2.3 and Theorem 4.3.2 and Chapter 7, Theorem 3.2.2 and Proposition 2.2.7 below).

The idea is that there are 'enough' of nil-closed maps to fix the behavior of our extension on all objects and morphisms.

0.1.4. The 2-categorical features are essential in the proof of Theorem 1.1.9. Namely, we perceive the datum of a functor

$$\mathrm{IndCoh}_{\mathrm{Corr}(\mathrm{indinfSch}_{\mathrm{aft}})^{\text{nil-closed}}_{\mathrm{all;all}}} : \mathrm{Corr}(\mathrm{indinfSch}_{\mathrm{aft}})^{\text{nil-closed}}_{\mathrm{all;all}} \to \mathrm{DGCat}_{\mathrm{cont}}$$

as that of simplicial functor between $(\infty,1)$-categories

$$'\mathbf{Grid}^{\geq \mathrm{dgnl}}_{\bullet}(\mathrm{indinfSch}_{\mathrm{aft}})^{\text{nil-closed}}_{\mathrm{all;all}} \to \mathrm{Seq}^{\mathrm{ext}}_{\bullet}(\mathbb{S}).$$

The corresponding functors

$$'\mathbf{Grid}^{\geq \mathrm{dgnl}}_{n}(\mathrm{indinfSch}_{\mathrm{aft}})^{\text{nil-closed}}_{\mathrm{all;all}} \to \mathrm{Seq}^{\mathrm{ext}}_{n}(\mathbb{S})$$

are obtained as *left Kan extensions* of the corresponding functors

$$'\mathbf{Grid}^{\geq \mathrm{dgnl}}_{n}(\mathrm{Sch}_{\mathrm{aft}})^{\text{nil-closed}}_{\mathrm{all;all}} \to \mathrm{Seq}^{\mathrm{ext}}_{n}(\mathbb{S}).$$

That is to say, here we use in an essential way the fact that

$$'\mathbf{Grid}^{\geq \mathrm{dgnl}}_{n}(\mathrm{indinfSch}_{\mathrm{aft}})^{\text{nil-closed}}_{\mathit{all;all}}$$

is a category, and not just a space: we approximate diagrams

$$\underline{\mathbf{d}} \in \mathrm{Grid}_n(\mathrm{indinfSch}_{\mathrm{aft}})^{\text{nil-closed}}_{\mathrm{all;all}}$$

by diagrams $\underline{\mathbf{c}} \in \mathrm{Grid}_n(\mathrm{indinfSch}_{\mathrm{aft}})^{\text{nil-closed}}_{\mathrm{all;all}}$, where the maps $\underline{\mathbf{c}} \to \underline{\mathbf{d}}$ are required to be term-wise nil-closed.

So, it would be impossible to prove an analog of Theorem 1.1.9, if instead of the $(\infty,2)$-categories

$$\mathrm{Corr}(\mathrm{Sch}_{\mathrm{aft}})^{\text{nil-closed}}_{\mathrm{all;all}} \text{ and } \mathrm{Corr}(\mathrm{indinfSch}_{\mathrm{aft}})^{\text{nil-closed}}_{\mathrm{all;all}}$$

we used the underlying $(\infty,1)$-categories

$$\mathrm{Corr}(\mathrm{Sch}_{\mathrm{aft}})_{\mathrm{all;all}} \text{ and } \mathrm{Corr}(\mathrm{indinfSch}_{\mathrm{aft}})_{\mathrm{all;all}}.$$

0.1.5. The bulk of the proof of Theorem 1.1.9 is concentrated in Proposition 1.2.5 that guarantees that the left Kan extension extensions of

$$'\mathbf{Grid}_n^{\geq \mathrm{dgnl}}(\mathrm{Sch}_{\mathrm{aft}})^{\text{nil-closed}}_{\mathrm{all;all}} \to \mathrm{Seq}_n^{\mathrm{ext}}(\mathbb{S})$$

along

$$'\mathbf{Grid}_n^{\geq \mathrm{dgnl}}(\mathrm{Sch}_{\mathrm{aft}})^{\text{nil-closed}}_{\mathrm{all;all}} \to {'\mathbf{Grid}}_n^{\geq \mathrm{dgnl}}(\mathrm{indinfSch}_{\mathrm{aft}})^{\text{nil-closed}}_{\mathrm{all;all}}$$

'does the right thing', i.e., produces the expected strings of 1-morphisms.

In the process of proving Proposition 1.2.5 we will need to make a digression and study the behavior of colimits in the categories $\mathrm{Seq}_n^{\mathrm{ext}}(\mathbb{S})$, and how these colimits behave with respect to restriction functors

$$\mathrm{Seq}_n^{\mathrm{ext}}(\mathbb{S}) \to \mathrm{Seq}_m^{\mathrm{ext}}(\mathbb{S}),$$

corresponding to maps $[m] \to [n]$ in the category $\boldsymbol{\Delta}$.

0.2. The horizontal extension procedure. Our second extension result, Theorem 6.1.5 is a general framework designed to treat the following situation. We start with the functor

$$\mathrm{IndCoh}_{\mathrm{Corr}(\mathrm{Sch}_{\mathrm{aft}})^{\mathrm{proper}}_{\mathrm{all;all}}} : \mathrm{Corr}(\mathrm{Sch}_{\mathrm{aft}})^{\mathrm{proper}}_{\mathrm{all;all}} \to \mathrm{DGCat}_{\mathrm{cont}},$$

and we want to extend it to a functor

$$\mathrm{IndCoh}_{\mathrm{Corr}(\mathrm{PreStk}_{\mathrm{laft}})^{\mathrm{sch}\ \&\ \mathrm{proper}}_{\mathrm{sch;all}}} : \mathrm{Corr}(\mathrm{PreStk}_{\mathrm{laft}})^{\mathrm{proper}}_{\mathrm{sch;all}} \to \mathrm{DGCat}_{\mathrm{cont}}.$$

0.2.1. The above extension procedure is an instance of the 2-categorical right Kan extension. We do not develop the general theory of right Kan extensions for $(\infty,2)$-categories in this book.

What saves the day here is the fact that the present situation has a feature that makes it particularly simple:

Whatever that right Kan

$$\mathrm{RKE}_{\mathrm{Corr}(\mathrm{Sch}_{\mathrm{aft}})^{\mathrm{proper}}_{\mathrm{all;all}} \to \mathrm{Corr}(\mathrm{PreStk}_{\mathrm{laft}})^{\mathrm{sch}\ \&\ \mathrm{proper}}_{\mathrm{sch;all}}}(\mathrm{IndCoh}_{\mathrm{Corr}(\mathrm{Sch}_{\mathrm{aft}})^{\mathrm{proper}}_{\mathrm{all;all}}})$$

is, its restriction to the underlying $(\infty,1)$-categories is given by the usual (i.e., 1-categorical) right Kan extension, i.e., the canonical map

$$\begin{aligned}&\mathrm{RKE}_{\mathrm{Corr}(\mathrm{Sch}_{\mathrm{aft}})^{\mathrm{proper}}_{\mathrm{all;all}} \to \mathrm{Corr}(\mathrm{PreStk}_{\mathrm{laft}})^{\mathrm{sch}\ \&\ \mathrm{proper}}_{\mathrm{sch;all}}}\\ &\quad \times (\mathrm{IndCoh}_{\mathrm{Corr}(\mathrm{Sch}_{\mathrm{aft}})^{\mathrm{proper}}_{\mathrm{all;all}}})|_{\mathrm{Corr}(\mathrm{PreStk}_{\mathrm{laft}})_{\mathrm{all;all}}}\\ &\qquad \to \mathrm{RKE}_{\mathrm{Corr}(\mathrm{Sch}_{\mathrm{aft}})_{\mathrm{all;all}} \to \mathrm{Corr}(\mathrm{PreStk}_{\mathrm{laft}})_{\mathrm{sch;all}}}(\mathrm{IndCoh}_{\mathrm{Corr}(\mathrm{Sch}_{\mathrm{aft}})_{\mathrm{all;all}}})\end{aligned}$$

is an isomorphism.

0.2.2. A general statement of when a functor $F : \mathbb{T}_1 \to \mathbb{T}_2$ between $(\infty, 2)$-categories has the property that for any $\Phi : \mathbb{T}_1 \to \mathbb{S}$, the map

$$\mathrm{RKE}_F(\Phi)|_{\mathbb{T}_2^{1\text{-Cat}}} \to \mathrm{RKE}_{F|_{\mathbb{T}_1^{1\text{-Cat}}}}(\Phi|_{\mathbb{T}_1^{1\text{-Cat}}})$$

is an isomorphism, is given in Lemma 6.3.3.

The idea of the condition of this lemma says that for any $t_1 \in \mathbb{T}_1$ and $t_2 \in \mathbb{T}_2$, the morphisms in the category

$$\mathbf{Maps}_{\mathbb{T}_2}(t_2, F(t_1))$$

'come' from 2-morphisms in $\mathbb{T}_1$.

This guarantees that the 1-categorical right Kan extension

$$\mathrm{RKE}_{F|_{\mathbb{T}_1^{1\text{-Cat}}}}(\Phi|_{\mathbb{T}_1^{1\text{-Cat}}}) : \mathbb{T}_2^{1\text{-Cat}} \to \mathbb{S}$$

can be canonically extended to 2-morphisms using the data of Φ itself.

1. Functors obtained by bivariant extension

In this section we will describe one of the two extension results of this chapter that allows to extend a functor from a given $(\infty, 2)$-category of correspondences to a larger one.

A typical situation in which the procedure described in this section will be applied is when we want to extend IndCoh as a functor out of the category of correspondences of schemes to that of ind-inf-schemes. I.e., we start with the functor

$$\mathrm{IndCoh}_{\mathrm{Corr}(\mathrm{Sch}_{\mathrm{aft}})^{\text{nil-closed}}_{\mathrm{all;all}}} : \mathrm{Corr}(\mathrm{Sch}_{\mathrm{aft}})^{\text{nil-closed}}_{\mathrm{all;all}} \to \mathrm{DGCat}_{\mathrm{cont}}$$

and we want to extend it to a functor

$$\mathrm{IndCoh}_{\mathrm{Corr}(\mathrm{indinfSch}_{\mathrm{aft}})^{\text{nil-closed}}_{\mathrm{all;all}}} : \mathrm{Corr}(\mathrm{indinfSch}_{\mathrm{aft}})^{\text{nil-closed}}_{\mathrm{all;all}} \to \mathrm{DGCat}_{\mathrm{cont}}.$$

1.1. Set-up for the bivariant extension. In this subsection we will describe the context of our extension procedure and state the main result of this section, Theorem 1.1.9.

1.1.1. Let $(\mathbf{C}, vert, horiz, adm)$ be as in Chapter 7, Sect. 1.1.1. We will also assume that all three classes $vert$, $horiz$ and adm satisfy the '2 out of 3' property.

Suppose we have a functor

$$\Phi^{adm}_{vert;horiz} : \mathrm{Corr}(\mathbf{C})^{adm}_{vert;horiz} \to \mathbb{S},$$

where $\mathbb{S} \in 2\text{-Cat}$.

Denote

$$\Phi := \Phi^{adm}_{vert;horiz}|_{\mathbf{C}_{vert}} \text{ and } \Phi^! := \Phi^{adm}_{vert;horiz}|_{(\mathbf{C}_{horiz})^{\mathrm{op}}};$$

$$\Phi_{adm} := \Phi|_{\mathbf{C}_{adm}} \simeq \Phi^{adm}_{vert;horiz}|_{\mathbf{C}_{adm}}$$

and

$$\Phi^!_{adm} := \Phi^!|_{(\mathbf{C}_{adm})^{\mathrm{op}}} \simeq \Phi^{adm}_{vert;horiz}|_{(\mathbf{C}_{adm})^{\mathrm{op}}}.$$

For an object $\mathbf{c} \in \mathbf{C}$ we will simply write $\Phi(\mathbf{c})$ for $\Phi^{adm}_{vert;horiz}(\mathbf{c})$.

In our main application we will take $\mathbb{S}$ to be $\mathrm{DGCat}_{\mathrm{cat}}$ and $\mathbf{C} = \mathrm{Sch}_{\mathrm{aft}}$ with $vert = horiz = \mathrm{all}$ and $adm = \text{nil-closed}$. We take $\Phi^{adm}_{vert;horiz}$ to be the functor

$$\mathrm{IndCoh}_{\mathrm{Corr}(\mathrm{Sch}_{\mathrm{aft}})^{\text{nil-closed}}_{\mathrm{all;all}}} : \mathrm{Corr}(\mathrm{Sch}_{\mathrm{aft}})^{\text{nil-closed}}_{\mathrm{all;all}} \to \mathrm{DGCat}_{\mathrm{cont}}.$$

1.1.2. Let $(\mathbf{D}, vert, horiz, adm)$ be another datum as above, and assume that $\mathbf{D}$ admits all fiber products. Let $F : \mathbf{C} \to \mathbf{D}$ be a functor that preserves the corresponding classes of 1-morphisms, i.e., that it gives rise to well-defined functors

$$F_{vert} : \mathbf{C}_{vert} \to \mathbf{D}_{vert}, \; F_{horiz} : \mathbf{C}_{horiz} \to \mathbf{D}_{horiz} \text{ and } F_{adm} : \mathbf{C}_{adm} \to \mathbf{D}_{adm},$$

and that each of the above functors (including F itself) is fully faithful.

We will assume that F takes Cartesian squares as in Chapter 7, Diagram (1.1) to Cartesian squares. Hence, F induces a functor

$$F^{adm}_{vert;horiz} : \mathrm{Corr}(\mathbf{C})^{adm}_{vert;horiz} \to \mathrm{Corr}(\mathbf{D})^{adm}_{vert;horiz}.$$

1.1.3. We will also assume that for every $\mathbf{d} \in \mathbf{D}$ there exists $\mathbf{c} \in \mathbf{C}$ equipped with a map $\mathbf{c} \to \mathbf{d}$ that belongs to *adm*.

1.1.4. The goal of this section is to extend the functor $\Phi^{adm}_{vert;horiz}$ to a functor

$$\Psi^{adm}_{vert;horiz} : \mathrm{Corr}(\mathbf{D})^{adm}_{vert;horiz} \to \mathbb{S},$$

under certain conditions on the $(\infty, 2)$-category $\mathbb{S}$ (see Sect. 1.1.5) and on the functor $\Phi^{adm}_{vert;horiz}$ (see Sect. 1.1.6).

1.1.5. *Conditions on the target* 2*-category.* Let $\mathbb{S}$ be a target $(\infty, 2)$-category. We will impose the following conditions on $\mathbb{S}$:

(1) For any $s', s'' \in \mathbb{S}$, the category $\mathbf{Maps}_{\mathbb{S}}(s', s'')$ is presentable;
(2) For any $s', s'' \in \mathbb{S}$, the category $\mathbf{Maps}_{\mathbb{S}}(s', s'')$ is pointed, i.e., the map from the initial object to the final object is an isomorphism;
(3) For any $s', s'' \in \mathbb{S}$, and a fixed $\widetilde{s}' \overset{\alpha}{\to} s'$ (resp., $s'' \overset{\beta}{\to} \widetilde{s}''$), the functors

$$\mathbf{Maps}_{\mathbb{S}}(s', s'') \overset{-\circ\alpha}{\longrightarrow} \mathbf{Maps}_{\mathbb{S}}(\widetilde{s}', s'') \text{ and } \mathbf{Maps}_{\mathbb{S}}(s', s'') \overset{\beta\circ -}{\longrightarrow} \mathbf{Maps}_{\mathbb{S}}(s', \widetilde{s}'')$$

preserve colimits;
(4) $\mathbb{S}^{1\text{-Cat}}$ is presentable;
(5) $\mathbb{S}^{1\text{-Cat}}$ is pointed;
(6) For $s \in \mathbb{S}$ there exist objects $[1] \otimes s$ and $s^{[1]}$ equipped with functorial identifications

$$\mathrm{Maps}([1] \otimes s, s') \simeq \mathrm{Maps}([1], \mathbf{Maps}_{\mathbb{S}}(s, s'))$$

and

$$\mathrm{Maps}(s', s^{[1]}) \simeq \mathrm{Maps}([1], \mathbf{Maps}_{\mathbb{S}}(s', s)),$$

respectively.

These conditions are satisfied, e.g., if $\mathbb{S}$ is $1\text{-}\mathrm{Cat}^{\mathrm{St,cocmpl}}_{\mathrm{cont}}$ ot $\mathrm{DGCat}_{\mathrm{cont}}$.

1.1.6. *Conditions on the functor* $\Phi^{adm}_{vert;horiz}$. Denote

$$\Psi_{adm} := \mathrm{LKE}_{F_{adm}}(\Phi_{adm}), \quad \Psi^!_{adm} := \mathrm{RKE}_{F^{\mathrm{op}}_{adm}}(\Phi^!_{adm}),$$

$$\Psi_{vert} := \mathrm{LKE}_{F_{vert}}(\Phi_{vert}), \quad \Psi^!_{horiz} := \mathrm{RKE}_{F^{\mathrm{op}}_{horiz}}(\Phi^!_{horiz}).$$

We impose the following conditions on the interaction of $\Phi^{adm}_{vert;horiz}$ and F:

(1) The functor Ψ_{vert} satisfies the left Beck-Chevalley condition with respect to $adm \subset vert$.
(2) The canonical map $\Psi_{adm} \to \Psi_{vert}|_{\mathbf{D}_{adm}}$ is an isomorphism.

(3) The functor $\Psi^!$ satisfies the right Beck-Chevalley condition with respect to $adm \subset horiz$.

(4) The canonical map $\Psi^!_{horiz}|_{(\mathbf{D}_{adm})^{\mathrm{op}}} \to \Psi^!_{adm}$ is an isomorphism.

1.1.7. Finally, we impose one more technical condition:

$(*)$ For every morphism $\mathbf{d}' \to \mathbf{d}$ in $\mathbf{D}$, the map

$$\underset{\mathbf{c} \in (\mathbf{D}_{adm})_{/\mathbf{d}} \underset{\mathbf{D}_{adm}}{\times} \mathbf{C}_{adm}}{\mathrm{colim}} \Psi_{adm}(\mathbf{d}' \underset{\mathbf{d}}{\times} \mathbf{c}) \to \Psi_{adm}(\mathbf{d}')$$

is an isomorphism.

1.1.8. The main result of the present section is the following:

THEOREM 1.1.9. *Under the above circumstances there exists a uniquely defined functor*

$$\Psi^{adm}_{vert;horiz} : \mathrm{Corr}(\mathbf{D})^{adm}_{vert;horiz} \to \mathbb{S},$$

equipped with an identification,

$$\Phi^{adm}_{vert;horiz} \simeq \Psi^{adm}_{vert;horiz} \circ F^{adm}_{vert;horiz},$$

such that the induced natural transformation

$$\mathrm{LKE}_{F_{adm}}(\Phi_{adm}) \to \Psi^{adm}_{vert;horiz}|_{\mathbf{D}_{adm}}$$

is an isomorphism.

In addition, the functor $\Psi^{adm}_{vert;horiz}$ has the following properties:

- *The induced natural transformation*

$$\mathrm{LKE}_{F_{vert}}(\Phi) \to \Psi^{adm}_{vert;horiz}|_{\mathbf{D}_{vert}}$$

is an isomorphism;

- *The induced natural transformation*

$$\Psi^{adm}_{vert;horiz}|_{(\mathbf{D}_{horiz})^{\mathrm{op}}} \to \mathrm{RKE}_{(F_{horiz})^{\mathrm{op}}}(\Phi^!)$$

is an isomorphism.

Note that we can reformulate the uniqueness part of Theorem 1.1.9 as follows:

COROLLARY 1.1.10. *Let*

$$\widetilde{\Psi}^{adm}_{vert;horiz} : \mathrm{Corr}(\mathbf{D})^{adm}_{vert;horiz} \to \mathbb{S}$$

be a functor such that the following maps are isomorphisms:

(i) $\mathrm{LKE}_{F_{vert}}(\widetilde{\Psi}^{adm}_{vert;horiz} \circ F_{vert}|_{\mathbf{C}_{vert}}) \to \widetilde{\Psi}^{adm}_{vert;horiz}|_{\mathbf{D}_{vert}}$;

(ii) $\Psi^{adm}_{vert;horiz}|_{\mathbf{D}^{\mathrm{op}}_{horiz}} \to \mathrm{LKE}_{F^{\mathrm{op}}_{horiz}}(\widetilde{\Psi}^{adm}_{vert;horiz} \circ F^{\mathrm{op}}_{horiz}|_{\mathbf{C}^{\mathrm{op}}_{horiz}})$;

(iii) $\mathrm{LKE}_{F_{adm}}(\widetilde{\Psi}^{adm}_{vert;horiz} \circ F_{adm}|_{\mathbf{C}_{adm}}) \to \widetilde{\Psi}^{adm}_{vert;horiz}|_{\mathbf{D}_{adm}}$

(iv) *The map*

$$\underset{\mathbf{c} \in (\mathbf{D}_{adm})_{/\mathbf{d}} \underset{\mathbf{D}_{adm}}{\times} \mathbf{C}_{adm}}{\mathrm{colim}} \widetilde{\Psi}_{adm}(\mathbf{d}' \underset{\mathbf{d}}{\times} \mathbf{c}) \to \widetilde{\Psi}_{adm}(\mathbf{d}')$$

for every morphism $\mathbf{d}' \to \mathbf{d}$ in $\mathbf{D}$.

Then the functor $\Phi^{adm}_{vert;horiz} := \widetilde{\Psi}^{adm}_{vert;horiz} \circ F^{adm}_{vert;horiz}$ satisfies the assumptions of Theorem 1.1.9, and $\widetilde{\Psi}^{adm}_{vert;horiz}$ identifies canonically with the extension given by that theorem.

1.2. Construction of the functor. The rest of this section and the following four sections are devoted to the proof of Theorem 1.1.9.

We will construct $\Psi^{adm}_{vert;horiz}$ as a simplicial functor

$$\Psi_\bullet : {}'\mathbf{Grid}^{\geq \mathrm{dgnl}}_\bullet(\mathbf{D})^{adm}_{vert;horiz} \to \mathrm{Seq}^{\mathrm{ext}}_\bullet(\mathbb{S}).$$

1.2.1. The functor $F : \mathbf{C} \to \mathbf{D}$ gives rise to a simplicial functor

$$F_n : {}'\mathbf{Grid}^{\geq \mathrm{dgnl}}_n(\mathbf{C})^{adm}_{vert;horiz} \to {}'\mathbf{Grid}^{\geq \mathrm{dgnl}}_n(\mathbf{D})^{adm}_{vert;horiz},$$

which is fully faithful for each n.

We define the functor

$$\Psi_n : {}'\mathbf{Grid}^{\geq \mathrm{dgnl}}_n(\mathbf{D})^{adm}_{vert;horiz} \to \mathrm{Seq}^{\mathrm{ext}}_n(\mathbb{S}),$$

by

$$\Psi_n := \mathrm{LKE}_{F_n}(\Phi_n),$$

where

$$\Phi_\bullet : {}'\mathbf{Grid}^{\geq \mathrm{dgnl}}_\bullet(\mathbf{C})^{adm}_{vert;horiz} \to \mathrm{Seq}^{\mathrm{ext}}_\bullet(\mathbb{S}),$$

is the simplicial functor corresponding to $\Phi^{adm}_{vert;horiz}$. (It will follow from Corollary 2.3.5 that this left Kan extension exists.)

1.2.2. Consider the following set up. Let $\mathbf{I}$ be an index category, and let $F_{\mathbf{I}} : \mathbf{C}_{\mathbf{I}} \to \mathbf{D}_{\mathbf{I}}$ be a map between co-Cartesian fibrations over $\mathbf{I}$ that takes co-Cartesian arrows to co-Cartesian ones. Let $\mathbf{S}_{\mathbf{I}}$ be another co-Cartesian fibration over $\mathbf{I}$, and let $\Phi_{\mathbf{I}} : \mathbf{C}_{\mathbf{I}} \to \mathbf{S}_{\mathbf{I}}$ be a functor over $\mathbf{I}$ that also takes co-Cartesian arrows to co-Cartesian arrows.

For each arrow $r : i \to j$ in $\mathbf{I}$ there are canonical natural transformations

$$\mathrm{LKE}_{F_i}(\Phi_j \circ r_{\mathbf{C}}) \to \mathrm{LKE}_{F_j}(\Phi_j) \circ r_{\mathbf{D}} \tag{1.1}$$

and

$$\mathrm{LKE}_{F_i}(r_{\mathbf{S}} \circ \Phi_i) \to r_{\mathbf{S}} \circ \mathrm{LKE}_{F_i}(\Phi_i) \tag{1.2}$$

as functors $\mathbf{D}_i \to \mathbf{S}_j$, while

$$F_j \circ r_{\mathbf{C}} \simeq r_{\mathbf{D}} \circ F_i \text{ and } r_{\mathbf{S}} \circ \Phi_i \simeq \Phi_j \circ r_{\mathbf{C}},$$

where

$$r_{\mathbf{C}} : \mathbf{C}_i \to \mathbf{C}_j,\ r_{\mathbf{D}} : \mathbf{D}_i \to \mathbf{D}_j,\ r_{\mathbf{S}} : \mathbf{S}_i \to \mathbf{S}_j$$

denote the corresponding functors.

We have:

LEMMA 1.2.3. *Assume that the maps (1.1) and (1.2) are isomorphisms. Then relative left Kan extension defines a functor $\Psi_{\mathbf{I}} : \mathbf{D}_{\mathbf{I}} \to \mathbf{S}_{\mathbf{I}}$, which has the property that it sends co-Cartesian arrows to co-Cartesian ones. Furthermore, for every $i \in \mathbf{I}$, the natural map*

$$\mathrm{LKE}_{F_i}(\Phi_i) \to \Psi_{\mathbf{I}}|_{\mathbf{C}_i}$$

is an isomorphism.

1.2.4. We will apply Lemma 1.2.3 to

$$\mathbf{I} := \mathbf{\Delta}^{\mathrm{op}}, \quad F_{\mathbf{I}} := F_\bullet, \quad \mathbf{S}_{\mathbf{I}} := \mathrm{Seq}^{\mathrm{ext}}_\bullet(\mathbb{S}), \quad \Phi_{\mathbf{I}} := \Phi_\bullet.$$

For a map $r : [m] \to [n]$ in $\mathbf{\Delta}$, let $r^*_{\mathbb{S}}$ (resp., $r^*_{\mathbf{C}}$, $r^*_{\mathbf{D}}$) denotes the functor $\mathrm{Seq}^{\mathrm{ext}}_n(\mathbb{S}) \to \mathrm{Seq}^{\mathrm{ext}}_m(\mathbb{S})$ (resp., $'\mathbf{Grid}^{\geq \mathrm{dgnl}}_n(\mathbf{C})^{adm}_{vert;horiz} \to {'\mathbf{Grid}}^{\geq \mathrm{dgnl}}_m(\mathbf{C})^{adm}_{vert;horiz}$ and similarly for $\mathbf{D}$).

Consider the resulting functors

$$\mathrm{LKE}_{F_n}(\Phi_m \circ r^*_{\mathbf{C}}) \to \mathrm{LKE}_{F_m}(\Phi_m) \circ r^*_{\mathbf{D}}. \tag{1.3}$$

and

$$\mathrm{LKE}_{F_n}(r^*_{\mathbb{S}} \circ \Phi_n) \to r^*_{\mathbb{S}} \circ \mathrm{LKE}_{F_n}(\Phi_n). \tag{1.4}$$

The bulk of the proof of Theorem 1.1.9 will amount to the proof of the next proposition:

PROPOSITION 1.2.5.
(a) *The maps* (1.3) *are isomorphisms.*
(b) *The maps* (1.4) *are isomorphisms.*

1.3. Proof of Theorem 1.1.9, existence. In this subsection we will assume Proposition 1.2.5 and will deduce the existence part of Theorem 1.1.9. The uniqueness assertion will be proved in Sect. 3.2.

1.3.1. First, assuming Proposition 1.2.5, and using Lemma 1.2.3, we obtain that that the functors

$$\Psi_n := \mathrm{LKE}_{F_n}(\Phi_n)$$

give rise to a simplicial functor

$$\Psi_\bullet : {'\mathbf{Grid}}^{\geq \mathrm{dgnl}}_\bullet(\mathbf{D})^{adm}_{vert;horiz} \to \mathrm{Seq}^{\mathrm{ext}}_\bullet(\mathbb{S}).$$

By the $(\mathfrak{L}^{\mathrm{ext}}, \mathrm{Seq}^{\mathrm{ext}})$-adjunction, from $\Psi_\bullet$ we obtain a functor

$$\mathfrak{L}^{\mathrm{ext}}({'\mathbf{Grid}}^{\geq \mathrm{dgnl}}_\bullet(\mathbf{D})^{adm}_{vert;horiz}) \to \mathbb{S},$$

and finally, using Chapter 7, Corollary 1.4.6, the sought-for functor

$$\Psi^{adm}_{vert;horiz} : \mathrm{Corr}(\mathbf{D})^{adm}_{vert;horiz} \to \mathbb{S}.$$

1.3.2. The composite

$$'\mathbf{Grid}^{\geq \mathrm{dgnl}}_\bullet(\mathbf{C})^{adm}_{vert;horiz} \to {'\mathbf{Grid}}^{\geq \mathrm{dgnl}}_\bullet(\mathbf{D})^{adm}_{vert;horiz} \xrightarrow{\Psi_\bullet} \mathrm{Seq}^{\mathrm{ext}}_\bullet(\mathbb{S})$$

identifies with $\Phi_\bullet$. This gives an identification

$$\Phi^{adm}_{vert;horiz} \simeq \Psi^{adm}_{vert;horiz} \circ F^{adm}_{vert;horiz}.$$

Denote

$$\Psi := \Psi^{adm}_{vert;horiz}|_{\mathbf{D}_{vert}} \text{ and } \Psi^! := \Psi^{adm}_{vert;horiz}|_{(\mathbf{D}_{horiz})^{\mathrm{op}}};$$

$$\Psi_{adm} := \Psi|_{\mathbf{D}_{adm}} \simeq \Psi^{adm}_{vert;horiz}|_{\mathbf{D}_{adm}}$$

and

$$\Psi^!_{adm} := \Psi^!|_{(\mathbf{D}_{adm})^{\mathrm{op}}} \simeq \Psi^{adm}_{vert;horiz}|_{(\mathbf{D}_{adm})^{\mathrm{op}}}.$$

For an object $\mathbf{d} \in \mathbf{D}$ we shall simply write $\Psi(\mathbf{d})$ for $\Psi^{adm}_{vert;horiz}(\mathbf{d})$.

1.3.3. Let us now show that that the induced natural transformation

$$\mathrm{LKE}_{F_{adm}}(\Phi_{adm}) \to \Psi_{adm} \tag{1.5}$$

is an isomorphism.

It suffices to show that this natural transformation is an isomorphism on objects. I.e., we have to show that for $\mathbf{d} \in \mathbf{D}$, the map

$$\underset{\gamma:\mathbf{c}\to\mathbf{d},\gamma\in adm}{\mathrm{colim}} \Phi(\mathbf{c}) \to \Psi(\mathbf{d})$$

is an isomorphism, where the above map is given by a compatible family of maps

$$\Phi(\mathbf{c}) \overset{\sim}{\to} \Psi(\mathbf{c}) \overset{\Psi(\gamma)}{\to} \Psi(\mathbf{d}). \tag{1.6}$$

By the definition of the functor Ψ_n for $n = 0$, we have a tautological isomorphism

$$\underset{\gamma:\mathbf{c}\to\mathbf{d},\gamma\in adm}{\mathrm{colim}} \Phi(\mathbf{c}) = \Psi_0(\mathbf{d}) = \Psi(\mathbf{d}). \tag{1.7}$$

Now, by the definition of Ψ_n for $n = 1$, the compatible family of maps

$$\Phi(\mathbf{c}) \to \Psi(\mathbf{d})$$

that comprise (1.6) identifies with that in (1.7).

1.3.4. Let us show that the natural transformation

$$\mathrm{LKE}_{F_{vert}}(\Phi) \to \Psi$$

is an isomorphism.

Again, it is enough to do so at the level of objects. Consider the commutative diagram

$$\begin{array}{ccc} \mathrm{LKE}_{F_{adm}}(\Phi_{adm}) & \longrightarrow & \Psi_{adm} \\ \downarrow & & \downarrow = \\ \mathrm{LKE}_{F_{vert}}(\Phi)|_{\mathbf{D}_{adm}} & \longrightarrow & \Psi|_{\mathbf{D}_{adm}}. \end{array}$$

The top horizontal arrow in this diagram is an isomorphism by Sect. 1.3.3 above. The left vertical arrow is an isomorphism by the second condition in Sect. 1.1.6. This implies that the bottom horizontal arrow is an isomorphism, as desired.

1.3.5. Let us now show that the natural transformation

$$\Psi^! \to \mathrm{RKE}_{(F_{horiz})^{\mathrm{op}}}(\Phi^!)$$

is an isomorphism.

As in Sect. 1.3.4, it suffices to show that the natural transformation

$$\Psi^!_{adm} \to \mathrm{RKE}_{(F_{adm})^{\mathrm{op}}}(\Phi^!_{adm})$$

is an isomorphism.

Since the functor $\Psi^!_{adm}$ is obtained from Ψ_{adm} by passing to right adjoints (by Chapter 7, Theorem 3.2.2), we need to show that for $\mathbf{d} \in \mathbf{D}$ the map

$$\Psi(\mathbf{d}) \to \underset{\gamma:\mathbf{c}\to\mathbf{d},\gamma\in adm}{\lim} \Phi(\mathbf{c}),$$

comprised of functors

$$\Psi(\mathbf{d}) \to \Phi(\mathbf{c}),$$

right adjoint to those in (1.7), is an isomorphism. However, this follows from Proposition 2.2.5 below.

2. Limits and colimits of sequences

In this section we prepare for the proof of Proposition 1.2.5 by making a digression on the behavior of limits and colimits in categories of the form $\mathrm{Seq}_n^{\mathrm{ext}}(\mathbb{S})$, where $\mathbb{S}$ is an $(\infty,2)$-category as in Sect. 1.1.5.

2.1. Limits and colimits of presentable categories. In this subsection we recall the behavior of limits and colimits in the $(\infty,1)$-category, whose objects are presentable categories, and whose 1-morphisms are colimit-preserving functors.

2.1.1. Recall from Chapter 1, Sect. 2.5 that $1\text{-Cat}_{\mathrm{Prs}}$ denotes the 1-full subcategory of 1-Cat, whose objects are presentable $(\infty,1)$-categories, and whose morphisms are continuous (i.e., *colimit-preserving*) functors. Recall that the embedding

$$1\text{-Cat}_{\mathrm{Prs}} \hookrightarrow 1\text{-Cat}$$

commutes with *limits*, see Chapter 1, Lemma 2.5.2(b).

Let us recall the setting of Chapter 1, Proposition 2.5.7, which we will extensively use. Let

$$\mathbf{I} \to 1\text{-Cat}_{\mathrm{Prs}}, \quad i \mapsto \mathbf{C}_i \tag{2.1}$$

be a functor, and consider the object

$$\mathbf{C}' := \underset{i\in\mathbf{I}}{\mathrm{colim}}\, \mathbf{C}_i \in 1\text{-Cat}_{\mathrm{Prs}}.$$

For an index i, let ins_i denote the tautological functor

$$\mathbf{C}_i \to \mathbf{C}'.$$

Assume now that for every arrow $i \to j$ in $\mathbf{I}$, the corresponding functor $\mathbf{C}_i \to \mathbf{C}_j$ admits a *continuous* right adjoint. Consider the resulting functor

$$\mathbf{I}^{\mathrm{op}} \to 1\text{-Cat}_{\mathrm{Prs}}, \quad i \mapsto \mathbf{C}_i,$$

obtained from (2.1) by passing to right adjoints. Consider the object

$$\mathbf{C}'' := \lim_{i\in\mathbf{I}^{\mathrm{op}}} \mathbf{C}_i \in 1\text{-Cat}_{\mathrm{Prs}}.$$

For an index i, let ev_i denote the tautological functor

$$\mathbf{C}'' \to \mathbf{C}_i.$$

Then Chapter 1, Proposition 2.5.7 says that each of the functors ev_i admits a left adjoint, and that the resulting functor

$$\mathbf{C}' \to \mathbf{C}'', \quad (\mathrm{ev}_i)^L : \mathbf{C}_i \to \mathbf{C}'' \tag{2.2}$$

is an equivalence. In other words, we have an equivalence $\mathbf{C}' \simeq \mathbf{C}''$, under which, the adjoint pair

$$\mathrm{ins}_i : \mathbf{C}_i \rightleftarrows \mathbf{C}' : (\mathrm{ins}_i)^R$$

identifies with

$$(\mathrm{ev}_i)^L : \mathbf{C}_i \rightleftarrows \mathbf{C}'' : \mathrm{ev}_i\,.$$

As a formal consequence, we obtain:

COROLLARY 2.1.2.

(a) *The natural transformation*

$$\underset{i\in\mathbf{I}}{\operatorname{colim}}\ \operatorname{ins}_i\circ(\operatorname{ins}_i)^R\to\operatorname{Id}_{\mathbf{C}'}$$

is an isomorphism, where the colimit is taken in either the category

$$\operatorname{Maps}_{1\text{-}\mathrm{Cat}_{\mathrm{Prs}}}(\mathbf{C}',\mathbf{C}')$$

or

$$\operatorname{Maps}_{1\text{-}\mathrm{Cat}}(\mathbf{C}',\mathbf{C}').$$

(b) *The functor* $\mathbf{C}''\to\mathbf{C}'$, *given by*

$$\underset{i\in\mathbf{I}}{\operatorname{colim}}\ \operatorname{ins}_i\circ\operatorname{ev}_i$$

provides an inverse to the functor $\mathbf{C}'\to\mathbf{C}''$ *from* (2.2), *where the colimit is taken in either the category* $\operatorname{Maps}_{1\text{-}\mathrm{Cat}_{\mathrm{Prs}}}(\mathbf{C}'',\mathbf{C}')$ *or* $\operatorname{Maps}_{1\text{-}\mathrm{Cat}}(\mathbf{C}'',\mathbf{C}')$.

(c) *For* $\mathbf{D}\in 1\text{-}\mathrm{Cat}_{\mathrm{Prs}}$, *the natural map*

$$\underset{i\in\mathbf{I}}{\operatorname{colim}}\ \operatorname{Funct}_{\mathrm{cont}}(\mathbf{D},\mathbf{C}_i)\to\operatorname{Funct}_{\mathrm{cont}}(\mathbf{D},\mathbf{C}')$$

is an isomorphism, where the colimit is taken in the category $1\text{-}\mathrm{Cat}_{\mathrm{Prs}}$.

PROOF. Let us prove point (a). It suffices to show that the natural transformation

$$\underset{i\in\mathbf{I}}{\operatorname{colim}}\ (\operatorname{ev}_i)^L\circ\operatorname{ev}_i\to\operatorname{Id}_{\mathbf{C}''}$$

is an equivalence. I.e., we have to show that for $\underline{\mathbf{c}},\underline{\widetilde{\mathbf{c}}}\in\mathbf{C}''$, given by compatible systems of objects $\mathbf{c}_i,\widetilde{\mathbf{c}}_i\in\mathbf{C}_i$, respectively, the map

$$\operatorname{Maps}_{\mathbf{C}''}(\underline{\mathbf{c}},\underline{\widetilde{\mathbf{c}}})\to\lim_{i\in\mathbf{I}}\operatorname{Maps}_{\mathbf{C}''}((\operatorname{ev}_i)^L\circ\operatorname{ev}_i(\underline{\mathbf{c}}),\underline{\widetilde{\mathbf{c}}})$$

is an isomorphism. We rewrite the right-hand side as

$$\lim_{i\in\mathbf{I}}\operatorname{Maps}_{\mathbf{C}_i}((\operatorname{ev}_i(\underline{\mathbf{c}}),\operatorname{ev}_i(\underline{\widetilde{\mathbf{c}}})),$$

and now the assertion becomes manifest.

Point (b) follows formally from (a). Point (c) follows formally from (b) and the commutative diagram

$$\begin{array}{ccc}\underset{i\in\mathbf{I}}{\operatorname{colim}}\ \operatorname{Funct}_{\mathrm{cont}}(\mathbf{D},\mathbf{C}_i) & \longrightarrow & \operatorname{Funct}_{\mathrm{cont}}(\mathbf{D},\mathbf{C}') \\ {\scriptstyle\sim}\uparrow & & \uparrow{\scriptstyle\sim} \\ \lim\limits_{i\in\mathbf{I}^{\mathrm{op}}}\operatorname{Funct}_{\mathrm{cont}}(\mathbf{D},\mathbf{C}_i) & \xrightarrow{\sim} & \operatorname{Funct}_{\mathrm{cont}}(\mathbf{D},\mathbf{C}'').\end{array}$$

□

2.2. Limits and colimits in $\mathbb{S}^{1\text{-}\mathrm{Cat}}$. Let

$$\mathbf{1}\text{-}\mathbf{Cat}_{\mathrm{Prs}}\subset\mathbf{1}\text{-}\mathbf{Cat}$$

be the 1-full subcategory, corresponding to $1\text{-}\mathrm{Cat}_{\mathrm{Prs}}\subset 1\text{-}\mathrm{Cat}$. We can view the equivalence (2.2) as a result about functors $\mathbf{I}\to\mathbf{1}\text{-}\mathbf{Cat}_{\mathrm{Prs}}$.

In this subsection we will generalize the equivalence (2.2) by replacing $\mathbf{1}\text{-}\mathbf{Cat}_{\mathrm{Prs}}$ by a more general $(\infty,2)$-category $\mathbb{S}$.

2.2.1. *Limits of mapping categories.* Let $\mathbb{S}$ be an $(\infty, 2)$-category satisfying assumptions (1), (3), (4) and (6) of Sect. 1.1.5.

LEMMA 2.2.2.
(a) *For* $\mathbf{I} \in 1\text{-Cat}$ *and a functor* $\mathbf{I} \to \mathbb{S}$, $i \mapsto s_i$ *with* $s := \underset{i \in \mathbf{I}}{\operatorname{colim}}\, s_i$, *for any* $s' \in \mathbb{S}$, *the resulting map*

$$\mathbf{Maps}_{\mathbb{S}}(s, s') \to \lim_{i \in \mathbf{I}^{\mathrm{op}}} \mathbf{Maps}_{\mathbb{S}}(s_i, s')$$

is an isomorphism in $1\text{-Cat}_{\mathrm{Prs}}$.
(b) *For* $\mathbf{I} \in \infty\text{-Cat}$ *and a functor* $I \to \mathbb{S}$, $i \mapsto s_i$ *with* $s := \lim_{i \in \mathbf{I}} s_i$, *for any* $s' \in \mathbb{S}$, *the resulting map*

$$\mathbf{Maps}_{\mathbb{S}}(s', s) \to \lim_{i \in \mathbf{I}} \mathbf{Maps}_{\mathbb{S}}(s', s_i)$$

is an isomorphism in $1\text{-Cat}_{\mathrm{Prs}}$.

REMARK 2.2.3. By definition, the maps in the lemma a priori induce isomorphisms of the underlying spaces.

PROOF. Follows from the above remark by replacing s' by the objects $s'^{[1]}$ and $[1] \otimes s'$, respectively. □

2.2.4. Let

$$\mathbf{I} \to \mathbb{S}, \ i \mapsto s_i \tag{2.3}$$

be a functor, and assume that for every arrow $i \to j$, the corresponding 1-morphism $s_i \to s_j$ admits a right adjoint. Consider the functor

$$\mathbf{I}^{\mathrm{op}} \to \mathbb{S}, \ i \mapsto s_i,$$

obtained from (2.3) by passing to right adjoints.

Denote

$$s' := \underset{i \in \mathbf{I}}{\operatorname{colim}}\, s_i \text{ and } s'' := \lim_{i \in \mathbf{I}^{\mathrm{op}}} s_i$$

and let

$$\mathrm{ins}_i : s_i \to s' \text{ and } \mathrm{ev}_i : s'' \to s_i$$

denote the corresponding 1-morphisms.

We are going to prove:

PROPOSITION 2.2.5.
(a) *Each of the 1-morphisms* ev_i *admits a left adjoint. The 1-morphism* $s' \to s''$, *given by the compatible family*

$$(\mathrm{ev}_i)^L : s_i \to s'',$$

is an isomorphism. Under this identification, the functors $(\mathrm{ev}_i)^L$ *correspond to* ins_i, *and the functors* ev_i *correspond to the the right adjoints* $(\mathrm{ins}_i)^R$ *of* ins_i.
(b) *The map*

$$\underset{i \in \mathbf{I}}{\operatorname{colim}}\ \mathrm{ins}_i \circ (\mathrm{ins}_i)^R \to \mathrm{id}_{s'}$$

is an isomorphism, where the colimit is taken in $\mathbf{Maps}_{\mathbb{S}}(s', s')$.

(c) *The inverse 1-morphism to one in point (a) is given by*

$$\underset{i\in\mathbf{I}}{\operatorname{colim}}\ \mathrm{ins}_i \circ \mathrm{ev}_i,$$

where the colimit is taken in $\mathbf{Maps}_{\mathbb{S}}(s'', s')$.

(d) *For any* $t \in \mathbb{S}$, *the natural functor*

$$\underset{i\in\mathbf{I}}{\operatorname{colim}}\,\mathbf{Maps}_{\mathbb{S}}(t, s_i) \to \mathbf{Maps}_{\mathbb{S}}(t, s')$$

is an equivalence, where the colimit is taken in $1\text{-}\mathrm{Cat}_{\mathrm{Prs}}$.

PROOF. To show that the 1-morphism ev_j admits a left adjoint, it is enough to show that for any $t \in \mathbb{S}$, the induced morphism

$$\mathbf{Maps}_{\mathbb{S}}(t, s'') \to \mathbf{Maps}_{\mathbb{S}}(t, s_j)$$

admits a left adjoint and that for a 1-morphism $t_0 \to t_1$, the diagram

$$\begin{array}{ccc}
\mathbf{Maps}_{\mathbb{S}}(t_1, s'') & \longleftarrow & \mathbf{Maps}_{\mathbb{S}}(t_1, s_j) \\
\downarrow & & \downarrow \\
\mathbf{Maps}_{\mathbb{S}}(t_0, s'') & \longleftarrow & \mathbf{Maps}_{\mathbb{S}}(t_0, s_j)
\end{array}$$

obtained by passing to left adjoints along the horizontal arrows in the commutative diagram

$$\begin{array}{ccc}
\mathbf{Maps}_{\mathbb{S}}(t_1, s'') & \longrightarrow & \mathbf{Maps}_{\mathbb{S}}(t_1, s_j) \\
\downarrow & & \downarrow \\
\mathbf{Maps}_{\mathbb{S}}(t_0, s'') & \longrightarrow & \mathbf{Maps}_{\mathbb{S}}(t_0, s_j),
\end{array}$$

that a priori *commutes up to a natural transformation*, actually commutes.

By Lemma 2.2.2, we have

$$\mathbf{Maps}_{\mathbb{S}}(t, s'') \simeq \lim_{i\in\mathbf{I}^{\mathrm{op}}} \mathbf{Maps}_{\mathbb{S}}(t, s_i),$$

and now the assertion follows from the equivalence (2.2).

We will now show that the composite

$$s' \to s'' \to s',$$

where the first arrow is the map point (a) and the second arrow is the map in point (c), is isomorphic to $\mathrm{id}_{s'}$. We need to show that for every $j \in \mathbf{I}$, the composition

$$s_j \xrightarrow{\mathrm{ins}_j} s' \to s'' \to s' \tag{2.4}$$

is canonically isomorphic to ins_j, in a way functorial in j.

Note now that for any $t \in \mathbb{S}$ we have a commutative diagram

$$\begin{array}{ccccc}
\mathbf{Maps}_{\mathbb{S}}(t, s') & \longrightarrow & \mathbf{Maps}_{\mathbb{S}}(t, s'') & \longrightarrow & \mathbf{Maps}_{\mathbb{S}}(t, s') \\
\uparrow & & \uparrow & & \uparrow \\
\underset{i\in\mathbf{I}}{\operatorname{colim}}\,\mathbf{Maps}_{\mathbb{S}}(t, s_i) & \longrightarrow & \lim\limits_{i\in\mathbf{I}^{\mathrm{op}}} \mathbf{Maps}_{\mathbb{S}}(t, s_i) & \longrightarrow & \underset{i\in\mathbf{I}}{\operatorname{colim}}\,\mathbf{Maps}_{\mathbb{S}}(t, s_i),
\end{array} \tag{2.5}$$

where the colimits in the bottom row are taken in $1\text{-}\mathrm{Cat}_{\mathrm{Prs}}$. Now, Corollary 2.1.2(b) asserts that the bottom composition is canonically isomorphic to the identity on $\underset{i\in\mathbf{I}}{\operatorname{colim}}\,\mathbf{Maps}_{\mathbb{S}}(t, s_i)$.

Applying this to $t = s_j$ and the tautological map

$$s_j \to \underset{i \in \mathbf{I}}{\operatorname{colim}} \mathbf{Maps}_{\mathbb{S}}(s_j, s_i),$$

we obtain that the composition (2.4) is indeed isomorphic to ins_j.

Thus, we obtain that the composite $s' \to s'' \to s'$ is indeed isomorphic to the identity map. In particular, the composite in the top row of (2.5) is also isomorphic to the identity map. Now, since the middle vertical map in (2.5) is an isomorphism (by Lemma 2.2.2(b)), we obtain that the left vertical map is an isomorphism as well.

This proves point (d) of the proposition. Now, since all maps in the bottom row of (2.5) are isomorphisms, we obtain that the same is true for the top row. This proves point (a) of the proposition.

Point (c) has been established already. Point (b) follows formally from point (c).

□

2.2.6. As an application of Proposition 2.2.5, we shall now prove the following. Let

$$F : \mathbf{C} \to \mathbf{D}$$

be a functor between $(\infty, 1)$-categories. Let

$$\Phi : \mathbf{C} \to \mathbb{S}$$

be a functor, and set $\Psi := \text{LKE}_F(\Phi)$.

Assume that for every arrow $\mathbf{c}_1 \to \mathbf{c}_2$ in $\mathbf{C}$, the resulting 1-morphism $\Phi(\mathbf{c}_1) \to \Phi(\mathbf{c}_2)$ admits a right adjoint. Let $\Phi^! : \mathbf{C}^{\text{op}} \to \mathbb{S}$ be the functor, obtained from Φ by passing to right adjoints.

We claim:

PROPOSITION 2.2.7. *Under the above circumstances, the functor*

$$\Psi^! := \text{RKE}_{F^{\text{op}}}(\Phi^!)$$

is obtained from Ψ by passing to right adjoints.

PROOF. Let us first show that for any arrow $\mathbf{d}_1 \to \mathbf{d}_2$ in $\mathbf{D}$, the 1-morphism

$$\Psi(\mathbf{d}_1) \to \Psi(\mathbf{d}_2)$$

admits a right adjoint.

We have:

$$\Psi(\mathbf{d}_1) = \underset{\mathbf{c}_1, F(\mathbf{c}_1) \to \mathbf{d}_1}{\operatorname{colim}} \Phi(\mathbf{c}_1) \text{ and } \Psi(\mathbf{d}_2) = \underset{\mathbf{c}_2, F(\mathbf{c}_2) \to \mathbf{d}_2}{\operatorname{colim}} \Phi(\mathbf{c}_2),$$

and the 1-morphism $\Psi(\mathbf{d}_1) \to \Psi(\mathbf{d}_2)$ is obtained from the map of index categories

$$(\mathbf{c}_1, F(\mathbf{c}_1) \to \mathbf{d}_1) \mapsto (\mathbf{c}_1, F(\mathbf{c}_1) \to \mathbf{d}_1 \to \mathbf{d}_2).$$

The right adjoint in question is given, in terms of the isomorphism of Proposition 2.2.5(a) by

$$\underset{\mathbf{c}_2, F(\mathbf{c}_2) \to \mathbf{d}_2}{\lim} \Phi(\mathbf{c}_2) \to \underset{\mathbf{c}_1, F(\mathbf{c}_1) \to \mathbf{d}_1}{\lim} \Phi(\mathbf{c}_1).$$

Let $\Psi^!$ denote the functor obtained from Ψ by passing to right adjoints. Let us construct a natural transformation

$$\Psi^! \to \mathrm{RKE}_{F^{\mathrm{op}}}(\Phi^!), \tag{2.6}$$

which by definition amounts to a natural transformation

$$\Psi^! \circ F^{\mathrm{op}} \to \Phi^!.$$

This natural transformation is obtained by passing to right adjoints in the canonical natural transformation

$$\Phi \to \Psi \circ F.$$

Let us now show that the natural transformation (2.6) is an isomorphism. It is enough to check this at the level of objects. We have

$$\Psi^!(\mathbf{d}) = \underset{\mathbf{c}, F(\mathbf{c}) \to \mathbf{d}}{\mathrm{colim}} \Psi(\mathbf{c}) \text{ and } \mathrm{RKE}_{F^{\mathrm{op}}}(\Phi^!)(\mathbf{d}) = \underset{\mathbf{c}, F(\mathbf{c}) \to \mathbf{d}}{\lim} \Phi(\mathbf{c}).$$

By unwinding the definitions, we obtain that the resulting map

$$\underset{\mathbf{c}, F(\mathbf{c}) \to \mathbf{d}}{\mathrm{colim}} \Psi(\mathbf{c}) \to \underset{\mathbf{c}, F(\mathbf{c}) \to \mathbf{d}}{\lim} \Phi(\mathbf{c})$$

is the map of Proposition 2.2.5(a).

□

2.3. Colimits in $\mathrm{Seq}_n^{\mathrm{ext}}(\mathbb{S})$. In this subsection we will record results pertaining to the behavior of colimits in the category $\mathrm{Seq}_n^{\mathrm{ext}}(\mathbb{S})$: these are Propositions 2.3.2, 2.3.4 and Corollary 2.3.7.

2.3.1. *Colimits of strings.* We begin with the following observation:

PROPOSITION 2.3.2. *For a morphism $r : [0,m] \to [0,n]$ in $\mathbf{\Delta}$, the functor*

$$r_{\mathbb{S}}^* : \mathrm{Seq}_n^{\mathrm{ext}}(\mathbb{S}) \to \mathrm{Seq}_m^{\mathrm{ext}}(\mathbb{S})$$

commutes with colimits in the following cases:

(a) $m = 0$.

(b) $m = 1$ *and* r *sends* $0 \mapsto i$ *and* $1 \mapsto i+1$ *for* $0 \le i < n$.

(c) *For* r *inert; i.e. of the form* $i \mapsto k + i$ *for some* $0 \le k \le n - m$.

(d) r *is a surjection.*

PROOF. We claim that in each of the above cases, the functor $r_{\mathbb{S}}^*$ admits a right adjoint. Clearly, (a) and (b) are particular cases of (c).

The right adjoint in question sends a string

$$s^0 \to s^1 \to \ldots \to s^{m-1} \to s^m$$

to

$$* \to \ldots \to * \to s^0 \to s^1 \to \ldots \to s^{m-1} \to s^m \to * \to \ldots \to *,$$

where $*$ denotes the initial/final object of $\mathbb{S}$. Here we are using the fact that for $s', s'' \in \mathbb{S}$, the map

$$\mathbf{Maps}(s', *) \times \mathbf{Maps}(*, s'') \to \mathbf{Maps}(s', s'')$$

sends the unique object in $\mathbf{Maps}(s', *) \times \mathbf{Maps}(*, s'')$ to the initial/final object in $\mathbf{Maps}(s', s'')$.

For r surjective, by transitivity, it suffices to consider the case when r collapses the interval $\{i-1\} \to \{i\}$ in $[n]$ to $\{i-1\} \in [n-1]$. In this case, the right adjoint in question sends

$$s_0 \to \ldots \to s^{i-2} \to s^{i-1} \to s^i \to s^{i+1} \to \ldots \to s^n$$

to

$$s^0 \to \ldots \to s^{i-2} \to s'^{i-1} \to s^{i+1} \to \ldots \to s^n,$$

where

$$s'^{i-1} = s^{i-1} \underset{s^i}{\times} (s^i)^{[1]},$$

where $(s^i)^{[1]} \to s^i$ corresponds to the map $[0] \to [1]$ given by $0 \mapsto 0$.

□

2.3.3. *Colimits of 1-morphisms.* We shall now describe explicitly how to compute colimits in $\mathrm{Seq}_1^{\mathrm{ext}}(\mathbb{S})$. Let $\mathbf{I}$ be an $(\infty,1)$-category, and let

$$i \mapsto \underline{s}_i := (s_i^0 \overset{\beta_i}{\to} s_i^1)$$

be a functor $\mathbf{I} \to \mathrm{Seq}_1^{\mathrm{ext}}(\mathbb{S})$.

Let us denote

$$s^0 \simeq \underset{i\in\mathbf{I}}{\mathrm{colim}}\, s_i^0 \text{ and } s^1 \simeq \underset{i\in\mathbf{I}}{\mathrm{colim}}\, s_i^1. \tag{2.7}$$

Let ins_i^0 and ins_i^1 denote the canonical 1-morphisms

$$s_i^0 \to s^0 \text{ and } s_i^1 \to s^1,$$

respectively.

Note that by Proposition 2.3.2(b), if the colimit

$$\underset{i\in\mathbf{I}}{\mathrm{colim}}\, \underline{s}_i \in \mathrm{Seq}_1^{\mathrm{ext}}(\mathbb{S})$$

exists, the 0-th (resp., 1-st) component of the corresponding object identifies with s_0 (resp., s_1).

Assume now that for every 1-morphism $i \to j$ in $\mathbf{I}$, the 1-morphism $s_i^0 \to s_j^0$ admits a right adjoint. In this case, the 1-morphism ins_i^0 also admits a right adjoint, which we denote by $(\mathrm{ins}_i^0)^R$.

PROPOSITION 2.3.4. *Under the above circumstances, the colimit* $\underline{s} := \underset{i\in\mathbf{I}}{\mathrm{colim}}\, \underline{s}_i \in$ $\mathrm{Seq}_1^{\mathrm{ext}}(\mathbb{S})$ *exists. The resulting* 1*-morphism*

$$s^0 \to s^1$$

identifies canonically with

$$\beta := \underset{i\in\mathbf{I}}{\mathrm{colim}}\, \mathrm{ins}_i^1 \circ \beta^i \circ (\mathrm{ins}_i^0)^R \in \mathbf{Maps}_{\mathbb{S}}(s_0, s_1).$$

PROOF. Let

$$s'^0 \overset{\beta'}{\to} s'^1$$

be an object in $\mathrm{Seq}_1^{\mathrm{ext}}(\mathbb{S})$. We need to show that a compatible system of data

(2.8)

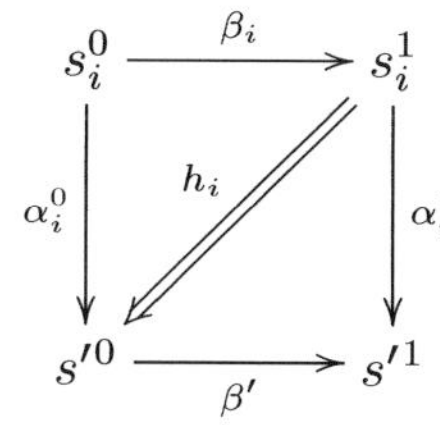

is equivalent to that of

(2.9)

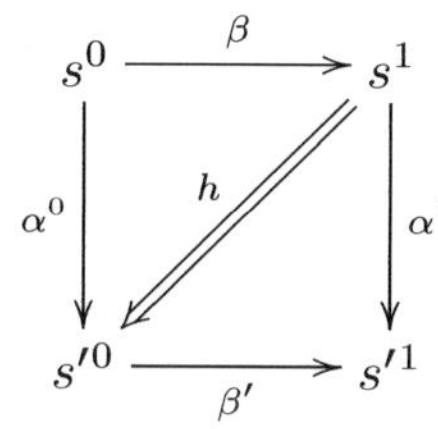

The 1-morphisms α^0 and α^1 are uniquely recovered from the compatible families α_i^0 and α_i^1 respectively, by (2.7).

Since compositions of 1-morphisms commute with colimits, by the construction of β, the data of h is equivalent to that of a compatible system of 2-morphisms

$$\alpha_i^1 \circ \beta_i \circ (\mathrm{ins}_i^0)^R \simeq \alpha^1 \circ \mathrm{ins}_i^1 \circ \beta_i \circ (\mathrm{ins}_i^0)^R \to \beta' \circ \alpha^0,$$

which by adjunction is equivalent to

$$\alpha_i^1 \circ \beta_i \to \beta' \circ \alpha^0 \circ \mathrm{ins}_i^0 \simeq \beta' \circ \alpha_i^0,$$

as desired. □

Combing with Proposition 2.3.2(b), we obtain:

COROLLARY 2.3.5. *Let $i \mapsto \underline{s}_i$ be a functor $\mathbf{I} \to \mathrm{Seq}_n^{\mathrm{ext}}(\mathbb{S})$, such that for every $k \in \{0, ..., n-1\}$, and every 1-morphism $i \to j$ in $\mathbf{I}$, the corresponding 1-morphism $s_i^k \to s_j^k$ admits a right adjoint. Then the colimit*

$$\underline{s} := \underset{i \in \mathbf{I}}{\mathrm{colim}}\, \underline{s}_i \in \mathrm{Seq}_n^{\mathrm{ext}}(\mathbb{S})$$

exists.

2.3.6. *The product situation.* Let $\underline{s}$

$$s^0 \to s^1 \to ... \to s^n \tag{2.10}$$

be an object of $\mathrm{Seq}_n^{\mathrm{ext}}(\mathbb{S})$.

Let us be given, for each $k = 0, ..., n$ an index category $\mathbf{I}_k$ and a colimit diagram

$$\underset{i_k \in \mathbf{I}_k}{\mathrm{colim}}\, s_{i_k} \to s^k,$$

such that each of the 1-morphisms $\mathrm{ins}_{i_k} : s_{i_k} \to s^k$ and $\alpha_{i_k,i'_k} : s_{i_k} \to s_{i'_k}$ (for $(i_k \to i'_k) \in \mathbf{I}_k$) admits a right adjoint.

Set $\mathbf{I} := \mathbf{I}_0 \times \ldots \times \mathbf{I}_n$. We define an $\mathbf{I}$-diagram $\underline{i} \mapsto \underline{s}_{\underline{i}}$ in $\mathrm{Seq}_n^{\mathrm{ext}}(\mathbb{S})$ by setting for $\underline{i} = (i_0, \ldots, i_k)$

$$s^k_{\underline{i}} = s_{i_k}, \quad s_{i_{k-1}} \overset{\mathrm{ins}_{i_{k-1}}}{\longrightarrow} s^{k-1} \to s^k \overset{(\mathrm{ins}_{i_k})^R}{\longrightarrow} s_{i_k}.$$

By construction, the above $\mathbf{I}$-family $\underline{s}_{\underline{i}}$ is equipped with a map to the object $\underline{s}$ of (2.10).

From Propositions 2.3.4 and 2.3.2(b), we obtain:

COROLLARY 2.3.7. *Assume that each of the index categories $\mathbf{I}_k$ is contractible. Then:*

(a) *The map*

$$\underset{\underline{i} \in \mathbf{I}}{\mathrm{colim}}\, \underline{s}_{\underline{i}} \to \underline{s}$$

is an isomorphism.

(b) *For any $r : [m] \to [n]$, the map*

$$r^*_{\mathbb{S}}(\underset{\underline{i} \in \mathbf{I}}{\mathrm{colim}}\, \underline{s}_{\underline{i}}) \to \underset{\underline{i} \in \mathbf{I}}{\mathrm{colim}}\, (r^*_{\mathbb{S}}(\underline{s}_{\underline{i}})).$$

is an isomorphism.

3. The core of the proof

3.1. Calculation of $\mathrm{LKE}_{F_1}(\Phi_1)$. Recall the notation

$$\Psi := \mathrm{LKE}_{F_{vert}}(\Phi) : \mathbf{D}_{vert} \to \mathbb{S}$$

and

$$\Psi^! := \mathrm{RKE}_{(F_{horiz})^{\mathrm{op}}}(\Phi^!) : (\mathbf{D}_{horiz})^{\mathrm{op}} \to \mathbb{S}.$$

Note that we can identify the values of Ψ and $\Psi^!$ on objects of $\mathbf{D}$ by Proposition 2.2.7 and the conditions (2) and (4) in Sect. 1.1.6, i.e., that

$$\mathrm{LKE}_{F_{adm}}(\Phi|_{\mathbf{C}_{adm}}) \simeq \Psi|_{\mathbf{D}_{adm}}$$

and

$$\Psi^!|_{(\mathbf{D}_{adm})^{\mathrm{op}}} \simeq \mathrm{RKE}_{(F_{adm})^{\mathrm{op}}}(\Phi^!|_{(\mathbf{C}_{adm})^{\mathrm{op}}}).$$

Let an object $\underline{\mathbf{d}} \in \mathrm{Seq}_1^{\mathrm{ext}}(\mathrm{Corr}(\mathbf{D})^{adm}_{vert;horiz})$ be given by a diagram

$$(3.1) \qquad \begin{array}{ccc} \mathbf{d}_{0,1} & \overset{\alpha_{\mathbf{d}}}{\longrightarrow} & \mathbf{d}_{0,0} \\ {\scriptstyle \beta_{\mathbf{d}}}\downarrow & & \\ \mathbf{d}_{1,1}. & & \end{array}$$

The goal of this subsection is to construct a canonical identification of

$$\mathrm{LKE}_{F_1}(\Psi_1)(\underline{\mathbf{d}}_1), W$$

which is a 1-morphism

$$\Psi(\mathbf{d}_{0,0}) \to \Psi(\mathbf{d}_{1,1}),$$

with $\Psi(\beta_{\mathbf{d}}) \circ \Psi^!(\alpha_{\mathbf{d}})$.

3.1.1. The 1-morphism

$$\mathrm{LKE}_{F_n}(r_{\mathbb{S}}^* \circ \Phi_n)(\underline{\mathbf{d}})$$

is the colimit in $\mathrm{Seq}_1^{\mathrm{ext}}(\mathbb{S})$ over

$$(\underline{\gamma} : \underline{\mathbf{c}} \to \underline{\mathbf{d}}) \in ({}'\mathbf{Grid}_1^{\geq \mathrm{dgnl}}(\mathbf{C})_{vert;horiz}^{adm})_{/\underline{\mathbf{d}}}$$

of

$$\Phi(\beta_{\mathbf{c}}) \circ \Phi^!(\alpha_{\mathbf{c}}),$$

for the morphisms

$$\begin{array}{ccc} \mathbf{c}_{0,1} & \xrightarrow{\alpha_{\mathbf{c}}} & \mathbf{c}_{0,0} \\ {\scriptstyle \beta_{\mathbf{c}}} \downarrow & & \\ \mathbf{c}_{1,1}. & & \end{array}$$

We calculate this colimit using Lemma 2.3.4, and we obtain

$$\underset{\underline{\gamma}:\underline{\mathbf{c}}\to\underline{\mathbf{d}}}{\mathrm{colim}}\, \Psi_{adm}(\gamma_{1,1}) \circ (\Phi(\beta_{\mathbf{c}}) \circ \Phi^!(\alpha_{\mathbf{c}})) \circ (\Psi_{adm}(\gamma_{0,0}))^R,$$

which can be rewritten as

$$\underset{\underline{\gamma}:\underline{\mathbf{c}}\to\underline{\mathbf{d}}}{\mathrm{colim}}\, \Psi(\gamma_{1,1}) \circ \Psi(\beta_{\mathbf{c}}) \circ \Psi^!(\alpha_{\mathbf{c}}) \circ \Psi^!(\gamma_{0,0}). \tag{3.2}$$

3.1.2. Consider the following diagram:

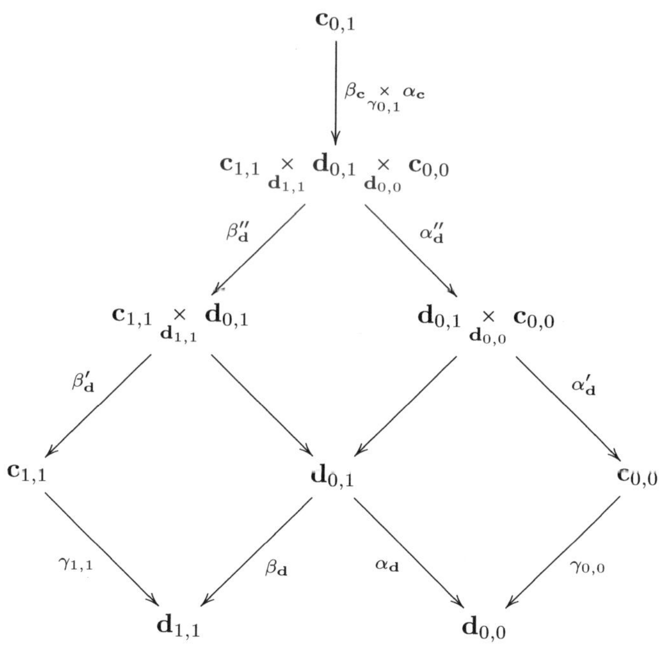

We rewrite the expression in (3.2) as the colimit over

$$\{\underline{\gamma} : \underline{\mathbf{c}} \to \underline{\mathbf{d}}\} \in (\mathbf{Grid}_1^{\geq \mathrm{dgnl}}(\mathbf{C})_{vert;horiz}^{adm})_{/\underline{\mathbf{d}}}$$

of

$$\Psi(\gamma_{1,1})\circ\Psi(\beta'_{\mathbf{d}})\circ\Psi(\beta''_{\mathbf{d}})\circ\Psi(\beta_{\mathbf{c}} \underset{\gamma_{0,1}}{\times} \alpha_{\mathbf{c}})\circ\Psi^!(\beta_{\mathbf{c}} \underset{\gamma_{0,1}}{\times} \alpha_{\mathbf{c}})\circ\Psi^!(\alpha''_{\mathbf{d}})\circ\Psi^!(\alpha'_{\mathbf{d}})\circ\Psi^!(\gamma_{0,0}). \tag{3.3}$$

3.1.3. The forgetful functor

$$(\mathbf{Grid}_1^{\geq \text{dgnl}}(\mathbf{C})^{adm}_{vert;horiz})_{/\underline{\mathbf{d}}} \to (\mathbf{C}_{adm})_{/\mathbf{d}_{0,0}} \times (\mathbf{C}_{adm})_{/\mathbf{d}_{1,1}}$$

is a co-Cartesian fibration. For a fixed object

$$\gamma_{0,0} : \mathbf{c}_{0,0} \to \mathbf{d}_{0,0} \text{ and } \gamma_{1,1} : \mathbf{c}_{1,1} \to \mathbf{d}_{1,1},$$

the fiber of $(\mathbf{Grid}_1^{\geq \text{dgnl}}(\mathbf{C})^{adm}_{vert;horiz})_{/\underline{\mathbf{d}}}$ over it is canonically equivalent to

$$(\mathbf{C}_{adm})_{/\mathbf{c}_{1,1} \underset{\mathbf{d}_{1,1}}{\times} \mathbf{d}_{0,1} \underset{\mathbf{d}_{0,0}}{\times} \mathbf{c}_{0,0}}.$$

Hence, by Proposition 2.2.5(b), the colimit of the expressions (3.3) over the above fiber is canonically isomorphic to

$$\Psi(\gamma_{1,1}) \circ \Psi(\beta'_{\mathbf{d}}) \circ \Psi(\beta''_{\mathbf{d}}) \circ \Psi^!(\alpha''_{\mathbf{d}}) \circ \Psi^!(\alpha'_{\mathbf{d}}) \circ \Psi^!(\gamma_{0,0}).$$

3.1.4. Applying the Beck-Chevalley isomorphisms (i.e., Conditions (1) and (3) from Sect. 1.1.6), we rewrite the latter expression as

$$\Psi(\gamma_{1,1}) \circ \Psi^!(\gamma_{1,1}) \circ \Psi(\beta_{\mathbf{d}}) \circ \Psi^!(\alpha_{\mathbf{d}}) \circ \Psi(\gamma_{0,0}) \circ \Psi^!(\gamma_{0,0}). \tag{3.4}$$

We obtain that the colimit over $\underline{\mathbf{c}} \in (\mathbf{Grid}_1^{\geq \text{dgnl}}(\mathbf{C})^{adm}_{vert;horiz})_{/\underline{\mathbf{d}}}$ is isomorphic to the colimit over

$$(\mathbf{c}_{0,0} \times \mathbf{c}_{1,1}) \in (\mathbf{C}_{adm})_{/\mathbf{d}_{0,0}} \times (\mathbf{C}_{adm})_{/\mathbf{d}_{1,1}}$$

of the expressions (3.4). Applying Proposition 2.2.5(b) again, we obtain that the latter is canonically isomorphic to

$$\Psi(\beta_{\mathbf{d}}) \circ \Psi^!(\alpha_{\mathbf{d}}),$$

as asserted.

3.2. Proof of uniqueness in Theorem 1.1.9. In this subsection we will continue to assume Proposition 1.2.5 and will deduce the uniqueness assertion in Theorem 1.1.9.

3.2.1. Let

$$\widetilde{\Psi}^{adm}_{vert;horiz} : \text{Corr}(\mathbf{D})^{adm}_{vert;horiz} \to \mathbb{S}$$

be a functor.

Denote

$$\widetilde{\Psi} := \widetilde{\Psi}^{adm}_{vert;horiz}|_{\mathbf{D}_{vert}}, \quad \widetilde{\Psi}^! := \widetilde{\Psi}^{adm}_{vert;horiz}|_{(\mathbf{D}_{horiz})^{\text{op}}}.$$

Let us be given a natural transformation

$$\Phi^{adm}_{vert;horiz} \to \widetilde{\Psi}^{adm}_{vert;horiz} \circ F^{adm}_{vert;horiz}. \tag{3.5}$$

Let $\widetilde{\Psi}_\bullet$ denote the functor of simplicial categories

$$'\mathbf{Grid}_\bullet^{\geq \text{dgnl}}(\mathbf{D})^{adm}_{vert;horiz} \xrightarrow{\widetilde{\Psi}_\bullet} \text{Seq}_\bullet^{\text{ext}}(\mathbb{S}),$$

corresponding to $\widetilde{\Psi}$.

By the construction of $\Psi^{adm}_{vert;horiz}$, we obtain a natural transformation between simplicial functors

$$\Psi_\bullet \to \widetilde{\Psi}_\bullet,$$

as functors from the simplicial category $'\mathbf{Grid}_\bullet^{\geq \text{dgnl}}(\mathbf{D})^{adm}_{vert;horiz}$ to the simplicial category $\text{Seq}_\bullet^{\text{ext}}(\mathbb{S})$.

3.2.2. Restricting along

$$''\mathbf{Grid}_\bullet^{\geq \mathrm{dgnl}}(\mathbf{D})^{adm}_{vert;horiz} \hookrightarrow {}'\mathbf{Grid}_\bullet^{\geq \mathrm{dgnl}}(\mathbf{D})^{adm}_{vert;horiz},$$

we obtain a natural transformation between simplicial functors

$$\Psi_\bullet|_{''\mathbf{Grid}_\bullet^{\geq \mathrm{dgnl}}(\mathbf{D})^{adm}_{vert;horiz}} \to \widetilde{\Psi}_\bullet|_{''\mathbf{Grid}_\bullet^{\geq \mathrm{dgnl}}(\mathbf{D})^{adm}_{vert;horiz}}, \tag{3.6}$$

as functors from the simplicial category

$$''\mathbf{Grid}_\bullet^{\geq \mathrm{dgnl}}(\mathbf{D})^{adm}_{vert;horiz} \simeq \mathrm{Seq}_\bullet(\mathrm{Corr}(\mathbf{D})^{adm}_{vert;horiz})$$

to the simplicial category $\mathrm{Seq}_\bullet^{\mathrm{ext}}(\mathbb{S})$, while both functors take values in

$$\mathrm{Seq}_\bullet(\mathbb{S}) \subset \mathrm{Seq}_\bullet^{\mathrm{ext}}(\mathbb{S}).$$

3.2.3. Suppose now that (3.5) is an isomorphism, and that the natural transformation

$$\mathrm{LKE}_{F_{adm}}(\Phi_{adm}) \to \widetilde{\Psi}^{adm}_{vert;horiz}|_{\mathbf{D}_{adm}}, \tag{3.7}$$

induced by (3.5), is also an isomorphism.

We will show that in this case, the natural transformation (3.6) is an isomorphism. By the Segal condition, it suffices to do so on 0-simplices and 1-simplices.

For 0-simplices, this is just the fact that the map (3.7) is an isomorphism. For 1-simplices, we need to show that for a 1-morphism $f : \mathbf{d} \to \mathbf{d}'$ in $\mathrm{Corr}(\mathbf{D})^{adm}_{vert;horiz}$ the 2-morphism in

$$\begin{array}{ccc} \Psi(\mathbf{d}) & \longrightarrow & \Psi(\mathbf{d}') \\ \sim\downarrow & \overset{h}{\swarrow} & \downarrow\sim \\ \widetilde{\Psi}(\mathbf{d}) & \longrightarrow & \widetilde{\Psi}(\mathbf{d}') \end{array} \tag{3.8}$$

is an isomorphism. It is enough to consider separately the cases when $f : \mathbf{d} \to \mathbf{d}'$ is vertical or horizontal.

3.2.4. Note that the natural transformation (3.5) gives rise to a natural transformation

$$\Phi \to \widetilde{\Psi} \circ F_{vert},$$

and hence to

$$\Psi := \mathrm{LKE}_{F_{vert}}(\Phi) \to \widetilde{\Psi}. \tag{3.9}$$

Restricting to $\mathbf{D}_{adm}$ we obtain the commutative diagram

$$\begin{array}{ccc} \mathrm{LKE}_{F_{adm}}(\Phi_{adm}) & \xrightarrow{\mathrm{id}} & \mathrm{LKE}_{F_{adm}}(\Phi_{adm}) \\ \downarrow & & \downarrow \\ \Psi|_{\mathbf{D}_{adm}} & \longrightarrow & \widetilde{\Psi}|_{\mathbf{D}_{adm}}. \end{array} \tag{3.10}$$

Parenthetically, note that the vertical arrows in (3.10) are isomorphisms. Hence, $\Psi|_{\mathbf{D}_{adm}} \to \widetilde{\Psi}|_{\mathbf{D}_{adm}}$ is an isomorphism. Therefore, (3.9) is also an isomorphism, because it is such on objects.

3.2.5. We claim that for $f : \mathbf{d} \to \mathbf{d}'$ being a vertical morphism $\mathbf{d} \overset{\beta}{\to} \mathbf{d}'$, the diagram (3.8) represents the natural transformation (3.9) evaluated on β.

Indeed, it follows from the calculation of Ψ_1 in Sect. 3.1 that $\Psi_1(\beta)$ can be written as a colimit of the category

$$\begin{array}{ccc} \mathbf{c} & \xrightarrow{\beta_{\mathbf{c}}} & \mathbf{c}' \\ {\scriptstyle\gamma}\downarrow & & \downarrow{\scriptstyle\gamma'} \\ \mathbf{d} & \xrightarrow{\beta} & \mathbf{d}' \end{array}$$

of the objects $(\Phi(\mathbf{c}) \overset{\Phi(\beta_{\mathbf{c}})}{\longrightarrow} \Phi(\mathbf{c}')) \in \mathrm{Seq}_1^{\mathrm{ext}}(\mathbb{S})$. Thus, to show that (3.8) is represented by

$$\begin{array}{ccc} \Psi(\mathbf{d}) & \xrightarrow{\Psi(\beta)} & \Psi(\mathbf{d}') \\ \downarrow & & \downarrow \\ \widetilde{\Psi}(\mathbf{d}) & \xrightarrow{\widetilde{\Psi}(\beta)} & \widetilde{\Psi}(\mathbf{d}'), \end{array}$$

we need to construct a compatible of diagrams

$$\begin{array}{ccc} \Phi(\mathbf{c}) & \xrightarrow{\Phi(\beta_{\mathbf{c}})} & \Phi(\mathbf{c}') \\ {\scriptstyle\Psi(\gamma)}\downarrow & & \downarrow{\scriptstyle\Psi(\gamma')} \\ \Psi(\mathbf{d}) & \xrightarrow{\Psi(\beta)} & \Psi(\mathbf{d}') \\ \downarrow & & \downarrow \\ \widetilde{\Psi}(\mathbf{d}) & \xrightarrow{\widetilde{\Psi}(\beta)} & \widetilde{\Psi}(\mathbf{d}'), \end{array}$$

in which the outer squares are identified with

$$\begin{array}{ccc} \Phi(\mathbf{c}) & \longrightarrow & \Phi(\mathbf{c}') \\ {\scriptstyle\widetilde{\Psi}(\gamma)}\downarrow & & \downarrow{\scriptstyle\widetilde{\Psi}(\gamma')} \\ \widetilde{\Psi}(\mathbf{d}) & \xrightarrow{\widetilde{\Psi}(\beta)} & \widetilde{\Psi}(\mathbf{d}'). \end{array}$$

However, this is given by the diagram (3.10).

3.2.6. As in Sect. 3.2.4, from the *isomorphism* (3.5), we obtain a natural transformation

$$\widetilde{\Psi}^! \circ (F_{horiz})^{\mathrm{op}} \to \Phi^!,$$

and hence a natural transformation

$$\widetilde{\Psi}^! \to \Psi^!, \tag{3.11}$$

which is also an isomorphism by the same logic.

We claim that for a 1-morphism $f : \mathbf{d} \to \mathbf{d}'$ in $\mathrm{Corr}(\mathbf{D})^{adm}_{vert;horiz}$, given by a horizontal morphism $\alpha : \mathbf{d}' \to \mathbf{d}$, the diagram (3.8) represents the *inverse* of the natural transformation (3.11) evaluated on α.

Indeed, using the calculation of Ψ_1 in Sect. 3.1, we obtain $\Psi_1(f)$ can be written as a colimit of the category

$$\begin{array}{ccc} \mathbf{c} & \xleftarrow{\alpha_{\mathbf{c}}} & \mathbf{c}' \\ {\scriptstyle \gamma}\downarrow & & \downarrow{\scriptstyle \gamma'} \\ \mathbf{d} & \xleftarrow{\alpha} & \mathbf{d}' \end{array}$$

of the objects $(\Phi(\mathbf{c}) \overset{\Phi^!(\alpha_{\mathbf{c}})}{\longrightarrow} \Phi(\mathbf{c}')) \in \mathrm{Seq}_1^{\mathrm{ext}}(\mathbb{S})$.

Thus, to show that (3.8) is represented by

$$\begin{array}{ccc} \Psi(\mathbf{d}) & \xrightarrow{\Psi^!(\alpha)} & \Psi(\mathbf{d}') \\ {\scriptstyle \sim}\downarrow & & \downarrow{\scriptstyle \sim} \\ \widetilde{\Psi}(\mathbf{d}) & \xrightarrow{\widetilde{\Psi}^!(\alpha)} & \widetilde{\Psi}(\mathbf{d}'), \end{array}$$

we need to construct a compatible family of diagrams

$$\begin{tikzcd} \Phi(\mathbf{c}) \arrow[r, "\Phi^!(\alpha_{\mathbf{c}})"] \arrow[d, "\widetilde{\Psi}(\gamma)"'] & \Phi(\mathbf{c}') \arrow[d, "\widetilde{\Psi}(\gamma')"] \arrow[dl, Rightarrow] \\ \widetilde{\Psi}(\mathbf{d}) \arrow[r, "\widetilde{\Psi}^!(\alpha)"'] \arrow[d, "\sim"'] & \widetilde{\Psi}(\mathbf{d}') \arrow[d, "\sim"] \\ \Psi(\mathbf{d}) \arrow[r, "\Psi^!(\alpha)"'] & \Psi(\mathbf{d}') \end{tikzcd}$$

where the outer square is

$$\begin{tikzcd} \Phi(\mathbf{c}) \arrow[r, "\Phi^!(\alpha_{\mathbf{c}})"] \arrow[d, "\Psi(\gamma)"'] & \Phi(\mathbf{c}') \arrow[d, "\Psi(\gamma')"] \arrow[dl, Rightarrow] \\ \Psi(\mathbf{d}). \arrow[r, "\Psi^!(\alpha)"'] & \Psi(\mathbf{d}') \end{tikzcd}$$

Now, the required family of diagrams is obtained by passing to left adjoints along the γ-arrows in the family of *commutative* diagrams

$$
\begin{CD}
\widetilde{\Psi}(\mathbf{d}) @>{\widetilde{\Psi}^!(\alpha)}>> \widetilde{\Psi}(\mathbf{d}') \\
@V{\sim}VV @VV{\sim}V \\
\Psi(\mathbf{d}) @>{\Psi^!(\alpha)}>> \Psi(\mathbf{d}') \\
@V{\Psi^!(\gamma)}VV @VV{\Psi^!(\gamma')}V \\
\Phi(\mathbf{c}) @>>{\Phi^!(\alpha_{\mathbf{c}})}> \Phi(\mathbf{c}')
\end{CD}
$$

4. Proof of Proposition 1.2.5: easy reduction steps

4.1. Reductions for Proposition 1.2.5(a).

4.1.1. We claim that the map (1.3) is an isomorphism if r is of the form $i \mapsto i + k$ for $0 \leq k \leq n - m$ (in particular, it is an isomorphism for $m = 0$).

In fact, we claim that in this case, the map

$$\mathrm{LKE}_{F_n}(G \circ r^*_{\mathbf{C}}) \to \mathrm{LKE}_{F_m}(G) \circ r^*_{\mathbf{D}}$$

is an isomorphism for any functor G out of $'\mathbf{Grid}^{\geq \mathrm{dgnl}}_m(\mathbf{C})^{adm}_{vert;horiz}$.

Note that if the latter statement holds for $r : [m] \to [n]$ and $q : [l] \to [m]$, then it holds for the composition $r \circ q : [l] \to [n]$. This reduces the assertion to the case of r being a map $[n-1] \to [n]$, which is either $i \mapsto i$ or $i \mapsto i+1$.

Fix an object $\underline{\mathbf{d}}_n \in {}'\mathbf{Grid}^{\geq \mathrm{dgnl}}_n(\mathbf{D})^{adm}_{vert;horiz}$. We need to show that the map

$$\underset{\underline{\mathbf{c}}_n, F_n(\underline{\mathbf{c}}_n) \to \underline{\mathbf{d}}_n}{\mathrm{colim}} G(r^*_{\mathbf{C}}(\underline{\mathbf{c}}_n)) \to \underset{\underline{\mathbf{c}}_m, F_m(\underline{\mathbf{c}}_m) \to r^*_{\mathbf{D}}(\underline{\mathbf{d}}_n)}{\mathrm{colim}} G(\underline{\mathbf{c}}_m)$$

is an isomorphism.

We will show that in each of the above cases, the functor of index categories, i.e.,

$$\{\underline{\mathbf{c}}_n, F_n(\underline{\mathbf{c}}_n) \to \underline{\mathbf{d}}_n\} \to \{\underline{\mathbf{c}}_m, F_m(\underline{\mathbf{c}}_m) \to r^*_{\mathbf{D}}(\underline{\mathbf{d}}_n)\}, \quad \underline{\mathbf{c}}_n \mapsto r^*_{\mathbf{C}}(\underline{\mathbf{c}}_n) \tag{4.1}$$

is cofinal. By symmetry, it is sufficient to consider the case of r being the map $i \mapsto i$.

The functor (4.1) is (obviously) a co-Cartesian fibration. Hence, it is enough to show that it has contractible fibers.

For a given

$$\underline{\mathbf{c}}_{n-1} \to r^*_{\mathbf{D}}(\underline{\mathbf{d}}),$$

the fiber over it is the category of diagrams

$$\{\mathbf{c}_{n,n} \to \mathbf{d}_{n,n}, \mathbf{c}_{n-1,n} \to \mathbf{c}_{n-1,n-1} \underset{\mathbf{d}_{n-1,n-1}}{\times} \mathbf{d}_{n-1,n} \underset{\mathbf{d}_{n,n}}{\times} \mathbf{c}_{n,n}\},$$

where both maps are in *adm*.

This category is a co-Cartesian fibration over the category of

$$\{\mathbf{c}_{n,n} \to \mathbf{d}_{n,n}\}.$$

We claim that this category is contractible. Indeed, this follows from Lemma 4.1.2 below.

So, it is enough to show that each fiber of this co-Cartesian fibration , i.e.,

$$\{\mathbf{c}_{n-1,n} \to \mathbf{c}_{n-1,n-1} \underset{\mathbf{d}_{n-1,n-1}}{\times} \mathbf{d}_{n-1,n} \underset{\mathbf{d}_{n,n}}{\times} \mathbf{c}_{n,n}\}$$

is contractible. However, this also follows from Lemma 4.1.2.

LEMMA 4.1.2. *For a given $\mathbf{d} \in \mathbf{D}$, the category $\mathbf{C}_{adm} \underset{\mathbf{D}_{adm}}{\times} (\mathbf{D}_{adm})_{/\mathbf{d}}$ is contractible.*

PROOF. We claim that $\mathbf{C}_{adm} \underset{\mathbf{D}_{adm}}{\times} (\mathbf{D}_{adm})_{/\mathbf{d}}$ is in fact cofiltered (i.e., its opposite category is filtered).

Indeed, the category $(\mathbf{D}_{adm})_{/\mathbf{d}}$ has products, and therefore is cofiltered. Note now that every object of $(\mathbf{D}_{adm})_{/\mathbf{d}}$ admits a map from an object in $\mathbf{C}_{adm} \underset{\mathbf{D}_{adm}}{\times} (\mathbf{D}_{adm})_{/\mathbf{d}}$, by Sect. 1.1.3.

Now, we have the following general assertion: let $\mathbf{E}' \to \mathbf{E}$ be a fully faithful embedding with $\mathbf{E}$ filtered. Assume that every object of $\mathbf{E}$ admits a morphism to an object of $\mathbf{E}'$. Then $\mathbf{E}'$ is also filtered (and its embedding into $\mathbf{E}$ is cofinal). □

4.1.3. We will now show that it is enough to prove that (1.3) is an isomorphism for $m = 1$. Indeed, in order to show that

$$\mathrm{LKE}_{F_n}(\Phi_m \circ r^*_{\mathbf{C}}) \to \mathrm{LKE}_{F_m}(\Phi_m) \circ r^*_{\mathbf{D}}$$

is an isomorphism, it is enough to show that the induced natural transformation

$$q^*_{\mathbb{S}} \circ \mathrm{LKE}_{F_n}(\Phi_m \circ r^*_{\mathbf{C}}) \to q^*_{\mathbb{S}} \circ \mathrm{LKE}_{F_m}(\Phi_m) \circ r^*_{\mathbf{D}}$$

is an isomorphism for every $m \geq 1$ and $q : [1] \to [m]$ of the form

$$0 \mapsto i,\ 1 \mapsto i+1.$$

We have a commutative diagram

$$\begin{array}{ccc}
q^*_{\mathbb{S}} \circ \mathrm{LKE}_{F_n}(\Phi_m \circ r^*_{\mathbf{C}}) & \longrightarrow & q^*_{\mathbb{S}} \circ \mathrm{LKE}_{F_m}(\Phi_m) \circ r^*_{\mathbf{D}} \\
\uparrow & & \uparrow \\
\mathrm{LKE}_{F_n}(q^*_{\mathbb{S}} \circ \Phi_m \circ r^*_{\mathbf{C}}) & \longrightarrow & \mathrm{LKE}_{F_m}(q^*_{\mathbb{S}} \circ \Phi_m) \circ r^*_{\mathbf{D}} \\
{\scriptstyle\simeq}\downarrow & & \downarrow{\scriptstyle\simeq} \\
\mathrm{LKE}_{F_n}(\Phi_1 \circ q^*_{\mathbf{C}} \circ r^*_{\mathbf{C}}) & & \mathrm{LKE}_{F_m}(\Phi_1 \circ q^*_{\mathbf{C}}) \circ r^*_{\mathbf{D}} \\
{\scriptstyle=}\downarrow & & \downarrow \\
\mathrm{LKE}_{F_n}(\Phi_1 \circ (r \circ q)^*_{\mathbf{C}}) & & \mathrm{LKE}_{F_m}(\Phi_1) \circ q^*_{\mathbf{D}} \circ r^*_{\mathbf{D}} \\
{\scriptstyle\mathrm{id}}\downarrow & & \downarrow{\scriptstyle=} \\
\mathrm{LKE}_{F_n}(\Phi_1 \circ (r \circ q)^*_{\mathbf{C}}) & \longrightarrow & \mathrm{LKE}_{F_1}(\Phi_1) \circ (r \circ q)^*_{\mathbf{D}}.
\end{array}$$

By assumption, the bottom horizontal arrow is an isomorphism. The second-from-the-bottom right vertical arrow is an isomorphism by Sect. 4.1.1. The upper left and upper right vertical arrows are isomorphisms by Proposition 2.3.2(b).

Hence, the top horizontal arrow is also an isomorphism, as required.

4.1.4. We will now show that it is enough to show that the map (1.3) is an isomorphism for $r : [1] \to [n]$ of the form $0 \mapsto 0$ and $1 \mapsto n$.

Indeed, given a map $r : [1] \to [n]$ decompose it as $p \circ q$, where $q : [1] \to [m]$ is of the form $0 \mapsto 0$ and $1 \mapsto m$ and p is of the form $i \mapsto i + k$ for $0 \le k \le n - m$.

We have a commutative diagram

$$\begin{array}{ccc} \mathrm{LKE}_{F_n}(\Phi_1 \circ (p \circ q)^*_{\mathbf{C}}) & \longrightarrow & \mathrm{LKE}_{F_1}(\Phi_1) \circ (p \circ q)^*_{\mathbf{D}} \\ {\scriptstyle =}\downarrow & & \downarrow{\scriptstyle =} \\ \mathrm{LKE}_{F_n}(\Phi_1 \circ q^*_{\mathbf{C}} \circ p^*_{\mathbf{C}}) & & \mathrm{LKE}_{F_1}(\Phi_1) \circ q^*_{\mathbf{D}} \circ p^*_{\mathbf{D}} \\ \downarrow & & \downarrow{\scriptstyle \mathrm{id}} \\ \mathrm{LKE}_{F_m}(\Phi_1 \circ q^*_{\mathbf{C}}) \circ p^*_{\mathbf{D}} & \longrightarrow & \mathrm{LKE}_{F_1}(\Phi_1) \circ q^*_{\mathbf{D}} \circ p^*_{\mathbf{D}}. \end{array}$$

By assumption, the bottom horizontal map is an isomorphism. The lower left vertical is an isomorphism by Sect. 4.1.1. This implies that the top horizontal map is an isomorphism, as required.

4.1.5. To summarize, in order to prove Proposition 1.2.5(a) it remains to consider the following two cases:

(I) r is the map $[1] \to [n]$ given by $0 \mapsto 0$ and $1 \mapsto n$ with $n > 1$.

(II) r is the degeneracy map $[1] \to [0]$.

4.2. Reductions for Proposition 1.2.5(b).

4.2.1. First, we note that the map (1.4) is an isomorphism for $m = 0$, by Proposition 2.3.2(a).

4.2.2. We will show that it is sufficient it is sufficient to prove that the map (1.4) is an isomorphism for $m = 1$.

Indeed, in order to show that (1.4) is an isomorphism, it is sufficient to show that the induced natural transformation

$$q^*_{\mathbb{S}} \circ \mathrm{LKE}_{F_n}(r^*_{\mathbb{S}} \circ \Phi_n) \to q^*_{\mathbb{S}} \circ r^*_{\mathbb{S}} \circ \mathrm{LKE}_{F_n}(\Phi_n)$$

is an isomorphism, for every $m \ge 1$ and $q : [1] \to [m]$ of the form

$$0 \mapsto i,\ 1 \mapsto i + 1.$$

We have a commutative diagram

$$\begin{array}{ccc}
q_{\mathbb{S}}^* \circ \mathrm{LKE}_{F_n}(r_{\mathbb{S}}^* \circ \Phi_n) & \longrightarrow & q_{\mathbb{S}}^* \circ r_{\mathbb{S}}^* \circ \mathrm{LKE}_{F_n}(\Phi_n) \\
\uparrow & & \Big\downarrow{\scriptstyle =} \\
\mathrm{LKE}_{F_n}(q_{\mathbb{S}}^* \circ r_{\mathbb{S}}^* \circ \Phi_n) & & \\
{\scriptstyle =}\Big\downarrow & & \\
\mathrm{LKE}_{F_n}((r \circ q)_{\mathbb{S}}^* \circ \Phi_n) & \longrightarrow & (r \circ q)_{\mathbb{S}}^* \circ \mathrm{LKE}_{F_n}(\Phi_n),
\end{array}$$

where the upper left vertical arrow is an isomorphism by Proposition 2.3.2(b).

This establishes the announced reduction step.

4.2.3. We will now further reduce the verification of the fact that (1.4) is an isomorphism to the case when $r : [1] \to [n]$ sends $0 \mapsto 0$ and $1 \mapsto n$.

Given a map $r : [1] \to [n]$ we can factor it as

$$[1] \overset{p}{\to} [m] \overset{q}{\to} [n],$$

where q is of the form $i \mapsto i + k$ for $0 \leq k \leq n - m$, and p sends $0 \mapsto 0$ and $1 \mapsto m$.

We claim that if the map

$$\mathrm{LKE}_{F_n}(p_{\mathbb{S}}^* \circ \Phi_m) \to p_{\mathbb{S}}^* \circ \mathrm{LKE}_{F_n}(\Phi_m)$$

is an isomorphism, then so is the map

$$\mathrm{LKE}_{F_n}(r_{\mathbb{S}}^* \circ \Phi_m) \to r_{\mathbb{S}}^* \circ \mathrm{LKE}_{F_n}(\Phi_m).$$

Indeed, we have a commutative diagram

$$\begin{array}{ccc}
\mathrm{LKE}_{F_n}(r_{\mathbb{S}}^* \circ \Phi_n) & \longrightarrow & r_{\mathbb{S}}^* \circ \mathrm{LKE}_{F_n}(\Phi_n) \\
{\scriptstyle =}\uparrow & & \uparrow{\scriptstyle =} \\
\mathrm{LKE}_{F_n}(p_{\mathbb{S}}^* \circ q_{\mathbb{S}}^* \circ \Phi_n) & & p_{\mathbb{S}}^* \circ q_{\mathbb{S}}^* \circ \mathrm{LKE}_{F_n}(\Phi_n) \\
{\scriptstyle \sim}\uparrow & & \uparrow \\
\mathrm{LKE}_{F_n}(p_{\mathbb{S}}^* \circ \Phi_m \circ q_{\mathbf{C}}^*) & \longrightarrow & p_{\mathbb{S}}^* \circ \mathrm{LKE}_{F_n}(q_{\mathbb{S}}^* \circ \Phi_n) \\
\downarrow & & \downarrow{\scriptstyle \sim} \\
\mathrm{LKE}_{F_m}(p_{\mathbb{S}}^* \circ \Phi_m) \circ q_{\mathbf{D}}^* & & p_{\mathbb{S}}^* \circ \mathrm{LKE}_{F_n}(\Phi_m \circ q_{\mathbf{C}}^*) \\
{\scriptstyle \mathrm{id}}\downarrow & & \downarrow \\
\mathrm{LKE}_{F_m}(p_{\mathbb{S}}^* \circ \Phi_m) \circ q_{\mathbf{D}}^* & \longrightarrow & p_{\mathbb{S}}^* \circ \mathrm{LKE}_{F_m}(\Phi_m) \circ q_{\mathbf{D}}^*
\end{array}$$

By assumption, the bottom horizontal map is an isomorphism, and we wish to deduce that so is the top horizontal map. We claim that all the vertical maps are isomorphisms.

The second-from-the-bottom left vertical map and the lower right vertical map are isomorphisms by Sect. 4.1.1. The second-from-the-top right vertical map is an isomorphism by Proposition 2.3.2(c).

5. End of the proof of Proposition 1.2.5

5.1. Proof of Proposition 1.2.5(b); the main case. We need to show that the map

$$\mathrm{LKE}_{F_n}(r_{\mathbb{S}}^* \circ \Phi_n) \to r_{\mathbb{S}}^* \circ \mathrm{LKE}_{F_n}(\Phi_n) \tag{5.1}$$

is an isomorphism for r being the map $[1] \to [n]$, $0 \mapsto 0$ and $1 \mapsto n$.

Fix an object $\underline{\mathbf{d}}_n \in {}'\mathbf{Grid}_n^{\geq \mathrm{dgnl}}(\mathbf{D})_{vert;horiz}^{adm}$. We need to show that the map

$$\operatorname*{colim}_{\underline{\mathbf{c}}_n} r_{\mathbb{S}}^* \circ \Phi_n(\underline{\mathbf{c}}_n) \to r_{\mathbb{S}}^*(\operatorname*{colim}_{\underline{\mathbf{c}}_n} \Phi_n(\underline{\mathbf{c}}_n)) \tag{5.2}$$

is an isomorphism, where $\underline{\mathbf{c}}_n$ runs over the category

$$({}'\mathbf{Grid}_n^{\geq \mathrm{dgnl}}(\mathbf{C})_{vert;horiz}^{adm})_{/\underline{\mathbf{d}}_n}.$$

5.1.1. We first calculate the colimit

$$\operatorname*{colim}_{\underline{\mathbf{c}}_n} \Phi_n(\underline{\mathbf{c}}_n) \in \mathrm{Seq}_n(\mathbb{S}), \tag{5.3}$$

which is a string

$$\Psi(\mathbf{d}_{0,0}) \to \Psi(\mathbf{d}_{1,1}) \to \ldots \to \Psi(\mathbf{d}_{n,n}).$$

We claim that (5.3) is given by

$$\Psi(\mathbf{d}_{i-1,i-1}) \overset{\Psi(\beta_{i-1,i})\circ\Psi^!(\alpha_{i,i-1})}{\longrightarrow} \Psi(\mathbf{d}_{i,i}) \tag{5.4}$$

$i = 1, \ldots, n$.

For $i = 1, \ldots, n$, let q denote the map $[1] \to [n]$, given by $0 \mapsto i-1$, $1 \mapsto i$. We will identify

$$q_{\mathbb{S}}^*(\operatorname*{colim}_{\underline{\mathbf{c}}_n} \Phi_n(\underline{\mathbf{c}}_n))$$

with (5.4).

Note that by Proposition 2.3.2(b), we have

$$q_{\mathbb{S}}^*(\operatorname*{colim}_{\underline{\mathbf{c}}_n} \Phi_n(\underline{\mathbf{c}}_n)) \simeq \operatorname*{colim}_{\underline{\mathbf{c}}_n} q_{\mathbb{S}}^* \circ \Phi_n(\underline{\mathbf{c}}_n). \tag{5.5}$$

For $i = 1, \ldots, n$, let $\underline{\mathbf{d}}_n^i$ be the object of ${}'\mathbf{Grid}_1^{\geq \mathrm{dgnl}}(\mathbf{D})_{vert;horiz}^{adm}$ given by

$$\begin{array}{ccc} \mathbf{d}_{i-1,i} & \longrightarrow & \mathbf{d}_{i-1,i-1} \\ \downarrow & & \\ \mathbf{d}_{i,i}. & & \end{array}$$

We have a natural restriction functor

$$r_{\mathbf{C}}^* : ({}'\mathbf{Grid}_n^{\geq \mathrm{dgnl}}(\mathbf{C})_{vert;horiz}^{adm})_{/\underline{\mathbf{d}}_n} \to ({}'\mathbf{Grid}_1^{\geq \mathrm{dgnl}}(\mathbf{C})_{vert;horiz}^{adm})_{/\underline{\mathbf{d}}_n^i}. \tag{5.6}$$

The functor

$$r_{\mathbb{S}}^* \circ \Phi_n : ({}'\mathbf{Grid}_n^{\geq \mathrm{dgnl}}(\mathbf{C})_{vert;horiz}^{adm})_{/\underline{\mathbf{d}}_n} \to \mathrm{Seq}_1(\mathbb{S})$$

identifies with $\Phi_1 \circ r_{\mathbf{C}}^*$.

Now, it is easy to see that the functor (5.6) is a co-Cartesian fibration. Moreover, it follows from Lemma 4.1.2 that its fibers are contractible. Hence, the functor (5.6) is cofinal.

Therefore, the colimit in (5.5) identifies with

$$\underset{\underline{\mathbf{c}}_1}{\operatorname{colim}} \Phi_1(\underline{\mathbf{c}}_1),$$

where $\underline{\mathbf{c}}_1$ runs over the category $({}'\mathbf{Grid}_1^{\geq \mathrm{dgnl}}(\mathbf{C})^{adm}_{vert;horiz})_{/\underline{\mathbf{d}}^i_n}$.

Now, the fact that the latter colimit identifies with (5.4) is the calculation carried out in Sect. 3.1.

5.1.2. Consider the index category in (5.2), i.e., $({}'\mathbf{Grid}_n^{\geq \mathrm{dgnl}}(\mathbf{C})^{adm}_{vert;horiz})_{/\underline{\mathbf{d}}_n}$. It is a co-Cartesian fibration over

$$\underline{\gamma} := (\mathbf{c}_{0,0} \overset{\gamma_{0,0}}{\longrightarrow} \mathbf{d}_{0,0}), ..., (\mathbf{c}_{n,n} \overset{\gamma_{n,n}}{\longrightarrow} \mathbf{d}_{n,n}).$$

Hence, the map in (5.2) can be written as a composite of

$$\underset{\underline{\gamma}}{\operatorname{colim}}\, \underset{\underline{\mathbf{c}}_n}{\operatorname{colim}}\, r^*_{\mathbb{S}} \circ \Phi_n(\underline{\mathbf{c}}_n) \to \underset{\underline{\gamma}}{\operatorname{colim}}\, r^*_{\mathbb{S}}(\underset{\underline{\mathbf{c}}_n}{\operatorname{colim}} \Phi_n(\underline{\mathbf{c}}_n)) \tag{5.7}$$

and

$$\underset{\underline{\gamma}}{\operatorname{colim}}\, r^*_{\mathbb{S}}(\underset{\underline{\mathbf{c}}_n}{\operatorname{colim}} \Phi_n(\underline{\mathbf{c}}_n)) \to r^*_{\mathbb{S}}(\underset{\underline{\gamma}}{\operatorname{colim}}\, \underset{\underline{\mathbf{c}}_n}{\operatorname{colim}} \Phi_n(\underline{\mathbf{c}}_n)), \tag{5.8}$$

where now $\underline{\mathbf{c}}_n$ runs over the fiber category

$$(({}'\mathbf{Grid}_n^{\geq \mathrm{dgnl}}(\mathbf{C})^{adm}_{vert;horiz})_{/\underline{\mathbf{d}}_n})_{\underline{\gamma}}.$$

We will show that each of the maps (5.7) and (5.8) is an isomorphism.

5.1.3. We start with the following observation:

For a given $\underline{\gamma}$, let $\underline{\mathbf{d}}'_n$ be the object of ${}'\mathbf{Grid}_n^{\geq \mathrm{dgnl}}(\mathbf{D})^{adm}_{vert;horiz}$ with

$$\mathbf{d}'_{i,i} = \mathbf{c}_{i,i} \text{ and } \mathbf{d}'_{i,i+1} = \mathbf{c}_{i+1,i+1} \underset{\mathbf{d}_{i+1,i+1}}{\times} \mathbf{d}_{i,i+1} \underset{\mathbf{d}_{i,i}}{\times} \mathbf{c}_{i,i};$$

the other coordinates of $\underline{\mathbf{d}}'$ are uniquely determined by the condition that the inner squares should be Cartesian.

Consider the category $({}'\mathbf{Grid}_n^{\geq \mathrm{dgnl}}(\mathbf{C})^{adm}_{vert;horiz})_{/\underline{\mathbf{d}}'_n}$. Note, however, that cofinal in this category is the full subcategory consisting of $\underline{\mathbf{c}}'_n \to \underline{\mathbf{d}}'_n$ with $\mathbf{c}'_{i,i} = \mathbf{c}_{i,i}$ for $i = 0, ..., n$. And note that this subcategory identifies tautologically with the fiber category $(({}'\mathbf{Grid}_n^{\geq \mathrm{dgnl}}(\mathbf{C})^{adm}_{vert;horiz})_{/\underline{\mathbf{d}}_n})_{\underline{\gamma}}$ appearing in the colimits (5.7) and (5.8).

Hence, we need to show that the maps

$$\underset{\underline{\gamma}}{\operatorname{colim}}\, \underset{\underline{\mathbf{c}}_n}{\operatorname{colim}}\, r^*_{\mathbb{S}} \circ \Phi_n(\underline{\mathbf{c}}_n) \to \underset{\underline{\gamma}}{\operatorname{colim}}\, r^*_{\mathbb{S}}(\underset{\underline{\mathbf{c}}_n}{\operatorname{colim}} \Phi_n(\underline{\mathbf{c}}_n)) \tag{5.9}$$

and

$$\underset{\underline{\gamma}}{\operatorname{colim}}\, r^*_{\mathbb{S}}(\underset{\underline{\mathbf{c}}_n}{\operatorname{colim}} \Phi_n(\underline{\mathbf{c}}'_n)) \to r^*_{\mathbb{S}}(\underset{\underline{\gamma}}{\operatorname{colim}}\, \underset{\underline{\mathbf{c}}_n}{\operatorname{colim}} \Phi_n(\underline{\mathbf{c}}'_n)), \tag{5.10}$$

are isomorphisms, where $\underline{\mathbf{c}}'_n$ runs over the category $({}'\mathbf{Grid}_n^{\geq \mathrm{dgnl}}(\mathbf{C})^{adm}_{vert;horiz})_{/\underline{\mathbf{d}}'_n}$.

5.1.4. We begin by showing that the map (5.10) is an isomorphism.

According to Sect. 5.1.1, the colimit $\underset{\underline{\mathbf{c}}_n}{\operatorname{colim}} \Phi_n(\underline{\mathbf{c}}_n))$, which is a string

$$\Psi(\mathbf{c}_{0,0}) \to \Psi(\mathbf{c}_{1,1}) \to \ldots \to \Psi(\mathbf{c}_{n,n}),$$

is given by

$$\Psi(\mathbf{d}'_{i-1,i-1}) \overset{\Psi(\beta'_{i-1,i}) \circ \Psi^!(\alpha'_{i,i-1})}{\longrightarrow} \Psi(\mathbf{d}'_{i,i}). \tag{5.11}$$

Now, as in Sect. 3.1.4, by the Beck-Chevalley conditions, for each $i = 1, ..., n$, the composition $\Psi(\beta'_{i-1,i}) \circ \Psi^!(\alpha'_{i,i-1})$ identifies with

$$\Psi^!(\gamma_i) \circ \Psi(\beta_{i-1,i}) \circ \Psi^!(\alpha_{i,i-1}) \circ \Psi(\gamma_{i-1}).$$

Hence, the fact that (5.10) is an isomorphism follows from Corollary 2.3.7(b). Here, the corresponding categories

$$(\mathbf{c}_{i,i} \overset{\gamma_{i,i}}{\to} \mathbf{d}_{i,i}) = (\mathbf{C}_{adm})_{/\mathbf{d}_{i,i}}$$

are contractible by Lemma 4.1.2.

5.1.5. We will now show that the map (5.9) is an isomorphism. In fact, we claim that the map

$$\underset{\underline{\mathbf{c}}_n}{\operatorname{colim}}\, r_{\mathbb{S}}^* \circ \Phi_n(\underline{\mathbf{c}}_n) \to r_{\mathbb{S}}^*(\underset{\underline{\mathbf{c}}_n}{\operatorname{colim}} \Phi_n(\underline{\mathbf{c}}_n)) \tag{5.12}$$

is already an isomorphism.

Note, however, that the map (5.12) is the map (5.2) for $\underline{\mathbf{d}}_n$ replaced by $\underline{\mathbf{d}}'_n$. I.e., we have reduced the original problem to the case when $\mathbf{d}_{i,i} \in \mathbf{C}$.

We note that in this case the category $('\mathbf{Grid}_n^{\geq \mathrm{dgnl}}(\mathbf{C})^{adm}_{vert;horiz})_{/\underline{\mathbf{d}}_n}$ identifies with the product

$$(\mathbf{C}_{adm})_{/\mathbf{d}_{0,1}} \times \ldots \times (\mathbf{C}_{adm})_{/\mathbf{d}_{i-1,i}} \times \ldots . \times (\mathbf{C}_{adm})_{/\mathbf{d}_{n-1,n}}.$$

For every $\underline{\mathbf{c}} \in ('\mathbf{Grid}_n^{\geq \mathrm{dgnl}}(\mathbf{C})^{adm}_{vert;horiz})_{/\underline{\mathbf{d}}_n}$, the object $r_{\mathbb{S}}^* \circ \Phi_n(\underline{\mathbf{c}}_n)$, which is a map

$$\Psi(\mathbf{d}_{0,0}) \to \Psi(\mathbf{d}_{n,n})$$

equals the composite

$$(\Psi(\beta_{n-1,n}) \circ \Psi(\gamma_{n-1,n}) \circ \Psi^!(\gamma_{n-1,n}) \circ \Psi^!(\alpha_{n,n-1})) \circ \ldots \circ \\ \circ \ldots \circ (\Psi(\beta_{0,1}) \circ \Psi(\gamma_{0,1}) \circ \Psi^!(\gamma_{0,1}) \circ \Psi^!(\alpha_{1,0})). \tag{5.13}$$

Hence, as in Sect. 3.1.3, the left-hand side in (5.12) identifies with

$$(\Psi(\beta_{n-1,n}) \circ \Psi^!(\alpha_{n,n-1})) \circ \ldots \circ (\Psi(\beta_{0,1}) \circ \Psi^!(\alpha_{1,0})). \tag{5.14}$$

Now, according to Sect. 5.1.1, the right-hand side in (5.12) also identifies with (5.14). By unwinding the constructions, it is easy to see that the map in (5.12) corresponds to the identity endomorphism on (5.14) in terms of the above identifications.

5.2. Proof of Proposition 1.2.5(a): the degeneracy map. To finish the proof of Proposition 1.2.5(a), we need to treat the cases specified in Sect. 4.1.5. In this subsection we will consider the degeneracy map $[1] \to [0]$.

5.2.1. Fix an object $\mathbf{d} \in \mathbf{D} = {}'\mathbf{Grid}_0^{\geq \mathrm{dgnl}}(\mathbf{C})_{vert;horiz}^{adm}$. We need to show that the map

$$\underset{\mathbf{c},\gamma:\mathbf{c}\to\mathbf{d}}{\mathrm{colim}}\, \Phi_1(r_{\mathbf{C}}^*(\mathbf{c})) \to \underset{\underline{\mathbf{c}}_1,\underline{\mathbf{c}}_1\to r_{\mathbf{D}}^*(\mathbf{d})}{\mathrm{colim}}\, \Phi_1(\underline{\mathbf{c}}_1)$$

is an isomorphism.

5.2.2. We compose the above map with the isomorphism

$$\underset{\underline{\mathbf{c}}_1,\underline{\mathbf{c}}_1\to r_{\mathbf{D}}^*(\mathbf{d})}{\mathrm{colim}}\, \Phi_1(\underline{\mathbf{c}}_1) \simeq \mathrm{id}_{\Psi(\mathbf{d})}$$

of Sect. 3.1.

So, we need to show that the map

$$\underset{\mathbf{c},\gamma:\mathbf{c}\to\mathbf{d}}{\mathrm{colim}}\, \Phi_1(r_{\mathbf{C}}^*(\mathbf{c})) \simeq \underset{\mathbf{c},\gamma:\mathbf{c}\to\mathbf{d}}{\mathrm{colim}}\, \mathrm{id}_{\Phi(\mathbf{c})} \to \mathrm{id}_{\Psi(\mathbf{d})}$$

is an isomorphism.

However, this follows from Proposition 2.3.4.

5.3. Proof of Proposition 1.2.5(a): the main case. The case we now need to consider is that of r being the map $[1] \to [n]$, $n > 1$, with $0 \mapsto 0$ and $1 \mapsto n$.

5.3.1. Fix an object $\underline{\mathbf{d}}_n \in {}'\mathbf{Grid}_n^{\geq \mathrm{dgnl}}(\mathbf{D})_{vert;horiz}^{adm}$. We need to show that the map

$$\underset{\underline{\mathbf{c}}_n,F_n(\underline{\mathbf{c}}_n)\to\underline{\mathbf{d}}_n}{\mathrm{colim}}\, \Phi_1(r_{\mathbf{C}}^*(\underline{\mathbf{c}}_n)) \to \underset{\underline{\mathbf{c}}_1,F_1(\underline{\mathbf{c}}_1)\to r_{\mathbf{D}}^*(\underline{\mathbf{d}}_n)}{\mathrm{colim}}\, \Phi_1(\underline{\mathbf{c}}_1) \tag{5.15}$$

is an isomorphism.

Consider the maps

$$\alpha_{n,0} : \mathbf{d}_{n,0} \to \mathbf{d}_{0,0} \text{ and } \beta_{0,n} : \mathbf{d}_{n,0} \to \mathbf{d}_{n,n}.$$

By Sect. 3.1, the right-hand side in (5.15), which is a 1-morphism $\Psi(\mathbf{d}_{0,0}) \to \Psi(\mathbf{d}_{n,n})$, identifies with

$$\Psi(\beta_{0,n}) \circ \Psi^!(\alpha_{n,0}).$$

By Proposition 2.3.4, the left-hand side in (5.15) identifies with

$$\underset{\underline{\mathbf{c}}_n,F_n(\underline{\mathbf{c}}_n)\to\underline{\mathbf{d}}_n}{\mathrm{colim}}\, \Psi(\beta_{0,n}) \circ \Psi(\gamma_{n,0}) \circ \Psi^!(\gamma_{n,0}) \circ \Psi^!(\alpha_{n,0}).$$

5.3.2. Thus, it suffices to show that the map

$$\underset{\underline{\mathbf{c}}_n,F_n(\underline{\mathbf{c}}_n)\to\underline{\mathbf{d}}_n}{\mathrm{colim}}\, \Psi(\gamma_{n,0}) \circ \Psi^!(\gamma_{n,0}) \to \mathrm{id}_{\Psi(\mathbf{d}_{n,0})}$$

is an isomorphism.

However, this follows by induction from Condition (*) in Sect. 1.1.6.

6. Functors obtained by horizontal extension

In this section we prove the second extension result in this chapter. A typical situation that it applies to is when we start with the functor

$$\mathrm{IndCoh}_{\mathrm{Corr}(\mathrm{Sch}_{\mathrm{aft}})_{\mathrm{all;all}}^{\mathrm{proper}}} : \mathrm{Corr}(\mathrm{Sch}_{\mathrm{aft}})_{\mathrm{all;all}}^{\mathrm{proper}} \to \mathrm{DGCat}_{\mathrm{cont}},$$

and we want to extend it to a functor

$$\mathrm{IndCoh}_{\mathrm{Corr}(\mathrm{PreStk}_{\mathrm{laft}})_{\mathrm{sch;all}}^{\mathrm{sch\ \&\ proper}}} : \mathrm{Corr}(\mathrm{PreStk}_{\mathrm{laft}})_{\mathrm{sch;all}}^{\mathrm{sch\ \&\ proper}} \to \mathrm{DGCat}_{\mathrm{cont}}.$$

6.1. Set up for the horizontal extension. In this subsection we formulate the main result of this section, Theorem 6.1.5.

6.1.1. Let $(\mathbf{C}, vert, horiz, adm)$ and $(\mathbf{D}, vert, horiz, adm)$ both be as in Chapter 7, Sect. 1.1.1, and let $F : \mathbf{C} \to \mathbf{D}$ be a functor that preserves the corresponding classes of 1-morphisms, i.e., that it gives rise to well-defined functors

$$F_{vert} : \mathbf{C}_{vert} \to \mathbf{D}_{vert},\ F_{horiz} : \mathbf{C}_{horiz} \to \mathbf{D}_{horiz} \text{ and } F_{adm} : \mathbf{C}_{adm} \to \mathbf{D}_{adm}.$$

Furthermore, suppose that F takes Cartesian squares in Chapter 7, Diagram (1.1) to Cartesian squares. Hence, F induces a functor

$$F^{adm}_{vert;horiz} : \mathrm{Corr}(\mathbf{C})^{adm}_{vert;horiz} \to \mathrm{Corr}(\mathbf{D})^{adm}_{vert;horiz}.$$

6.1.2. Now, suppose we have a functor

$$\Phi^{adm}_{vert;horiz} : \mathrm{Corr}(\mathbf{C})^{adm}_{vert;horiz} \to \mathbb{S},$$

where $\mathbb{S}$ is a $(\infty, 2)$-category, and let

$$\Phi^! := \Phi^{adm}_{vert;horiz}|_{(\mathbf{C}_{horiz})^{\mathrm{op}}} : (\mathbf{C}_{horiz})^{\mathrm{op}} \to \mathbb{S}^{1\text{-Cat}}.$$

Our interest in this section the *right Kan extension* of $\Phi^{adm}_{vert;horiz}$ under $F^{adm}_{vert;horiz}$.

By definition, such a right Kan extension is a functor (if it exists)

$$\Psi^{adm}_{vert;horiz} : \mathrm{Corr}(\mathbf{D})^{adm}_{vert;horiz} \to \mathbb{S},$$

universal with respect to the property of being endowed with a natural transformation

$$\Psi^{adm}_{vert;horiz} \circ F^{adm}_{vert;horiz} \to \Phi^{adm}_{vert;horiz}.$$

6.1.3. We make the following assumptions on $\mathbb{S}$:

- The $(\infty, 1)$-category $\mathbb{S}^{1\text{-Cat}}$ admits limits;
- For every $s \in \mathbb{S}$ there exists and object $[1] \otimes s \in \mathbb{S}$ equipped with a functorial identification

$$\mathrm{Maps}([1] \otimes s, s') \simeq \mathrm{Maps}([1], \mathbf{Maps}_{\mathbb{S}}(s, s')).$$

respectively.

Note that in this case the conclusion of Lemma 2.2.2(b) is applicable to $\mathbb{S}$.

6.1.4. We do not intend to develop the general theory of right Kan extensions in the 2-categorical context. However, we will prove the following result:

THEOREM 6.1.5. *Assume that for any* $\mathbf{c} \in \mathbf{C}$ *the functor* F *induces an equivalence*

$$(\mathbf{C}_{vert})_{/\mathbf{c}} \to (\mathbf{D}_{vert})_{/F(\mathbf{c})}.$$

Then the right Kan extension

$$\mathrm{RKE}_{F^{adm}_{vert;horiz}}(\Phi^{adm}_{vert;horiz}) : \mathrm{Corr}(\mathbf{D})^{adm}_{vert;horiz} \to \mathbb{S}$$

exists and the natural maps

$$\mathrm{RKE}_{F^{adm}_{vert;horiz}}(\Phi^{adm}_{vert;horiz})|_{\mathrm{Corr}(\mathbf{D})_{vert;horiz}} \to \mathrm{RKE}_{F_{vert;horiz}}(\Phi_{vert;horiz}) \tag{6.1}$$

and

$$\mathrm{RKE}_{F_{vert;horiz}}(\Phi_{vert;horiz})|_{(\mathbf{D}_{horiz})^{\mathrm{op}}} \to \mathrm{RKE}_{(F_{horiz})^{\mathrm{op}}}(\Phi^!) \tag{6.2}$$

are isomorphisms.

The rest of this section is devoted to the proof of Theorem 6.1.5.

6.2. Proof of Theorem 6.1.5: the easy case. As a warm-up, we shall first prove the easy case of Theorem 6.1.5, namely, when adm = isom. In this case, the assertion of the theorem amounts to the isomorphism (6.2).

6.2.1. Fix an object $\mathbf{d} \in \mathbf{D}$. We need to show that the map

$$\lim_{\mathbf{c}\in(\mathrm{Corr}(\mathbf{C})_{vert;horiz})_{\mathbf{d}/}} \Phi(\mathbf{c}) \to \lim_{\mathbf{c}\in((\mathbf{C}_{horiz})_{/\mathbf{d}})^{\mathrm{op}}} \Phi(\mathbf{c})$$

is an isomorphism.

We claim that the functor of index categories, i.e.,

$$(\mathbf{C}_{horiz})_{/\mathbf{d}} \to ((\mathrm{Corr}(\mathbf{C})_{vert;horiz})_{\mathbf{d}/})^{\mathrm{op}}$$

is cofinal.

6.2.2. We claim that the above functor admits a left adjoint. Note that the objects of the category $(\mathrm{Corr}(\mathbf{C})_{vert;horiz})_{\mathbf{d}/}$ are diagrams

$$\begin{array}{ccc} \widetilde{\mathbf{d}} & \xrightarrow{\alpha} & \mathbf{d} \\ {\scriptstyle\beta}\big\downarrow & & \\ F(\mathbf{c}) & & \end{array}$$

with $\mathbf{c} \in \mathbf{C}$, $\alpha \in horiz$ and $\beta \in vert$.

However, the condition of the theorem implies that the vertical arrow

$$\widetilde{\mathbf{d}} \overset{\beta}{\to} F(\mathbf{c})$$

is of the form

$$F(\widetilde{\mathbf{c}}) \to F(\mathbf{c})$$

for a canonically defined $\widetilde{\mathbf{c}} \overset{\beta'}{\to} \mathbf{c}$ in $\mathbf{C}$. So, the above diagram has the form

$$\begin{array}{ccc} F(\widetilde{\mathbf{c}}) & \xrightarrow{\alpha} & \mathbf{d} \\ {\scriptstyle F(\beta')}\big\downarrow & & \\ F(\mathbf{c}). & & \end{array}$$

The left adjoint in question sends such a diagram to

$$F(\widetilde{\mathbf{c}}) \to \mathbf{d}.$$

6.3. Proof of Theorem 6.1.5: the principle. We do not intend to tackle the general theory of right Kan extensions in $(\infty, 2)$-categories. However, it turns out that under the condition of the theorem, the 2-categorical right Kan extension amounts to the 1-categorical one.

6.3.1. We consider the following general paradigm. Let

$$F : \mathbb{T}_1 \to \mathbb{T}_2$$

be a functor between $(\infty, 2)$-categories, and let

$$\Phi : \mathbb{T}_1 \to \mathbb{S}$$

be another functor, where $\mathbb{S}$ satisfies the assumptions of Sect. 6.1.3.

6.3.2. We now make the following assumption on the functor F. For $t_2 \in \mathbb{T}_2$ let $\mathbf{I}$ denote the index category, whose objects are pairs

$$(t'_1 \in \mathbb{T}_1, g : t_2 \to F(t'_1)),$$

and whose morphisms are *commutative diagrams* (i.e., we only allow invertible 2-morphisms).

For an object $t_1 \in \mathbb{T}_1$, we have an $\mathbf{I}^{\mathrm{op}}$-diagram of categories

$$(t'_1 \in \mathbb{T}_1, g : t_2 \to F(t'_1)) \mapsto \mathrm{Maps}_{\mathbb{T}_1}(t'_1, t_1).$$

We have a naturally defined functor

$$\underset{(t'_1,g)\in\mathbf{I}^{\mathrm{op}}}{\mathrm{colim}} \mathrm{Maps}_{\mathbb{T}_1}(t'_1, t_1) \to \mathrm{Maps}_{\mathbb{T}_2}(t_2, F(t_1)), \quad (t'_1 \overset{f}{\to} t_1) \mapsto F(f) \circ g, \tag{6.3}$$

where the colmit is taken in 1-Cat.

We claim:

LEMMA 6.3.3. *Suppose that the functor* (6.3) *is an equivalence. Then*

$$\mathrm{RKE}_F(\Phi) : \mathbb{T}_2 \to \mathbb{S}$$

exists and the canonical map

$$\mathrm{RKE}_F(\Phi)|_{\mathbb{T}_2^{1\text{-Cat}}} \to \mathrm{RKE}_{F|_{\mathbb{T}_1^{1\text{-Cat}}}}(\Phi|_{\mathbb{T}_1^{1\text{-Cat}}})$$

is an isomorphism.

REMARK 6.3.4. Intuitively, the lemma says that 2-functoriality of

$$\mathrm{RKE}_{F|_{\mathbb{T}_1^{1\text{-Cat}}}}(\Phi|_{\mathbb{T}_1^{1\text{-Cat}}}) : \mathbb{T}_2^{1\text{-Cat}} \to \mathbb{S}^{1\text{-Cat}}$$

is already built in, because 2-morphisms between arrows

$$t_2 \to F(t_1)$$

all come from 2-morphisms in $\mathbb{T}_1$, and thus are encoded by the 2-functoriality of Φ.

6.4. Proof of Theorem 6.1.5: the general case. We will prove that (6.1) by applying Lemma 6.3.3 to our functor

$$F^{adm}_{vert;horiz} : \mathrm{Corr}(\mathbf{C})^{adm}_{vert;horiz} \to \mathrm{Corr}(\mathbf{D})^{adm}_{vert;horiz}.$$

6.4.1. Fix an object $\mathbf{d} \in \mathbf{D}$ and $\mathbf{c} \in \mathbf{C}$. Consider the corresponding category $\mathbf{I}$. First we note that as in Sect. 6.2, cofinal in $\mathbf{I}^{\mathrm{op}}$ is its full subcategory $\mathbf{I}'^{\mathrm{op}}$, consisting of horizontal morphisms $F(\mathbf{c}') \to \mathbf{D}$.

6.4.2. Thus, we need to show that the functor

$$\underset{\mathbf{c}', F(\mathbf{c}') \to \mathbf{d}}{\operatorname{colim}} \operatorname{Maps}_{\operatorname{Corr}(\mathbf{C})^{adm}_{vert;horiz}}(\mathbf{c}', \mathbf{c}) \to \operatorname{Maps}_{\operatorname{Corr}(\mathbf{C})^{adm}_{vert;horiz}}(\mathbf{d}, F(\mathbf{c}))$$

is an equivalence.

However, we claim that the above functor admits an explicit inverse: it sends

$$\begin{array}{ccc} \mathbf{d}' & \longrightarrow & \mathbf{d} \\ {\scriptstyle \beta_{\mathbf{d}}}\downarrow & & \\ F(\mathbf{c}) & & \end{array}$$

to the object $(F(\mathbf{c}') \to \mathbf{d}) \in \mathbf{I}'$ and

$$\begin{array}{ccc} \mathbf{c}' & \longrightarrow & \mathbf{c}' \\ {\scriptstyle \beta_{\mathbf{c}}}\downarrow & & \\ \mathbf{c}, & & \end{array}$$

where $\mathbf{d}' = F(\mathbf{c}')$, $\beta_{\mathbf{D}} = F(\beta_{\mathbf{c}})$. □

CHAPTER 9

The (symmetric) monoidal structure on the category of correspondences

Introduction

0.1. Why do we want it? The goal of this chapter is to provide a general framework for theorems along the lines that the functor

$$\text{IndCoh}_{\text{Corr}(\text{Sch}_{\text{aft}})} : \text{Corr}(\text{Sch}_{\text{aft}}) \to \text{DGCat}_{\text{cont}} \tag{0.1}$$

is symmetric monoidal.

Why do we care about this? There are at least two applications for having such a formalism.

0.1.1. The first application is the following. We show that a symmetric monoidal structure on (0.1) encodes the duality on IndCoh; in this case, Serre self-duality of $\text{IndCoh}(X)$ for $X \in \text{Sch}_{\text{aft}}$.

Namely, let $\mathbf{C}$ be an arbitrary category closed under finite products, and let

$$\Phi : \text{Corr}(\mathbf{C}) \to \mathbf{O}$$

be a functor, where $\mathbf{O}$ be a symmetric monoidal category.

In Sect. 2.1.4 we show that the symmetric monoidal structure on $\mathbf{C}$, given by Cartesian products, gives rise to a symmetric monoidal structure on $\text{Corr}(\mathbf{C})$. Assume now that the functor Φ is endowed with a symmetric monoidal structure.

It then follows from Proposition 2.3.4 that for every $\mathbf{c} \in \mathbf{C}$, the object $\Phi(\mathbf{c}) \in \mathbf{O}$ is canonically self-dual, so that the duality datum is provided by applying Φ to the 1-morphisms

$$\begin{array}{ccc} \mathbf{c} & \longrightarrow & \mathbf{c} \times \mathbf{c} \\ \downarrow & & \\ * & & \end{array}$$

and

$$\begin{array}{ccc} \mathbf{c} & \longrightarrow & * \\ \downarrow & & \\ \mathbf{c} \times \mathbf{c}. & & \end{array}$$

In particular, for map $\mathbf{c}_1 \xrightarrow{f} \mathbf{c}_2$ in $\mathbf{C}$, the maps

$$\Phi(\mathbf{c}_1) \rightleftarrows \Phi(\mathbf{c}_2)$$

in $\mathbf{O}$, given by the diagrams

$$\begin{array}{ccc} \mathbf{c}_1 & \xrightarrow{\text{id}} & \mathbf{c}_1 \\ {\scriptstyle f}\downarrow & & \\ \mathbf{c}_2 & & \end{array}$$

and

$$\begin{array}{ccc} \mathbf{c}_1 & \xrightarrow{f} & \mathbf{c}_2 \\ {\scriptstyle \text{id}}\downarrow & & \\ \mathbf{c}_1 & & \end{array}$$

are the duals of each other.

Going back to the example of

$$\mathbf{C} = \text{Sch}_{\text{aft}},\ \mathbf{O} = \text{DGCat}_{\text{cont}} \text{ and } \Phi = \text{IndCoh}_{\text{Corr}(\text{Sch}_{\text{aft}})},$$

the resulting identification

$$\text{IndCoh}(X)^\vee \simeq \text{IndCoh}(X)$$

is the ind-extension of the Serre duality anti-equivalence $\text{Coh}(X)^{\text{op}} \to \text{Coh}(X)$.

0.1.2. Another application is the following. We show that a monoidal structure on (0.1) encodes the formation of *convolution categories.*

Let $\mathbf{C}$ be a category that admits finite limits. Let $\mathbf{c} \in \mathbf{C}$ be an object, and let $\mathbf{c}^\bullet \in \mathbf{C}^{\mathbf{\Delta}^{\text{op}}}$ be a *Segal object* [1] acting on $\mathbf{c}$. I.e., we have an identification $\mathbf{c}^0 = \mathbf{c}$ and we require that for any $n \geq 2$, the map

$$\mathbf{c}^1 \underset{\mathbf{c}}{\times} \ldots \underset{\mathbf{c}}{\times} \mathbf{c}^1,$$

given by the product of the maps

$$[1] \to [n], \quad 0 \mapsto i, 1 \mapsto i+1, \quad i = 0, \ldots, n-1,$$

be an isomorphism.

In Theorem 4.4.2 we show that $\mathbf{c}^1$, regarded as an object of $\text{Corr}(\mathbf{C})$, has a natural structure of associative algebra (with respect to the (symmetric) monoidal structure on $\text{Corr}(\mathbf{C})$), where the binary operation on $\mathbf{c}^1$ is given by the diagram

$$\begin{array}{ccc} \mathbf{c}^2 & \longrightarrow & \mathbf{c}^1 \times \mathbf{c}^1 \\ \downarrow & & \\ \mathbf{c}^1, & & \end{array}$$

in which the vertical map is given by the active map $[1] \to [2]$, and the horizontal map is given by the product of the two inert maps $[1] \to [2]$.

In particular, taking $\mathbf{C} = \text{Sch}_{\text{aft}}$ and applying the (symmetric) monoidal functor

$$\text{IndCoh}_{\text{Corr}(\text{Sch}_{\text{aft}})} : \text{Corr}(\text{Sch}_{\text{aft}}) \to \text{DGCat}_{\text{cont}},$$

[1] Alternative terminology: *category-object.*

we obtain that for a Segal object $X^\bullet$ in the category of schemes, the category $\text{IndCoh}(X^1)$ is endowed with a monoidal structure, given by *convolution*. I.e., it is given by pull-push along the diagram

$$\begin{array}{ccc} X_1 \underset{X_0}{\times} X_1 & \longrightarrow & X^1 \times X^1 \\ \downarrow & & \\ X^1. & & \end{array}$$

0.2. What is done in this Chapter?

0.2.1. In Sect. 1 we make a general review, following [**Lu2**], of 'what it means to be (symmetric) monoidal'.

First, we define the notion of (commutative) monoid in an ∞-category.

As a result we obtain the notions of (symmetric) monoidal $(\infty,1)$-category and $(\infty,2)$-category.

We also review the notions of right-lax and left-lax (symmetric) monoidal functors between (symmetric) monoidal $(\infty,1)$-categories and $(\infty,2)$-categories. The latter leads to the notion of (commutative) algebra object in a (symmetric) monoidal $(\infty,1)$-category.

0.2.2. In Sect. 2, we show that if an $(\infty,1)$-category $\mathbf{C}$ has a (symmetric) monoidal structure, and $vert, horiz, adm$ are three classes of objects, preserved by the monoidal operation, then the $(\infty,2)$-category $\text{Corr}(\mathbf{C})^{adm}_{vert;horiz}$ acquires a (symmetric) monoidal structure.

In the applications, we will take $\mathbf{C}$ endowed with the *Cartesian* symmetric monoidal structure.

In Sect. 2.2 we show that the $(\infty,1)$-category

$$\text{Corr}(\mathbf{C}) := \text{Corr}(\mathbf{C})^{\text{isom}}_{\text{all;all}}$$

is endowed with a canonical anti-involution, given by swapping the roles of the vertical and horizontal arrows.

We prove that this anti-involution is canonically isomorphic to the *dualization* functor, when $\text{Corr}(\mathbf{C})$ is considered as a symmetric monoidal category.

0.2.3. In Sect. 3 we show that the extension results of Chapter 7, Sects. 4 and 5 and Chapter 8 carry through to the (symmetric) monoidal world.

0.2.4. In Sect. 4 we give the following two constructions, starting from a Segal object $\mathbf{c}^\bullet$ in an $(\infty,1)$-category $\mathbf{C}$, acting on $\mathbf{c} = \mathbf{c}^0 \in \mathbf{C}$.

In the first construction (which does *not* appeal to the symmetric monoidal structure on the category of correspondences), we show that a Segal object as above defines an algebra object in the monoidal category

$$\mathbf{Maps}_{\text{Corr}(\mathbf{C})^{adm}_{vert;horiz}}(\mathbf{c},\mathbf{c})$$

of endomorphisms of $\mathbf{c}$ in the $(\infty,2)$-category $\text{Corr}(\mathbf{C})^{adm}_{vert;horiz}$.

In the second construction (which does talk about the symmetric monoidal structure on the category of correspondences), we show that the object $\mathbf{c}^1$ has a natural structure of associative algebra in $\text{Corr}(\mathbf{C})_{vert;horiz}$.

1. (Symmetric) monoidal structures: recollections

In this section we review, mostly following [**Lu2**], and partly repeating the material of Chapter 1, Sect. 3, the notions of (commutative) monoid (in a given $(\infty,1)$-category), (symmetric) monoidal $(\infty,1)$-category, and (symmetric) monoidal $(\infty,2)$-category.

For a usual category $\mathcal{C}$, a monoid in it is an object $c \in \mathcal{C}$ equipped with a product operation $c \times c \to c$ and a unit map $* \to c$ that satisfy the usual axioms (in fact, three altogether).

The main feature in the ∞-setting is that if we were to imitate this definition when $\mathcal{C}$ is an $(\infty,1)$-category, in addition to the above binary operation we will need to supply a whole tail of higher operations (e.g., a homotopy between the two tertiary operations $c^{\times 3} \rightrightarrows c$), and axioms on the compatibilities between them. The problem is that this becomes too unwieldy to work with.

The main idea is that when defining monoids, instead of specifying just one object c, we specify the entire datum of its products and maps between them. Such a data is encoded by just one functor $\mathbf{\Delta}^{\mathrm{op}} \to \mathcal{C}$, where the original c is the value of our functor on $[1] \in \mathbf{\Delta}^{\mathrm{op}}$. This functor must satisfy some obvious condition (that expresses the fact that its value on $[n] \in \mathbf{\Delta}^{\mathrm{op}}$ is the the product of n copies of c). This description was first formulated in [**Seg**].

The above approach to the definition creates a very convenient framework for working with monoids (and the related notions of monoidal categories, algebras in them, etc.). It also allows for an immediate generalization in the world of $(\infty,2)$-categories.

1.1. Monoids and commutative monoids.

In this subsection we recall the notions of monoid and commutative monoid in the setting of $(\infty,1)$-categories.

1.1.1. Let $\mathcal{C}$ be an $(\infty,1)$-category with finite products (including the empty finite product, i.e., a final object). One can then talk about *monoids* in $\mathcal{C}$. By definition, they form a full subcategory, denoted $\mathrm{Monoid}(\mathcal{C})$, in $\mathcal{C}^{\mathbf{\Delta}^{\mathrm{op}}}$, consisting of objects $\mathcal{R}^\bullet$, such that $\mathcal{R}^0 = *_{\mathcal{C}}$, and such that for any $n \geq 2$, the map

$$\mathcal{R}^n \to \mathcal{R}^1 \times ... \times \mathcal{R}^1, \tag{1.1}$$

given by the product of the maps

$$[1] \to [n], \quad 0 \mapsto i, 1 \mapsto i+1, \quad i = 0, ..., n-1,$$

is an isomorphism.

1.1.2. Similarly, one can talk about *commutative monoids*, denoted $\mathrm{ComMonoid}(\mathcal{C})$ in $\mathcal{C}$. Instead of $\mathbf{\Delta}^{\mathrm{op}}$ we use the category Fin_* of pointed finite sets. The condition now is that $\mathcal{R}^{\{*\}} = *_{\mathcal{C}}$ and for every $(* \in I) \in \mathrm{Fin}_*$ the map

$$\mathcal{R}^I \to \prod_{i \in I - \{*\}} \mathcal{R}^{(* \in \{* \sqcup i\})} \tag{1.2}$$

is an isomorphism.

1.1.3. Recall that we have a canonically defined functor $\mathbf{\Delta}^{\mathrm{op}} \to \mathrm{Fin}_*$, see Chapter 1, Sect. 3.3.2.

Pre-composing, we obtain the forgetful functor

$$\mathrm{ComMonoid}(\mathcal{C}) \to \mathrm{Monoid}(\mathcal{C}).$$

In what follows we will focus on the symmetric monoidal case, while the monoidal case can be treated similarly.

1.2. Symmetric monoidal categories. In this subsection we recall the notion of symmetric monoidal $(\infty,1)$-category, see Chapter 1, Sect. 3.3, from a slightly different perspective.

1.2.1. Applying Sect. 1.1 to $\mathcal{C} = 1\text{-Cat}$, we obtain the notion of *symmetric monoidal category*. Unstraightening associates to a symmetric monoidal category a coCartesian fibration

$$\mathcal{C}^{\otimes,\mathrm{Fin}_*} \to \mathrm{Fin}_* \tag{1.3}$$

and also a Cartesian fibration

$$\mathcal{C}^{\otimes,\mathrm{Fin}_*^{\mathrm{op}}} \to \mathrm{Fin}_*^{\mathrm{op}}. \tag{1.4}$$

The data of a symmetric monoidal category is equivalent to either (1.3) and (1.4) satisfying the condition that $\mathcal{C}^{\{*\}} = *$ and for every $(* \in I) \in \mathrm{Fin}_*$ the corresponding functor

$$\mathcal{C}^I \to \prod_{i \in I - \{*\}} \mathcal{C}^{(* \in \{* \sqcup i\})} \tag{1.5}$$

is an equivalence.

1.2.2. By a symmetric monoidal functor we shall mean a 1-morphism in the category

$$\mathrm{ComMonoid}(1\text{-Cat}).$$

Equivalently, this is a functor over Fin_*

$$\mathcal{C}_1^{\otimes,\mathrm{Fin}_*} \to \mathcal{C}_2^{\otimes,\mathrm{Fin}_*} \tag{1.6}$$

that sends coCartesian arrows to coCartesian arrows, and still equivalently, a functor over $(\mathrm{Fin}_*)^{\mathrm{op}}$

$$\mathcal{C}_1^{\otimes,\mathrm{Fin}_*^{\mathrm{op}}} \to \mathcal{C}_2^{\otimes,\mathrm{Fin}_*^{\mathrm{op}}} \tag{1.7}$$

that sends Cartesian arrows to Cartesian arrows.

1.2.3. We shall say that a map $(* \in I) \to (* \in J)$ is *inert* (resp., *idle*) if any element in $J - \{*\}$ has exactly (resp. at most) one preimage.

By a *right-lax* symmetric monoidal functor between $\mathcal{C}_1$ and $\mathcal{C}_2$ we shall mean a functor as in (1.6) that is only required to send coCartesian arrows that lie over *idle* maps in Fin_* to coCartesian arrows.

By a *non-unital right-lax* symmetric monoidal functor between $\mathcal{C}_1$ and $\mathcal{C}_2$ we shall mean a functor as in (1.6) that is only required to send coCartesian arrows that lie over *inert* maps in Fin_* to coCartesian arrows.

Similarly, we obtain that notions of *left-lax* and *non-unital left-lax* symmetric monoidal functors: use (1.7) instead of (1.6) and 'Cartesian' instead of 'coCartesian'.

1.2.4. By a *commutative algebra* in a symmetric monoidal category $\mathcal{C}$ we shall mean a non-unital right-lax symmetric monoidal functor $* \to \mathcal{C}$.

We let $\text{ComAlg}(\mathcal{C})$ denote the category of commutative algebras in $\mathcal{C}$. (Note that these are the *unital* commutative algebras!)

By construction, a right-lax monoidal functor $\mathcal{C}_1 \to \mathcal{C}_2$ induces a functor

$$\text{ComAlg}(\mathcal{C}_1) \to \text{ComAlg}(\mathcal{C}_2).$$

1.3. The Cartesian symmetric monoidal structure. Let $\mathcal{C}$ be again an $(\infty, 1)$-category with finite products.

It is then intuitively clear that the operation of Cartesian product defines on $\mathcal{C}$ a symmetric monoidal structure, called the *Cartesian symmetric monoidal structure.* We will formalize this in the present subsection, following [**Lu2**, Sect. 2.4.1].

1.3.1. We start with the functor

$$\text{Fin}_*^{\text{op}} \to 1\text{-Cat}, \quad (* \in I) \mapsto \mathcal{C}^{I-\{*\}}, \tag{1.8}$$

see Chapter 1, Sect. 3.3.3.

Straightening defines a Cartesian fibration

$$\mathcal{C}^{\times,\text{Fin}_*} \to \text{Fin}_*. \tag{1.9}$$

However, the condition that $\mathcal{C}$ admits finite products implies that (1.9) is also a coCartesian fibration, thereby giving rise to the datum as in (1.3).

It is clear that the functors in (1.5) are equivalences. Hence, (1.9) corresponds to a canonically defined symmetric monoidal structure on $\mathcal{C}$. This is the Cartesian symmetric monoidal structure.

1.3.2. Note that any functor $\mathcal{C}_1 \to \mathcal{C}_2$ gives rise to a functor over Fin_*

$$\mathcal{C}_1^{\times,\text{Fin}_*} \to \mathcal{C}_2^{\times,\text{Fin}_*}$$

and thus defines a left-lax functor from $\mathcal{C}_1$ to $\mathcal{C}_2$, when both are considered as equipped with the Cartesian symmetric monoidal structure.

This functor is (strictly) symmetric monoidal if and only if the initial functor $\mathcal{C}_1 \to \mathcal{C}_2$ commutes with finite products.

1.3.3. Note now that on the one hand, we have the notion of commutative monoid in $\mathcal{C}$, and on the other hand, we have the notion of commutative algebra in $\mathcal{C}$, considered as a symmetric monoidal category.

However, that these two notions coincide, i.e., the categories $\text{ComMonoid}(\mathcal{C})$ and $\text{ComAlg}(\mathcal{C})$ are canonically equivalent, see [**Lu2**, Proposition 2.4.2.5].

Let us see explicitly the functor

$$\text{ComAlg}(\mathcal{C}) \to \text{ComMonoid}(\mathcal{C}).$$

Indeed, the operation of assigning to $(* \in I) \in \text{Fin}_*$, the functor of Cartesian product along $I - \{*\}$

$$\mathcal{C}^{I-\{*\}} \to \mathcal{C}$$

defines a functor

$$\mathcal{C}^{\times,\text{Fin}_*} \to \mathcal{C} \times \text{Fin}_*$$

over Fin_*.

Given a section $\mathrm{Fin}_* \to \mathcal{C}^{\times,\mathrm{Fin}_*}$, we thus obtain a functor $\mathrm{Fin}_* \to \mathcal{C}$. It is easy to see that the requirement on the above section to be right-lax symmetric monoidal implies that the resulting object of $\mathcal{C}^{\mathrm{Fin}_*}$ is a commutative monoid.

1.4. Symmetric monoidal $(\infty,2)$-categories. In this subsection we introduce symmetric monoidal $(\infty,2)$-categories. We refer the reader to Chapter 10 for our conventions regarding $(\infty,2)$-categories.

1.4.1. The $(\infty,1)$-category 2-Cat has finite products. Hence, we can talk about commutative monoids in 2-Cat (or, equivalently, according to Sect. 1.3.3 above, about commutative algebras in 2-Cat with respect to the Cartesian symmetric monoidal structure).

Thus, we obtain the notion of *symmetric monoidal* $(\infty,2)$-*category.*

1.4.2. Applying the 2-categorical unstraigtening (see Chapter 11, Theorem 2.1.8), we can encode the datum of a symmetric monoidal $(\infty,2)$-category $\mathbb{S}$ by a 2-coCartesian fibration

$$\mathbb{S}^{\otimes,\mathrm{Fin}_*} \to \mathrm{Fin}_*, \tag{1.10}$$

or equivalently a 2-Cartesian fibration

$$\mathbb{S}^{\otimes,\mathrm{Fin}_*^{\mathrm{op}}} \to \mathrm{Fin}_*^{\mathrm{op}}. \tag{1.11}$$

As in the case of $(\infty,1)$-categories, the datum of a symmetric monoidal $(\infty,2)$-category is equivalent to that of (1.10) (or (1.11)) such that the corresponding functors

$$\mathbb{S}^I \to \prod_{i\in I-\{*\}} \mathbb{S}^{(*\in\{*\sqcup i\})}$$

are equivalences.

1.4.3. As in the case of $(\infty,1)$-categories, this leads to the notion of (resp., non-unital) right-lax symmetric monoidal functor $\mathbb{S}_1 \to \mathbb{S}_2$ between symmetric monoidal $(\infty,2)$-categories $\mathbb{S}_1$ and $\mathbb{S}_2$: this is a functor

$$\mathbb{S}_1^{\otimes,\mathrm{Fin}_*} \to \mathbb{S}_2^{\otimes,\mathrm{Fin}_*}$$

satisfying the same condition for idle (resp., inert) arrows in Fin_*.

1.4.4. Applying the above for $\mathbb{S}_1 = *$ and $\mathbb{S}_2 = \mathbb{S}$, we obtain the notion of commutative algebra object in $\mathbb{S}$.

Note, however, that since $*$ is a 1-category, commutative algebras in $\mathbb{S}$ (and homomorphisms between them) are the same as the corresponding notions for $\mathbb{S}^{1\text{-}\mathrm{Cat}}$

1.4.5. The feature of the 2-categorical situation is that, given right-lax (resp., non-unital) symmetric monoidal functors

$$\Phi', \Phi'' : \mathbb{S}_1 \to \mathbb{S}_2,$$

in addition to usual natural transformations between them equipped with a symmetric monoidal structure, one can talk about natural transformations between them equipped with a *right-lax* (resp., *left-lax*) symmetric monoidal structure.

By definition those are right-lax (resp., left-lax) natural transformations (see Chapter 10, Sect. 3.2.7 for what this means) between the corresponding functors

$$\Phi'^{\otimes,\mathrm{Fin}_*}, \Phi''^{\otimes,\mathrm{Fin}_*} : \mathbb{S}_1^{\otimes,\mathrm{Fin}_*} \rightrightarrows \mathbb{S}_2^{\otimes,\mathrm{Fin}_*},$$

that are strict over idle arrows in Fin_*.

One can also talk about natural transformations between them equipped with a *non-unital* right-lax (resp., left-lax) symmetric monoidal structure: replace the word 'idle' by 'inert' in the above definition.

Thus, given a symmetric monoidal $(\infty,2)$-category $\mathbb{S}$, we obtain the notion of *right-lax* (resp., *left-lax*) homomorphism

$$s' \to s''$$

between two commutative algebra objects in $\mathbb{S}$.

Similarly, we obtain the notion of *non-unital* right-lax (resp., left-lax) homomorphism.

1.4.6. As in the case of 1-Cat, if an $(\infty,2)$-category $\mathbb{S}$ has finite products, it acquires the Cartesian symmetric monoidal structure.

By Sects. 1.4.4 and 1.3.3, commutative algebra objects in $\mathbb{S}$ in the Cartesian symmetric monoidal structure are the same as commutative monads in $\mathbb{S}^{1\text{-Cat}}$.

1.4.7. Let us take $\mathbb{S} = \mathbf{1\text{-}Cat}$. By Sect. 1.4.6, we obtain that $\mathbf{1\text{-}Cat}$ is a symmetric monoidal $(\infty,2)$-category. Commutative algebra objects in it are the same as symmetric monoidal categories.

By unwinding the definitions, we obtain that given two symmetric monoidal categories $\mathcal{C}_1,\mathcal{C}_2$, regarded as commutative algebra objects in $\mathbf{1\text{-}Cat}$, *right-lax* (resp., *left-lax*) homomorphisms between them are the same as *right-lax* (resp., *left-lax*) symmetric monoidal functors

$$\mathcal{C}_1 \to \mathcal{C}_2$$

as defined in Sect. 1.2.3. Indeed, both identify with right-lax (resp., left-lax) natural transformations between functors

$$\mathrm{Fin}_* \to \mathbf{1\text{-}Cat}.$$

2. (Symmetric) monoidal structures and correspondences

Recall that in Chapter 7, Sect. 1 we associated to an $(\infty,1)$-category equipped with three classes of morphisms $vert, horiz, adm$ (satisfying some natural conditions) an $(\infty,2)$-category $\mathrm{Corr}(\mathbf{C})^{adm}_{vert;horiz}$.

The first observation in that a Cartesian symmetric monoidal structure on $\mathbf{C}$ induces a symmetric monoidal structure on $\mathrm{Corr}(\mathbf{C})^{adm}_{vert;horiz}$.

We will now show that for the $(\infty,1)$-category

$$\mathrm{Corr}(\mathbf{C}) = \mathrm{Corr}(\mathbf{C})^{\mathrm{isom}}_{\mathrm{all;all}},$$

the operation of *dualization* with respect to its symmetric monoidal structure (induced by the Cartesian symmetric monoidal structure on $\mathbf{C}$) can be interpreted as the anti-involution that swaps the roles of vertical and horizontal arrows.

2.1. The (symmetric) monoidal structure on the functor Corr.

Let $\mathbf{C}$ be an $(\infty,1)$-category, equipped with three classes of morphisms *vert*, *horiz*, *adm* as in Chapter 7, Sect. 1.1, so that we can form the $(\infty,2)$-category $\mathrm{Corr}(\mathbf{C})^{adm}_{vert;horiz}$.

Assume that $\mathbf{C}$ has finite products, and each of the above classes of morphisms is preserved by finite products.

In this subsection we show (which is completely tautological) that the $(\infty,2)$-category $\mathrm{Corr}(\mathbf{C})^{adm}_{vert;horiz}$ acquires a symmetric monoidal structure.

2.1.1. Let Trpl be the $(\infty,1)$-category whose objects are given by $(\infty,1)$-categories $\mathbf{C}$ together with three classes of 1-morphisms *vert*, *horiz* and *adm* as in Chapter 7, Sect. 1.1.

A 1-morphism $(\mathbf{C}_1, vert_1, horiz_1, adm_1) \to (\mathbf{C}_2, vert_2, horiz_2, adm_2)$ is a functor from $\mathbf{C}_1$ to $\mathbf{C}_2$ that preserves each of the three classes of 1-morphisms as well as the Cartesian squares from Chapter 7, Sect. 1.1.

We endow Trpl with the Cartesian symmetric monoidal structure.

2.1.2. The assignment

$$(\mathbf{C}, vert, horiz, adm) \mapsto {}''\mathbf{Grid}^{\geq \mathrm{dgnl}}(\mathbf{C})^{adm}_{vert;horiz} \tag{2.1}$$

is clearly a functor

$$\mathrm{Trpl} \to 1\text{-}\mathrm{Cat}^{\mathbf{\Delta}^{\mathrm{op}}},$$

whose essential image belongs to the essnetial image of the functor $\mathrm{Seq}_\bullet$.

Hence, we obtain that the assignment

$$(\mathbf{C}, vert, horiz, adm) \mapsto \mathrm{Corr}(\mathbf{C})^{adm}_{vert;horiz}$$

is a functor

$$\mathrm{Corr} : \mathrm{Trpl} \to 2\text{-}\mathrm{Cat}. \tag{2.2}$$

By Sect. 1.3.2, the above functor Corr carries a left-lax symmetric monoidal structure. However, it is easy to see that this left-lax symmetric structure is actually strict, e.g., because this is the case for the functors (2.1) and $\mathrm{Seq}_\bullet$.

Thus, we obtain that (2.2) has a natural symmetric monoidal structure.

2.1.3. Let $(\mathbf{C}, vert, horiz, adm)$ be an object of Trpl, and let $\mathbf{C}$ be endowed with a symmetric monoidal structure. Assume that each of the classes of the morphisms *vert*, *horiz*, and *adm* is preserved by the tensor product functor

$$\mathbf{C} \times \mathbf{C} \to \mathbf{C}.$$

Since the forgetful functor

$$\mathrm{Trpl} \to 1\text{-}\mathrm{Cat}$$

that remembers the underlying category $\mathbf{C}$ is 1-fully faithful, by Chapter 4, Lemma 2.2.7, the symmetric monoidal structure on $\mathbf{C} \in 1\text{-}\mathrm{Cat}$ gives rise to a symmetric monoidal structure on $(\mathbf{C}, vert, horiz, adm) \in \mathrm{Trpl}$.

Hence, we obtain that

$$\mathrm{Corr}(\mathbf{C})^{adm}_{vert;horiz} \in 2\text{-}\mathrm{Cat}$$

acquires a structure of symmetric monoidal $(\infty,2)$-category.

2.1.4. In our main application, we will work with the *Cartesian* symmetric monoidal structure on $\mathbf{C}$.

Thus, in this case we assume that the classes of 1-morphisms $(vert, horiz, adm)$ are preserved by finite products, and we obtain that

$$\mathrm{Corr}(\mathbf{C})^{adm}_{vert;horiz} \in 2\text{-Cat}$$

acquires a structure of symmetric monoidal $(\infty, 2)$-category.

In the sequel, unless explicitly stated otherwise, when discussing a symmetric monoidal structure on the $\mathrm{Corr}(\mathbf{C})^{adm}_{vert;horiz}$, we shall mean the one, coming from the Cartesian symmetric monoidal structure on $\mathbf{C}$.

2.2. The canonical anti-involution on the category of correspondences. In this subsection we show that the $(\infty, 1)$-category $\mathrm{Corr}(\mathbf{C})$ carries a canonical anti-involution, given swapping the roles of vertical and horizontal arrows.

We will also show that this anti-involution is canonically isomorphic to the dualization functor on $\mathrm{Corr}(\mathbf{C})$, when the latter is regarded as a symmetric monoidal $(\infty, 1)$-category.

2.2.1. Let us take $adm = \mathrm{isom}$, so that $\mathrm{Corr}(\mathbf{C})^{adm}_{vert;horiz} = \mathrm{Corr}(\mathbf{C})_{vert;horiz}$ is an $(\infty, 1)$-category.

Let us also take $vert = horiz = \mathrm{all}$. Note that in this case,

$$\mathrm{Corr}(\mathbf{C}) := \mathrm{Corr}(\mathbf{C})_{\mathrm{all,all}}$$

carries a canonical anti-involution, denoted ϖ.

At the level of objects ϖ acts as identity. At the level of 1-morphisms it sends

$$\begin{array}{ccc} \mathbf{c}_{1,0} & \xrightarrow{\alpha_0} & \mathbf{c}_0 \\ {\scriptstyle\alpha_1}\downarrow & & \\ \mathbf{c}_1 & & \end{array}$$

to

$$\begin{array}{ccc} \mathbf{c}_{1,0} & \xrightarrow{\alpha_1} & \mathbf{c}_1 \\ {\scriptstyle\alpha_0}\downarrow & & \\ \mathbf{c}_0. & & \end{array}$$

2.2.2. The formal definition is as follows. To define ϖ, we need to construct an involutive identification

$$\mathrm{Seq}_\bullet(\mathrm{Corr}(\mathbf{C})) \simeq \mathrm{Seq}_\bullet(\mathrm{Corr}(\mathbf{C})) \circ (\mathrm{rev})^{\mathrm{op}}, \tag{2.3}$$

where

$$\mathrm{rev} : \mathbf{\Delta} \to \mathbf{\Delta}$$

is the reversal involution on $\mathbf{\Delta}$, see Chapter 1, Sect. 1.1.9.

By definition,

$$\mathrm{Seq}_\bullet(\mathrm{Corr}(\mathbf{C})) = \mathrm{Grid}^{\geq \mathrm{dgnl}}_\bullet(\mathbf{C}),$$

and (2.3) comes from the involutive identification

$$([\bullet] \times [\bullet])^{\geq \mathrm{dgnl}} \simeq ([\bullet] \times [\bullet])^{\geq \mathrm{dgnl}} \circ \mathrm{rev}, \tag{2.4}$$

as functors $\mathbf{\Delta} \to 1\text{-Cat}_{\mathrm{ordn}}$, given by reflecting half-grids over the NW-SE diagonal.

I.e., for every $[n] \in \mathbf{\Delta}$, the corresponding involution on

$$([n] \times [n])^{\geq \mathrm{dgnl}}$$

is $(i,j) \mapsto (n-j, n-i)$.

2.2.3. Suppose now that the $(\infty,1)$-category $\mathbf{C}$, in addition, admits finite products.

In this case, by Sect. 2.1.4, the Cartesian symmetric monoidal structure on $\mathbf{C}$ induces a symmetric monoidal structure on $\mathrm{Corr}(\mathbf{C})$. The unit for this symmetric monoidal structure is the final object $* \in \mathbf{C}$, viewed as an object of $\mathrm{Corr}(\mathbf{C})$. (Note, however, that $*$ is *not* the final object in $\mathrm{Corr}(\mathbf{C})$.)

Unwinding the definitions, we see that in this case, ϖ has a natural structure of symmetric monoidal functor.

2.3. Relationship to the dualization functor. In this subsection we will show that the anti-involution ϖ, defined above, is the dualization functor on $\mathrm{Corr}(\mathbf{C})$, when the latter is regarded as a symmetric monoidal category.

2.3.1. Recall that if a symmetric monoidal category $\mathbf{O}$ is such that every object $\mathbf{o} \in \mathbf{O}$ is dualizable, there is a canonical anti-involution

$$\mathbf{O} \overset{\sim}{\to} \mathbf{O}^{\mathrm{op}} \tag{2.5}$$

that takes $\mathbf{o} \in \mathbf{O}$ to its monoidal dual $\mathbf{o}^\vee$, see Chapter 1, Sect. 4.1.4.

This dualization functor is characterized by an isomorphism

$$\mathrm{Maps}(\mathbf{o}, \mathbf{o}') \overset{\sim}{\to} \mathrm{Maps}(\mathbf{1}_{\mathbf{O}}, \mathbf{o}^\vee \otimes \mathbf{o}') \tag{2.6}$$

as functors $\mathbf{O}^{\mathrm{op}} \times \mathbf{O} \to \mathrm{Spc}$, see Chapter 1, Sect. 4.1.2.

2.3.2. We take $\mathbf{O} := \mathrm{Corr}(\mathbf{C})$, where the latter is regarded as a symmetric monoidal category by the procedure of Sect. 2.1.4.

Let us observe that every object $\mathbf{c} \in \mathrm{Corr}(\mathbf{C})$ is dualizable and in fact self-dual. The duality data is supplied by the 1-morphisms in $\mathrm{Corr}(\mathbf{C})$:

$$\begin{array}{ccc} \mathbf{c} & \longrightarrow & * \\ \downarrow & & \\ \mathbf{c} \times \mathbf{c} & & \end{array} \quad \text{and} \quad \begin{array}{ccc} \mathbf{c} & \longrightarrow & \mathbf{c} \times \mathbf{c} \\ \downarrow & & \\ * & & \end{array} \tag{2.7}$$

2.3.3. We will prove:

PROPOSITION 2.3.4. *The anti-involution*

$$\varpi : \mathrm{Corr}(\mathbf{C}) \to \mathrm{Corr}(\mathbf{C})^{\mathrm{op}}$$

is canonically isomorphic to the dualization functor.

2.3.5. *Variant.* Let $(\mathbf{C}, vert, horiz, adm)$ be an object of Trpl as in Sect. 2.1.4, with adm = isom and $vert = horiz$. Then the involution ϖ restricts to an involution on $\mathrm{Corr}(\mathbf{C})_{vert,horiz}$.

Assume now that for every $\mathbf{c} \in \mathbf{C}$, the diagonal map $\mathbf{c} \to \mathbf{c} \times \mathbf{c}$ and the tautological maps $\mathbf{c} \to *$ belong to $vert = horiz$. In this case every object of $\mathrm{Corr}(\mathbf{C})_{vert,horiz}$ is dualizable, and the assertion of Proposition 2.3.4 holds verbatim (indeed, replace the original $\mathbf{C}$ by $\mathbf{C}_{vert} = \mathbf{C}_{horiz}$).

2.4. A digression: the twisted arrows category.

2.4.1. For an integer n let tw_n denote the (ordinary) category

$$-n \to ... \to -1 \to -0 \to 0 \to 1 \to ... \to n.$$

We have the natural functors

$$[n]^{\mathrm{op}} \to \mathrm{tw}_n \leftarrow [n].$$

The assignment $n \rightsquigarrow \mathrm{tw}_n$ is naturally a functor

$$\mathrm{tw}_\bullet : \mathbf{\Delta} \to 1\text{-}\mathrm{Cat}_{\mathrm{ordn}} \subset 1\text{-}\mathrm{Cat},$$

equipped with the natural transformations

$$[\bullet]^{\mathrm{op}} \to \mathrm{tw}_\bullet \leftarrow [\bullet].$$

2.4.2. For a $(\infty,1)$-category $\mathbf{D}$, set

$$\mathrm{Tw}_n(\mathbf{D}) := \mathrm{Maps}_{1\text{-}\mathrm{Cat}}(\mathrm{tw}_n, \mathbf{D}).$$

Thus, $\mathrm{Tw}_\bullet(\mathbf{D})$ is an object of

$$\mathrm{Funct}(\mathbf{\Delta}^{\mathrm{op}}, \mathrm{Spc}) = \mathrm{Spc}^{\mathbf{\Delta}^{\mathrm{op}}},$$

which is easily seen to be a complete Segal space, equipped with the maps

$$\mathrm{Seq}_\bullet(\mathbf{D}^{\mathrm{op}}) \leftarrow \mathrm{Tw}_\bullet(\mathbf{D}) \to \mathrm{Seq}_\bullet(\mathbf{D}). \tag{2.8}$$

2.4.3. We define the *twisted arrow category* of $\mathbf{D}$, denoted $\mathrm{Tw}(\mathbf{D})$ so that

$$\mathrm{Seq}_\bullet(\mathrm{Tw}(\mathbf{D})) = \mathrm{Tw}_\bullet(\mathbf{D}).$$

The maps (2.8) give rise to a functor

$$\mathrm{Tw}(\mathbf{D}) \to \mathbf{D}^{\mathrm{op}} \times \mathbf{D}. \tag{2.9}$$

It is not difficult to see (see [**Lu6**, Proposition 4.2.5]) that the functor (2.9) is a co-Cartesian fibration, which is the unstraightening of the Yoneda functor

$$\mathbf{D}^{\mathrm{op}} \times \mathbf{D} \to \mathrm{Spc}.$$

2.5. Proof of Proposition 2.3.4.

2.5.1. We will prove Proposition 2.3.4 by exhibiting a canonical isomorphism between the functors

$$\mathrm{Corr}(\mathbf{C})^{\mathrm{op}} \times \mathrm{Corr}(\mathbf{C}) \to \mathrm{Spc}$$

given by

$$\mathrm{Maps}_{\mathrm{Corr}(\mathbf{C})}(-,-) \text{ and } \mathrm{Maps}_{\mathrm{Corr}(\mathbf{C})}(*, \varpi(-) \otimes -), \tag{2.10}$$

respectively.

Unstraightening, we need to construct an isomorphism between the coCartesian fibrations in spaces that correspond to the two functors in (2.10).

2.5.2. By Sect. 2.4, the functor $\mathrm{Maps}_{\mathrm{Corr}(\mathbf{C})}(-,-)$ corresponds to the coCartesian fibration

$$\mathrm{Tw}(\mathrm{Corr}(\mathbf{C})) \to (\mathrm{Corr}(\mathbf{C}))^{\mathrm{op}} \times \mathrm{Corr}(\mathbf{C}).$$

The functor

$$\mathrm{Maps}_{\mathrm{Corr}(\mathbf{C})}(*,-) : \mathrm{Corr}(\mathbf{C}) \to \mathrm{Spc}$$

is given by the coCartesian fibration $\mathrm{Corr}(\mathbf{C})_{*/}$.

Thus, in order to construct an isomorphism between the functors (2.10) we need to construct a functor

$$\mathrm{Tw}(\mathrm{Corr}(\mathbf{C})) \to \mathrm{Corr}(\mathbf{C})_{*/} \tag{2.11}$$

that fits into a pullback diagram

$$\begin{array}{ccc} \mathrm{Tw}(\mathrm{Corr}(\mathbf{C})) & \longrightarrow & \mathrm{Corr}(\mathbf{C})_{*/} \\ \downarrow & & \downarrow \\ (\mathrm{Corr}(\mathbf{C}))^{\mathrm{op}} \times \mathrm{Corr}(\mathbf{C}) & \xrightarrow{\mathrm{mult}\circ(\varpi\times\mathrm{Id})} & \mathrm{Corr}(\mathbf{C}) \end{array} \tag{2.12}$$

where $\mathrm{mult} : \mathrm{Corr}(\mathbf{C}) \times \mathrm{Corr}(\mathbf{C}) \to \mathrm{Corr}(\mathbf{C})$ is the functor of tensor product.

2.5.3. We have

$$\mathrm{Seq}_n(\mathrm{Corr}(\mathbf{C})_{*/}) \simeq \mathrm{Grid}_{n+1}^{\geq\mathrm{dgnl}}(\mathbf{C}) \underset{\mathbf{C}}{\times} *,$$

where $\mathrm{Grid}_{n+1}^{\geq\mathrm{dgnl}}(\mathbf{C}) \to \mathbf{C}$ is evaluation on the object

$$(0,0) \in ([n+1] \times [n+1])^{\geq\mathrm{dgnl}}.$$

The sought-for maps

$$\mathrm{Seq}_n(\mathrm{Tw}(\mathrm{Corr}(\mathbf{C}))) \to \mathrm{Seq}_n(\mathrm{Corr}(\mathbf{C})_{*/})$$

are defined as follows.

An object

$$\underline{\mathbf{c}} \in \mathrm{Maps}((\mathrm{tw}_n \times \mathrm{tw}_n)^{\geq\mathrm{dgnl}}, \mathbf{C})$$

gets sent to

$$\underline{\mathbf{c}}' \in \mathrm{Maps}(([n+1] \times [n+1])^{\geq\mathrm{dgnl}}, \mathbf{C}),$$

given by the formula:

$$\underline{\mathbf{c}}'_{i,j} = \begin{cases} * \text{ if } i = j = 0 \\ \underline{\mathbf{c}}_{-(j-1),j-1} \text{ if } i = 0, j \geq 1 \\ \underline{\mathbf{c}}_{i-1,j-1} \times \underline{\mathbf{c}}_{-(j-1),-(1-i)} \text{ if } i \geq 1. \end{cases}$$

In other words, we fold the half-grid over the NW-SE diagonal. For example, for $n = 0$, this map sends a diagram

$$\begin{array}{ccc} \mathbf{c}_{0,-0} & \longrightarrow & \mathbf{c}_{-0,-0} \\ \downarrow & & \\ \mathbf{c}_{0,0} & & \end{array}$$

to

$$
\begin{CD}
\mathbf{c}_{0,-0} @>>> * \\
@VVV \\
\mathbf{c}_{0,0} \times \mathbf{c}_{-0,-0}
\end{CD}
$$

and for $n = 1$ a diagram

$$
\begin{CD}
\mathbf{c}_{-1,1} @>>> \mathbf{c}_{-1,0} @>>> \mathbf{c}_{-1,-0} @>>> \mathbf{c}_{-1,-1} \\
@VVV @VVV @VVV \\
\mathbf{c}_{-0,1} @>>> \mathbf{c}_{-0,0} @>>> \mathbf{c}_{-0,-0} \\
@VVV @VVV \\
\mathbf{c}_{0,1} @>>> \mathbf{c}_{0,0} \\
@VVV \\
\mathbf{c}_{1,1}
\end{CD}
$$

to

$$
\begin{CD}
\mathbf{c}_{-1,1} @>>> \mathbf{c}_{-0,0} @>>> * \\
@VVV @VVV \\
\mathbf{c}_{0,1} \times \mathbf{c}_{-1,-0} @>>> \mathbf{c}_{0,0} \times \mathbf{c}_{-0,-0} \\
@VVV \\
\mathbf{c}_{1,1} \times \mathbf{c}_{-1,-1}.
\end{CD}
$$

The fact that (2.12) is a pullback diagram is an easy verification.

3. Extension results in the symmetric monoidal context

In Chapter 7, Sects. 4 and 5 and in Chapter 8 we proved several theorems that say that a functor out of a certain $(\infty, 2)$-category of correspondences can be uniquely extended to a functor out of another $(\infty, 2)$-category of correspondences.

In this section we will show that these extension procedures are compatible with symmetric monoidal structures.

3.1. 'No cost' and factorization extensions. In this subsection, we let $\mathbf{C}$ be a symmetric monoidal category, equipped with three classes of morphisms $vert$, $horiz$, adm as in Sect. 2.1.3[2].

We will study how the extension paradigm in Chapter 7, Sects. 4 and 5 interacts with the symmetric monoidal structures.

[2]We *do not* need the symmetric monoidal structure on $\mathbf{C}$ to be Cartesian.

3.1.1. Recall the setting of Chapter 7, Sect. 4. I.e., we start with a $\mathbf{C}$, equipped with four classes of morphisms $vert, horiz, adm, adm'$, satisfying the assumptions of Chapter 7, Sect. 4.1.1.

Let $\mathbb{S}$ be an $(\infty, 2)$-category and let us be given a functor

$$\Phi^{adm'}_{vert,horiz} : \mathrm{Corr}(\mathbf{C})^{adm'}_{vert,horiz} \to \mathbb{S}.$$

Assume that all four classes of morphisms are preserved by the functor of tensor product $\mathbf{C} \times \mathbf{C} \to \mathbf{C}$, so that

$$\mathrm{Corr}(\mathbf{C})^{adm}_{vert,horiz} \text{ and } \mathrm{Corr}(\mathbf{C})^{adm'}_{vert,horiz}$$

acquire symmetric monoidal structures by Sect. 2.1.3.

Assume also that $\mathbb{S}$ is equipped with a symmetric monoidal structure. Assume also that

$$\Phi^{adm}_{vert,horiz} := \Phi^{adm'}_{vert,horiz}|_{\mathrm{Corr}(\mathbf{C})^{adm}_{vert,horiz}}$$

is equipped with a symmetric monoidal structure.

We claim:

PROPOSITION 3.1.2. *The symmetric monoidal structure on $\Phi^{adm}_{vert,horiz}$ extends uniquely to one on $\Phi^{adm'}_{vert,horiz}$.*

PROOF. Indeed, Chapter 7, Theorem 4.1.3 implies that for any n, the functor

$$(\mathrm{Corr}(\mathbf{C})^{adm}_{vert,horiz})^{\times n} \to (\mathrm{Corr}(\mathbf{C})^{adm'}_{vert,horiz})^{\times n}$$

is a categorical epimorphism.

Now, our assertion follows from the next observation:

LEMMA 3.1.3. *Let $\mathbf{O}$ be a symmetric monoidal category, and let $\mathbf{o}_1 \to \mathbf{o}_2$ be a homomrphism between commutative algebras on $\mathbf{O}$, such that for any n, the map $(\mathbf{o}_1)^{\otimes n} \to (\mathbf{o}_2)^{\otimes n}$ is a categorical epimorphism. Then for another commutative algebra $\mathbf{o}'$ in $\mathbf{O}$, restriction defines an isomorphism from the space of homomorphisms $\mathbf{o}_2 \to \mathbf{o}'$ and the subspace of homomorphisms $\mathbf{o}_1 \to \mathbf{o}'$ that factor through $\mathbf{o}_2$ as maps of plain objects of $\mathbf{O}$.*

□

3.1.4. The above discussion applies verbatim to the setting of Chapter 7, Sect. 5. I.e., we start with $\mathbf{C}$, equipped with four classes of morphisms $vert, horiz, adm$ and $co\text{-}adm$, satisfying the assumptions of Chapter 7, Sect. 5.1.

Let $\mathbb{S}$ be an $(\infty, 2)$-category and let us be given a functor

$$\Phi^{adm}_{vert,horiz} : \mathrm{Corr}(\mathbf{C})^{adm}_{vert,horiz} \to \mathbb{S}.$$

Assume now that $\mathbf{C}$ all four classes of morphisms are preserved by the functor of tensor product $\mathbf{C} \times \mathbf{C} \to \mathbf{C}$, so that

$$\mathrm{Corr}(\mathbf{C})^{\mathrm{isom}}_{vert,co\text{-}adm} \text{ and } \mathrm{Corr}(\mathbf{C})^{adm}_{vert,horiz}$$

acquire symmetric monoidal structures by Sect. 2.1.3.

Assume also that

$$\Phi^{\mathrm{isom}}_{vert,co\text{-}adm} := \Phi^{adm}_{vert,horiz}|_{\mathrm{Corr}(\mathbf{C})^{\mathrm{isom}}_{vert,co\text{-}adm}}$$

is equipped with a symmetric monoidal structure.

We have (with the same proof as above):

PROPOSITION 3.1.5. *The symmetric monoidal structure on* $\Phi^{\text{isom}}_{vert,co\text{-}adm}$ *extends uniquely to one on* $\Phi^{adm}_{vert,horiz}$.

3.2. Right Kan extensions and symmetric monoidal structures. In this subsection we will review the notion of right Kan extension of 1-morphisms in $(\infty,2)$-categories, and how it interacts with symmetric monoidal structures.

As an application we will show that the extension procedure in Chapter 8, Sect. 6 is (lax!) compatible with symmetric monoidal structures.

3.2.1. Let $\mathbb{S}$ be an $(\infty,2)$-category, and let $\alpha : s_1 \to s_2$ be a 1-morphism in $\mathbb{S}$. Given another object $s' \in \mathbb{S}$, restriction defines a functor

$$\mathbf{Maps}_{\mathbb{S}}(s_2, s') \to \mathbf{Maps}_{\mathbb{S}}(s_1, s').$$

The (partially defined) right adjoint functor to the above restriction functor is called the functor of *right Kan extension*, and is denoted by RKE_α.

3.2.2. Suppose now that $\mathbb{S}$ has a symmetric monoidal structure. Let s_1 and s_2 be commutative algebra objects in $\mathbb{S}$, and let

$$\alpha : s_1 \to s_2$$

be a homomorphism.

Let now s' be another commutative algebra object in $\mathbb{S}$, and let

$$\phi_1 : s_1 \to s'$$

be a *right-lax* homomorphism (see Sect. 1.4.5 for what this means).

Suppose that RKE_α is defined on ϕ_1 regarded as a plain object of $\mathbf{Maps}_{\mathbb{S}}(s_1, s')$. Suppose, moreover, that for every n, the canonical map from the composition

$$s_2^{\otimes n} \overset{\otimes}{\to} s_2 \overset{\mathrm{RKE}_\alpha(\phi_1)}{\longrightarrow} s'$$

to the right Kan extension along $\alpha^{\otimes n} : s_1^{\otimes n} \to s_2^{\otimes n}$ of the composition

$$s_1^{\otimes n} \overset{\otimes}{\to} s_1 \overset{\phi_1}{\longrightarrow} s'$$

is an isomorphism (in particular, the latter right Kan extension is also defined).

By unwinding the construction, we obtain that in this case

$$\phi_2 := \mathrm{RKE}_\alpha(\phi_1)$$

has a unique structure of *right-lax* homomorphism $s_2 \to s'$ so that the co-unit of the adjunction

$$\phi_2 \circ \alpha \to \phi_1$$

has the structure of a map between right-lax homomorphisms.

3.2.3. Consider now the situation of Chapter 8, Theorem 6.1.5, where

$$(\mathbf{C}, vert, horiz, adm)$$

and

$$(\mathbf{D}, vert, horiz, adm)$$

are as in Sect. 2.1.4. Assume also that the functor F takes products to products, so that it induces a symmetric monoidal functor

$$F^{adm}_{vert,horiz} : \operatorname{Corr}(\mathbf{C})^{adm}_{vert,horiz} \to \operatorname{Corr}(\mathbf{D})^{adm}_{vert,horiz}.$$

Let $\mathbb{S}$ be a symmetric monoidal $(\infty, 2)$-category, and assume that

$$\Phi^{adm}_{vert,horiz} : \operatorname{Corr}(\mathbf{C})^{adm}_{vert,horiz} \to \mathbb{S}$$

is endowed with a right-lax symmetric monoidal structure.

We claim:

PROPOSITION 3.2.4. *Suppose that for any n and any map*

$$\mathbf{c} \to \mathbf{d}_1 \times ... \times \mathbf{d}_n$$

that is in horiz (here $\mathbf{c} \in \mathbf{C}$, $\mathbf{d}_i \in \mathbf{D}$), each of the projections $\mathbf{c} \to \mathbf{d}_i$ is also in horiz. Then

$$\Psi^{adm}_{vert,horiz} := \operatorname{RKE}_{F^{adm}_{vert,horiz}}(\Phi^{adm}_{vert,horiz}) : \operatorname{Corr}(\mathbf{D})^{adm}_{vert,horiz} \to \mathbb{S}$$

acquires a uniquely defined right-lax symmetric monoidal structure, for which the natural transformation

$$\Psi^{adm}_{vert,horiz} \circ F^{adm}_{vert,horiz} \to \Phi^{adm}_{vert,horiz}$$

has the structure of a map between right-lax symmetric monoidal functors.

PROOF. We need to show that the isomorphism condition from Sect. 3.2.2 is satisfied.

By Chapter 8, Theorem 6.1.5, it is enough to check the corresponding 1-categorical statement. Thus, we need to show that for an integer n and an n-tuple of objects $\mathbf{d}_1, ..., \mathbf{d}_n$, the map

$$\lim_{\mathbf{c} \in \mathbf{C}, \alpha:\mathbf{c} \to \mathbf{d}_1 \times ... \times \mathbf{d}_n, \alpha \in horiz} \Phi(\mathbf{c}) \to \lim_{\mathbf{c}_i \in \mathbf{C}, \alpha_i:\mathbf{c}_i \to \mathbf{d}_i, \alpha_i \in horiz} \Phi(\mathbf{c}_1 \times ... \times \mathbf{c}_n)$$

is an isomorphism.

However, the condition of the proposition implies that the corresponding map of index categories is cofinal.

□

3.3. Symmetric monoidal structure on the bivariant extension. In this subsection we show that the extension procedure of Chapter 8, Sect. 1 is compatible with symmetric monoidal structures.

3.3.1. Let us now be in the setting of Chapter 8, Sect. 1, where both

$$(\mathbf{C}, vert, horiz, adm)$$

and

$$(\mathbf{D}, vert, horiz, adm)$$

are as in Sect. 2.1.4. We let the functor

$$F : \mathbf{C} \to \mathbf{D}$$

be endowed with a symmetric monoidal structure.

Let the target $(\infty, 2)$-category $\mathbb{S}$ be equipped with a symmetric monoidal structure. Assume also that the tensor product functor

$$\mathbb{S} \times \mathbb{S} \to \mathbb{S}$$

is such that the underlying functor $\mathbb{S}^{1\text{-Cat}} \times \mathbb{S}^{1\text{-Cat}} \to \mathbb{S}^{1\text{-Cat}}$ commutes with *colimits in each variable.*

3.3.2. Let us be given a functor

$$\Phi^{adm}_{vert,horiz} : \mathrm{Corr}(\mathbf{C})^{adm}_{vert,horiz} \to \mathbb{S},$$

satisfying the assumptions of Chapter 8, Sect. 1.1.6, and let

$$\Psi^{adm}_{vert,horiz} : \mathrm{Corr}(\mathbf{D})^{adm}_{vert,horiz} \to \mathbb{S}$$

be its unique extension, satisfying the requirements of Chapter 8, Theorem 1.1.9.

Assume that $\Phi^{adm}_{vert,horiz}$ is equipped with a symmetric monoidal structure. We claim:

PROPOSITION 3.3.3. *Suppose that for any n and any map*

$$\mathbf{c} \to \mathbf{d}_1 \times \ldots \times \mathbf{d}_n$$

that is in horiz (resp., vert, adm), each of the projections $\mathbf{c} \to \mathbf{d}_i$ is also in horiz (resp., vert, adm). Assume also that for every $\mathbf{d} \in \mathbf{D}$, the maps

$$* \to \mathbf{d} \text{ and } \mathbf{d} \to \mathbf{d} \times \mathbf{d}$$

are in adm, and that the functor

$$\mathrm{Maps}_{\mathbb{S}}(\mathbf{1}_{\mathbb{S}}, -) : \mathbb{S}^{1\text{-Cat}} \to \mathrm{Spc}$$

is conservative. Then then functor $\Psi^{adm}_{vert,horiz}$ carries a unique symmetric monoidal structure, which induces the given one on

$$\Phi^{adm}_{vert,horiz} \simeq \Psi^{adm}_{vert,horiz}|_{\mathrm{Corr}(\mathbf{C})^{adm}_{vert,horiz}}.$$

PROOF. It is enough to show that for any integer n, both circuits of the diagram

$$\begin{array}{ccc} (\mathrm{Corr}(\mathbf{D})^{adm}_{vert,horiz})^{\times n} & \xrightarrow{(\Psi^{adm}_{vert,horiz})^{\times n}} & \mathbb{S}^{\times n} \\ \text{product map} \big\downarrow & & \big\downarrow \text{product map} \\ \mathrm{Corr}(\mathbf{D})^{adm}_{vert,horiz} & \xrightarrow{\Psi^{adm}_{vert,horiz}} & \mathbb{S} \end{array} \tag{3.1}$$

identify with the *canonical* extension, given by Chapter 8, Theorem 1.1.9, of the functor given by the (canonically identified) two circuits of the diagram

$$\begin{array}{ccc} (\mathrm{Corr}(\mathbf{C})^{adm}_{vert,horiz})^{\times n} & \xrightarrow{(\Phi^{adm}_{vert,horiz})^{\times n}} & \mathbb{S}^{\times n} \\ \text{product map}\downarrow & & \downarrow\text{product map} \\ \mathrm{Corr}(\mathbf{C})^{adm}_{vert,horiz} & \xrightarrow{\Phi^{adm}_{vert,horiz}} & \mathbb{S}. \end{array}$$

We claim that the two circuits in (3.1) satisfy the assumptions of Chapter 8, Corollary 1.1.10.

We first check condition (i) for the anti-clockwise circuit. We need to show that for $\mathbf{d}_1,..,\mathbf{d}_n \in \mathbf{D}$, the map

$$\underset{\mathbf{c}_i\in\mathbf{C},\beta_i:\mathbf{c}_i\to\mathbf{d}_i,\beta_i\in vert}{\mathrm{colim}} \Phi(\mathbf{c}_1\times...\times\mathbf{c}_n) \to \underset{\mathbf{c}\in\mathbf{C},\beta:\mathbf{c}\to\mathbf{d}_1\times...\times\mathbf{d}_n,\beta\in vert}{\mathrm{colim}} \Phi(\mathbf{c}) =: \Psi(\mathbf{c}_1\times...\times\mathbf{c}_n)$$

is an isomorphism. However, this follows from the fact that the corresponding map of index categories is cofinal. Condition (ii) for the anti-clockwise circuit follows in the same way as condition (i).

Let us check (i) for the clockwise circuit. We need to show that for $\mathbf{d}_1,..,\mathbf{d}_n \in \mathbf{D}$, the map

$$\underset{\mathbf{c}_i\in\mathbf{C},\beta_i:\mathbf{c}_i\to\mathbf{d}_i,\beta_i\in vert}{\mathrm{colim}} \Phi(\mathbf{c}_1)\otimes...\otimes\Phi(\mathbf{c}_n) \to \bigotimes_i \left(\underset{\mathbf{c}_i\in\mathbf{C},\beta_i:\mathbf{c}_i\to\mathbf{d}_i,\beta_i\in vert}{\mathrm{colim}} \Phi(\mathbf{c}_i)\right)$$

is an isomorphism. However, this follows from the commutation of the tensor product on $\mathbb{S}^{1\text{-Cat}}$ with colimits in each variable.

In particular, we obtain that the restrictions of the two circuits in (3.1) to

$$(\mathbf{D}_{vert})^{\times n} \subset (\mathrm{Corr}(\mathbf{D})^{adm}_{vert,horiz})^{\times n},$$

and further to

$$(\mathbf{D}_{adm})^{\times n} \subset (\mathrm{Corr}(\mathbf{D})^{adm}_{vert,horiz})^{\times n},$$

are canonically identified.

Conditions (iii) and (iv) for the clockwise circuit follows by the same argument as condition (i). Hence, we obtain that they also hold for the anti-clockwise circuit, since the two functors are identified on $(\mathbf{D}_{adm})^{\times n}$.

Note that this, in particular, establishes the assertion of the proposition in the case when $horiz = adm = vert$. Thus, the restriction of $\Psi^{adm}_{vert,horiz}$ to

$$\mathrm{Corr}(\mathbf{D})^{adm}_{adm,adm}$$

has a symmetric monoidal structure. By further restricting to $\mathrm{Corr}(\mathbf{D})_{adm,adm}$, and applying Proposition 2.3.4, we obtain that for every $\mathbf{d}\in\mathbf{D}$, the object $\Psi(\mathbf{d})\in\mathbb{S}$ is *dualizable*.

It remains to check condition (ii) for the clockwise circuit. I.e., we need to show that the map

$$\bigotimes_i \left(\underset{\mathbf{c}_i\in\mathbf{C},\alpha_i:\mathbf{c}_i\to\mathbf{d}_i,\alpha_i\in horiz}{\lim} \Phi^!(\mathbf{c}_i)\right) \to \underset{\mathbf{c}_i\in\mathbf{C},\alpha_i:\mathbf{c}_i\to\mathbf{d}_i,\alpha_i\in horiz}{\lim} \Phi^!(\mathbf{c}_1)\otimes...\otimes\Phi^!(\mathbf{c}_n)$$

is an isomorphism. However, this follows as in Chapter 3, Proposition 3.1.7 from the fact that each $\Phi^!(\mathbf{c}_i)$ and each

$$\lim_{\mathbf{c}_i \in \mathbf{C}, \alpha_i : \mathbf{c}_i \to \mathbf{d}_i, \alpha_i \in horiz} \Phi^!(\mathbf{c}_i) \simeq \Psi^!(\mathbf{d}_i)$$

is dualizable, using Chapter 1, Lemma 4.1.6(a).

□

4. Monads and associative algebras in the category of correspondences

It turns out that the category of correspondences is well adapted to the formalism of convolution algebras and convolution categories.

Let $\mathbf{c}^\bullet$ be a *Segal object* of $\mathbf{C}$ (such data are also called *category-objects*); see Sect. 4.1.3 for the definition. Let $p_s, p_t : \mathbf{c}^1 \to \mathbf{c}^0 = \mathbf{c}$ be the source and target maps, respectively, corresponding to the two maps $[0] \to [1]$.

Consider the 1-morphism

$$\begin{array}{ccc} \mathbf{c}^1 & \xrightarrow{p_s} & \mathbf{c} \\ {\scriptstyle p_t}\downarrow & & \\ \mathbf{c} & & \end{array}$$

as an object of the monoidal $(\infty, 1)$-category $\mathbf{Maps}_{\mathrm{Corr}(\mathbf{C})^{\mathrm{all}}_{\mathrm{all;all}}}(\mathbf{c}, \mathbf{c})$.

When $\mathbf{C}$ is an ordinary category, it is clear that the above 1-morphism is an associative algebra object in $\mathbf{Maps}_{\mathrm{Corr}(\mathbf{C})^{\mathrm{all}}_{\mathrm{all;all}}}(\mathbf{c}, \mathbf{c})$, and that all associative algebra objects are obtained in this way. The first result of this section, Proposition 4.1.5 shows that the same is true in the ∞-setting.

As a corollary, we obtain that if Φ is a functor from $\mathrm{Corr}(\mathbf{C})^{adm}_{vert;horiz}$ with values in $\mathbf{1}\text{-}\mathbf{Cat}$, a Segal object $\mathbf{c}^\bullet$ (under appropriate conditions) defines a monad on the category $\Phi(\mathbf{c})$.

The second result of this section, Theorem 4.4.2 is (essentially) the following. It says that for a Segal object $\mathbf{c}^\bullet$, the object $\mathbf{c}^1 \in \mathrm{Corr}(\mathbf{C})_{\mathrm{all;all}}$ has a natural structure of associative algebra in the (symmetric) monoidal category $\mathrm{Corr}(\mathbf{C})_{\mathrm{all;all}}$.

As a consequence, we obtain that if Φ is a *monoidal functor* from $\mathrm{Corr}(\mathbf{C})_{\mathrm{all;all}}$ to 1-Cat, the category $\Phi(\mathbf{c}^1)$ acquires a canonical monoidal structure, given by convolution.

4.1. Monads and Segal objects.

In this subsection we will articulate the following idea:

The category of algebras in the monoidal category of endomorphisms of an object $\mathbf{c} \in \mathrm{Corr}(\mathbf{C})^{\mathrm{all}}_{\mathrm{all;all}}$ is canonically equivalent to the category if Segal objects acting on $\mathbf{c}$, i.e., simplicial objects $\mathbf{c}^\bullet$ with $\mathbf{c}^0 = \mathbf{c}$.

4.1.1. Note that if $\mathbb{S}$ is an $(\infty,2)$-category and $s \in \mathbb{S}$ is an object, then the $(\infty,1)$-category $\mathbf{Maps}_{\mathbb{S}}(s,s)$ acquires a natural monoidal structure. Indeed, the corresponding functor

$$\mathbf{\Delta}^{\mathrm{op}} \to 1\text{-Cat}$$

is given by

$$\mathrm{Seq}_n(\mathbb{S}) \underset{\mathrm{Seq}_0(\mathbb{S})\times\ldots\times\mathrm{Seq}_0(\mathbb{S})}{\times} \{s \times \ldots \times s\},$$

where $\mathrm{Seq}_n(\mathbb{S}) \to \mathrm{Seq}_0(\mathbb{S}) \times \ldots \times \mathrm{Seq}_0(\mathbb{S})$ corresponds to the map

$$[0] \sqcup \ldots \sqcup [0] \simeq [n]^{\mathrm{Spc}} \to [n].$$

By definition, a *monad* acting on s is an associative algebra in the monoidal $(\infty,1)$-category $\mathbf{Maps}_{\mathbb{S}}(s,s)$.

4.1.2. Let $\mathbf{C}$ be an $(\infty,1)$-category with finite limits, and take $vert = horiz = adm =$ all. Take $\mathbb{S} = \mathrm{Corr}(\mathbf{C})^{\mathrm{all}}_{\mathrm{all;all}}$.

We will be interested in the category of monads in $\mathrm{Corr}(\mathbf{C})^{\mathrm{all}}_{\mathrm{all;all}}$ acting on a given $\mathbf{c} \in \mathbf{C}$.

4.1.3. Recall now that if $\mathbf{C}$ is any $(\infty,1)$-category with finite limits, one can talk about Segal objects[3] in $\mathbf{C}$, acting on a given $\mathbf{c} \in \mathbf{C}$. Denote this category by $\mathrm{Seg}(\mathbf{c})$.

By definition, this is a full subcategory $\mathbf{C}^{\mathbf{\Delta}^{\mathrm{op}}}$ of simplicial objects $\mathbf{c}^\bullet$ equipped with an identification $\mathbf{c}^0 = \mathbf{c}$, consisting of objects for which for every $n \geq 2$, the map

$$\mathbf{c}^n \to \mathbf{c}^1 \underset{\mathbf{c}}{\times} \ldots \underset{\mathbf{c}}{\times} \mathbf{c}^1,$$

given by the product of the maps

$$[1] \to [n], \quad 0 \mapsto i, 1 \mapsto i+1, \quad i = 0, \ldots, n-1,$$

is an isomorphism.

This condition can be equivalently formulated as saying that for any $\mathbf{c}' \in \mathbf{C}$, the simplicial space $\mathrm{Maps}_{\mathbf{C}}(\mathbf{c}', \mathbf{c}^\bullet)$ is a Segal space (but not necessarily a complete Segal space).

4.1.4. We will prove:

PROPOSITION-CONSTRUCTION 4.1.5. *There exists a canonical equivalence between* $\mathrm{Seg}(\mathbf{c})$ *and the category* $\mathrm{AssocAlg}(\mathbf{Maps}_{\mathrm{Corr}(\mathbf{C})^{\mathrm{all}}_{\mathrm{all;all}}}(\mathbf{c},\mathbf{c}))$.

4.1.6. *Variant.* Let $(\mathbf{C}, vert, horiz, adm)$ be an object of Trpl. Let $\mathbf{c}^\bullet$ be an object of $\mathrm{Seg}(\mathbf{c})$. Suppose that:

- The 'source' map $\mathbf{c}^1 \to \mathbf{c}^0$ (i.e., one corresponding to $(0 \in [0]) \mapsto (0 \in [1])$) belongs to *horiz*;
- The 'target' map $\mathbf{c}^1 \to \mathbf{c}^0$ (i.e., one corresponding to $(0 \in [0]) \mapsto (1 \in [1])$) belongs to *vert*;
- The multiplication map $\mathbf{c}^2 \to \mathbf{c}^1$ (i.e., one corresponding to the active map $[1] \to [2]$) belongs to *adm*.

[3] An alternative terminology for this is 'category-objects' in $\mathbf{C}$, acting on $\mathbf{c} \in \mathbf{C}$.

In this case, we obtain that the algebra object in $\mathbf{Maps}_{\mathrm{Corr}(\mathbf{C})^{\mathrm{all}}_{\mathrm{all;all}}}(\mathbf{c},\mathbf{c})$ corresponding to $\mathbf{c}^\bullet$ by Proposition 4.1.5, defines an algebra object in the 1-full monoidal subcategory

$$\mathbf{Maps}_{\mathrm{Corr}(\mathbf{C})^{adm}_{vert;horiz}}(\mathbf{c},\mathbf{c}) \subset \mathbf{Maps}_{\mathrm{Corr}(\mathbf{C})^{\mathrm{all}}_{\mathrm{all;all}}}(\mathbf{c},\mathbf{c}).$$

4.2. Proof of Proposition 4.1.5. The equivalence stated in the proposition is completely evident when $\mathbf{C}$ is an ordinary category, and the reader should check it before proceeding to the ∞-case.

The proof in the latter case will use (a little bit) of diagram manipulation.

4.2.1. For $(\infty,2)$-category and $s \in \mathbb{S}$, the monoidal $(\infty,1)$-category $\mathbf{Maps}_{\mathbb{S}}(s,s)$ is described as follows: the corresponding functor $\mathbf{\Delta}^{\mathrm{op}} \to 1\text{-Cat}$ sends $[n]$ to

$$\mathrm{Seq}_n(\mathbb{S}) \underset{\mathrm{Seq}_0(\mathbb{S})^{\times n+1}}{\times} *,$$

where the maps $* \to \mathrm{Seq}_0(\mathbb{S})$ are given by $s \in \mathbb{S}$.

Applying this to $\mathbb{S} = \mathrm{Corr}(\mathbf{C})^{\mathrm{all}}_{\mathrm{all;all}}$ and $s = \mathbf{C}$, we obtain that the monoidal category $\mathbf{Maps}_{\mathrm{Corr}(\mathbf{C})^{\mathrm{all}}_{\mathrm{all;all}}}(\mathbf{c},\mathbf{c})$ is given by the functor $\mathbf{\Delta}^{\mathrm{op}} \to 1\text{-Cat}$ sends $[n]$ to

$$\mathbf{Grid}^{\geq\,\mathrm{dgnl}}_n(\mathbf{C}) \underset{\mathbf{C}^{n+1}}{\times} *.$$

I.e., this is the category whose objects are half-grids with all squares being Cartesian, and with the diagonal entries identified with $\mathbf{c}$.

4.2.2. Thus, associative algebras in $\mathbf{Maps}_{\mathrm{Corr}(\mathbf{C})^{\mathrm{all}}_{\mathrm{all;all}}}(\mathbf{c},\mathbf{c})$ are right-lax natural transformations

$$[n] \mapsto \left(* \Rightarrow \mathbf{Grid}^{\geq\,\mathrm{dgnl}}_n(\mathbf{C}) \underset{\mathbf{C}^{n+1}}{\times} \{\mathbf{c},...,\mathbf{c}\}\right), \tag{4.1}$$

with the corresponding natural transformations being isomorphisms over *inert* arrows in $\mathbf{\Delta}^{\mathrm{op}}$.

We will now describe the above category of natural transformations (4.1) slightly differently.

4.2.3. Consider the functor

$$\mathbf{\Delta} \to 1\text{-Cat}_{\mathrm{ordn}}, \quad [n] \mapsto ([n] \times [n]^{\mathrm{op}})^{\geq\,\mathrm{dgnl}}.$$

Let $\mathbf{I}$ denote the corresponding *Cartesian* fibration over $\mathbf{\Delta}^{\mathrm{op}}$. Let $\mathbf{I}_n$ denote the fiber of $\mathbf{I}$ over $[n] \in \mathbf{\Delta}^{\mathrm{op}}$, i.e., $\mathbf{I}_n = ([n] \times [n]^{\mathrm{op}})^{\geq\,\mathrm{dgnl}}$.

Let $\mathbf{I}' \subset \mathbf{I}$ be the full subcategory, such that for each n, the subcategory $\mathbf{I}'_n \subset \mathbf{I}_n$ corresponds to the element $(0,n) \in ([n] \times [n]^{\mathrm{op}})^{\geq\,\mathrm{dgnl}}$, i.e., the NW corner of the half-grid. It is easy to see that the projection

$$\mathbf{I}' \to \mathbf{\Delta}^{\mathrm{op}}$$

is an equivalence.

Let $\mathbf{I}'' \subset \mathbf{I}$ be the full subcategory, such that for every n, the subcategory $\mathbf{I}''_n \subset \mathbf{I}_n$ corresponds to the union of the elements (i,i), i.e., the diagonal entries.

4.2.4. Then the category of natural transformations (4.1) is a full subcategory in the category of functors

$$\mathbf{I} \to \mathbf{C},$$

such that:

- The restriction of the functor to $\mathbf{I}''$ is identified with the constant functor with value $\mathbf{c}$;
- For every n and every square in $\mathbf{I}_n$, the resulting square in $\mathbf{C}$ is Cartesian;
- Arrows in $\mathbf{I}$ that are Cartesian over inert arrows in $\mathbf{\Delta}^{\text{op}}$ get sent to isomorphisms in $\mathbf{C}$.

4.2.5. We note now that restriction along $\mathbf{I}' \to \mathbf{I}$ associates to a functor $\mathbf{I} \to \mathbf{C}$ a simplicial object in $\mathbf{C}$. Functors satisfying the assumptions of Sect. 4.2.4 are easily seen to give rise to functors $\mathbf{\Delta}^{\text{op}} \to \mathbf{C}$ that satisfy the conditions of being a Segal object.

Vice versa, starting from a functor

$$\mathbf{\Delta}^{\text{op}} \simeq \mathbf{I}' \to \mathbf{C}, \tag{4.2}$$

we apply right Kan extension along the embedding $\mathbf{I}' \to \mathbf{I}$, and thus obtain a functor $\mathbf{I} \to \mathbf{C}$.

It is easy to see that if (4.2) is a Segal object, then the resulting functor $\mathbf{I} \to \mathbf{C}$ satisfies the assumptions of Sect. 4.2.4.

This defines the desired equivalence

$$\text{AssocAlg}(\mathbf{Maps}_{\text{Corr}(\mathbf{C})^{\text{all}}_{\text{all;all}}}(\mathbf{c}, \mathbf{c})) \to \text{Seg}(\mathbf{c}).$$

4.3. Action on a module: first version. In this subsection we will add a few remarks concerning the action of the monoidal category $\mathbf{Maps}_{\text{Corr}(\mathbf{C})^{\text{all}}_{\text{all;all}}}(\mathbf{c}, \mathbf{c})$ on the categories of the form $\mathbf{Maps}_{\text{Corr}(\mathbf{C})^{\text{all}}_{\text{all;all}}}(\mathbf{c}', \mathbf{c})$ for $\mathbf{c}' \in \mathbf{C}$.

4.3.1. Let s and s' be objects in an $(\infty, 2)$-category $\mathbb{S}$. Then the category $\mathbf{Maps}_{\mathbb{S}}(s', s)$ is naturally a module for the monoidal category $\mathbf{Maps}_{\mathbb{S}}(s, s)$.

Applying this to $\mathbb{S} = \text{Corr}(\mathbf{C})^{\text{all}}_{\text{all;all}}$, we obtain that for $\mathbf{c}' \in \mathbf{C}$, the monoidal category $\mathbf{Maps}_{\text{Corr}(\mathbf{C})^{\text{all}}_{\text{all;all}}}(\mathbf{c}, \mathbf{c})$ acts naturally on the category

$$\mathbf{Maps}_{\text{Corr}(\mathbf{C})^{\text{all}}_{\text{all;all}}}(\mathbf{c}', \mathbf{c}).$$

4.3.2. Let now $\mathbf{c}^\bullet$ be a monad acting on $\mathbf{c}$ *over* an object $\mathbf{c}'$, i.e., is a monad in the category $\mathbf{C}_{/\mathbf{c}'}$. Let γ denote the morphism $\mathbf{c} \to \mathbf{c}'$.

Then it follows from the construction that the object

$$(\mathbf{c}' \to \mathbf{c}) \in \mathbf{Maps}_{\text{Corr}(\mathbf{C})^{\text{all}}_{\text{all;all}}}(\mathbf{c}', \mathbf{c})$$

, corresponding to the diagram

$$\begin{array}{ccc} \mathbf{c} & \xrightarrow{\gamma} & \mathbf{c}' \\ {\scriptstyle \text{id}}\big\downarrow & & \\ \mathbf{c} & & \end{array} \tag{4.3}$$

is a module over the algebra in $\mathbf{Maps}_{\text{Corr}(\mathbf{C})^{\text{all}}_{\text{all;all}}}(\mathbf{c}, \mathbf{c})$, corresponding to $\mathbf{c}^\bullet$.

4.3.3. Note now that if $s' \xrightarrow{f} s$ is a 1-morphism in an $(\infty, 2)$-category $\mathbb{S}$ such that f admits a left adjoint, then the composition $(f \circ f^L) \in \mathbf{Maps}_{\mathbb{S}}(s,s)$ has a natural structure of algebra (this follows, e.g., from the description of the procedure of passage to the adjoint morphism given in Chapter 12, Theorem 1.2.4). In fact, this is a universal algebra object in $\mathbf{Maps}_{\mathbb{S}}(s,s)$ that acts on $f \in \mathbf{Maps}_{\mathbb{S}}(s',s)$.

Note that the 1-morphism (4.3) admits a left adjoint given by

$$\begin{array}{ccc} \mathbf{c} & \xrightarrow{\mathrm{id}} & \mathbf{c} \\ {\scriptstyle\gamma}\downarrow & & \\ \mathbf{c}'. & & \end{array} \tag{4.4}$$

Hence, we obtain that the algebra in $\mathbf{Maps}_{\mathrm{Corr}(\mathbf{C})^{\mathrm{all}}_{\mathrm{all;all}}}(\mathbf{c},\mathbf{c})$, corresponding to $\mathbf{c}^\bullet$, admits a canonical homomorphism to the algebra corresponding to the composition of (4.3) and (4.4). This map corresponds to the morphism

$$\mathbf{c}^1 \to \mathbf{c} \underset{\mathbf{c}'}{\times} \mathbf{c},$$

expressing the fact that $\mathbf{c}^1$ acts on $\mathbf{c}$ *over* $\mathbf{c}'$.

4.3.4. Let $\mathbf{c}^\bullet$ now be the Čech nerve of the map $\beta : \mathbf{c} \to \mathbf{c}'$. In this case we claim that the above homomorphism of algebras is an isomorphism. Indeed, it suffices to check this fact at the level of the underlying objects of $\mathbf{Maps}_{\mathrm{Corr}(\mathbf{C})^{\mathrm{all}}_{\mathrm{all;all}}}(\mathbf{c},\mathbf{c})$, and the required isomorphism follows from the Cartesian diagram

$$\begin{array}{ccc} \mathbf{c}^1 & \xrightarrow{p_s} & \mathbf{c} \\ {\scriptstyle p_t}\downarrow & & \downarrow{\scriptstyle\gamma} \\ \mathbf{c} & \xrightarrow{\gamma} & \mathbf{c}'. \end{array}$$

4.3.5. *Variant.* Retaining the assumptions of Sect. 4.1.6, assume that the map $\gamma : \mathbf{c} \to \mathbf{c}'$ belongs to *horiz*, and that the target map $p_s : \mathbf{c}^1 \to \mathbf{c}$ belongs to *adm*.

Then we obtain that the object of $\mathbf{Maps}_{\mathrm{Corr}(\mathbf{C})^{adm}_{vert;horiz}}(\mathbf{c}',\mathbf{c})$, given by (4.4), is a module for the algebra in $\mathbf{Maps}_{\mathrm{Corr}(\mathbf{C})^{adm}_{vert;horiz}}(\mathbf{c},\mathbf{c})$, corresponding to $\mathbf{c}^\bullet$.

Furthermore, if $\gamma \in adm$, in which case the corresponding morphism

$$(\mathbf{c}' \to \mathbf{c}) \in \mathrm{Corr}(\mathbf{C})^{adm}_{vert;horiz}$$

admits a left adjoint, the isomorphism from Sect. 4.3.4 holds at the level of algebras in the monoidal category $\mathbf{Maps}_{\mathrm{Corr}(\mathbf{C})^{adm}_{vert;horiz}}(\mathbf{c},\mathbf{c})$.

4.4. From monads/Segal objects to algebras. In this subsection we formulate the main result of this section, Theorem 4.4.2. It says that if $\mathbf{c}^\bullet$ is a Segal object of $\mathbf{C}$ acting on $\mathbf{c}$, then its first term $\mathbf{c}^1$ acquires a natural algebra structure in $\mathrm{Corr}(\mathbf{C})$.

4.4.1. Let $\mathbf{C}$ be an $(\infty,1)$-category with finite limits, and take $vert = horiz = adm =$ all. We let

$$\operatorname{Corr}(\mathbf{C}) := \operatorname{Corr}(\mathbf{C})^{\text{isom}}_{\text{all;all}}$$

be endowed with a symmetric monoidal structure as in Sect. 2.1.4.

We will prove the following:

THEOREM-CONSTRUCTION 4.4.2. *There exists a canonical right-lax homomorphism of monoidal* $(\infty,1)$*-categories*

$$\mathbf{Maps}_{\operatorname{Corr}(\mathbf{C})^{\text{all}}_{\text{all;all}}}(\mathbf{c},\mathbf{c}) \to \operatorname{Corr}(\mathbf{C}).$$

4.4.3. Let us explain the content of Theorem 4.4.2 when $\mathbf{C}$ is an ordinary category. In this case the sought-for functor

$$\mathbf{Maps}_{\operatorname{Corr}(\mathbf{C})^{\text{all}}_{\text{all;all}}}(\mathbf{c},\mathbf{c}) \to \operatorname{Corr}(\mathbf{C})$$

is explicitly defined as follows:

It sends an object of $\mathbf{Maps}_{\operatorname{Corr}(\mathbf{C})^{\text{all}}_{\text{all;all}}}(\mathbf{c},\mathbf{c})$, given by

$$\begin{array}{ccc} \widetilde{\mathbf{c}} & \longrightarrow & \mathbf{c} \\ \downarrow & & \\ \mathbf{c} & & \end{array}$$

to $\widetilde{\mathbf{c}} \in \operatorname{Corr}(\mathbf{C})$. The monoidal structure is defined as follows: for a pair of objects

$$\begin{array}{ccc} \widetilde{\mathbf{c}}_1 & \longrightarrow & \mathbf{c} \\ \downarrow & & \\ \mathbf{c} & & \end{array}$$

and

$$\begin{array}{ccc} \widetilde{\mathbf{c}}_2 & \longrightarrow & \mathbf{c} \\ \downarrow & & \\ \mathbf{c} & & \end{array}$$

their tensor product in $\mathbf{Maps}_{\operatorname{Corr}(\mathbf{C})^{\text{all}}_{\text{all;all}}}(\mathbf{c},\mathbf{c})$ is given by

$$\begin{array}{ccc} \widetilde{\mathbf{c}}_1 \underset{\mathbf{c}}{\times} \widetilde{\mathbf{c}}_2 & \longrightarrow & \mathbf{c} \\ \downarrow & & \\ \mathbf{c} & & \end{array}$$

and the corresponding 1-morphism $\widetilde{\mathbf{c}}_1 \times \widetilde{\mathbf{c}}_2 \to \widetilde{\mathbf{c}}_1 \underset{\mathbf{c}}{\times} \widetilde{\mathbf{c}}_2$ (note the direction of the arrow!) in $\operatorname{Corr}(\mathbf{C})$ is given by the diagram

$$\begin{array}{ccc} \widetilde{\mathbf{c}}_1 \underset{\mathbf{c}}{\times} \widetilde{\mathbf{c}}_2 & \longrightarrow & \widetilde{\mathbf{c}}_1 \times \widetilde{\mathbf{c}}_2 \\ {\scriptstyle \text{id}}\downarrow & & \\ \widetilde{\mathbf{c}}_1 \underset{\mathbf{c}}{\times} \widetilde{\mathbf{c}}_2 . & & \end{array}$$

4.4.4. As a formal consequence of Theorem 4.4.2, combined with Proposition 4.1.5, we obtain:

COROLLARY 4.4.5. *For any* $\mathbf{c} \in \mathbf{C}$, *there is a canonically defined functor*

$$\operatorname{Seg}(\mathbf{c}) \to \operatorname{AssocAlg}(\operatorname{Corr}(\mathbf{C})).$$

4.4.6. We note that if $\mathbf{c}^\bullet$ is an object of $\operatorname{Seg}(\mathbf{c})$, the object of $\operatorname{Corr}(\mathbf{C})$, underlying the corresponding algebra is $\mathbf{c}^1$, and the product map is given by the diagram

$$\begin{array}{ccc} \mathbf{c}^2 & \longrightarrow & \mathbf{c}^1 \times \mathbf{c}^1 \\ \downarrow & & \\ \mathbf{c}^1, & & \end{array}$$

where the horizontal arrow corresponds to the two inert maps $[1] \to [2]$.

4.4.7. *Variant.* Let $(\mathbf{C}, vert, horiz, \text{isom})$ be an object of Trpl. Assume that $vert \subset horiz$. Let $\mathbf{c}^\bullet$ be an object of $\operatorname{Seg}(\mathbf{c})$. Assume that:

- The 'source' map $\mathbf{c}^1 \to \mathbf{c}^0$ belongs to *horiz*;
- The 'target' map $\mathbf{c}^1 \to \mathbf{c}^0$ belongs to *vert*;
- The multiplication map $\mathbf{c}^2 \to \mathbf{c}^1$ belongs to *vert*.

In this case, Corollary 4.4.5 implies that to $\mathbf{c}^\bullet$ there corresponds a canonically defined algebra object in $\operatorname{Corr}(\mathbf{C})_{vert;horiz}$.

4.5. Action on a module: second version. In this subsection we will add some comments on how the construction in Theorem 4.4.2 interacts with an action on modules. The upshot is that the associative algebra in $\operatorname{Corr}(\mathbf{C})$, corresponding to a Segal object $\mathbf{c}^\bullet$, acts on $\mathbf{c}^0$.

4.5.1. Let us return to the setting of Sect. 4.3. For a pair of objects $\mathbf{c}, \mathbf{c}' \in \mathbf{C}$ we consider the monoidal category $\mathbf{Maps}_{\operatorname{Corr}(\mathbf{C})^{\text{all}}_{\text{all;all}}}(\mathbf{c}, \mathbf{c})$ and its module category $\mathbf{Maps}_{\operatorname{Corr}(\mathbf{C})^{\text{all}}_{\text{all;all}}}(\mathbf{c}', \mathbf{c})$.

It will follow from the construction in Theorem 4.4.2 that the right-lax monoidal functor

$$\mathbf{Maps}_{\operatorname{Corr}(\mathbf{C})^{\text{all}}_{\text{all;all}}}(\mathbf{c}, \mathbf{c}) \to \operatorname{Corr}(\mathbf{C})$$

extends to a right-lax map between module categories

$$\mathbf{Maps}_{\operatorname{Corr}(\mathbf{C})^{\text{all}}_{\text{all;all}}}(\mathbf{c}', \mathbf{c}) \to \operatorname{Corr}(\mathbf{C}),$$

which at the level of objects sends

$$\begin{array}{ccc} \widetilde{\mathbf{c}}' & \longrightarrow & \mathbf{c}' \\ \downarrow & & \\ \mathbf{c} & & \end{array} \tag{4.5}$$

to $\widetilde{\mathbf{c}}'$.

At the level of objects, this right-lax monoidal structure is defined as follows. For an object of $\mathbf{Maps}_{\mathrm{Corr}(\mathbf{C})^{\mathrm{all}}_{\mathrm{all;all}}}(\mathbf{c},\mathbf{c})$, given by the diagram

$$\begin{array}{ccc} \widetilde{\mathbf{c}} & \longrightarrow & \mathbf{c} \\ \downarrow & & \\ \mathbf{c} & & \end{array}$$

and the object of $\mathbf{Maps}_{\mathrm{Corr}(\mathbf{C})^{\mathrm{all}}_{\mathrm{all;all}}}(\mathbf{c}',\mathbf{c})$ given by the diagram (4.5), the corresponding 1-morphism in $\mathrm{Corr}(\mathbf{C})$ is given by

$$\begin{array}{ccc} \widetilde{\mathbf{c}} \underset{\mathbf{c}}{\times} \widetilde{\mathbf{c}}' & \longrightarrow & \widetilde{\mathbf{c}} \times \widetilde{\mathbf{c}}' \\ {\scriptstyle \mathrm{id}}\downarrow & & \\ \widetilde{\mathbf{c}} \underset{\mathbf{c}}{\times} \widetilde{\mathbf{c}}'. & & \end{array}$$

4.5.2. Let us take $\mathbf{c}' = *$, and consider the object of $\mathbf{Maps}_{\mathrm{Corr}(\mathbf{C})^{\mathrm{all}}_{\mathrm{all;all}}}(*,\mathbf{c})$, given by

$$\begin{array}{ccc} \mathbf{c} & \longrightarrow & * \\ {\scriptstyle \mathrm{id}}\downarrow & & \\ \mathbf{c}. & & \end{array} \tag{4.6}$$

By Sect. 4.3.2, we obtain that for any Segal object $\mathbf{c}^\bullet$ acting on $\mathbf{c}$, the object (4.6) is naturally a module over the corresponding algebra in $\mathbf{Maps}_{\mathrm{Corr}(\mathbf{C})^{\mathrm{all}}_{\mathrm{all;all}}}(\mathbf{c},\mathbf{c})$.

Hence, applying Sect. 4.5.1, we obtain that the corresponding algebra $\mathbf{c}^1 \in \mathrm{Corr}(\mathbf{C})$ acts on the object $\mathbf{c} \in \mathrm{Corr}(\mathbf{C})$.

Explicitly, the corresponding action map is given by the diagram

$$\begin{array}{ccc} \mathbf{c}^1 & \xrightarrow{\mathrm{id} \times p_s} & \mathbf{c}^1 \times \mathbf{c} \\ {\scriptstyle p_t}\downarrow & & \\ \mathbf{c}. & & \end{array}$$

4.5.3. Note now that if an object $\mathbf{o}$ in the (symmetric) monoidal category $\mathbf{O}$ is dualizable, there exists a universal associative algebra, denoted $\underline{\mathrm{End}}_{\mathbf{O}}(\mathbf{o})$ in $\mathbf{O}$, acting on $\mathbf{o}$. The object of $\mathbf{O}$ underlying $\underline{\mathrm{End}}_{\mathbf{O}}(\mathbf{o})$ is $\mathbf{o}^\vee \otimes \mathbf{o}$.

Applying this to $\mathbf{O} = \mathrm{Corr}(\mathbf{C})$ and $\mathbf{o} = \mathbf{c}$, we obtain a canonically defined homomorphism of algebras

$$\mathbf{c}^1 \to \underline{\mathrm{End}}_{\mathrm{Corr}(\mathbf{C})}(\mathbf{c}). \tag{4.7}$$

Identifying $\mathbf{c}^\vee \simeq \mathbf{c}$ (see Proposition 2.3.4), the map in $\mathrm{Corr}(\mathbf{C})$, underlying the above homomorphism is

$$\mathbf{c}^1 \xrightarrow{p_s \times p_t} \mathbf{c} \times \mathbf{c}. \tag{4.8}$$

4.5.4. Let us now take $\mathbf{c}^\bullet$ to be the Čech nerve of the map $\mathbf{c} \to *$, i.e., $\mathbf{c}^n = \mathbf{c}^{\times(n+1)}$. Note that in this case we obtain that the homomorphism (4.7) is an isomorphism. Indeed, this is so because the map of the underlying objects of $\mathrm{Corr}(\mathbf{C})$, i.e., (4.7) is an isomorphism.

4.5.5. *Variant.* Let us be in the situation of Sect. 4.4.7. Assume in addition that $\mathbf{c}$ is such that the diagonal map $\mathbf{c} \to \mathbf{c} \times \mathbf{c}$ and the tautological map $\mathbf{c} \to *$ belong to *horiz*.

In this case we still obtain that $\mathbf{c} \in \mathrm{Corr}(\mathbf{C})_{vert;horiz}$ is a module over $\mathbf{c}^1$, where the latter is viewed as an associative algebra.

If we assume that the diagonal map $\mathbf{c} \to \mathbf{c} \times \mathbf{c}$ and the tautological map $\mathbf{c} \to *$ belong to *vert* as well, then $\mathbf{c}$ is dualizable in $\mathrm{Corr}(\mathbf{C})_{vert;horiz}$, and the isomorphism of Sect. 4.5.4 happens at the level of associative algebras in $\mathrm{Corr}(\mathbf{C})_{vert;horiz}$.

4.6. Proof of Theorem 4.4.2: introduction.

4.6.1. Let

$$\mathbf{E} := \mathbf{Maps}_{\mathrm{Corr}(\mathbf{C})^{\mathrm{all}}_{\mathrm{all;all}}}(\mathbf{c}, \mathbf{c})^{\otimes, \mathbf{\Delta}^{\mathrm{op}}} \text{ and } \mathbf{F} := \mathrm{Corr}(\mathbf{C})^{\otimes, \mathbf{\Delta}^{\mathrm{op}}}$$

be the coCartesian fibrations over $\mathbf{\Delta}^{\mathrm{op}}$ corresponding to $\mathbf{Maps}_{\mathrm{Corr}(\mathbf{C})^{\mathrm{all}}_{\mathrm{all;all}}}(\mathbf{c}, \mathbf{c})$ and $\mathrm{Corr}(\mathbf{C})$, respectively.

We need to construct a functor $\mathbf{E} \to \mathbf{F}$ over $\mathbf{\Delta}^{\mathrm{op}}$.

4.6.2. Recall (see Sect. 4.2.1) that $\mathbf{E}$ is the coCartesian fibration attached to the functor $\mathbf{\Delta}^{\mathrm{op}} \to 1\text{-Cat}$, given by

$$[n] \mapsto \mathbf{Grid}^{\geq \mathrm{dgnl}}_n(\mathbf{C}) \underset{\mathbf{C}^{n+1}}{\times} \{\mathbf{c}, ..., \mathbf{c}\}. \tag{4.9}$$

Consider another functor $\mathbf{\Delta}^{\mathrm{op}} \to 1\text{-Cat}$,

$$[n] \mapsto \mathbf{Grid}^{> \mathrm{dgnl}}_n(\mathbf{C}), \tag{4.10}$$

where $\mathbf{Grid}^{> \mathrm{dgnl}}_n$ is defined in the same way as as $\mathbf{Grid}^{\geq \mathrm{dgnl}}$, with the difference that we use $([n] \times [n]^{\mathrm{op}})^{> \mathrm{dgnl}}$ instead of $([n] \times [n]^{\mathrm{op}})^{\geq \mathrm{dgnl}}$.

Restriction defines a natural transformation (4.9) $\Rightarrow$ (4.10). Let

$$\mathbf{E}' \to \mathbf{\Delta}^{\mathrm{op}}$$

be the coCartesian fibration corresponding to the functor (4.10). We obtain a functor

$$\mathbf{E} \to \mathbf{E}'.$$

4.6.3. The sought-for functor $\mathbf{E} \to \mathbf{F}$ will be obtained as a composition of the above functor $\mathbf{E} \to \mathbf{E}'$ and a functor $\mathbf{E}' \to \mathbf{F}$ that we will now proceed to define.

Fix an integer k and an object $\alpha \in \mathrm{Seq}_k(\mathbf{\Delta}^{\mathrm{op}})$. We will construct a map

$$\mathrm{Seq}_k(\mathbf{E}') \underset{\mathrm{Seq}_k(\mathbf{\Delta}^{\mathrm{op}})}{\times} \{\alpha\} \to \mathrm{Seq}_k(\mathbf{F}) \underset{\mathrm{Seq}_k(\mathbf{\Delta}^{\mathrm{op}})}{\times} \{\alpha\}, \tag{4.11}$$

functorially in $[k] \in \mathbf{\Delta}^{\mathrm{op}}$ and α.

4.7. Proof of Theorem 4.4.2: Step 1. We shall first give an explicit description of the space

$$\mathrm{Seq}_k(\mathbf{E}') \underset{\mathrm{Seq}_k(\mathbf{\Delta}^{\mathrm{op}})}{\times} \{\alpha\}.$$

4.7.1. We begin with the following general observation. Let $\mathbf{I}$ be an index $(\infty,1)$-category, and let $\mathbf{D} \to \mathbf{I}^{\mathrm{op}}$ be a Cartesian fibration, corresponding to a functor $\mathbf{I} \to 1\text{-Cat}$. Let $\mathbf{D}' \to \mathbf{I}$ be the coCartesian fibration, corresponding to the same functor. I.e., $\mathbf{D}'$ and $\mathbf{D}$ have the same fibers over objects of $\mathbf{I}$.

Note that the space $\mathrm{Seq}_k(\mathbf{D}')$ can be described as follows in terms of $\mathbf{D}$. Namely, $\mathrm{Seq}_k(\mathbf{D}')$ consists of functors

$$([k] \times [k]^{\mathrm{op}})^{\geq \mathrm{dgnl}} \to \mathbf{D}$$

with the property that all the vertical arrows in $([k] \times [k]^{\mathrm{op}})^{\geq \mathrm{dgnl}}$ map to isomorphisms in $\mathbf{I}^{\mathrm{op}}$, and all the horizontal arrows map to arrows in $\mathbf{D}'$ that are Cartesian over $\mathbf{I}^{\mathrm{op}}$.

4.7.2. Note that assignment

$$(i,j) \mapsto ([\alpha(j)] \times [\alpha(j)]^{\mathrm{op}})^{> \mathrm{dgnl}}$$

gives a functor

$$([k] \times [k]^{\mathrm{op}})^{\geq \mathrm{dgnl}} \to 1\text{-Cat}_{\mathrm{ordn}}.$$

Let $\mathbf{I}_\alpha$ denote the corresponding coCartesian fibration over $([k] \times [k]^{\mathrm{op}})^{\geq \mathrm{dgnl}}$.

Consider the space

$$\mathrm{Maps}(\mathbf{I}_\alpha, \mathbf{C}).$$

4.7.3. It follows from Sect. 4.7.1 that the space $\mathrm{Seq}_k(\mathbf{E}') \underset{\mathrm{Seq}_k(\mathbf{\Delta}^{\mathrm{op}})}{\times} \{\alpha\}$ is a full subspace

$$'\mathrm{Maps}(\mathbf{I}_\alpha, \mathbf{C}) \subset \mathrm{Maps}(\mathbf{I}_\alpha, \mathbf{C}),$$

consisting of functors that satisfy the following conditions:

- For every $(i,j) \in ([k] \times [k]^{\mathrm{op}})^{\geq \mathrm{dgnl}}$, the resulting functor
$$([\alpha(j)] \times [\alpha(j)]^{\mathrm{op}})^{> \mathrm{dgnl}} \to \mathbf{C}$$
sends squares to Cartesian squares in $\mathbf{C}$.
- Every arrow in $\mathbf{I}_\alpha$, which is coCartesian over a horizontal arrow in $([k] \times [k]^{\mathrm{op}})^{\geq \mathrm{dgnl}}$, gets sent to an isomorphism in $\mathbf{C}$.

4.8. Proof of Theorem 4.4.2: Step 2. We shall now describe the space

$$\mathrm{Seq}_k(\mathbf{F}) \underset{\mathrm{Seq}_k(\mathbf{\Delta}^{\mathrm{op}})}{\times} \{\alpha\}.$$

4.8.1. Let $\mathbf{C}'$ be a monoidal $(\infty,1)$-category with finite limits, such that the monoidal operation commutes with finite limits. Then $(\mathbf{C}', \mathrm{all}, \mathrm{all}, \mathrm{isom})$ is naturally an associative algebra object in Trpl.

Hence, by Sect. 2.1.2, the $(\infty,1)$-category $\mathrm{Corr}(\mathbf{C}')$ acquires a monoidal structure. Consider the corresponding coCartesian fibration $\mathrm{Corr}(\mathbf{C}')^{\otimes,\mathbf{\Delta}^{\mathrm{op}}}$ over $\mathbf{\Delta}^{\mathrm{op}}$.

Let $\mathbf{C}'^{\otimes,\mathbf{\Delta}} \to \mathbf{\Delta}$ be the *Cartesian* fibration corresponding to the monoidal structure on $\mathbf{C}'$.

Then the space $\mathrm{Seq}_k(\mathrm{Corr}(\mathbf{C}')^{\otimes,\mathbf{\Delta}^{\mathrm{op}}})$ admits the following description. It consists of functors

$$([k] \times [k]^{\mathrm{op}})^{\geq \mathrm{dgnl}} \to \mathbf{C}'^{\otimes,\mathbf{\Delta}},$$

such that vertical arrows in $([k] \times [k]^{\mathrm{op}})^{\geq \mathrm{dgnl}}$ get sent to arrows in $\mathbf{C}'^{\otimes,\mathbf{\Delta}}$ that project to isomorphisms in $\mathbf{\Delta}$;

4.8.2. We apply this to $\mathbf{C}' = \mathbf{C}$ equipped with the Cartesian monoidal structure. We obtain that the space

$$\operatorname{Seq}_k(\mathbf{F}) \underset{\operatorname{Seq}_k(\mathbf{\Delta}^{\mathrm{op}})}{\times} \{\alpha\} \tag{4.12}$$

can be described as follows.

Recall the coCartesian fibration $\mathbf{I}_\alpha \to ([k] \times [k]^{\mathrm{op}})^{\geq \mathrm{dgnl}}$, see Sect. 4.7.2. Then (4.12) identifies with the subspace

$$''\mathrm{Maps}(\mathbf{I}_\alpha, \mathbf{C}) \subset \mathrm{Maps}(\mathbf{I}_\alpha, \mathbf{C}),$$

consisting of functors that satisfy the following condition:

- For every $(i,j) \in ([k] \times [k]^{\mathrm{op}})^{\geq \mathrm{dgnl}}$, the tautological extension of the corresponding functor
$$([\alpha(j)] \times [\alpha(j)]^{\mathrm{op}})^{> \mathrm{dgnl}} \to \mathbf{C}$$
to a functor
$$([\alpha(j)] \times [\alpha(j)]^{\mathrm{op}})^{\geq \mathrm{dgnl}} \to \mathbf{C},$$
where we send the diagonal entries to $* \in \mathbf{C}$ has the property that it sends squares to Cartesian squares in $\mathbf{C}$.

4.8.3. Let us decipher the above condition. Set $n = \alpha(j)$. Let the functor

$$([n] \times [n]^{\mathrm{op}})^{> \mathrm{dgnl}} \to \mathbf{C} \tag{4.13}$$

be given by $\underline{\mathbf{c}} \in \mathbf{Grid}_n^{> \mathrm{dgnl}}(\mathbf{C})$.

Our condition says that all the squares in $\underline{\mathbf{c}}$ must be Cartesian (so it is the same as the corresponding condition for $'\mathrm{Maps}(\mathbf{I}_\alpha, \mathbf{C})$, see Sect. 4.7.3). In addition, we require that for every $m = 0, ..., n-2$, the map

$$\mathbf{c}_{m,m+2} \to \mathbf{c}_{m+1,m+2} \times \mathbf{c}_{m,m+1}$$

be an isomorphism.

Note that, the datum of a functor (4.13) as above is equivalent to that of an object of $\mathbf{C}^{\times n}$ (as it should be).

4.9. Proof of Theorem 4.4.2: Step 3. We shall now complete the construction of a map (4.11) by constructing a map

$$'\mathrm{Maps}(\mathbf{I}_\alpha, \mathbf{C}) \to {''\mathrm{Maps}}(\mathbf{I}_\alpha, \mathbf{C}).$$

4.9.1. Consider the *category*

$$\mathbf{Maps}(\mathbf{I}_\alpha, \mathbf{C}),$$

and the corresponding full subcategories

$$'\mathbf{Maps}(\mathbf{I}_\alpha, \mathbf{C}) \hookrightarrow \mathbf{Maps}(\mathbf{I}_\alpha, \mathbf{C}) \hookleftarrow {''\mathbf{Maps}}(\mathbf{I}_\alpha, \mathbf{C}).$$

It is easy to see, however, that the embedding

$$''\mathbf{Maps}(\mathbf{I}_\alpha, \mathbf{C}) \hookrightarrow \mathbf{Maps}(\mathbf{I}_\alpha, \mathbf{C})$$

admits a left adjoint.

4.9.2. Composing with this left adjoint, we obtain a functor

$$'\mathbf{Maps}(\mathbf{I}_\alpha, \mathbf{C}) \hookrightarrow \mathbf{Maps}(\mathbf{I}_\alpha, \mathbf{C}) \to {}''\mathbf{Maps}(\mathbf{I}_\alpha, \mathbf{C}).$$

Passing to the underlying spaces, we obtain the desired map

$$'\mathrm{Maps}(\mathbf{I}_\alpha, \mathbf{C}) \to {}''\mathrm{Maps}(\mathbf{I}_\alpha, \mathbf{C}).$$

□

Appendix. $(\infty, 2)$-categories

Introduction

1. Why do we need them?

This part plays a service role for Part III, in which we develop the formalism of categories of correspondences.

1.1. As was explained before, an adequate framework to encode the information carried by the assignment

$$S \in \mathrm{Sch}_{\mathrm{aft}} \rightsquigarrow \mathrm{IndCoh}(S) \in \mathrm{DGCat}_{\mathrm{cont}}$$

is in terms of the functor

$$\mathrm{IndCoh}_{\mathrm{Corr}(\mathrm{Sch}_{\mathrm{aft}})^{\mathrm{proper}}_{\mathrm{all,all}}} : \mathrm{Corr}(\mathrm{Sch}_{\mathrm{aft}})^{\mathrm{proper}}_{\mathrm{all,all}} \to (\mathrm{DGCat}_{\mathrm{cont}})^{2\text{-Cat}}. \tag{1.1}$$

Now, the construction of the above functor is such that even if one is ultimately interested only in the 1-categorical data, i.e., the corresponding functor of $(\infty, 1)$-categories

$$\mathrm{IndCoh}_{\mathrm{Corr}(\mathrm{Sch}_{\mathrm{aft}})} : \mathrm{Corr}(\mathrm{Sch}_{\mathrm{aft}}) \to \mathrm{DGCat}_{\mathrm{cont}},$$

in order to produce it, one needs to construct (1.1).

So, $(\infty, 2)$-categories are necessary in order to get IndCoh off the ground.

1.2. Now, one possible approach would be to believe that there exists a reasonable notion of $(\infty, 2)$-category (and companion notions of functor, natural transformation, etc.) and not worry about the details. For example, just imagine that a $(\infty, 2)$-category is a $(\infty, 1)$-category enriched over the monoidal $(\infty, 1)$-category 1-Cat. (The actual definition is indeed along these lines.)

The problem with that is that we need more than just the existence of these notions. We will actually need to perform some pretty non-trivial operations with them. Let us explain what these operations are.

1.3. First off, let us be given an $(\infty, 2)$-category $\mathbb{S}$, equipped with a class of 1-morphisms $\mathbf{C}$ (closed under compositions and containing all isomorphisms). To this data we need to be able to associate a *bi-simplicial* space, denoted $\mathrm{Sq}^{\mathrm{Pair}}_{\bullet,\bullet}(\mathbb{S}, \mathbf{C})$.

The corresponding space of (m, n)-simplices is that of diagrams

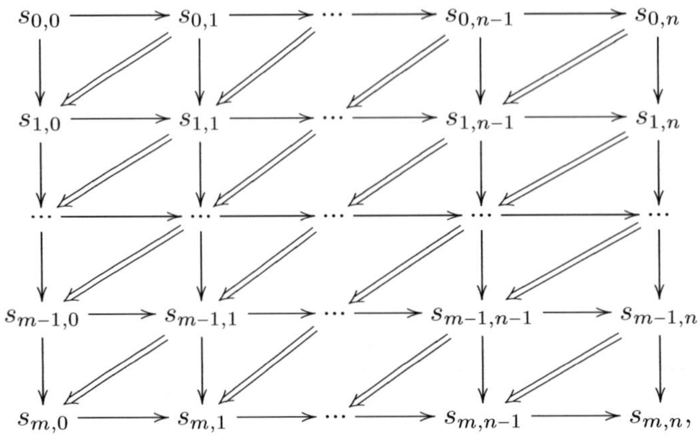

where the horizontal arrows are arbitrary 1-morphisms in $\mathbb{S}$, and the vertical arrows are 1-morphisms that belong to $\mathbf{C}$, and each square represents a (not necessarily invertible) 2-morphism.

Moreover, we need the assignment

$$(\mathbb{S}, \mathbf{C}) \rightsquigarrow \mathrm{Sq}^{\mathrm{Pair}}_{\bullet,\bullet}(\mathbb{S}, \mathbf{C}),$$

viewed as a functor from the category $2\text{-}\mathrm{Cat}^{\mathrm{Pair}}$ of pairs $(\mathbb{S}, \mathbf{C})$ to the category $\mathrm{Spc}^{\mathbf{\Delta}^{\mathrm{op}} \times \mathbf{\Delta}^{\mathrm{op}}}$ of bi-simplicial spaces, to be *fully faithful* with essential image given by some explicit conditions (see Chapter 10, Theorem 5.2.3 for the latter).

1.4. Secondly, for a pair of $(\infty, 2)$-categories $\mathbb{S}$ and $\mathbb{T}$, in addition to the $(\infty, 2)$-category $\mathrm{Funct}(\mathbb{S}, \mathbb{T})$ of functors $\mathbb{S} \to \mathbb{T}$, we need to be able to form its two enlargements, denoted

$$\mathrm{Funct}(\mathbb{S}, \mathbb{T})_{\text{right-lax}} \text{ and } \mathrm{Funct}(\mathbb{S}, \mathbb{T})_{\text{left-lax}},$$

respectively, that have the same class of objects, but where we allow as 1-morphisms right-lax (resp., left-lax) natural transformations (see Chapter 10, Sect. 3.2.7 for the definition).

1.5. While the previous two properties of the sought-for notion of $(\infty, 2)$-category can still be taken for granted, the next one cannot. We will need to be able to perform the following manipulation:

For a 1-morphism in an $(\infty, 2)$-category, it makes sense to ask whether this 1-morphism admits a left or right adjoint (these are notions that take place in the underlying *ordinary* 2-category). Given a functor $\mathbb{S} \to \mathbb{T}$, we shall say that it is right (resp., left) *adjointable* if for every 1-morphism in $\mathbb{S}$, its image in $\mathbb{T}$ admits a right (resp., left) adjoint.

Let

$$\mathrm{Funct}(\mathbb{S}, \mathbb{T})^{R}_{\text{right-lax}} \subset \mathrm{Funct}(\mathbb{S}, \mathbb{T})_{\text{right-lax}}$$

and

$$\mathrm{Funct}(\mathbb{S}, \mathbb{T})^{L}_{\text{left-lax}} \subset \mathrm{Funct}(\mathbb{S}, \mathbb{T})_{\text{left-lax}}$$

be the full $(\infty, 2)$-subcategories that correspond to functors that are left (resp., right) adjointable.

What we need is to have a canonical equivalence

$$\text{Funct}(\mathbb{S}, \mathbb{T})^R_{\text{right-lax}} \simeq \text{Funct}(\mathbb{S}^{1\&2\text{-op}}, \mathbb{T})^L_{\text{left-lax}}, \tag{1.2}$$

given by *passage to adjoint 1-morphisms.*

The construction of this equivalence will be the subject of Chapter 12.

1.6. That said, it is a sensible strategy to get the idea of how we approach $(\infty, 2)$-categories by reading the rest of this introduction, and skipping the bulk of the Appendix on the first pass.

2. Setting up the theory of $(\infty, 2)$-categories

2.1. In Chapter 10 we define what we mean by $(\infty, 2)$-categories.

The idea is to mimic the approach to $(\infty, 1)$-categories via *complete Segal spaces.* And this is what one obtains if one wants to express the idea is that an $(\infty, 2)$-category is just an $(\infty, 1)$-category, *enriched over* 1-Cat: we upgrade the *spaces* of morphisms to $(\infty, 1)$*-categories.*

2.2. So, for us the datum of an $(\infty, 2)$-category $\mathbb{S}$ is that of a *simplicial* $(\infty, 1)$*-category* that we denote $\text{Seq}_\bullet(\mathbb{S})$ (here "Seq" stands for sequences).

Namely, the $(\infty, 1)$-category $\text{Seq}_0(\mathbb{S})$ is actually a *space*, formed by objects of $\mathbb{S}$. I.e., it is the same as one of the underlying $(\infty, 1)$-category $\mathbb{S}^{1\text{-Cat}}$, obtained by discarding non-invertible 1-morphisms in $\mathbb{S}$.

The $(\infty, 1)$-category $\text{Seq}_1(\mathbb{S})$ has as objects 1-morphisms in $\mathbb{S}$. Again, these are the same as objects of $\text{Seq}_1(\mathbb{S}^{1\text{-Cat}})$.

However, whereas the latter is a *space* (i.e.., we only allow homotopies between 1-morphisms in $\mathbb{S}^{1\text{-Cat}}$), in the case of $\text{Seq}_1(\mathbb{S})$, we have non-invertible morphisms. Namely, morphisms between two objects $s_0 \overset{\alpha}{\to} s_1$ and $s_0 \overset{\beta}{\to} s_1$ are 2-morphisms $\alpha \Rightarrow \beta$ in $\mathbb{S}$.

The higher $\text{Seq}_n(\mathbb{S})$ have as objects sequences

$$s_0 \to s_1 \to \ldots \to s_{n-1} \to s_n$$

of objects of $\mathbb{S}$, and as morphisms sequences of 2-morphisms

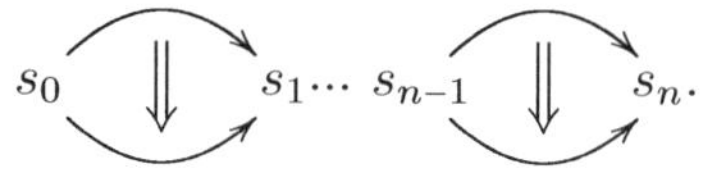

2.3. Formally, we define the $(\infty, 1)$-category of $(\infty, 2)$-categories 2-Cat to be a full subcategory in $1\text{-Cat}^{\mathbf{\Delta}^{\text{op}}}$, given by explicit conditions that are analogous to the condition on an object of $\text{Spc}^{\mathbf{\Delta}^{\text{op}}}$ to be a complete Segal space.

In Chapter 10, Sect. 2.4 we introduce the main example of $(\infty, 2)$-category: this is the $(\infty, 2)$-category of $(\infty, 1)$-categories, denoted **1 - Cat**.

The $(\infty, 2)$-category **1 - Cat** plays the same role vis-à-vis 2-Cat as the $(\infty, 1)$-category Spc vis-à-vis 1-Cat. In particular, it is the recipient of the 2-categorical Yoneda functor, discussed below.

2.4. The above definition $(\infty,2)$-categories is amenable to introducing the notion of *right-lax* functor $\mathbb{S} \dashrightarrow \mathbb{T}$ between two $(\infty,2)$-categories $\mathbb{S}$ to $\mathbb{T}$. The idea of right-lax functors is that they do not strictly preserve compositions of 1-morphisms, but only do so up to (not necessarily invertible) 2-morphisms.

By definition, right-lax functors $\mathbb{S} \dashrightarrow \mathbb{T}$ are functors (subject to a certain non-degeneracy conditions) between the coCartesian fibrations

$$\mathbb{S}^{\oint} \to \mathbf{\Delta}^{\mathrm{op}} \text{ and } \mathbb{T}^{\oint} \to \mathbf{\Delta}^{\mathrm{op}}$$

corresponding to

$$\mathrm{Seq}_\bullet(\mathbb{S}) : \mathbf{\Delta}^{\mathrm{op}} \to 1\text{-Cat} \text{ and } \mathrm{Seq}_\bullet(\mathbb{T}) : \mathbf{\Delta}^{\mathrm{op}} \to 1\text{-Cat},$$

respectively, see Chapter 10, Sect. 3.1.

2.5. Having defined right-lax functors, we can now define the $(\infty,2)$-category

$$\mathrm{Funct}(\mathbb{S},\mathbb{T})_{\text{right-lax}}.$$

Namely, for a test $(\infty,2)$-category $\mathbb{X}$, the space of maps

$$\mathbb{X} \to \mathrm{Funct}(\mathbb{S},\mathbb{T})_{\text{right-lax}}$$

is a certain full subspace in the space of right-lax functors $\mathbb{X} \times \mathbb{S} \dashrightarrow \mathbb{T}$, see Chapter 10, Sect. 3.2.7. Namely, we take those right-lax functors that:

- For every $x \in \mathbb{X}$ the corresponding right-lax functor $\{x\} \times \mathbb{S} \to \mathbb{T}$ is strict;
- For every $s \in \mathbb{S}$ the corresponding right-lax functor $\mathbb{X} \times \{s\} \to \mathbb{T}$ is strict;
- For every $x_0 \overset{\alpha}{\to} x_1$ and $s_0 \overset{\beta}{\to} s_1$, the 2-morphism in $\mathbb{T}$, corresponding to the composition

  $$(x_0,s_0) \overset{(\alpha,\mathrm{id})}{\longrightarrow} (x_1,s_0) \overset{(\mathrm{id},\beta)}{\longrightarrow} (x_1,s_1)$$

  is invertible.

2.6. Having defined the $(\infty,2)$-categories $\mathrm{Funct}(\mathbb{S},\mathbb{T})_{\text{right-lax}}$, we can define the functor

$$\mathrm{Sq}^{\mathrm{Pair}} : 2\text{-Cat}^{\mathrm{Pair}} \to \mathrm{Spc}^{\mathbf{\Delta}^{\mathrm{op}} \times \mathbf{\Delta}^{\mathrm{op}}}, \tag{2.1}$$

mentioned in Sect. 1.3.

Namely, given a pair $(\mathbb{S},\mathbf{C})$, we let the space of (m,n)-simplices in $\mathrm{Sq}^{\mathrm{Pair}}_{\bullet,\bullet}(\mathbb{S},\mathbf{C})$ be the subspace of the space of functors

$$[m] \to \mathrm{Funct}([n],\mathbb{S})_{\text{right-lax}},$$

such that for every $i \in [n]$, the corresponding functor $[m] \to \mathbb{S}$ factors through $\mathbf{C}$.

2.7. We made the decision to leave some statements in Chapter 10 without proof. The majority of the these have to do with the notion of *Gray product*. The most important of them is the theorem that says that the functor (2.1) is fully faithful. The missing proofs will be supplied elsewhere.

3. The rest of the Appendix

3.1. We start Chapter 11 by upgrading the structure of $(\infty,1)$-category on the totality of $(\infty,2)$-categories to that of $(\infty,2)$-category. We denote the latter by **2-Cat**, so that

$$(\mathbf{2}\text{-}\mathbf{Cat})^{1\text{-Cat}} = 2\text{-Cat}.$$

3.2. In Chapter 11, Sect. 2, we introduce the notions of what it means for a functor $\mathbb{T} \to \mathbb{S}$ to be a 1-Cartesian and 2-Cartesian fibration. Both of these notion are obtained by imposing certain *conditions* (as opposed to additional pieces of structure).

The main result of Chapter 11 is the straightening/unstraightening theorem. It says that the $(\infty, 2)$-category of 2-Cartesian (resp., 1-Cartesian) fibrations over $\mathbb{S}$ (with 1-morphisms being functors preserving Cartesian arrows) is equivalent to $(\infty, 2)$-category of functors

$$\mathbb{S}^{1\text{-op}} \to \mathbf{2\text{-}Cat} \quad (\text{resp., } \mathbb{S}^{1\text{-op}} \to \mathbf{1\text{-}Cat}).$$

3.3. Having at our disposal the straightening theorem, starting from the $(\infty, 2)$-category

$$\text{Funct}([1], \mathbb{S})_{\text{right-lax}},$$

projecting to $\mathbb{S} \times \mathbb{S}$ (by evaluation on the two ends of $[1]$), we obtain the 2-categorical Yoneda functor

$$\mathbb{S} \hookrightarrow \text{Funct}(\mathbb{S}^{1\text{-op}}, \mathbf{1\text{-}Cat})$$

that we prove to be a fully faithful embedding.

3.4. Having developed the basics of $(\infty, 2)$-categories, in Chapter 12 we finally address the construction of functors obtained by *passing to adjoints along 1-morphisms*, mentioned in Sect. 1.5.

The main construction of Chapter 12 is given in Sects. 2.2 and 2.3. Namely, given an $(\infty, 2)$-category $\mathbb{S}$, we explicitly describe another $(\infty, 2)$-category, denoted $\mathbb{S}^R$, equipped with a functor

$$\mathbb{S} \to \mathbb{S}^R,$$

which is *universal* with respect to the property of being *left adjointable.*

The construction of $\mathbb{S}^R$ is given in terms of the functor $\text{Sq}_{\bullet,\bullet}$, mentioned in Sect. 1.3, and its left adjoint, denoted $\mathfrak{L}^{Sq}$.

Having this explicit description of $\mathbb{S}^R$ allows to to establish the desired equivalence (1.2).

CHAPTER 10

Basics of 2-Categories

Introduction

0.1. What are $(\infty,2)$-categories? There are multiple definitions of $(\infty,2)$-categories. In this Chapter we adopt the one that mimics the approach to $(\infty,1)$-categories via complete Segal spaces. Let us first recall the latter.

0.1.1. Given an $(\infty,1)$-category $\mathbf{C}$, we attach to it the simplicial space, denoted $\mathrm{Seq}_\bullet(\mathbf{C})$, whose space $\mathrm{Seq}_n(\mathbf{C})$ of n-simplices is the space of n-fold compositions in $\mathbf{C}$, i.e.,

$$\mathrm{Seq}_n(\mathbf{C}) = \mathrm{Maps}_{1\text{-Cat}}([n],\mathbf{C}),$$

where $[n]$ is the category $0 \to 1 \to \dots \to n$.

It turns out that the functor

$$\mathrm{Seq}_\bullet : 1\text{-Cat} \to \mathrm{Spc}^{\mathbf{\Delta}^{\mathrm{op}}}$$

is fully faithful, and one can explicitly (and concisely) describe its essential image: it consists of *complete Segal spaces*, see Sect. 1.2 for the definition.

0.1.2. We *define* $(\infty,2)$-categories by a similar procedure: we let the datum of an $(\infty,2)$-category $\mathbb{S}$ to be a simplicial $(\infty,1)$-*category* $\mathrm{Seq}_\bullet(\mathbb{S})$, subject to conditions analogous to those that single out complete Segal spaces among simplicial spaces.

Thus, we obtain an $(\infty,1)$-category 2-Cat, which is a full subcategory in $\mathrm{Spc}^{\mathbf{\Delta}^{\mathrm{op}}}$.

0.1.3. The idea of $\mathrm{Seq}_\bullet(\mathbb{S})$ is the following. For $n=0$, the $(\infty,1)$-category $\mathrm{Seq}_0(\mathbb{S})$ is the *space* of objects in $\mathbb{S}$, denoted $\mathbb{S}^{\mathrm{Spc}}$.

For $n=1$, the category $\mathrm{Seq}_1(\mathbb{S})$ has as objects 1-morphisms $\alpha : s_0 \to s_1$. Now, morphisms in $\mathrm{Seq}_1(\mathbb{S})$ are diagrams

(0.1)

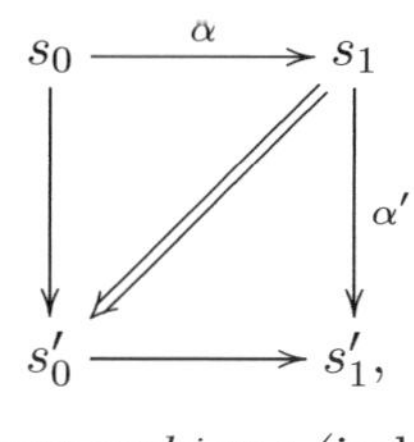

where the vertical arrows are *isomorphisms* (indeed, they must be such, because each column in the above diagram must restrict to a morphism in $\mathrm{Seq}_0(\mathbb{S})$).

Thus, $\mathrm{Seq}_1(\mathbb{S})$ splits as a disjoint union according to $\pi_0(\mathrm{Seq}_0(\mathbb{S}))\times\pi_0(\mathrm{Seq}_0(\mathbb{S}))$. For fixed $(s_0,s_1) \in \mathrm{Seq}_0(\mathbb{S}) \times \mathrm{Seq}_0(\mathbb{S})$, the category

$$\mathrm{Seq}_1(\mathbb{S}) \underset{\mathrm{Seq}_0(\mathbb{S})\times\mathrm{Seq}_0(\mathbb{S})}{\times} \{(s_0,s_1)\}$$

has as morphisms diagrams

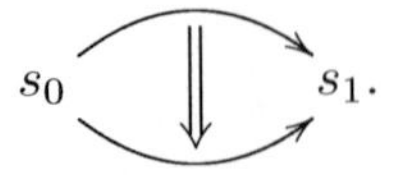

0.1.4. We note that this approach is morally close to the 'enriched ideology' (although we do not attempt to pursue the latter): for each $s_0, s_1 \in \mathbb{S}^{\mathrm{Spc}}$ we upgrade the *space*

$$\mathrm{Maps}_{\mathbb{S}}(s_0, s_1)$$

(which records the structure of the $(\infty,1)$-category underlying $\mathbb{S}$) to an $(\infty,1)$-category

$$\mathbf{Maps}_{\mathbb{S}}(s_0, s_1),$$

the latter being $\mathrm{Seq}_1(\mathbb{S}) \underset{\mathrm{Seq}_0(\mathbb{S})\times\mathrm{Seq}_0(\mathbb{S})}{\times} \{(s_0, s_1)\}$. I.e.,

$$\mathrm{Maps}_{\mathbb{S}}(s_0, s_1) = (\mathbf{Maps}_{\mathbb{S}}(s_0, s_1))^{\mathrm{Spc}}.$$

0.1.5. Having defined $(\infty,2)$-categories, we can now explain the main example of one such: the $(\infty,2)$-category of $(\infty,1)$-categories, denoted $\mathbf{1\text{-}Cat}$, so that

$$(\mathbf{1\text{-}Cat})^{1\text{-}\mathrm{Cat}} = 1\text{-}\mathrm{Cat}.$$

Here is its definition: the corresponding $(\infty,1)$-category $\mathrm{Seq}_n(\mathbf{1\text{-}Cat})$ has as objects *Cartesian* fibrations over $[n]^{\mathrm{op}}$.

Morphisms in this $(\infty,1)$-category are functors over $[n]^{\mathrm{op}}$ (that do *not necessarily* take Cartesian edges to Cartesian edges) but ones that induce an equivalence over each $i \in [n]$.

0.2. What if did not insist that the vertical arrows be isomorphisms?

0.2.1. Having defined $(\infty,2)$-categories, it is natural to ask the following question. Let us attach to an $(\infty,2)$-category $\mathbb{S}$ another simplicial category (denote it by $\mathrm{Seq}^{\mathrm{ext}}_{\bullet}(\mathbb{S})$), as follows:

We let $\mathrm{Seq}^{\mathrm{ext}}_0(\mathbb{S})$ be the $(\infty,1)$-*category* $\mathbb{S}^{1\text{-}\mathrm{Cat}}$ underlying $\mathbb{S}$ (i.e., $\mathbb{S}^{1\text{-}\mathrm{Cat}}$ is obtained from $\mathbb{S}$ by removing non-invertible 2-morphisms). Recall, by contrast, that $\mathrm{Seq}_0(\mathbb{S})$ was the space $\mathbb{S}^{\mathrm{Spc}}$.

The category $\mathrm{Seq}^{\mathrm{ext}}_1(\mathbb{S})$ will still have as objects 1-morphisms $s_0 \to s_1$. But morphisms in $\mathrm{Seq}^{\mathrm{ext}}_1(\mathbb{S})$ will be diagrams (0.1), where we allow *arbitrary* 1-morphisms along the vertical edges (i.e., we no longer require that these 1-morphisms be isomorphisms).

0.2.2. The above construction indeed defines a functor

$$\mathrm{Seq}^{\mathrm{ext}}_{\bullet} : 2\text{-}\mathrm{Cat} \to 1\text{-}\mathrm{Cat}^{\mathbf{\Delta}^{\mathrm{op}}},$$

and this functor also happens to be fully faithful (this is one of the results that are left unproved in this book).

Thus, the $(\infty,1)$-category 2-Cat admits two *different* realizations as a full subcategory in $1\text{-}\mathrm{Cat}^{\mathbf{\Delta}^{\mathrm{op}}}$.

0.2.3. Of course, it is nice to know that the functor $\mathrm{Seq}^{\mathrm{ext}}_\bullet$ is fully faithful. But do we actually need this in order to develop the theory?

The answer is 'yes', and that is mainly for the following reason: we will use the $\mathrm{Seq}^{\mathrm{ext}}_\bullet$ realization of 2-Cat in order to talk about adjunctions.

In more detail, for any $(\infty,2)$-category $\mathbb{T}$, it makes sense to ask whether a given 1-morphism $t_0 \to t_1$ admits a left or right adjoint. Now, let

$$F : \mathbb{S} \to \mathbb{T}$$

be a functor, such that for every 1-morphism $s_0 \overset{\alpha}{\to} s_1$, the corresponding 1-morphism

$$F(s_0) \overset{F(\alpha)}{\longrightarrow} F(s_1)$$

admits a left (resp., right) adjoint.

Then it is natural to expect that in this case, we will be able to canonically construct a functor

$$F^L : \mathbb{S}^{1\&2\text{-op}} \to \mathbb{T} \text{ or } F^R : \mathbb{S}^{1\&2\text{-op}} \to \mathbb{T}$$

(here $\mathbb{S}^{1\&2\text{-op}}$ is the $(\infty,2)$-category obtained from $\mathbb{S}$ by inverting 1- and 2-morphisms), which is the same as F at the level of objects, and at the level of 1-morphisms replaces each $F(\alpha)$ by its left (resp., right) adjoint.

Such a construction is indeed possible, and the functor $\mathrm{Seq}^{\mathrm{ext}}_\bullet$ will be the main tool for carrying it out.

0.2.4. Finally, one can ask the following question: if $\mathrm{Seq}^{\mathrm{ext}}_\bullet$ is so good, why do we not use that instead of $\mathrm{Seq}_\bullet$ in the definition of $(\infty,2)$-categories?

The answer is that we need the $\mathrm{Seq}_\bullet$-realization in order to define the notion of *lax functor* between $(\infty,2)$-categories, see Sect. 0.3.2 below.

So, to summarize, we need both realizations $\mathrm{Seq}_\bullet$ and $\mathrm{Seq}^{\mathrm{ext}}_\bullet$.

0.3. What else is done in this chapter?

0.3.1. In Sect. 1 we recall the realization of 1-Cat via complete Segal spaces, and in Sect. 2 we introduce $(\infty,2)$-categories according to the recipe explained above.

Skipping Sect. 3 for a second, in Sect. 4 we explain the approach to $(\infty,2)$-categories via the functor $\mathrm{Seq}^{\mathrm{ext}}_\bullet$, and in Sect. 5 we describe its essential image (rather, that of its variant $\mathrm{Sq}^{\mathrm{Pair}}_{\bullet,\bullet}$).

In Sect. 6 we upgrade the $(\infty,1)$-category 2-Cat to an $(\infty,2)$-category **2-Cat**.

0.3.2. Let us now return to Sect. 3. In this section we introduce the notion of *right-lax* functor between $(\infty,2)$-categories. Morally, a right-lax functor

$$F : \mathbb{S} \dashrightarrow \mathbb{T}$$

is the same as a functor, with the difference that it only respects composition *up to a not necessarily invertible* 2-morphism.

I.e., for a string

$$s_0 \overset{\alpha}{\to} s_1 \overset{\beta}{\to} s_2$$

in $\mathbb{S}$, we are supposed to be given a 2-morphism in $\mathbb{T}$

$$F(\beta) \circ F(\alpha) \to F(\beta \circ \alpha).$$

0.3.3. Formally, the definition is given as follows. Let $\mathbb{S}^{\oint}$ and $\mathbb{T}^{\oint}$ be the coCartesian fibrations over $\mathbf{\Delta}^{\mathrm{op}}$, corresponding to the functors

$$\mathrm{Seq}_\bullet(\mathbb{S}), \mathrm{Seq}_\bullet(\mathbb{T}) : \mathbf{\Delta}^{\mathrm{op}} \to 1\text{-Cat},$$

respectively.

A genuine (i.e., strict) functor $\mathbb{S} \to \mathbb{T}$ is the same as a functor $\mathbb{S}^{\oint} \to \mathbb{T}^{\oint}$ over $\mathbf{\Delta}^{\mathrm{op}}$ that takes coCartesian edges to coCartesian edges.

A right-lax functor $\mathbb{S} \dashrightarrow \mathbb{T}$ is, by definition, a functor $\mathbb{S}^{\oint} \to \mathbb{T}^{\oint}$ over $\mathbf{\Delta}^{\mathrm{op}}$ that takes coCartesian edges *that lie over idle arrows in* $\mathbf{\Delta}^{\mathrm{op}}$ to coCartesian edges (we refer the reader to Sect. 3.1.2 where the notion of idle arrow in $\mathbf{\Delta}^{\mathrm{op}}$ is defined).

0.3.4. The notion of right-lax functor allows us to introduce the notion of *Gray product* of $(\infty, 2)$-categories. Given, $\mathbb{S}, \mathbb{T} \in 2\text{-Cat}$, their Gray product, denoted

$$\mathbb{S} \circledast \mathbb{T}$$

is a $(\infty, 2)$-category, equipped with a *right-lax* functor

$$\mathbb{S} \times \mathbb{T} \dashrightarrow \mathbb{S} \circledast \mathbb{T},$$

universal with respect to the following property:

For a pair of 1-morphisms

$$s_0 \overset{\phi}{\to} s_1 \text{ and } t_0 \overset{\psi}{\to} t_1,$$

the diagram

$$\begin{array}{ccc} (s_0, t_0) & \xrightarrow{(\mathrm{id},\psi)} & (s_0, t_1) \\ {\scriptstyle (\phi,\mathrm{id})}\downarrow & & \downarrow{\scriptstyle (\phi,\mathrm{id})} \\ (s_1, t_0) & \xrightarrow{(\mathrm{id},\psi)} & (s_1, t_1) \end{array}$$

in $\mathbb{S} \circledast \mathbb{T}$ no longer commutes, but only does so up to a *non-invertible* 2-morphism

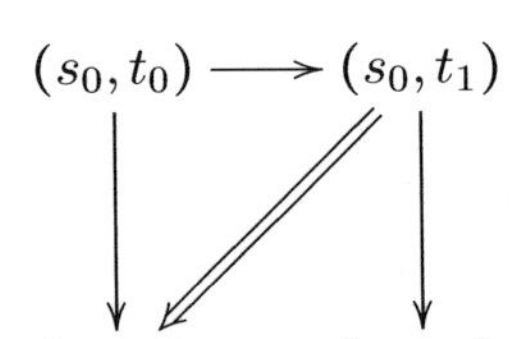

0.3.5. The Gray product produces something non-trivial even if $\mathbb{S} = \mathbf{I}$ and $\mathbb{T} = \mathbf{J}$ are $(\infty, 1)$-categories. Consider the simplest example of $\mathbf{I} = \mathbf{J} = [1]$. In this case, the $(\infty, 2)$-category

$$[1] \circledast [1] =: [1, 1]$$

can be depicted as

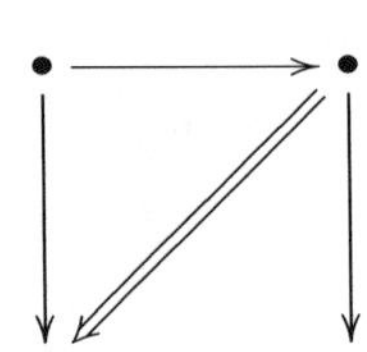

0.3.6. The notion of Gray product allows to introduce the notion of *right-lax natural transformation* between functors. In general, for $\mathbb{S},\mathbb{T}\in 2\text{-Cat}$, we introduce the $(\infty,2)$-category

$$\mathrm{Funct}(\mathbb{S},\mathbb{T})_{\text{right-lax}}$$

(of genuine functors, but where we allow right-lax natural transformations) by

$$\mathrm{Maps}_{2\text{-Cat}}(\mathbb{X},\mathrm{Funct}(\mathbb{S},\mathbb{T})_{\text{right-lax}}) = \mathrm{Maps}_{2\text{-Cat}}(\mathbb{X}\circledast\mathbb{S},\mathbb{T}).$$

The $(\infty,2)$-category contains as a 1-full subcategory the usual $(\infty,2)$-category of functors, denoted $\mathrm{Funct}(\mathbb{S},\mathbb{T})$, and defined by

$$\mathrm{Maps}_{2\text{-Cat}}(\mathbb{X},\mathrm{Funct}(\mathbb{S},\mathbb{T})) = \mathrm{Maps}_{2\text{-Cat}}(\mathbb{X}\times\mathbb{S},\mathbb{T}).$$

0.3.7. For example, we have:

$$\mathrm{Seq}_n^{\mathrm{ext}}(\mathbb{S}) = \mathrm{Funct}([n],\mathbb{T})_{\text{right-lax}}.$$

0.4. Status of the assertions.

0.4.1. Unfortunately, the existing literature on $(\infty,2)$-categories does not contain the proofs of all the statements that we need. We decided to leave some of the statements unproved, and supply the corresponding proofs elsewhere (including the proofs here would have altered the order of the exposition, and would have come at the expense of clarity).

0.4.2. Here is the list of the unproved statements:

Proposition 3.2.6 says that the formation of Gray product commutes with colimits in each variable.

Proposition 3.2.9 asserts the associativity of the Gray product.

Theorems 4.1.3, Theorem 4.3.5, Theorem 4.6.3 and Theorem 5.2.3 are all generalizations of the assertion that the functor $\mathrm{Seq}_\bullet^{\mathrm{ext}}$ (or, rather, its variant $\mathrm{Sq}_{\bullet,\bullet}^{\mathrm{Pair}}$) is fully faithful with specified essential image.

Proposition 4.5.4 gives an explicit description of the Gray product in terms of the functor $\mathrm{Sq}_{\bullet,\bullet}^{\mathrm{Pair}}$.

It is quite possible that references for (some of) the above statements do exist, and we would be grateful if the reader could point them out to us.

1. Recollections: $(\infty,1)$-categories via complete Segal spaces

As a warm-up to the definition of $(\infty,2)$-categories, in this section we will recall the description of $(\infty,1)$-categories as *complete Segal spaces*.

In the case of $(\infty,2)$-categories, we will follow the same route, but with its inherent complications.

1.1. The category $\boldsymbol{\Delta}$. Let us recall the following from Chapter 1:

1.1.1. For an integer n we let $[n]$ denote the ordinary 1-category symbolically represented as

$$0\to 1\to\ldots\to n.$$

1.1.2. We let $\mathbf{\Delta}$ be the (ordinary) category, whose objects are $[n]$ for $n \in \mathbf{N}$ and whose morphisms are functors $[n_1] \to [n_2]$.

By constrution, $\mathbf{\Delta}$ comes equipped with a fully faithful functor to 1-Cat.

1.1.3. The category $\mathbf{\Delta}$ carries a canonical involution, denoted rev. It acts as identity on objects, and on morphisms it is defined via the commutative diagrams

$$\begin{array}{ccc} [n_1] & \xrightarrow{\text{rev}(\alpha)} & [n_2] \\ \downarrow & & \downarrow \\ [n_1]^{\text{op}} & \xrightarrow{\alpha} & [n_2]^{\text{op}}, \end{array}$$

where the vertical arrows are the canonical equivalences

$$[n] \to [n]^{\text{op}}, \quad i \mapsto n - i.$$

1.2. (Complete) Segal spaces.

1.2.1. Consider the $(\infty,1)$-category of simplicial spaces, i.e., $\text{Spc}^{\mathbf{\Delta}^{\text{op}}}$. Let us recall that an object

$$E_\bullet \in \text{Spc}^{\mathbf{\Delta}^{\text{op}}}$$

is said to be a *Segal* space if the following condition is satisfied:

For any $n = n_1 + n_2$, the natural map

$$E_n \to E_{n_1} \underset{E_0}{\times} E_{n_2},$$

is an isomorphism in Spc.

In the above formula, the maps $E_{n_1} \to E_0 \leftarrow E_{n_2}$ are given by

$$0 \in [0] \mapsto n_1 \in [n_1] \text{ and } 0 \in [0] \mapsto 0 \in [n_2],$$

respectively.

1.2.2. We leave it to the reader to work out that for a Segal space, the set $\pi_0(E_1)$ acquires an associative binary operation

$$\pi_0(E_1) \underset{\pi_0(E_0)}{\times} \pi_0(E_1) \to \pi_0(E_1),$$

so that it makes sense to talk about the subset of *invertible* arrows

$$(\pi_0(E_1))^{\text{invert}} \subset \pi_0(E_1).$$

We let $(E_1)^{\text{invert}} \subset E_1$ be the corresponding full subspace (i.e., union of connected components).

It follows from the definitions that the degeneracy map $E_0 \to E_1$ factors through a map

$$E_0 \to (E_1)^{\text{invert}}.$$

1.2.3. Recall that a Segal space $E_\bullet$ is said to be *complete* if the above map $E_0 \to (E_1)^{\text{invert}}$ is an isomorphism in Spc.

1.3. The functor $\text{Seq}_\bullet$**.** The idea of the functor $\text{Seq}_\bullet$ is very simple: we want to record the datum of an ∞-category by keeping track of *spaces* of n-fold compositions, for every n, along with its simplicial structure as we vary n.

1.3.1. We construct the functor of $(\infty,1)$-categories

$$\mathrm{Seq}_\bullet : 1\text{-Cat} \to \mathrm{Spc}^{\mathbf{\Delta}^{\mathrm{op}}}$$

by sending $\mathbf{C} \in 1\text{-Cat}$ to the simplicial space, whose n-simplices is the space

$$\mathrm{Maps}_{1\text{-Cat}}([n], \mathbf{C})$$

of functors $[n] \to \mathbf{C}$.

1.3.2. The functor $\mathrm{Seq}_\bullet$ admits a left adjoint, denoted $\mathfrak{L}$. Tautologically, $\mathfrak{L}$ is the left-Kan extension along the Yoneda embedding $\mathbf{\Delta} \hookrightarrow \mathrm{Spc}^{\mathbf{\Delta}^{\mathrm{op}}}$ of the functor tautological functor

$$\mathbf{\Delta} \to 1\text{-Cat},$$

i.e., the functor that sends an ordered finite set to itself, viewed as an ordinary category.

1.3.3. We now quote the following fundamental fact ([**Rezk1, JT**]):

THEOREM 1.3.4. *The above functor* $\mathrm{Seq}_\bullet$ *is fully faithful. Its essential image is the full subcategory of* $\mathrm{Spc}^{\mathbf{\Delta}^{\mathrm{op}}}$ *that consists of complete Segal spaces.*

1.3.5. The category 1-Cat carries a natural involution, denoted

$$\mathbf{C} \mapsto \mathbf{C}^{\mathrm{op}}.$$

It is uniquely characterized by the property that the the functor $\mathrm{Seq}_\bullet$ intertwines this involution with one on $\mathrm{Spc}^{\mathbf{\Delta}^{\mathrm{op}}}$, induced by the functor $\mathrm{rev} : \mathbf{\Delta} \to \mathbf{\Delta}$.

1.4. Properties of categories and functors in terms of $\mathrm{Seq}_\bullet$**.** In this subsection we will show how to translate various properties of ∞-categories (such as the property of being ordinary) or functors (such as the property of being fully faithful) into properties of the corresponding simplicial space.

1.4.1. First, let us observe that an $(\infty,1)$-category $\mathbf{C}$ is ordinary if for any $\mathbf{c}_0, \mathbf{c}_1 \in \mathrm{Seq}_0(\mathbf{c})$, the space

$$\mathrm{Seq}_1(\mathbf{C}) \underset{\mathrm{Seq}_0(\mathbf{c}) \times \mathrm{Seq}_0(\mathbf{c})}{\times} \{(\mathbf{c}_0, \mathbf{c}_1)\},$$

is *discrete*.

Recall that

$$1\text{-Cat}_{\mathrm{ordn}} \subset 1\text{-Cat}$$

denotes the full subcategory that consists of ordinary categories, and that the above inclusion admits a left adjoint, denoted

$$\mathbf{C} \mapsto \mathbf{C}^{\mathrm{ordn}}.$$

Sometimes, $\mathbf{C}^{\mathrm{ordn}}$ is called the *homotopy category* of $\mathbf{C}$, and is denoted $\mathrm{Ho}(\mathbf{C})$.

1.4.2. We have a fully faithful inclusion

$$\mathrm{Spc} \hookrightarrow 1\text{-Cat}.$$

Namely, $\mathbf{C} \in 1\text{-Cat}$ is a space if and only if $\mathrm{Seq}_\bullet(\mathbf{C})$ is degenerate, i.e., the degeneracy map $\mathrm{Seq}_0(\mathbf{C}) \to \mathrm{Seq}_n(\mathbf{C})$ is an isomorphism for every n. Note that given the Segal condition, it is enough to check this for $n = 1$.

The inclusion $\mathrm{Spc} \hookrightarrow 1\text{-Cat}$ admits a right adjoint, given by

$$\mathbf{C} \mapsto \mathbf{C}^{\mathrm{Spc}}.$$

We have

$$\mathrm{Seq}_n(\mathbf{C}^{\mathrm{Spc}}) \simeq \mathrm{Seq}_0(\mathbf{C}), \quad \forall\, n.$$

1.4.3. It follows from the definitions that a functor between $(\infty,1)$-categories $F : \mathbf{C} \to \mathbf{D}$ is *fully faithful* if and only if the corresponding map of spaces

$$\mathrm{Seq}_1(\mathbf{C}) \to \mathrm{Seq}_1(\mathbf{D}) \underset{\mathrm{Seq}_0(\mathbf{D})\times\mathrm{Seq}_0(\mathbf{D})}{\times} (\mathrm{Seq}_0(\mathbf{C}) \times \mathrm{Seq}_0(\mathbf{C})) \tag{1.1}$$

is an isomorphism (in Spc).

Note that if $\mathbf{C}$ and $\mathbf{D}$ are both spaces, then F is fully faithful if and only if it is a *monomorphism*, i.e., the inclusion of a union of connected components. Indeed, the above condition is equivalent to

$$\mathrm{Seq}_0(\mathbf{C}) \to \mathrm{Seq}_0(\mathbf{C}) \underset{\mathrm{Seq}_0(\mathbf{D})}{\times} \mathrm{Seq}_0(\mathbf{C})$$

being an isomorphism.

1.4.4. Recall that notion of a functor being *1-fully faithful*, see Chapter 1, Sect. 1.2.4 (for a functor between ordinary categories '1-fully faithful' is what is usually called 'faithful').

It is easy to see that $F : \mathbf{C} \to \mathbf{D}$ is 1-fully faithful if and only if the map (1.1) is a monomorphism.

1.4.5. Recall also the notion of a *1-replete* functor, see Chapter 1, Sect. 1.2.5. It is not difficult to see that this is equivalent to the condition that the functor

$$\mathrm{Seq}_1(\mathbf{C}) \to \mathrm{Seq}_1(\mathbf{D})$$

should be fully faithful.

1.5. The $(\infty,1)$-category 1-Cat. In this subsection we will describe the object

$$\mathrm{Seq}_\bullet(1\text{-Cat}) \in 1\text{-Cat}^{\mathbf{\Delta}^{\mathrm{op}}}.$$

1.5.1. For an $(\infty,1)$-category $\mathbf{I}$, recall that

$$\text{coCart}_{/\mathbf{I}} \subset 1\text{-Cat}_{/\mathbf{I}}$$

denotes the full subcategory consisting of coCartesian fibrations.

Recall that

$$(\text{coCart}_{/\mathbf{I}})_{\text{strict}} \subset \text{coCart}_{/\mathbf{I}}$$

denotes the 1-full subcategory with the same objects, but where 1-morphisms are functors that send arrows that are coCartesian over $\mathbf{I}$ to arrows that are coCartesian over $\mathbf{I}$.

Recall also that we denote by

$$0\text{-coCart}_{/\mathbf{I}} \subset \text{coCart}_{/\mathbf{I}}$$

the full subcategory consisting of coCartesian fibrations in spaces.

The above notation carries over *mutatis mutandis* to the case of *Cartesian* fibrations.

1.5.2. By definition,

$$\text{Seq}_\bullet(1\text{-Cat}) \simeq \text{Maps}_{1\text{-Cat}}([\bullet], 1\text{-Cat}).$$

Applying Chapter 1, Sect. 1.4.2, we obtain that

$$\text{Seq}_\bullet(1\text{-Cat}) \simeq (\text{coCart}_{/[\bullet]})^{\text{Spc}}.$$

Under this identification, the full subcategory $\text{Spc} \subset 1\text{-Cat}$ corresponds to

$$(0\text{-coCart}_{/[\bullet]})^{\text{Spc}} \subset (\text{coCart}_{/[\bullet]})^{\text{Spc}}.$$

1.6. Unstraightening.

1.6.1. The incarnation of $(\infty,1)$-categories as complete Segal spaces gives an explicit description of the unstraightening functor

$$\text{Maps}_{1\text{-Cat}}(\mathbf{C}, 1\text{-Cat}) \to (\text{coCart}_{/\mathbf{C}})^{\text{Spc}}.$$

Let us be given a functor $F : \mathbf{C} \to 1\text{-Cat}$. Let us describe the complete Segal space of the corresponding coCartesian fibration $\widetilde{\mathbf{C}} \to \mathbf{C}$.

1.6.2. For each n consider the category

$$([n] \times [n])^{\geq \text{diag}},$$

equipped[1] with the projection on the second coordinate

$$([n] \times [n])^{\geq \text{diag}} \to [n].$$

This is a coCartesian fibration of *ordinary* categories, and consider the corresponding functor

$$\iota_n : [n] \to 1\text{-Cat}.$$

1.6.3. Now, the space $\text{Seq}_n(\widetilde{\mathbf{C}})$ is described as follows: it is the space of pairs consisting of a functor

$$[n] \to \mathbf{C},$$

and a natural transformation from ι_n to the composite functor

$$[n] \to \mathbf{C} \xrightarrow{F} 1\text{-Cat}.$$

[1]This is the full subcategory of $[n] \times [n]$ consisting of objects (i, j) with $i \leq j$.

2. The notion of $(\infty,2)$-category

In this section we give the definition of an $(\infty,2)$-category. In doing so, we will follow C. Barwick in approaching $(\infty,2)$-categories via complete Segal spaces.

Namely, the datum of a $(\infty,2)$-category will consist of an assignment for every n of the $(\infty,1)$-category whose objects are n-fold compositions, and whose morphisms are strings of 2-morphisms.

2.1. Definition of the $(\infty,1)$-category of $(\infty,2)$-categories. In this subsection we introduce the $(\infty,1)$-category of $(\infty,2)$-categories, to be denoted 2-Cat.

2.1.1. We define 2-Cat as a full subcategory of $1\text{-Cat}^{\boldsymbol{\Delta}^{\mathrm{op}}}$, defined by the following three conditions:

Condition 0: We require that $\mathbf{E}_0 \in 1\text{-Cat}$ be a space.

Condition 1: We require that for any $n = n_1 + n_2$, the map (i.e., functor between $(\infty,1)$-categories)

$$\mathbf{E}_n \to \mathbf{E}_{n_1} \underset{\mathbf{E}_0}{\times} \mathbf{E}_{n_2}$$

be an isomorphism in 1-Cat (i.e., an equivalence of $(\infty,1)$-categories).

To formulate Condition 2 we note that given Condition 1, the ordinary category $(\mathbf{E}_1)^{\mathrm{ordn}}$ contains a 1-full subcategory $((\mathbf{E}_1)^{\mathrm{ordn}})^{\mathrm{invert}}$. We let $(\mathbf{E}_1)^{\mathrm{invert}}$ to be the corresponding 1-full subcategory of $\mathbf{E}_1$.

It is easy to see that the degeneracy functor $\mathbf{E}_0 \to \mathbf{E}_1$ (automatically, uniquely) factors as $\mathbf{E}_0 \to (\mathbf{E}_1)^{\mathrm{invert}}$.

Condition 2. We require that the above map (i.e., a functor of $(\infty,1)$-categories)

$$\mathbf{E}_0 \to (\mathbf{E}_1)^{\mathrm{invert}}$$

be an isomorphism in 1-Cat (i.e., be an equivalence of $(\infty,1)$-categories).

REMARK 2.1.2. One can show that, given Conditions 0 and 1, the $(\infty,1)$-category $(\mathbf{E}_1)^{\mathrm{invert}}$ is actually a space. In fact, Conditions 1 and 2 imply that $(\mathbf{E}_\bullet)^{\mathrm{Spc}}$ is a Segal space and the natural map $((\mathbf{E}_1)^{\mathrm{Spc}})^{\mathrm{invert}} \to (\mathbf{E}_1)^{\mathrm{invert}}$ is an isomorphism (of spaces).

From here one deduces that Condition 2 can be replaced by a seemingly weaker condition:

Condition 2': We require that the Segal space $(\mathbf{E}_\bullet)^{\mathrm{Spc}}$ be complete.

2.1.3. We will denote the tautological fully faithful embedding

$$2\text{-Cat} \to 1\text{-Cat}^{\boldsymbol{\Delta}^{\mathrm{op}}}$$

by $\mathrm{Seq}_\bullet$.

2.1.4. Note that the category 2-Cat has limits, and the functor $\mathrm{Seq}_\bullet$ commutes with limits. Indeed, it suffices to observe that Conditions 0, 1 and 2 above are all stable under taking limits in $1\text{-Cat}^{\boldsymbol{\Delta}^{\mathrm{op}}}$.

For future reference we record:

LEMMA 2.1.5. *The category* 2-Cat *is presentable (in particular, contains colimits).*

2.1.6. The category 2-Cat carries a pair of mutually commuting involutions, denoted

$$\mathbb{S} \mapsto \mathbb{S}^{1\text{-op}} \text{ and } \mathbb{S} \mapsto \mathbb{S}^{2\text{-op}},$$

respectively.

The involution $\mathbb{S} \mapsto \mathbb{S}^{1\text{-op}}$ is uniquely characterized by the property that the functor $\mathrm{Seq}_\bullet$ intertwines it with the involution on $1\text{-Cat}^{\mathbf{\Delta}^{\mathrm{op}}}$, induced by the involution $\mathrm{rev} : \mathbf{\Delta} \to \mathbf{\Delta}$.

The involution $\mathbb{S} \mapsto \mathbb{S}^{1\text{-op}}$ is uniquely characterized by the property that the functor $\mathrm{Seq}_\bullet$ intertwines it with the involution on $1\text{-Cat}^{\mathbf{\Delta}^{\mathrm{op}}}$, induced by the involution

$$1\text{-Cat} \to 1\text{-Cat}, \quad \mathbf{C} \mapsto \mathbf{C}^{\mathrm{op}}.$$

In what follows, we shall denote by

$$\mathbb{S} \mapsto \mathbb{S}^{1\&2\text{-op}} \simeq \mathbb{S}^{2\&1\text{-op}}$$

the composition of the above two involutions.

2.2. Basic properties of $(\infty, 2)$-categories. Since $(\infty, 2)$-categories were defined via simplicial $(\infty, 1)$-categories, their properties (such as being ordinary) are expressed in such terms.

2.2.1. We have a fully faithful embedding

$$1\text{-Cat} \to 2\text{-Cat} \tag{2.1}$$

that makes the diagram

$$\begin{array}{ccc} 1\text{-Cat} & \xrightarrow{\mathrm{Seq}_\bullet} & \mathrm{Spc}^{\mathbf{\Delta}^{\mathrm{op}}} \\ \downarrow & & \downarrow \\ 2\text{-Cat} & \xrightarrow{\mathrm{Seq}_\bullet} & 1\text{-Cat}^{\mathbf{\Delta}^{\mathrm{op}}} \end{array} \tag{2.2}$$

commute.

2.2.2. The embedding (2.1) admits a right adjoint, to be denoted

$$\mathbb{S} \mapsto \mathbb{S}^{1\text{-Cat}}.$$

This right adjoint can be characterized by the fact that the natural transformation in the diagram

$$\begin{array}{ccc} 1\text{-Cat} & \xrightarrow{\mathrm{Seq}_\bullet} & \mathrm{Spc}^{\mathbf{\Delta}^{\mathrm{op}}} \\ {\scriptstyle \mathbb{S} \mapsto \mathbb{S}^{1\text{-Cat}}} \uparrow & & \uparrow {\scriptstyle \mathbf{E}_\bullet \mapsto (\mathbf{E}_\bullet)^{\mathrm{Spc}}} \\ 2\text{-Cat} & \xrightarrow{\mathrm{Seq}_\bullet} & 1\text{-Cat}^{\mathbf{\Delta}^{\mathrm{op}}}, \end{array}$$

obtained by passing to right adjoints along the vertical arrows in (2.2), is an isomorphism.

We denote

$$\mathbb{S}^{\mathrm{Spc}} := (\mathbb{S}^{1\text{-Cat}})^{\mathrm{Spc}}.$$

2.2.3. We shall say that an $(\infty, 2)$-category $\mathbb{S}$ is *ordinary* if the $(\infty, 1)$-category $\mathrm{Seq}_1(\mathbb{S})$ is ordinary.

Let $2\text{-Cat}_{\mathrm{ordn}}$ denote the full subcategory of 2-Cat consisting of ordinary 2-categories. It is easy to see that $2\text{-Cat}_{\mathrm{ordn}}$ identifies with the category of *ordinary* (a.k.a., usual, classical) 2-categories.

REMARK 2.2.4. We always work up to coherent homotopy and what we call an ordinary 2-category is what is often called a 'bicategory' in the literature.

2.2.5. The embedding

$$2\text{-Cat}_{\mathrm{ordn}} \hookrightarrow 2\text{-Cat}$$

admits a left adjoint, to be denoted

$$\mathbb{S} \mapsto \mathbb{S}^{\mathrm{ordn}}.$$

It is not difficult to see that this left adjoint is given by sending the corresponding $\mathbf{E}_\bullet \in 1\text{-Cat}^{\mathbf{\Delta}^{\mathrm{op}}}$ to the ordinary simplicial category $(\mathbf{E}_\bullet)^{\mathrm{ordn}}$.

REMARK 2.2.6. Note, however, that the diagram

$$\begin{array}{ccc} 1\text{-Cat} & \longrightarrow & 1\text{-Cat}_{\mathrm{ordn}} \\ \downarrow & & \downarrow \\ 2\text{-Cat} & \longrightarrow & 2\text{-Cat}_{\mathrm{ordn}} \end{array}$$

does not commute.

I.e.,, an $(\infty, 2)$-category may be ordinary, only have invertible 2-morphisms, thus being an $(\infty, 1)$-category, but not an *ordinary* category.

2.2.7. In what follows we shall refer to points of $\mathrm{Seq}_0(\mathbb{S})$ as *objects* of $\mathbb{S}$. For $s', s'' \in \mathbb{S}$, we consider the category

$$\mathrm{Seq}_1(\mathbb{S}) \underset{\mathrm{Seq}_0(\mathbb{S}) \times \mathrm{Seq}_0(\mathbb{S})}{\times} \{s' \times s''\}.$$

We shall refer to it as the *category of morphisms* from s' to s'' and denote it by $\mathbf{Maps}_{\mathbb{S}}(s', s'')$.

We shall use the notation $\mathrm{Maps}_{\mathbb{S}}(s', s'')$ for

$$\mathbf{Maps}_{\mathbb{S}}(s', s'')^{\mathrm{Spc}} \simeq \mathrm{Maps}_{\mathbb{S}^{1\text{-Cat}}}(s', s'').$$

2.3. Properties of functors between $(\infty, 2)$-categories.

2.3.1. Let $F : \mathbb{S} \to \mathbb{T}$ be a functor between $(\infty, 2)$-categories. We shall say that is *fully faithful* if the resulting functor if $(\infty, 1)$-categories

$$\mathrm{Seq}_1(\mathbb{S}) \to \mathrm{Seq}_1(\mathbb{T}) \underset{\mathrm{Seq}_0(\mathbb{T}) \times \mathrm{Seq}_0(\mathbb{T})}{\times} (\mathrm{Seq}_0(\mathbb{S}) \times \mathrm{Seq}_0(\mathbb{S})) \tag{2.3}$$

is an equivalence.

Equivalently, this means that for every $s', s'' \in \mathbb{S}$, the functor

$$\mathbf{Maps}_{\mathbb{S}}(s', s'') \to \mathbf{Maps}_{\mathbb{T}}(F(s'), F(s'')) \tag{2.4}$$

is an equivalence.

2.3.2. We shall say that F is *1-fully faithful* if the functor (2.3) is fully faithful. Equivalently, this means the functor (2.4) is fully faithful for any $s', s'' \in \mathbb{S}$.

2.3.3. We shall say that F is *1-replete* if the functor

$$\mathrm{Seq}_1(\mathbb{S}) \to \mathrm{Seq}_1(\mathbb{T})$$

is fully faithful.

A functor which is 1-replete is an equivalence onto what we call a *1-full subcategory.* For an $(\infty, 2)$-category $\mathbb{S}$, its 1-full subcategories are in bijection with those in $\mathbb{S}^{\mathrm{ordn}}$, and also with those of $\mathbb{S}^{1\text{-}\mathrm{Cat}}$.

2.3.4. We shall say that a functor F is *2-fully faithful* if the functor (2.3) is 1-fully faithful (equivalently, if the functor (2.4) is 1-fully faithful for any $s', s'' \in \mathbb{S}$).

2.3.5. We shall say that F is *2-replete* if the functor (2.3) is 1-replete (equivalently, if the functor (2.4) is 1-replete for any $s', s'' \in \mathbb{S}$).

2.3.6. A functor which is 2-replete is an equivalence onto what we call a *2-full subcategory.*

For an $(\infty, 2)$-category $\mathbb{S}$, its 2-full subcategories are in bijection with those in $\mathbb{S}^{\mathrm{ordn}}$. Each such is determined by a subset of $\pi_0(\mathrm{Seq}_1(\mathrm{Seq}_1(\mathbb{S})))$, closed under compositions.

2.4. The $(\infty, 2)$-category $\mathbf{1\text{-}Cat}$. The main example of an $(\infty, 2)$-category is the $(\infty, 2)$-category of $(\infty, 1)$-categories, denoted $\mathbf{1\text{-}Cat}$. In this subsection we define what it is.

2.4.1. We introduce the $(\infty, 2)$-category $\mathbf{1\text{-}Cat}$ as follows. We let $\mathrm{Seq}_n(\mathbf{1\text{-}Cat})$ be the 1-full subcategory of $\mathrm{Cart}_{/[n]^{\mathrm{op}}}$, where we restrict 1-morphisms to functors that induce an equivalence over each $i \in [n]$.

The assignment

$$n \mapsto \mathrm{Seq}_n(\mathbf{1\text{-}Cat})$$

clearly defines an object of $1\text{-}\mathrm{Cat}^{\mathbf{\Delta}^{\mathrm{op}}}$.

PROPOSITION 2.4.2.

(a) *The above object* $\mathrm{Seq}_\bullet(\mathbf{1\text{-}Cat})$ *lies in the essential image of the functor*

$$\mathrm{Seq}_\bullet : 2\text{-}\mathrm{Cat} \to 1\text{-}\mathrm{Cat}^{\mathbf{\Delta}^{\mathrm{op}}}.$$

(b) *The resulting object* $\mathbf{1\text{-}Cat} \in 2\text{-}\mathrm{Cat}$ *is equipped with a canonical identification*

$$(\mathbf{1\text{-}Cat})^{1\text{-}\mathrm{Cat}} \simeq 1\text{-}\mathrm{Cat}.$$

PROOF. First, we note that by construction, the $(\infty, 1)$-category $\mathrm{Seq}_0(\mathbf{1\text{-}Cat})$ tautologically identifies with $(1\text{-}\mathrm{Cat})^{\mathrm{Spc}}$. Similarly, we have a canonical identification

$$\mathrm{Seq}_\bullet(1\text{-}\mathrm{Cat}) := \mathrm{Maps}([\bullet], 1\text{-}\mathrm{Cat}) \simeq (\mathrm{Cart}_{/[\bullet]^{\mathrm{op}}})^{\mathrm{Spc}} = (\mathrm{Seq}_\bullet(\mathbf{1\text{-}Cat}))^{\mathrm{Spc}},$$

where the second isomorphism is given by Chapter 1, Sect. 1.4.5.

It remains to show that the simplicial category $\mathrm{Seq}_\bullet(\mathbf{1\text{-}Cat})$ satisfies the Segal condition. Indeed, this follows from the fact that for $n = n_1 + n_2$, the functor

$$\mathrm{Cart}_{/[n]^{\mathrm{op}}} \to \mathrm{Cart}_{/[n_1]^{\mathrm{op}}} \underset{1\text{-}\mathrm{Cat}}{\times} \mathrm{Cart}_{/[n_2]^{\mathrm{op}}}$$

is an equivalence.

□

2.4.3. We also have:

COROLLARY 2.4.4. *For* $\mathbf{S}, \mathbf{T} \in \mathbf{1}\text{-}\mathbf{Cat}$, *there is a canonical equivalence*

$$\mathbf{Maps}_{\mathbf{1}\text{-}\mathbf{Cat}}(\mathbf{S}, \mathbf{T}) \simeq \mathrm{Funct}(\mathbf{S}, \mathbf{T}).$$

PROOF. It suffices to show that for each $[n] \in \Delta^{\mathrm{op}}$, there is a natural equivalence

$$\mathrm{Maps}([n], \mathrm{Funct}(\mathbf{S}, \mathbf{T})) \simeq \mathrm{Maps}([n], \mathrm{Cart}_{/[1]^{\mathrm{op}}} \underset{1\text{-}\mathrm{Cat} \times 1\text{-}\mathrm{Cat}}{\times} \{(\mathbf{S}, \mathbf{T})\}).$$

Tautologically, we have

$$\mathrm{Maps}([n], \mathrm{Funct}(\mathbf{S}, \mathbf{T})) \simeq \mathrm{Maps}_{\mathrm{coCart}_{/[n]}}(\mathbf{S} \times [n], \mathbf{T} \times [n])$$

$$\simeq \mathrm{Maps}([1], \mathrm{coCart}_{/[n]}) \underset{\mathrm{coCart}_{/[n]} \times \mathrm{coCart}_{/[n]}}{\times} (\mathbf{S} \times [n], \mathbf{T} \times [n]).$$

By Chapter 12, Proposition 2.1.3, the latter is naturally equivalent to

$$\mathrm{Maps}([n], \mathrm{Cart}_{/[1]^{\mathrm{op}}}) \underset{\mathrm{Maps}([n], 1\text{-}\mathrm{Cat}) \times \mathrm{Maps}([n], 1\text{-}\mathrm{Cat})}{\times} (\mathbf{S}, \mathbf{T})$$

$$\simeq \mathrm{Maps}([n], \mathrm{Cart}_{/[1]^{\mathrm{op}}} \underset{1\text{-}\mathrm{Cat} \times 1\text{-}\mathrm{Cat}}{\times} \{(\mathbf{S}, \mathbf{T})\}),$$

as desired. □

REMARK 2.4.5. Suppose in the above definition of $\mathbf{1}\text{-}\mathbf{Cat}$, we replace $\mathrm{Cart}_{/[n]^{\mathrm{op}}}$ by $\mathrm{coCart}_{/[n]}$. (Note that the underlying simplicial spaces are both identified with $\mathrm{Maps}([\bullet], 1\text{-}\mathrm{Cat})$.)

The latter simplicial category also lies in the essential image of the functor $\mathrm{Seq}_\bullet$, and the resulting $(\infty, 2)$-category identifies with $(\mathbf{1}\text{-}\mathbf{Cat})^{2\text{-}\mathrm{op}}$.

2.5. The $(\infty, 2)$-category of functors. So far, we have defined on the totality of $(\infty, 2)$-categories a structure of $(\infty, 1)$-category. In particular, for $\mathbb{S}, \mathbb{T}$ we have a well-defined *space*

$$\mathrm{Maps}_{2\text{-}\mathrm{Cat}}(\mathbb{S}, \mathbb{T}).$$

We claim, however, that the above space lifts, in a natural way to an object of 2-Cat.

2.5.1. We have the following basic result:

THEOREM 2.5.2 ([**Rezk2, BarS**]). *For* $\mathbb{S}, \mathbb{T} \in 2\text{-}\mathrm{Cat}$, *the functor*

$$\mathbb{X} \mapsto \mathrm{Maps}_{2\text{-}\mathrm{Cat}}(\mathbb{X} \times \mathbb{S}, \mathbb{T})$$

is representable.

Note that Theorem 2.5.2 is not at all tautological. By the Adjoint Functor Theorem, it is equivalent to the following one:

THEOREM 2.5.3. *The functor*

$$2\text{-}\mathrm{Cat} \times 2\text{-}\mathrm{Cat} \to 2\text{-}\mathrm{Cat}, \quad \mathbb{S}, \mathbb{T} \mapsto \mathbb{S} \times \mathbb{T}$$

commutes with colimits in each variable.

2.5.4. In what follows we shall denote the object representing the functor in Theorem 2.5.2 by

$$\text{Funct}(\mathbb{S}, \mathbb{T}) \in 2\text{-Cat}.$$

Note that by definition

$$(\text{Funct}(\mathbb{S}, \mathbb{T}))^{\text{Spc}} \simeq \text{Maps}_{2\text{-Cat}}(\mathbb{S}, \mathbb{T}).$$

2.6. $(\infty, 2)$-categories via bi-simplicial spaces.

2.6.1. The category $1\text{-Cat}^{\mathbf{\Delta}^{\text{op}}}$ admits a fully faithful embedding into

$$(\text{Spc}^{\mathbf{\Delta}^{\text{op}}})^{\mathbf{\Delta}^{\text{op}}} \simeq \text{Spc}^{\mathbf{\Delta}^{\text{op}} \times \mathbf{\Delta}^{\text{op}}},$$

given by $(\text{Seq}_\bullet)^{\mathbf{\Delta}^{\text{op}}}$, i.e., apply the functor $\text{Seq}_\bullet : 1\text{-Cat} \to \text{Spc}^{\mathbf{\Delta}^{\text{op}}}$ simplex-wise.

Hence, we obtain a fully faithful embedding

$$\text{Sq}^{\sim}_{\bullet,\bullet} = (\text{Seq}_\bullet)^{\mathbf{\Delta}^{\text{op}}} \circ \text{Seq}_\bullet : 2\text{-Cat}^{\mathbf{\Delta}^{\text{op}}} \to \text{Spc}^{\mathbf{\Delta}^{\text{op}} \times \mathbf{\Delta}^{\text{op}}}. \tag{2.5}$$

2.6.2. For a pair of integers $m, n \geq 0$, consider the functor $\text{Sq}^{\sim}_{m,n}$. Since this functor commutes with limits, it is co-representable. We let $[m,n]^{\sim} \in 2\text{-Cat}$ denote the co-representing object.

It is easy to see that $[0,n]^{\sim} \simeq [n]$ and $[m,0]^{\sim} \simeq *$. Note that Segal condition implies that the natural maps

$$[m, n_1]^{\sim} \underset{*}{\sqcup} [m, n_2]^{\sim} \to [m, n_1 + n_2]^{\sim} \tag{2.6}$$

and

$$[m_1, n]^{\sim} \underset{[n]}{\sqcup} [m_2, n]^{\sim} \to [m_1 + m_2, n]^{\sim} \tag{2.7}$$

are isomorphisms. Pictorially, one could think of $[m,n]^{\sim}$ as the diagram

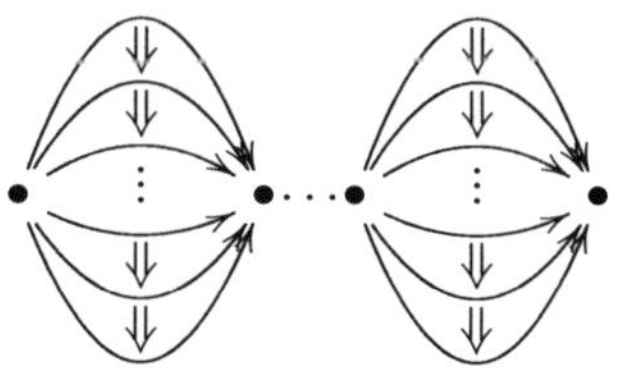

2.6.3. The material in the rest of this subsection is included for the sake of completeness.

Note that we have the following explicit description of the *ordinary* 2-category

$$([m,n]^{\sim})^{\text{ordn}}.$$

Its objects are integers $0, ..., n$. For $0 \leq i, j \leq n$, the 1-category of morphisms $i \to j$ is described as follows:

(a) It is empty if $i > j$;

(b) It is $\{*\}$ if $i = j$;

(c) For $i < j$ it is the poset of sequences

$$(f_{i,i+1}, ..., f_{j-1,j}) \in \{0, ..., m\}$$

The order relation is

$$(f_{i,i+1},...,f_{j-1,j}) \leq (f'_{i,i+1},...,f'_{j-1,j}) \Leftrightarrow \forall i \leq k \leq j-1,\ f_{k,k+1} \leq f'_{k,k+1}.$$

2.6.4. We have the following result:

THEOREM 2.6.5 ([**BarS**, Lemma 12.5]). *For any (m,n), the tautological functor*

$$[m,n]^{\sim} \to ([m,n]^{\sim})^{\mathrm{ordn}}$$

is an equivalence.

In other words, this theorem says that for every m,n the object of 2-Cat, co-representing the functor

$$\mathbb{S} \mapsto \mathrm{Sq}^{\sim}_{m,n}(\mathbb{S}) \simeq \mathrm{Seq}_m(\mathrm{Seq}_n(\mathbb{S})), \quad 2\text{-Cat} \to \mathrm{Spc}$$

is an *ordinary* 2-category.

3. Lax functors and the Gray product

One of the key new features of $(\infty,2)$-categories, as compared to $(\infty,1)$-categories, is the notion of *right-lax* (or *left-lax*) functor. We will introduce these notions in the present section.

3.1. Lax functors. The idea of a lax functor is that the composition of 1-morphisms does not have to go to the composition, but rather be connected to it by a (not necessarily invertible) 2-morphism.

3.1.1. Let $\mathbb{S}$ be an $(\infty,2)$-category, thought of as a functor $\mathbf{\Delta}^{\mathrm{op}} \to 1\text{-Cat}$. Let $\mathbb{S}^{\oint}$ be the total space of the corresponding coCartesian fibration over $\mathbf{\Delta}^{\mathrm{op}}$.

By definition, a functor between a pair of $(\infty,2)$-categories $\mathbb{S}$ and $\mathbb{T}$ is a functor

$$\mathbb{S}^{\oint} \to \mathbb{T}^{\oint}$$

that sends *coCartesian arrows to coCartesian* arrows.

3.1.2. We give the following definitions: a map

$$\alpha : [m] \to [n]$$

is said to be:

- *active* if $\alpha(0) = 0$ and $\alpha(m) = n$.
- *idle* if for all $0 \leq j \leq n$ for which there exist $0 \leq i_1, i_2 \leq m$ with $\alpha(i_i) \leq j \leq \alpha(i_2)$, there exists $0 \leq i \leq m$ such that $\alpha(i) = j$;
- *inert* if for all $0 \leq j \leq n$ for which there exist $0 \leq i_1, i_2 \leq m$ with $\alpha(i_i) \leq j \leq \alpha(i_2)$, there exists a *unique* $0 \leq i \leq m$ such that $\alpha(i) = j$;

In other words, an inert map sends $i \mapsto i + k$ for some $0 \leq k \leq n - m$.

3.1.3. We define a *non-unital right-lax functor* from $\mathbb{S}$ to $\mathbb{T}$ to be a functor

$$\mathbb{S}^{\oint} \to \mathbb{T}^{\oint}$$

which takes coCartesian edges over *inert* morphisms of Δ to coCartesian edges.

We define a *right-lax functor* from $\mathbb{S}$ to $\mathbb{T}$ to be a functor

$$\mathbb{S}^{\oint} \to \mathbb{T}^{\oint}$$

which takes coCartesian edges over *idle* morphisms of Δ to coCartesian edges.

We define a *non-unital left-lax functor* (resp., *left-lax functor*) from $\mathbb{S}$ to $\mathbb{T}$ to be a *non-unital right-lax functor* (resp., *right-lax functor*) from $\mathbb{S}^{2\text{-op}}$ to $\mathbb{T}^{2\text{-op}}$.

REMARK 3.1.4. At the level of ordinary 2-categories, the notion of 'non-unital right-lax functor' differs from what is called a lax functor in the literature on ordinary 2-categories. In particular, unlike the classical notion, the notion of a non-unital right-lax functor is invariant with respect to equivalence of 2-categories.

On the other hand, the notion of 'right-lax functor', at the level of ordinary 2-categories, agrees with what is usually called a 'normal lax functor' in the literature on ordinary 2-categories.

3.1.5. In what follows, given a pair of $(\infty,2)$-categories $\mathbb{S}$ and $\mathbb{T}$ we shall symbolically denote right-lax and left-lax functors from $\mathbb{S}$ to $\mathbb{T}$ by

$$\mathbb{S} \dashrightarrow \mathbb{T},$$

to distinguish them from actual (i.e., strict) functors, which we denote by $\mathbb{S} \to \mathbb{T}$.

3.1.6. Let $F : \mathbb{S} \dashrightarrow \mathbb{T}$ be a right-lax functor. Then, for a string of objects $s_0 \overset{\phi}{\to} s_1 \overset{\psi}{\to} s_2$ in $\mathbb{S}$, we are given a (not necessarily invertible) 2-morphism

$$F(\psi) \circ F(\phi) \to F(\psi \circ \phi), \tag{3.1}$$

i.e., a 1-morphism in $\mathbf{Maps}_{\mathbb{T}}(F(s_0), F(s_1))$.

For a left-lax functor, we have a map in the opposite direction: $F(\psi \circ \phi) \to F(\psi) \circ F(\phi)$.

3.1.7. We can introduce 1-full subcategories

$$2\text{-Cat} \hookrightarrow 2\text{-Cat}_{\text{right-lax}} \hookrightarrow 2\text{-Cat}_{\text{right-lax}_{\text{non-untl}}}$$

of the full subcategory of $1\text{-Cat}^{\mathbf{\Delta}^{\text{op}}}$ formed by coCartesian fibrations, by imposing increasingly weaker conditions on 1-morphisms.

Thus, for a pair of objects $\mathbb{S}, \mathbb{T} \in 2\text{-Cat}$ we have the well-defined spaces

$$\text{Maps}_{2\text{-Cat}}(\mathbb{S},\mathbb{T}) \subset \text{Maps}_{2\text{-Cat}_{\text{right-lax}}}(\mathbb{S},\mathbb{T}) \subset \text{Maps}_{2\text{-Cat}_{\text{right-lax}_{\text{non-untl}}}}(\mathbb{S},\mathbb{T}).$$

All of the above categories contain limits, and the above embeddings commute with limits.

3.2. The Gray tensor product. The notion of right-lax functor allows one to introduce the notion of *Gray product* of $(\infty,2)$-categories. Given $(\infty,2)$-categories $\mathbb{S}$ and $\mathbb{T}$, the Gray product $\mathbb{S}\circledast\mathbb{T}$ has the same objects as $\mathbb{S}\times\mathbb{T}$. However, for $s_0 \overset{\phi}{\to} s_1$ and $t_0 \overset{\psi}{\to} t_1$, the diagram

$$\begin{array}{ccc} (s_0,t_0) & \xrightarrow{(\mathrm{id},\psi)} & (s_0,t_1) \\ {\scriptstyle(\phi,\mathrm{id})}\big\downarrow & & \big\downarrow{\scriptstyle(\phi,\mathrm{id})} \\ (s_1,t_0) & \xrightarrow{(\mathrm{id},\psi)} & (s_1,t_1) \end{array}$$

will *no longer commute*, but will do so up to a *non-invertible* 2-morphism.

The formation of the Gray product will allow us to talk about right-lax natural transformation between functors.

3.2.1. For an n-tuple of $(\infty,2)$-categories $\mathbb{S}_1,\mathbb{S}_2,...,\mathbb{S}_n$ and another $(\infty,2)$-category $\mathbb{T}$, let

$$\mathrm{Maps}_{2\text{-Cat}}(\mathbb{S}_1\circledast...\circledast\mathbb{S}_n,\mathbb{T}) \subset \mathrm{Maps}_{2\text{-Cat}^{\mathrm{lax}}}(\mathbb{S}_1\times...\times\mathbb{S}_n,\mathbb{T})$$

to be the subspace given by right-lax functors such that:

(1) For each i and an object $\hat{s}_i \in \prod\limits_{j\neq i}\mathbb{S}_j$, the composite lax functor

$$\mathbb{S}_i \overset{\mathrm{id}\times\hat{s}_i}{\longrightarrow} \mathbb{S}_1\times...\times\mathbb{S}_n \overset{F}{\dashrightarrow} \mathbb{T}$$

is a strict functor.

(2) For any morphism

$$f=(f_i):(s_1,...,s_n)\to(s_1',...,s_n')$$

in $\mathbb{S}_1\times...\times\mathbb{S}_n$ and $1\leq k\leq n-1$, the 2-morphism in $\mathbb{T}$, corresponding to splitting f as a composition (see (3.1))

$$(s_1,...,s_k,s_{k+1},...,s_n) \overset{(f_1,...,f_k,\mathrm{id},...,\mathrm{id})}{\longrightarrow}$$
$$\to (s_1',...,s_k',s_{k+1},...,s_n) \overset{(\mathrm{id},...,\mathrm{id},f_{k+1},...,f_n)}{\longrightarrow} (s_1',...,s_k',s_{k+1}',...,s_n'),$$

is an isomorphism.

3.2.2. For example, if $n=2$ and F is an object of $\mathrm{Maps}_{2\text{-Cat}}(\mathbb{S}_1\circledast\mathbb{S}_2,\mathbb{T})$, for any

$$(s_1,s_2)\overset{(f_1,f_2)}{\to}(s_1',s_2')$$

in $\mathbb{S}_1\times\mathbb{S}_2$, we obtain a 2-morphism in $\mathbb{T}$

$$F(f_1,\mathrm{id}_{s_2'})\circ F(\mathrm{id}_{s_1},f_2)\to F(f_1,f_2)\simeq F(\mathrm{id}_{s_1'},f_2)\circ F(f_1,\mathrm{id}_{s_2}).$$

3.2.3. Since the functor $\mathrm{Maps}_{2\text{-Cat}}(\mathbb{S}_1 \circledast \ldots \circledast \mathbb{S}_n, -)$ commutes with limits, it is co-represented by an $(\infty, 2)$-category, to be denoted $\mathbb{S}_1 \circledast \ldots \circledast \mathbb{S}_n$, and called the *Gray tensor product.*

By definition, we have a tautological projection

$$\mathbb{S}_1 \circledast \ldots \circledast \mathbb{S}_n \to \mathbb{S}_1 \times \ldots \times \mathbb{S}_n,$$

and a canonically defined lax functor

$$\mathbb{S}_1 \times \ldots \times \mathbb{S}_n \dashrightarrow \mathbb{S}_1 \circledast \ldots \circledast \mathbb{S}_n,$$

such that the composition

$$\mathbb{S}_1 \times \ldots \times \mathbb{S}_n \dashrightarrow \mathbb{S}_1 \circledast \ldots \circledast \mathbb{S}_n \to \mathbb{S}_1 \times \ldots \times \mathbb{S}_n$$

is the identity functor.

3.2.4. Note also that by construction, we have a canonical identification

$$(\mathbb{S}_n \circledast \ldots \circledast \mathbb{S}_1)^{2\text{-op}} \simeq (\mathbb{S}_1^{2\text{-op}} \circledast \ldots \circledast \mathbb{S}_n^{2\text{-op}}). \tag{3.2}$$

3.2.5. We have the following basic fact[2]

PROPOSITION 3.2.6. *The functor*

$$2\text{-Cat} \times \ldots \times 2\text{-Cat} \to 2\text{-Cat}, \quad \mathbb{S}_1, \ldots, \mathbb{S}_n \mapsto \mathbb{S}_1 \circledast \ldots \circledast \mathbb{S}_n$$

commutes with colimits in each variable.

3.2.7. For a pair of $(\infty, 2)$-categories $\mathbb{S}$ and $\mathbb{T}$, recall the $(\infty,2)$-category $\mathrm{Funct}(\mathbb{S}, \mathbb{T})$. We introduce its enlargement $\mathrm{Funct}(\mathbb{S}, \mathbb{T})_{\text{right-lax}}$ (which has the same underlying space, but more 1-morphisms) as follows:

We set

$$\mathrm{Maps}_{2\text{-Cat}}(\mathbb{X}, \mathrm{Funct}(\mathbb{S}, \mathbb{T})_{\text{right-lax}}) = \mathrm{Maps}_{2\text{-Cat}_{\text{right-lax}}}(\mathbb{X} \circledast \mathbb{S}, \mathbb{T}),$$

where the co-representability is insured by Proposition 3.2.6.

We call 1-morphisms in $\mathrm{Funct}(\mathbb{S}, \mathbb{T})_{\text{right-lax}}$ *right-lax natural transformations.* By definition, given a right-lax natural transformation $\alpha : F_1 \dashrightarrow F_2$, for an object $s \in \mathbb{S}$ we have a 1-morphism

$$\alpha(s) : F_1(s) \to F_2(s)$$

in $\mathbb{T}$, and for a 1-morphism $\phi : s \to s'$, we have a 2-morphism

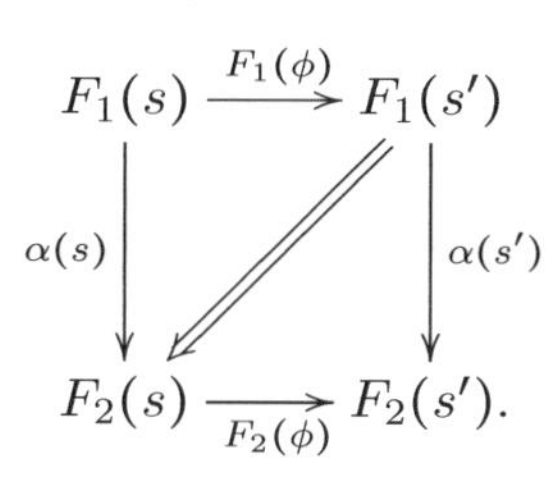

Similarly, we introduce the $(\infty, 2)$-category $\mathrm{Funct}(\mathbb{S}, \mathbb{T})_{\text{left-lax}}$.

[2]We do not prove it, and we were not able to find a reference.

3.2.8. For $n = n_1 + n_2$ and an n-tuple of $(\infty, 2)$-categories $\mathbb{S}_1, ..., \mathbb{S}_n$ consider the right-lax functor

$$\begin{aligned} \mathbb{S}_1 \times ... \times \mathbb{S}_n \simeq (\mathbb{S}_1 \times ... \times \mathbb{S}_{n_1}) \times (\mathbb{S}_{n_1+1} \times ... \times \mathbb{S}_{n_1+n_2}) \dashrightarrow \\ \dashrightarrow (\mathbb{S}_1 \circledast ... \circledast \mathbb{S}_{n_1}) \times (\mathbb{S}_{n_1+1} \circledast ... \circledast \mathbb{S}_{n_1+n_2}) \dashrightarrow \\ \dashrightarrow (\mathbb{S}_1 \circledast ... \circledast \mathbb{S}_{n_1}) \circledast (\mathbb{S}_{n_1+1} \circledast ... \circledast \mathbb{S}_{n_1+n_2}) \end{aligned}$$

It follows from the definitions that the above right-lax functor gives rise to a functor

$$\mathbb{S}_1 \circledast ... \circledast \mathbb{S}_{n_1} \circledast \mathbb{S}_{n_1+1} \circledast ... \circledast \mathbb{S}_{n_1+n_2} \dashrightarrow (\mathbb{S}_1 \circledast ... \circledast \mathbb{S}_{n_1}) \circledast (\mathbb{S}_{n_1+1} \circledast ... \circledast \mathbb{S}_{n_1+n_2}). \tag{3.3}$$

We have the following proposition[3]:

PROPOSITION 3.2.9. *The functor* (3.3) *is an equivalence.*

REMARK 3.2.10. It is easy to see that Proposition 3.2.9 implies that the Cartesian monoidal structure on 2-Cat induces a monoidal structure on 2-Cat, given by the Gray product.

3.3. Cubical 2-categories.

3.3.1. For an integer k and a k-tuple $n_1, ..., n_k$ we let

$$[n_1, ..., n_k] \in 2\text{-Cat}$$

denote the $(\infty, 2)$-category

$$[n_1] \circledast ... \circledast [n_k].$$

3.3.2. From Proposition 3.2.9 we obtain that for $k = k_1 + k_2$, the natural functor

$$[n_1, ..., n_k] \to [n_1, ..., n_{k_1}] \circledast [n_{k_1+1}, ..., n_{k_1+k_2}]$$

is an equivalence.

3.3.3. In addition, from Proposition 3.2.6 we obtain that for $1 \leq i \leq k$ and $n_i = n_i' + n_i''$, the natural functor

$$\begin{aligned} [n_1, ..., n_{i-1}, n_i', n_{i+1}, ..., n_k] \underset{[n_1,...,n_{i-1},0,n_{i+1},...,n_k]}{\sqcup} [n_1, ..., n_{i-1}, n_i'', n_{i+1}, ..., n_k] \to \\ \to [n_1, ..., n_{i-1}, n_i, n_{i+1}, ..., n_k] \end{aligned} \tag{3.4}$$

is an equivalence, where we note that

$$[n_1, ..., n_{i-1}, 0, n_{i+1}, ..., n_k] \simeq [n_1, ..., n_{i-1}, n_{i+1}, ..., n_k].$$

3.3.4. The following proposition is quoted for the sake of completeness[4]:

PROPOSITION 3.3.5. *The* $(\infty, 2)$*-categories* $[n_1, ..., n_k]$ *are ordinary.*

3.4. Squares. Our primary interest will be the $(\infty, 2)$-categories $[m, n]$.

---

[3] We do not prove it, and we were not able to find a reference.

[4] We do not prove it, and we were not able to find a reference.

3.4.1. We consider first the case $m = n = 1$. It follows from the definitions that for $\mathbb{S} \in 2\text{-Cat}$, the space

$$\mathrm{Maps}_{2\text{-Cat}}([1,1], \mathbb{S}) \tag{3.5}$$

identifies canonically with the space of diagrams

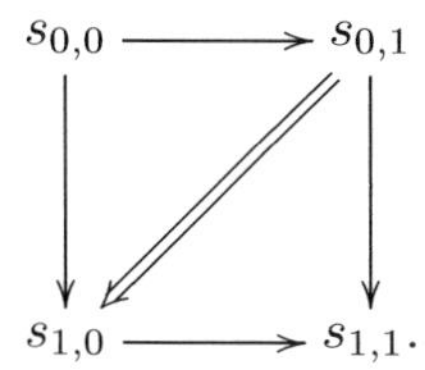

In other words,

$$\mathrm{Maps}_{2\text{-Cat}}([1,1], \mathbb{S}) \simeq (\mathrm{Sq}^{\sim}_{0,2} \times \mathrm{Sq}^{\sim}_{0,2}) \underset{\mathrm{Sq}^{\sim}_{0,1} \times \mathrm{Sq}^{\sim}_{0,1}}{\times} \mathrm{Sq}^{\sim}_{1,1},$$

where both maps $\mathrm{Sq}^{\sim}_{0,2} \to \mathrm{Sq}^{\sim}_{0,1}$ correspond to the unique active map $[1] \to [2]$.

REMARK 3.4.2. Using (3.4), for any m, n we can depict $[m, n]$ by the diagram

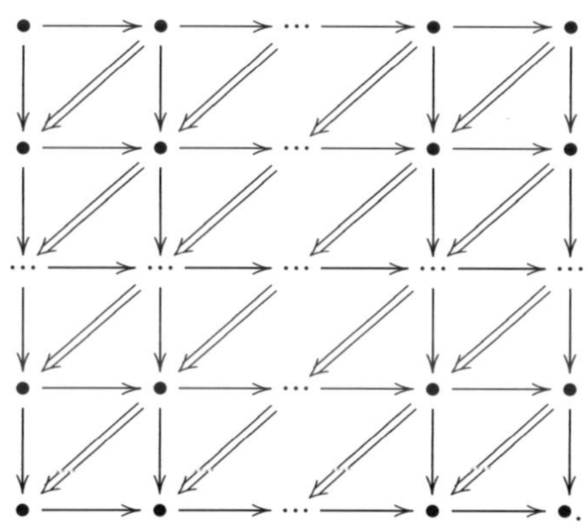

3.4.3. Recall the notation $[m, n]^{\sim}$, see Sect. 2.6.2.

Using the description of the space (3.5), we obtain that there is a canonical identification

$$[1,1] \underset{[1] \sqcup [1]}{\sqcup} (* \sqcup *) \simeq [1,1]^{\sim}, \tag{3.6}$$

where the two maps $[1] \to [1,1]$ are

$$[1] \simeq [1,0] \rightrightarrows [1,1],$$

corresponding to the two maps $[0] \rightrightarrows [1]$.

Pictorially, $[1,1]^{\sim}$, which is

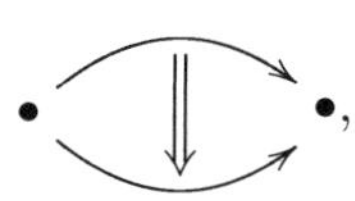

is obtained from $[1,1]$, which is

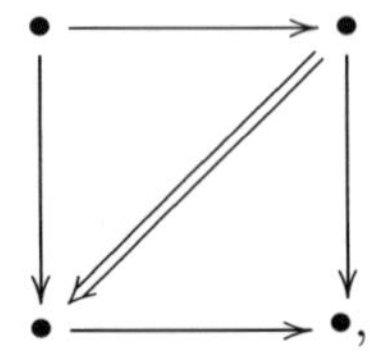

by collapsing the vertical edges.

3.4.4. We claim that

$$[m,n] \underset{[m]\sqcup...\sqcup[m]}{\sqcup} (* \sqcup .... \sqcup *) \simeq [m,n]^{\sim}, \tag{3.7}$$

functorial in $[m],[n] \in \mathbf{\Delta}$.

Indeed, unwinding the definitions, to specify a map $\to$ in (3.7), we need to specify a functorial assigment for each $[i],[j] \in \mathbf{\Delta}$ and a pair of maps

$$[i] \to [m] \times [n], [j] \to [m] \times [n] \tag{3.8}$$

a point in $\mathrm{Sq}^{\sim}_{i,j}([m,n]^{\sim})$.

The sought-for point is obtained from the tautological point of $\mathrm{Sq}^{\sim}_{i,j}([i,j]^{\sim})$ by composing with

$$([i],[j]) \to ([m],[n]),$$

where the latter is obtained from (3.8) by projection.

3.4.5. It follows from the construction, that the map (3.7) canonically factors through a map

$$[m,n] \underset{[m]\sqcup...\sqcup[m]}{\sqcup} (* \sqcup .... \sqcup *) \to [m,n]^{\sim}, \tag{3.9}$$

functorial in $[m],[n] \in \mathbf{\Delta}$, where the $n+1$ maps $[m] \to [m,n]$ are given by

$$[m] \simeq [m,0] \to [m,n],$$

corresponding to the $n+1$ maps $[0] \to [n]$.

PROPOSITION 3.4.6. *The map* (3.9) *is an isomorphism.*

PROOF. Using (3.4) and (2.6), we obtain that it is sufficient to consider the cases when m and n are equal to 0 or 1.

When m or n are equal to 0, there is nothing to prove. The case of $m = n = 1$ follows from the isomorphism (3.6).

□

3.4.7. Note that by applying Proposition 3.4.6 in the case $n = 1$, we obtain:

COROLLARY 3.4.8. *For $\mathbb{S} \in$ 2-Cat and $s_0, s_1 \in \mathbb{S}$, there exists a canonical isomorphism*

$$\mathrm{Funct}([1],\mathbb{S})_{\text{right-lax}} \underset{\mathbb{S}\times\mathbb{S}}{\times} \{(s_0,s_1)\} \simeq \mathbf{Maps}_{\mathbb{S}}(s_0,s_1).$$

4. $(\infty, 2)$-categories via squares

Recall that we originally realized 2-Cat as a full subcategory in $1\text{-Cat}^{\mathbf{\Delta}^{\mathrm{op}}}$ via the functor $\mathrm{Seq}_\bullet$.

In this subsection, we will discuss a *different* realization of 2-Cat as a full subcategory in $1\text{-Cat}^{\mathbf{\Delta}^{\mathrm{op}}}$, this time via the functor that we denote $\mathrm{Seq}_\bullet^{\mathrm{ext}}$.

Recall that for an $(\infty,2)$-category $\mathbb{S}$, for $n = 0$, the $(\infty,1)$-category $\mathrm{Seq}_0(\mathbb{S})$ recorded the space $\mathbb{S}^{\mathrm{Spc}}$. For $n = 1$, $(\infty,1)$-category $\mathrm{Seq}_1(\mathbb{S})$ had as objects 1-morphisms $s_0 \to s_1$ and as morphisms 2-morphisms, i.e., diagrams

$$s_0 \underset{\longrightarrow}{\overset{\longrightarrow}{\Downarrow}} s_1.$$

The idea of $\mathrm{Seq}_\bullet^{\mathrm{ext}}$ is the following. The $(\infty,1)$-category $\mathrm{Seq}_0^{\mathrm{ext}}(\mathbb{S})$ will be $\mathbb{S}^{1\text{-Cat}}$. Now, for $n = 1$, the category $\mathrm{Seq}_1^{\mathrm{ext}}(\mathbb{S})$ will have as objects 1-morphisms $s_0 \to s_1$ as before, but as morphisms diagrams

$$\begin{array}{ccc} s_0 & \longrightarrow & s_1 \\ \Big\downarrow & \swarrow\!\!\swarrow & \Big\downarrow \\ s_0' & \longrightarrow & s_1'. \end{array}$$

I.e., $\mathrm{Seq}_1^{\mathrm{ext}}(\mathbb{S})$ will be the $(\infty,1)$-category $\mathrm{Funct}([1],\mathbb{S})_{\text{right-lax}}$.

4.1. The functor $\mathrm{Sq}_{\bullet,\bullet}$. Before introducing the functor $\mathrm{Seq}_\bullet^{\mathrm{ext}}$, we introduce the corresponding version, denoted $\mathrm{Sq}_{\bullet,\bullet}$, of the functor $\mathrm{Sq}_{\bullet,\bullet}^{\sim}$. It will have the advantage of respecting more symmetries of the target category, i.e., $\mathrm{Spc}^{\mathbf{\Delta}^{\mathrm{op}}\times\mathbf{\Delta}^{\mathrm{op}}}$.

4.1.1. Consider the functor

$$\mathrm{Sq}_{\bullet,\bullet} : 2\text{-Cat} \to \mathrm{Spc}^{\mathbf{\Delta}^{\mathrm{op}}\times\mathbf{\Delta}^{\mathrm{op}}},$$

defined by

$$\mathbb{S} \mapsto ([m],[n] \mapsto \mathrm{Maps}_{2\text{-Cat}}([m,n],\mathbb{S}))\,.$$

4.1.2. This section is devoted to the discussion of the generalizations of the following fundamental result[5]:

THEOREM 4.1.3. *The functor* $\mathrm{Sq}_{\bullet,\bullet}$ *is fully faithful.*

4.1.4. Note from (3.9) and Proposition 3.4.6 we obtain a natural transformation

$$\mathrm{Sq}_{\bullet,\bullet}^{\sim} \to \mathrm{Sq}_{\bullet,\bullet},$$

such that for $\mathbb{S} \in 2\text{-Cat}$ and any m, n, the corresponding map

$$\mathrm{Sq}_{\bullet,\bullet}^{\sim}(\mathbb{S}) \to \mathrm{Sq}_{\bullet,\bullet}(\mathbb{S}) \tag{4.1}$$

is a full embedding.

Indeed, if we denote $E_{\bullet,\bullet} = \mathrm{Sq}_{\bullet,\bullet}(\mathbb{S})$, the essential image of (4.1) is the full subspace of $E_{m,n}$ consisting of points, for which for every $[1] \to [m]$ and $[0] \to [n]$,

[5] We do not prove it, and we were not able to find a reference.

the corresponding point in $E_{1,0}$ is degenerate, i.e., lies in the essential image of $E_{0,0} \to E_{1,0}$.

4.1.5. Let

$$E_{\bullet,\bullet} \mapsto (E_{\bullet,\bullet})^{\text{reflect}}$$

denote the involution on $\text{Spc}^{\mathbf{\Delta}^{\text{op}} \times \mathbf{\Delta}^{\text{op}}}$, corresponding to swapping the two factors of $\mathbf{\Delta}^{\text{op}}$.

It follows from (3.2), that there is a canonical identification

$$\text{Sq}_{\bullet,\bullet}(\mathbb{S}^{2\text{-op}}) \simeq (\text{Sq}_{\bullet,\bullet}(\mathbb{S}))^{\text{reflect}}.$$

4.1.6. Let

$$E_{\bullet,\bullet} \mapsto (E_{\bullet,\bullet})^{\text{vert-op}} \text{ and } (E_{\bullet,\bullet})^{\text{horiz-op}}$$

be the involutions on $\text{Spc}^{\mathbf{\Delta}^{\text{op}} \times \mathbf{\Delta}^{\text{op}}}$ induced by the involution rev along the first and second copy of $\mathbf{\Delta}^{\text{op}}$, respectively.

Let

$$E_{\bullet,\bullet} \mapsto (E_{\bullet,\bullet})^{\text{vert\&horiz-op}}$$

denote their composition.

It follows that we have a canonical identification

$$\text{Sq}_{\bullet,\bullet}(\mathbb{S}^{1\&2\text{-op}}) \simeq (\text{Sq}_{\bullet,\bullet}(\mathbb{S}))^{\text{vert\&horiz-op}}.$$

4.1.7. For $m = 0$, we have a canonical identification

$$\text{Sq}_{0,n}(\mathbb{S}) \simeq \text{Maps}_{2\text{-Cat}}([n], \mathbb{S}) \simeq (\text{Seq}_n(\mathbb{S}))^{\text{Spc}},$$

and similarly for $n = 0$, we have

$$\text{Sq}_{m,0}(\mathbb{S}) \simeq \text{Maps}_{2\text{-Cat}}([m], \mathbb{S}) \simeq (\text{Seq}_m(\mathbb{S}))^{\text{Spc}},$$

Note that (3.4) implies that for $n = n_1 + n_2$, and $\mathbb{S} \in 2\text{-Cat}$, the natural map

$$\text{Sq}_{m,n}(\mathbb{S}) \to \text{Sq}_{m,n_1}(\mathbb{S}) \underset{\text{Sq}_{m,0}(\mathbb{S})}{\times} \text{Sq}_{m,n_2}(\mathbb{S})$$

is an isomorphism and for $m = m_1 + m_2$, the natural map

$$\text{Sq}_{m,n}(\mathbb{S}) \to \text{Sq}_{m_1,n}(\mathbb{S}) \underset{\text{Sq}_{0,n}(\mathbb{S})}{\times} \text{Sq}_{m_2,n}(\mathbb{S}) \tag{4.2}$$

is an isomorphism.

4.2. The functor $\mathrm{Seq}^{\mathrm{ext}}_\bullet$.
4.2.1. The isomorphism (4.2) implies that for a fixed n, the functor

$$2\text{-Cat} \to \mathrm{Spc}^{\mathbf{\Delta}^{\mathrm{op}}}, \quad \mathbb{S} \mapsto \mathrm{Sq}_{\bullet,n}(\mathbb{S})$$

lands in the subcategory of Segal spaces. Moreover, it is easy to see that it actually lands in the full subcategory of *complete Segal spaces.*

Hence, we obtain a well-defined functor, to be denoted $\mathrm{Seq}^{\mathrm{ext}}_\bullet$

$$2\text{-Cat} \to 1\text{-Cat}^{\mathbf{\Delta}^{\mathrm{op}}}$$

so that

$$\mathrm{Sq}_{\bullet,\bullet} = (\mathrm{Seq}_\bullet)^{\mathbf{\Delta}^{\mathrm{op}}} \circ \mathrm{Seq}^{\mathrm{ext}}_\bullet .$$

In fact, by definition,

$$\mathrm{Seq}^{\mathrm{ext}}_\bullet(\mathbb{S}) \simeq \mathrm{Funct}([\bullet], \mathbb{S})_{\text{right-lax}}.$$

4.2.2. From Theorem 4.1.3, combined with Theorem 1.3.4 we obtain:

COROLLARY 4.2.3. *The functor* $\mathrm{Seq}^{\mathrm{ext}}_\bullet$ *is fully faithful.*

4.2.4. Note that the functor $\mathrm{Seq}^{\mathrm{ext}}_\bullet$ is *different* from the functor $\mathrm{Seq}_\bullet$. For example,

$$\mathrm{Seq}^{\mathrm{ext}}_0(\mathbb{S}) \simeq \mathbb{S}^{1\text{-Cat}},$$

whereas

$$\mathrm{Seq}_0(\mathbb{S}) = \mathbb{S}^{\mathrm{Spc}}.$$

Note that we have a natural transformation between the functors

$$\mathrm{Seq}_\bullet \to \mathrm{Seq}^{\mathrm{ext}}_\bullet,$$

and for every $\mathbb{S} \in 2\text{-Cat}$ and $n \in \mathbf{\Delta}$, we have a *replete* embedding

$$\mathrm{Seq}_n(\mathbb{S}) \hookrightarrow \mathrm{Seq}^{\mathrm{ext}}_n(\mathbb{S}).$$

Indeed,

$$\mathrm{Seq}_n(\mathbb{S}) = \mathrm{Seq}^{\mathrm{ext}}_n(\mathbb{S}) \underset{\mathbb{S}^{1\text{-Cat}} \times \ldots \times \mathbb{S}^{1\text{-Cat}}}{\times} (\mathbb{S}^{\mathrm{Spc}} \times \ldots \times \mathbb{S}^{\mathrm{Spc}}).$$

4.3. The category of pairs. As we have seen above, to $\mathbb{S} \in 2\text{-Cat}$, and m, n we can assign two spaces

$$\mathrm{Sq}^{\sim}_{m,n} \subset \mathrm{Sq}_{m,n} .$$

In this subsection, we will see that one can produce an entire array of intermediate spaces, one for each 1-full subcategory $\mathbf{C} \in \mathbb{S}^{1\text{-Cat}}$ with the same space of objects.

4.3.1. Let $2\text{-Cat}^{\text{Pair}}$ be the following $(\infty,1)$-category. Its objects are pairs $(\mathbb{S},\mathbf{C})$, where $\mathbb{S}\in 2\text{-Cat}$, and $\mathbf{C}$ is a 1-full subcategory in $\mathbb{S}^{1\text{-Cat}}$ such that $\mathbf{C}^{\text{Spc}}=\mathbb{S}^{\text{Spc}}$.

For a pair of objects $(\mathbb{S}_1,\mathbf{C}_1)$ and $(\mathbb{S}_2,\mathbf{C}_2)$, the space of morphisms between them consists of functors $F:\mathbb{S}_1\to\mathbb{S}_2$, such that the induced functor $\mathbf{C}_1\to\mathbb{S}_2$ factors (automatically uniquely) via $\mathbf{C}_2$.

The ∞-categorical structure on $2\text{-Cat}^{\text{Pair}}$ is uniquely determined by the requirement that the forgetful functor

$$\text{OblvSubcat}:2\text{-Cat}^{\text{Pair}}\to 2\text{-Cat},\quad (\mathbb{S},\mathbf{C})\mapsto\mathbb{S}$$

should be 1-fully faithful.

4.3.2. The above functor OblvSubcat admits a left and a right adjoints, given by

$$\mathbb{S}\mapsto(\mathbb{S},\mathbb{S}^{\text{Spc}})\text{ and }\text{ and }\mathbb{S}\mapsto(\mathbb{S},\mathbb{S}^{1\text{-Cat}}),$$

respectively.

4.3.3. We define the functor

$$\text{Sq}^{\text{Pair}}_{\bullet,\bullet}:2\text{-Cat}^{\text{Pair}}\to\text{Spc}^{\mathbf{\Delta}^{\text{op}}\times\mathbf{\Delta}^{\text{op}}}$$

as follows.

For $(\mathbb{S},\mathbf{C})\in\text{Pair}$ we let $\text{Sq}^{\text{Pair}}_{m,n}(\mathbb{S},\mathbf{C})$ be the full subspace of $\text{Sq}_{m,n}(\mathbb{S})$, consisting of points such that for every $[1]\to[m]$ and $[0]\to[n]$, the resulting point of

$$\text{Sq}_{1,0}(\mathbb{S})\simeq(\text{Seq}_1(\mathbb{S}))^{\text{Spc}}$$

belongs to

$$\text{Seq}_1(\mathbf{C})\subset(\text{Seq}_1(\mathbb{S}))^{\text{Spc}}.$$

The sought-for functor

$$\text{Sq}^{\text{Pair}}_{\bullet,\bullet}(\mathbb{S},\mathbf{C}):\mathbf{\Delta}^{\text{op}}\times\mathbf{\Delta}^{\text{op}}\to\text{Spc}$$

is uniquely determined by the requirement that the embeddings

$$\text{Sq}^{\text{Pair}}_{m,n}(\mathbb{S},\mathbf{C})\hookrightarrow\text{Sq}_{m,n}(\mathbb{S})$$

upgrade to a natural transformation

$$\text{Sq}^{\text{Pair}}_{\bullet,\bullet}(\mathbb{S},\mathbf{C})\to\text{Sq}_{\bullet,\bullet}\circ\text{OblvSubcat}.$$

Note that we have

$$\text{Sq}^{\text{Pair}}_{\bullet,\bullet}(\mathbb{S},\mathbb{S}^{1\text{-Cat}})\simeq\text{Sq}_{\bullet,\bullet}(\mathbb{S})\text{ and }\text{Sq}^{\text{Pair}}_{\bullet,\bullet}(\mathbb{S},\mathbb{S}^{\text{Spc}})\simeq\text{Sq}^{\sim}_{\bullet,\bullet}(\mathbb{S}).$$

4.3.4. We have the following generalization of Theorem 4.1.3[6]:

THEOREM 4.3.5. *The functor* $\text{Sq}^{\text{Pair}}_{\bullet,\bullet}$ *is fully faithful.*

[6]We do not prove it, and we were not able to find a reference.

4.3.6. Note that for a given $(\mathbb{S}, \mathbf{C}) \in 2\text{-Cat}^{\text{Pair}}$, we have

$$\mathrm{Sq}^{\text{Pair}}_{0,\bullet}(\mathbb{S}, \mathbf{C}) \simeq \mathrm{Sq}_{0,\bullet}(\mathbb{S}) \simeq \mathrm{Seq}_\bullet(\mathbb{S}^{1\text{-Cat}}),$$

while

$$\mathrm{Sq}^{\text{Pair}}_{\bullet,0}(\mathbb{S}, \mathbf{C}) \simeq \mathrm{Seq}_\bullet(\mathbf{C}).$$

In addition, for $n = n_1 + n_2$, the natural map

$$\mathrm{Sq}^{\text{Pair}}_{m,n}(\mathbb{S}, \mathbf{C}) \to \mathrm{Sq}^{\text{Pair}}_{m,n_1}(\mathbb{S}, \mathbf{C}) \underset{\mathrm{Sq}^{\text{Pair}}_{m,0}(\mathbb{S}, \mathbf{C})}{\times} \mathrm{Sq}^{\text{Pair}}_{m,n_2}(\mathbb{S}, \mathbf{C})$$

is an isomorphism and for $m = m_1 + m_2$, the natural map

$$\mathrm{Sq}^{\text{Pair}}_{m,n}(\mathbb{S}, \mathbf{C}) \to \mathrm{Sq}^{\text{Pair}}_{m_1,n}(\mathbb{S}, \mathbf{C}) \underset{\mathrm{Sq}^{\text{Pair}}_{0,n}(\mathbb{S}, \mathbf{C})}{\times} \mathrm{Sq}^{\text{Pair}}_{m_2,n}(\mathbb{S}, \mathbf{C}) \tag{4.3}$$

is an isomorphism.

4.3.7. As in Sect. 4.2.1, we obtain that there exists a well-defined functor

$$\mathrm{Seq}^{\text{Pair}}_\bullet : 2\text{-Cat}^{\text{Pair}} \to 1\text{-Cat}^{\mathbf{\Delta}^{\text{op}}}$$

so that

$$\mathrm{Sq}^{\text{Pair}}_{\bullet,\bullet} = (\mathrm{Seq}_\bullet)^{\mathbf{\Delta}^{\text{op}}} \circ \mathrm{Seq}^{\text{Pair}}_\bullet .$$

From Theorem 4.3.5, combined with Theorem 1.3.4, we obtain:

COROLLARY 4.3.8. *The functor* $\mathrm{Seq}^{\text{Pair}}_\bullet$ *is fully faithful.*

4.4. Left adjoint functors.

4.4.1. By construction, the functor $\mathrm{Seq}^{\text{ext}}_\bullet$ commutes with limits. Hence, by the Adjoint Functor Theorem, it admits a left adjoint, to be denoted $\mathfrak{L}^{\text{ext}}$.

Similarly, the functor

$$\mathrm{Seq}_\bullet : 2\text{-Cat} \to 1\text{-Cat}^{\mathbf{\Delta}^{\text{op}}}$$

admits a left adjoint, to be denoted $\mathfrak{L}$.

It is clear that when we restrict $\mathfrak{L}$ to the full subcategory $\mathrm{Spc}^{\mathbf{\Delta}^{\text{op}}} \subset 1\text{-Cat}^{\mathbf{\Delta}^{\text{op}}}$, the resulting functor lands in $1\text{-Cat} \subset 2\text{-Cat}$, thereby providing the left adjoint to the functor

$$\mathrm{Seq}_\bullet : 1\text{-Cat} \to \mathrm{Spc}^{\mathbf{\Delta}^{\text{op}}}.$$

4.4.2. We have the natural transformations of functors $2\text{-Cat}^{\text{Pair}} \to 1\text{-Cat}^{\mathbf{\Delta}^{\text{op}}}$

$$\mathrm{Seq}_\bullet \circ \mathrm{OblvSubcat} \to \mathrm{Seq}^{\text{Pair}}_\bullet \to \mathrm{Seq}^{\text{ext}}_\bullet \circ \mathrm{OblvSubcat}.$$

Composing with $\mathfrak{L}^{\text{ext}}$ and the co-unit of the adjunction, we obtain the natural transformations

$$\mathfrak{L}^{\text{ext}} \circ \mathrm{Seq}_\bullet \circ \mathrm{OblvSubcat} \to \mathfrak{L}^{\text{ext}} \circ \mathrm{Seq}^{\text{Pair}}_\bullet \to \mathfrak{L}^{\text{ext}} \circ \mathrm{Seq}^{\text{ext}}_\bullet \circ \mathrm{OblvSubcat} \to \mathrm{OblvSubcat},$$

where the last arrow is an isomorphism by Corollary 4.2.3.

We claim:

PROPOSITION 4.4.3. *The natural transformations*

$$\mathfrak{L}^{\text{ext}} \circ \mathrm{Seq}_\bullet \circ \mathrm{OblvSubcat} \to \mathfrak{L}^{\text{ext}} \circ \mathrm{Seq}^{\text{Pair}}_\bullet \to \mathfrak{L}^{\text{ext}} \circ \mathrm{Seq}^{\text{ext}}_\bullet \circ \mathrm{OblvSubcat} \to \mathrm{OblvSubcat}$$

are isomorphisms.

PROOF. By adjunction, we need to show that for $(\mathbb{S}, \mathbf{C}) \in 2\text{-Cat}^{\mathrm{Pair}}$ and $\mathbb{T} \in 2\text{-Cat}$, the natural map

$$\mathrm{Maps}_{2\text{-Cat}}(\mathrm{Seq}^{\mathrm{ext}}_{\bullet}(\mathbb{S}), \mathrm{Seq}^{\mathrm{ext}}_{\bullet}(\mathbb{T})) \to \mathrm{Maps}_{2\text{-Cat}}(\mathrm{Seq}^{\mathrm{Pair}}_{\bullet}(\mathbb{S}, \mathbf{C}), \mathrm{Seq}^{\mathrm{ext}}_{\bullet}(\mathbb{T})),$$

given by the inclusion $\mathrm{Seq}^{\mathrm{Pair}}_{\bullet}(\mathbb{S}, \mathbf{C}) \hookrightarrow \mathrm{Seq}^{\mathrm{ext}}_{\bullet}(\mathbb{S})$, is an isomorphism.

Note, however, that the above map fits into a commutative diagram

$$\begin{array}{ccc}
\mathrm{Maps}_{2\text{-Cat}}(\mathrm{Seq}^{\mathrm{ext}}_{\bullet}(\mathbb{S}), \mathrm{Seq}^{\mathrm{ext}}_{\bullet}(\mathbb{T})) & \longrightarrow & \mathrm{Maps}_{2\text{-Cat}}(\mathrm{Seq}^{\mathrm{Pair}}_{\bullet}(\mathbb{S}, \mathbf{C}), \mathrm{Seq}^{\mathrm{ext}}_{\bullet}(\mathbb{T})) \\
\uparrow & & \uparrow \\
\mathrm{Maps}_{2\text{-Cat}^{\mathrm{Pair}}}((\mathbb{S}, \mathbb{S}^{1\text{-Cat}}), (\mathbb{T}, \mathbb{T}^{1\text{-Cat}})) & \longrightarrow & \mathrm{Maps}_{2\text{-Cat}^{\mathrm{Pair}}}((\mathbb{S}, \mathbf{C}), (\mathbb{T}, \mathbb{T}^{1\text{-Cat}})),
\end{array}$$

where the vertical arrows are isomorphisms by Corollary 4.3.8. Now, the bottom horizontal arrow is an isomorphism, since the functor

$$\mathbb{T} \mapsto (\mathbb{T}, \mathbb{T}^{1\text{-Cat}}), \quad 2\text{-Cat} \to 2\text{-Cat}^{\mathrm{Pair}}$$

is the right adjoint to OblvSubcat.

□

4.4.4. Let

$$\mathfrak{L}^{\mathrm{Sq}} : \mathrm{Spc}^{\mathbf{\Delta}^{\mathrm{op}} \times \mathbf{\Delta}^{\mathrm{op}}} \to 2\text{-Cat},$$

denote the left adjoint of the functor $\mathrm{Sq}_{\bullet,\bullet}$. Tautologically, we have:

$$\mathfrak{L}^{\mathrm{Sq}} = \mathfrak{L}^{\mathrm{ext}} \circ \mathfrak{L}^{\mathbf{\Delta}^{\mathrm{op}}}.$$

From Proposition 4.4.3 we obtain:

COROLLARY 4.4.5. *The natural transformations*

$$\mathfrak{L}^{\mathrm{Sq}} \circ \mathrm{Sq}^{\sim}_{\bullet,\bullet} \circ \mathrm{OblvSubcat} \to \mathfrak{L}^{\mathrm{Sq}} \circ \mathrm{Sq}^{\mathrm{Pair}}_{\bullet,\bullet} \to \mathfrak{L}^{\mathrm{Sq}} \circ \mathrm{Sq}_{\bullet,\bullet} \circ \mathrm{OblvSubcat} \to \mathrm{OblvSubcat}$$

are isomorphisms.

4.5. The Gray product via Squares.

4.5.1. Let $\mathbb{S}$ and $\mathbb{T}$ be a pair of $(\infty, 2)$-categories. Consider the following object of $\mathrm{Spc}^{\mathbf{\Delta}^{\mathrm{op}} \times \mathbf{\Delta}^{\mathrm{op}}}$

$$(\mathrm{Sq}^{\sim}_{\bullet,\bullet}(\mathbb{S}^{2\text{-op}}))^{\mathrm{reflect}} \times \mathrm{Sq}^{\sim}_{\bullet,\bullet}(\mathbb{T}). \tag{4.4}$$

I.e., its space of m,n-simplices is

$$\mathrm{Sq}^{\sim}_{n,m}(\mathbb{S}^{2\text{-op}}) \times \mathrm{Sq}^{\sim}_{m,n}(\mathbb{T}).$$

We claim that we have a canonically defined map

$$\mathfrak{L}^{\mathrm{Sq}}\big((\mathrm{Sq}^{\sim}_{\bullet,\bullet}(\mathbb{S}^{2\text{-op}}))^{\mathrm{reflect}} \times \mathrm{Sq}^{\sim}_{\bullet,\bullet}(\mathbb{T})\big) \to \mathbb{S} \circledast \mathbb{T}, \tag{4.5}$$

functorial in $\mathbb{S}$ and $\mathbb{T}$.

4.5.2. Indeed, the datum of a map (4.5) is equivalent to that of a map in $\mathrm{Spc}^{\mathbf{\Delta}^{\mathrm{op}}\times\mathbf{\Delta}^{\mathrm{op}}}$

$$(\mathrm{Sq}^{\sim}_{\bullet,\bullet}(\mathbb{S}^{2\text{-op}}))^{\mathrm{reflect}} \times \mathrm{Sq}^{\sim}_{\bullet,\bullet}(\mathbb{T}) \to \mathrm{Sq}_{\bullet,\bullet}(\mathbb{S}\circledast\mathbb{T}). \tag{4.6}$$

The datum of the map (4.6) (when we require functoriality in $\mathbb{S}$ and $\mathbb{T}$) is equivalent to that of a map of bi-cosimplicial objects in 2-Cat

$$[m,n] \to ([n,m]^{\sim})^{2\text{-op}} \circledast [m,n]^{\sim}. \tag{4.7}$$

The functors (4.7) are constructed as follows. Consider the composition of (right-lax) functors

$$\begin{aligned}(4.8)\quad [m]\times[n] &\overset{\mathrm{diag}}{\longrightarrow} [m]\times[n]\times[m]\times[n] \dashrightarrow [m]\circledast[n]\circledast[m]\circledast[n] \to \\ &\to ([m]\circledast[n])\circledast([m]\circledast[n]) \simeq ([n]^{2\text{-op}}\circledast[m]^{2\text{-op}})^{2\text{-op}}\circledast([m]\circledast[n]) \simeq \\ &\simeq ([n]\circledast[m])^{2\text{-op}}\circledast([m]\circledast[n]) = ([n,m])^{2\text{-op}}\circledast[m,n] \to ([n,m]^{\sim})^{2\text{-op}}\circledast[m,n]^{\sim}\end{aligned}$$

Now, unwinding the construction, one checks that the composite map in (4.8) indeed gives rise to a map (4.7) (the corresponding 2-morphisms are isomorphisms).

4.5.3. We have the following result[7]:

PROPOSITION 4.5.4. *The map* (4.5) *is an equivalence.*

4.6. Cubes.

4.6.1. Let k be an integer ≥ 1. The assignment

$$(n_1,...,n_k) \mapsto [n_1,...,n_k]$$

gives a functor

$$\mathbf{\Delta}^{\times k} \to 2\text{-Cat}.$$

Hence, we obtain a well-defined functor

$$\mathbf{Cu} : 2\text{-Cat} \to \mathrm{Spc}^{(\mathbf{\Delta}^{\mathrm{op}})^{\times k}}$$

that sends $\mathbb{S}$ to

$$(n_1,...,n_k) \mapsto \mathrm{Maps}_{2\text{-Cat}}([n_1,...,n_k],\mathbb{S}).$$

4.6.2. We have the following generalization of Theorem 4.1.3[8]:

THEOREM 4.6.3. *For any $k \geq 2$, the corresponding functor* **Cu** *is fully faithful.*

5. Essential image of the functor $\mathrm{Sq}_{\bullet,\bullet}$

The goal of this section is to describe the essential image of the functors $\mathrm{Sq}_{\bullet,\bullet}$ and $\mathrm{Sq}^{\mathrm{Pair}}_{\bullet,\bullet}$.

5.1. Invertible angles.

5.1.1. Let $E_{\bullet,\bullet}$ be an object of $\mathrm{Spc}^{\mathbf{\Delta}^{\mathrm{op}}\times\mathbf{\Delta}^{\mathrm{op}}}$. We shall say that $E_{\bullet,\bullet}$ is a double category if for every n, the objects $E_{\bullet,n}$ and $E_{n,\bullet}$ of $\mathrm{Spc}^{\mathbf{\Delta}^{\mathrm{op}}}$ are complete Segal spaces.

[7]We do not prove it, and we were not able to find a reference.

[8]We do not prove it, and we were not able to find a reference.

5.1.2. Let $E_{\bullet,\bullet}$ be a double category. Let $(E_{1,1})^{\ulcorner} \subset E_{1,1}$ be the full subspace consisting of squares in which the right vertical side and the bottom horizontal side are degenerate. I.e., these are diagrams

$$\begin{array}{ccc} x & \xrightarrow{\alpha} & y \\ {\scriptstyle\beta}\downarrow & & \downarrow{\scriptstyle\mathrm{id}} \\ y & \xrightarrow{\mathrm{id}} & y. \end{array} \tag{5.1}$$

Let $(E_{1,1})^{\lrcorner} \subset E_{1,1}$ be the full subspace consisting of squares in which the left vertical side and the top horizontal side are degenerate. I.e., these are diagrams

$$\begin{array}{ccc} x & \xrightarrow{\mathrm{id}} & x \\ {\scriptstyle\mathrm{id}}\downarrow & & \downarrow{\scriptstyle\beta} \\ x & \xrightarrow{\alpha} & y. \end{array}$$

5.1.3. We define the full subspace

$$(E_{1,1})^{\ulcorner,\mathrm{invert}} \subset (E_{1,1})^{\ulcorner}$$

of *invertible angles* as follows. A point (5.1) is invertible if the following two conditions hold:

(I) There exists a point in $E_{2,1}$

$$\begin{array}{ccc} x & \xrightarrow{\mathrm{id}} & x \\ {\scriptstyle\mathrm{id}}\downarrow & & \downarrow{\scriptstyle\beta'} \\ x & \xrightarrow{\alpha} & y \\ {\scriptstyle\beta}\downarrow & & \downarrow{\scriptstyle\mathrm{id}} \\ y & \xrightarrow{\mathrm{id}} & y, \end{array} \tag{5.2}$$

in which the lower square is the original (5.1), the top square is in $(E_{1,1})^{\lrcorner}$, and the outer square is degenerate (i.e., pulled back via $[1] \times [1] \to [1] \times [0]$).

(II) There exists a point in $E_{1,2}$

$$\begin{array}{ccccc} x & \xrightarrow{\mathrm{id}} & x & \xrightarrow{\alpha} & y \\ {\scriptstyle\mathrm{id}}\downarrow & & {\scriptstyle\beta}\downarrow & & \downarrow{\scriptstyle\mathrm{id}} \\ x & \xrightarrow{\alpha'} & y & \xrightarrow{\mathrm{id}} & y, \end{array} \tag{5.3}$$

in which the right square is the original (5.1), the left square is in $(E_{1,1})^{\lrcorner}$, and the outer square is degenerate (i.e., pulled back via $[1] \times [1] \to [0] \times [1]$).

5.1.4. The following is an elementary check:

LEMMA 5.1.5. *The restriction maps*

$$(E_{1,1})^{\ulcorner,\mathrm{invert}} \to E_{1,0} \text{ and } (E_{1,1})^{\ulcorner,\mathrm{invert}} \to E_{0,1},$$

given by taking (5.1) *to its left vertical side and its top horizontal side, are* monomorphisms.

5.2. Description of the essential image. Let $(\mathbb{S},\mathbf{C})$ be an object of $2\text{-Cat}^{\mathrm{Pair}}$, and consider a point $\beta \in \mathrm{Sq}^{\mathrm{Pair}}_{1,0}(\mathbb{S},\mathbf{C})$. This point represents an 1-morphism in $\mathbf{C}$, and the same point can be represented by an element in $\alpha \in \mathrm{Sq}^{\mathrm{Pair}}_{0,1}(\mathbb{S},\mathbf{C})$, completing β to a point (5.1) in $(\mathrm{Sq}^{\mathrm{Pair}}_{1,1}(\mathbb{S},\mathbf{C}))^{\ulcorner,\mathrm{invert}}$.

It turns out that the property that one can complete a point in $\mathrm{Sq}^{\mathrm{Pair}}_{1,0}(\mathbb{S},\mathbf{C})$ to a point in $(\mathrm{Sq}^{\mathrm{Pair}}_{1,1}(\mathbb{S},\mathbf{C}))^{\ulcorner,\mathrm{invert}}$ characterizes the essential image of $2\text{-Cat}^{\mathrm{Pair}}$ in $\mathrm{Spc}^{\mathbf{\Delta}^{\mathrm{op}}\times\mathbf{\Delta}^{\mathrm{op}}}$.

5.2.1. Let $E_{\bullet,\bullet}$ be again a double category. We shall say that $E_{\bullet,\bullet}$ is *anti-clockwise reversible* if the map

$$(E_{1,1})^{\ulcorner,\mathrm{invert}} \to E_{1,0}$$

from Lemma 5.1.5 is an isomorphism.

We shall say that $E_{\bullet,\bullet}$ is *reversible* if both maps in Lemma 5.1.5 are isomorphisms.

5.2.2. We now give the following sharpening of Theorems 4.1.3 and 4.3.5[9]:

THEOREM 5.2.3.
(a) *The essential image of the functor*

$$\mathrm{Sq}_{\bullet,\bullet} : 2\text{-Cat} \to \mathrm{Spc}^{\mathbf{\Delta}^{\mathrm{op}}\times\mathbf{\Delta}^{\mathrm{op}}}$$

is the full subcategory consisting of reversible double categories.
(b) *The essential image of the functor*

$$\mathrm{Sq}^{\mathrm{Pair}}_{\bullet,\bullet} : 2\text{-Cat}^{\mathrm{Pair}} \to \mathrm{Spc}^{\mathbf{\Delta}^{\mathrm{op}}\times\mathbf{\Delta}^{\mathrm{op}}}$$

is the full subcategory consisting of anti-clockwise reversible double categories.

5.3. The $(\infty,2)$-category $\mathbf{1}\text{-}\mathbf{Cat}$ via squares. In Sect. 2.4 we introduced the $(\infty,2)$-category $\mathbf{1}\text{-}\mathbf{Cat}$. In this subsection we will describe its essential image under the functor $\mathrm{Seq}^{\mathrm{ext}}_{\bullet}$.

5.3.1. Consider the following object, denoted $\mathrm{Seq}^{\mathrm{ext}}_{\bullet}(\mathbf{1}\text{-}\mathbf{Cat})$ of $1\text{-Cat}^{\mathbf{\Delta}^{\mathrm{op}}}$. Namely,

$$\mathrm{Seq}^{\mathrm{ext}}_{n}(\mathbf{1}\text{-}\mathbf{Cat}) := \mathrm{Cart}_{/[n]^{\mathrm{op}}}.$$

Note the difference between $\mathrm{Seq}^{\mathrm{ext}}_{n}(\mathbf{1}\text{-}\mathbf{Cat})$ and $\mathrm{Seq}_{n}(\mathbf{1}\text{-}\mathbf{Cat})$: the two have the same objects, while the latter has fewer morphisms.

5.3.2. We claim:

PROPOSITION 5.3.3.
(a) *The object* $\mathrm{Seq}^{\mathrm{ext}}_{\bullet}(\mathbf{1}\text{-}\mathbf{Cat})$ *lies in the essential image of the functor*

$$\mathrm{Seq}^{\mathrm{ext}} : 2\text{-Cat} \to 1\text{-Cat}^{\mathbf{\Delta}^{\mathrm{op}}}.$$

(b) *The resulting object of* 2-Cat *identifies canonically with* $\mathbf{1}\text{-}\mathbf{Cat}$.

[9]We do not prove it, and we were not able to find a reference.

Proof. First, it is clear that simplicial category $\mathrm{Seq}_\bullet(\mathbf{1\text{-}Cat})$ is obtained from $\mathrm{Seq}^{\mathrm{ext}}_\bullet(\mathbf{1\text{-}Cat})$ by the procedure of Sect. 4.2.4.

The fact that the simplicial category $\mathrm{Seq}^{\mathrm{ext}}_\bullet(\mathbf{1\text{-}Cat})$ satisfies the Segal condition follows in the same way as in the case of $\mathrm{Seq}_\bullet(\mathbf{1\text{-}Cat})$.

Consider the bi-simplicial space

$$\mathrm{Sq}_{\bullet,\bullet}(\mathbf{1\text{-}Cat}) := \mathrm{Seq}_\bullet(\mathrm{Seq}^{\mathrm{ext}}_\bullet(\mathbf{1\text{-}Cat})).$$

It is easy to see that it is a *complete* Segal space along each row and column, so it is a double category.

By Theorem 5.2.3(a), in order to prove the proposition, we need to show that $\mathrm{Sq}_{\bullet,\bullet}(\mathbf{1\text{-}Cat})$ is reversible.

For every n, consider the 1-full subcategory

$$\mathrm{Funct}([n], 1\text{-Cat}) \simeq (\mathrm{Cart}_{/[n]^{\mathrm{op}}})_{\mathrm{strict}} \subset \mathrm{Cart}_{/[n]^{\mathrm{op}}},$$

and the corresponding full bi-simplicial subspace

$$\mathrm{Seq}_\bullet((\mathrm{Cart}_{/[\bullet]^{\mathrm{op}}})_{\mathrm{strict}}) =: \mathrm{Sq}'_{\bullet,\bullet}(\mathbf{1\text{-}Cat}) \subset \mathrm{Sq}_{\bullet,\bullet}(\mathbf{1\text{-}Cat}).$$

Note that

$$\mathrm{Sq}'_{m,n}(\mathbf{1\text{-}Cat}) \subset \mathrm{Sq}_{m,n}(1\text{-Cat})$$

is an isomorphism when either m or n equals 0. Hence, its is enough to show that $\mathrm{Sq}'_{\bullet,\bullet}(\mathbf{1\text{-}Cat})$ is reversible.

However, by construction, $\mathrm{Sq}'_{\bullet,\bullet}(\mathbf{1\text{-}Cat})$ identifies with $\mathrm{Sq}_{\bullet,\bullet}(1\text{-Cat})$, where $1\text{-Cat} \in 1\text{-Cat}$ is regarded as an $(\infty,2)$-category. In particular, it is reversible. □

6. The $(\infty, 2)$-category of $(\infty, 2)$-categories

In this section we upgrade the structure of $(\infty,1)$-category on the totality of $(\infty,2)$-categories to a structure of $(\infty,2)$-category. I.e., we will define an $(\infty,2)$-category $\mathbf{2\text{-}Cat}$ equipped with an identification

$$(\mathbf{2\text{-}Cat})^{1\text{-Cat}} \simeq 2\text{-Cat}.$$

We will show that for $\mathbb{S}, \mathbb{T} \in \mathbf{2\text{-}Cat}$, the $(\infty,1)$-category $\mathbf{Maps}_{\mathbf{2\text{-}Cat}}(\mathbb{S},\mathbb{T})$ is canonically equivalent to $(\mathrm{Funct}(\mathbb{S},\mathbb{T}))^{1\text{-Cat}}$, where $\mathrm{Funct}(\mathbb{S},\mathbb{T})$ is the $(\infty,2)$-category of functors defined in Sect. 2.5.4.

We will also show that the $(\infty,2)$-category $\mathbf{1\text{-}Cat}$ sits inside $\mathbf{2\text{-}Cat}$ as a full subcategory.

Note, however, that the structure of $(\infty,2)$-category on the totality of $(\infty,2)$-categories is not the end of the story: the latter must in fact form an $(\infty,3)$-category. However, we will not pursue this here.

6.1. The $\mathrm{Seq}^{\mathrm{ext}}_\bullet$ model for $\mathbf{2\text{-}Cat}$.

6.1.1. We introduce the $(\infty, 2)$-category $\mathbf{2\text{-}Cat}$ to be the full subcategory in

$$\operatorname{Funct}(\mathbf{\Delta}^{\mathrm{op}}, \mathbf{1\text{-}Cat}),$$

whose objects are functors $\mathbf{\Delta}^{\mathrm{op}} \to \mathbf{1\text{-}Cat}$ such that, when regarded as functor $\mathbf{\Delta}^{\mathrm{op}} \to 1\text{-Cat}$, they belong to

$$2\text{-Cat} \overset{\mathrm{Seq}_\bullet^{\mathrm{ext}}}{\hookrightarrow} \operatorname{Funct}(\mathbf{\Delta}^{\mathrm{op}}, 1\text{-Cat}).$$

I.e., we take the $(\infty, 1)$-category 2-Cat realized as a full subcategory of

$$\operatorname{Funct}(\mathbf{\Delta}^{\mathrm{op}}, 1\text{-Cat})$$

via the functor $\mathrm{Seq}_\bullet^{\mathrm{ext}}$ and extend it to an $(\infty, 2)$-category by addinng non-invertible 2-morphisms to be those given by extending the target $(\infty, 1)$-category 1-Cat to the $(\infty, 2)$-category $\mathbf{1\text{-}Cat}$.

By construction, we have

$$(\mathbf{2\text{-}Cat})^{1\text{-Cat}} \simeq 2\text{-Cat}.$$

REMARK 6.1.2. Note that in giving the above definition, it is important that we are dealing with the functor $\mathrm{Seq}_\bullet^{\mathrm{ext}}$, rather than $\mathrm{Seq}_\bullet$.

6.1.3. By unwinding the definition, we obtain the following description of the functor

$$\mathrm{Seq}_\bullet^{\mathrm{ext}}(\mathbf{2\text{-}Cat}) : \mathbf{\Delta}^{\mathrm{op}} \to 1\text{-Cat}.$$

Namely, $\mathrm{Seq}_n^{\mathrm{ext}}(\mathbf{2\text{-}Cat})$ is the full subcategory in

$$\Big(\operatorname{Funct}\big([n], \operatorname{Funct}(\mathbf{\Delta}^{\mathrm{op}}, \mathbf{1\text{-}Cat})\big)_{\text{right-lax}}\Big)^{1\text{-Cat}} \subset$$
$$\subset \Big(\operatorname{Funct}\big(\mathbf{\Delta}^{\mathrm{op}}, \operatorname{Funct}([n], \mathbf{1\text{-}Cat})_{\text{right-lax}}\big)\Big)^{1\text{-Cat}} \simeq$$
$$\simeq \operatorname{Funct}\Big(\mathbf{\Delta}^{\mathrm{op}}, \big(\operatorname{Funct}([n], \mathbf{1\text{-}Cat})_{\text{right-lax}}\big)^{1\text{-Cat}}\Big) = \operatorname{Funct}(\mathbf{\Delta}^{\mathrm{op}}, \mathrm{Cart}_{/[n]^{\mathrm{op}}}),$$

consisting of objects $\mathbf{E}_\bullet$ that satisfy the following:

- As an object of $(\operatorname{Funct}(\mathbf{\Delta}^{\mathrm{op}}, \mathrm{Cart}_{/[n]^{\mathrm{op}}}))^{\mathrm{Spc}}$, we require that $\mathbf{E}_\bullet$ belong to

$$\mathrm{Maps}_{1\text{-Cat}}(\mathbf{\Delta}^{\mathrm{op}} \times [n], 1\text{-Cat}) \simeq \mathrm{Maps}_{1\text{-Cat}}(\mathbf{\Delta}^{\mathrm{op}}, \operatorname{Funct}([n], 1\text{-Cat})) \simeq$$
$$\simeq \mathrm{Maps}_{1\text{-Cat}}(\mathbf{\Delta}^{\mathrm{op}}, (\mathrm{Cart}_{/[n]^{\mathrm{op}}})_{\mathrm{strict}}) \subset$$
$$\subset \mathrm{Maps}_{1\text{-Cat}}(\mathbf{\Delta}^{\mathrm{op}}, \mathrm{Cart}_{/[n]^{\mathrm{op}}}) = (\operatorname{Funct}(\mathbf{\Delta}^{\mathrm{op}}, \mathrm{Cart}_{/[n]^{\mathrm{op}}}))^{\mathrm{Spc}};$$

- For every $i \in [n]$, we require that the resulting object

$$\mathbf{E}_{\bullet, i} \in \operatorname{Funct}(\mathbf{\Delta}^{\mathrm{op}}, 1\text{-Cat})$$

lie in the essential image of the functor

$$\mathrm{Seq}_\bullet^{\mathrm{ext}} : 2\text{-Cat} \to \operatorname{Funct}(\mathbf{\Delta}^{\mathrm{op}}, 1\text{-Cat}) = 1\text{-Cat}^{\mathbf{\Delta}^{\mathrm{op}}}.$$

6.2. Identifying the categories of maps. Let $\mathbb{S}$ and $\mathbb{T}$ be two objects of $\mathbf{2\text{-}Cat}$. The first test on whether the above definition of $\mathbf{2\text{-}Cat}$ is reasonable, is whether or not the $(\infty,1)$-category $\mathbf{Maps}_{\mathbf{2\text{-}Cat}}(\mathbb{S},\mathbb{T})$ indeed recovers the $(\infty,1)$-category $(\mathrm{Funct}(\mathbb{S},\mathbb{T}))^{1\text{-Cat}}$ of functors from $\mathbb{S}$ to $\mathbb{T}$, defined in Sect. 2.5.4.

6.2.1. We claim:

PROPOSITION-CONSTRUCTION 6.2.2. *For* $\mathbb{S},\mathbb{T}\in\mathbf{2\text{-}Cat}$, *the* $(\infty,1)$-*category* $\mathbf{Maps}_{\mathbf{2\text{-}Cat}}(\mathbb{S},\mathbb{T})$ *identifies canonically with*

$$(\mathrm{Funct}(\mathbb{S},\mathbb{T}))^{1\text{-Cat}}.$$

The rest of this subsection is devoted to the proof of this proposition.

6.2.3. Unwinding the definition of $\mathbf{Maps}_{\mathbf{2\text{-}Cat}}(\mathbb{S},\mathbb{T})$, and using Chapter 1, Sect. 1.4.5, we obtain that for $\mathbf{I}\in 1\text{-Cat}$, we have a canonical isomorphism

$$\mathrm{Maps}_{1\text{-Cat}}(\mathbf{I},\mathbf{Maps}_{\mathbf{2\text{-}Cat}}(\mathbb{S},\mathbb{T}))\simeq \mathrm{Maps}_{\mathrm{Funct}(\mathbf{\Delta}^{\mathrm{op}},1\text{-Cat})}(\mathbf{I}\times \mathrm{Seq}^{\mathrm{ext}}_\bullet(\mathbb{S}),\mathrm{Seq}^{\mathrm{ext}}_\bullet(\mathbb{T})),$$

functorial in $\mathbf{I}$.

Thus, in order to prove the proposition, we need to construct an identification

$$\mathrm{Maps}_{2\text{-Cat}}(\mathbf{I}\times\mathbb{S},\mathbb{T})\simeq \mathrm{Maps}_{\mathrm{Funct}(\mathbf{\Delta}^{\mathrm{op}},1\text{-Cat})}(\mathbf{I}\times \mathrm{Seq}^{\mathrm{ext}}_\bullet(\mathbb{S}),\mathrm{Seq}^{\mathrm{ext}}_\bullet(\mathbb{T})). \tag{6.1}$$

functorial in $\mathbf{I}$.

6.2.4. Recall the $(\infty,2)$-category $\mathrm{Funct}(\mathbf{I},\mathbb{T})_{\text{left-lax}}$, defined so that

$$\mathrm{Maps}(\mathbb{S}',\mathrm{Funct}(\mathbf{I},\mathbb{T})_{\text{left-lax}}):=\mathrm{Maps}(\mathbf{I}\circledast\mathbb{S}',\mathbb{T}).$$

Note that for every n we have a canonical fully faithful embedding

$$\mathbf{Maps}(\mathbf{I},\mathrm{Seq}^{\mathrm{ext}}_n(\mathbb{T}))\hookrightarrow \mathrm{Seq}^{\mathrm{ext}}_n(\mathrm{Funct}(\mathbf{I},\mathbb{T})_{\text{left-lax}}).$$

Indeed, for every m we have

$$\mathrm{Seq}_m(\mathbf{Maps}(\mathbf{I},\mathrm{Seq}^{\mathrm{ext}}_n(\mathbb{T})))=\mathrm{Maps}(\mathbf{I}\times[m],\mathrm{Seq}^{\mathrm{ext}}_n(\mathbb{T}))\simeq\mathrm{Maps}((\mathbf{I}\times[m])\circledast[n],\mathbb{T}),$$

while

$$\mathrm{Seq}_m(\mathrm{Seq}^{\mathrm{ext}}_n(\mathrm{Funct}(\mathbf{I},\mathbb{T})_{\text{left-lax}}))=$$
$$=\mathrm{Maps}([m]\circledast[n],\mathrm{Funct}(\mathbf{I},\mathbb{T})_{\text{left-lax}})=\mathrm{Maps}(\mathbf{I}\circledast([m]\circledast[n]),\mathbb{T}),$$

and the embedding in question comes from the projection

$$\mathbf{I}\circledast([m]\circledast[n])\simeq\mathbf{I}\circledast[m]\circledast[n]\simeq(\mathbf{I}\circledast[m])\circledast[n]\to(\mathbf{I}\times[m])\circledast[n].$$

6.2.5. Thus, we obtain that the right-hand side in (6.1), interpreted as

$$\mathrm{Maps}_{\mathrm{Funct}(\mathbf{\Delta}^{\mathrm{op}},1\text{-Cat})}(\mathrm{Seq}^{\mathrm{ext}}_\bullet(\mathbb{S}),\mathbf{Maps}(\mathbf{I},\mathrm{Seq}^{\mathrm{ext}}_\bullet(\mathbb{T}))),$$

admits a fully faithful embedding into

$$\mathrm{Maps}_{\mathrm{Funct}(\mathbf{\Delta}^{\mathrm{op}},1\text{-Cat})}(\mathrm{Seq}^{\mathrm{ext}}_\bullet(\mathbb{S}),\mathrm{Seq}^{\mathrm{ext}}_\bullet(\mathrm{Funct}(\mathbf{I},\mathbb{T})_{\text{left-lax}}))\simeq$$
$$\simeq\mathrm{Maps}(\mathbb{S},\mathrm{Funct}(\mathbf{I},\mathbb{T})_{\text{left-lax}})=\mathrm{Maps}(\mathbf{I}\circledast\mathbb{S},\mathbb{T}).$$

Furthermore, it is easy to see that the essential image of the right-hand side in (6.1) in $\mathrm{Maps}(\mathbf{I}\circledast\mathbb{S},\mathbb{T})$ equals that of the fully faithful embedding

$$\mathrm{Maps}(\mathbf{I}\times\mathbb{S},\mathbb{T})\hookrightarrow\mathrm{Maps}(\mathbf{I}\circledast\mathbb{S},\mathbb{T}), \tag{6.2}$$

thereby giving rise to the sought-for isomorphism (6.1).

6.3. Another interpretation for $\mathbf{1\text{-}Cat}$. We will now show that $\mathbf{1\text{-}Cat}$, as defined in Sect. 2.4, embeds fully faithfully into $\mathbf{2\text{-}Cat}$.

6.3.1. Let us (temporarily) denote by

$$\mathbf{1\text{-}Cat}' \subset \mathbf{2\text{-}Cat}$$

the full subcategory, defined so that $(\mathbf{1\text{-}Cat}')^{\mathrm{Spc}}$, viewed as a subspace of $(\mathbf{2\text{-}Cat})^{\mathrm{Spc}}$, equals

$$(1\text{-Cat})^{\mathrm{Spc}} \subset (2\text{-Cat})^{\mathrm{Spc}} \simeq (\mathbf{2\text{-}Cat})^{\mathrm{Spc}}.$$

We are going to prove:

PROPOSITION-CONSTRUCTION 6.3.2. *There is a canonical equivalence of $(\infty, 2)$-categories*

$$\mathbf{1\text{-}Cat} \simeq \mathbf{1\text{-}Cat}',$$

extending the identification

$$(\mathbf{1\text{-}Cat})^{1\text{-Cat}} \simeq 1\text{-Cat} \simeq (\mathbf{1\text{-}Cat}')^{1\text{-Cat}}.$$

The rest of this subsection is devoted to the proof of this proposition.

6.3.3. We construct the functor $\mathbf{1\text{-}Cat} \to \mathbf{1\text{-}Cat}'$ in the guise of a map of simplicial categories:

$$\mathrm{Seq}^{\mathrm{ext}}_\bullet(\mathbf{1\text{-}Cat}) \to \mathrm{Seq}^{\mathrm{ext}}_\bullet(\mathbf{1\text{-}Cat}'). \tag{6.3}$$

We recall that

$$\mathrm{Seq}^{\mathrm{ext}}_n(\mathbf{1\text{-}Cat}) = \mathrm{Cart}_{/[n]^{\mathrm{op}}}$$

and

$$\mathrm{Seq}^{\mathrm{ext}}_n(\mathbf{1\text{-}Cat}') \subset \mathrm{Funct}(\mathbf{\Delta}^{\mathrm{op}}, \mathrm{Cart}_{/[n]^{\mathrm{op}}}) \subset \mathrm{Funct}(\mathbf{\Delta}^{\mathrm{op}}, 1\text{-Cat}_{/[n]^{\mathrm{op}}}).$$

The sought-for functor in (6.3) is given by

$$\mathrm{Cart}_{/[n]^{\mathrm{op}}} \to \mathrm{Funct}(\mathbf{\Delta}^{\mathrm{op}}, 1\text{-Cat}_{/[n]^{\mathrm{op}}}), \quad (\mathbf{E} \to [n]) \mapsto \left([m] \mapsto \mathrm{Funct}([m], \mathbf{E}) \underset{\mathrm{Funct}([m],[n]^{\mathrm{op}})}{\times} [n]^{\mathrm{op}}\right),$$

where $[n]^{\mathrm{op}} \to \mathrm{Funct}([m], [n]^{\mathrm{op}})$ is given by

$$[n]^{\mathrm{op}} = \mathrm{Funct}(\{*\}, [n]^{\mathrm{op}}) \to \mathrm{Funct}([m], [n]^{\mathrm{op}}).$$

It is easy to see that the image of the above map indeed lands in $\mathrm{Seq}^{\mathrm{ext}}_n(\mathbf{1\text{-}Cat}')$.

6.3.4. It follows from the construction that the resulting functor $\mathbf{1\text{-}Cat} \to \mathbf{1\text{-}Cat}'$ makes the diagram

$$\begin{array}{ccc}
(\mathbf{1\text{-}Cat})^{1\text{-Cat}} & \longrightarrow & (\mathbf{1\text{-}Cat}')^{1\text{-Cat}} \\
\sim\uparrow & & \uparrow\sim \\
1\text{-Cat} & \xrightarrow{\mathrm{Id}} & 1\text{-Cat}
\end{array}$$

commute.

Hence, it remains to show that it is fully faithful. However, it follows from the construction that for $\mathbf{S}, \mathbf{T} \in 1\text{-Cat}$, the diagram

$$\begin{array}{ccc} \mathbf{Maps}_{\mathbf{1\text{-}Cat}}(\mathbf{S},\mathbf{T}) & \longrightarrow & \mathbf{Maps}_{\mathbf{1\text{-}Cat}'}(\mathbf{S},\mathbf{T}) \\ {\scriptstyle \text{Corollary 2.4.4}}\uparrow{\scriptstyle \sim} & & {\scriptstyle \sim}\uparrow{\scriptstyle \text{Proposition 6.2.2}} \\ \mathrm{Funct}(\mathbf{S},\mathbf{T}) & \xrightarrow{\mathrm{Id}} & \mathrm{Funct}(\mathbf{S},\mathbf{T}) \end{array}$$

commutes, establishing the required fully-faithfulness.

CHAPTER 11

Straightening and Yoneda for $(\infty,2)$-categories

Introduction

0.1. What is done in this Chapter? The goal of this Chapter is to construct the 2-categorical Yoneda embedding

$$\text{Yon}_{\mathbb{S}} : \mathbb{S} \hookrightarrow \text{Funct}(\mathbb{S}^{1\text{-op}}, \mathbf{1}\text{-}\mathbf{Cat}), \quad \mathbb{S} \in 2\text{-Cat}, \tag{0.1}$$

which will, in turn, be needed for the proof of the Adjunction Theorem in Chapter 12.

As in the case of $(\infty,1)$-categories, in the present 2-categorical context, a natural approach to the construction of the functor $\text{Yon}_{\mathbb{S}}$ is via the straightening/unstraightening procedure.

The latter is an equivalence between the $(\infty,2)$-category of functors $\mathbb{S}^{1\text{-op}} \to \mathbf{1}\text{-}\mathbf{Cat}$ and the $(\infty,2)$-category of *1-Cartesian* fibrations over $\mathbb{S}$.

0.1.1. Let us comment on the notion of 1-Cartesian fibration over a given $\mathbb{S} \in 2\text{-Cat}$.

The space of such will be a full subspace in $(\mathbf{2}\text{-}\mathbf{Cat}_{/\mathbb{S}})^{\text{Spc}}$, and it is singled out by certain explicit conditions; the actual definition is given in Sect. 1.2.1. The definition is rigged so that the datum of a 1-Cartesian fibration over $\mathbb{S}$ is equivalent to that of a functor $\mathbb{S}^{1\text{-op}} \to \mathbf{1}\text{-}\mathbf{Cat}$.

As to the 2-categorical structure, there are actually two natural $(\infty,2)$-categories

$$(\mathbf{1}\text{-}\mathbf{Cart}_{/\mathbb{S}})_{\text{strict}} \subset (\mathbf{1}\text{-}\mathbf{Cart}_{/\mathbb{S}})_{2\text{-strict}},$$

one being a 1-full subcategory in the other.

In Sect. 1 we state the sub-main result of this Chapter, Corollary 1.2.6, that says that there is a canonical 'straightening/unstraightening' equivalence

$$(\mathbf{1}\text{-}\mathbf{Cart}_{/\mathbb{S}})_{2\text{-strict}} \simeq \text{Funct}(\mathbb{S}^{1\text{-op}}, \mathbf{1}\text{-}\mathbf{Cat})_{\text{right-lax}}, \tag{0.2}$$

which induces an equivalence

$$(\mathbf{1}\text{-}\mathbf{Cart}_{/\mathbb{S}})_{\text{strict}} \simeq \text{Funct}(\mathbb{S}^{1\text{-op}}, \mathbf{1}\text{-}\mathbf{Cat}).$$

0.1.2. Here is, however, a catch: the above straightening/unstraightening assertion (i.e., the equivalence (0.2)) is *too weak* to be amenable to a natural proof.

Namely, the equivalence (0.2) does not contain enough functoriality (the mechanics of how this happens can be seen by tracing through the proof of the main theorem of this Chapter, Theorem 1.1.8; see also Sect. 0.1.5 below).

0.1.3. To remedy this, we engage a more ambitious straightening/unstraightening procedure.

Namely, in Sect. 1 we introduce the notion of 2-Cartesian fibration (over a given $(\infty, 2)$-category $\mathbb{S}$). Again, the space of such is a full subspace of $(\mathbf{2\text{-}Cat}_{/\mathbb{S}})^{\mathrm{Spc}}$, and it is singled out by certain explicit conditions specified in Sect. 1.1.1.

As in the case of 1-Cartesian fibrations, there two natural $(\infty, 2)$-categories

$$(\mathbf{2\text{-}Cart}_{/\mathbb{S}})_{\mathrm{strict}} \subset (\mathbf{2\text{-}Cart}_{/\mathbb{S}})_{\text{2-strict}},$$

one being a 1-full subcategory in the other.

The 2-categorical straightening/unstraightening assertion, Theorem 1.1.8, which is the main result of this Chapter, says that there exists a canonical equivalence

$$(\mathbf{2\text{-}Cart}_{/\mathbb{S}})_{\text{2-strict}} \simeq \mathrm{Funct}(\mathbb{S}^{1\text{-op}}, \mathbf{2\text{-}Cat})_{\text{right-lax}}, \tag{0.3}$$

which induces an equivalence

$$(\mathbf{2\text{-}Cart}_{/\mathbb{S}})_{\mathrm{strict}} \simeq \mathrm{Funct}(\mathbb{S}^{1\text{-op}}, \mathbf{2\text{-}Cat}).$$

0.1.4. The proof of Theorem 1.1.8 is spread over Sects. 2-4. Let us indicate its main steps.

In Sect. 2 we establish the particular case of the isomorphism (0.3), when $\mathbb{S}$ is the interval $[n]$. This is done by a combinatorial procedure, which essentially amounts to unwinding the definitions.

In Sect. 3 we realize 2-Cartesian fibrations over the Gray product $\mathbb{S}_1 \circledast \mathbb{S}_2$ as an explicit full subspace in $2\text{-Cat}_{/\mathbb{S}_1 \times \mathbb{S}_2}$.

In Sect. 4 we use the results of the previous two sections to establish the isomorphism (0.3) *at the level of spaces underlying the* $(\infty, 2)$*-categories on both sides*, in the case when $\mathbb{S} = [m] \circledast [n]$.

Using Chapter 10, Theorems 4.1.3 and 5.2.3 we deduce from this that the isomorphism (0.3) holds *at the level of spaces* for any $\mathbb{S} \in 2\text{-Cat}$.

0.1.5. So, far, the same strategy would have worked if we worked with 1-Cartesian fibrations and $\mathbf{1\text{-}Cat}$ instead of $\mathbf{2\text{-}Cat}$ as a target.

However, now, in the 2-Cartesian context, we observe that the statement that we are trying to prove has enough functoriality, that it allows to formally deduce the equivalence (0.3) from just knowing it at the level of the underlying spaces.

0.2. What else is done in this Chapter?

0.2.1. As was mentioned before, our actual goal is to construct the Yoneda embedding (0.1) (and prove its fully faithfulness).

Having proved the 2-categorical straightening theorem in the earlier sections, the construction of the Yoneda embedding and the proof of its properties is carried out in Sect. 5.

0.2.2. In addition, this Chapter contains two sections in the Appendix.

In Sect. A, given $\mathbb{S} \in 2\text{-Cat}$, we give an explicit description of the universal non-unital right-lax functor out of $\mathbb{S}$:

$$\mathbb{S} \overset{\iota_{\mathbb{S}}}{\dashrightarrow} \mathrm{RLax}_{\text{non-untl}}(\mathbb{S}),$$

so that any non-unital right-lax functor $F : \mathbb{S} \dashrightarrow \mathbb{T}$ is obtained as

$$\widetilde{F} \circ \iota_{\mathbb{S}},$$

for a canonically defined *strict* functor $\widetilde{F} : \mathrm{RLax}_{\text{non-untl}}(\mathbb{S}) \to \mathbb{T}$.

The explicit description of $\mathrm{RLax}_{\text{non-untl}}(\mathbb{S})$ is used in Sect. 3.

0.2.3. In Sect. B we discuss the condition on a functor $\mathbb{S} \to \mathbb{T}$ between $(\infty,2)$-categories to be a *localization on 1-morphisms*. Informally, this means that $\mathbb{T}$ is obtained from $\mathbb{S}$ by inverting certain 2-morphisms.

This notion is used in the description of 2-Cartesian fibrations over Gray products, also in Sect. 3.

1. Straightening for $(\infty,2)$-categories

In this section we define the notion of a 2-Cartesian fibration of $(\infty,2)$-categories and formulate the main result in this Chapter: this is the *straightening theorem* that says that 2-Cartesian fibrations over a given $(\infty,2)$-category $\mathbb{S}$ are equivalent to functors $\mathbb{S}^{1\text{-op}} \to \mathbf{2\text{-}Cat}$.

1.1. The notion of 2-Cartesian fibration. In this subsection we will introduce the notion of 2-Cartesian fibration between $(\infty,2)$-categories.

When defining it, one should basically 'follow one's nose', keeping in mind that a 2-Cartesian fibration over $\mathbb{S}$ should be the same as a functor $\mathbb{S}^{1\text{-op}} \to \mathbf{2\text{-}Cat}$, while adapting the definition of Cartesian fibration in the context of $(\infty,1)$-categories.

1.1.1. Let $F : \mathbb{T} \to \mathbb{S}$ be a functor between $(\infty,2)$-categories. We shall say that a 1-morphism $t_0 \overset{\alpha}{\to} t_1$ is 2-Cartesian over $\mathbb{S}$ if for every $t \in \mathbb{T}$, the functor

$$\mathbf{Maps}_{\mathbb{T}}(t,t_0) \to \mathbf{Maps}_{\mathbb{T}}(t,t_1) \underset{\mathbf{Maps}_{\mathbb{S}}(F(t),F(t_1))}{\times} \mathbf{Maps}_{\mathbb{S}}(F(t),F(t_0)),$$

given by composition with α, is an equivalence of $(\infty,1)$-categories.

DEFINITION 1.1.2. *We shall say that F is a 2-Cartesian fibration if the following conditions hold:*

(1) *For every $t \in \mathbb{T}$ and a 1-morphism $s' \overset{\beta}{\to} F(t)$ there exists a 2-Cartesian 1-morphism $t' \overset{\alpha}{\to} t$ with $F(\alpha) \simeq \beta$.*

(2) *For every $t', t \in \mathbb{T}$, the functor*

$$\mathbf{Maps}(t',t) \to \mathbf{Maps}(F(t'),F(t))$$

is a coCartesian fibration (of $(\infty,1)$-categories), and for any $\widetilde{t}' \to t'$ and $t \to \widetilde{t}$, the corresponding functors

$$\mathbf{Maps}(t',t) \to \mathbf{Maps}(\widetilde{t}',t) \text{ and } \mathbf{Maps}(t',t) \to \mathbf{Maps}(t',\widetilde{t}),$$

given by composition, send arrows that are coCartesian over

$$\mathbf{Maps}(F(t'),F(t))$$

to arrows that are coCartesian over

$$\mathbf{Maps}(F(\widetilde{t}'),F(t)) and \mathbf{Maps}(F(t'),F(\widetilde{t})),$$

respectivelly.

1.1.3. Let us assume that condition (1) above holds, and let us write down the second condition in more explicit terms.

Let $\alpha_{\mathbb{S}}: s' \to s$ be a 1-morphism in $\mathbb{S}$, and let t be an object of $\mathbb{T}$ that lies over s. Then condition (1) implies that there exists a canonically defined object $t' \in \mathbb{T}$ that lies over s' and a 1 -morphism

$$\alpha_{\mathbb{T}}: t' \to t$$

that covers $\alpha_{\mathbb{S}}$.

Suppose now that we are given a pair of 1-morphisms

$$\alpha^1_{\mathbb{S}}, \alpha^2_{\mathbb{S}}: s' \rightrightarrows s$$

and a 2-morphism $\alpha^1_{\mathbb{S}} \xrightarrow{\phi_{\mathbb{S}}} \alpha^2_{\mathbb{S}}$. Then the second condition says that there exists a 1-morphism

$$\beta: t^{1\prime} \to t^{2\prime}$$

and a 2-morphism

$$\alpha^1_{\mathbb{T}} \xrightarrow{\phi_{\mathbb{T}}} \alpha^2_{\mathbb{T}} \circ \beta,$$

with the following property: for any $\widetilde{t}'$ in the fiber of $\mathbb{T}$ over s', and a pair of morphisms

$$\gamma_1 : \widetilde{t}' \to t^{1\prime} \text{ and } \gamma_2 : \widetilde{t}' \to t^{2\prime},$$

composition with $\phi_{\mathbb{T}}$ defines an isomorphism from the space of 2-morphisms

$$\beta \circ \gamma_1 \to \gamma_2$$

to the space of 2-morphisms

$$\alpha^1_{\mathbb{T}} \circ \gamma_1 \to \alpha^2_{\mathbb{T}} \circ \gamma_2$$

covering $\phi_{\mathbb{S}}$.

Furthermore, the formation of β is compatible in the natural sense with compositions

$$(\alpha^1_{\mathbb{S}}, \alpha^2_{\mathbb{S}}) \mapsto (\widetilde{\alpha}_{\mathbb{S}} \circ \alpha^1_{\mathbb{S}}, \widetilde{\alpha}_{\mathbb{S}} \circ \alpha^2_{\mathbb{S}}), \quad \widetilde{\alpha}_{\mathbb{S}}: s \to \widetilde{s}$$

and

$$(\alpha^1_{\mathbb{S}}, \alpha^2_{\mathbb{S}}) \mapsto (\alpha^1_{\mathbb{S}} \circ \widetilde{\alpha}'_{\mathbb{S}}, \alpha^2_{\mathbb{S}} \circ \widetilde{\alpha}'_{\mathbb{S}}), \quad \widetilde{\alpha}'_{\mathbb{S}}: \widetilde{s}' \to s'.$$

1.1.4. Let $\mathbf{2}\text{-}\mathbf{Cart}_{/\mathbb{S}} \subset \mathbf{2}\text{-}\mathbf{Cat}_{/\mathbb{S}}$ denote the full subcategory spanned by 2-Cartesian fibrations.

Let $(\mathbf{2}\text{-}\mathbf{Cart}_{/\mathbb{S}})_{\text{1-strict}} \subset \mathbf{2}\text{-}\mathbf{Cart}_{/\mathbb{S}}$ be the 1-full subcategory, where we allow as 1-morphisms those functors $\mathbb{T}_1 \to \mathbb{T}_2$ over $\mathbb{S}$ that send 1-morphisms in $\mathbb{T}_1$ that are 2-Cartesian over $\mathbb{S}$ to 1-morphisms in $\mathbb{T}_2$ that are 2-Cartesian over $\mathbb{S}$.

1.1.5. Let $(\mathbf{2}\text{-}\mathbf{Cart}_{/\mathbb{S}})_{\text{2-strict}} \subset \mathbf{2}\text{-}\mathbf{Cart}_{/\mathbb{S}}$ be the 1-full subcategory, where we impose the following condition on 1-morphisms:

Given $F_1 : \mathbb{T}_1 \to \mathbb{S}$ and $F_2 : \mathbb{T}_2 \to \mathbb{S}$, we consider those functors $G : \mathbb{T}_1 \to \mathbb{T}_2$ over $\mathbb{S}$ such that the corresponding functors

$$\mathbf{Maps}_{\mathbb{T}_1}(t_1', t) \to \mathbf{Maps}_{\mathbb{T}_2}(G(t_1'), G(t))$$

send arrows that are coCartesian over $\mathbf{Maps}_{\mathbb{S}}(F_1(t_1'), F_1(t))$ to arrows that are coCartesian over

$$\mathbf{Maps}_{\mathbb{S}}(F_2 \circ G(t_1'), F_2 \circ G(t)) \simeq \mathbf{Maps}_{\mathbb{S}}(F_1(t_1'), F_1(t)).$$

1.1.6. Let $(\mathbf{2}\text{-}\mathbf{Cart}_{/\mathbb{S}})_{\text{strict}} \subset \mathbf{2}\text{-}\mathbf{Cart}_{/\mathbb{S}}$ be the 1-full subcategory equal to

$$(\mathbf{2}\text{-}\mathbf{Cart}_{/\mathbb{S}})_{\text{1-strict}} \cap (\mathbf{2}\text{-}\mathbf{Cart}_{/\mathbb{S}})_{\text{2-strict}}.$$

Denote also

$$\text{2-Cart}_{/\mathbb{S}} := (\mathbf{2}\text{-}\mathbf{Cart}_{/\mathbb{S}})^{\text{1-Cat}}, \ (\text{2-Cart}_{/\mathbb{S}})_{\text{2-strict}} := ((\mathbf{2}\text{-}\mathbf{Cart}_{/\mathbb{S}})_{\text{2-strict}})^{\text{1-Cat}}$$

and

$$(\text{2-Cart}_{/\mathbb{S}})_{\text{strict}} := ((\mathbf{2}\text{-}\mathbf{Cart}_{/\mathbb{S}})_{\text{strict}})^{\text{1-Cat}}.$$

1.1.7. Our goal in the next few sections will be to prove:

THEOREM-CONSTRUCTION 1.1.8.

(a) *There exists a canonical equivalence*

$$(\mathbf{2}\text{-}\mathbf{Cart}_{/\mathbb{S}})_{\text{2-strict}} \simeq \text{Funct}(\mathbb{S}^{\text{1-op}}, \mathbf{2}\text{-}\mathbf{Cat})_{\text{right-lax}},$$

functorial in $\mathbb{S}$.

(b) *Under the equivalence of point (a), the 1-full subcategories*

$$(\mathbf{2}\text{-}\mathbf{Cart}_{/\mathbb{S}})_{\text{strict}} \subset (\mathbf{2}\text{-}\mathbf{Cart}_{/\mathbb{S}})_{\text{2-strict}}$$

and

$$\text{Funct}(\mathbb{S}^{1\text{-op}}, \mathbf{2}\text{-}\mathbf{Cat}) \subset \text{Funct}(\mathbb{S}^{1\text{-op}}, \mathbf{2}\text{-}\mathbf{Cat})_{\text{right-lax}}$$

correspond to one another.

1.2. The notion of 1-Cartesian fibration. According to Theorem 1.1.8, 2-Cartesian fibrations over $\mathbb{S}$ correspond to functors $\mathbb{S}^{1\text{-op}} \to \mathbf{2}\text{-}\mathbf{Cat}$.

In this subsection we will define the notion of *1-Cartesian fibration.* Those will form a full subcategory among 2-Cartesian fibrations, and they will correspond to functors $\mathbb{S}^{1\text{-op}} \to \mathbf{1}\text{-}\mathbf{Cat}$.

1.2.1. Let $F : \mathbb{T} \to \mathbb{S}$ be a functor between $(\infty,2)$-categories.

DEFINITION 1.2.2. *We shall say that F is a 1-Cartesian fibration if the following conditions hold:*

(1) *The induced functor*

$$\mathbb{T}^{1\text{-Cat}} \to \mathbb{S}^{1\text{-Cat}}$$

is a Cartesian fibration;

(2) *For every $t', t \in \mathbb{T}$, the functor*

$$\mathbf{Maps}_{\mathbb{T}}(t', t) \to \mathbf{Maps}_{\mathbb{S}}(F(t'), F(t))$$

is a coCartesian fibration in spaces.

If $F : \mathbb{T} \to \mathbb{S}$ is a 1-Cartesian fibration, we will say that a 1-morphism in $\mathbb{T}$ is *Cartesian* if the corresponding morphism in $\mathbb{T}^{1\text{-Cat}}$ is Cartesian over $\mathbb{S}^{1\text{-Cat}}$.

1.2.3. Let $\mathbf{1}\text{-}\mathbf{Cart}_{/\mathbb{S}}$ denote the full subcategory of $\mathbf{2}\text{-}\mathbf{Cat}_{/\mathbb{S}}$ formed by 1-Cartesian fibrations.

We let $(\mathbf{1}\text{-}\mathbf{Cart}_{/\mathbb{S}})_{\text{strict}}$ be the 1-full subcategory of $\mathbf{1}\text{-}\mathbf{Cart}_{/\mathbb{S}}$, where we restrict morphisms to those functors $\mathbb{T}_1 \to \mathbb{T}_2$ over $\mathbb{S}$, such that send arrows in $(\mathbb{T}_1)^{1\text{-Cat}}$ Cartesian over $\mathbb{S}^{1\text{-Cat}}$ to arrows in $(\mathbb{T}_2)^{1\text{-Cat}}$ with the same property.

Denote also

$$\text{1-Cart}_{/\mathbb{S}} := (\mathbf{1}\text{-}\mathbf{Cart}_{/\mathbb{S}})^{1\text{-Cat}} \text{ and } (\text{1-Cart}_{/\mathbb{S}})_{\text{strict}} := ((\mathbf{1}\text{-}\mathbf{Cart}_{/\mathbb{S}})_{\text{strict}})^{1\text{-Cat}}.$$

1.2.4. We claim:

LEMMA 1.2.5.

(a) *For a functor $F : \mathbb{T} \to \mathbb{S}$ the following conditions are equivalent:*

(i) *F is a 1-Cartesian fibration;*

(ii) *F is a 2-Cartesian fibration and the fiber of F over every $s \in \mathbb{S}$ is an $(\infty,1)$-category.*

(b) *If $\mathbb{T} \to \mathbb{S}$ is a 1-Cartesian fibration, then a 1-morphism in $\mathbb{T}$ is 2-Cartesian over $\mathbb{S}$ if and only if it is Cartesian.*

Hence, combining this lemma with Theorem 1.1.8 and Chapter 10, Proposition 6.3.2, we obtain:

COROLLARY 1.2.6.

(a) *There exists a canonical equivalence*

$$\mathbf{1}\text{-}\mathbf{Cart}_{/\mathbb{S}} \simeq \operatorname{Funct}(\mathbb{S}^{1\text{-op}}, \mathbf{1}\text{-}\mathbf{Cat})_{\text{right-lax}},$$

functorial in $\mathbb{S} \in 2\text{-Cat}$.

(b) *Under the equivalence of point (a), the 1-full subcategories*

$$(\mathbf{1}\text{-}\mathbf{Cart}_{/\mathbb{S}})_{\text{strict}} \subset \mathbf{1}\text{-}\mathbf{Cart}_{/\mathbb{S}}$$

and

$$\operatorname{Funct}(\mathbb{S}^{1\text{-op}}, \mathbf{1}\text{-}\mathbf{Cat}) \subset \operatorname{Funct}(\mathbb{S}^{1\text{-op}}, \mathbf{1}\text{-}\mathbf{Cat})_{\text{right-lax}}$$

correspond to one another.

1.2.7. Let $\mathbb{S} = \mathbf{S}$ be an $(\infty, 1)$-category. We note:

LEMMA 1.2.8. *A functor* $\mathbb{T} \to \mathbf{S}$ *is a 1-Cartesian fibration if and only if the following conditions hold:*

- $\mathbb{T} = \mathbf{T} \in 1\text{-Cat}$*;*
- *The resulting functor* $\mathbf{T} \to \mathbf{S}$ *is a Cartesian fibration.*

I.e., we obtain that in the above case, the notion of 1-Cartesian fibration reduces to the usual notion of 1-Cartesian fibration on $(\infty, 1)$-categries.

It will follow from the construction that the equivalence of Corollary 1.2.6(b) in this case, i.e.,

$$(\mathbf{1}\text{-}\mathbf{Cart}_{/\mathbf{S}})_{\text{strict}} \simeq \operatorname{Funct}(\mathbf{S}^{1\text{-op}}, \mathbf{1}\text{-}\mathbf{Cat}),$$

induces at the level of the underlying $(\infty, 1)$-categories, i.e.,

$$(\mathrm{Cart}_{/\mathbf{S}})_{\text{strict}} \text{ and } \operatorname{Maps}(\mathbf{S}^{1\text{-op}}, \mathbf{1}\text{-}\mathbf{Cat}),$$

the equivalence of Chapter 1, Sect. 1.4.5.

REMARK 1.2.9. Let us take $\mathbb{S} = \mathbf{S} = [n]^{\text{op}}$. We obtain that in this case the equivalence of Corollary 1.2.6(a) at the level of the underlying $(\infty, 1)$-categories amounts to the *definition* of the $(\infty, 1)$-category $\operatorname{Seq}_n^{\text{ext}}(\mathbf{1}\text{-}\mathbf{Cat})$, see Chapter 10, Sect. 5.3.

The idea of the proof of Theorem 1.1.8 is to give a similar interpretation of $\operatorname{Seq}_n^{\text{ext}}(\mathbf{2}\text{-}\mathbf{Cat})$, namely, as 2-Cartesian fibrations over $[n]^{\text{op}}$. This will be furnished by Theorem 2.0.1.

The rest of the proof of Theorem 1.1.8 will amount to bootstrapping the statement for any $\mathbb{S} \in 2\text{-Cat}$ from the case $\mathbb{S} = [n]^{\text{op}}$, and lifting the 1-categorical equivalence to a 2-categorical one.

1.3. Variants. In this subsection we will introduce the companion notions of *2-coCartesian* and *1-coCartesian* fibrations over an $(\infty, 2)$-category.

1.3.1. We shall say that a functor between $(\infty,2)$-categories $\mathbb{T}\to\mathbb{S}$ is *2-coCartesian* (resp., *1-coCartesian*) fibration if the corresponding functor $\mathbb{T}^{1\&2\text{-op}}\to\mathbb{S}^{1\&2\text{-op}}$ is a 2-Cartesian (resp., 1-Cartesian) fibration.

Similarly, we introduce the 1-full subcategories

$$(\mathbf{2}\text{-}\mathbf{coCart}_{/\mathbb{S}})_{\text{strict}}\subset(\mathbf{2}\text{-}\mathbf{coCart}_{/\mathbb{S}})_{\text{2-strict}}\subset\mathbf{2}\text{-}\mathbf{coCart}_{/\mathbb{S}}\subset\mathbf{2}\text{-}\mathbf{Cat}_{/\mathbb{S}}$$

and

$$(\mathbf{1}\text{-}\mathbf{coCart}_{/\mathbb{S}})_{\text{strict}}\subset(\mathbf{1}\text{-}\mathbf{coCart}_{/\mathbb{S}})_{\text{2-strict}}\subset\mathbf{1}\text{-}\mathbf{coCart}_{/\mathbb{S}}\subset\mathbf{2}\text{-}\mathbf{Cat}_{/\mathbb{S}}.$$

1.3.2. From Theorem 1.1.8 we obtain:

COROLLARY 1.3.3.
(a) *There exists a canonical equivalence*

$$(\mathbf{2}\text{-}\mathbf{coCart}_{/\mathbb{S}})_{\text{2-strict}}\simeq\text{Funct}(\mathbb{S},\mathbf{2}\text{-}\mathbf{Cat})_{\text{left-lax}},$$

functorial in $\mathbb{S}$.
(b) *Under the equivalence of point (a) the 1-full subcategories*

$$(\mathbf{2}\text{-}\mathbf{coCart}_{/\mathbb{S}})_{\text{strict}}\subset(\mathbf{2}\text{-}\mathbf{coCart}_{/\mathbb{S}})_{\text{2-strict}}$$

and

$$\text{Funct}(\mathbb{S},\mathbf{2}\text{-}\mathbf{Cat})\subset\text{Funct}(\mathbb{S},\mathbf{2}\text{-}\mathbf{Cat})_{\text{left-lax}}$$

correspond to one another.

Similarly, from Corollary 1.2.6 we obtain:

COROLLARY 1.3.4.
(a) *There exists a canonical equivalence*

$$\mathbf{1}\text{-}\mathbf{coCart}_{/\mathbb{S}}\simeq\text{Funct}(\mathbb{S},\mathbf{1}\text{-}\mathbf{Cat})_{\text{left-lax}},$$

functorial in $\mathbb{S}\in 2\text{-Cat}$.
(b) *Under the equivalence of point (a), the 1-full subcategories*

$$(\mathbf{1}\text{-}\mathbf{coCart}_{/\mathbb{S}})_{\text{strict}}\subset\mathbf{1}\text{-}\mathbf{coCart}_{/\mathbb{S}}\ \textit{and}\ \text{Funct}(\mathbb{S},\mathbf{1}\text{-}\mathbf{Cat})\subset\text{Funct}(\mathbb{S},\mathbf{1}\text{-}\mathbf{Cat})_{\text{left-lax}}$$

correspond to one another.

1.3.5. We note that in addition to the notions of 2-Cartesian and 2-coCartesian (resp., 1-Cartesian and 1-coCartesian) fibration, there exist two more notions, induced by the involution $\mathbb{S}\mapsto\mathbb{S}^{2\text{-op}}$ on 2-Cat.

These notions correspond to functors from $\mathbb{S}^{1\text{-op}}$ and $\mathbb{S}^{2\text{-op}}$ with values in **2 - Cat** and **1 - Cat**, respectively.

2. Straightening over intervals

In this section we will establish the following particular case of Theorem 1.1.8:

We will take the base $\mathbb{S}$ to be the interval $[n]$, and we will identify the $(\infty,1)$-categories underlying the $(\infty,2)$-categories appearing on the two sides in Theorem 1.1.8.

More precisely, our goal is to prove the following:

THEOREM-CONSTRUCTION 2.0.1.

(a) *There exists a canonical equivalence of simplicial categories*

$$\mathrm{Seq}_\bullet^{\mathrm{ext}}(\mathbf{2}\text{-}\mathbf{Cat}) \simeq (2\text{-Cart}_{/[\bullet]^{\mathrm{op}}})_{2\text{-strict}}.$$

(b) *For an individual n, under the equivalence*

$$\mathrm{Seq}_n^{\mathrm{ext}}(\mathbf{2}\text{-}\mathbf{Cat}) \simeq (2\text{-Cart}_{/[n]^{\mathrm{op}}})_{2\text{-strict}},$$

the 1-full subcategories

$$\mathrm{Seq}_n^{\mathrm{ext}}(2\text{-Cat}) \subset \mathrm{Seq}_n^{\mathrm{ext}}(\mathbf{2}\text{-}\mathbf{Cat}) \text{ and } (2\text{-coCart}_{/[n]^{\mathrm{op}}})_{\mathrm{strict}} \subset (2\text{-Cart}_{/[n]^{\mathrm{op}}})_{2\text{-strict}}$$

correspond to one another.

REMARK 2.0.2. Note that since $[n]^{\mathrm{op}}$ is a 1-category, the inclusion

$$(2\text{-Cart}_{/[n]^{\mathrm{op}}})_{2\text{-strict}} \subset 2\text{-Cart}_{/[n]^{\mathrm{op}}}$$

is an equivalence.

2.1. The main construction. We now proceed to defining the functor in one direction

$$2\text{-Cart}_{/[n]^{\mathrm{op}}} \to \mathrm{Seq}_n^{\mathrm{ext}}(\mathbf{2}\text{-}\mathbf{Cat}).$$

The idea of the construction is pretty straightforward: we think of an object of $\mathrm{Seq}_n^{\mathrm{ext}}(\mathbf{2}\text{-}\mathbf{Cat})$ as a string

$$\mathbb{T}^0 \to \mathbb{T}^1 \to \ldots \to \mathbb{T}^n$$

of $(\infty,2)$-categories, which we encode by means of a functor

$$\mathbf{\Delta}^{\mathrm{op}} \to \mathrm{Cart}_{/[n]^{\mathrm{op}}},$$

see Chapter 10, Sect. 6.1.3.

The value of this functor on $[m] \in \mathbf{\Delta}^{\mathrm{op}}$ is the category of strings $t_0^i \to t_1^i \to \ldots \to t_m^i$ in $\mathbb{T}^i$, where i varies along $[n]$. We interpret such strings as strings in the 'total' $(\infty,2)$-category over $[n]$ that project to a single object in $[n]$.

The total category in question is precisely the object $\mathbb{T} \in 2\text{-Cart}_{/[n]^{\mathrm{op}}}$ that we start from. We will now make turn this idea into an actual construction.

2.1.1. Given $(\mathbb{T} \to [n]^{\mathrm{op}}) \in 2\text{-Cart}_{/[n]^{\mathrm{op}}}$ we define an object

$$\mathbf{E}_{\bullet,n} \in \mathrm{Funct}(\mathbf{\Delta}^{\mathrm{op}}, \mathrm{Cart}_{/[n]^{\mathrm{op}}})$$

as follows:

We let

$$\mathbf{E}_{m,n} := \mathrm{Seq}_m^{\mathrm{ext}}(\mathbb{T}) \underset{\mathrm{Seq}_m^{\mathrm{ext}}([n]^{\mathrm{op}})}{\times} [n]^{\mathrm{op}},$$

where $[n]^{\mathrm{op}} \to \mathrm{Seq}_m^{\mathrm{ext}}([n]^{\mathrm{op}})$ is the functor

$$[n]^{\mathrm{op}} = \mathrm{Funct}(\{*\}, [n]^{\mathrm{op}}) \to \mathrm{Funct}([m], [n]^{\mathrm{op}}) = \mathrm{Seq}_m^{\mathrm{ext}}([n]^{\mathrm{op}}).$$

It is straightforward to check that $\mathbf{E}_{m,n}$, viewed as a category over $[n]^{\mathrm{op}}$, is a Cartesian fibration, and that the object $\mathbf{E}_{\bullet,n}$ thus constructed defines an object of $\mathrm{Seq}_n^{\mathrm{ext}}(\mathbf{2\text{-}Cat})$.

Furthermore, this construction is clearly functorial in $\mathbb{T}$, thereby giving rise to a functor

$$2\text{-Cart}_{/[n]^{\mathrm{op}}} \to \mathrm{Seq}_n^{\mathrm{ext}}(\mathbf{2\text{-}Cat}). \tag{2.1}$$

Furthermore, it is clear that the above functor sends the 1-full subcategory

$$(2\text{-Cart}_{/[n]^{\mathrm{op}}})_{\mathrm{strict}} \subset 2\text{-Cart}_{/[n]^{\mathrm{op}}}$$

to the 1-full subcategory

$$\mathrm{Seq}_n^{\mathrm{ext}}(2\text{-Cat}) \subset \mathrm{Seq}_n^{\mathrm{ext}}(\mathbf{2\text{-}Cat}).$$

REMARK 2.1.2. Note that the construction presented above is a generalization of the construction in Chapter 10, Proposition 6.3.2. The reason that we cannot finish the proof of Theorem 2.0.1 as easily as in the case of Chapter 10, Proposition 6.3.2 is that we do not yet know that for given $\mathbb{S}_0, \mathbb{S}_1 \in 2\text{-Cat}$, the category

$$2\text{-Cart}_{/[1]^{\mathrm{op}}} \underset{2\text{-Cat} \times 2\text{-Cat}}{\times} \{\mathbb{S}_0 \times \mathbb{S}_1\}$$

identifies with

$$\mathbf{Maps}_{\mathbf{2\text{-}Cat}}(\mathbb{S}_0, \mathbb{S}_1) \simeq \mathrm{Funct}(\mathbb{S}_0, \mathbb{S}_1)^{1\text{-Cat}}.$$

2.2. Proof of Theorem 2.0.1: the inverse map. We will define a functor

$$\mathrm{Seq}_n^{\mathrm{ext}}(\mathbf{2\text{-}Cat}) \to 2\text{-Cart}_{/[n]^{\mathrm{op}}} \tag{2.2}$$

inverse to (2.1).

We now want to recover the 'total' $(\infty, 2)$-category $\mathbb{T}$ over $[n]$, i.e., for each m, we want to recover the corresponding category of strings

$$t_0 \to t_1 \to \dots \to t_n,$$

while we know the category of strings that project to a single element in $[n]$.

We will recover all strings by a variant of the construction used in Chapter 10, Sect. 1.6 to define the unstraightening procedure for $(\infty, 1)$-categories.

2.2.1. In order to define the functor (2.2), we will need the following combinatorial construction. Let $\mathrm{Tot}(\mathbf{\Delta})$ be the coCartesian fibration over $\mathbf{\Delta}$ corresponding to the tautological functor

$$\mathbf{\Delta} \to 1\text{-Cat}.$$

Note that $\mathrm{Tot}(\mathbf{\Delta})$ is an ordinary category, whose objects are pairs $([n] \in \mathbf{\Delta}, i \in [n])$, and such that the set of morphisms $([n_0], i_0) \to ([n_1], i_1)$ is the set of morphisms $\phi : [n_0] \to [n_1]$ such that $\phi(i_0) = i_1$.

We let $p : \mathrm{Tot}(\mathbf{\Delta}) \to \mathbf{\Delta}$ the tautological projection $([n], i) \mapsto [n]$. We let $\mathrm{Tot}(\mathbf{\Delta})_{[m]}$ the fiber of $\mathrm{Tot}(\mathbf{\Delta})$ over $[m] \in \mathbf{\Delta}$; tautologically $\mathrm{Tot}(\mathbf{\Delta})_{[m]} \simeq [m]$.

We note now that in addition to p, there is another canonically defined functor

$$q : \mathrm{Tot}(\mathbf{\Delta}) \to \mathbf{\Delta}.$$

Namely, we set

$$q([n], i) := [i], \quad q(([n_0], i_0) \overset{\phi}{\to} ([n_1], i_1)) = ([i_0] \overset{\phi|_{[i_0]}}{\longrightarrow} [i_1]).$$

In particular, restricting to $\mathrm{Tot}(\mathbf{\Delta})_{[m]}$, we obtain the functor

$$q_{[m]} : [m] \to \mathbf{\Delta}, \quad i \mapsto [i].$$

2.2.2. Going back to the desired functor (2.2), let $\mathbf{E}_{\bullet,n}$ be an object of

$$\mathrm{Seq}_n^{\mathrm{ext}}(\mathbf{2}\text{-}\mathbf{Cat}),$$

thought of as a functor

$$\mathbf{\Delta}^{\mathrm{op}} \to \mathrm{Cart}_{/[n]^{\mathrm{op}}}.$$

We can view[1] the data of $\mathbf{E}_{\bullet,n}$ as an $(\infty,1)$-category $\mathbf{E}^{\oint}$ over $\mathbf{\Delta}^{\mathrm{op}} \times [n]^{\mathrm{op}}$, such that:

- The composition $\mathbf{E}^{\oint} \to \mathbf{\Delta}^{\mathrm{op}} \times [n]^{\mathrm{op}} \to \mathbf{\Delta}^{\mathrm{op}}$ is a coCartesian fibration;
- The composition $\mathbf{E}^{\oint} \to \mathbf{\Delta}^{\mathrm{op}} \times [n]^{\mathrm{op}} \to [n]^{\mathrm{op}}$ is a Cartesian fibration;
- The functor $\mathbf{E}^{\oint} \to \mathbf{\Delta}^{\mathrm{op}} \times [n]^{\mathrm{op}}$, viewed as a functor between coCartesian fibrations over $\mathbf{\Delta}^{\mathrm{op}}$, belongs to $(\mathrm{coCart}_{/\mathbf{\Delta}^{\mathrm{op}}})_{\mathrm{strict}}$;
- The functor $\mathbf{E}^{\oint} \to \mathbf{\Delta}^{\mathrm{op}} \times [n]^{\mathrm{op}}$, viewed as a functor between Cartesian fibrations over $[n]^{\mathrm{op}}$, belongs to $(\mathrm{Cart}_{/[n]^{\mathrm{op}}})_{\mathrm{strict}}$.

2.2.3. We construct the object $\mathbb{T} \in 2\text{-}\mathrm{Cart}_{/[n]^{\mathrm{op}}}$ corresponding to $\mathbf{E}_{\bullet,n}$ as follows. We define the category

$$\mathrm{Funct}([m]^{\mathrm{op}}, \mathbb{T})_{\text{right-lax}}$$

(which will be the same as $\mathrm{Seq}_m^{\mathrm{ext}}(\mathbb{T})$, up to the involution rev) to be a certain full subcategory in the $(\infty,1)$-category of pairs (ϕ, Φ), where ϕ is a functor $[m] \to [n]$, and Φ is a lift of the functor

$$(\phi^{\mathrm{op}}, (\mathrm{rev} \circ q_{[m]})^{\mathrm{op}}) : [m]^{\mathrm{op}} \to [n]^{\mathrm{op}} \times \mathbf{\Delta}^{\mathrm{op}}$$

to a functor

$$[m]^{\mathrm{op}} \to \mathbf{E}^{\oint}.$$

[1] See the elementary Chapter 12, Proposition 2.1.3 for a general assertion to this effect.

2.2.4. We single out $\mathrm{Funct}([m]^{\mathrm{op}}, \mathbb{T})_{\text{right-lax}}$ by imposing the following condition on objects.

Fix $i = 1, ..., m$. Consider a coCartesian lift in $\mathbf{E}^{\oint}$

$$\Phi(i) \to e'$$

of the 1-morphism

$$(\mathrm{rev} \circ q_{[m]}(i) \to \mathrm{rev} \circ q_{[m]}(i-1)) \in \mathbf{\Delta}^{\mathrm{op}}.$$

Consider a Cartesian lift in $\mathbf{E}^{\oint}$

$$e'' \to \Phi(i-1)$$

of the 1-morphism

$$(\phi(i) \to \phi(i-1)) \in [n]^{\mathrm{op}}.$$

Note that by the last two properties of $\mathbf{E}^{\oint}$ listed in Sect. 2.2.2, we have a canonical map

$$e' \to e''.$$

We require that this map be an isomorphism.

2.2.5. Clearly, the assignment

$$m \mapsto \mathrm{Funct}([m]^{\mathrm{op}}, \mathbb{T})_{\text{right-lax}}$$

extends to a functor $\mathbf{\Delta}^{\mathrm{op}} \to 1\text{-Cat}$.

We set

$$\mathrm{Seq}^{\mathrm{ext}}_{\bullet}(\mathbb{T}) := \mathrm{Funct}([\bullet]^{\mathrm{op}}, \mathbb{T})_{\text{right-lax}} \circ \mathrm{rev},$$

where rev is the reversal involution on $\mathbf{\Delta}^{\mathrm{op}}$.

Using Chapter 10, Theorem 5.2.3(a), we show:

LEMMA 2.2.6. *The simplicial category* $\mathrm{Seq}^{\mathrm{ext}}_{\bullet}(\mathbb{T})$ *belongs to the essential image of the functor*

$$\mathrm{Seq}^{\mathrm{ext}}_{\bullet} : 2\text{-Cat} \to \mathrm{Funct}(\mathbf{\Delta}^{\mathrm{op}}, 1\text{-Cat}).$$

Let $\mathbb{T}$ denote the resulting object of 2-Cat.

2.2.7. By construction, the simplicial category $\mathrm{Seq}^{\mathrm{ext}}_{\bullet}(\mathbb{T})$ maps to the simplicial category

$$m \mapsto \mathrm{Funct}([m], [n]^{\mathrm{op}}).$$

Hence, the $(\infty, 2)$-category $\mathbb{T}$, constructed above, comes equipped with a functor

$$\mathbb{T} \to [n]^{\mathrm{op}}.$$

It is a straightforward verification that the above functor $\mathbb{T} \to [n]^{\mathrm{op}}$ is a 2-Cartesian fibration.

2.2.8. Thus, we have constructed a functor

$$\mathrm{Seq}^{\mathrm{ext}}_{n}(\mathbf{2\text{-}Cat}) \to 2\text{-}\mathrm{Cart}_{/[n]^{\mathrm{op}}}.$$

It is again a straightforward verification that this functor is inverse to (2.1).

3. Locally 2-Cartesian and 2-Cartesian fibrations over Gray products

As was mentioned before, the assertion of Theorem 1.1.8 will be deduced from that of Theorem 2.0.1 by a certain bootstrapping procedure.

However, in order to do so, we will need to enlarge the entities that appear in both the left-hand and the right-hand side. For the left-hand side, the corresponding notion is that of *locally 2-Cartesian fibration.*

3.1. The notion of locally 2-Cartesian fibration. The idea of the notion of locally 2-Cartesian fibration is the following: whereas 2-Cartesian fibrations over $\mathbb{S}$ correspond to functors

$$\mathbb{S}^{1\text{-op}} \to \mathbf{2\text{-}Cat},$$

locally 2-Cartesian fibrations correspond to *right-lax* functors

$$\mathbb{S}^{1\text{-op}} \dashrightarrow \mathbf{2\text{-}Cat}.$$

3.1.1. Let $F : \mathbb{T} \to \mathbb{S}$ be a functor between $(\infty, 2)$-categories. We shall say that a 1-morphism α in $\mathbb{T}$ is locally 2-Cartesian over $\mathbb{S}$, if the resulting 1-morphism in $[1] \underset{F(\alpha),\mathbb{S}}{\times} \mathbb{T}$ is 2-Cartesian with respect to the projection

$$[1] \underset{F(\alpha),\mathbb{S}}{\times} \mathbb{T} \to [1].$$

DEFINITION 3.1.2. *We shall say that F is a locally 2-Cartesian fibration if the following conditions hold:*

(1) *For every $t \in \mathbb{T}$ and a 1-morphism $s' \overset{\beta}{\to} F(t)$ there exists a locally 2-Cartesian 1-morphism $t' \overset{\alpha}{\to} t$ with $F(\alpha) \simeq \beta$.*
(2) *Condition (2) in Definition* 1.1.2 *holds.*

3.1.3. Note that if $F : \mathbb{T} \to \mathbb{S}$ is a a locally 2-Cartesian fibration, then for every 1-morphism $s_0 \to s_1$, the functor

$$[1] \underset{\mathbb{S}}{\times} \mathbb{T} \to [1]$$

is a 2-Cartesian fibration. In particular, by Theorem 2.0.1 and Chapter 10, Proposition 6.2.2, it gives rise to a well-defined functor

$$\mathbb{T}_{s_1} \to \mathbb{T}_{s_0}.$$

We will refer to it as the *pullback functor* along the given 1-morphism.

3.1.4. The next assertion follows from the definitions:

LEMMA 3.1.5.

(a) *A functor $F : \mathbb{T} \to \mathbb{S}$ is a 2-Cartesian fibration if and only it is a locally 2-Cartesian fibration and the induced functor $\mathbb{T}^{1\text{-Cat}} \to \mathbb{S}^{1\text{-Cat}}$ is a Cartesian fibration of $(\infty,1)$-categories.*

(b) *If $F : \mathbb{T} \to \mathbb{S}$ is a 2-Cartesian fibration, then any 1-morphism in $\mathbb{T}$ that is locally 2-Cartesian over $\mathbb{S}$ is automatically 2-Cartesian.*

(c) *If $F : \mathbb{T} \to \mathbb{S}$ is a locally 2-Cartesian fibration, then a 1-morphism in $\mathbb{T}$ is locally 2-Cartesian over $\mathbb{S}$ if and only if the corresponding 1-morphism in $\mathbb{T}^{\text{ordn}}$ is is locally 2-Cartesian over $\mathbb{S}^{\text{ordn}}$.*

(d) *If $F:\mathbb{T}\to\mathbb{S}$ is a locally 2-Cartesian fibration, then it is 2-Cartesian if and only if the corresponding functor $\mathbb{T}^{\mathrm{ordn}}\to\mathbb{S}^{\mathrm{ordn}}$ is.*

3.1.6. Let $2\text{-}\mathrm{Cart}^{\mathrm{loc}}_{/\mathbb{S}}$ denote the the full subcategory of $2\text{-}\mathrm{Cat}_{/\mathbb{S}}$ formed by locally 2-Cartesian fibrations in $(\infty,1)$-categories. Let

$$(2\text{-}\mathrm{Cart}^{\mathrm{loc}}_{/\mathbb{S}})_{1\text{-strict}}\supset(2\text{-}\mathrm{Cart}^{\mathrm{loc}}_{/\mathbb{S}})_{\mathrm{strict}}\subset(2\text{-}\mathrm{Cart}^{\mathrm{loc}}_{/\mathbb{S}})_{2\text{-strict}}$$

be the 1-full subcategories, defined by the same conditions as in Sects. 1.1.4-1.1.6.

3.2. Locally 2-Cartesian fibrations vs 2-Cartesian fibrations over $\mathrm{RLax}_{\text{non-untl}}(\mathbb{S})$. In this subsection we will formulate, and begin the proof of, the main assertion of this section, Theorem 3.2.2.

3.2.1. Let

$$\mathbb{S}\dashrightarrow\mathrm{RLax}_{\text{non-untl}}(\mathbb{S}),$$

be the universal non-unital right-lax functor, see Sect. A.

We are going to prove:

THEOREM-CONSTRUCTION 3.2.2. *There exists a canonical fully faithful embedding*

$$(2\text{-}\mathrm{Cart}^{\mathrm{loc}}_{/\mathbb{S}})^{\mathrm{Spc}}\overset{\Phi}{\hookrightarrow}(2\text{-}\mathrm{Cart}_{/\mathrm{RLax}_{\text{non-untl}}(\mathbb{S})})^{\mathrm{Spc}},$$

functorial in $\mathbb{S}$, whose essential image consists of those 2-Cartesian fibrations for which the pullback functors along quasi-invertible arrows (see Sect. A.3*) are equivalences.*

REMARK 3.2.3. The above proposition is stated as an isomorphism of spaces. However, it will follow from the construction that this equivalence extends to one between the corresponding $(\infty,2)$-categories (both 2-strict and strict versions).

REMARK 3.2.4. If we assume Theorem 1.1.8, then Theorem 3.2.2 implies that the space

$$(2\text{-}\mathrm{Cart}^{\mathrm{loc}}_{/\mathbb{S}})^{\mathrm{Spc}}$$

is isomorphic to space of right-lax functors

$$\mathbb{S}^{1\text{-op}}\dashrightarrow\mathbf{2\text{-}Cat}.$$

3.2.5. In the rest of this subsection we will construct the map in the easy direction, i.e.,

$$(2\text{-}\mathrm{Cart}^{\mathrm{loc}}_{/\mathbb{S}})^{\mathrm{Spc}}\overset{\Psi}{\leftarrow}(2\text{-}\mathrm{Cart}_{/\mathrm{RLax}_{\text{non-untl}}(\mathbb{S})})^{\mathrm{Spc}}.$$

Consider the coCartesian fibrations

$$\mathbb{S}^{\oint}\to\mathbf{\Delta}^{\mathrm{op}}\text{ and }\mathrm{RLax}_{\text{non-untl}}(\mathbb{S})^{\oint}\to\mathbf{\Delta}^{\mathrm{op}},$$

and the adjoint functors

$$\iota^{\oint}_{\mathbb{S}}:\mathbb{S}^{\oint}\rightleftarrows\mathrm{RLax}_{\text{non-untl}}(\mathbb{S})^{\oint}:\rho^{\oint}_{\mathbb{S}},$$

see Sect. A.

3.2.6. Starting from a 2-Cartesian fibration $\widetilde{\mathbb{T}} \to \mathrm{RLax}_{\text{non-untl}}(\mathbb{S})$, define

$$\Psi(\widetilde{\mathbb{T}})^{\oint} := \widetilde{\mathbb{T}}^{\oint} \underset{\mathrm{RLax}_{\text{non-untl}}(\mathbb{S})^{\oint}}{\times} \mathbb{S}^{\oint},$$

where the functor $\mathbb{S}^{\oint} \to \mathrm{RLax}_{\text{non-untl}}(\mathbb{S})^{\oint}$ is $\iota_{\mathbb{S}}^{\oint}$.

We have:

LEMMA 3.2.7.

(a) *The composite functor*

$$\Psi(\widetilde{\mathbb{T}})^{\oint} \to \mathbb{S}^{\oint} \to \boldsymbol{\Delta}^{\text{op}}$$

is a coCartesian fibration.

(b) *The functor $\boldsymbol{\Delta}^{\text{op}} \to 1\text{-Cat}$, corresponding to the coCartesian fibration of point (a) lies in the essential image of the functor $\mathrm{Seq}_\bullet : 2\text{-Cat} \to \mathrm{Funct}(\boldsymbol{\Delta}^{\text{op}}, 1\text{-Cat})$. Denote the resulting $(\infty,2)$-category by $\Psi(\widetilde{\mathbb{T}})$.*

(c) *The functor*

$$\Psi(\widetilde{\mathbb{T}})^{\oint} \to \mathbb{S}^{\oint}$$

maps arrows that are coCartesian over $\boldsymbol{\Delta}^{\text{op}}$ to arrows that are coCartesian over $\boldsymbol{\Delta}^{\text{op}}$.

(d) *The functor $\Psi(\widetilde{\mathbb{T}}) \to \mathbb{S}$ arising from point (c) is a locally 2-Cartesian fibration.*

3.3. Proof of Theorem 3.2.2, the inverse map. In this subsection we define the sought-for map

$$(2\text{-Cart}_{/\mathbb{S}}^{\text{loc}})^{\text{Spc}} \overset{\Phi}{\hookrightarrow} (2\text{-Cart}_{/\mathrm{RLax}_{\text{non-untl}}(\mathbb{S})})^{\text{Spc}}.$$

3.3.1. Given $\mathbb{T} \in 2\text{-Cart}_{/\mathbb{S}}^{\text{loc}}$, consider the corresponding functor

$$\mathbb{T}^{\oint} \to \mathbb{S}^{\oint}.$$

We define

$$'\Phi(\mathbb{T})^{\oint} := \mathbb{T}^{\oint} \underset{\mathbb{S}^{\oint}}{\times} \mathrm{RLax}_{\text{non-untl}}(\mathbb{S})^{\oint},$$

where the functor $\mathrm{RLax}_{\text{non-untl}}(\mathbb{S})^{\oint} \to \mathbb{S}^{\oint}$ is $\rho_{\mathbb{S}}^{\oint}$.

We will define the sought-for $(\infty,1)$-category $\Phi(\mathbb{T})^{\oint}$ is a certain full subcategory of $'\Phi(\mathbb{T})^{\oint}$.

3.3.2. Fix an object of

$$\{\gamma\} \underset{((\boldsymbol{\Delta}_{\text{actv}})_{[m]/})^{\text{op}}}{\times} \left(\mathrm{RLax}_{\text{non-untl}}(\mathbb{S})^{\oint} \underset{\boldsymbol{\Delta}^{\text{op}}}{\times} \{[m]\}\right) \simeq \mathrm{Seq}_n(\mathbb{S}), \quad \gamma : [m] \to [n],$$

given by

$$\underline{s} = s_0 \to s_1 \to \ldots \to s_n,$$

see Sect. A.1.3 for the notation.

The fiber of $'\Phi(\mathbb{T})^{\oint}$ over the above object of $\mathrm{RLax}_{\text{non-untl}}(\mathbb{S})^{\oint}$ is by definition

$$\mathrm{Seq}_n(\mathbb{T}) \underset{\mathrm{Seq}_n(\mathbb{S})}{\times} \{s_0 \to s_1 \to \ldots \to s_n\}, \tag{3.1}$$

i.e., this is the category of strings

$$\underline{t} = t_0 \to t_1 \to \ldots \to t_n$$

in $\mathbb{T}$ that project to $\underline{s}$.

3.3.3. The full subcategory of (3.1), corresponding to $\Phi(\mathbb{T})^{\oint} \subset {}'\Phi(\mathbb{T})^{\oint}$ consists of those $\underline{t}$, for which for every $i \in 1,\ldots,n$ for which there exists a $j \in 1,\ldots,m$ with

$$\gamma(j-1) \leq i-1 < i \leq \gamma(j),$$

the corresponding 1-morphism $t_{i-1} \to t_i$ in $\mathbb{T}$ is locally 2-Cartesian over $s_{i-1} \to s_i$.

3.3.4. We claim:

LEMMA 3.3.5.
(a) *The composite functor*

$$\Phi(\mathbb{T})^{\oint} \to \mathrm{RLax}_{\text{non-untl}}(\mathbb{S})^{\oint} \to \mathbf{\Delta}^{\mathrm{op}}$$

is a coCartesian fibration.

(b) *The functor* $\mathbf{\Delta}^{\mathrm{op}} \to 1\text{-Cat}$, *corresponding to the coCartesian fibration of point (a) lies in the essential image of the functor* $\mathrm{Seq}_\bullet : 2\text{-Cat} \to \mathrm{Funct}(\mathbf{\Delta}^{\mathrm{op}}, 1\text{-Cat})$. *Denote the resulting* $(\infty,2)$*-category by* $\Phi(\mathbb{T})$.

(c) *The functor*

$$\Phi(\mathbb{T})^{\oint} \to \mathrm{RLax}_{\text{non-untl}}(\mathbb{S})^{\oint}$$

maps arrows that are coCartesian over $\mathbf{\Delta}^{\mathrm{op}}$ *to arrows that are coCartesian over* $\mathbf{\Delta}^{\mathrm{op}}$.

(d) *The functor* $\Phi(\mathbb{T}) \to \mathrm{RLax}_{\text{non-untl}}(\mathbb{S})$ *arising from point (c) is a 2-Cartesian fibration, for which the pullback functors along quasi-invertible arrows are equivalences.*

3.4. Proof of Theorem 3.2.2, computation of the compositions. In this subsection we will conclude the proof of Theorem 3.2.2 by showing that the maps Φ and Ψ constructed above are mutually inverse.

3.4.1. Since the composition $\rho_{\mathbb{S}}^{\oint} \circ \iota_{\mathbb{S}}^{\oint}$ is isomorphic to $\mathrm{Id}_{\mathbb{S}^{\oint}}$, we obtain immediately that the composition

$$(2\text{-Cart}^{\text{loc}}_{/\mathbb{S}})^{\text{Spc}} \overset{\Psi}{\to} (2\text{-Cart}_{/\mathrm{RLax}_{\text{non-untl}}(\mathbb{S})})^{\text{Spc}} \overset{\Phi}{\to} (2\text{-Cart}^{\text{loc}}_{/\mathbb{S}})^{\text{Spc}}$$

is canonically isomorphic to the identity map.

3.4.2. Let now $\widetilde{\mathbb{T}} \to \mathrm{RLax}_{\text{non-untl}}(\mathbb{S})$ be a 2-Cartesian fibration. We will now construct a functor

$$\widetilde{\mathbb{T}} \to {}'\Phi(\Psi(\widetilde{\mathbb{T}})).$$

The datum of such a functor is equivalent to that of a functor

$$\widetilde{\mathbb{T}}^{\oint} \to \widetilde{\mathbb{T}}^{\oint} \tag{3.2}$$

that fits into the commutative diagram

$$\begin{array}{ccc} \widetilde{\mathbb{T}}^{\oint} & \xrightarrow{(3.2)} & \widetilde{\mathbb{T}}^{\oint} \\ \downarrow & & \downarrow \\ \mathrm{RLax}_{\text{non-untl}}(\mathbb{S})^{\oint} & \xrightarrow{\iota_{\mathbb{S}}^{\oint} \circ \rho_{\mathbb{S}}^{\oint}} & \mathrm{RLax}_{\text{non-untl}}(\mathbb{S})^{\oint}. \end{array} \tag{3.3}$$

3.4.3. The construction of the functor (3.2) is based on the following lemma:

LEMMA 3.4.4. *Let $F : \mathbf{C} \to \mathbf{D}$ be a functor between $(\infty, 1)$-categories, and let $\Gamma_{\mathbf{D}} : [1] \times \mathbf{D} \to \mathbf{D}$ be a functor such that $\Gamma_{\mathbf{D}}|_{\{1\}\times\mathbf{D}} = \mathrm{Id}_{\mathbf{D}}$. Suppose that for every $\mathbf{c} \in \mathbf{C}$ there exists a Cartesian arrow $\mathbf{c}' \to \mathbf{c}$ that covers the 1-morphism $\Gamma_{\mathbf{D}}|_{[1]\times\{F(\mathbf{c})\}}$ in $\mathbf{D}$. Then there exists a uniquely defined functor $\Gamma_{\mathbf{C}} : [1] \times \mathbf{C} \to \mathbf{C}$, such that:*

- *$\Gamma_{\mathbf{C}}$ is equipped with an identification $\Gamma_{\mathbf{C}}|_{\{1\}\times\mathbf{C}} = \mathrm{Id}_{\mathbf{C}}$;*
- *The diagram*

$$\begin{array}{ccc} [1] \times \mathbf{C} & \xrightarrow{\Gamma_{\mathbf{C}}} & \mathbf{C} \\ {\scriptstyle \mathrm{Id}_{[1]}\times F}\downarrow & & \downarrow{\scriptstyle F} \\ [1] \times \mathbf{D} & \xrightarrow{\Gamma_{\mathbf{D}}} & \mathbf{D} \end{array} \tag{3.4}$$

  commutes
- *For any $\mathbf{c} \in \mathbf{C}$, the 1-morphism given by $\Gamma_{\mathbf{C}}|_{[1]\times\{\mathbf{c}\}}$ is Cartesian over $\mathbf{D}$.*

3.4.5. We apply the above lemma to $\mathbf{D} := \mathrm{RLax}_{\text{non-untl}}(\mathbb{S})^{\oint}$,

$$\mathbf{C} := \widetilde{\mathbb{T}}^{\mathrm{ext},\oint} \underset{\mathrm{RLax}_{\text{non-untl}}(\mathbb{S})^{\mathrm{ext},\oint}}{\times} \mathrm{RLax}_{\text{non-untl}}(\mathbb{S})^{\mathrm{ext},\oint},$$

with F induced by the projection $\widetilde{\mathbb{T}} \to \mathrm{RLax}_{\text{non-untl}}(\mathbb{S})$. We let $\Gamma_{\mathbf{D}}$ be given by the natural transformation

$$\iota_{\mathbb{S}}^{\oint} \circ \rho_{\mathbb{S}}^{\oint} \to \mathrm{Id}_{\mathrm{RLax}_{\text{non-untl}}(\mathbb{S})^{\oint}},$$

corresponding to the $(\iota_{\mathbb{S}}^{\oint}, \rho_{\mathbb{S}}^{\oint})$-adjunction.

Applying Lemma 3.4.4 we obtain a functor

$$[1] \times \widetilde{\mathbb{T}}^{\mathrm{ext},\oint} \underset{\mathrm{RLax}_{\text{non-untl}}(\mathbb{S})^{\mathrm{ext},\oint}}{\times} \mathrm{RLax}_{\text{non-untl}}(\mathbb{S})^{\mathrm{ext},\oint} \to$$
$$\to \widetilde{\mathbb{T}}^{\mathrm{ext},\oint} \underset{\mathrm{RLax}_{\text{non-untl}}(\mathbb{S})^{\mathrm{ext},\oint}}{\times} \mathrm{RLax}_{\text{non-untl}}(\mathbb{S})^{\mathrm{ext},\oint}.$$

Restricting to $\{0\} \in [1]$, and composing with

$$\widetilde{\mathbb{T}}^{\oint} \hookrightarrow \widetilde{\mathbb{T}}^{\mathrm{ext},\oint} \underset{\mathrm{RLax}_{\text{non-untl}}(\mathbb{S})^{\mathrm{ext},\oint}}{\times} \mathrm{RLax}_{\text{non-untl}}(\mathbb{S})^{\mathrm{ext},\oint},$$

we obtain a functor

$$\widetilde{\mathbb{T}}^{\oint} \to \widetilde{\mathbb{T}}^{\mathrm{ext},\oint} \underset{\mathrm{RLax}_{\text{non-untl}}(\mathbb{S})^{\mathrm{ext},\oint}}{\times} \mathrm{RLax}_{\text{non-untl}}(\mathbb{S})^{\mathrm{ext},\oint}. \tag{3.5}$$

Now, by unwinding the definitions, we obtain that the above functor (3.5) factors through the 1-full subcategory

$$\widetilde{\mathbb{T}}^{\oint} \subset \widetilde{\mathbb{T}}^{\mathrm{ext},\oint} \underset{\mathrm{RLax}_{\text{non-untl}}(\mathbb{S})^{\mathrm{ext},\oint}}{\times} \mathrm{RLax}_{\text{non-untl}}(\mathbb{S})^{\mathrm{ext},\oint}.$$

The resulting functor

$$\widetilde{\mathbb{T}}^{\oint} \to \widetilde{\mathbb{T}}^{\oint}$$

is the desired functor (3.2).

3.4.6. By further unwinding the definitions, we obtain that the essential image of the functor

$$\widetilde{\mathbb{T}} \to {}'\Phi(\Psi(\widetilde{\mathbb{T}}))$$

constructed above, belongs to $\Phi(\Psi(\widetilde{\mathbb{T}})) \subset {}'\Phi(\Psi(\widetilde{\mathbb{T}}))$.

Finally, if $\widetilde{\mathbb{T}} \to \text{RLax}_{\text{non-untl}}(\mathbb{S})$ is such that the pullback functors along quasi-invertible arrows (see Sect. A.3) are equivalences, then the resulting functor

$$\widetilde{\mathbb{T}} \to \Phi(\Psi(\widetilde{\mathbb{T}}))$$

is an equivalence.

□(Theorem 3.2.2)

3.5. Gray products and 2-Cartesian fibrations. In this subsection we will use Theorem 3.2.2 to give an explicit description of 2-Cartesian fibrations over Gray products.

3.5.1. Recall the condition on a functor between $(\infty, 2)$-categories to be a *localization on 1-morphisms*, see Sect. B.1. The following is straightforward:

LEMMA 3.5.2. *Let $\widetilde{\mathbb{S}} \to \mathbb{S}$ be a localization on 1-morphisms. Then the map*

$$\mathbf{2\text{-}Cart}_{/\mathbb{S}} \to \mathbf{2\text{-}Cart}_{/\widetilde{\mathbb{S}}},$$

defined by pullback, is fully faithful. Its essential image consists of those $F : \widetilde{\mathbb{T}} \to \widetilde{\mathbb{S}}$ that satisfy the following condition:
For every $t \in \widetilde{\mathbb{T}}$, a pair of 1-morphisms $\beta_0, \beta_1 : s' \to F(t)$ and a 2-morphism

$$\phi \in \text{Maps}_{\mathbf{Maps}_{\widetilde{\mathbb{S}}}(s', F(t))}(\beta_0, \beta_1),$$

if we denote by $\alpha_0 : t' \to t$ the 2-Cartesian lift of β_0 and by $\psi \in \text{Maps}_{\mathbf{Maps}_{\widetilde{\mathbb{T}}}(t',t)}(\alpha_0, \alpha_1)$ the coCartesian lift of ϕ, if the image of ϕ in $\mathbb{S}$ is invertible, *then the 1-morphism $\alpha_1 : t' \to t$ is 2-Cartesian over β_1.*

3.5.3. We fix $\mathbb{S}_1, \mathbb{S}_2 \in$ 2-Cat. We shall now describe the space

$$(\text{2-Cart}_{/\mathbb{S}_1 \circledast \mathbb{S}_2})^{\text{Spc}}$$

in a way functorial in $\mathbb{S}_1$ and $\mathbb{S}_2$. Indeed, combining Lemma 3.5.2 applied to

$$\text{RLax}_{\text{non-untl}}(\mathbb{S}_1 \times \mathbb{S}_2) \to \mathbb{S}_1 \circledast \mathbb{S}_2,$$

with Proposition 3.2.2, we obtain:

COROLLARY 3.5.4. *There exists a canonically defined fully faithful embedding*

$$(\text{2-Cart}_{/\mathbb{S}_1 \circledast \mathbb{S}_2})^{\text{Spc}} \hookrightarrow (\text{2-Cart}^{\text{loc}}_{/\mathbb{S}_1 \times \mathbb{S}_2})^{\text{Spc}}.$$

Its essential image consists of those $\mathbb{T} \to \mathbb{S}_1 \times \mathbb{S}_2$ that satisfy:

- *For every pair of composable 1-morphisms in $\mathbb{T}$, locally Cartesian over $\mathbb{S}_1 \times \mathbb{S}_2$, that cover two morphisms in $\mathbb{S}_1 \times \mathbb{S}_2$ both of which project to isomorphisms under $\mathbb{S}_1 \times \mathbb{S}_2 \to \mathbb{S}_1$, their composition is locally Cartesian over $\mathbb{S}_1 \times \mathbb{S}_2$.*
- *For every pair of composable 1-morphisms in $\mathbb{T}$, locally Cartesian over $\mathbb{S}_1 \times \mathbb{S}_2$, that cover two morphisms in $\mathbb{S}_1 \times \mathbb{S}_2$ both of which project to isomorphisms under $\mathbb{S}_1 \times \mathbb{S}_2 \to \mathbb{S}_2$, their composition is locally Cartesian over $\mathbb{S}_1 \times \mathbb{S}_2$.*

- *For every pair of 1-morphisms* $(s_1' \overset{\alpha_1}{\to} s_1) \in \mathbb{S}_1$ *and* $(s_2' \overset{\alpha_2}{\to} s_2) \in \mathbb{S}_2$ *and locally Cartesian 1-morphisms* $t'' \overset{\beta}{\to} t'$ *and* $t' \overset{\gamma}{\to} t$ *covering* $(\alpha_1, \mathrm{id}_{s_2'})$ *and* $(\mathrm{id}_{s_1}, \alpha_2)$*, respectively, the 1-morphism* $\gamma \circ \beta$ *is locally Cartesian over* (α_1, α_2).

COROLLARY 3.5.5.

(a) *The essential image of the (fully faithful) map*

$$(2\text{-Cart}_{/\mathbb{S}_1 \circledast \mathbb{S}_2})^{\mathrm{Spc}} \to (2\text{-Cart}^{\mathrm{loc}}_{/\mathbb{S}_1 \times \mathbb{S}_2})^{\mathrm{Spc}} \subset (2\text{-Cat}_{/\mathbb{S}_1 \times \mathbb{S}_2})^{\mathrm{Spc}}$$

consists of those

$$\mathbb{T} \to \mathbb{S}_1 \times \mathbb{S}_2$$

such that:

1. *The composition* $\mathbb{T} \to \mathbb{S}_1 \times \mathbb{S}_2 \to \mathbb{S}_1$ *is a 2-Cartesian fibration;*
2. *The functor* $\mathbb{T} \to \mathbb{S}_1 \times \mathbb{S}_2$*, viewed as a map in* $2\text{-Cart}_{/\mathbb{S}_1}$*, belongs to* $(2\text{-Cart}_{/\mathbb{S}_1})_{\text{strict}}$*;*
3. *For every* $s_1 \in \mathbb{S}_1$*, the resulting functor* $\mathbb{T}_{s_1} \to \mathbb{S}_2$ *is a 2-Cartesian fibration.*
4. *For every 1-morphism* $s_1 \to s_1'$ *in* $\mathbb{S}_1$*, the pullback functor* $\mathbb{T}_{s_1'} \to \mathbb{T}_{s_1}$*, which by the previous point is a 1-morphism in* $2\text{-Cart}_{/\mathbb{S}_2}$*, belongs* $(2\text{-Cart}_{/\mathbb{S}_2})_{2\text{-strict}}$.

(b) *The subspace*

$$(2\text{-Cart}_{/\mathbb{S}_1 \times \mathbb{S}_2})^{\mathrm{Spc}} \subset (2\text{-Cart}_{/\mathbb{S}_1 \circledast \mathbb{S}_2})^{\mathrm{Spc}}$$

corresponds to replacing in condition (4) the category $(2\text{-Cart}_{/\mathbb{S}_2})_{2\text{-strict}}$ *by its 1-full subcategory*

$$(2\text{-Cart}_{/\mathbb{S}_2})_{\text{strict}} \subset (2\text{-Cart}_{/\mathbb{S}_2})_{2\text{-strict}}.$$

4. Proof of Theorem 1.1.8

4.1. Proof of Theorem 1.1.8, Step 1: identifying the underlying spaces. In this subsection we will establish the assertion of Theorem 1.1.8 at the level of the underlying spaces.

4.1.1. First, we notice that Theorem 2.0.1 can be reformulated as follows:

COROLLARY 4.1.2. *There exists a canonical equivalence of bi-simplicial spaces that send* m, n *to*

$$\mathrm{Sq}_{m,n}(\mathbf{2\text{-}Cat}) \text{ and } \mathrm{Maps}([m], 2\text{-Cart}_{/[n]^{\mathrm{op}}}),$$

respectively.

4.1.3. Applying the 1-fully faithful embedding

(4.1)
$$\mathrm{Funct}([m], 2\text{-Cat}) \simeq \mathrm{Seq}^{\mathrm{ext}}_m(2\text{-Cat}) \simeq (2\text{-Cart}_{/[m]^{\mathrm{op}}})_{\text{strict}} \hookrightarrow 2\text{-Cart}_{/[m]^{\mathrm{op}}} \hookrightarrow 2\text{-Cat}_{/[m]^{\mathrm{op}}}$$

(where the second isomorphism is Theorem 2.0.1(b)), we obtain a fully faithful map

$$\mathrm{Maps}([m], 2\text{-Cat}_{/[n]^{\mathrm{op}}}) \to ((2\text{-Cat}_{/[m]^{\mathrm{op}}})_{/[n]^{\mathrm{op}} \times [m]^{\mathrm{op}}})^{\mathrm{Spc}} \simeq (2\text{-Cat}_{/[n]^{\mathrm{op}} \times [m]^{\mathrm{op}}})^{\mathrm{Spc}}.$$

Restricting to $2\text{-Cart}_{/[n]^{\mathrm{op}}} \hookrightarrow 2\text{-Cat}_{/[n]^{\mathrm{op}}}$, we obtain a fully faithful map

$$\mathrm{Maps}([m], 2\text{-Cart}_{/[n]^{\mathrm{op}}}) \to (2\text{-Cat}_{/[n]^{\mathrm{op}} \times [m]^{\mathrm{op}}})^{\mathrm{Spc}}. \tag{4.2}$$

LEMMA 4.1.4. *The essential image of the map* (4.2) *lies in*

$$(2\text{-Cart}^{\text{loc}}_{/[n]^{\text{op}}\times[m]^{\text{op}}})^{\text{Spc}} \subset (2\text{-Cat}_{/[n]^{\text{op}}\times[m]^{\text{op}}})^{\text{Spc}}$$

and coincides with the essential image of fully faithful embedding

$$(2\text{-Cart}_{/[n]^{\text{op}}\circledast[m]^{\text{op}}})^{\text{Spc}} \hookrightarrow (2\text{-Cart}^{\text{loc}}_{/[n]^{\text{op}}\times[m]^{\text{op}}})^{\text{Spc}}$$

of Corollary 3.5.4.

PROOF. Follows from Corollary 3.5.5(a). □

4.1.5. Thus, combining Lemma 4.1.4 and Corollary 4.1.2 we obtain a canonical identification of bi-simplicial spaces

$$\text{Sq}_{m,n}(\mathbf{2\text{-}Cat}) \simeq (2\text{-Cart}_{/[n]^{\text{op}}\circledast[m]^{\text{op}}})^{\text{Spc}}. \tag{4.3}$$

We can now establish the assertion of Theorem 1.1.8 at the level of the underlying spaces:

COROLLARY 4.1.6. *For* $\mathbb{S} \in 2\text{-Cat}$, *there exists a canonical equivalence*

$$(2\text{-Cart}_{/\mathbb{S}})^{\text{Spc}} \simeq \text{Maps}_{2\text{-Cat}}(\mathbb{S}^{1\text{-op}}, \mathbf{2\text{-}Cat}),$$

functorial in $\mathbb{S}$.

PROOF. It follows from Chapter 10, Theorems 4.1.3 and 5.2.3(a) that for $\mathbb{S} \in 2\text{-Cat}$, the restriction map

$$(2\text{-Cat}_{/\mathbb{S}})^{\text{Spc}} \to \text{Maps}_{\text{Funct}(\mathbf{\Delta}^{\text{op}}\times\mathbf{\Delta}^{\text{op}},\text{Spc})}(\text{Sq}_{\bullet,\bullet}(\mathbb{S}), (2\text{-Cat}_{/[\bullet]\circledast[\bullet]})^{\text{Spc}})$$

is an isomorphism, and under this isomorphism the subspaces

$$(2\text{-Cart}_{/\mathbb{S}})^{\text{Spc}} \subset (2\text{-Cat}_{/\mathbb{S}})^{\text{Spc}}$$

and

$$\text{Maps}_{\text{Funct}(\mathbf{\Delta}^{\text{op}}\times\mathbf{\Delta}^{\text{op}},\text{Spc})}(\text{Sq}_{\bullet,\bullet}(\mathbb{S}), (2\text{-Cart}_{/[\bullet]\circledast[\bullet]})^{\text{Spc}}) \subset$$
$$\subset \text{Maps}_{\text{Funct}(\mathbf{\Delta}^{\text{op}}\times\mathbf{\Delta}^{\text{op}},\text{Spc})}(\text{Sq}_{\bullet,\bullet}(\mathbb{S}), (2\text{-Cat}_{/[\bullet]\circledast[\bullet]})^{\text{Spc}})$$

correspond to one another.

Hence, the assertion of the corollary follows from the isomorphism (4.3) using the canonical identification of bi-cosimplicial objects of 2-Cat:

$$([m]\circledast[n])^{1\text{-op}} \simeq [n]^{\text{op}} \circledast [m]^{\text{op}}.$$

□

4.2. Proof of Theorem 1.1.8, Step 2: identifying the underlying $(\infty,1)$-categories. In this subsection we will construct the identification of the $(\infty,1)$-categories underlying the two sides in Theorem 1.1.8(b).

4.2.1. We need to construct an isomorphism of simplicial spaces

$$\text{Maps}_{1\text{-Cat}}([m], (2\text{-Cart}_{/\mathbb{S}})_{\text{strict}}) \simeq \text{Maps}_{2\text{-Cat}}(\mathbb{S}^{1\text{-op}} \times [m], \mathbf{2\text{-}Cat}), \quad [m] \in \mathbf{\Delta}.$$

Taking into account Corollary 4.1.6, we need to construct an isomorphism of simplicial spaces

$$\text{Maps}([m], (2\text{-Cart}_{/\mathbb{S}})_{\text{strict}}) \simeq (2\text{-Cart}_{/\mathbb{S}\times[m]^{\text{op}}})^{\text{Spc}}, \quad [m] \in \mathbf{\Delta}. \tag{4.4}$$

4.2.2. Given $[m] \in \mathbf{\Delta}$, using the 1-fully faithful embedding

$$(4.5)\quad \text{Funct}([m], 2\text{-Cat}) \simeq \text{Seq}_m^{\text{ext}}(2\text{-Cat}) \simeq \\ \simeq (2\text{-Cart}_{/[m]^{\text{op}}})_{\text{strict}} \hookrightarrow 2\text{-Cart}_{/[m]^{\text{op}}} \hookrightarrow 2\text{-Cat}_{/[m]^{\text{op}}}$$

of (4.1) (which, we note, uses the statement of Theorem 2.0.1), we obtain a fully faithful map

$$\text{Maps}([m], 2\text{-Cat}_{/\mathbb{S}}) \hookrightarrow ((2\text{-Cat}_{/[m]^{\text{op}}})_{/\mathbb{S}\times[m]^{\text{op}}})^{\text{Spc}} \simeq (2\text{-Cat}_{/\mathbb{S}\times[m]^{\text{op}}})^{\text{Spc}}.$$

Composing with the embedding

$$\text{Maps}([m], (2\text{-Cart}_{/\mathbb{S}})_{\text{strict}}) \hookrightarrow \text{Maps}([m], 2\text{-Cart}_{/\mathbb{S}}) \hookrightarrow \text{Maps}([m], 2\text{-Cat}_{/\mathbb{S}}),$$

we obtain a fully faithful map

$$(4.6)\quad \text{Maps}([m], (2\text{-Cart}_{/\mathbb{S}})_{\text{strict}}) \to (2\text{-Cat}_{/\mathbb{S}\times[m]^{\text{op}}})^{\text{Spc}}.$$

LEMMA 4.2.3. *The essential image of the map* (4.6) *equals*

$$(2\text{-Cart}_{/\mathbb{S}\times[m]^{\text{op}}})^{\text{Spc}} \subset (2\text{-Cat}_{/\mathbb{S}\times[m]^{\text{op}}})^{\text{Spc}}.$$

PROOF. Follows from Corollary 3.5.5(b). □

Thus, we obtain the required identification (4.4).

4.3. Proof of Theorem 1.1.8, Step 3: end of the argument.

4.3.1. Given $\mathbb{T} \in 2\text{-Cat}$, we need to construct an isomorphism of spaces

$$\text{Maps}(\mathbb{T}, (\mathbf{2\text{-}coCart}_{/\mathbb{S}})_{2\text{-strict}}) \simeq \text{Maps}_{2\text{-Cat}}(\mathbb{T} \circledast \mathbb{S}^{1\text{-op}}, \mathbf{2\text{-}Cat}),$$

functorial in $\mathbb{T}$ and $\mathbb{S}$, so that the subspaces

$$\text{Maps}(\mathbb{T}, (\mathbf{2\text{-}coCart}_{/\mathbb{S}})_{\text{strict}}) \subset \text{Maps}(\mathbb{T}, (\mathbf{2\text{-}coCart}_{/\mathbb{S}})_{2\text{-strict}})$$

and

$$\text{Maps}_{2\text{-Cat}}(\mathbb{T} \times \mathbb{S}^{1\text{-op}}, \mathbf{2\text{-}Cat}) \subset \text{Maps}_{2\text{-Cat}}(\mathbb{T} \circledast \mathbb{S}^{1\text{-op}}, \mathbf{2\text{-}Cat})$$

correspond to one another.

Taking into account Corollary 4.1.6, we need to construct an isomorphism of spaces

$$(4.7)\quad \text{Maps}(\mathbb{T}, (\mathbf{2\text{-}coCart}_{/\mathbb{S}})_{2\text{-strict}}) \simeq (2\text{-coCart}_{/\mathbb{S}\circledast\mathbb{T}^{1\text{-op}}})^{\text{Spc}},$$

so that

$$\text{Maps}(\mathbb{T}, (\mathbf{2\text{-}coCart}_{/\mathbb{S}})_{\text{strict}}) \subset \text{Maps}(\mathbb{T}, (\mathbf{2\text{-}coCart}_{/\mathbb{S}})_{2\text{-strict}})$$

maps to

$$(2\text{-coCart}_{/\mathbb{S}\times\mathbb{T}^{1\text{-op}}})^{\text{Spc}} \subset (2\text{-coCart}_{/\mathbb{S}\circledast\mathbb{T}^{1\text{-op}}})^{\text{Spc}}.$$

4.3.2. The equivalence of $(\infty,1)$-categories

$$\mathbf{Maps}_{2\text{-Cat}}(\mathbb{T}, \mathbf{2\text{-}Cat}) \simeq (2\text{-coCart}_{/\mathbb{T}^{1\text{-op}}})_{\text{strict}}$$

established in Step 2, gives rise to a 1-fully faithful embedding

$$\mathbf{Maps}_{2\text{-Cat}}(\mathbb{T}, \mathbf{2\text{-}Cat}) \hookrightarrow 2\text{-Cat}_{/\mathbb{T}^{1\text{-op}}}.$$

From here, we obtain a fully faithful embedding

$$\text{Maps}(\mathbb{T}, \mathbf{2\text{-}Cat}_{/\mathbb{S}}) \hookrightarrow ((2\text{-Cat}_{/\mathbb{T}^{1\text{-op}}})_{/\mathbb{S}\times\mathbb{T}^{1\text{-op}}})^{\text{Spc}} = (2\text{-Cat}_{/\mathbb{S}\times\mathbb{T}^{1\text{-op}}})^{\text{Spc}}.$$

Composing with

$$\text{Maps}(\mathbb{T}, (\mathbf{2\text{-}coCart}_{/\mathbb{S}})_{2\text{-strict}}) \subset \text{Maps}(\mathbb{T}, \mathbf{2\text{-}Cat}_{/\mathbb{S}}),$$

we obtain a fully faithful map

$$\text{Maps}(\mathbb{T}, (\mathbf{2\text{-}coCart}_{/\mathbb{S}})_{2\text{-strict}}) \hookrightarrow (2\text{-Cat}_{/\mathbb{S}\times\mathbb{T}^{1\text{-op}}})^{\text{Spc}}. \tag{4.8}$$

We claim:

LEMMA 4.3.3.
(a) *The essential image of the map* (4.8) *is contained in* $(2\text{-Cat}^{\text{loc}}_{/\mathbb{S}\times\mathbb{T}^{1\text{-op}}})^{\text{Spc}}$ *and equals the essential image of the fully faithful embedding*

$$(2\text{-coCart}_{/\mathbb{S}\circledast\mathbb{T}^{1\text{-op}}})^{\text{Spc}} \hookrightarrow (2\text{-Cat}^{\text{loc}}_{/\mathbb{S}\times\mathbb{T}^{1\text{-op}}})^{\text{Spc}}$$

of Corollary 3.5.4.
(b) *Under the resulting isomorphism*

$$\text{Maps}(\mathbb{T}, (\mathbf{2\text{-}coCart}_{/\mathbb{S}})_{2\text{-strict}}) \simeq (2\text{-coCart}_{/\mathbb{S}\circledast\mathbb{T}^{1\text{-op}}})^{\text{Spc}}$$

the subspace

$$\text{Maps}(\mathbb{T}, (\mathbf{2\text{-}coCart}_{/\mathbb{S}})_{\text{strict}}) \subset \text{Maps}(\mathbb{T}, (\mathbf{2\text{-}coCart}_{/\mathbb{S}})_{2\text{-strict}})$$

maps to

$$(2\text{-coCart}_{/\mathbb{S}\times\mathbb{T}^{1\text{-op}}})^{\text{Spc}} \subset (2\text{-coCart}_{/\mathbb{S}\circledast\mathbb{T}^{1\text{-op}}})^{\text{Spc}}.$$

PROOF. Follows from Corollary 3.5.5. □

The last lemma establishes the desired isomorphism (4.7).

□(Theorem 1.1.8)

5. The Yoneda embedding

The goal of this section is to discuss the several incarnations of what can be called the Yoneda lemma in the context of $(\infty,2)$-categories.

For example, we will show that to $s \in \mathbb{S}$ there corresponds a Yoneda functor

$$h_s : \mathbb{S} \to \mathbf{1\text{-}Cat}, \quad h_s(s') = \mathbf{Maps}_{\mathbb{S}}(s, s'), \tag{5.1}$$

and for any $\mathbb{S} \xrightarrow{F} \mathbf{1\text{-}Cat}$ we have an equivalence

$$\mathbf{Maps}_{\text{Funct}(\mathbb{S},\mathbf{1\text{-}Cat})}(h_s, F) \simeq F(s). \tag{5.2}$$

By letting s vary, we will construct the Yoneda embedding

$$\text{Yon} : \mathbb{S} \hookrightarrow \text{Funct}(\mathbb{S}^{1\text{-op}}, \mathbf{1\text{-}Cat}).$$

5.1. The right-lax slice construction. In order to construct the Yoneda functors, we will use Corollary 1.3.4 in order to interpret the datum of a functor $\mathbb{S} \to \mathbf{1\text{-}Cat}$ as a 1-coCartesian fibration.

In this subsection we will construct the corresponding 1-coCartesian fibrations (up to reversing the arrows).

5.1.1. Let $\mathbb{S}$ be an $(\infty, 2)$-category, and $s \in \mathbb{S}$ an object. We define the $(\infty, 2)$-category $\mathbb{S}_{//s}$ to be

$$\text{Funct}([1], \mathbb{S})_{\text{right-lax}} \underset{\mathbb{S}}{\times} \{s\},$$

where the fiber product is formed using functor

$$\text{Funct}([1], \mathbb{S})_{\text{right-lax}} \to \mathbb{S}.$$

given by evaluation at $1 \in [1]$.

5.1.2. Let $p_s : \mathbb{S}_{//s} \to \mathbb{S}$ be the functor given by evaluation at $0 \in [1]$. By definition, the fiber of p over $s' \in \mathbb{S}$ is an $(\infty, 2)$-category

$$\text{Funct}([1], \mathbb{S})_{\text{right-lax}} \underset{\mathbb{S}\times\mathbb{S}}{\times} \{(s', s)\},$$

which by Chapter 10, Corollary 3.4.8 is an $(\infty, 1)$-category, equipped with a canonical identification with $\mathbf{Maps}_{\mathbb{S}}(s', s)$.

5.1.3. We claim:

LEMMA 5.1.4. *The functor $p_s : \mathbb{S}_{//s} \to \mathbb{S}$ is a 1-Cartesian fibration.*

PROOF. Let $\alpha : s_0' \to s$ be an object of $\mathbb{S}_{//s}$, and and let $\beta : s_1' \to s_0'$ is a 1-morphism in $\mathbb{S}$. Then it is easy to see that the commutative diagram

$$\begin{array}{ccc} s_0' & \xrightarrow{\alpha} & s \\ {\scriptstyle\beta}\uparrow & & \uparrow{\scriptstyle \text{id}_s} \\ s_1' & \xrightarrow{\alpha\circ\beta} & s \end{array}$$

represents a Cartesian arrow in $(\mathbb{S}_{//s})^{1\text{-Cat}}$ over β: indeed this is an assertion at the level of the underlying $(\infty, 1)$-categories.

To finish the proof of the lemma, given a pair of objects

$$\underline{s}_0 = (\alpha_0 : s_0' \to s) \text{ and } \underline{s}_1 = (\alpha_1 : s_1' \to s)$$

of $\mathbb{S}_{/s}$, we need to show that the functor

$$\mathbf{Maps}_{\mathbb{S}_{//s}}(\underline{s}_0, \underline{s}_1) \to \mathbf{Maps}_{\mathbb{S}}(s_0, s_1)$$

is a coCartesian fibration in spaces.

The category $\mathbf{Maps}_{\mathbb{S}_{//s}}(\underline{s}_0, \underline{s}_1)$ has as objects pairs (β, ϕ), where $\beta : s_0' \to s_1'$ and ϕ is a 2-morphism $\alpha_0 \to \alpha_1 \circ \beta$. Morphisms from (β, ϕ) to $(\widetilde{\beta}, \widetilde{\phi})$ is the space of 2-morphisms $\psi : \beta \to \widetilde{\beta}$, equipped with an identification $\widetilde{\phi} \simeq \alpha_1(\psi) \circ \phi$. This makes it clear that the assignment

$$(\beta, \phi) \mapsto \beta$$

is coCartesian fibration in spaces.

□

5.1.5. Applying Corollary 1.2.6, from the 1-Cartesian fibration $\mathbb{S}_{//s} \to \mathbb{S}$ we obtain a functor

$$\widetilde{h}_s : \mathbb{S}^{1\text{-op}} \to \mathbf{1\text{-}Cat}.$$

The value of this functor on a given $s' \in \mathbb{S}$ is

$$(\mathbb{S}_{//s})_{s'} \simeq \mathbf{Maps}_{\mathbb{S}}(s', s).$$

5.2. The 2-categorical Yoneda lemma. In this subsection we will establish the isomorphism (5.2).

5.2.1. For a pair of 1-Cartesian fibrations in $(\infty,1)$-categories $\mathbb{T}_0, \mathbb{T}_1$ over $\mathbb{S}$, let us denote by

$$\mathbf{Maps}^{\text{strict}}_{\mathbb{S}}(\mathbb{T}_0, \mathbb{T}_1) := \mathbf{Maps}_{(\mathbf{1\text{-}Cart}_{/\mathbb{S}})_{\text{strict}}}(\mathbb{T}_0, \mathbb{T}_1),$$

where the notation $(\mathbf{1\text{-}Cart}_{/\mathbb{S}})_{\text{strict}}$ is as in Sect. 1.2.3.

I.e., $\mathbf{Maps}^{\text{strict}}_{\mathbb{S}}(\mathbb{T}_0, \mathbb{T}_1)$ is the full subcategory of $\mathbf{Maps}_{\mathbb{S}}(\mathbb{T}_0, \mathbb{T}_1)$ that consists of those functors that map 1-morphisms in $\mathbb{T}_0$ that are Cartesian over $\mathbb{S}$ to 1-morphisms in $\mathbb{T}_1$ with the same property.

5.2.2. We claim:

PROPOSITION 5.2.3. *For a 1-Cartesian fibration $F : \mathbb{T} \to \mathbb{S}$, evaluation at $(s \overset{\text{id}_s}{\to} s) \in \mathbb{S}_{//s}$ defines an equivalence*

$$\mathbf{Maps}^{\text{strict}}_{\mathbb{S}}(\mathbb{S}_{//s}, \mathbb{T}) \to \mathbb{T}_s.$$

PROOF. Let

$$(\text{Funct}([1], \mathbb{T})_{\text{right-lax}})^{\text{Cart}_{/\mathbb{S}}} \subset \text{Funct}([1], \mathbb{T})_{\text{right-lax}}$$

denote the full subcategory whose objects are 1-morphisms Cartesian over $\mathbb{S}$.

Evaluation defines functors

$$\text{ev}_0, \text{ev}_1 : (\text{Funct}([1], \mathbb{T})_{\text{right-lax}})^{\text{Cart}_{/\mathbb{S}}} \to \mathbb{T}.$$

Consider the fiber product

$$(\text{Funct}([1], \mathbb{T})_{\text{right-lax}})^{\text{Cart}_{/\mathbb{S}}} \underset{\text{ev}_1, \mathbb{T}}{\times} \mathbb{T}_s \simeq (\text{Funct}([1], \mathbb{T})_{\text{right-lax}})^{\text{Cart}_{/\mathbb{S}}} \underset{F \circ \text{ev}_1, \mathbb{S}}{\times} \{s\}.$$

It is easy to see that the functor (between $(\infty,2)$-categories over $\mathbb{S}$)

$$(\text{Funct}([1], \mathbb{T})_{\text{right-lax}})^{\text{Cart}_{/\mathbb{S}}} \underset{\text{ev}_1, \mathbb{T}}{\times} \mathbb{T}_s \to \mathbb{S}_{//s} \times \mathbb{T}_s$$

is an equivalence.

Hence, we obtain a functor (between $(\infty,2)$-categories over $\mathbb{S}$)

$$\mathbb{S}_{//s} \times \mathbb{T}_s \to (\text{Funct}([1], \mathbb{T})_{\text{lax}})^{\text{Cart}_{/\mathbb{S}}} \underset{\text{ev}_1, \mathbb{T}}{\times} \mathbb{T}_s \to (\text{Funct}([1], \mathbb{T})_{\text{lax}})^{\text{Cart}_{/\mathbb{S}}} \overset{\text{ev}_0}{\longrightarrow} \mathbb{T}.$$

The latter gives rise to a functor

$$\mathbb{T}_s \to \mathbf{Maps}_{\mathbb{S}}(\mathbb{S}_{//s}, \mathbb{T}).$$

It is easy to see that the latter functor takes values in $\mathbf{Maps}^{\text{strict}}_{\mathbb{S}}(\mathbb{S}_{//s}, \mathbb{T})$ and provides an inverse to one in the statement of the proposition.

□

5.2.4. Applying Corollary 1.2.6, from Proposition 5.2.3 we obtain:

COROLLARY 5.2.5. *For $F : \mathbb{S}^{1\text{-op}} \to \mathbf{1}\text{-}\mathbf{Cat}$, evaluation at $s \in \mathbb{S}$ defines an equivalence*

$$\mathbf{Maps}_{\mathrm{Funct}(\mathbb{S}^{1\text{-op}},\mathbf{1}\text{-}\mathbf{Cat})}(\widetilde{h}_s, F) \simeq F(s).$$

5.3. The 2-categorical Yoneda embedding. We will now show how to turn $s \in \mathbb{S}$ into a parameter and thus obtain the Yoneda functor

$$\mathrm{Yon}_{\mathbb{S}} : \mathbb{S} \to \mathrm{Funct}(\mathbb{S}^{1\text{-op}}, \mathbf{1}\text{-}\mathbf{Cat}).$$

We will then show that $\mathrm{Yon}_{\mathbb{S}}$ is fully faithful.

5.3.1. For $\mathbb{S} \in 2\text{-Cat}$, consider the $(\infty, 2)$-category

$$\mathrm{Funct}([1], \mathbb{S})_{\text{right-lax}}.$$

Evaluation on $0, 1 \in [1]$ defines two functors

$$\mathrm{ev}_0, \mathrm{ev}_1 : \mathrm{Funct}([1], \mathbb{S})_{\text{right-lax}} \rightrightarrows \mathbb{S}.$$

As in Lemma 5.1.4 one shows:

LEMMA 5.3.2.
(a) *The functor* $\mathrm{ev}_1 : \mathrm{Funct}([1], \mathbb{S})_{\text{right-lax}} \to \mathbb{S}$ *is a 2-coCartesian fibration of* $(\infty, 2)$-*categories.*
(b) *The functor*

$$(\mathrm{ev}_0 \times \mathrm{ev}_1) : \mathrm{Funct}([1], \mathbb{S})_{\text{right-lax}} \to \mathbb{S} \times \mathbb{S}$$

is a strict functor between 2-coCartesian fibrations over $\mathbb{S}$.

5.3.3. Applying Corollary 1.3.3, from the functor $\mathrm{ev}_0 \times \mathrm{ev}_1$ we obtain a functor

$$\mathbb{S} \to \mathbf{2}\text{-}\mathbf{Cat},$$

equipped with a natural transformation to the constant functor with value $\mathbb{S}$.

I.e., we obtain a functor

$$\mathbb{S} \to \mathbf{2}\text{-}\mathbf{Cat}_{/\mathbb{S}} \tag{5.3}$$

5.3.4. Note, however, that by Lemma 5.1.4, the functor (5.3) takes values in the full subcategory

$$\mathbf{1}\text{-}\mathbf{Cart}_{/\mathbb{S}} \subset \mathbf{2}\text{-}\mathbf{Cat}_{/\mathbb{S}}.$$

Moreover, the functor (5.3) factors through the 1-full subcategory

$$(\mathbf{1}\text{-}\mathbf{Cart}_{/\mathbb{S}})_{\text{strict}} \subset \mathbf{1}\text{-}\mathbf{Cart}_{/\mathbb{S}}.$$

I.e., we have a functor

$$\mathbb{S} \to (\mathbf{1}\text{-}\mathbf{Cart}_{/\mathbb{S}})_{\text{strict}}. \tag{5.4}$$

5.3.5. Applying the equivalence $(\mathbf{1}\text{-}\mathbf{Cart}_{/\mathbb{S}})_{\text{strict}} \simeq \text{Funct}(\mathbb{S}^{1\text{-op}}, \mathbf{1}\text{-}\mathbf{Cat})$, from (5.4), we obtain a functor

$$\text{Yon}_{\mathbb{S}} : \mathbb{S} \to \text{Funct}(\mathbb{S}^{1\text{-op}}, \mathbf{1}\text{-}\mathbf{Cat}), \tag{5.5}$$

or, equivalently, a functor

$$\mathbb{S}^{1\text{-op}} \times \mathbb{S} \to \mathbf{1}\text{-}\mathbf{Cat}. \tag{5.6}$$

We will refer to the functor $\text{Yon}_{\mathbb{S}}$ of (5.5) as the *2-categorical Yoneda functor.*

5.3.6. We claim:

PROPOSITION 5.3.7. *The functor* (5.5) *is fully faithful.*

PROOF. We need to show that for $s, s' \in \mathbb{S}$, the functor

$$\mathbf{Maps}_{\mathbb{S}}(s, s') \to \mathbf{Maps}_{\text{Funct}(\mathbb{S}^{1\text{-op}}, \mathbf{1}\text{-}\mathbf{Cat})}(\text{Yon}_{\mathbb{S}}(s), \text{Yon}_{\mathbb{S}}(s'))$$

is an equivalence.

Equivalently (by virtue of Corollary 1.2.6), we need to show that the composite functor

$$\mathbf{Maps}_{\mathbb{S}}(s, s') \to \mathbf{Maps}_{\text{Funct}(\mathbb{S}^{1\text{-op}}, \mathbf{1}\text{-}\mathbf{Cat})}(\text{Yon}(s), \text{Yon}(s')) \to \mathbf{Maps}^{\text{strict}}_{\mathbb{S}}(\mathbb{S}_{//s}, \mathbb{S}_{//s'}) \tag{5.7}$$

is an equivalence.

By construction, the above map (5.7) has the property that for any $t \in \mathbb{S}$ the induced map

$$\begin{aligned} &\mathbf{Maps}_{\mathbb{S}}(s, s') \to \mathbf{Maps}^{\text{strict}}_{\mathbb{S}}(\mathbb{S}_{//s}, \mathbb{S}_{//s'}) \to \\ &\quad \to \mathbf{Maps}((\mathbb{S}_{//s})_t, (\mathbb{S}_{//s'})_t) \simeq \mathbf{Maps}(\mathbf{Maps}_{\mathbb{S}}(t, s), \mathbf{Maps}_{\mathbb{S}}(t, s')) \end{aligned}$$

is the map

$$\mathbf{Maps}_{\mathbb{S}}(s, s') \to \mathbf{Maps}(\mathbf{Maps}_{\mathbb{S}}(t, s), \mathbf{Maps}_{\mathbb{S}}(t, s')),$$

given by composition of 1-morphisms.

Taking $t = s$ and evaluating at id_s, we obtain that the composition

$$\begin{aligned} &\mathbf{Maps}_{\mathbb{S}}(s, s') \to \mathbf{Maps}^{\text{strict}}_{\mathbb{S}}(\mathbb{S}_{//s}, \mathbb{S}_{//s'}) \to \\ &\to \mathbf{Maps}((\mathbb{S}_{//s})_s, (\mathbb{S}_{//s'})_s) \to \mathbf{Maps}(\mathbf{Maps}_{\mathbb{S}}(s, s), \mathbf{Maps}_{\mathbb{S}}(s, s')) \to \mathbf{Maps}_{\mathbb{S}}(s, s') \end{aligned}$$

is the identity map.

Now, according to Proposition 5.2.3, the composition

$$\begin{aligned} &\mathbf{Maps}^{\text{strict}}_{\mathbb{S}}(\mathbb{S}_{//s}, \mathbb{S}_{//s'}) \to \mathbf{Maps}((\mathbb{S}_{//s})_s, (\mathbb{S}_{//s'})_s) \to \\ &\quad \to \mathbf{Maps}(\mathbf{Maps}_{\mathbb{S}}(s, s), \mathbf{Maps}_{\mathbb{S}}(s, s')) \to \mathbf{Maps}_{\mathbb{S}}(s, s') \end{aligned}$$

is an isomorphism, implying that (5.7) is an equivalence as well.

$\square$

A. The universal right-lax functor

A.1. The construction.

A.1.1. Consider the 1-fully faithful functor

$$2\text{-Cat} \to 2\text{-Cat}_{\text{right-lax}_{\text{non-untl}}},$$

see Chapter 10, Sect. 3.1.5.

This functor is easily seen to commute with limits. Hence, it admits a left adjoint, to be denoted

$$\mathbb{S} \mapsto \text{RLax}_{\text{non-untl}}(\mathbb{S}).$$

It turns out that this functor can be described rather explicitly, and this description is useful.

A.1.2. Recall the notation $\mathbb{S}^{\oint}$, see Chapter 10, Sect. 3.1.1.

Starting from $\mathbb{S} \in 2\text{-Cat}$, consider the following $(\infty,1)$-category:

$$\text{RLax}_{\text{non-untl}}(\mathbb{S})^{\oint} := \mathbb{S}^{\oint} \underset{\mathbf{\Delta}^{\text{op}}}{\times} \text{Actv},$$

where Actv is the full subcategory of $\text{Funct}([1], \mathbf{\Delta}^{\text{op}})$, spanned by active morphisms, and $\text{Actv} \to \mathbf{\Delta}^{\text{op}}$ is the functor of evaluation at $0 \in [1]$.

Evaluation on $1 \in [1]$ defines a functor

$$\text{RLax}_{\text{non-untl}}(\mathbb{S})^{\oint} \to \mathbf{\Delta}^{\text{op}}. \tag{A.1}$$

A.1.3. For example,

$$\text{RLax}_{\text{non-untl}}(\mathbb{S})^{\oint} \underset{\mathbf{\Delta}^{\text{op}}}{\times} \{[0]\} \simeq \text{Seq}_0(\mathbb{S}).$$

The category $\text{RLax}_{\text{non-untl}}(\mathbb{S})^{\oint} \underset{\mathbf{\Delta}^{\text{op}}}{\times} \{[1]\}$ is described as follows. It is a co-Cartesian fibration over $\mathbf{\Delta}^{\text{op}}_{\text{actv}}$ (where $\mathbf{\Delta}_{\text{actv}}$ is the 1-full subcategory of $\mathbf{\Delta}$ where we restrict the arrows to active morphisms). We have

$$\{[n]\} \underset{\mathbf{\Delta}_{\text{actv}}}{\times} \left(\text{RLax}_{\text{non-untl}}(\mathbb{S})^{\oint} \underset{\mathbf{\Delta}^{\text{op}}}{\times} \{[1]\}\right) \simeq \text{Seq}_n(\mathbb{S}).$$

For an active map $\alpha : [m] \to [n]$ the corresponding functor between the fibers identifies with the functor

$$\text{Seq}_n(\mathbb{S}) \to \text{Seq}_m(\mathbb{S}),$$

induced by α.

A.1.4. The projection $[1] \to [0]$ defines a functor $\mathbf{\Delta}^{\text{op}} \to \text{Actv}$, which in turn gives rise to a functor

$$\iota^{\oint}_{\mathbb{S}} : \mathbb{S}^{\oint} \to \text{RLax}_{\text{non-untl}}(\mathbb{S})^{\oint} := \mathbb{S}^{\oint} \underset{\mathbf{\Delta}^{\text{op}}}{\times} \text{Actv},$$

compatible with projections to $\mathbf{\Delta}^{\text{op}}$.

We will prove:

THEOREM A.1.5.
(i) *The functor* $\mathrm{RLax}_{\text{non-untl}}(\mathbb{S})^{\oint} \to \mathbf{\Delta}^{\mathrm{op}}$ *of* (A.1) *is a coCartesian fibration, and the resulting functor* $\mathbf{\Delta}^{\mathrm{op}} \to 1\text{-Cat}$ *lies in the essential image of the functor* $\mathrm{Seq}_\bullet$*; denote the resulting* $(\infty,2)$*-category by* $\mathrm{RLax}_{\text{non-untl}}(\mathbb{S})$.

(ii) *The functor* $\iota_{\mathbb{S}}^{\oint}$ *sends coCartesian arrows over inert morphisms in* $\mathbf{\Delta}^{\mathrm{op}}$ *to coCartesian arrows. Denote the resulting lax functor* $\mathbb{S} \dashrightarrow \mathrm{RLax}_{\text{non-untl}}(\mathbb{S})$ *by* $\iota_{\mathbb{S}}$.

(iii) *For any* $\mathbb{T} \in 2\text{-Cat}$*, the composite map*

$$\mathrm{Maps}_{2\text{-Cat}}(\mathrm{RLax}_{\text{non-untl}}(\mathbb{S}),\mathbb{T}) \to \mathrm{Maps}_{2\text{-Cat}_{\text{right-lax}_{\text{non-untl}}}}(\mathrm{RLax}_{\text{non-untl}}(\mathbb{S}),\mathbb{T}) \to \\ \to \mathrm{Maps}_{2\text{-Cat}_{\text{right-lax}_{\text{non-untl}}}}(\mathbb{S},\mathbb{T}),$$

where the second arrow is given by precomposition with $\iota_{\mathbb{S}}$*, is an isomorphism.*

A.1.6. Note that the functor

$$\iota_{\mathbb{S}}^{\oint} : \mathbb{S}^{\oint} \to \mathrm{RLax}_{\text{non-untl}}(\mathbb{S})^{\oint}$$

admits a left adjoint; to be denoted $\lambda_{\mathbb{S}}^{\oint}$. This is a functor between categories over $\mathbf{\Delta}^{\mathrm{op}}$ that sends coCartesian edges to coCartesian edges.

For example, the corresponding functor

$$\mathrm{RLax}_{\text{non-untl}}(\mathbb{S})^{\oint} \underset{\mathbf{\Delta}^{\mathrm{op}}}{\times} [1] \to \mathrm{Seq}_1(\mathbb{S})$$

is given, in terms of the description in Sect. A.1.3 by the compatible family of functors

$$\mathrm{Seq}_n(\mathbb{S}) \to \mathrm{Seq}_1(\mathbb{S}),$$

each corresponding to the unique active map $[1] \to [n]$.

Hence, we obtain that the functor $\lambda_{\mathbb{S}}^{\oint}$ corresponds to a functor

$$\lambda_{\mathbb{S}} : \mathrm{RLax}_{\text{non-untl}}(\mathbb{S}) \to \mathbb{S}.$$

We claim:

PROPOSITION A.1.7. *The functor* $\lambda_{\mathbb{S}} : \mathrm{RLax}_{\text{non-untl}}(\mathbb{S}) \to \mathbb{S}$ *is the counit of the adjunction, i.e., corresponds to the identity functor on* $\mathbb{S}$*, considered as a non-unital right-lax functor.*

PROOF. We need to show that the composite lax functor

$$\mathbb{S} \overset{\iota_{\mathbb{S}}}{\dashrightarrow} \mathrm{RLax}_{\text{non-untl}}(\mathbb{S}) \overset{\lambda_{\mathbb{S}}}{\to} \mathbb{S}$$

identifies with the identity functor on $\mathbb{S}$.

For that we need to show that the composite functor

$$\lambda_{\mathbb{S}}^{\oint} \circ \iota_{\mathbb{S}}^{\oint} : \mathbb{S}^{\oint} \to \mathbb{S}^{\oint}$$

is the identity functor. But this follows from the fact that the functor $\iota_{\mathbb{S}}^{\oint}$ is fully faithful.

□

A.2. Proof of Theorem A.1.5.

A.2.1. To prove point (i) of the theorem, let us explicitly describe the functor

$$\mathbf{\Delta}^{\mathrm{op}} \to 1\text{-Cat}$$

corresponding to the projection

$$\mathrm{RLax}_{\text{non-untl}}(\mathbb{S})^{\oint} \to \mathbf{\Delta}^{\mathrm{op}}.$$

Namely, this functor sends $[m]$ to a coCartesian fibration over $((\mathbf{\Delta}_{\mathrm{actv}})_{[m]/})^{\mathrm{op}}$, whose fiber over an active map $\gamma : [m] \to [n]$ is

$$\{\gamma\} \underset{((\mathbf{\Delta}_{\mathrm{actv}})_{[m]/})^{\mathrm{op}}}{\times} \left(\mathrm{RLax}_{\text{non-untl}}(\mathbb{S})^{\oint} \underset{\mathbf{\Delta}^{\mathrm{op}}}{\times} \{[m]\}\right) = \mathrm{Seq}_n(\mathbb{S}),$$

and where for active map $\alpha : [n_1] \to [n_2]$ the corresponding functor

$$\mathrm{Seq}_{n_2}(\mathbb{S}) \to \mathrm{Seq}_{n_1}(\mathbb{S})$$

is induced by α.

For a map $\beta : [m_1] \to [m_2]$, the corresponding functor

$$\mathrm{RLax}_{\text{non-untl}}(\mathbb{S})^{\oint} \underset{\mathbf{\Delta}^{\mathrm{op}}}{\times} \{[m_2]\} \to \mathrm{RLax}_{\text{non-untl}}(\mathbb{S})^{\oint} \underset{\mathbf{\Delta}^{\mathrm{op}}}{\times} \{[m_1]\} \tag{A.2}$$

is described as follows.

For an active map $\gamma_2 : [m_2] \to [n_2]$, the category of factorizations of $\gamma_2 \circ \beta$ as

$$[m_1] \overset{\gamma_1'}{\to} [n_1'] \overset{\alpha'}{\to} [n_2]$$

has a final object

$$[m_1] \overset{\gamma_1}{\to} [n_1] \overset{\alpha}{\to} [n_2]$$

In fact, α is the injection of the sub-segment with the smallest element $\gamma_2 \circ \beta(0)$ and the largest element $\gamma_2 \circ \beta(m_1)$.

The corresponding functor in (A.2) sends

$$\{\gamma_2\} \underset{((\mathbf{\Delta}_{\mathrm{actv}})_{[m_2]/})^{\mathrm{op}}}{\times} \left(\mathrm{RLax}_{\text{non-untl}}(\mathbb{S})^{\oint} \underset{\mathbf{\Delta}^{\mathrm{op}}}{\times} \{[m_2]\}\right) \to$$
$$\to \{\gamma_1\} \underset{((\mathbf{\Delta}_{\mathrm{actv}})_{[m_1]/})^{\mathrm{op}}}{\times} \left(\mathrm{RLax}_{\text{non-untl}}(\mathbb{S})^{\oint} \underset{\mathbf{\Delta}^{\mathrm{op}}}{\times} \{[m_1]\}\right)$$

and equals the functor

$$\mathrm{Seq}_{n_2}(\mathbb{S}) \to \mathrm{Seq}_{n_1}(\mathbb{S})$$

is induced by α.

The verification of Conditions (0)-(2) for being an $(\infty, 2)$-category is now straightforward. It is equally easy to see that the functor $\iota_{\mathbb{S}}^{\oint}$ sends coCartesian arrows over inert arrows in $\mathbf{\Delta}$ to coCartesian arrows.

A.2.2. Let us now show that the map

$$\mathrm{Maps}_{2\text{-Cat}}(\mathrm{RLax}_{\text{non-untl}}(\mathbb{S}),\mathbb{T}) \to \mathrm{Maps}_{2\text{-Cat}_{\text{right-lax}_{\text{non-untl}}}}(\mathbb{S},\mathbb{T})$$

is an isomorphism.

Given $\mathbb{T} \in 2\text{-Cat}$, the operation of relative left Kan extension along $\iota_{\mathbb{S}}^{\oint}$ gives rise to a fully faithful embedding of spaces

$$\mathrm{Maps}_{1\text{-Cat}_{/\mathbf{\Delta}^{\mathrm{op}}}}(\mathbb{S}^{\oint},\mathbb{T}^{\oint}) \to \mathrm{Maps}_{1\text{-Cat}_{/\mathbf{\Delta}^{\mathrm{op}}}}(\mathrm{RLax}_{\text{non-untl}}(\mathbb{S})^{\oint},\mathbb{T}^{\oint}). \tag{A.3}$$

Let

$$\mathrm{Maps}'_{1\text{-Cat}_{/\mathbf{\Delta}^{\mathrm{op}}}}(\mathbb{S}^{\oint},\mathbb{T}^{\oint}) \subset \mathrm{Maps}_{1\text{-Cat}_{/\mathbf{\Delta}^{\mathrm{op}}}}(\mathbb{S}^{\oint},\mathbb{T}^{\oint})$$

be the subspace consisting of functors that send coCartesian arrows over inert morphisms in $\mathbf{\Delta}^{\mathrm{op}}$ to coCartesian morphisms. Let

$$\mathrm{Maps}'_{1\text{-Cat}_{/\mathbf{\Delta}^{\mathrm{op}}}}(\mathrm{RLax}_{\text{non-untl}}(\mathbb{S})^{\oint},\mathbb{T}^{\oint}) \subset \mathrm{Maps}_{1\text{-Cat}_{/\mathbf{\Delta}^{\mathrm{op}}}}(\mathrm{RLax}_{\text{non-untl}}(\mathbb{S})^{\oint},\mathbb{T}^{\oint})$$

be the subspace consisting of functors that send all coCartesian arrows to coCartesian morphisms. We will show that the map (A.3) defines an isomorphism

$$\mathrm{Maps}'_{1\text{-Cat}_{/\mathbf{\Delta}^{\mathrm{op}}}}(\mathbb{S}^{\oint},\mathbb{T}^{\oint}) \to \mathrm{Maps}'_{1\text{-Cat}_{/\mathbf{\Delta}^{\mathrm{op}}}}(\mathrm{RLax}_{\text{non-untl}}(\mathbb{S})^{\oint},\mathbb{T}^{\oint}). \tag{A.4}$$

A.2.3. Note that the functor

$$\iota_{\mathbb{S}}^{\oint} : \mathbb{S}^{\oint} \to \mathrm{RLax}_{\text{non-untl}}(\mathbb{S})^{\oint}$$

admits a right adjoint, to be denoted $\rho_{\mathbb{S}}^{\oint}$. Explicitly, for every m and $\gamma : [m] \to [n]$, the functor $\rho_{\mathbb{S}}^{\oint}$ makes the following diagram commutative

$$\begin{array}{ccc}
\{\gamma\} \underset{((\mathbf{\Delta}_{\mathrm{actv}})_{[m]/})^{\mathrm{op}}}{\times} \left(\mathrm{RLax}_{\text{non-untl}}(\mathbb{S})^{\oint} \underset{\mathbf{\Delta}^{\mathrm{op}}}{\times} \{[m]\}\right) & \longrightarrow & \mathrm{RLax}_{\text{non-untl}}(\mathbb{S})^{\oint} \\
\sim\downarrow & & \downarrow \rho_{\mathbb{S}}^{\oint} \\
\mathrm{Seq}_n(\mathbb{S}) & \longrightarrow & \mathbb{S}^{\oint}.
\end{array}$$

In particular, we note that $\rho_{\mathbb{S}}^{\oint}$ does *not* respect the projections

$$\mathrm{RLax}_{\text{non-untl}}(\mathbb{S})^{\oint} \to \mathbf{\Delta}^{\mathrm{op}} \text{ and } \mathbb{S}^{\oint} \to \mathbf{\Delta}^{\mathrm{op}}.$$

We have the following general assertion:

LEMMA A.2.4. *Suppose we have a diagram of $(\infty,1)$-categories*

$$\begin{array}{ccc}
\mathbf{C}' & \xrightarrow{\iota} & \mathbf{C} \\
& \searrow \quad \swarrow & \\
& \mathbf{I} &
\end{array}$$

such that ι is fully faithful and admits a right adjoint ρ. Then for any coCartesian fibration $\mathbf{D} \to \mathbf{I}$, relative left Kan extension gives a fully faithful embedding

$$\mathrm{Maps}_{1\text{-Cat}_{/\mathbf{I}}}(\mathbf{C}',\mathbf{D}) \hookrightarrow \mathrm{Maps}_{1\text{-Cat}_{/\mathbf{I}}}(\mathbf{C},\mathbf{D})$$

with the image consisting of functors $F : \mathbf{C} \to \mathbf{D}$ over $\mathbf{I}$ such that for every $c \in \mathbf{C}$, the counit of the adjunction $\iota \circ \rho(c) \to c$ induces the arrow

$$F(\iota \circ \rho(c)) \to F(c)$$

in $\mathbf{D}$ that is coCartesian over $\mathbf{I}$.

A.2.5. Applying this lemma, we need to show that for a functor

$$F : \mathrm{RLax}_{\text{non-untl}}(\mathbb{S})^{\oint} \to \mathbb{T}^{\oint}$$

the following conditions are equivalent:

(1) F takes coCartesian arrows to coCartesian arrows;

(2) F takes the arrows coming from the counit of the adjunction $\iota_{\mathbb{S}}^{\oint} \circ \rho_{\mathbb{S}}^{\oint} \to \mathrm{id}$ and also arrows of the form $\iota_{\mathbb{S}}^{\oint}(f)$, where f is a coCartesian arrow in $\mathbb{S}^{\oint}$ lying over an inert map in $\mathbf{\Delta}^{\mathrm{op}}$, to coCartesian arrows.

We have the following general observation:

LEMMA A.2.6. *Let $\mathbf{D} \to \mathbf{I}$ be a coCartesian fibration of $(\infty,1)$-categories. Then an arrow in $\mathbf{D}$ is coCartesian over $\mathbf{I}$ if and only if its image in $\mathbf{D}^{\mathrm{ordn}}$ is coCartesian over $\mathbf{I}^{\mathrm{ordn}}$.*

This lemma allows to replace the verification of the equivalence of conditions (1) and (2) above to the case when $\mathbb{T}$ (and hence also $\mathbb{S}$) is an ordinary 2-category. In this case the assertion is straightforward.

A.3. Quasi-invertible 1-morphisms.

A.3.1. Since

$$\mathrm{Seq}_0(\mathrm{RLax}_{\text{non-untl}}(\mathbb{S})) \simeq \mathrm{Seq}_0(\mathbb{S}),$$

the categories $\mathbb{S}_0$ and $\mathrm{RLax}_{\text{non-untl}}(\mathbb{S})$ have the same spaces of objects.

Note that the subcategory

$$(\mathrm{Seq}_1(\mathrm{RLax}_{\text{non-untl}}(\mathbb{S})))^{\mathrm{invert}} \subset \mathrm{Seq}_1(\mathrm{RLax}_{\text{non-untl}}(\mathbb{S}))$$

identifies with

$$\mathrm{Seq}_0(\mathbb{S}) \simeq \{[0]\} \underset{\mathbf{\Delta}_{\mathrm{actv}}}{\times} \left(\mathrm{RLax}_{\text{non-untl}}(\mathbb{S})^{\oint} \underset{\mathbf{\Delta}^{\mathrm{op}}}{\times} \{[1]\}\right) \subset$$
$$\subset \mathrm{RLax}_{\text{non-untl}}(\mathbb{S})^{\oint} \underset{\mathbf{\Delta}^{\mathrm{op}}}{\times} \{[1]\} = \mathrm{Seq}_1(\mathrm{RLax}_{\text{non-untl}}(\mathbb{S})).$$

A.3.2. We shall say that a 1-morphism is *quasi-invertible* if it belongs to the full subcategory, to be denoted $(\mathrm{Seq}_1(\mathrm{RLax}_{\text{non-untl}}(\mathbb{S})))^{\text{q-invert}}$, and equal to

$$\mathrm{Seq}_0(\mathbb{S}) \simeq (\mathrm{Seq}_1(\mathbb{S}))^{\mathrm{invert}} \subset \mathrm{Seq}_1(\mathbb{S}) \simeq \{[1]\} \underset{\mathbf{\Delta}_{\mathrm{actv}}}{\times} \left(\mathrm{RLax}_{\text{non-untl}}(\mathbb{S})^{\oint} \underset{\mathbf{\Delta}^{\mathrm{op}}}{\times} \{[1]\}\right) \subset$$
$$\subset \mathrm{RLax}_{\text{non-untl}}(\mathbb{S})^{\oint} \underset{\mathbf{\Delta}^{\mathrm{op}}}{\times} \{[1]\} = \mathrm{Seq}_1(\mathrm{RLax}_{\text{non-untl}}(\mathbb{S})).$$

REMARK A.3.3. Note that we thus obtain two *different* fully faithful functors

$$\mathrm{Seq}_0(\mathbb{S}) \simeq (\mathrm{Seq}_1(\mathrm{RLax}_{\text{non-untl}}(\mathbb{S})))^{\mathrm{invert}} \hookrightarrow \mathrm{Seq}_1(\mathrm{RLax}_{\text{non-untl}}(\mathbb{S}))$$

and

$$\mathrm{Seq}_0(\mathbb{S}) \simeq (\mathrm{Seq}_1(\mathrm{RLax}_{\text{non-untl}}(\mathbb{S})))^{\text{q-invert}} \hookrightarrow \mathrm{Seq}_1(\mathrm{RLax}_{\text{non-untl}}(\mathbb{S})).$$

By construction, these functors are connected by a natural transformation (from the former to the latter).

A.3.4. We observe:

LEMMA A.3.5. *A non-unital right-lax functor* $\mathbb{S} \dashrightarrow \mathbb{T}$ *is unital if and only if the corresponding functor*

$$\mathrm{RLax}_{\text{non-untl}}(\mathbb{S}) \to \mathbb{T}$$

sends quasi-invertible 1-morphisms to isomorphisms.

B. Localizations on 1-morphisms

B.1. The notion of localization on 1-morphisms.

B.1.1. Let $\mathbf{C}$ be an $(\infty, 1)$-category, and let $\mathbf{C}' \subset \mathbf{C}$ be a 1-full subcategory with the same class of objects. (I.e., the datum of $\mathbf{C}$ amounts to specifying a class of 1-morphisms containing all isomorphisms and closed under compositions).

Recall that the localization of $\mathbf{C}$ with respect to $\mathbf{C}'$ is a pair

$$(\mathbf{C}, F_{\text{can}} : \mathbf{C} \to \widetilde{\mathbf{C}}_{\text{can}}),$$

universal with respect to functors $F : \mathbf{C} \to \widetilde{\mathbf{C}}$ that map 1-morphisms from $\mathbf{C}'$ to isomorphisms.

B.1.2. Let $F : \mathbb{S} \to \mathbb{T}$ be a functor between $(\infty, 2)$-categories.

DEFINITION B.1.3. *We shall say that* F *is a* localization on 1-morphisms *if:*

(1) *The functor* $\mathrm{Seq}_0(\mathbb{S}) \to \mathrm{Seq}(\mathbb{T})_0$ *is an isomorphism (in* Spc*);*

(2) *The functor* $\mathrm{Seq}_1(\mathbb{S}) \to \mathrm{Seq}(\mathbb{T})_1$ *is a localization.*

B.1.4. We claim:

PROPOSITION B.1.5. *For a functor* $F : \mathbb{S} \to \mathbb{T}$*, the following are equivalent:*

(1) F *is a localization on 1-morphisms;*

(2) *The corresponding functor* $\mathbb{S}^{\oint} \to \mathbb{T}^{\oint}$ *is a localization.*

PROOF. Follows from the next general lemma:

LEMMA B.1.6. *Let* $\mathbf{C} \to \mathbf{I}$ *and* $\mathbf{D} \to \mathbf{I}$ *be coCartesian fibrations, and let* $F : \mathbf{C} \to \mathbf{D}$ *be a functor compatible with the projections to* $\mathbf{I}$ *such that* F *sends coCartesian arrows to coCartesian arrows. Then* F *is a localization if and only if for every* $\mathbf{i} \in \mathbf{I}$ *the corresponding functor* $\mathbf{C} \underset{\mathbf{I}}{\times} \{\mathbf{i}\} \to \mathbf{D} \underset{\mathbf{I}}{\times} \{\mathbf{i}\}$ *is a localization.*

□

As a corollary, we obtain:

COROLLARY B.1.7. *Let* $\mathbb{S} \to \mathbb{T}$ *be a localization on 1-morphisms. Then for any* $\mathbb{X} \in 2\text{-Cat}$*, the maps*

$$\mathrm{Maps}_{2\text{-Cat}}(\mathbb{T}, \mathbb{X}) \to \mathrm{Maps}_{2\text{-Cat}}(\mathbb{S}, \mathbb{X}),$$

$$\mathrm{Maps}_{2\text{-Cat}_{\text{right-lax}}}(\mathbb{T}, \mathbb{X}) \to \mathrm{Maps}_{2\text{-Cat}_{\text{right-lax}}}(\mathbb{S}, \mathbb{X})$$

and

$$\mathrm{Maps}_{2\text{-Cat}_{\text{right-lax}_{\text{non-untl}}}}(\mathbb{T}, \mathbb{X}) \to \mathrm{Maps}_{2\text{-Cat}_{\text{right-lax}_{\text{non-untl}}}}(\mathbb{S}, \mathbb{X})$$

are fully faithful.

B.1.8. It is easy to see that if $\mathbb{S} \to \mathbb{T}$ is a localization on 1-morphisms, then for any $\mathbb{X} \in 2\text{-Cat}$, so is the functor

$$\mathbb{S} \times \mathbb{X} \to \mathbb{S} \times \mathbb{T}.$$

From here we obtain:

COROLLARY B.1.9. *Let $\mathbb{S} \to \mathbb{T}$ be a localization on 1-morphisms. Then for any $\mathbb{X} \in 2\text{-Cat}$, the functor*

$$\mathrm{Funct}(\mathbb{T}, \mathbb{X}) \to \mathrm{Funct}(\mathbb{S}, \mathbb{X})$$

is fully faithful.

B.2. Description of localizations.

B.2.1. We have:

PROPOSITION B.2.2. *Let $\mathbb{S}$ be an $(\infty, 2)$-category. The following pieces of data are equivalent:*

(i) *The datum of a functor $\mathbb{S} \to \mathbb{T}$, which is a localization on 1-morphisms.*

(ii) *The datum of a functor $\mathbb{S}^{\mathrm{ordn}} \to \mathbb{T}^{\mathrm{ordn}}$, which is a localization on 1-morphisms.*

(iii) *The datum of a subset of isomorphism classes of morphisms in* $\mathrm{Seq}_1(\mathbb{S})$ *that contains all isomorphisms and is closed under the composition operation*

$$\pi_0(\mathrm{Seq}_1(\mathbb{S})) \underset{\pi_0(\mathrm{Sq}(\mathbb{S})_0)}{\times} \pi_0(\mathrm{Seq}_1(\mathbb{S})) \to \pi_0(\mathrm{Seq}_1(\mathbb{S})).$$

PROOF. Follows from the next general lemma:

LEMMA B.2.3. *Let $\mathbf{C} \to \mathbf{I}$ be a coCartesian fibration. Then the datum of a localization $F : \mathbf{C} \to \mathbf{D}$, such that $\mathbf{D}$ is also a coCartesian fibration over $\mathbf{I}$ and F sends coCartesian arrows to coCartesian arrows is equivalent to the datum of a localization $\mathbf{C} \underset{\mathbf{I}}{\times} \{\mathbf{i}\} \to \mathbf{D}_{\mathbf{i}}$ for each $\mathbf{i} \in \mathbf{I}$, such that for every 1-morphism $\mathbf{i}_1 \to \mathbf{i}_2$ in $\mathbf{I}$ the corresponding functor*

$$\mathbf{C} \underset{\mathbf{I}}{\times} \{\mathbf{i}_1\} \to \mathbf{C} \underset{\mathbf{I}}{\times} \{\mathbf{i}_2\}$$

sends the 1-morphisms that become isomorphisms on $\mathbf{D}_{\mathbf{i}_1}$ to 1-morphisms that become isomorphisms on $\mathbf{D}_{\mathbf{i}_2}$.

□

B.2.4. As a corollary we obtain:

COROLLARY B.2.5. *For $\mathbb{S} \in 2\text{-Cat}$, the canonical functors*

$$\lambda_{\mathbb{S}} : \mathrm{RLax}_{\text{non-untl}}(\mathbb{S}) \to \mathbb{S},$$

$$\mathrm{RLax}_{\text{non-untl}}(\mathbb{S} \times \mathbb{T}) \to \mathbb{S} \circledast \mathbb{T} \text{ and } \mathbb{S} \circledast \mathbb{T} \to \mathbb{S} \times \mathbb{T}$$

are localizations on 1-morphisms.

CHAPTER 12

Adjunctions in $(\infty, 2)$-categories

Introduction

0.1. What is done in this Chapter? This Chapter contains the only piece of original mathematics pertaining to $(\infty, 2)$-categories that we develop in this book. It has to do with pairs of functors obtained from one another by the procedure of *passage adjoints.*

0.1.1. First, we note that if $\mathbb{T}$ is an ordinary 2-category, and $t \overset{\alpha}{\to} t'$ is a 1-morphism, there exists an (elementary) notion of right (resp., left) adjoint 1-morphism. For example, if $\mathbb{T} = (\mathbf{1}\text{-}\mathbf{Cat})^{\text{ordn}}$, this is the usual notion of adjunction for functors between $(\infty, 1)$-categories.

If $\mathbb{T}$ is a $(\infty, 2)$-category, we will say that a 1-morphism admits a right (resp., left) adjoint, if it does so when considered as a 1-morphism in $\mathbb{T}^{\text{ordn}}$.

0.1.2. Consider the following situation: let $\mathbb{S}$ and $\mathbb{T}$ be $(\infty, 2)$-categories, and let $F : \mathbb{S} \to \mathbb{T}$.

We shall say that F is right-adjointable (resp., left-adjointable) if for every 1-morphism $s \overset{\alpha}{\to} s'$ in $\mathbb{S}$, the corresponding 1-morphism $F(\alpha)$ admits a right (resp., left) adjoint.

We let

$$\text{Maps}_{2\text{-Cat}}(\mathbb{S}, \mathbb{T})^R \subset \text{Maps}_{2\text{-Cat}}(\mathbb{S}, \mathbb{T})$$

denote the full subspace spanned by left-adjointable functors (the superscript "R" is because 1-morphisms generated by these these functors are right adjoints).

Let

$$\text{Maps}_{2\text{-Cat}}(\mathbb{S}, \mathbb{T})^L \subset \text{Maps}_{2\text{-Cat}}(\mathbb{S}, \mathbb{T})$$

denote the full subspace spanned by right-adjointable functors.

0.1.3. Assume for a moment that $\mathbb{T}$ is an ordinary 2-category. In this case, the procedure of *passage to adjoint 1-morphisms* defines a canonical isomorphism

$$\text{Maps}_{2\text{-Cat}}(\mathbb{S}, \mathbb{T})^R \simeq \text{Maps}_{2\text{-Cat}}(\mathbb{S}^{1\&2\text{-op}}, \mathbb{T})^L. \tag{0.1}$$

The main result of this Chapter (stated in Corollary 1.3.4) is that the isomorphism (0.1) holds for an arbitrary $(\infty, 2)$-category.

0.1.4. Moreover, we note that the spaces

$$\mathrm{Maps}_{2\text{-Cat}}(\mathbb{S}, \mathbb{T})^R$$

and

$$\mathrm{Maps}_{2\text{-Cat}}(\mathbb{S}^{1\&2\text{-op}}, \mathbb{T})^L$$

can naturally be extended to $(\infty, 2)$-categories

$$\mathrm{Funct}(\mathbb{S}, \mathbb{T})^R_{\text{right-lax}} \text{ and } \mathrm{Funct}(\mathbb{S}^{1\&2\text{-op}}, \mathbb{T})^L_{\text{left-lax}},$$

where we allow right-lax (resp., left-lax) natural transformations.

We will show (see Corollary 3.1.9) that the isomorphism (0.1) extends to an equivalence of $(\infty, 2)$-categories

$$\mathrm{Funct}(\mathbb{S}, \mathbb{T})^R_{\text{right-lax}} \simeq \mathrm{Funct}(\mathbb{S}^{1\&2\text{-op}}, \mathbb{T})^L_{\text{left-lax}}.$$

0.2. How is this done? To simplify the discussion, we will focus on the construction of the isomorphism of spaces (0.1).

0.2.1. It is not difficult to see that for a given $(\infty, 2)$-category $\mathbb{S}$, there exists a *universal left-adjointable functor*

$$F_{\text{univ}} : \mathbb{S} \to \mathbb{S}^R.$$

I.e., any left-adjointable functor $F : \mathbb{S} \to \mathbb{T}$ uniquely factors as

$$G \circ F_{\text{univ}}, \quad G : \mathbb{S}^R \to \mathbb{T}.$$

Similarly, we have the universal right-adjointable functor $\mathbb{S} \to \mathbb{S}^L$.

0.2.2. The isomorphism (0.1), functorial in $\mathbb{T}$, is equivalent to the existence of a canonical equivalence of $(\infty, 2)$-categories

$$(\mathbb{S}^{2\text{-op}})^R \simeq (\mathbb{S}^{1\text{-op}})^L. \tag{0.2}$$

The construction of the equivalence (0.2) is based on the explicit description of the $(\infty, 2)$-category $\mathbb{S}^R$ (resp., $\mathbb{S}^L$). Such a description is given by Theorem 1.2.4 and is the key idea of this Chapter.

0.2.3. Namely, we prove that the $(\infty, 2)$-category $\mathbb{S}^R$ is obtained by applying the functor $\mathfrak{L}^{\mathrm{Sq}}$, left adjoint to

$$\mathrm{Sq}_{\bullet,\bullet} : 2\text{-Cat} \to \mathrm{Spc}^{\mathbf{\Delta}^{\mathrm{op}} \times \mathbf{\Delta}^{\mathrm{op}}},$$

to a particular object of $\mathrm{Spc}^{\mathbf{\Delta}^{\mathrm{op}} \times \mathbf{\Delta}^{\mathrm{op}}}$.

That object of $\mathrm{Spc}^{\mathbf{\Delta}^{\mathrm{op}} \times \mathbf{\Delta}^{\mathrm{op}}}$ is the following: we consider

$$\mathrm{Sq}_{\bullet,\bullet}(\mathbb{S}^{2\text{-op}}),$$

and then we *invert the vertical arrows*, i.e., apply the involution

$$\mathrm{rev} : \mathbf{\Delta}^{\mathrm{op}} \to \mathbf{\Delta}^{\mathrm{op}}$$

along the *first factor* in $\mathbf{\Delta}^{\mathrm{op}} \times \mathbf{\Delta}^{\mathrm{op}}$.

We note that this object *does not* belong to the essential image of the functor $\mathrm{Sq}_{\bullet,\bullet}$, so the above procedure is doing something non-trivial.

0.2.4. By unwinding the construction, we see that the procedures for obtaining $(\mathbb{S}^{2\text{-op}})^R$ and $(\mathbb{S}^{1\text{-op}})^L$ are exactly the same, thereby leading to the isomorphism (0.2).

1. Adjunctions

In this section we study the following situation: let $\mathbb{S}$ be a $(\infty, 2)$-category equipped with a 1-full subcategory $\mathbf{C} \subset \mathbb{S}^{1\text{-Cat}}$ with the same class of objects. We will consider functors $F : \mathbb{S} \to \mathbb{T}$ such that for every 1-morphism α in $\mathbf{C}$, the corresponding 1-morphism $F(\alpha)$ in $\mathbb{T}$ *admits a left adjoint*. We will state a theorem to the effect that there exists a universal such functor

$$F_{\text{univ}} : \mathbb{S} \to \mathbb{S}^{R_{\mathbf{C}}},$$

i.e., any functor $F : \mathbb{S} \to \mathbb{T}$ with the above property uniquely factors as

$$G \circ F_{\text{univ}}, \quad G : \mathbb{S}^{R_{\mathbf{C}}} \to \mathbb{T}.$$

However more importantly, the $(\infty, 2)$-category $\mathbb{S}^{R_{\mathbf{C}}}$ can be described explicitly.

This explicit description will allow us to show that to a functor $F : \mathbb{S} \to \mathbb{T}$ that maps *all* 1-morphisms in $\mathbb{T}$ to left-adjointable 1-morphisms in $\mathbb{T}$, there canonically corresponds a functor

$$\mathbb{S}^{1\&2\text{-op}} \to \mathbb{T}$$

obtained from F by *passing to left adjoints* along 1-morphisms.

1.1. Adjointable arrows and functors. In this subsection we will define what it means for a 1-morphism in an $(\infty, 2)$-category to admit an adjoint, and the related notion of a functor to be *adjointable*.

The main feature of these notions is that they *do not* depend on the ∞-categorical structure, i.e., these are conditions on arrows/functors between the underlying *ordinary* 2-categories.

1.1.1. Let $\mathbb{T}$ be an *ordinary* 2-category, and let $t \overset{\alpha}{\to} t'$ be a 1-morphism. In this case there exists the notion of *right adjoint 1-morphism*.

Namely, a 1-morphism $t' \overset{\beta}{\to} t$ is said to be the *right adjoint* of α if we are given 2-morphisms

$$\text{co-unit} : \alpha \circ \beta \to \text{id}_{t'} \text{ and unit} : \text{id}_t \to \beta \circ \alpha$$

such that the compositions

$$\alpha \overset{\alpha \circ \text{unit}}{\longrightarrow} \alpha \circ \beta \circ \alpha \overset{\text{co-unit} \circ \alpha}{\longrightarrow} \alpha \text{ and } \beta \overset{\text{unit} \circ \beta}{\longrightarrow} \beta \circ \alpha \circ \beta \overset{\beta \circ \text{co-unit}}{\longrightarrow} \beta$$

are the identity 2-morphisms.

It is easy to see that if a right adjoint 1-morphism exists, it is defined up to a unique isomorphism.

1.1.2. Replacing $\mathbb{T}$ by $\mathbb{T}^{2\text{-op}}$, we obtain the notion of *left adjoint* 1-morphism.

It is easy to see that the data defining β as a right adjoint of α is equivalent to the data defining α as a left adjoint of β.

1.1.3. Let now $\mathbb{T}$ be an $(\infty,2)$-category, and let $t \overset{\alpha}{\to} t'$ be a 1-morphism.

DEFINITION 1.1.4. *We shall say that α admits a right (resp., left) adjoint, if it does so in the ordinary 2-category $\mathbb{T}^{\text{ordn}}$.*

Let $\text{Seq}_1(\mathbb{T})^R \subset \text{Seq}_1(\mathbb{T})$ be the full subcategory spanned by 1-morphisms that admit a *left adjoint*[1].

Let $\text{Seq}_1(\mathbb{T})^L \subset \text{Seq}_1(\mathbb{T})$ be the full subcategory spanned by 1-morphisms that admit a *right adjoint.*

The procedure of passage to the adjoint 1-morphism defines an equivalence

$$(\text{Seq}_1(\mathbb{T})^R)^{\text{ordn}} \simeq \big(((\text{Seq}_1(\mathbb{T})^L)^{\text{ordn}})\big)^{\text{op}}.$$

REMARK 1.1.5. In Corollary 3.1.9 we will see that the above equivalence of ordinary categories in fact lifts to an equivalence of $(\infty,1)$-categories

$$\text{Seq}_1(\mathbb{T})^R \simeq \big((\text{Seq}_1(\mathbb{T})^L\big)^{\text{op}}.$$

1.1.6. Let $\mathbb{S}$ be an $(\infty,2)$-category, and $\mathbf{C} \subset \mathbb{S}^{1\text{-Cat}}$ be a 1-full subcategory with the same class of objects. Let

$$F : \mathbb{S} \to \mathbb{T}$$

be a functor, where $\mathbb{T}$ is another $(\infty,2)$-category.

DEFINITION 1.1.7. *We shall say that F is* right adjointable with respect to $\mathbf{C}$, *if for every 1-morphism $s \overset{\alpha}{\to} s'$ in $\mathbf{C}$, the 1-morphism*

$$F(s) \overset{F(\alpha)}{\longrightarrow} F(s')$$

admits a right adjoint.

In a similar way we define the notion of functor *left* adjointable with respect to $\mathbf{C}$.

1.1.8. We denote by

$$\text{Funct}(\mathbb{S},\mathbb{T})^{R_{\mathbf{C}}}_{\text{right-lax}} \subset \text{Funct}(\mathbb{S},\mathbb{T})_{\text{right-lax}}$$

the full subcategory corresponding to functors that are *left* adjointable with respect to $\mathbf{C}$.

Let

$$\text{Funct}(\mathbb{S},\mathbb{T})^{L_{\mathbf{C}}}_{\text{left-lax}} \subset \text{Funct}(\mathbb{S},\mathbb{T})_{\text{right-lax}}$$

denote the full subcategory corresponding to functors *right* adjointable with respect to $\mathbf{C}$.

Let

$$\text{Maps}_{2\text{-Cat}}(\mathbb{S},\mathbb{T})^{R_{\mathbf{C}}} \subset \text{Maps}_{2\text{-Cat}}(\mathbb{S},\mathbb{T}) \supset \text{Maps}_{2\text{-Cat}}(\mathbb{S},\mathbb{T})^{L_{\mathbf{C}}}$$

the be the corresponding full subspaces.

Clearly, under the isomorphism

$$\text{Funct}(\mathbb{S},\mathbb{T})_{\text{left-lax}} \simeq \text{Funct}(\mathbb{S}^{2\text{-op}},\mathbb{T}^{2\text{-op}})_{\text{right-lax}}$$

---

[1]The superscript "R" means the 1-morphisms in question are themselves right adjoint to something.

we have:

$$\mathrm{Funct}(\mathbb{S},\mathbb{T})^{L_{\mathbf{C}}}_{\text{left-lax}} \simeq \mathrm{Funct}(\mathbb{S}^{2\text{-op}},\mathbb{T}^{2\text{-op}})^{R_{\mathbf{C}}}_{\text{right-lax}}.$$

Consider the particular case when $\mathbf{C}$ is all of $\mathbb{S}^{1\text{-Cat}}$. In this case we will simply write

$$\mathrm{Funct}(\mathbb{S},\mathbb{T})^{R}_{\text{right-lax}} \text{ and } \mathrm{Funct}(\mathbb{S},\mathbb{T})^{L}_{\text{left-lax}}$$

and

$$\mathrm{Maps}_{2\text{-Cat}}(\mathbb{S},\mathbb{T})^{R} \text{ and } \mathrm{Maps}_{2\text{-Cat}}(\mathbb{S},\mathbb{T})^{L},$$

respectively.

1.1.9. Assume now that $\mathbb{T}$ is ordinary. In this case one shows that there is a canonical equivalence of (ordinary) 2-categories

$$\mathrm{Funct}(\mathbb{S},\mathbb{T})^{R}_{\text{right-lax}} \simeq \mathrm{Funct}(\mathbb{S}^{1\&2\text{-op}},\mathbb{T})^{L}_{\text{left-lax}}, \tag{1.1}$$

given by *passage* to left adjoint 1-morphisms.

The goal of this chapter is to generalize the equivalence (1.1) to the case when $\mathbb{T}$ is an $(\infty,2)$-category. This will ultimately be achieved in Corollary 3.1.9.

1.2. The universal adjointable functor. It is fairly easy to see that, given a pair $(\mathbb{S},\mathbf{C})$, where $\mathbb{S}$ is an $(\infty,2)$-category and $\mathbf{C}\subset\mathbb{S}^{1\text{-Cat}}$ is a 1-full subcategory with the same class of objects, there exists a *unversal* recipient, denoted $\mathbb{S}^{R_{\mathbf{C}}}$, of functors left-adjointable with respect to $\mathbf{C}$.

The point is that this $(\infty,2)$-category $\mathbb{S}^{R_{\mathbf{C}}}$ can be described explicitly in terms of the adjoint functors

$$\mathrm{Sq}_{\bullet,\bullet}: 2\text{-Cat} \rightleftarrows \mathrm{Spc}^{\mathbf{\Delta}^{\mathrm{op}}\times\mathbf{\Delta}^{\mathrm{op}}} : \mathfrak{L}^{\mathrm{Sq}}.$$

This description is given by Theorem 1.2.4. In fact, the $(\infty,2)$-category $\mathbb{S}^{R_{\mathbf{C}}}$ is obtained from $\mathbb{S}$ as a combination of the following three steps:

- Starting from $\mathbb{S}$, we pass to $\mathbb{S}^{2\text{-op}}$ and form the bi-simplicial groupoid $\mathrm{Sq}^{\mathrm{Pair}}_{\bullet,\bullet}(\mathbb{S}^{2\text{-op}},\mathbf{C})$, see Chapter 10, Sect. 4.3 for the notation.
- We take $\mathrm{Sq}^{\mathrm{Pair}}_{\bullet,\bullet}(\mathbb{S}^{2\text{-op}},\mathbf{C})$ and *reverse* its vertical arrows. Note that the resulting bi-simplicial groupoid will *not* be in the essential image of the functor $\mathrm{Sq}_{\bullet,\bullet}$.
- We take $\left(\mathrm{Sq}^{\mathrm{Pair}}_{\bullet,\bullet}(\mathbb{S}^{2\text{-op}},\mathbf{C})\right)^{\text{vert-op}}$ and apply to it the functor $\mathfrak{L}^{\mathrm{Sq}}$.

The idea of this construction is that the reversed vertical arrows will supply the data of left adjoints for 1 morphisms in $\mathbf{C}$.

1.2.1. Let $\mathbb{S}$ be an $(\infty,2)$-category, and $\mathbf{C}\subset\mathbb{S}^{1\text{-Cat}}$ be a 1-full subcategory with the same space of objects.

Consider the bi-simplicial category $(\mathrm{Sq}^{\mathrm{Pair}}_{\bullet,\bullet}(\mathbb{S}^{2\text{-op}},\mathbf{C}))^{\text{vert-op}}$, where the notation $(-)^{\text{vert-op}}$ is as in Chapter 10, Sect. 4.1.5, and $\mathrm{Sq}^{\mathrm{Pair}}_{\bullet,\bullet}$ is as in Chapter 10, Sect. 4.3.3.

We define the $(\infty,2)$-category $\mathbb{S}^{R_{\mathbf{C}}}$ to be

$$\mathfrak{L}^{\mathrm{Sq}}((\mathrm{Sq}^{\mathrm{Pair}}_{\bullet,\bullet}(\mathbb{S}^{2\text{-op}},\mathbf{C}))^{\text{vert-op}}),$$

where $\mathfrak{L}^{\mathrm{Sq}}$ is as in Chapter 10, Secr. 4.4.4.

I.e., by definition, for $\mathbb{T}\in 2\text{-Cat}$,

$$\mathrm{Maps}_{2\text{-Cat}}(\mathbb{S}^{R_{\mathbf{C}}},\mathbb{T}) = \mathrm{Maps}_{\mathrm{Funct}(\mathbf{\Delta}^{\mathrm{op}},1\text{-Cat})}((\mathrm{Sq}^{\mathrm{Pair}}_{\bullet,\bullet}(\mathbb{S}^{2\text{-op}},\mathbf{C}))^{\text{vert-op}},\mathrm{Sq}_{\bullet,\bullet}(\mathbb{T})).$$

1.2.2. We claim that we have a canonically defined functor

$$\mathbb{S} \to \mathbb{S}^{R_{\mathbf{C}}}. \tag{1.2}$$

Namely, it is obtained via the isomorphism

$$\mathbb{S} \simeq \mathfrak{L}^{\mathrm{Sq}} \circ \mathrm{Sq}^{\sim}_{\bullet,\bullet}(\mathbb{S})$$

(of Chapter 10, Corollary 4.4.5) by applying $\mathfrak{L}^{\mathrm{Sq}}$ to the tautological bi-simplicial functor

$$\mathrm{Sq}^{\sim}_{\bullet,\bullet}(\mathbb{S}) \simeq (\mathrm{Sq}^{\sim}_{\bullet,\bullet}(\mathbb{S}^{2\text{-op}}))^{\text{vert-op}} \hookrightarrow (\mathrm{Sq}^{\mathrm{Pair}}_{\bullet,\bullet}(\mathbb{S}^{2\text{-op}}, \mathbf{C}))^{\text{vert-op}}. \tag{1.3}$$

1.2.3. Recall the notation

$$\mathrm{Maps}_{2\text{-Cat}}(\mathbb{S}, \mathbb{T})^{R_{\mathbf{C}}} \subset \mathrm{Maps}_{2\text{-Cat}}(\mathbb{S}, \mathbb{T}),$$

see Sect. 1.1.8.

We will prove the following result:

THEOREM 1.2.4. *Restriction along* (1.2) *defines an isomorphism*

$$\mathrm{Maps}_{2\text{-Cat}}(\mathbb{S}^{R_{\mathbf{C}}}, \mathbb{T}) \to \mathrm{Maps}_{2\text{-Cat}}(\mathbb{S}, \mathbb{T})^{R_{\mathbf{C}}}.$$

1.3. The case $\mathbf{C} = \mathbb{S}^{1\text{-Cat}}$. Let us consider a particular case of Theorem 1.2.4 when $\mathbf{C} = \mathbb{S}^{1\text{-Cat}}$. In this case we shall simply write $\mathbb{S}^{R}$. The point is that in this case we will have a canonical equivalence

$$(\mathbb{S}^{R})^{2\text{-op}} \simeq (\mathbb{S}^{1\text{-op}})^{R},$$

which will allow to realize the *passage to adjoints* construction.

1.3.1. Recall that for $\mathbb{X} \in 2\text{-Cat}$ we have

$$(\mathrm{Sq}_{\bullet,\bullet}(\mathbb{X}))^{\text{vert\&horiz-op}} \simeq \mathrm{Sq}_{\bullet,\bullet}(\mathbb{X}^{1\&2\text{-op}}),$$

see Chapter 10, Sect. 4.1.6.

Recall also the involution reflect on $\mathrm{Spc}^{\mathbf{\Delta}^{\mathrm{op}} \times \mathbf{\Delta}^{\mathrm{op}}}$, see Chapter 10, Sect. 4.1.5. For $\mathbb{X} \in 2\text{-Cat}$ we have

$$(\mathrm{Sq}_{\bullet,\bullet}(\mathbb{X}))^{\text{reflect}} \simeq \mathrm{Sq}_{\bullet,\bullet}(\mathbb{X}^{2\text{-op}}). \tag{1.4}$$

Hence, for $\mathbb{S} \in 2\text{-Cat}$ we have obtain a canonical isomorphism

(1.5)

$$\Big(\big(\mathrm{Sq}_{\bullet,\bullet}(\mathbb{S}^{2\text{-op}})\big)^{\text{vert-op}}\Big)^{\text{reflect}} \simeq \Big(\big(\mathrm{Sq}_{\bullet,\bullet}(\mathbb{S}^{2\text{-op}})\big)^{\text{reflect}}\Big)^{\text{horiz-op}} \simeq \big(\mathrm{Sq}_{\bullet,\bullet}(\mathbb{S})\big)^{\text{horiz-op}} \simeq$$
$$\simeq \Big(\big(\mathrm{Sq}_{\bullet,\bullet}(\mathbb{S})\big)^{\text{vert\&horiz-op}}\Big)^{\text{vert-op}} \simeq \big(\big(\mathrm{Sq}_{\bullet,\bullet}(\mathbb{S}^{1\&2\text{-op}})\big)\big)^{\text{vert-op}}.$$

1.3.2. Note that from (1.4) it follows that for $E_{\bullet,\bullet} \in \mathrm{Spc}^{\mathbf{\Delta}^{\mathrm{op}} \times \mathbf{\Delta}^{\mathrm{op}}}$ we have

$$\mathfrak{L}^{\mathrm{Sq}}\left((E_{\bullet,\bullet})^{\mathrm{reflect}}\right) \simeq (\mathfrak{L}^{\mathrm{Sq}}(E_{\bullet,\bullet}))^{2\text{-op}}. \tag{1.6}$$

Hence, we obtain an identification

$$\begin{aligned}(\mathbb{S}^R)^{2\text{-op}} &= \left(\mathfrak{L}^{\mathrm{Sq}}\left((\mathrm{Sq}_{\bullet,\bullet}(\mathbb{S}^{2\text{-op}}))^{\mathrm{vert\text{-}op}}\right)\right)^{2\text{-op}} \\ &\overset{(1.6)}{\simeq} \mathfrak{L}^{\mathrm{Sq}}\left(\left((\mathrm{Sq}_{\bullet,\bullet}(\mathbb{S}^{2\text{-op}}))^{\mathrm{vert\text{-}op}}\right)^{\mathrm{reflect}}\right) \\ &\overset{(1.5)}{\simeq} \mathfrak{L}^{\mathrm{Sq}}\left((\mathrm{Sq}_{\bullet,\bullet}(\mathbb{S}^{1\&2\text{-op}}))^{\mathrm{vert\text{-}op}}\right) = (\mathbb{S}^{1\text{-op}})^R.\end{aligned} \tag{1.7}$$

1.3.3. Combining (1.7) and (1.2), we obtain a canonically defined map

$$\mathbb{S}^{1\text{-op}} \to (\mathbb{S}^R)^{2\text{-op}} \tag{1.8}$$

and hence a map

$$\mathbb{S}^{1\&2\text{-op}} \to \mathbb{S}^R \tag{1.9}$$

Applying Theorem 1.2.4 we obtain:

COROLLARY 1.3.4. *The functors*

$$\mathbb{S} \to \mathbb{S}^R \text{ and } \mathbb{S}^{1\&2\text{-op}} \to \mathbb{S}^R$$

define isomorphisms

$$\mathrm{Maps}_{2\text{-Cat}}(\mathbb{S}^R, \mathbb{T}) \to \mathrm{Maps}_{2\text{-Cat}}(\mathbb{S}, \mathbb{T})^R$$

and

$$\begin{aligned}\mathrm{Maps}_{2\text{-Cat}}(\mathbb{S}^R, \mathbb{T}) &\simeq \mathrm{Maps}_{2\text{-Cat}}(((\mathbb{S}^{1\text{-op}})^R)^{2\text{-op}}, \mathbb{T}) \\ &\simeq \mathrm{Maps}_{2\text{-Cat}}((\mathbb{S}^{1\text{-op}})^R, \mathbb{T}^{2\text{-op}}) \to \mathrm{Maps}_{2\text{-Cat}}(\mathbb{S}^{1\text{-op}}, \mathbb{T}^{2\text{-op}})^R \\ &\simeq \mathrm{Maps}_{2\text{-Cat}}(\mathbb{S}^{1\&2\text{-op}}, \mathbb{T})^L.\end{aligned}$$

In particular, we obtain a canonical identification

$$\mathrm{Maps}_{2\text{-Cat}}(\mathbb{S}, \mathbb{T})^R \simeq \mathrm{Maps}_{2\text{-Cat}}(\mathbb{S}^{1\&2\text{-op}}, \mathbb{T})^L. \tag{1.10}$$

We will refer to the isomorphism of (1.10) as the procedure of *passing to right adjoints*.

1.3.5. *The adjoint 1-morphism.* Let us specialize the above discussion further to the case $\mathbb{S} = [1]$. For $\mathbb{T} \in 2\text{-Cat}$, let

$$((\mathrm{Seq}_1(\mathbb{T}))^{\mathrm{Spc}})^R \subset (\mathrm{Seq}_1(\mathbb{T}))^{\mathrm{Spc}} \supset ((\mathrm{Seq}_1(\mathbb{T}))^{\mathrm{Spc}})^L,$$

be the subspaces of 1-morphisms that admit right and left adjoints, respectively.

As a particular case of Corollary 1.3.4 we obtain:

COROLLARY 1.3.6. *There exists a canonical isomorphism of spaces*

$$((\mathrm{Seq}_1(\mathbb{T}))^{\mathrm{Spc}})^R \simeq ((\mathrm{Seq}_1(\mathbb{T}))^{\mathrm{Spc}})^L$$

that induces the isomorphism

$$\pi_0\left(((\mathrm{Seq}_1(\mathbb{T}))^{\mathrm{Spc}})^R\right) \simeq \pi_0\left(((\mathrm{Seq}_1(\mathbb{T}))^{\mathrm{Spc}})^L\right),$$

given by passage to the adjoint 1-morphism.

1.4. Proof of Theorem 1.2.4 for ordinary 2-categories. In this subsection we take $\mathbb{T}$ to be an *ordinary* 2-category. In this case Theorem 1.2.4 can be proved by explicit analysis.

1.4.1. Let us first show that the image of the restriction functor

$$\text{Maps}_{2\text{-Cat}}(\mathbb{S}^{R_{\mathbf{C}}}, \mathbb{T}) \to \text{Maps}_{2\text{-Cat}}(\mathbb{S}, \mathbb{T})$$

belongs to $\text{Maps}_{2\text{-Cat}}(\mathbb{S}, \mathbb{T})^{R_{\mathbf{C}}} \subset \text{Maps}_{2\text{-Cat}}(\mathbb{S}, \mathbb{T})$.

Let $F : \mathbb{S} \to \mathbb{T}$ be a functor, such that the map

$$\text{Sq}^{\sim}_{\bullet,\bullet}(\mathbb{S}) \to \text{Sq}_{\bullet,\bullet}(\mathbb{T})$$

has been extended to a map

$$F_{\bullet,\bullet} : (\text{Sq}^{\text{Pair}}_{\bullet,\bullet}(\mathbb{S}^{2\text{-op}}, \mathbf{C}))^{\text{vert-op}} \to \text{Sq}_{\bullet,\bullet}(\mathbb{T}).$$

Let $s \overset{\alpha}{\to} s'$ be a morphism in $\mathbf{C}$, and let $F(s) \overset{F(\alpha)}{\longrightarrow} F(s')$ be its image in $\mathbb{T}$. We wish to show that $F(\alpha)$ admits a left adjoint.

Consider α as a $(1,0)$-simplex in $\text{Sq}^{\text{Pair}}_{\bullet,\bullet}(\mathbb{S}, \mathbf{C})$. Let us now vertically invert it, and thus consider it as a $(1,0)$-simplex in $(\text{Sq}^{\text{Pair}}_{\bullet,\bullet}(\mathbb{S}^{2\text{-op}}, \mathbf{C}))^{\text{vert-op}}$. The image of the latter under $F_{\bullet,\bullet}$ is a $(1,0)$-simplex in $\text{Sq}_{\bullet,\bullet}(\mathbb{T})$, i.e., a 1-morphism

$$F(s') \overset{\beta}{\to} F(s)$$

(note the direction of the arrow!).

Let us show that β is the left adjoint of $F(\alpha)$.

1.4.2. Consider the following point in $\text{Sq}^{\text{Pair}}_{1,1}(\mathbb{S}^{2\text{-op}}, \mathbf{C})$:

(1.11)

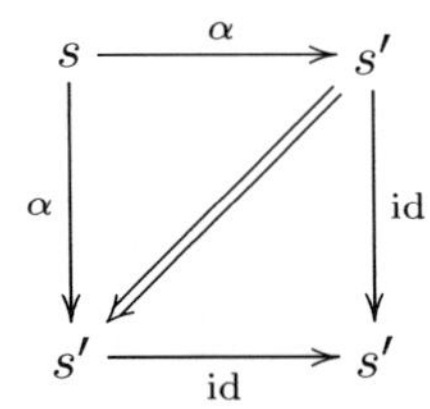

where the 2-morphism is the identity map $\alpha \Rightarrow \alpha$.

Let us vertically invert it, and thus consider it as a $(1,1)$-simplex in

$$(\text{Sq}^{\text{Pair}}_{\bullet,\bullet}(\mathbb{S}^{2\text{-op}}, \mathbf{C}))^{\text{vert-op}}$$

. Take the image of the latter under $F_{\bullet,\bullet}$. We obtain a $(1,1)$-simplex in

$$\text{Sq}_{\bullet,\bullet}(\mathbb{T}),$$

i.e., a diagram

(1.12)

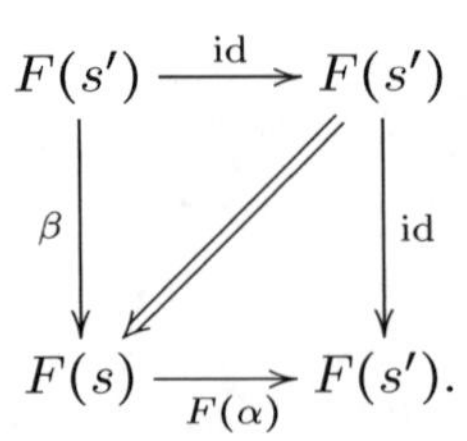

The $(1,1)$-simplex (1.12) represents a 2-morphism

$$\mathrm{id} \to F(\alpha) \circ \beta, \tag{1.13}$$

which will be the unit of the adjunction $(\beta, F(\alpha))$-adjunction.

1.4.3. Consider the following point in $\mathrm{Sq}^{\mathrm{Pair}}_{1,1}(\mathbb{S}^{2\text{-op}}, \mathbf{C})$:

(1.14)

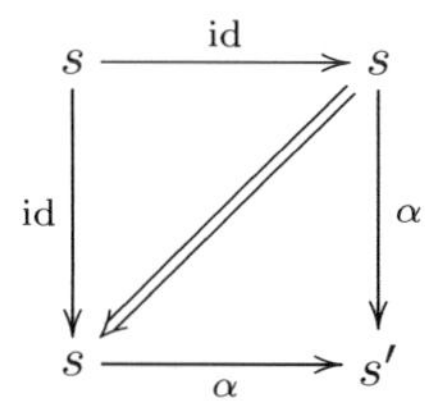

where the 2-morphism is the identity map $\alpha \Rightarrow \alpha$.

Let us vertically invert it, and thus consider it as a $(1,1)$-simplex in

$$(\mathrm{Sq}^{\mathrm{Pair}}_{\bullet,\bullet}(\mathbb{S}^{2\text{-op}}, \mathbf{C}))^{\text{vert-op}}.$$

Take the image of the latter under $F_{\bullet,\bullet}$. We obtain a $(1,1)$-simplex in $\mathrm{Sq}_{\bullet,\bullet}(\mathbb{T})$, i.e., a diagram

(1.15)

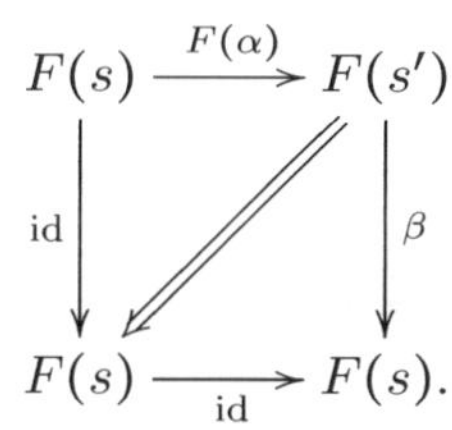

The $(1,1)$-simplex (1.12) represents a 2-morphism

$$\beta \circ F(\alpha) \to \mathrm{id}, \tag{1.16}$$

which will be the co-unit of the adjunction $(\beta, F(\alpha))$-adjunction.

1.4.4. The adjunction identities for (1.16) and (1.13) follow by concatenating the diagrams (1.11) and (1.14) first vertically, and then horizontally.

1.4.5. Let us now be given a functor $F : \mathbb{S} \to \mathbb{T}$ such that for each arrow $s \overset{\alpha}{\to} s'$ in $\mathbf{C}$, the corresponding 1-morphism $F(\alpha)$ admits a left adjoint. Let us construct the corresponding map

$$F_{\bullet,\bullet} : (\mathrm{Sq}^{\mathrm{Pair}}_{\bullet,\bullet}(\mathbb{S}^{2\text{-op}}, \mathbf{C}))^{\text{vert-op}} \to \mathrm{Sq}_{\bullet,\bullet}(\mathbb{T}).$$

With no restriction of generality, we can assume that $\mathbb{S}$ is also ordinary. We will define the map in question for $(0,0)$, $(0,1)$, $(1,0)$ and $(1,1)$ simplices, and it will be clear that it extends to a map of bi-simplicial *sets* by associativity.

1.4.6. At the level of $(0,0)$ simplices, $F_{\bullet,\bullet}$ sends a vertex

$$s \in \mathbb{S}^{\mathrm{Spc}},$$

thought of the space of $(0,0)$-simplices in $(\mathrm{Sq}^{\mathrm{Pair}}_{\bullet,\bullet}(\mathbb{S}^{2\text{-op}}, \mathbf{C}))^{\text{vert-op}}$, to

$$F(s) \in \mathbb{T}^{\mathrm{Spc}} = \mathrm{Sq}_{0,0}(\mathbb{T}).$$

1.4.7. At the level of $(0,1)$ simplices, $F_{\bullet,\bullet}$ sends

$$(s_0 \overset{\alpha}{\to} s_1) \in (\mathrm{Seq}_1(\mathbb{S}))^{\mathrm{Spc}},$$

thought of the space of $(0,1)$-simplices in $(\mathrm{Sq}^{\mathrm{Pair}}_{\bullet,\bullet}(\mathbb{S}^{2\text{-op}},\mathbf{C}))^{\mathrm{vert\text{-}op}}$, to

$$\left(F(s_0) \overset{F(\alpha)}{\longrightarrow} F(s_1)\right) \in (\mathrm{Seq}_1(\mathbb{T}))^{\mathrm{Spc}} = \mathrm{Sq}_{0,1}(\mathbb{T}).$$

1.4.8. At the level of $(1,0)$ simplices, $F_{\bullet,\bullet}$ sends

$$(s_0 \overset{\alpha}{\to} s_1) \in \mathrm{Seq}_1(\mathbf{C}),$$

thought of the space of $(1,0)$-simplices in $(\mathrm{Sq}^{\mathrm{Pair}}_{\bullet,\bullet}(\mathbb{S}^{2\text{-op}},\mathbf{C}))^{\mathrm{vert\text{-}op}}$, to

$$\left(F(s_1) \overset{F(\alpha)^L}{\longrightarrow} F(s_0)\right) \in (\mathrm{Seq}_1(\mathbb{T}))^{\mathrm{Spc}} = \mathrm{Sq}_{1,0}(\mathbb{T}),$$

where $F(\alpha)^L$ is the left adjoint of $F(\alpha)$.

1.4.9. At the level of $(1,1)$ simplices, $F_{\bullet,\bullet}$ sends a point

$$\begin{array}{ccc} s_{0,0} & \xrightarrow{\alpha_0} & s_{0,1} \\ {\scriptstyle \beta_0}\downarrow & \overset{\phi}{\swarrow} & \downarrow{\scriptstyle \beta_1} \\ s_{1,0} & \xrightarrow[\alpha_1]{} & s_{1,1} \end{array} \tag{1.17}$$

in $\mathrm{Sq}^{\mathrm{Pair}}_{1,1}(\mathbb{S}^{2\text{-op}},\mathbf{C})$, thought of as a $(1,1)$-simplex in $(\mathrm{Sq}^{\mathrm{Pair}}_{\bullet,\bullet}(\mathbb{S}^{2\text{-op}},\mathbf{C}))^{\mathrm{vert\text{-}op}}$ to the element of $\mathrm{Sq}_{1,1}(\mathbb{T})$, given by the diagram

$$\begin{array}{ccc} F(s_{1,0}) & \xrightarrow{F(\alpha_1)} & F(s_{1,1}) \\ {\scriptstyle F(\beta_0)^L}\downarrow & \overset{\psi}{\swarrow} & \downarrow{\scriptstyle F(\beta_1)^L} \\ F(s_{0,0}) & \xrightarrow[F(\alpha_0)]{} & F(s_{0,1}), \end{array}$$

where the 2-morphism

$$\psi : F(\beta_1)^L \circ F(\alpha_1) \to F(\alpha_0) \circ F(\beta_0)^L,$$

is obtained from

$$\phi : F(\alpha_1) \circ F(\beta_0) \to F(\beta_1) \circ F(\alpha_0)$$

by adjunction. (Note the direction in which ϕ goes–this is due to the fact that (1.17) was a $(1,1)$-simplex in $\mathrm{Sq}^{\mathrm{Pair}}_{1,1}(\mathbb{S}^{2\text{-op}},\mathbf{C})$, i.e., we inverted the 2-morphisms in $\mathbb{S}$.)

2. Proof of Theorem 1.2.4

We will first prove Theorem 1.2.4 in the case when the target $(\infty, 2)$-category is $\mathbf{1}\text{-}\mathbf{Cat}$, and then reduce to this case using the Yoneda embedding of Chapter 11, Sect. 6.3.

The idea of the proof in the case of $\mathbb{T} = \mathbf{1}\text{-}\mathbf{Cat}$ is the following.

Consider the 2-category $[m, n] := [m] \circledast [n]$. Let $\mathbf{C}$ be the 1-full subcategory in $[m, n]^{1\text{-Cat}}$, corresponding to the horizontal direction (i.e., the 2nd coordinate). The main observation is that the corresponding space $\text{Maps}_{2\text{-Cat}}(\mathbb{S}, \mathbf{1}\text{-}\mathbf{Cat})^{R\mathbf{C}}$ can be described very explicitly.

Namely, this is the space of functors

$$[m] \to \text{biCart}_{/[n]^{\text{op}}},$$

where $\text{biCart}_{/\mathbf{I}}$ denoted the category of *bi-Caretsian fibrations* over a given $(\infty, 1)$-category $\mathbf{I}$, i.e.,

$$\text{biCart}_{/\mathbf{I}} = \text{Cart}_{/\mathbf{I}} \cap \text{coCart}_{/\mathbf{I}} \subset 1\text{-Cat}_{/\mathbf{I}}.$$

2.1. The swapping procedure. In this subsection we will make a (relatively elementary) observation pertaining to $(\infty, 1)$-categories that lies in the heart of the proof of Theorem 1.2.4 for the target $\mathbb{T} = \mathbf{1}\text{-}\mathbf{Cat}$.

2.1.1. Let $\mathbf{I}$ and $\mathbf{J}$ be $(\infty, 1)$-categories. Consider the following $(\infty, 1)$-category, denoted

$$\text{Cart-coCart}_{\mathbf{I},\mathbf{J}}.$$

This is the full subcategory of $1\text{-Cat}_{/\mathbf{I}\times\mathbf{J}}$, that consists of those $(\infty, 1)$-categories $\mathbf{C}$ over $\mathbf{I} \times \mathbf{J}$ that satisfy:

- The composite functor $\mathbf{C} \to \mathbf{I} \times \mathbf{J} \to \mathbf{I}$ is a Cartesian fibration;
- The functor $\mathbf{C} \to \mathbf{I} \times \mathbf{J}$, viewed as a functor between Cartesian fibrations over $\mathbf{I}$, sends Cartesian arrows to Cartesian arrows (i.e., belongs to $(\text{Cart}_{/\mathbf{I}})_{\text{strict}}$);
- The composite functor $\mathbf{C} \to \mathbf{I} \times \mathbf{J} \to \mathbf{J}$ is a coCartesian fibration;
- The functor $\mathbf{C} \to \mathbf{I} \times \mathbf{J}$, viewed as a functor between coCartesian fibrations over $\mathbf{J}$, sends coCartesian arrows to coCartesian arrows (i.e., belongs to $(\text{coCart}_{/\mathbf{J}})_{\text{strict}}$);.

2.1.2. The first two conditions define a 1-fully faithful embedding

$$\text{Cart-coCart}_{\mathbf{I},\mathbf{J}} \hookrightarrow \text{Funct}(\mathbf{I}^{\text{op}}, 1\text{-Cat}_{/\mathbf{J}}),$$

and the second two conditions imply that it factors as

$$\text{Cart-coCart}_{\mathbf{I},\mathbf{J}} \hookrightarrow \text{Funct}(\mathbf{I}^{\text{op}}, \text{coCart}_{/\mathbf{J}}).$$

Similarly, we have a 1-fully faithful embedding

$$\text{Cart-coCart}_{\mathbf{I},\mathbf{J}} \hookrightarrow \text{Funct}(\mathbf{J}, \text{Cart}_{/\mathbf{I}}).$$

We claim:

PROPOSITION 2.1.3. *The induced maps*

$$(\text{Cart-coCart}_{\mathbf{I},\mathbf{J}})^{\text{Spc}} \to \text{Maps}(\mathbf{I}^{\text{op}}, \text{coCart}_{/\mathbf{J}})$$

and

$$(\text{Cart-coCart}_{\mathbf{I},\mathbf{J}})^{\text{Spc}} \to \text{Maps}(\mathbf{J}, \text{Cart}_{/\mathbf{I}})$$

are isomorphisms.

PROOF. We will prove the first isomorphism, the second being similar. The inverse map is constructed as follows: given

$$\mathbf{I}^{\text{op}} \to 1\text{-Cat}_{/\mathbf{J}},$$

we tautologically construct a Cartesian fibration $\mathbf{C} \to \mathbf{I}$, equipped with a functor $\mathbf{C} \to \mathbf{I} \times \mathbf{J}$ that takes Cartesian arrows to Cartesian arrows.

We need to show that if the initial map takes values in $\text{coCart}_{/\mathbf{J}} \subset 1\text{-Cat}_{/\mathbf{J}}$, then the resulting functor $\mathbf{C} \to \mathbf{J}$ is a coCartesian fibration, and the functor $\mathbf{C} \to \mathbf{I} \times \mathbf{J}$, viewed as a functor between coCartesian fibrations over $\mathbf{J}$, sends coCartesian arrows to coCartesian arrows. This is a straightforward verification.

□

COROLLARY 2.1.4. *There exists a canonical isomorphism*

$$\text{Maps}_{1\text{-Cat}}(\mathbf{I}^{\text{op}}, \text{coCart}_{/\mathbf{J}}) \simeq \text{Maps}_{1\text{-Cat}}(\mathbf{J}, \text{Cart}_{/\mathbf{I}}).$$

2.2. Proof of Theorem 1.2.4 for $\mathbb{T} = \mathbf{1}\text{-}\mathbf{Cat}$.

2.2.1. The datum of a functor

$$\mathbb{S} \to \mathbf{1}\text{-}\mathbf{Cat},$$

is equivalent to that of a functor

$$\mathbb{S}^{2\text{-op}} \to (\mathbf{1}\text{-}\mathbf{Cat})^{2\text{-op}},$$

which by Chapter 10, Corollary 4.4.5, is equivalent to the datum of a bi-simplicial map

$$\text{Sq}^{\text{Pair}}_{\bullet,\bullet}(\mathbb{S}^{2\text{-op}}, \mathbf{C}) \to \text{Sq}_{\bullet,\bullet}((\mathbf{1}\text{-}\mathbf{Cat})^{2\text{-op}}),$$

and finally a map

$$\text{Sq}^{\text{Pair}}_{\bullet,\bullet}(\mathbb{S}^{2\text{-op}}, \mathbf{C}) \to (\text{Sq}_{\bullet,\bullet}(\mathbf{1}\text{-}\mathbf{Cat}))^{\text{reflect}}. \tag{2.1}$$

2.2.2. The datum of a functor

$$\mathbb{S}^{R\text{c}} \to \mathbf{1}\text{-}\mathbf{Cat}$$

is equivalent to that of a bi-simplicial map

$$\text{Sq}^{\text{Pair}}_{\bullet,\bullet}(\mathbb{S}^{2\text{-op}}, \mathbf{C}) \to (\text{Sq}_{\bullet,\bullet}(\mathbf{1}\text{-}\mathbf{Cat}))^{\text{vert-op}}. \tag{2.2}$$

2.2.3. Recall that the space $\mathrm{Sq}_{m,n}(\mathbf{1}\text{-}\mathbf{Cat})$ is described as

$$\mathrm{Maps}_{1\text{-Cat}}([m], \mathrm{Cart}_{/[n]^{\mathrm{op}}}).$$

Hence, the space of (m,n)-simplices of $(\mathrm{Sq}_{\bullet,\bullet}(\mathbf{1}\text{-}\mathbf{Cat}))^{\mathrm{reflect}}$ is described as

$$\mathrm{Maps}_{1\text{-Cat}}([n], \mathrm{Cart}_{/[m]^{\mathrm{op}}}).$$

We claim that a functor

$$\mathbb{S} \to \mathbf{1}\text{-}\mathbf{Cat},$$

belongs to $\mathrm{Maps}(\mathbb{S}, \mathbf{1}\text{-}\mathbf{Cat})^{R_{\mathbf{C}}}$ if and only if each of the maps

$$\mathrm{Sq}^{\mathrm{Pair}}_{m,n}(\mathbb{S}^{2\text{-op}}, \mathbf{C})) \to \mathrm{Maps}_{1\text{-Cat}}([n], \mathrm{Cart}_{/[m]^{\mathrm{op}}})$$

takes values in

$$\mathrm{Maps}_{1\text{-Cat}}([n], \mathrm{biCart}_{/[m]^{\mathrm{op}}}) \subset \mathrm{Maps}_{1\text{-Cat}}([n], \mathrm{Cart}_{/[m]^{\mathrm{op}}}),$$

where for $\mathbf{I} \in 1\text{-Cat}$, we let $\mathrm{biCart}_{\mathbf{I}}$ denote the full subcategory of $1\text{-Cat}_{/\mathbf{I}}$ equal to

$$\mathrm{Cart}_{/\mathbf{I}} \cap \mathrm{coCart}_{/\mathbf{I}}.$$

Indeed, this assertion can be checked at the level of the underlying ordinary 1-categories, in which case it is a straightforward verification.

2.2.4. The space of (m,n)-simplices of $(\mathrm{Sq}_{\bullet,\bullet}(\mathbf{1}\text{-}\mathbf{Cat}))^{\text{vert-op}}$ is described as

$$\mathrm{Maps}_{1\text{-Cat}}([m]^{\mathrm{op}}, \mathrm{Cart}_{/[n]^{\mathrm{op}}}).$$

We claim that for every functor

$$\mathbb{S}^{R_{\mathbf{C}}} \to \mathbf{1}\text{-}\mathbf{Cat},$$

then each of the maps

$$\mathrm{Sq}^{\mathrm{Pair}}_{m,n}(\mathbb{S}^{2\text{-op}}, \mathbf{C}) \to \mathrm{Maps}_{1\text{-Cat}}([m]^{\mathrm{op}}, \mathrm{Cart}_{/[n]^{\mathrm{op}}}) \overset{\text{Corollary 2.1.4}}{\simeq} \mathrm{Maps}_{1\text{-Cat}}([n], \mathrm{coCart}_{/[m]^{\mathrm{op}}})$$

takes values in

$$\mathrm{Maps}_{1\text{-Cat}}([n], \mathrm{biCart}_{/[m]^{\mathrm{op}}}) \subset \mathrm{Maps}_{1\text{-Cat}}([n], \mathrm{coCart}_{/[m]^{\mathrm{op}}}).$$

Indeed, this assertion can be checked at the level of the underlying ordinary 1-categories, in which case it follows from the validity of Theorem 1.2.4 with values in $\mathbf{1}\text{-}\mathbf{Cat}_{\mathrm{ordn}}$.

2.2.5. Hence, when considering the bi-simplicial maps (2.1) and (2.2) we can replace the bi-simplicial spaces

$$(\mathrm{Sq}_{\bullet,\bullet}(\mathbf{1}\text{-}\mathbf{Cat}))^{\mathrm{reflect}} \text{ and } (\mathrm{Sq}_{\bullet,\bullet}(\mathbf{1}\text{-}\mathbf{Cat}))^{\text{vert-op}}$$

by their common full bi-simplicial subspace

$$(2.3)\quad '(\mathrm{Sq}_{\bullet,\bullet}(\mathbf{1}\text{-}\mathbf{Cat}))^{\mathrm{reflect}} \simeq {}'(\mathrm{Sq}_{\bullet,\bullet}(\mathbf{1}\text{-}\mathbf{Cat}))^{\text{vert-op}}, \\ (m,n) \mapsto \mathrm{Maps}_{1\text{-}\mathrm{Cat}}([n], \mathrm{biCart}_{/[m]^{\mathrm{op}}}).$$

Thus, we obtain that the datum of a bi-simplicial map in (2.1) is equivalent to the datum of a bi-simplicial map in (2.2), thereby establishing an isomorphism

$$\mathrm{Maps}_{2\text{-}\mathrm{Cat}}(\mathbb{S}, \mathbf{1}\text{-}\mathbf{Cat})^{R\mathrm{c}} \simeq \mathrm{Maps}_{2\text{-}\mathrm{Cat}}(\mathbb{S}^{R\mathrm{c}}, \mathbf{1}\text{-}\mathbf{Cat}).$$

2.2.6. Finally, it follows from the construction, that the composed map

$$\mathrm{Maps}_{2\text{-}\mathrm{Cat}}(\mathbb{S}, \mathbf{1}\text{-}\mathbf{Cat})^{R\mathrm{c}} \simeq \mathrm{Maps}_{2\text{-}\mathrm{Cat}}(\mathbb{S}^{R\mathrm{c}}, \mathbf{1}\text{-}\mathbf{Cat}) \overset{(1.2)}{\longrightarrow} \mathrm{Maps}_{2\text{-}\mathrm{Cat}}(\mathbb{S}, \mathbf{1}\text{-}\mathbf{Cat})$$

is the tautological embedding

$$\mathrm{Maps}_{2\text{-}\mathrm{Cat}}(\mathbb{S}, \mathbf{1}\text{-}\mathbf{Cat})^{R\mathrm{c}} \hookrightarrow \mathrm{Maps}_{2\text{-}\mathrm{Cat}}(\mathbb{S}, \mathbf{1}\text{-}\mathbf{Cat}).$$

This shows that the map

$$\mathrm{Maps}_{2\text{-}\mathrm{Cat}}(\mathbb{S}^{R\mathrm{c}}, \mathbf{1}\text{-}\mathbf{Cat}) \overset{(1.2)}{\longrightarrow} \mathrm{Maps}_{2\text{-}\mathrm{Cat}}(\mathbb{S}, \mathbf{1}\text{-}\mathbf{Cat})$$

is an isomorphism onto $\mathrm{Maps}_{2\text{-}\mathrm{Cat}}(\mathbb{S}, \mathbf{1}\text{-}\mathbf{Cat})^{R\mathrm{c}}$, as required.

2.3. Swapping procedure: relative version. The contents of this subsection are needed in order to generalize the contents of Sect. 2.2 to the case when instead of the target $(\infty,2)$-category $\mathbf{1}\text{-}\mathbf{Cat}$, we are dealing with

$$\mathrm{Funct}(\mathbf{I} \circledast \mathbf{J}, \mathbf{1}\text{-}\mathbf{Cat}), \quad \mathbf{I}, \mathbf{J} \in 1\text{-}\mathrm{Cat}.$$

2.3.1. For $\mathbf{I}, \mathbf{J} \in 1\text{-}\mathrm{Cat}$ we let

$$\text{Cart-Cart}_{\mathbf{I},\mathbf{J}}$$

be the full subcategory of $1\text{-}\mathrm{Cat}_{/\mathbf{I}\times\mathbf{J}}$ that consists of objects $\mathbf{C} \to \mathbf{I} \times \mathbf{J}$ satisfying the following conditions:

- The composite functor $\mathbf{C} \to \mathbf{I} \times \mathbf{J} \to \mathbf{I}$ is a Cartesian fibration;
- The functor $\mathbf{C} \to \mathbf{I} \times \mathbf{J}$, viewed as a functor between Cartesian fibrations over $\mathbf{I}$ sends coCartesian arrows to Cartesian ones;
- For every $i \in \mathbf{I}$, the resulting functor $\mathbf{C}_i \to \mathbf{J}$ is a Cartesian fibration.

Unstraightening over $\mathbf{I}$ defines an equivalence

$$\mathrm{Maps}_{1\text{-}\mathrm{Cat}}(\mathbf{I}^{\mathrm{op}}, \mathrm{Cart}_{/\mathbf{J}}) \simeq (\text{Cart-Cart}_{\mathbf{I},\mathbf{J}})^{\mathrm{Spc}}.$$

2.3.2. For a triplet of $(\infty, 1)$-categories $\mathbf{J}, \mathbf{K}, \mathbf{L}$ let

$$\text{Cart-Cart-Cart}_{\mathbf{J},\mathbf{K},\mathbf{L}}$$

denote the full subcategory of $1\text{-Cat}_{/\mathbf{J}\times\mathbf{K}\times\mathbf{L}}$ that consists of objects $\mathbf{C} \to \mathbf{J} \times \mathbf{K} \times \mathbf{L}$ satisfying the following conditions:

- When viewed as a category over $\mathbf{K} \times (\mathbf{J} \times \mathbf{L})$, it belongs to $\text{Cart}_{\mathbf{K},\mathbf{J}\times\mathbf{L}}$;
- For every fixed $l \in \mathbf{L}$, the category $\mathbf{C}_l$ is a Cartesian fibration over $\mathbf{J} \times \mathbf{K}$.

Let

$$(\text{Cart-Cart-Cart}_{\mathbf{J},\mathbf{K},\mathbf{L}})_{\text{strict}} \subset \text{Cart-Cart-Cart}_{\mathbf{J},\mathbf{K},\mathbf{L}}$$

be the following 1-full subcategory: Given two objects $\mathbf{C}, \mathbf{C}' \in \text{Cart-Cart-Cart}_{\mathbf{J},\mathbf{K},\mathbf{L}}$, we restrict 1-morphisms to those functors $F : \mathbf{C} \to \mathbf{C}'$ over $\mathbf{J} \times \mathbf{K} \times \mathbf{L}$ that:

- For every fixed $j \in \mathbf{J}$ and $l \in \mathbf{L}$, the corresponding functor $F_{j,l} : \mathbf{C}_{j,l} \to \mathbf{C}'_{j,l}$ carries arrows Cartesian over $\mathbf{K}$ to arrows Cartesian over $\mathbf{K}$.
- For every fixed $j \in \mathbf{J}$ and $k \in \mathbf{K}$, the corresponding functor $F_{j,k} : \mathbf{C}_{j,k} \to \mathbf{C}'_{j,k}$ carries arrows Cartesian over $\mathbf{L}$ to arrows Cartesian over $\mathbf{L}$.

For another category $\mathbf{I}$, consider the goupoid

$$\text{Maps}(\mathbf{I}, (\text{Cart-Cart-Cart}_{\mathbf{J},\mathbf{K},\mathbf{L}})_{\text{strict}}).$$

We shall now describe it in several different ways.

2.3.3. Let $\text{coCart-Cart-Cart}_{\mathbf{J},\mathbf{K},\mathbf{L}}$ denote the full subcategory of $1\text{-Cat}_{/\mathbf{J}\times\mathbf{K}\times\mathbf{L}}$ that consists of objects $\mathbf{C} \to \mathbf{J} \times \mathbf{K} \times \mathbf{L}$ satisfying the following conditions:

- The composite functor $\mathbf{C} \to \mathbf{J} \times \mathbf{K} \times \mathbf{L} \to \mathbf{J}$ is a coCartesian fibration;
- The functor $\mathbf{C} \to \mathbf{J} \times \mathbf{K} \times \mathbf{L}$, when viewed as a map between coCartesian fibrations over $\mathbf{J}$, sends coCartesian arrows to coCartesian ones;
- For every $j \in \mathbf{J}$, the resulting object $\mathbf{C}_j \in 1\text{-Cat}_{\mathbf{K}\times\mathbf{L}}$ belongs to $\text{Cart-Cart}_{/\mathbf{K}\times\mathbf{L}}$;
- For every fixed $k \in \mathbf{K}$ and an arrow $j_0 \to j_1$ in $\mathbf{J}$, the resulting functor $\mathbf{C}_{j_0,k} \to \mathbf{C}_{j_1,k}$ carries arrows Cartesian over $\mathbf{L}$ to arrows Cartesian over $\mathbf{L}$.
- For every fixed $l \in \mathbf{L}$ and an arrow $j_0 \to j_1$ in $\mathbf{J}$, the resulting functor $\mathbf{C}_{j_0,l} \to \mathbf{C}_{j_1,l}$ carries arrows Cartesian over $\mathbf{K}$ to arrows Cartesian over $\mathbf{K}$.

Let

$$(\text{coCart-Cart-Cart}_{\mathbf{J},\mathbf{K},\mathbf{L}})_{\text{strict}} \subset \text{coCart-Cart-Cart}_{\mathbf{J},\mathbf{K},\mathbf{L}}$$

be the following 1-full subcategory: Given two objects $\mathbf{C}, \mathbf{C}' \in \text{Cart-Cart-Cart}_{\mathbf{J},\mathbf{K},\mathbf{L}}$, we restrict 1-morphisms to those functors $F : \mathbf{C} \to \mathbf{C}'$ over $\mathbf{J} \times \mathbf{K} \times \mathbf{L}$ that:

- For any fixed $j \in \mathbf{J}$ and $l \in \mathbf{L}$, the corresponding functor $F_{j,l} : \mathbf{C}_{j,l} \to \mathbf{C}'_{j,l}$ carries arrows Cartesian over $\mathbf{K}$ to arrows Cartesian over $\mathbf{K}$.
- For any fixed $j \in \mathbf{J}$ and $k \in \mathbf{K}$ the corresponding functor $F_{j,k} : \mathbf{C}_{j,k} \to \mathbf{C}'_{j,k}$ carries arrows Cartesian over $\mathbf{L}$ to arrows Cartesian over $\mathbf{L}$.

For another category $\mathbf{I}$, consider the goupoid

$$\text{Maps}(\mathbf{I}, (\text{coCart-Cart-Cart}_{\mathbf{J},\mathbf{K},\mathbf{L}})_{\text{strict}}).$$

2.3.4. Now, we claim that as in Corollary 2.1.4 we have a canonical isomorphism:

(2.4)
$$\mathrm{Maps}(\mathbf{J}, (\text{Cart-Cart-Cart}_{\mathbf{I},\mathbf{K},\mathbf{L}})_{\mathrm{strict}}) \simeq \mathrm{Maps}(\mathbf{I}^{\mathrm{op}}, (\text{coCart-Cart-Cart}_{\mathbf{J},\mathbf{K},\mathbf{L}})_{\mathrm{strict}}).$$

This is obtained by identifying both sides with the *space* of the full subcategory

$$\text{Cart-coCart-Cart-Cart}_{\mathbf{I},\mathbf{J},\mathbf{K},\mathbf{L}} \subset 1\text{-Cat}_{\mathbf{I}\times\mathbf{J}\times\mathbf{K}\times\mathbf{L}},$$

consisting of $\mathbf{C}$ over $\mathbf{I}\times\mathbf{J}\times\mathbf{K}\times\mathbf{L}$ that satisfy:

- The composite functor $\mathbf{C}\to\mathbf{I}\times\mathbf{J}\times\mathbf{K}\times\mathbf{L}\to\mathbf{J}$ is a coCartesian fibration;
- The functor $\mathbf{C}\to\mathbf{I}\times\mathbf{J}\times\mathbf{K}\times\mathbf{L}$, when viewed as a map between coCartesian fibrations over $\mathbf{J}$, sends coCartesian arrows to coCartesian ones;
- For every $j\in\mathbf{J}$, the resulting object $\mathbf{C}_j\in 1\text{-Cat}_{\mathbf{I}\times\mathbf{K}\times\mathbf{L}}$ belongs to
$$\text{Cart-Cart-Cart}_{/\mathbf{I}\times\mathbf{K}\times\mathbf{L}};$$
- For every fixed $k\in\mathbf{K}$, $i\in\mathbf{I}$ and an arrow $j_0\to j_1$ in $\mathbf{J}$, the resulting functor $\mathbf{C}_{i,j_0,k}\to\mathbf{C}_{i,j_1,k}$ carries arrows Cartesian over $\mathbf{L}$ to arrows Cartesian over $\mathbf{L}$.
- For every fixed $l\in\mathbf{L}$, $i\in\mathbf{I}$ and an arrow $j_0\to j_1$ in $\mathbf{J}$, the resulting functor $\mathbf{C}_{i,j_0,l}\to\mathbf{C}_{i,j_1,l}$ carries arrows Cartesian over $\mathbf{K}$ to arrows Cartesian over $\mathbf{K}$.

2.3.5. Finally, we claim that the space

$$\mathrm{Maps}(\mathbf{I}, (\text{Cart-Cart-Cart}_{\mathbf{J},\mathbf{K},\mathbf{L}})_{\mathrm{strict}})$$

can be also identified with the *space* of the following full subcategory

$$\text{Cart-Cart-Cart-Cart}_{\mathbf{I}^{\mathrm{op}},\mathbf{J},\mathbf{K},\mathbf{L}} \subset 1\text{-Cat}_{\mathbf{I}^{\mathrm{op}}\times\mathbf{J}\times\mathbf{K}\times\mathbf{L}}.$$

Namely, for quadruplet of $(\infty,1)$-categories $\mathbf{I},\mathbf{J},\mathbf{K},\mathbf{L}$, we define

$$\text{Cart-Cart-Cart-Cart}_{\mathbf{I},\mathbf{J},\mathbf{K},\mathbf{L}}$$

to consist of those $\mathbf{C}$ over $\mathbf{I}\times\mathbf{J}\times\mathbf{K}\times\mathbf{L}$ that satisfy:

- When viewed as a category over $(\mathbf{I}\times\mathbf{K})\times(\mathbf{J}\times\mathbf{L})$, it belongs to
$$\text{Cart-Cart}_{\mathbf{I}\times\mathbf{K},\mathbf{J}\times\mathbf{L}};$$
- For every fixed $i\in\mathbf{I}$ and $l\in\mathbf{L}$, the resulting functor $\mathbf{C}_{i,l}\to\mathbf{J}\times\mathbf{K}$ is a Cartesian fibration;
- For every fixed $j\in\mathbf{J}$ and $k\in\mathbf{K}$, the resulting functor $\mathbf{C}_{j,k}\to\mathbf{I}\times\mathbf{L}$ is a Cartesian fibration.

2.4. Adjunctions: relative version. Let us fix another pair of $(\infty,1)$-categories $\mathbf{K}$ and $\mathbf{L}$. We will now explain the modifications needed to adapt the above proof of Theorem 1.2.4 for $\mathbb{T}=\mathbf{1}\text{-}\mathbf{Cat}$ to the case when $\mathbb{T}=\mathrm{Funct}(\mathbf{K}\circledast\mathbf{L},\mathbf{1}\text{-}\mathbf{Cat})$.

2.4.1. First, we claim that the bi-simplicial space $\mathrm{Sq}(\mathrm{Funct}(\mathbf{K}\circledast\mathbf{L},\mathbf{1}\text{-}\mathbf{Cat}))$ is described as follows:
(2.5)
$$\mathrm{Sq}_{m,n}(\mathrm{Funct}(\mathbf{K}\circledast\mathbf{L},\mathbf{1}\text{-}\mathbf{Cat}))\simeq \mathrm{Maps}([m],(\text{Cart-Cart-Cart}_{[n]^{\mathrm{op}},\mathbf{K}^{\mathrm{op}},\mathbf{L}^{\mathrm{op}}})_{\mathrm{strict}}),$$
where the notation $(\text{Cart-Cart-Cart}_{[n]^{\mathrm{op}},\mathbf{K}^{\mathrm{op}},\mathbf{L}^{\mathrm{op}}})_{\mathrm{strict}}$ is as in Sect. 2.3.2.

Assuming that, the proof in Sect. 2.2 goes through, once we substitute the isomorphism of Corollary 2.1.4 by that of (2.4).

2.4.2. To establish (2.5), we proceed as follows. We rewrite
$$\mathrm{Sq}_{m,n}(\mathrm{Funct}(\mathbf{K}\circledast\mathbf{L},\mathbf{1}\text{-}\mathbf{Cat}))=\mathrm{Maps}(([m]\circledast[n])\times(\mathbf{K}\circledast\mathbf{L}),\mathbf{1}\text{-}\mathbf{Cat}),$$
and further by Chapter 11, Corollary 2.2.6 as
$$(\mathbf{1}\text{-}\mathbf{Cart}_{/(([m]\circledast[n])\times(\mathbf{K}\circledast\mathbf{L}))^{1\text{-op}}})^{\mathrm{Spc}}\simeq(\mathbf{1}\text{-}\mathbf{Cart}_{/([n]^{\mathrm{op}}\circledast[m]^{\mathrm{op}})\times(\mathbf{L}^{\mathrm{op}}\circledast\mathbf{K}^{\mathrm{op}})})^{\mathrm{Spc}}.$$

Now, using Chapter 11, Lemmas 2.2.5 and 2.2.8 and Corollary 4.6.5, we obtain
$$(\mathbf{1}\text{-}\mathbf{Cart}_{/([n]^{\mathrm{op}}\circledast[m]^{\mathrm{op}})\times(\mathbf{L}^{\mathrm{op}}\circledast\mathbf{K}^{\mathrm{op}})})^{\mathrm{Spc}}\simeq(\text{Cart-Cart-Cart-Cart}_{[m]^{\mathrm{op}},[n]^{\mathrm{op}},\mathbf{K}^{\mathrm{op}},\mathbf{L}^{\mathrm{op}}})^{\mathrm{Spc}},$$
and finally, using Sect. 2.3.5, we identify
$$(\text{Cart-Cart-Cart-Cart}_{[m]^{\mathrm{op}},[n]^{\mathrm{op}},\mathbf{K}^{\mathrm{op}},\mathbf{L}^{\mathrm{op}}})^{\mathrm{Spc}}\simeq$$
$$\simeq\mathrm{Maps}([m],(\text{Cart-Cart-Cart}_{[n]^{\mathrm{op}},\mathbf{K}^{\mathrm{op}},\mathbf{L}^{\mathrm{op}}})_{\mathrm{strict}}),$$
as required.

2.5. Proofs of Theorem 1.2.4, the general case. The proof will amount to deducing Theorem 1.2.4 from the particular case of $\mathbb{T}=\mathrm{Funct}([m]\circledast[n],\mathbf{1}\text{-}\mathbf{Cat})$ using the 2-categorical Yoneda embedding.

2.5.1. For a target category $\mathbb{T}$ we consider its Yoneda embedding
$$\mathbb{T}\overset{\mathrm{Yon}_{\mathbb{T}}}{\hookrightarrow}\mathrm{Funct}(\mathbb{T}^{1\text{-op}},\mathbf{1}\text{-}\mathbf{Cat}).$$

Consider the commutative diagram
$$\begin{array}{ccc}\mathrm{Maps}_{2\text{-Cat}}(\mathbb{S}^{R_{\mathbf{C}}},\mathbb{T}) & \longrightarrow & \mathrm{Maps}_{2\text{-Cat}}(\mathbb{S},\mathbb{T})^{R_{\mathbf{C}}}\\ \downarrow & & \downarrow\\ \mathrm{Maps}_{2\text{-Cat}}\left(\mathbb{S}^{R_{\mathbf{C}}},\mathrm{Funct}(\mathbb{T}^{1\text{-op}},\mathbf{1}\text{-}\mathbf{Cat})\right) & \longrightarrow & \mathrm{Maps}_{2\text{-Cat}}\left(\mathbb{S},\mathrm{Funct}(\mathbb{T}^{1\text{-op}},\mathbf{1}\text{-}\mathbf{Cat})\right)^{R_{\mathbf{C}}},\end{array}$$
where the vertical arrows are fully faithful embeddings.

We claim that it is sufficient to show that the bottom arrow in the above diagram, i.e.,
(2.6)
$$\mathrm{Maps}_{2\text{-Cat}}\left(\mathbb{S}^{R_{\mathbf{C}}},\mathrm{Funct}(\mathbb{T}^{1\text{-op}},\mathbf{1}\text{-}\mathbf{Cat})\right)\to\mathrm{Maps}_{2\text{-Cat}}\left(\mathbb{S},\mathrm{Funct}(\mathbb{T}^{1\text{-op}},\mathbf{1}\text{-}\mathbf{Cat})\right)^{R_{\mathbf{C}}},$$
is an isomorphism.

2.5.2. Indeed, if this is the case, we obtain that the functor

$$\mathrm{Maps}_{2\text{-Cat}}(\mathbb{S}^{R_{\mathbf{C}}},\mathbb{T}) \to \mathrm{Maps}_{2\text{-Cat}}(\mathbb{S},\mathbb{T})^{R_{\mathbf{C}}}$$

is fully faithful, and it remains to show that it is essentially surjective.

This follows from the next general assertion:

LEMMA 2.5.3. *Let $\mathbb{T}_1 \hookrightarrow \mathbb{T}_2$ be a fully faithful functor. Then the diagram of spaces*

$$\begin{array}{ccc} \mathrm{Maps}_{2\text{-Cat}}(\mathbb{S}^{R_{\mathbf{C}}},\mathbb{T}_1) & \longrightarrow & \mathrm{Maps}_{2\text{-Cat}}(\mathbb{S}^{R_{\mathbf{C}}},\mathbb{T}_2) \\ \downarrow & & \downarrow \\ \mathrm{Maps}_{2\text{-Cat}}(\mathbb{S},\mathbb{T}_1) & \longrightarrow & \mathrm{Maps}_{2\text{-Cat}}(\mathbb{S},\mathbb{T}_2) \end{array}$$

is a pull-back square.

PROOF. Let us be given a functor $\mathbb{S}^{R_{\mathbf{C}}} \to \mathbb{T}_2$, such that the composition

$$\mathbb{S} \to \mathbb{S}^{R_{\mathbf{C}}} \to \mathbb{T}_2$$

factors through $\mathbb{T}_1 \subset \mathbb{T}_2$. We wish to show that the initial functor also factors through $\mathbb{T}_1 \subset \mathbb{T}_2$.

Consider the corresponding map of bi-simplicial spaces

$$\left(\mathrm{Sq}^{\mathrm{Pair}}_{\bullet,\bullet}(\mathbb{S}^{2\text{-op}},\mathbf{C})\right)^{\text{vert-op}} \to \mathrm{Sq}_{\bullet,\bullet}(\mathbb{T}_2).$$

We wish to show that it takes values in $\mathrm{Sq}_{\bullet,\bullet}(\mathbb{T}_1) \subset \mathrm{Sq}_{\bullet,\bullet}(\mathbb{T}_2)$.

The latter is enough to check on $(0,0)$-simplicies, and the assertion follows from the assumption as

$$\mathrm{Sq}^{\sim}_{0,0}(\mathbb{S}) \to \left(\mathrm{Sq}^{\mathrm{Pair}}_{0,0}(\mathbb{S}^{2\text{-op}},\mathbf{C})\right)^{\text{vert-op}}$$

is an isomorphism.

□

2.5.4. Thus, we wish to show that for $\mathbb{S},\mathbb{T} \in 2\text{-Cat}$, the map

$$\mathrm{Maps}_{2\text{-Cat}}\left(\mathbb{S}^{R_{\mathbf{C}}},\mathrm{Funct}(\mathbb{T},\mathbf{1}\text{-}\mathbf{Cat})\right) \to \mathrm{Maps}_{2\text{-Cat}}\left(\mathbb{S},\mathrm{Funct}(\mathbb{T},\mathbf{1}\text{-}\mathbf{Cat})\right)$$

is an isomorphism with essential image

$$\mathrm{Maps}_{2\text{-Cat}}\left(\mathbb{S},\mathrm{Funct}(\mathbb{T},\mathbf{1}\text{-}\mathbf{Cat})\right)^{R_{\mathbf{C}}} \subset \mathrm{Maps}_{2\text{-Cat}}\left(\mathbb{S},\mathrm{Funct}(\mathbb{T},\mathbf{1}\text{-}\mathbf{Cat})\right).$$

2.5.5. For any $\mathbb{S}' \in 2\text{-Cat}$, the space $\mathrm{Maps}_{2\text{-Cat}}\left(\mathbb{S}',\mathrm{Funct}(\mathbb{T},\mathbf{1}\text{-}\mathbf{Cat})\right)$ can be described as that of bi-simplicial functors

$$\mathrm{Sq}_{\bullet,\bullet}(\mathbb{T}) \to \mathrm{Sq}_{\bullet,\bullet}\left(\mathrm{Funct}(\mathbb{S}',\mathbf{1}\text{-}\mathbf{Cat})\right).$$

Note that the bi-simplicial space $\mathrm{Sq}_{\bullet,\bullet}\left(\mathrm{Funct}(\mathbb{S}',\mathbf{1}\text{-}\mathbf{Cat})\right)$ identifies with

$$\mathrm{Maps}_{2\text{-Cat}}\left(\mathbb{S}',\mathrm{Funct}([\bullet]\circledast[\bullet],\mathbf{1}\text{-}\mathbf{Cat})\right),$$

where the bi-simplicial $(\infty,2)$-category $\mathrm{Funct}([\bullet]\circledast[\bullet],\mathbf{1}\text{-}\mathbf{Cat})$ attaches to m,n the $(\infty,2)$-category

$$\mathrm{Funct}([m]\circledast[n],\mathbf{1}\text{-}\mathbf{Cat}).$$

2.5.6. Note that under the above identification

$$\mathrm{Maps}_{2\text{-Cat}}(\mathbb{S}, \mathrm{Funct}(\mathbb{T}, \mathbf{1}\text{-}\mathbf{Cat})) \simeq$$
$$\simeq \mathrm{Maps}_{\mathrm{Spc}^{\mathbf{\Delta}^{\mathrm{op}}\times\mathbf{\Delta}^{\mathrm{op}}}}\left(\mathrm{Sq}_{\bullet,\bullet}(\mathbb{T}), \mathrm{Maps}_{2\text{-Cat}}(\mathbb{S}, \mathrm{Funct}([\bullet]\circledast[\bullet], \mathbf{1}\text{-}\mathbf{Cat}))\right),$$

the subspace

$$\mathrm{Maps}_{2\text{-Cat}}(\mathbb{S}, \mathrm{Funct}(\mathbb{T}, \mathbf{1}\text{-}\mathbf{Cat}))^{R_{\mathrm{C}}} \subset \mathrm{Maps}(\mathbb{S}, \mathrm{Funct}(\mathbb{T}, \mathbf{1}\text{-}\mathbf{Cat}))$$

maps to

$$\mathrm{Maps}_{\mathrm{Spc}^{\mathbf{\Delta}^{\mathrm{op}}\times\mathbf{\Delta}^{\mathrm{op}}}}\left(\mathrm{Sq}_{\bullet,\bullet}(\mathbb{T}), \mathrm{Maps}_{2\text{-Cat}}(\mathbb{S}, \mathrm{Funct}([\bullet]\circledast[\bullet], \mathbf{1}\text{-}\mathbf{Cat}))^{R_{\mathrm{C}}}\right) \subset$$
$$\subset \mathrm{Maps}_{\mathrm{Spc}^{\mathbf{\Delta}^{\mathrm{op}}\times\mathbf{\Delta}^{\mathrm{op}}}}\left(\mathrm{Sq}_{\bullet,\bullet}(\mathbb{T}), \mathrm{Maps}_{2\text{-Cat}}(\mathbb{S}, \mathrm{Funct}([\bullet]\circledast[\bullet], \mathbf{1}\text{-}\mathbf{Cat}))\right). \tag{2.7}$$

Hence, we obtain that it is enough to show that the map

$$\mathrm{Maps}_{\mathrm{Spc}^{\mathbf{\Delta}^{\mathrm{op}}\times\mathbf{\Delta}^{\mathrm{op}}}}\left(\mathrm{Sq}_{\bullet,\bullet}(\mathbb{T}), \mathrm{Maps}_{2\text{-Cat}}\left(\mathbb{S}^{R_{\mathrm{C}}}, \mathrm{Funct}([\bullet]\circledast[\bullet], \mathbf{1}\text{-}\mathbf{Cat})\right)\right) \to$$
$$\to \mathrm{Maps}_{\mathrm{Spc}^{\mathbf{\Delta}^{\mathrm{op}}\times\mathbf{\Delta}^{\mathrm{op}}}}\left(\mathrm{Sq}_{\bullet,\bullet}(\mathbb{T}), \mathrm{Maps}_{2\text{-Cat}}(\mathbb{S}, \mathrm{Funct}([\bullet]\circledast[\bullet], \mathbf{1}\text{-}\mathbf{Cat}))\right))$$

is an isomorphism onto the subspace (2.7).

2.5.7. To prove the latter, it is sufficient to show that for every m, n, the map

$$\mathrm{Maps}_{2\text{-Cat}}\left(\mathbb{S}^{R_{\mathrm{C}}}, \mathrm{Funct}([m]\circledast[n], \mathbf{1}\text{-}\mathbf{Cat})\right) \to$$
$$\to \mathrm{Maps}_{2\text{-Cat}}(\mathbb{S}, \mathrm{Funct}([m]\circledast[n], \mathbf{1}\text{-}\mathbf{Cat})))$$

is an isomorphism onto

$$\mathrm{Maps}_{2\text{-Cat}}(\mathbb{S}, \mathrm{Funct}([m]\circledast[n], \mathbf{1}\text{-}\mathbf{Cat}))^{R_{\mathrm{C}}} \subset$$
$$\subset \mathrm{Maps}_{2\text{-Cat}}(\mathbb{S}, \mathrm{Funct}([m]\circledast[n], \mathbf{1}\text{-}\mathbf{Cat})).$$

2.5.8. However, the latter statement is the assertion of Theorem 1.2.4 for the target category $\mathrm{Funct}([m]\circledast[n], \mathbf{1}\text{-}\mathbf{Cat})$, and it holds due to Sect. 2.4.

3. Adjunction with parameters

Our current goal is to lift the isomorphism of spaces

$$\mathrm{Maps}_{2\text{-Cat}}(\mathbb{S}, \mathbb{T})^R \simeq \mathrm{Maps}_{2\text{-Cat}}(\mathbb{S}^{1\&2\text{-op}}, \mathbb{T})^L,$$

which is part of the statement of Corollary 1.3.4 to an equivalence of $(\infty, 2)$-categories

$$\mathrm{Funct}(\mathbb{S}, \mathbb{T})^R_{\text{right-lax}} \simeq \mathrm{Funct}(\mathbb{S}^{1\&2\text{-op}}, \mathbb{T})^L_{\text{left-lax}}. \tag{3.1}$$

3.1. The set-up for adjunction with parameters. By definition, for an $(\infty, 2)$-category $\mathbb{X}$, we have

$$\mathrm{Maps}_{2\text{-Cat}}(\mathbb{X}, \mathrm{Funct}(\mathbb{S}, \mathbb{T})_{\text{right-lax}}) = \mathrm{Maps}_{2\text{-Cat}}(\mathbb{X}\circledast\mathbb{S}, \mathbb{T})$$

and

$$\mathrm{Maps}_{2\text{-Cat}}(\mathbb{X}, \mathrm{Funct}(\mathbb{S}^{1\&2\text{-op}}, \mathbb{T})_{\text{left-lax}}) = \mathrm{Maps}_{2\text{-Cat}}(\mathbb{S}^{1\&2\text{-op}}\circledast\mathbb{X}, \mathbb{T}).$$

Hence, we have the subspaces

$$\mathrm{Maps}_{2\text{-Cat}}(\mathbb{X}\circledast\mathbb{S}, \mathbb{T})^{\mathbb{S}_R} \subset \mathrm{Maps}_{2\text{-Cat}}(\mathbb{X}\circledast\mathbb{S}, \mathbb{T}) \tag{3.2}$$

and

$$\text{Maps}_{2\text{-Cat}}(\mathbb{S}^{1\&2\text{-op}} \circledast \mathbb{X}, \mathbb{T})^{\mathbb{S}_L} \subset \text{Maps}_{2\text{-Cat}}(\mathbb{S}^{1\&2\text{-op}} \circledast \mathbb{X}, \mathbb{T}), \tag{3.3}$$

corresponding to

$$\text{Maps}_{2\text{-Cat}}(\mathbb{X}, \text{Funct}(\mathbb{S}, \mathbb{T})^R_{\text{right-lax}}) \subset \text{Maps}_{2\text{-Cat}}(\mathbb{X}, \text{Funct}(\mathbb{S}, \mathbb{T})_{\text{right-lax}})$$

and

$$\text{Maps}_{2\text{-Cat}}(\mathbb{X}, \text{Funct}(\mathbb{S}^{1\&2\text{-op}}, \mathbb{T})^L_{\text{left-lax}}) \subset \text{Maps}_{2\text{-Cat}}(\mathbb{X}, \text{Funct}(\mathbb{S}^{1\&2\text{-op}}, \mathbb{T})_{\text{left-lax}}),$$

respectively.

We will interpret the subspaces (3.2) and (3.3) via the *universal adjointable functors* of Theorem 1.2.4, which will allow to construct the desired equivalence

$$\text{Maps}_{2\text{-Cat}}(\mathbb{X} \circledast \mathbb{S}, \mathbb{T})^{\mathbb{S}_R} \simeq \text{Maps}_{2\text{-Cat}}(\mathbb{S}^{1\&2\text{-op}} \circledast \mathbb{X}, \mathbb{T})^{\mathbb{S}_L}.$$

3.1.1. We start with a pair of $(\infty,2)$-categories $\mathbb{S}_1$ and $\mathbb{S}_2$, and consider their Gray product $\mathbb{S}_1 \circledast \mathbb{S}_2$

We let $\mathbf{C}_1$ be the 1-full subcategory in $(\mathbb{S}_1 \circledast \mathbb{S}_2)^{1\text{-Cat}}$, corresponding to 1-morphisms of the form

$$(s_1, s_2) \overset{(\alpha,\text{id})}{\longrightarrow} (s_1', s_2), \quad \alpha \in \text{Maps}_{\mathbb{S}}(s_1, s_1').$$

We let $\mathbf{C}_2$ be the 1-full subcategory in $(\mathbb{S}_1 \circledast \mathbb{S}_2)^{1\text{-Cat}}$, corresponding to 1-morphisms of the form

$$(s_1, s_2) \overset{(\text{id},\beta)}{\longrightarrow} (s_1, s_2'), \quad \beta \in \text{Maps}_{\mathbb{S}}(s_2, s_2').$$

Consider the corresponding $(\infty,2)$-categories

$$(\mathbb{S}_1 \circledast \mathbb{S}_2)^{R_{\mathbb{S}_2}} := (\mathbb{S}_1 \circledast \mathbb{S}_2)^{R_{\mathbf{C}_2}} \text{ and } (\mathbb{S}_1 \circledast \mathbb{S}_2, \mathbb{T})^{L_{\mathbb{S}_1}} := (\mathbb{S}_1 \circledast \mathbb{S}_2)^{R_{\mathbf{C}_1}},$$

see Sect. 1.2.1.

3.1.2. For another $(\infty,2)$-category, let

$$\text{Maps}(\mathbb{S}_1 \circledast \mathbb{S}_2, \mathbb{T})^{R_{\mathbb{S}_2}} := \text{Maps}(\mathbb{S}_1 \circledast \mathbb{S}_2, \mathbb{T})^{R_{\mathbf{C}_2}}$$

and

$$\text{Maps}(\mathbb{S}_1 \circledast \mathbb{S}_2, \mathbb{T})^{L_{\mathbb{S}_1}} := \text{Maps}(\mathbb{S}_1 \circledast \mathbb{S}_2, \mathbb{T})^{L_{\mathbf{C}_1}}$$

denote the corresponding full subspaces, see Sect. 1.1.8.

3.1.3. Consider the bi-simplicial space

$$(\text{Sq}^{\sim}_{\bullet,\bullet}(\mathbb{S}_2^{1\text{-op}}))^{\text{reflect}} \times \text{Sq}^{\sim}_{\bullet,\bullet}(\mathbb{S}_1).$$

Recall (see Chapter 10, Formula (4.6)), there is a canonically defined map

$$(\text{Sq}^{\sim}_{\bullet,\bullet}(\mathbb{S}_2^{1\text{-op}}))^{\text{reflect}} \times \text{Sq}^{\sim}_{\bullet,\bullet}(\mathbb{S}_1) \to \text{Sq}_{\bullet,\bullet}(\mathbb{S}_2^{1\&2\text{-op}} \circledast \mathbb{S}_1). \tag{3.4}$$

Moreover, by Chapter 10, Proposition 4.5.4, the map

$$\mathfrak{L}^{\text{Sq}}\left((\text{Sq}^{\sim}_{\bullet,\bullet}(\mathbb{S}_2^{1\text{-op}}))^{\text{reflect}} \times \text{Sq}^{\sim}_{\bullet,\bullet}(\mathbb{S}_1)\right) \to \mathbb{S}_2^{1\&2\text{-op}} \circledast \mathbb{S}_1, \tag{3.5}$$

obtained from (3.4) by adjunction, is an isomorphism.

3.1.4. We claim that there is a canonically defined map

$$\mathfrak{L}^{\mathrm{Sq}}\left((\mathrm{Sq}^{\sim}_{\bullet,\bullet}(\mathbb{S}_2^{1\text{-op}}))^{\mathrm{reflect}} \times \mathrm{Sq}^{\sim}_{\bullet,\bullet}(\mathbb{S}_1)\right) \to (\mathbb{S}_1 \circledast \mathbb{S}_2)^{R_{\mathbb{S}_2}}. \tag{3.6}$$

Indeed, using Chapter 10, Formula (4.6) again, we obtain a map

$$(\mathrm{Sq}^{\sim}_{\bullet,\bullet}(\mathbb{S}_2))^{\mathrm{reflect}} \times \mathrm{Sq}^{\sim}_{\bullet,\bullet}(\mathbb{S}_1^{2\text{-op}}) \to \mathrm{Sq}_{\bullet,\bullet}(\mathbb{S}_2^{2\text{-op}} \circledast \mathbb{S}_1^{2\text{-op}}).$$

However, it follows by unwinding the construction, that the essential image of the latter map belongs to the full subspace

$$\mathrm{Sq}^{\mathrm{Pair}}_{\bullet,\bullet}(\mathbb{S}_2^{2\text{-op}} \circledast \mathbb{S}_1^{2\text{-op}}, \mathbf{C}_2) \subset \mathrm{Sq}_{\bullet,\bullet}(\mathbb{S}_2^{2\text{-op}} \circledast \mathbb{S}_1^{2\text{-op}})$$

thereby giving rise to a map

$$(\mathrm{Sq}^{\sim}_{\bullet,\bullet}(\mathbb{S}_2))^{\mathrm{reflect}} \times \mathrm{Sq}^{\sim}_{\bullet,\bullet}(\mathbb{S}_1^{2\text{-op}}) \to \mathrm{Sq}^{\mathrm{Pair}}_{\bullet,\bullet}(\mathbb{S}_2^{2\text{-op}} \circledast \mathbb{S}_1^{2\text{-op}}, \mathbf{C}_2) \tag{3.7}$$

Now, the sought-for map (3.6) is given by applying $\mathfrak{L}^{\mathrm{Sq}}$ to the composite map

(3.8)

$$\begin{aligned}(\mathrm{Sq}^{\sim}_{\bullet,\bullet}(\mathbb{S}_2^{1\text{-op}}))^{\mathrm{reflect}} \times \mathrm{Sq}^{\sim}_{\bullet,\bullet}(\mathbb{S}_1) \simeq \left((\mathrm{Sq}^{\sim}_{\bullet,\bullet}(\mathbb{S}_2))^{\mathrm{horiz\text{-}op}}\right)^{\mathrm{reflect}} \times (\mathrm{Sq}^{\sim}_{\bullet,\bullet}(\mathbb{S}_1^{2\text{-op}}))^{\mathrm{vert\text{-}op}} \simeq \\ \simeq \left((\mathrm{Sq}^{\sim}_{\bullet,\bullet}(\mathbb{S}_2))^{\mathrm{reflect}} \times \mathrm{Sq}^{\sim}_{\bullet,\bullet}(\mathbb{S}_1^{2\text{-op}})\right)^{\mathrm{vert\text{-}op}} \overset{(3.7)}{\longrightarrow} \left((\mathrm{Sq}^{\mathrm{Pair}}_{\bullet,\bullet}(\mathbb{S}_2^{2\text{-op}} \circledast \mathbb{S}_1^{2\text{-op}}, \mathbf{C}_2)\right)^{\mathrm{vert\text{-}op}} \simeq \\ \simeq \left((\mathrm{Sq}^{\mathrm{Pair}}_{\bullet,\bullet}((\mathbb{S}_1 \circledast \mathbb{S}_2)^{2\text{-op}}, \mathbf{C}_2)\right)^{\mathrm{vert\text{-}op}}.\end{aligned}$$

3.1.5. We claim:

THEOREM 3.1.6. *The composition*

$$\mathrm{Maps}(\mathbb{S}_1 \circledast \mathbb{S}_2, \mathbb{T})^{R_{\mathbb{S}_2}} \simeq \mathrm{Maps}((\mathbb{S}_1 \circledast \mathbb{S}_2)^{R_{\mathbb{S}_2}}, \mathbb{T}) \overset{(3.6)}{\longrightarrow}$$

$$\to \mathrm{Maps}\left(\mathfrak{L}^{\mathrm{Sq}}\left((\mathrm{Sq}^{\sim}_{\bullet,\bullet}(\mathbb{S}_2^{1\text{-op}}))^{\mathrm{reflect}} \times \mathrm{Sq}^{\sim}_{\bullet,\bullet}(\mathbb{S}_1)\right), \mathbb{T})\right) \overset{(3.5)}{\simeq} \mathrm{Maps}(\mathbb{S}_2^{1\&2\text{-op}} \circledast \mathbb{S}_1, \mathbb{T})$$

is fully faithful with essential image equal to

$$\mathrm{Maps}(\mathbb{S}_2^{1\&2\text{-op}} \circledast \mathbb{S}_1, \mathbb{T})^{L_{\mathbb{S}_2^{1\&2\text{-op}}}} \subset \mathrm{Maps}(\mathbb{S}_2^{1\&2\text{-op}} \circledast \mathbb{S}_1, \mathbb{T}).$$

As a corollary, we obtain:

COROLLARY 3.1.7. *There exists a canonical isomorphism*

$$\mathrm{Maps}(\mathbb{S}_1 \circledast \mathbb{S}_2, \mathbb{T})^{R_{\mathbb{S}_2}} \simeq \mathrm{Maps}(\mathbb{S}_2^{1\&2\text{-op}} \circledast \mathbb{S}_1, \mathbb{T})^{L_{\mathbb{S}_2^{1\&2\text{-op}}}}.$$

3.1.8. Since the equivalence of Corollary 3.1.7 is by construction functorial in $\mathbb{S}_1 \in$ 2-Cat, we obtain:

COROLLARY 3.1.9. *For* $\mathbb{S}, \mathbb{T} \in$ 2-Cat, *the isomorphism of* (1.10) *upgrades to an equivalence of* $(\infty, 2)$*-categories*

$$\mathrm{Funct}(\mathbb{S}, \mathbb{T})^{R}_{\mathrm{right\text{-}lax}} \simeq \mathrm{Funct}(\mathbb{S}^{1\&2\text{-op}}, \mathbb{T})^{L}_{\mathrm{left\text{-}lax}}.$$

3.2. Proof of Theorem 3.1.6. We will give a proof in the particular case when the target category $\mathbb{T}$ is **1 - Cat**. The general case is deduced by the same procedure as one employed in Sect. 2.5.

3.2.1. First, we claim that the assertion holds for **1 - Cat** replaced by $\mathbf{1\text{-}Cat}_{\mathrm{ordn}}$, i.e., when we consider functors from $\mathbb{S}^{2\text{-op}}$ (resp., $\mathbb{S}^L$) with values in ordinary 2-categories. This can be checked directly as in Sect. 1.4.

3.2.2. The datum of a map

$$\mathfrak{L}^{\mathrm{Sq}}\left((\mathrm{Sq}^{\sim}_{\bullet,\bullet}(\mathbb{S}_2^{1\text{-op}}))^{\mathrm{reflect}} \times \mathrm{Sq}^{\sim}_{\bullet,\bullet}(\mathbb{S}_1)\right) \to \mathbf{1}\text{-}\mathbf{Cat}$$

is equivalent to the datum of a map of bi-simplicial spaces

$$(\mathrm{Sq}^{\sim}_{\bullet,\bullet}(\mathbb{S}_2^{1\text{-op}}))^{\mathrm{reflect}} \times \mathrm{Sq}^{\sim}_{\bullet,\bullet}(\mathbb{S}_1) \to \mathrm{Sq}_{\bullet,\bullet}(\mathbf{1}\text{-}\mathbf{Cat}),$$

or equivalently

$$\left((\mathrm{Sq}^{\sim}_{\bullet,\bullet}(\mathbb{S}_2^{1\text{-op}}))^{\mathrm{reflect}} \times \mathrm{Sq}^{\sim}_{\bullet,\bullet}(\mathbb{S}_1)\right)^{\mathrm{vert\text{-}op}} \to \left(\mathrm{Sq}_{\bullet,\bullet}(\mathbf{1}\text{-}\mathbf{Cat})\right)^{\mathrm{vert\text{-}op}}.$$

3.2.3. Let us describe the subspace

$$\mathrm{Maps}_{\mathrm{Spc}^{\mathbf{\Delta}^{\mathrm{op}}\times\mathbf{\Delta}^{\mathrm{op}}}}\Big(\left((\mathrm{Sq}^{\sim}_{\bullet,\bullet}(\mathbb{S}_2^{1\text{-op}}))^{\mathrm{reflect}} \times \mathrm{Sq}^{\sim}_{\bullet,\bullet}(\mathbb{S}_1)\right)^{\mathrm{vert\text{-}op}}, \left(\mathrm{Sq}_{\bullet,\bullet}(\mathbf{1}\text{-}\mathbf{Cat})\right)^{\mathrm{vert\text{-}op}}\Big)^{L_{\mathbb{S}_2^{1\&2\text{-op}}}} \subset$$

$$\subset \mathrm{Maps}_{\mathrm{Spc}^{\mathbf{\Delta}^{\mathrm{op}}\times\mathbf{\Delta}^{\mathrm{op}}}}\Big(\left((\mathrm{Sq}^{\sim}_{\bullet,\bullet}(\mathbb{S}_2^{1\text{-op}}))^{\mathrm{reflect}} \times \mathrm{Sq}^{\sim}_{\bullet,\bullet}(\mathbb{S}_1)\right)^{\mathrm{vert\text{-}op}}, \left(\mathrm{Sq}_{\bullet,\bullet}(\mathbf{1}\text{-}\mathbf{Cat})\right)^{\mathrm{vert\text{-}op}}\Big)$$

that corresponds to

$$\mathrm{Maps}_{2\text{-}\mathrm{Cat}}(\mathbb{S}_2^{1\&2\text{-op}} \circledast \mathbb{S}_1, \mathbf{1}\text{-}\mathbf{Cat})^{L_{\mathbb{S}_2^{1\&2\text{-op}}}} \subset \mathrm{Maps}_{2\text{-}\mathrm{Cat}}(\mathbb{S}_2^{1\&2\text{-op}} \circledast \mathbb{S}_1, \mathbf{1}\text{-}\mathbf{Cat}) \overset{(3.5)}{\simeq}$$

$$\overset{(3.5)}{\simeq} \mathrm{Maps}_{\mathrm{Spc}^{\mathbf{\Delta}^{\mathrm{op}}\times\mathbf{\Delta}^{\mathrm{op}}}}\left((\mathrm{Sq}^{\sim}_{\bullet,\bullet}(\mathbb{S}_2^{1\text{-op}}))^{\mathrm{reflect}} \times \mathrm{Sq}^{\sim}_{\bullet,\bullet}(\mathbb{S}_1), \mathrm{Sq}_{\bullet,\bullet}(\mathbf{1}\text{-}\mathbf{Cat})\right) \simeq$$

$$\simeq \mathrm{Maps}_{\mathrm{Spc}^{\mathbf{\Delta}^{\mathrm{op}}\times\mathbf{\Delta}^{\mathrm{op}}}}\Big(\left((\mathrm{Sq}^{\sim}_{\bullet,\bullet}(\mathbb{S}_2^{1\text{-op}}))^{\mathrm{reflect}} \times \mathrm{Sq}^{\sim}_{\bullet,\bullet}(\mathbb{S}_1)\right)^{\mathrm{vert\text{-}op}}, \left(\mathrm{Sq}_{\bullet,\bullet}(\mathbf{1}\text{-}\mathbf{Cat})\right)^{\mathrm{vert\text{-}op}}\Big)$$

Namely, we claim that

$$\mathrm{Maps}_{\mathrm{Spc}^{\mathbf{\Delta}^{\mathrm{op}}\times\mathbf{\Delta}^{\mathrm{op}}}}\Big(\left((\mathrm{Sq}^{\sim}_{\bullet,\bullet}(\mathbb{S}_2^{1\text{-op}}))^{\mathrm{reflect}} \times \mathrm{Sq}^{\sim}_{\bullet,\bullet}(\mathbb{S}_1)\right)^{\mathrm{vert\text{-}op}}, \left(\mathrm{Sq}_{\bullet,\bullet}(\mathbf{1}\text{-}\mathbf{Cat})\right)^{\mathrm{vert\text{-}op}}\Big)^{L_{\mathbb{S}_2^{1\&2\text{-op}}}}$$

$$= \mathrm{Maps}_{\mathrm{Spc}^{\mathbf{\Delta}^{\mathrm{op}}\times\mathbf{\Delta}^{\mathrm{op}}}}\Big(\left((\mathrm{Sq}^{\sim}_{\bullet,\bullet}(\mathbb{S}_2^{1\text{-op}}))^{\mathrm{reflect}} \times \mathrm{Sq}^{\sim}_{\bullet,\bullet}(\mathbb{S}_1)\right)^{\mathrm{vert\text{-}op}}, {}'\left(\mathrm{Sq}_{\bullet,\bullet}(\mathbf{1}\text{-}\mathbf{Cat})\right)^{\mathrm{vert\text{-}op}}\Big),$$

where

$${}'\left(\mathrm{Sq}_{\bullet,\bullet}(\mathbf{1}\text{-}\mathbf{Cat})\right)^{\mathrm{vert\text{-}op}} \subset \left(\mathrm{Sq}_{\bullet,\bullet}(\mathbf{1}\text{-}\mathbf{Cat})\right)^{\mathrm{vert\text{-}op}}$$

is as in (2.3).

Indeed, this assertion can be checked at the level of of ordinary categories, where it is a straightforward verification.

3.2.4. To summarize, we obtain a canonical identification

$$\mathrm{Maps}_{2\text{-}\mathrm{Cat}}(\mathbb{S}_2^{1\&2\text{-op}} \circledast \mathbb{S}_1, \mathbf{1}\text{-}\mathbf{Cat})^{L_{\mathbb{S}_2^{1\&2\text{-op}}}} \simeq \tag{3.9}$$

$$\simeq \mathrm{Maps}_{\mathrm{Spc}^{\mathbf{\Delta}^{\mathrm{op}}\times\mathbf{\Delta}^{\mathrm{op}}}}\Big(\left((\mathrm{Sq}^{\sim}_{\bullet,\bullet}(\mathbb{S}_2^{1\text{-op}}))^{\mathrm{reflect}} \times \mathrm{Sq}^{\sim}_{\bullet,\bullet}(\mathbb{S}_1)\right)^{\mathrm{vert\text{-}op}}, \tag{3.10}$$

$${}'\left(\mathrm{Sq}_{\bullet,\bullet}(\mathbf{1}\text{-}\mathbf{Cat})\right)^{\mathrm{vert\text{-}op}}\Big). \tag{3.11}$$

3.2.5. According to Chapter 10, Proposition 4.5.4, the datum of a map

$$\mathbb{S}_1 \circledast \mathbb{S}_2 \to \mathbf{1}\text{-}\mathbf{Cat}$$

is equivalent to the datum of a bi-simplicial map

$$(\mathrm{Sq}^{\sim}_{\bullet,\bullet}(\mathbb{S}_1^{2\text{-op}}))^{\mathrm{reflect}} \times \mathrm{Sq}^{\sim}_{\bullet,\bullet}(\mathbb{S}_2) \to \mathrm{Sq}_{\bullet,\bullet}(\mathbf{1}\text{-}\mathbf{Cat}),$$

or, equivalently,

$$\left((\mathrm{Sq}^{\sim}_{\bullet,\bullet}(\mathbb{S}_1^{2\text{-op}}))^{\mathrm{reflect}} \times \mathrm{Sq}^{\sim}_{\bullet,\bullet}(\mathbb{S}_2)\right)^{\mathrm{reflect}} \to \left(\mathrm{Sq}_{\bullet,\bullet}(\mathbf{1}\text{-}\mathbf{Cat})\right)^{\mathrm{reflect}}.$$

3.2.6. Let us describe the subspace

$$\mathrm{Maps}_{\mathrm{Spc}^{\mathbf{\Delta}^{\mathrm{op}}\times\mathbf{\Delta}^{\mathrm{op}}}}\left(\left((\mathrm{Sq}^{\sim}_{\bullet,\bullet}(\mathbb{S}_1^{2\text{-op}}))^{\mathrm{reflect}} \times \mathrm{Sq}^{\sim}_{\bullet,\bullet}(\mathbb{S}_2)\right)^{\mathrm{reflect}}, \left(\mathrm{Sq}_{\bullet,\bullet}(\mathbf{1}\text{-}\mathbf{Cat})\right)^{\mathrm{reflect}}\right)^{R_{\mathbb{S}_2}} \subset$$
$$\subset \mathrm{Maps}_{\mathrm{Spc}^{\mathbf{\Delta}^{\mathrm{op}}\times\mathbf{\Delta}^{\mathrm{op}}}}\left(\left((\mathrm{Sq}^{\sim}_{\bullet,\bullet}(\mathbb{S}_1^{2\text{-op}}))^{\mathrm{reflect}} \times \mathrm{Sq}^{\sim}_{\bullet,\bullet}(\mathbb{S}_2)\right)^{\mathrm{reflect}}, \left(\mathrm{Sq}_{\bullet,\bullet}(\mathbf{1}\text{-}\mathbf{Cat})\right)^{\mathrm{reflect}}\right)$$

that corresponds to

$$\mathrm{Maps}_{2\text{-Cat}}(\mathbb{S}_1 \circledast \mathbb{S}_2, \mathbf{1}\text{-}\mathbf{Cat})^{R_{\mathbb{S}_2}} \subset \mathrm{Maps}_{2\text{-Cat}}(\mathbb{S}_1 \circledast \mathbb{S}_2, \mathbf{1}\text{-}\mathbf{Cat}).$$

Namely, we claim that

$$\mathrm{Maps}_{\mathrm{Spc}^{\mathbf{\Delta}^{\mathrm{op}}\times\mathbf{\Delta}^{\mathrm{op}}}}\left(\left((\mathrm{Sq}^{\sim}_{\bullet,\bullet}(\mathbb{S}_1^{2\text{-op}}))^{\mathrm{reflect}} \times \mathrm{Sq}^{\sim}_{\bullet,\bullet}(\mathbb{S}_2)\right)^{\mathrm{reflect}}, \left(\mathrm{Sq}_{\bullet,\bullet}(\mathbf{1}\text{-}\mathbf{Cat})\right)^{\mathrm{reflect}}\right)^{R_{\mathbb{S}_2}} =$$
$$= \mathrm{Maps}_{\mathrm{Spc}^{\mathbf{\Delta}^{\mathrm{op}}\times\mathbf{\Delta}^{\mathrm{op}}}}\left(\left((\mathrm{Sq}^{\sim}_{\bullet,\bullet}(\mathbb{S}_1^{2\text{-op}}))^{\mathrm{reflect}} \times \mathrm{Sq}^{\sim}_{\bullet,\bullet}(\mathbb{S}_2)\right)^{\mathrm{reflect}}, {}'\left(\mathrm{Sq}_{\bullet,\bullet}(\mathbf{1}\text{-}\mathbf{Cat})\right)^{\mathrm{reflect}}\right),$$

where

$${}'\left(\mathrm{Sq}_{\bullet,\bullet}(\mathbf{1}\text{-}\mathbf{Cat})\right)^{\mathrm{reflect}} \subset \left(\mathrm{Sq}_{\bullet,\bullet}(\mathbf{1}\text{-}\mathbf{Cat})\right)^{\mathrm{reflect}}$$

is an in (2.3).

Indeed, this assertion can be checked at the level of of ordinary categories, where it is a straightforward verification.

3.2.7. To summarize, we obtain a canonical identification

$$\mathrm{Maps}_{2\text{-Cat}}(\mathbb{S}_1 \circledast \mathbb{S}_2, \mathbf{1}\text{-}\mathbf{Cat})^{R_{\mathbb{S}_2}} \simeq \tag{3.12}$$
$$\simeq \mathrm{Maps}_{\mathrm{Spc}^{\mathbf{\Delta}^{\mathrm{op}}\times\mathbf{\Delta}^{\mathrm{op}}}}\left(\left((\mathrm{Sq}^{\sim}_{\bullet,\bullet}(\mathbb{S}_1^{2\text{-op}}))^{\mathrm{reflect}} \times \mathrm{Sq}^{\sim}_{\bullet,\bullet}(\mathbb{S}_2)\right)^{\mathrm{reflect}}, {}'\left(\mathrm{Sq}_{\bullet,\bullet}(\mathbf{1}\text{-}\mathbf{Cat})\right)^{\mathrm{reflect}}\right)$$

3.2.8. Note that we have a tautological identification:

$$\left((\mathrm{Sq}^{\sim}_{\bullet,\bullet}(\mathbb{S}_2^{1\text{-op}}))^{\mathrm{reflect}} \times \mathrm{Sq}^{\sim}_{\bullet,\bullet}(\mathbb{S}_1)\right)^{\mathrm{vert\text{-}op}} \simeq \left((\mathrm{Sq}^{\sim}_{\bullet,\bullet}(\mathbb{S}_1^{2\text{-op}}))^{\mathrm{reflect}} \times \mathrm{Sq}^{\sim}_{\bullet,\bullet}(\mathbb{S}_2)\right)^{\mathrm{reflect}}.$$

Hence, from the isomorphisms (3.9) and (3.12) and the isomorphism

$$'\left(\mathrm{Sq}_{\bullet,\bullet}(\mathbf{1}\text{-}\mathbf{Cat})\right)^{\mathrm{vert\text{-}op}} \simeq {}'\left(\mathrm{Sq}_{\bullet,\bullet}(\mathbf{1}\text{-}\mathbf{Cat})\right)^{\mathrm{reflect}},$$

we obtain an identification

$$\mathrm{Maps}_{2\text{-}\mathrm{Cat}}(\mathbb{S}_2^{1\&2\text{-op}} \circledast \mathbb{S}_1, \mathbf{1}\text{-}\mathbf{Cat})^{L_{\mathbb{S}_2^{1\&2\text{-op}}}} \simeq \mathrm{Maps}_{2\text{-}\mathrm{Cat}}(\mathbb{S}_1 \circledast \mathbb{S}_2, \mathbf{1}\text{-}\mathbf{Cat})^{R_{\mathbb{S}_2}}. \tag{3.13}$$

3.2.9. Consider now the map

$$\mathrm{Maps}_{2\text{-}\mathrm{Cat}}(\mathbb{S}_1 \circledast \mathbb{S}_2, \mathbf{1}\text{-}\mathbf{Cat})^{R_{\mathbb{S}_2}} \overset{(1.2)}{\longrightarrow} \mathrm{Maps}_{2\text{-}\mathrm{Cat}}((\mathbb{S}_1 \circledast \mathbb{S}_2)^{R_{\mathbb{S}_2}}, \mathbf{1}\text{-}\mathbf{Cat}) \overset{(3.6)}{\longrightarrow}$$
$$\to \mathrm{Maps}_{\mathrm{Spc}^{\mathbf{\Delta}^{\mathrm{op}} \times \mathbf{\Delta}^{\mathrm{op}}}}\left((\mathrm{Sq}^{\sim}_{\bullet,\bullet}(\mathbb{S}_2^{1\text{-op}}))^{\mathrm{reflect}} \times \mathrm{Sq}^{\sim}_{\bullet,\bullet}(\mathbb{S}_1), \mathbf{1}\text{-}\mathbf{Cat}\right)$$

that appears in Theorem 3.1.6.

It follows from the construction that it equals the composition

$$\mathrm{Maps}_{2\text{-}\mathrm{Cat}}(\mathbb{S}_1 \circledast \mathbb{S}_2, \mathbf{1}\text{-}\mathbf{Cat})^{R_{\mathbb{S}_2}} \overset{(3.12)}{\simeq}$$
$$\overset{(3.12)}{\simeq} \mathrm{Maps}_{\mathrm{Spc}^{\mathbf{\Delta}^{\mathrm{op}} \times \mathbf{\Delta}^{\mathrm{op}}}}\Big(\left((\mathrm{Sq}^{\sim}_{\bullet,\bullet}(\mathbb{S}_1^{2\text{-op}}))^{\mathrm{reflect}} \times \mathrm{Sq}^{\sim}_{\bullet,\bullet}(\mathbb{S}_2)\right)^{\mathrm{reflect}},$$
$${}'\left(\mathrm{Sq}_{\bullet,\bullet}(\mathbf{1}\text{-}\mathbf{Cat})\right)^{\mathrm{reflect}}\Big) =$$
$$= \mathrm{Maps}_{\mathrm{Spc}^{\mathbf{\Delta}^{\mathrm{op}} \times \mathbf{\Delta}^{\mathrm{op}}}}\Big(\left((\mathrm{Sq}^{\sim}_{\bullet,\bullet}(\mathbb{S}_1^{2\text{-op}}))^{\mathrm{reflect}} \times \mathrm{Sq}^{\sim}_{\bullet,\bullet}(\mathbb{S}_2)\right)^{\mathrm{reflect}},$$
$${}'\left(\mathrm{Sq}_{\bullet,\bullet}(\mathbf{1}\text{-}\mathbf{Cat})\right)^{\mathrm{vert\text{-}op}}\Big) \hookrightarrow$$
$$\hookrightarrow \mathrm{Maps}_{\mathrm{Spc}^{\mathbf{\Delta}^{\mathrm{op}} \times \mathbf{\Delta}^{\mathrm{op}}}}\Big(\left((\mathrm{Sq}^{\sim}_{\bullet,\bullet}(\mathbb{S}_1^{2\text{-op}}))^{\mathrm{reflect}} \times \mathrm{Sq}^{\sim}_{\bullet,\bullet}(\mathbb{S}_2)\right)^{\mathrm{reflect}},$$
$$\left(\mathrm{Sq}_{\bullet,\bullet}(\mathbf{1}\text{-}\mathbf{Cat})\right)^{\mathrm{vert\text{-}op}}\Big) \simeq$$
$$\simeq \mathrm{Maps}_{\mathrm{Spc}^{\mathbf{\Delta}^{\mathrm{op}} \times \mathbf{\Delta}^{\mathrm{op}}}}\left(\left(\left((\mathrm{Sq}^{\sim}_{\bullet,\bullet}(\mathbb{S}_1^{2\text{-op}}))^{\mathrm{reflect}} \times \mathrm{Sq}^{\sim}_{\bullet,\bullet}(\mathbb{S}_2)\right)^{\mathrm{reflect}}\right)^{\mathrm{vert\text{-}op}},\right.$$
$$\left.\mathrm{Sq}_{\bullet,\bullet}(\mathbf{1}\text{-}\mathbf{Cat})\right) \simeq$$
$$\simeq \mathrm{Maps}_{\mathrm{Spc}^{\mathbf{\Delta}^{\mathrm{op}} \times \mathbf{\Delta}^{\mathrm{op}}}}\left((\mathrm{Sq}^{\sim}_{\bullet,\bullet}(\mathbb{S}_2^{1\text{-op}}))^{\mathrm{reflect}} \times \mathrm{Sq}^{\sim}_{\bullet,\bullet}(\mathbb{S}_1), \mathrm{Sq}_{\bullet,\bullet}(\mathbf{1}\text{-}\mathbf{Cat})\right),$$

thus implying the assertion of the theorem.

4. An alternative proof

In this section we will give an alternative proof of Theorem 1.2.4 and Corollary 3.1.7.

4.1. An alternative proof of Corollary 3.1.7 for $\mathbb{T} = \mathbf{1\text{-}Cat}$. This proof will make a substantial use of the unstraightening equivalence of Chapter 11, Corollary 1.2.6.

We need to show that for $\mathbb{S}_1, \mathbb{S}_2 \in \mathbf{2\text{-}Cat}$ there exists a canonical isomorphism

$$\text{Maps}(\mathbb{S}_1 \circledast \mathbb{S}_2, \mathbf{1\text{-}Cat})^{R_{\mathbb{S}_2}} \simeq \text{Maps}(\mathbb{S}_2^{1\&2\text{-op}} \circledast \mathbb{S}_1, \mathbf{1\text{-}Cat})^{L_{\mathbb{S}_2^{1\&2\text{-op}}}}. \tag{4.1}$$

4.1.1. For a given $(\infty,2)$-category $\mathbb{S}$, recall the full subcategory

$$\mathbf{1\text{-}Cart}_{/\mathbb{S}} \subset \mathbf{2\text{-}Cat}_{/\mathbb{S}},$$

see Chapter 11, Sect. 1.2.3.

We let $\mathbf{1\text{-}biCart}_{/\mathbb{S}}$ be the full subcategory of $\mathbf{1\text{-}Cart}_{/\mathbb{S}}$ that consists of objects that under the equivalence

$$\mathbf{2\text{-}Cat}_{/\mathbb{S}} \simeq (\mathbf{2\text{-}Cat}_{/\mathbb{S}^{1\text{-op}}})^{2\text{-op}}, \quad \mathbb{T} \mapsto \mathbb{T}^{1\text{-op}}$$

corresponds to

$$\mathbf{1\text{-}Cart}_{/\mathbb{S}^{1\text{-op}}} \subset \mathbf{2\text{-}Cat}_{/\mathbb{S}^{1\text{-op}}}.$$

We have a tautological fully faithful embedding

$$\mathbf{1\text{-}biCart}_{/\mathbb{S}} \hookrightarrow (\mathbf{1\text{-}Cart}_{/\mathbb{S}^{1\text{-op}}})^{2\text{-op}}. \tag{4.2}$$

4.1.2. By definition, the left-hand side in (4.1) is a full subspace in

$$\text{Maps}(\mathbb{S}_1 \circledast \mathbb{S}_2, \mathbf{1\text{-}Cat}) \simeq \text{Maps}(\mathbb{S}_1, \text{Funct}(\mathbb{S}_2, \mathbf{1\text{-}Cat})_{\text{right-lax}}),$$

which we rewrite, using Chapter 11, Corollary 1.2.6, as

$$\text{Maps}(\mathbb{S}_1, \mathbf{1\text{-}Cart}_{/\mathbb{S}_2^{1\text{-op}}}).$$

Now, it follows from the definitions that the full subspace in question corresponds to

$$\text{Maps}(\mathbb{S}_1, \mathbf{1\text{-}biCart}_{/\mathbb{S}_2^{1\text{-op}}}).$$

4.1.3. Similarly, the right-hand side in (4.1) is a full subspace in

$$\text{Maps}(\mathbb{S}_2^{1\&2\text{-op}} \circledast \mathbb{S}_1, \mathbf{1\text{-}Cat}) \simeq \text{Maps}(\mathbb{S}_1^{2\text{-op}} \circledast \mathbb{S}_2^{1\text{-op}}, \mathbf{1\text{-}Cat}^{2\text{-op}}) \simeq$$
$$\simeq \text{Maps}(\mathbb{S}_1^{2\text{-op}} \circledast \mathbb{S}_2^{1\text{-op}}, \mathbf{1\text{-}Cat}),$$

where the last isomorphism comes from the identification $\mathbf{1\text{-}Cat} \simeq \mathbf{1\text{-}Cat}^{2\text{-op}}$ given by $\mathbb{T} \mapsto \mathbb{T}^{1\text{-op}}$.

We rewrite

$$\text{Maps}(\mathbb{S}_1^{2\text{-op}} \circledast \mathbb{S}_2^{1\text{-op}}, \mathbf{1\text{-}Cat}) \simeq \text{Maps}(\mathbb{S}_1^{2\text{-op}}, \text{Funct}(\mathbb{S}_2^{1\text{-op}}, \mathbf{1\text{-}Cat})_{\text{right-lax}}),$$

and further, using Chapter 11, Corollary 1.2.6, as

$$\text{Maps}(\mathbb{S}_1^{2\text{-op}}, \mathbf{1\text{-}Cart}_{/\mathbb{S}_2}) \simeq \text{Maps}(\mathbb{S}_1, (\mathbf{1\text{-}Cart}_{/\mathbb{S}_2})^{2\text{-op}}).$$

It again follows from the definitions that the right-hand side of (4.1), viewed as a full subspace in $\text{Maps}(\mathbb{S}_1, (\mathbf{1\text{-}Cart}_{/\mathbb{S}_2})^{2\text{-op}})$ equals to

$$\text{Maps}(\mathbb{S}_1, (\mathbf{1\text{-}biCart}_{/\mathbb{S}_2^{1\text{-op}}})) \subset \text{Maps}(\mathbb{S}_1, (\mathbf{1\text{-}Cart}_{/\mathbb{S}_2})^{2\text{-op}})$$

with respect to the fully faithful embedding (4.2).

4.1.4. Comparing the descriptions of the left-hand side and the right-hand side of (4.1), given in Sects. 4.1.2 and 4.1.3 above, we obtain the desired isomorphism. □

4.2. An alternative proof of Corollary 3.1.7 for a general $\mathbb{T}$. We need to show that for $\mathbb{S}_1, \mathbb{S}_2, \mathbb{T} \in \mathbf{2}\text{-}\mathbf{Cat}$ there exists a canonical isomorphism

$$\mathrm{Maps}(\mathbb{S}_1 \circledast \mathbb{S}_2, \mathbb{T})^{R_{\mathbb{S}_2}} \simeq \mathrm{Maps}(\mathbb{S}_2^{1\&2\text{-op}} \circledast \mathbb{S}_1, \mathbb{T})^{L_{\mathbb{S}_2^{1\&2\text{-op}}}}. \tag{4.3}$$

4.2.1. Note that the above isomorphism is equivalent to Corollary 3.1.9:

$$\mathrm{Funct}(\mathbb{S}, \mathbb{T})^R_{\text{right-lax}} \simeq \mathrm{Funct}(\mathbb{S}^{1\&2\text{-op}}, \mathbb{T})^L_{\text{left-lax}}.$$

In particular, we recover the isomorphism of spaces

$$\mathrm{Maps}_{2\text{-Cat}}(\mathbb{S}, \mathbb{T})^R_{\text{right-lax}} \simeq \mathrm{Maps}_{2\text{-Cat}}(\mathbb{S}^{1\&2\text{-op}}, \mathbb{T})^L_{\text{left-lax}}$$

of (1.10).

4.2.2. We start with the following lemma:

LEMMA 4.2.3. *Let $\mathbb{T}$ be an $(\infty,2)$-category, and let $\alpha : t_0 \to t_1$ be a 1-morphism. Then α admits a left adjoint if and only if the following two conditions hold:*
(i) *For every $s \in \mathbb{T}$, the resulting functor of $(\infty,1)$-categories*

$$\mathbf{Maps}_{\mathbb{T}}(s,t) \xrightarrow{\alpha\circ -} \mathbf{Maps}_{\mathbb{T}}(s,t')$$

admits a left adjoint;
(ii) *The Beck-Chevalley condition is satisfied. I.e., for every 1-morphism $\beta : s_0 \to s_1$, the corresponding natural transformation*

$$(\alpha \circ -)^L \circ (- \circ \beta) \to (- \circ \beta) \circ (\alpha \circ -)^L,$$

arising by adjunction from the isomorphism

$$(- \circ \beta) \circ (\alpha \circ -) \simeq (\alpha \circ -) \circ (- \circ \beta),$$

is an isomorphism.

PROOF. The assertion reduces to the case when $\mathbb{T}$ is an *ordinary* 2-category, and the latter is a straightforward verification. □

4.2.4. The Yoneda embedding for $\mathbb{T}$ gives rise to a fully faithful map

$$\mathrm{Maps}(\mathbb{S}_1 \circledast \mathbb{S}_2, \mathbb{T}) \hookrightarrow \mathrm{Maps}(\mathbb{S}_1 \circledast \mathbb{S}_2, \mathrm{Funct}(\mathbb{T}^{1\text{-op}}, \mathbf{1}\text{-}\mathbf{Cat})) \simeq$$
$$\simeq \mathrm{Maps}((\mathbb{S}_1 \circledast \mathbb{S}_2) \times \mathbb{T}^{1\text{-op}}), \mathbf{1}\text{-}\mathbf{Cat}),$$

which we further compose with the fully faithful embedding

$$\mathrm{Maps}((\mathbb{S}_1 \circledast \mathbb{S}_2) \times \mathbb{T}^{1\text{-op}}), \mathbf{1}\text{-}\mathbf{Cat}) \hookrightarrow \mathrm{Maps}(\mathbb{S}_1 \circledast \mathbb{T}^{1\text{-op}} \circledast \mathbb{S}_2, \mathbf{1}\text{-}\mathbf{Cat}) \simeq$$
$$\simeq \mathrm{Maps}((\mathbb{S}_1 \circledast \mathbb{T}^{1\text{-op}}) \circledast \mathbb{S}_2, \mathbf{1}\text{-}\mathbf{Cat}).$$

It is easy to see that the image of

$$\mathrm{Maps}(\mathbb{S}_1 \circledast \mathbb{S}_2, \mathbb{T})^{R_{\mathbb{S}_2}} \subset \mathrm{Maps}(\mathbb{S}_1 \circledast \mathbb{S}_2, \mathbb{T})$$

belongs to

$$\mathrm{Maps}((\mathbb{S}_1 \circledast \mathbb{T}^{1\text{-op}}) \circledast \mathbb{S}_2, \mathbf{1}\text{-}\mathbf{Cat})^{R_{\mathbb{S}_2}} \subset \mathrm{Maps}((\mathbb{S}_1 \circledast \mathbb{T}^{1\text{-op}}) \circledast \mathbb{S}_2, \mathbf{1}\text{-}\mathbf{Cat}).$$

Applying the isomorphism (4.1), we rewrite

$$\mathrm{Maps}((\mathbb{S}_1 \circledast \mathbb{T}^{1\text{-op}}) \circledast \mathbb{S}_2, \mathbf{1\text{-}Cat})^{R_{\mathbb{S}_2}}$$

as

$$\mathrm{Maps}(\mathbb{S}_2^{1\&2\text{-op}} \circledast (\mathbb{S}_1 \circledast \mathbb{T}^{1\text{-op}}), \mathbf{1\text{-}Cat})^{L_{\mathbb{S}_2^{1\&2\text{-op}}}}.$$

4.2.5. Consider the resulting fully faithful embedding

$$\mathrm{Maps}(\mathbb{S}_1 \circledast \mathbb{S}_2, \mathbb{T})^{R_{\mathbb{S}_2}} \hookrightarrow \mathrm{Maps}(\mathbb{S}_2^{1\&2\text{-op}} \circledast (\mathbb{S}_1 \circledast \mathbb{T}^{1\text{-op}}), \mathbf{1\text{-}Cat}) \simeq$$
$$\simeq \mathrm{Maps}(\mathbb{S}_2^{1\&2\text{-op}} \circledast \mathbb{S}_1 \circledast \mathbb{T}^{1\text{-op}}, \mathbf{1\text{-}Cat}).$$

It follows from Lemma 4.2.3 that the essential image of the latter map belongs to the full subspace

$$\mathrm{Maps}((\mathbb{S}_2^{1\&2\text{-op}} \circledast \mathbb{S}_1) \times \mathbb{T}^{1\text{-op}}, \mathbf{1\text{-}Cat}) \subset \mathrm{Maps}((\mathbb{S}_2^{1\&2\text{-op}} \circledast \mathbb{S}_1) \circledast \mathbb{T}^{1\text{-op}}, \mathbf{1\text{-}Cat}) \simeq$$
$$\simeq \mathrm{Maps}(\mathbb{S}_2^{1\&2\text{-op}} \circledast \mathbb{S}_1 \circledast \mathbb{T}^{1\text{-op}}, \mathbf{1\text{-}Cat}).$$

4.2.6. Finally, it is easy to see that the essential image of the resulting fully faithful map

$$\mathrm{Maps}(\mathbb{S}_1 \circledast \mathbb{S}_2, \mathbb{T})^{R_{\mathbb{S}_2}} \to \mathrm{Maps}((\mathbb{S}_2^{1\&2\text{-op}} \circledast \mathbb{S}_1) \times \mathbb{T}^{1\text{-op}}, \mathbf{1\text{-}Cat})$$

equals the essential image of

$$\mathrm{Maps}(\mathbb{S}_2^{1\&2\text{-op}} \circledast \mathbb{S}_1, \mathbb{T})^{L_{\mathbb{S}_2^{1\&2\text{-op}}}} \hookrightarrow \mathrm{Maps}(\mathbb{S}_2^{1\&2\text{-op}} \circledast \mathbb{S}_1, \mathbb{T}) \hookrightarrow$$
$$\hookrightarrow \mathrm{Maps}(\mathbb{S}_2^{1\&2\text{-op}} \circledast \mathbb{S}_1, \mathrm{Funct}(\mathbb{T}^{1\text{-op}}, \mathbf{1\text{-}Cat})) \simeq \mathrm{Maps}((\mathbb{S}_2^{1\&2\text{-op}} \circledast \mathbb{S}_1) \times \mathbb{T}^{1\text{-op}}, \mathbf{1\text{-}Cat}),$$

as desired.

□

4.3. An alternative proof of Theorem 1.2.4. Suppose we have a pair $(\mathbb{S}, \mathbf{C})$ and target $(\infty, 2)$-category $\mathbb{T}$. We will establish a canonical isomorphism

$$\mathrm{Maps}_{2\text{-Cat}}(\mathbb{S}, \mathbb{T})^{R_{\mathbf{C}}} \simeq \mathrm{Maps}_{2\text{-Cat}}(\mathbb{S}^{R_{\mathbf{C}}}, \mathbb{T}). \tag{4.4}$$

It will follow (see Sect. 4.3.3 below) that the map $\to$ in (4.4) is the same as the one given by restriction along (1.2).

4.3.1. Let $\mathbf{D} \subset \mathbb{T}^{1\text{-Cat}}$ be the 1-full subcategory consisting of 1-morphisms that admit a left adjoint. We have

$$\mathrm{Maps}_{2\text{-Cat}}(\mathbb{S}, \mathbb{T})^{R_{\mathbf{C}}} \simeq \mathrm{Maps}_{2\text{-Cat}^{\mathrm{Pair}}}((\mathbb{S}, \mathbf{C}), (\mathbb{T}, \mathbf{D})). \tag{4.5}$$

Since the functor $\mathrm{Sq}^{\mathrm{Pair}}_{\bullet,\bullet}$ is fully faithful, we can rewrite the right-hand side in (4.5) as

$$\mathrm{Maps}_{\mathrm{Spc}^{\boldsymbol{\Delta}^{\mathrm{op}} \times \boldsymbol{\Delta}^{\mathrm{op}}}}(\mathrm{Sq}^{\mathrm{Pair}}_{\bullet,\bullet}(\mathbb{S}, \mathbf{C}), \mathrm{Sq}^{\mathrm{Pair}}_{\bullet,\bullet}(\mathbb{T}, \mathbf{D})). \tag{4.6}$$

4.3.2. It is easy to see that the full subspace

$$\mathrm{Sq}^{\mathrm{Pair}}_{m,n}(\mathbb{T},\mathbf{D}) \subset \mathrm{Sq}_{m,n}(\mathbb{T}) = \mathrm{Maps}_{2\text{-Cat}}([m,n],\mathbb{S}) = \mathrm{Maps}_{2\text{-Cat}}([m]\circledast[n],\mathbb{T})$$

identifies with

$$\mathrm{Maps}_{2\text{-Cat}}([m]\circledast[n],\mathbb{T})^{R_{[m]}} \subset \mathrm{Maps}_{2\text{-Cat}}([m]\circledast[n],\mathbb{T}),$$

where the superscript $R_{[m]}$ follows the notational convention of (3.2).

Applying the isomorphism of Corollary 3.1.7, we rewrite

$$\mathrm{Maps}_{2\text{-Cat}}([m]\circledast[n],\mathbb{T})^{R_{[m]}} \simeq \mathrm{Maps}_{2\text{-Cat}}([n]^{\mathrm{op}}\circledast[m],\mathbb{T})^{L_{[m]}}.$$

Thus, $\mathrm{Maps}_{2\text{-Cat}}(\mathbb{S},\mathbb{T})^{R_{\mathbf{C}}}$ identifies with the full subspace of

$$\mathrm{Maps}_{\mathrm{Spc}^{\mathbf{\Delta}^{\mathrm{op}}\times\mathbf{\Delta}^{\mathrm{op}}}}(\mathrm{Sq}^{\mathrm{Pair}}_{\bullet,\bullet}(\mathbb{S},\mathbf{C}),((\mathrm{Sq}_{\bullet,\bullet}(\mathbb{T}))^{\text{vert-op}})^{\mathrm{reflect}}), \tag{4.7}$$

that corresponds to the bi-simplicial subspace

$$'((\mathrm{Sq}_{\bullet,\bullet}(\mathbb{T}))^{\text{vert-op}})^{\mathrm{reflect}} \subset ((\mathrm{Sq}_{\bullet,\bullet}(\mathbb{T}))^{\text{vert-op}})^{\mathrm{reflect}}$$

given by

$$\mathrm{Maps}_{2\text{-Cat}}([n]^{\mathrm{op}}\circledast[m],\mathbb{T})^{L_{[m]}} \subset \mathrm{Maps}_{2\text{-Cat}}([n]^{\mathrm{op}}\circledast[m],\mathbb{T}).$$

4.3.3. Note that the expression in (4.7) identifies tautologically with

$$\mathrm{Maps}_{\mathrm{Spc}^{\mathbf{\Delta}^{\mathrm{op}}\times\mathbf{\Delta}^{\mathrm{op}}}}(((\mathrm{Sq}^{\mathrm{Pair}}_{\bullet,\bullet}(\mathbb{S},\mathbf{C}))^{\mathrm{reflect}})^{\text{vert-op}},\mathrm{Sq}_{\bullet,\bullet}(\mathbb{T})) \simeq$$
$$\simeq \mathrm{Maps}_{2\text{-Cat}}(\mathfrak{L}^{\mathrm{Sq}}(((\mathrm{Sq}^{\mathrm{Pair}}_{\bullet,\bullet}(\mathbb{S},\mathbf{C}))^{\mathrm{reflect}})^{\text{vert-op}}),\mathbb{T}),$$

while the latter is

$$\mathrm{Maps}_{2\text{-Cat}}(\mathbb{S}^{R_{\mathbf{C}}},\mathbb{T}),$$

by the construction of $\mathbb{S}^{R_{\mathbf{C}}}$.

Thus, we have obtained a fully faithful embedding

$$\mathrm{Maps}_{2\text{-Cat}}(\mathbb{S},\mathbb{T})^{R_{\mathbf{C}}} \hookrightarrow \mathrm{Maps}_{2\text{-Cat}}(\mathbb{S}^{R_{\mathbf{C}}},\mathbb{T}). \tag{4.8}$$

It follows from the construction that the composite map

$$\mathrm{Maps}_{2\text{-Cat}}(\mathbb{S},\mathbb{T})^{R_{\mathbf{C}}} \hookrightarrow \mathrm{Maps}_{2\text{-Cat}}(\mathbb{S}^{R_{\mathbf{C}}},\mathbb{T}) \overset{(1.2)}{\longrightarrow} \mathrm{Maps}_{2\text{-Cat}}(\mathbb{S},\mathbb{T}) \tag{4.9}$$

is the tautological embedding $\mathrm{Maps}_{2\text{-Cat}}(\mathbb{S},\mathbb{T})^{R_{\mathbf{C}}} \hookrightarrow \mathrm{Maps}_{2\text{-Cat}}(\mathbb{S},\mathbb{T})$.

4.3.4. It remains to show that the essential image of (4.8) is everything. I.e., we need to show that for any functor

$$\mathbb{S}^{R_{\mathbf{C}}} \to \mathbb{T},$$

the corresponding map of bi-simplicial spaces

$$\mathrm{Sq}^{\mathrm{Pair}}_{\bullet,\bullet}(\mathbb{S},\mathbf{C}) \to ((\mathrm{Sq}_{\bullet,\bullet}(\mathbb{T}))^{\text{vert-op}})^{\mathrm{reflect}}$$

has the property that its essential image belongs to

$$'((\mathrm{Sq}_{\bullet,\bullet}(\mathbb{T}))^{\text{vert-op}})^{\mathrm{reflect}} \subset ((\mathrm{Sq}_{\bullet,\bullet}(\mathbb{T}))^{\text{vert-op}})^{\mathrm{reflect}}.$$

However, the latter assertion can be checked at the level of the *ordinary* 2-category underlying $\mathbb{T}$. And in the latter case, the assertion follows from Sect. 1.4:

Indeed we already know that the second map in (4.9) is an isomorphism onto $\mathrm{Maps}_{2\text{-Cat}}(\mathbb{S},\mathbb{T})^{R_{\mathbf{C}}} \subset \mathrm{Maps}_{2\text{-Cat}}(\mathbb{S},\mathbb{T})$.

□

Bibliography

[AG] D. Arinkin and D. Gaitsgory, *Singular support of coherent sheaves, and the geometric Langlands conjecture*, Selecta Math. N.S. **21** (2015), 1–199.

[BarS] C. Barwick and C. Schommer-Pries, *On the unicity of homotopy theory of higher categories*, arXiv: 1112.0040.

[BB] A. Beilinson and J. Bernstein, *A proof of Jantzen's conjectures*, Advances in Soviet Mathematics **16**, Part I (1993), 1–50.

[BD] A. Beilinson and V. Drinfeld, *Quantization of Hitchin?s integrable system and Hecke eigensheaves*, available at http://www.math.harvard.edu/~gaitsgde/grad_2009/.

[Bez] R. Bezrukavnikov, *On two geometric realizations of the affine Hecke algebra*, arXiv:1209.0403.

[BFN] D. Benzvi, J. Francis and D. Nadler, *Integral transforms and Drinfeld centers in derived algebraic geometry*, J. Amer. Math. Soc. **23** (2010), no. 4, 909–966.

[Bezr] R. Bezrukavnikov, *On two geometric realizations of the affine Hecke algebra*, arXiv:1209.0403.

[BoV] J. M. Boardman and J. M. Vogt *Homotopy Invariant Structures on Topological Spaces*, Lecture Notes in Mathematics **347**, Springer-Verlag, Berlin and New York (1973).

[De] P. Deligne, *Catégories tannakiennes*, in: "The Grothendieck Festschrift", Vol. II, 111–195, Progr. Math. **87**, Birkhäuser Boston, Boston, MA, 1990.

[Dr] V. Drinfeld, *DG Quotients of DG Categories*, J. Algebra **272** (2004), no. 2, 643–691.

[DrGa1] V. Drinfeld and D. Gaitsgory, *On some finiteness questions for algebraic stacks*, GAFA **23** (2013), 149–294.

[DrGa2] V. Drinfeld and D. Gaitsgory, *Compact generation of the category of D-modules on the stack of G-bundles on a curve*, joint with V. Drinfeld, arXiv:1112.2402, Cambridge Math Journal, **3** (2015), 19–125.

[Fra] J. Francis, *The tangent complex and Hochschild cohomology of E_n-rings*, Compos. Math. **149** (2013), no. 3, 430–480.

[FraG] J. Francis and D. Gaitsgory, *Chiral Koszul duality*, Selecta Math. (N.S.) **18** (2012), 27–87.

[FG] E. Frenkel and D. Gaitsgory, *D-modules on the affine flag variety and representations of affine Kac-Moody algebras*, Represent. Theory **13** (2009), 470–608.

[Ga1] D. Gaitsgory, *Ind-coherent sheaves*, arXiv:1105.4857.

[Ga2] D. Gaitsgory, *The Atiyah-Bott formula for the cohomology of the moduli space of bundles on a curve*, arXiv: 1505.02331.

[Ga3] D. Gaitsgory, *Sheaves of categories and the notion of 1-affineness*, Contemporary Mathematics **643** (2015), 1–99.

[Ga4] Notes on Geometric Langlands, *Generalities on DG categories*, available at http://www.math.harvard.edu/~gaitsgde/GL/.

[GaRo1] D. Gaitsgory and N. Rozenblyum, *DG indschemes*, Contemporary Mathematics **610** (2014), 139–251.

[GaRo2] D. Gaitsgory and N. Rozenblyum, *D-modules and crystals*, PAMQ **10**, no. 1 (2014), 57–155.

[Jo] A. Joyal, *Quasi-categories and Kan complexes*, (in Special volume celebrating the 70th birthday of Prof. Max Kelly) J. Pure Appl. Algebra **175** (2002), no. 1-3, 207–222.

[JT] A. Joyal, M. Tierney, *Quasi-categories vs Segal spaces*, Categories in algebra, geometry and mathematical physics, Contemp. Math. **431**, Amer. Math. Soc., Providence, RI (2007), 277–326.

[Kr] H. Krause, *The stable derived category of a Noetherian scheme*, Compos. Math., **141**(5) (2005), 1128–1162.

[LM] G. Laumon, L. Morret-Baily, *Champs algébriques*, Ergebnisse der Mathematik und ihrer Grenzgebiete (3 Folge, A Series of Modern Surveys in Mathematics), **39**, Springer-Verlag, Berlin, 2000.

[May] P. May, *The geometry of iterated loop spaces*, Lecture Notes in Mathematics **271**, Springer-Verlag, Berlin and New York (1972).

[LZ1] Y. Liu and W. Zheng, *Enhanced six operations and base change theorem for Artin stacks*, arXiv: 1211.5948.

[LZ2] Y. Liu and W. Zheng, *Gluing restricted nerves of infinity categories*, arXiv: 1211.5294..

[Lu1] J. Lurie, *Higher Topos Theory*, Annals of Mathematics Studies, **170**, Princeton University Press, Princeton, NJ, 2009.

[Lu2] J. Lurie, *Higher Algebra*, available at http://www.math.harvard.edu/~lurie.

[Lu3] J. Lurie, *$(\infty, 2)$-categories and Goodwillie calculus-I*, available at http://www.math.harvard.edu/~lurie.

[Lu4] J. Lurie, *DAG-VII, Spectral schemes*, available at http://www.math.harvard.edu/~lurie.

[Lu5] J. Lurie, *DAG-VIII*, available at http://www.math.harvard.edu/~lurie.

[Lu6] J. Lurie, *DAG-X*, available at http://www.math.harvard.edu/~lurie.

[Ne] A. Neeman, *The Grothendieck duality theorem via Bousfield techniques and Brown representability*, J. Amer. Math. Soc. **9** (1996), no. 1, 205–236.

[Rezk1] C. Rezk, *A model for the homotopy theory of homotopy theory*, Trans. Amer. Math. Soc. **353** (2001), no. 3, 973–1007.

[Rezk2] C. Rezk, *A Cartesian presentation of weak n-categories*, Geom. Topol. **14** (2010), no. 1, 521–571.

[Seg] G. Segal, *Categories and cohomology theories*, Topology **13**, (1974), 293–312.

[Sim] C. Simpson, *Algebraic (geometric) n-stacks*, arXiv: 9609014.

[To] B. Toen, *Descente fidèlement plate pour les n-champs d'Artin.*

[TV1] B. Toen and G. Vezzosi, *Homotopical Algebraic Geometry-I.*

[TV2] B. Toen and G. Vezzosi, *Homotopical Algebraic Geometry-II.*

[TT] R. Thomason and T. Trobaugh, *Higher algebraic K -theory of schemes and of derived categories*, The Grothendieck Festschrift, Vol. III, 247–435, Progr. Math., **88** (1990).

[YZ] A. Yekutieli and J. Zhang, *Dualizing Complexes and Perverse Sheaves on Noncommutative Ringed Schemes*, Selecta Math. **12** (2006), 137–177.

Index of Notations

Index

Selected Published Titles in This Series

221 **Dennis Gaitsgory and Nick Rozenblyum,** A Study in Derived Algebraic Geometry, 2017

219 **Richard Evan Schwartz,** The Projective Heat Map, 2017

218 **Tushar Das, David Simmons, and Mariusz Urbański,** Geometry and Dynamics in Gromov Hyperbolic Metric Spaces, 2017

217 **Benoit Fresse,** Homotopy of Operads and Grothendieck–Teichmüller Groups, 2017

216 **Frederick W. Gehring, Gaven J. Martin, and Bruce P. Palka,** An Introduction to the Theory of Higher-Dimensional Quasiconformal Mappings, 2017

215 **Robert Bieri and Ralph Strebel,** On Groups of PL-homeomorphisms of the Real Line, 2016

214 **Jared Speck,** Shock Formation in Small-Data Solutions to 3D Quasilinear Wave Equations, 2016

213 **Harold G. Diamond and Wen-Bin Zhang (Cheung Man Ping),** Beurling Generalized Numbers, 2016

212 **Pandelis Dodos and Vassilis Kanellopoulos,** Ramsey Theory for Product Spaces, 2016

211 **Charlotte Hardouin, Jacques Sauloy, and Michael F. Singer,** Galois Theories of Linear Difference Equations: An Introduction, 2016

210 **Jason P. Bell, Dragos Ghioca, and Thomas J. Tucker,** The Dynamical Mordell–Lang Conjecture, 2016

209 **Steve Y. Oudot,** Persistence Theory: From Quiver Representations to Data Analysis, 2015

208 **Peter S. Ozsváth, András I. Stipsicz, and Zoltán Szabó,** Grid Homology for Knots and Links, 2015

207 **Vladimir I. Bogachev, Nicolai V. Krylov, Michael Röckner, and Stanislav V. Shaposhnikov,** Fokker–Planck–Kolmogorov Equations, 2015

206 **Bennett Chow, Sun-Chin Chu, David Glickenstein, Christine Guenther, James Isenberg, Tom Ivey, Dan Knopf, Peng Lu, Feng Luo, and Lei Ni,** The Ricci Flow: Techniques and Applications: Part IV: Long-Time Solutions and Related Topics, 2015

205 **Pavel Etingof, Shlomo Gelaki, Dmitri Nikshych, and Victor Ostrik,** Tensor Categories, 2015

204 **Victor M. Buchstaber and Taras E. Panov,** Toric Topology, 2015

203 **Donald Yau and Mark W. Johnson,** A Foundation for PROPs, Algebras, and Modules, 2015

202 **Shiri Artstein-Avidan, Apostolos Giannopoulos, and Vitali D. Milman,** Asymptotic Geometric Analysis, Part I, 2015

201 **Christopher L. Douglas, John Francis, André G. Henriques, and Michael A. Hill, Editors,** Topological Modular Forms, 2014

200 **Nikolai Nadirashvili, Vladimir Tkachev, and Serge Vlăduţ,** Nonlinear Elliptic Equations and Nonassociative Algebras, 2014

199 **Dmitry S. Kaliuzhnyi-Verbovetskyi and Victor Vinnikov,** Foundations of Free Noncommutative Function Theory, 2014

198 **Jörg Jahnel,** Brauer Groups, Tamagawa Measures, and Rational Points on Algebraic Varieties, 2014

197 **Richard Evan Schwartz,** The Octagonal PETs, 2014

196 **Silouanos Brazitikos, Apostolos Giannopoulos, Petros Valettas, and Beatrice-Helen Vritsiou,** Geometry of Isotropic Convex Bodies, 2014

For a complete list of titles in this series, visit the AMS Bookstore at **www.ams.org/bookstore/survseries/**.